MW01200510

ACTIVE NETWORK ANALYSIS
Feedback Amplifier Theory

Second Edition

ADVANCED SERIES IN ELECTRICAL AND COMPUTER ENGINEERING

Editor: W. K. Chen *(University of Illinois, Chicago, USA)*

*For the complete list of titles in this series, please visit
http://www.worldscientific.com/series/asece

Advanced Series in Electrical and Computer Engineering – Vol. 15

ACTIVE NETWORK ANALYSIS
Feedback Amplifier Theory

Second Edition

Wai-Kai Chen
University of Illinois, Chicago, USA

 World Scientific

NEW JERSEY · LONDON · SINGAPORE · BEIJING · SHANGHAI · HONG KONG · TAIPEI · CHENNAI

Published by

World Scientific Publishing Co. Pte. Ltd.

5 Toh Tuck Link, Singapore 596224

USA office: 27 Warren Street, Suite 401-402, Hackensack, NJ 07601

UK office: 57 Shelton Street, Covent Garden, London WC2H 9HE

Library of Congress Cataloging-in-Publication Data
Names: Chen, Wai-Kai, 1936–   author.
Title: Active network analysis : feedback amplifier theory / Wai-Kai Chen
    (University of Illinois, Chicago, USA).
Description: 2nd edition. | New Jersey : World Scientific, 2016. | Series: Advanced series in
    electrical and computer engineering | Includes bibliographical references.
Identifiers: LCCN 2016000002| ISBN 9789814675888 (hc : alk. paper) |
    ISBN 9789814704465 (pbk : alk. paper)
Subjects: LCSH: Feedback amplifiers. | Electric network analysis. | Electric networks, Active.
Classification: LCC TK7871.58.F4 C4723 2016 | DDC 621.3815/35--dc23
LC record available at http://lccn.loc.gov/2016000002

British Library Cataloguing-in-Publication Data
A catalogue record for this book is available from the British Library.

Desk Editor: V. Vishnu Mohan

Typeset by Stallion Press
Email: enquiries@stallionpress.com

Printed in Singapore by B & Jo Enterprise Pte Ltd

PREFACE TO FIRST EDITION

Since Bode published his classical text "Network Analysis and Feedback Amplifier Design" in 1945, very few books have been written that treat the subject in any reasonable depth. The purpose of this book is to bridge this gap by providing an in-depth, up-to-date, unified, and comprehensive treatment of the fundamentals of the theory of active networks and its applications to feedback amplifier design. The guiding light throughout has been to extract the essence of the theory and to discuss the topics that are of fundamental importance and that will transcend the advent of new devices and design tools. Intended primarily as a text in network theory in electrical engineering for first-year graduate students, the book is also suitable as a reference for researchers and practicing engineers in industry. In selecting the level of presentation, considerable attention has been given to the fact that many readers may be encountering some of these topics for the first time. Thus, basic introductory material has been included. The background required is the usual undergraduate basic courses in circuits and electronics as well as the ability to handle matrices.

The book can be conveniently divided into three parts. The first part, comprising the first three chapters, deals with general network analysis. The second part, composed of the next four chapters, is concerned with feedback amplifier theory. The third part, consisting of the last two chapters, discusses the state-space and topological analyses of active networks and their relations to feedback theory.

Chapter 1 introduces many fundamental concepts used in the study of linear active networks. We start by dealing with general n-port networks and define passivity in terms of the universally encountered physical quantities *time* and *energy*. We then translate the time-domain passivity criteria into the equivalent frequency-domain passivity conditions. Chapter 2 presents a useful description of the external behavior of a multiterminal network in terms of the indefinite-admittance matrix and demonstrates how it can be employed effectively for the computation of network functions. The significance of this approach is that the indefinite-admittance matrix can usually be written down directly from the network by inspection and that the transfer functions can be expressed compactly as the ratios of the first-and/or second-order cofactors of the elements of the indefinite-admittance matrix. In Chapter 3 we consider the specialization of the general passivity condition for n-port networks in terms of the more immediately useful two-port parameters. We introduce various types of power gains, sensitivity, and the notion absolute stability as opposed to potential instability.

Chapters 4 and 5 are devoted to a study of single-loop feedback amplifiers. We begin the discussion by considering the conventional treatment of feedback amplifiers based on the ideal feedback model and analyzing several simple feedback networks. We then present in detail Bode's feedback theory, which is based on the concepts of return difference and null return difference. Bode's theory is formulated elegantly and compactly in terms of the first- and second-order cofactors of the elements of the indefinite-admittance matrix, and it is applicable to both simple and complicated networks, where the analysis by conventional method for the latter breaks down. We show that feedback may be employed to make the gain of an amplifier less sensitive to variations in the parameters of the active components, to control its transmission and driving-point properties, to reduce the effects of noise and nonlinear distortion, and to affect the stability or instability of the network. The fact that return difference can be measured experimentally for many practical amplifiers indicates that we can include all the parasitic effects in the stability study and that stability problems can be reduced to Nyquist plots.

The application of negative feedback in an amplifier improves its overall performance. However, we are faced with the stability problem in that, for sufficient amount of feedback, at some frequency the amplifier tends to oscillate and becomes unstable. Chapter 6 discusses various stability criteria and investigates several approaches to the stabilization of feedback amplifiers. The Nyquist stability criteria, the Bode plot, the root-locus technique, and root sensitivity are presented. The relationship between gain and phase shift and Bode's design theory is elaborated. Chapter 7 studies the multiple-loop feedback amplifiers that contain a multiplicity of inputs, outputs, and feedback loops. The concepts of return difference and null return difference for a single controlled source are now generalized to the notions of return difference matrix and null return difference matrix for a multiplicity of controlled sources. Likewise, the scalar sensitivity function is generalized to the sensitivity matrix, and formulas for computing multiparameter sensitivity functions are derived.

In Chapter 8, we formulate the network equations in the time domain as a system of first-order differential equations that govern the dynamic behavior of a network. The advantages of representing the network equations in this form are numerous. First of all, such a system has been widely studied in mathematics and its solution, both analytical and numerical, is known and readily available. Secondly, the representation can easily and naturally be extended to time-varying and nonlinear networks. In fact, nearly all time-varying and nonlinear networks are characterized by this approach. Finally, the first-order differential equations are easily programmed for a digital computer or simulated on an analog computer. We then formulate the general feedback theory in terms of the coefficient matrices of the state equations of a multiple-input, multiple-output and multiple-loop feedback amplifier, and derive expressions relating the zeros and poles of the determinants of the return difference matrix and the null return difference matrix to the eigenvalues of the coefficient matrices of the state equations under certain conditions. Finally, in Chapter 9 we study topological analysis of active networks and conditions under which there is a unique solution. These conditions are especially useful in computer-aided network analysis when a numerical solution does not converge. They help distinguish those cases where a network does not possess a unique solution from those where the fault lies with the integration technique. Thus, when a numerical solution does not converge, it is important to distinguish network instability, divergence due to improper numerical integration, and divergence due to lack of the existence of a unique solution.

The book is an outgrowth of notes developed over the past twenty-five years while teaching courses on active network theory at the graduate level at Ohio University and University of Illinois at Chicago. There is little difficulty in fitting the book into a one-semester or two-quarter course in active network theory. For example, the first four chapters plus some sections of Chapters 5, 6 and 8 would be ideal for a one- semester course, whereas the entire book can be covered adequately in a two-quarter course.

A special feature of the book is that it bridges the gap between theory and practice, with abundant examples showing how theory solves problems. These examples are actual practical problems, not idealized illustrations of the theory. A rich variety of problems has been presented at the end of each chapter, some of which are routine applications of results derived in the text. Others, however, require considerable extension of the text material. In all there are 286 problems.

Much of the material in the book was developed from my research. It is a pleasure to acknowledge publicly the research support of the National Science Foundation and the University of Illinois at Chicago through the Senior University Scholar Program. I am indebted to many graduate students who have made valuable contributions to this book. Special thanks are due to my doctoral student Hui Tang, who helped proofread Chapters 8 and 9, and to my secretary, Ms. Barbara Wehner, who assisted me in preparing the index. Finally, I express my appreciation to my wife, Shiao-Ling, for her patience and understanding during the preparation of the book.

Wai-Kai Chen
Naperville, Illinois
January 1, 1991

PREFACE TO SECOND EDITION

We are most gratified to find that the first edition of *Active Network Analysis* was well received and is widely used. Thus, we feel that our original goal of providing an in-depth, unified and comprehensive treatment of the fundamentals of the theory of active networks and its applications to feedback amplifier design was, indeed, worthwhile. Since then many changes have occurred, necessitating not only the updating of some of the material, but more startling, the addition and expansion of many topics.

The purpose of the book is to provide in a single volume a comprehensive reference work covering the broad spectrum of active networks and feedback amplifiers. It is written and developed for the practicing electrical engineers in industry, government, and academia. The goal is to provide the most up-to-date information in the classical fields of circuit theory, circuit components and their models, and feedback networks.

The new edition can again be conveniently divided into three parts. The first part, comprising the first three chapters, deals with fundamentals of general network analysis. The second part, composed of the next four chapters, is concerned with feedback amplifier theory and its design. In this part, we also included compact formulas expressing various feedback quantities of a linear multivariable and multiloop feedback network in terms of the first- and the second-order cofactors of the elements of its indefinite-admittance matrix. They are useful in computing the feedback matrices in that they do not require any matrix inversion in computing some of these quantities. Furthermore, they are suitable for symbolical analysis. The third part, consisting of the last four chapters, discusses the general formulations of multiloop feedback systems. In addition to the two original chapters on state-space and topological analyses of active networks and their relations to feedback theory, we added two new chapters. One is on generalization of topological feedback amplifier theory, in which topological formulas are derived. Extensions of topology and the summations of the products of all transmittances and their associated transfer immittances are also considered. The other chapter is on the indefinite-impedance matrix formulation of feedback amplifier theory. This dual concept as opposed to the more familiar indefinite-admittance matrix is rarely considered in the literature. Perhaps this is due to the fact that measuring the branch voltage is easier than measuring the branch current. However, advances in integrated op-amp circuits have made it possible to measure the branch current on line without opening any branch.

As before, the book stresses fundamental theory behind professional applications. In order to do so, it is reinforced with frequent examples. The reader is assumed to have a certain degree of sophistication and experience. However, brief reviews of theories, principles and mathematics of some subject areas are given. These reviews have been done concisely with perception. The prerequisite knowledge is a typical undergraduate mathematics background of calculus, complex variables, and simple matrix algebra plus a working knowledge in Laplace transform technique.

I am indebted to many of my students over the years who participated in testing the material of this book, and to my colleagues at the University of Illinois at Chicago for providing a stimulating milieu for discussions. Special thanks are due to my graduate students and visiting scholars Jiajian Lu, Jia-Long Lan, Mao-Da Tong, Hui-Yun Wang, and Yi Sheng Zhu, who made significant contributions to the field. In fact, some of the new materials included in the book are based on our joint research.

Wai-Kai Chen
Fremont, California
March 7, 2016

CONTENTS

CHARACTERIZATIONS OF NETWORKS

Over the past two decades, we have witnessed a rapid development of solid-state technology with its apparently unending proliferation of new devices. Presently available solid-state devices such as the transistor, the tunnel diode, the Zener diode, and the varactor diode have already replaced the old vacuum tube in most practical network applications. Moreover, the emerging field of integrated circuit technology threatens to push these relatively recent inventions into obsolescence. In order to understand fully the network properties and limitations of solid-state devices and to be able to cope with the applications of the new devices yet to come, it has become increasingly necessary to emphasize the fundamentals of active network theory that will transcend the advent of new devices and design tools.

The purpose of this chapter is to introduce many fundamental concepts used in the study of linear active networks. We first introduce the concepts of portwise linearity and time invariance. Then we define passivity in terms of the universally encountered physical quantities *time* and *energy*, and show that causality is a consequence of linearity and passivity. This is followed by a brief review of the general characterizations of *n*-port networks in the frequency-domain. The translation of the time-domain passivity criteria into the equivalent frequency-domain passivity conditions is taken up next. Finally, we introduce the discrete-frequency concepts of passivity. The significance of passivity in the study of active networks is that passivity is the formal negation of activity.

1.1 LINEARITY AND NONLINEARITY

A network is a structure comprised of a finite number of interconnected elements with a set of accessible terminal pairs called *ports* at which voltages and currents may be measured and the transfer of electromagnetic energy into or out of the structure can be made. Fundamental to the concept of a port is the assumption that the instantaneous current entering one terminal of the port is always equal to the instantaneous current leaving the other terminal of the port. A network with n such accessible ports is called an *n-port network* or simply an *n-port*, as depicted symbolically in Fig. 1.1. In this section we review briefly the concepts of linearity and nonlinearity and introduce the notion of port wise linearity and nonlinearity.

Refer to the general representation of an n-port network N of Fig. 1.1. The port voltages $v_k(t)$ and currents $i_k(t)$ (where $k = 1, 2, \ldots, n$) can be conveniently represented by the *port-voltage* and *port-current vectors* as

$$\mathbf{v}(t) = [v_1(t), v_2(t), \ldots, v_n(t)]' \qquad (1.1a)$$

$$\mathbf{i}(t) = [i_1(t), i_2(t), \ldots, i_n(t)]' \qquad (1.1b)$$

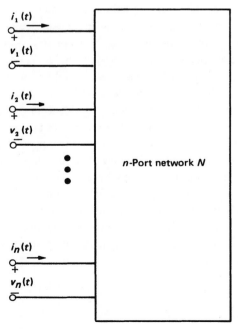

Figure 1.1 The general symbolic representation of an n-port network.

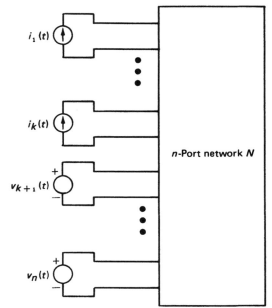

Figure 1.2 A specific input excitation of n-port network.

respectively, where the prime denotes the matrix transpose. There are $2n$ port signals, n port-voltage signals $v_k(t)$, and n port-current signals $i_k(t)$, and each port is associated with two signals $v_k(t)$ and $i_k(t)$. The port vectors $\mathbf{v}(t)$ and $\mathbf{i}(t)$ that can be supported by the n-port network N are said to constitute an *admissible signal pair* for the n-port network. Any n independent functions of these $2n$ port signals, taking one from each of the n ports, may be regarded as the *input* or *excitation* and the remaining n signals as the *output* or *response* of the n-port network. In Fig. 1.1 we may take, for example, $i_1(t), i_2(t), \ldots, i_k(t), v_{k+1}(t), \ldots, v_n(t)$ to be the input or excitation signals. Then $v_1(t), v_2(t), \ldots, v_k(t), i_{k+1}(t), \ldots, i_n(t)$ are the output or response signals. This input-output or excitation-response situation is shown in Fig. 1.2. To facilitate our discussion, let $\mathbf{u}(t)$ be the *excitation vector* associated with the excitation signals, and $\mathbf{y}(t)$ the *response vector* associated with the response signals. For the excitation-response situation of Fig. 1.2, the excitation and response vectors are given by

$$\mathbf{u}(t) = [i_1(t), i_2(t), \ldots, i_k(t), v_{k+1}(t), \ldots, v_n(t)]' \tag{1.2a}$$

$$\mathbf{y}(t) = [v_1(t), v_2(t), \ldots, v_k(t), i_{k+l}(t), \ldots, i_n(t)]' \tag{1.2b}$$

respectively. When we speak of zero excitation of an n-port, we mean that every excitation signal is zero; that is, $\mathbf{u}(t) = \mathbf{0}$. On the other hand, a nonzero excitation is meant a set of n excitation signals, not all of them being zero; that is, $\mathbf{u}(t) \neq \mathbf{0}$.

Generally speaking, a network is said to be *linear* if the superposition principle holds. This implies that the response resulting from all independent sources acting simultaneously is equal to the sum of the responses resulting from each independent source acting one at a time. In this sense, any network comprised of linear network elements (linear resistors, linear inductors, linear capacitors, linear transformers, or linear controlled sources) and independent sources is a linear network. Thus, to verify the linearity of a network by this definition, we must have the complete knowledge of the internal structure of the network. For an n-port, the accessible part of the network may be only at its n ports. For this reason, the above definition of linearity may not be adequate for an n-port. For our purposes, we introduce, in addition to the above definition, the notation of portwise linearity.

Definition 1.1: Linearity and nonlinearity An n-port network is said to be *portwise linear* or simply *linear* if the superposition principle holds at its n ports. An n-port is *portwise nonlinear* or simply *nonlinear* if it is not portwise linear.

In other words, if $\mathbf{y}_a(t)$ and $\mathbf{y}_b(t)$ are the responses of the excitations $\mathbf{u}_a(t)$ and $\mathbf{u}_b(t)$ of an n-port, respectively, then the n-port is portwise linear if and only if for any choice of real scalars α and β, the vector $\mathbf{y}(t) = \alpha \mathbf{y}_a(t) + \beta \mathbf{y}_b(t)$ represents the response of the excitation $\mathbf{u}(t) = \alpha \mathbf{u}_a(t) + \beta \mathbf{u}_b(t)$.

The network of Fig. 1.3 is linear in the usual sense. Let us form a one-port from this network as shown in Fig. 1.4. The port voltage and current are described by the equation

$$v(t) = \left(R_1 + \frac{R_2 R_3}{R_2 + R_3} \right) i(t) + \frac{E R_2}{R_2 + R_3} \tag{1.3}$$

Suppose that we take $i(t)$ to be the excitation and let $i_a(t) = i_b(t) = 1$ A be two excitations. Assume, for simplicity, that $\alpha = \beta = 1$. Then the corresponding

Figure 1.3 A linear network in the usual sense.

Figure 1.4   A nonlinear one-port network with $E \neq 0$.

Figure 1.5   A linear two-port network.

responses $v_a(t)$, $v_b(t)$, and $v_{a+b}(t)$ of the excitations $i_a(t)$, $i_b(t)$, and $i_a(t) + i_b(t)$ are given by

$$v_a(t) = v_b(t) = R_1 + \frac{R_2 R_3}{R_2 + R_3} + \frac{E R_2}{R_2 + R_3} \qquad (1.4a)$$

$$v_{a+b}(t) = 2R_1 + \frac{2R_2 R_3}{R_2 + R_3} + \frac{E R_2}{R_2 + R_3} \qquad (1.4b)$$

Since $v_{a+b}(t) \neq v_a(t) + v_b(t)$, the one-port is nonlinear in the portwise sense. Instead of forming a one-port, suppose that we form a two-port from the network of Fig. 1.3. The resulting two-port network is shown in Fig. 1.5; its port voltages and currents are characterized by

$$\begin{bmatrix} v_1(t) \\ v_2(t) \end{bmatrix} = \begin{bmatrix} R_1 + R_2 & R_2 \\ R_2 & R_2 + R_3 \end{bmatrix} \begin{bmatrix} i_1(t) \\ i_2(t) \end{bmatrix} \qquad (1.5)$$

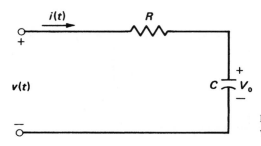

Figure 1.6 A nonlinear one-port network with nonzero initial capacitor voltage.

It is straightforward to demonstrate that the two-port is now portwise linear. Thus, a portwise nonlinear network need not contain any nonlinear network elements and can often be rendered portwise linear by extracting internal sources at newly formed ports.

As another example, consider the one-port of Fig. 1.6, in which the capacitor is initially charged to a voltage $v_C(0+) = V_0$. The terminal relation of the one-port is given by

$$v(t) = Ri(t) + \frac{1}{C} \int_0^t i(x)dx + V_0 \tag{1.6}$$

By following Eqs. (1.4), it is easy to confirm that the one-port is portwise nonlinear. Indeed, the presence of any independent sources or any initial conditions on the energy-storing elements in an n-port would render the n-port portwise nonlinear. On the other hand, an n-port network comprised of linear network elements with zero initial conditions and devoid of any independent sources is always portwise linear. For example, in the one-port networks of Figs. 1.4 and 1.6, if the independent source E and the initial voltage V_0 are set to zero, the resulting one-ports become portwise linear.

From the examples discussed above, it is clear that a portwise nonlinear n-port need not contain any nonlinear elements, but the presence of nonlinear elements does not necessarily imply that the n-port is portwise nonlinear. Figure 1.7 is a one-port comprised of two nonlinear resistors connected in series. The nonlinear resistors are characterized by the equations

$$v_\alpha(t) = i_\alpha(t) - i_\alpha^2(t) \tag{1.7a}$$

$$v_\beta(t) = i_\beta^2(t) \tag{1.7b}$$

The port voltage and current are related by the equation

$$v(t) = i(t) \tag{1.8}$$

Figure 1.7 A linear one-port network comprised of two nonlinear resistors.

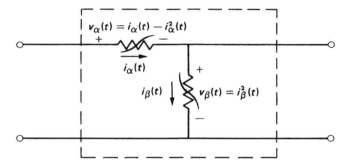

Figure 1.8 A nonlinear two-port network.

showing that this one-port is equivalent to a 1-Ω resistor and thus is portwise linear. Suppose that a two-port is formed from this one-port by connecting two wires across one of the resistors as shown in Fig. 1.8. The resulting two-port becomes portwise nonlinear.

We emphasize the difference between the linearity of a network and the portwise linearity of an n-port. Throughout the remainder of this book, we are concerned mainly with portwise linearity. For simplicity, the word portwise will usually be dropped, as also indicated in Definition 1.1, and will be used only for emphasis.

1.2 TIME INVARIANCE AND TIME VARIANCE

A network is said to be *time-invariant* if it contains no time-varying network elements. Otherwise, it is called a *time-varying network*. Like those discussed in the preceding section, if the port behavior of a network is the major concern, the above definition may not be adequate for an n-port. For this reason, we define portwise time-invariance.

Definition 1.2: Portwise time-invariance and time variance An n-port network is said to be *portwise time-invariant* or simply *time-invariant* if, for every real finite constant τ,

$$\mathbf{u}_a(t) = \mathbf{u}_b(t - \tau) \tag{1.9a}$$

then

$$\mathbf{y}_a(t) = \mathbf{y}_b(t - \tau) \tag{1.9b}$$

where $\mathbf{y}_a(t)$ and $\mathbf{y}_b(t)$ are the responses of arbitrary excitations $\mathbf{u}_a(t)$ and $\mathbf{u}_b(t)$, respectively. An n-port network is *portwise time-varying* or simply *time-varying* if it is not portwise time-invariant.

In other words, an n-port network is time-invariant if its port behavior is invariant to a shift in the time origin. Putting it differently, a time-invariant n-port should produce the same response no matter when a given excitation is applied. Thus, any n-port network comprised only of time-invariant network elements and devoid of any initial conditions is always portwise time-invariant. The converse, however, is not necessarily true. It is quite easy to conceive of an n-port with time-varying elements that exhibits a port behavior that is time-invariant. Figure 1.9 shows a one-port comprised of a series connection of two time-varying resistors, whose input impedance is $2R$ ohms. Thus, the one-port is portwise time-invariant. As before, if a two-port is formed from this one-port by connecting two wires across one of the time-varying resistors, as shown in Fig. 1.10, the resulting two-port becomes portwise time-varying. Also, in general, n-ports with initial stored energies that affect port behavior must be considered to be portwise time-varying.

It is important to distinguish between a time-invariant network and a portwise time-invariant n-port. The network of Fig. 1.9 is time-varying, but it is a portwise time-invariant one-port.

Figure 1.9 A time-invariant one-port network comprised of two time-varying resistors.

Figure 1.10 A time-varying two-port network.

In the remainder of this book, we shall mostly consider time-invariant networks. However, time-varying networks are important in modern engineering applications, having been used successfully in the design of parametric amplifiers and switching networks.

1.3 PASSIVITY AND ACTIVITY

The concept of passivity is not quite so well understood. For many, it simply means the absence of internal sources. A poorly designed transistor circuit may be passive in the sense that it is incapable of delivering energy at any time. A better definition of passivity, which is now widely accepted, is obtained by considering the energy delivery capability of a network.

Definition 1.3: Passivity and activity An n-port network is said to be *passive* if, for *all* admissible signal pairs $\mathbf{v}(t)$ and $\mathbf{i}(t)$, the total energy stored in and delivered to the n-port is nonnegative for all time t; that is,

$$\varepsilon(t) = \varepsilon(t_0) + \int_{t_0}^{t} \mathbf{v}'(x)\mathbf{i}(x)dx \geq 0 \tag{1.10}$$

for *all* initial time t_0 and *all* time $t \geq t_0$, where $\varepsilon(t_0)$ is the finite energy stored in the n-port at the initial time t_0. An n-port network is *active* if it is not passive. An n-port network is *strictly passive* if the equality is not attained in condition (1.10) for all nonzero admissible signal pairs $\mathbf{v}(t)$ and $\mathbf{i}(t)$.

In other words, a passive n-port is incapable of delivering energy at any time. To demonstrate activity, we need only find one excitation such that the condition (1.10) is violated for at least one time t. Thus, an n-port is active if and only if for *some* excitation

$$\varepsilon(t) = \varepsilon(t_0) + \int_{t_0}^{t} \mathbf{v}'(x)\mathbf{i}(x)dx < 0 \tag{1.11}$$

for *some* initial time t_0 and *some* time $t \geqslant t_0$.

The linear time-invariant resistor, inductor, and capacitor with nonnegative element values are simple examples of passive one-ports. The two-port network of Fig. 1.5 is, as expected, also passive. To demonstrate this, we compute its terminal voltages in terms of its terminal currents, which are given by the equations

$$v_1(t) = (R_1 + R_2)i_1(t) + R_2 i_2(t) \tag{1.12a}$$

$$v_2(t) = R_2 i_1(t) + (R_2 + R_3)i_2(t) \tag{1.12b}$$

Substituting these in Eq. (1.10) yields

$$\varepsilon(t) = \varepsilon(t_0) + \int_{t_0}^{t} [v_1(x)i_1(x) + v_2(x)i_2(x)]dx$$

$$= \int_{t_0}^{t} [R_2(i_1 + i_2)^2 + R_1 i_1^2 + R_3 i_2^2]dx \geqslant 0 \tag{1.13}$$

since the integrand is nonnegative, where $\varepsilon(t_0) = 0$. Thus, the two-port network of Fig. 1.5 is passive.

The absence of internal sources does not always imply passivity. For example, a nonlinear one-port resistor characterized by the equation

$$v(t) = i(t) + i^2(t) \tag{1.14}$$

is active, because we can demonstrate that (1.11) can be fulfilled for some excitation, some initial time t_0, and some time t. To see this, let us apply a current source $i(t) = -e^t$ to the nonlinear resistor. Then from (1.14), (1.11) can be expressed as

$$\varepsilon(t) = \varepsilon(t_0) + \int_{t_0}^{t} (i + i^2)i\,dx = \frac{1}{2}e^{2t} - \frac{e^{3t}}{3} - \frac{1}{2}e^{2t_0} + \frac{e^{3t_0}}{3} \tag{1.15}$$

where $\varepsilon(t_0) = 0$. For $t > \ln 1.5$ and $t_0 \leqslant \ln 1.5$, $\varepsilon(t)$ is negative, thus showing the nonlinear resistor to be active. In fact, any resistor, linear or otherwise, is passive if and only if its vi-characteristic lies solely in the first and third quadrants of the vi-plane. An unbiased tunnel diode, whose terminal VI-characteristic is as shown in Fig. 1.11b, is therefore a passive device. However, a properly biased tunnel diode of Fig. 1.12a, with its vi-characteristic as shown in Fig. 1.12b, is active in

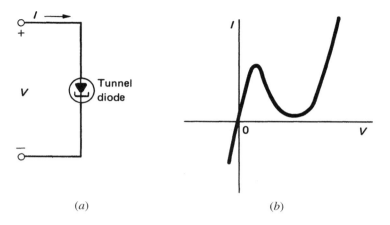

Figure 1.11 (a) An unbiased tunnel diode and (b) its terminal VI-characteristic.

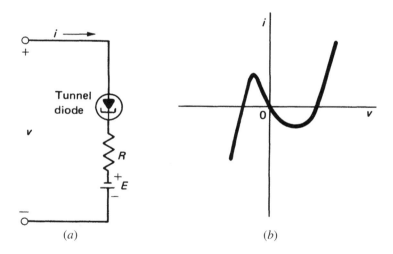

Figure 1.12 (a) A biased tunnel diode and (b) its terminal vi-characteristic.

that it is capable of delivering energy for small signal levels; its justification is left as an exercise (see Prob. 1.1).

The definition of passivity as defined above cannot be stated solely in terms of the port variables $\mathbf{v}(t)$ and $\mathbf{i}(t)$; the initial stored energy $\varepsilon(t_0)$ must be taken into account. Consider, for example, a linear time-invariant capacitor of capacitance C.

To test passivity, we compute

$$\varepsilon(t) = \varepsilon(t_0) + \int_{t_0}^{t} v(x)i(x)dx = \varepsilon(t_0) + C\int_{v(t_0)}^{v(t)} v\,dv$$

$$= \varepsilon(t_0) + \frac{1}{2}Cv^2(t) - \frac{1}{2}Cv^2(t_0) = \frac{1}{2}Cv^2(t) \qquad (1.16)$$

where $\varepsilon(t_0) = \frac{1}{2}Cv^2(t_0)$, showing that a linear time-invariant capacitor is passive if its capacitance is nonnegative and active if its capacitance is negative. Of course, negative capacitance is not physically realizable. However, it is possible to design an electronic circuit to attain a negative capacitance within a small operating range and a narrow frequency band. Now if the term corresponding to the initial stored energy $\varepsilon(t_0)$ is not included in Definition 1.3, Eq. (1.16) becomes $\varepsilon(t) = \frac{1}{2}Cv^2(t) - \frac{1}{2}Cv^2(t_0)$ and can then be made negative for some excitation, say, $v(t) = e^{-t}$ for positive C. This is totally unacceptable, and no one expects a linear time-invariant capacitor with positive capacitance to be active.

The above discussion of the capacitor brings out an important variation on the passivity of an n-port network. As is well known, the capacitor stores energy during part of the time and then returns it to the sources later. Thus, the total energy delivered to the capacitor from the very early time when the capacitor is completely quiescent starting at $t_0 = -\infty$ to the time $t = \infty$ must be identically zero. This gives rise to the concept of losslessness. An n-port network is lossless if all of the energy delivered into its ports is ultimately returned from the ports. More formally, we have the following definition.

Definition 1.4: Losslessness An n-port network is said to be *lossless* if, for *all finite* admissible port signals $\mathbf{v}(t)$ and $\mathbf{i}(t)$ that are square-integrable from t_0 to ∞, that is,

$$\int_{t_0}^{\infty} \mathbf{v}'(x)\mathbf{v}(x)dx < \infty \qquad (1.17a)$$

$$\int_{t_0}^{\infty} \mathbf{i}'(x)\mathbf{i}(x)dx < \infty \qquad (1.17b)$$

then

$$\varepsilon = \varepsilon(t_0) + \int_{t_0}^{\infty} \mathbf{v}'(x)\mathbf{i}(x)dx = 0 \qquad (1.17c)$$

for *all* initial time t_0, where $\varepsilon(t_0)$ denotes the finite energy stored in the n-port at the initial time t_0.

The requirement that the port signals be square-integrable is necessary. Consider, for example, a linear time-invariant inductor of inductance $L = 1$ H. Clearly, $v(t) = e^t$ and $i(t) = e^t$ are admissible port signals for the inductor, but they are not square-integrable. The total energy stored in the inductor at any time t is $\varepsilon(t) = \frac{1}{2}e^{2t}$. As t approaches infinity, $\varepsilon(t)$ becomes unbounded and Eq. (1.17c) is not satisfied. Thus, in testing for losslessness, Eq. (1.17c) should be checked for all finite admissible port signals $\mathbf{v}(t)$ and $\mathbf{i}(t)$ that are square-integrable from any initial time t_0 to $t = \infty$. For the integrals in (1.17) to exist, it is necessary that $v(\infty) = 0$ and $i(\infty) = 0$. Also, for $t_0 = -\infty$, we must have $v(-\infty) = 0$ and $i(-\infty) = 0$. In the case of the inductor considered above, the total energy stored at any time t is $\varepsilon(t) = \frac{1}{2}Li^2(t)$, which approaches zero as t approaches infinity, showing that a linear time-invariant inductor with nonnegative inductance is lossless.

When an n-port is completely quiescent with no stored energy at some very early time starting $t_0 = -\infty$, an alternative and more compact definition of passivity becomes

$$\varepsilon(t) = \int_{-\infty}^{t} \mathbf{v}'(x)\mathbf{i}(x)dx \geqslant 0 \tag{1.18}$$

for *all* excitations and for *all* time t. Likewise, an initially relaxed n-port network is active if for *some* excitation

$$\varepsilon(t) = \int_{-\infty}^{t} \mathbf{v}'(x)\mathbf{i}(x)dx < 0 \tag{1.19}$$

for *some* time t, and is lossless if for *all finite* admissible port signals $\mathbf{v}(t)$ and $\mathbf{i}(t)$ that are square-integrable, as defined in (1.17a) and (1.17b) with t_0 being replaced by $-\infty$,

$$\varepsilon = \int_{-\infty}^{\infty} \mathbf{v}'(x)\mathbf{i}(x)dx = 0 \tag{1.20}$$

Example 1.1 Figure 1.13 represents a gyrator of gyration resistance r. The gyrator is a two-port device whose terminal voltages and currents are characterized by the equations

$$v_1(t) = -ri_2(t) \tag{1.21a}$$
$$v_2(t) = ri_1(t) \tag{1.21b}$$

The total energy delivered to the gyrator from $-\infty$ to t is given by

$$\varepsilon(t) = \int_{-\infty}^{t} (v_1 i_1 + v_2 i_2)dx = \int_{-\infty}^{t} (ri_1 i_2 - ri_1 i_2)dx = 0 \tag{1.22}$$

showing that the gyrator is a lossless two-port network.

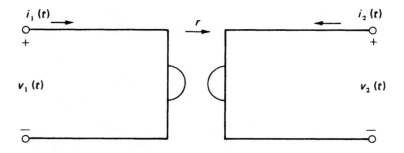

Figure 1.13 The diagrammatic symbol of the gyrator with gyration resistance r.

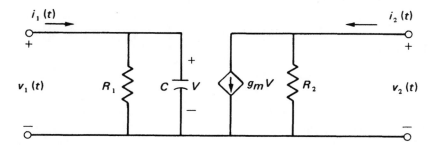

Figure 1.14 An equivalent network of a transistor.

Earlier, we demonstrated that the absence of internal sources in an n-port does not necessarily mean that the n-port is passive. However, the presence of internal sources, dependent or independent, does not guarantee that the n-port is always active. This is illustrated by the following example.

Example 1.2 Consider the equivalent network of a transistor as shown in Fig. 1.14. The port currents and voltages are related by the equations

$$i_1(t) = C\frac{dv_1(t)}{dt} + \frac{v_1(t)}{R_1} \tag{1.23a}$$

$$i_2(t) = g_m v_1(t) + \frac{v_2(t)}{R_2} \tag{1.23b}$$

Substituting these in $\varepsilon(t)$ of (1.18) or (1.19) gives

$$\varepsilon(t) = \int_{-\infty}^{t} \left(Cv_1\frac{dv_1}{dx} + \frac{v_1^2}{R_1} + \frac{v_2^2}{R_2} + g_m v_1 v_2 \right) dx \tag{1.24}$$

Choose port voltages $v_1(t)$ and $v_2(t)$ as the excitation and set

$$v_2(t) = -\frac{g_m R_2 v_1(t)}{2} \tag{1.25}$$

yielding the total energy delivered to the transistor at time t as

$$\varepsilon(t) = \frac{1}{2}Cv_1^2(t) - \frac{1}{4}R_2\left(g_m^2 - \frac{4}{R_1 R_2}\right)\int_{-\infty}^{t} v_1^2(x)dx \tag{1.26}$$

The first term on the right-hand side of Eq. (1.26) denotes the energy stored in the capacitor C at the time t. Since the capacitor is lossless, the first term reduces to zero as t approaches infinity. Equation (1.26) shows that if

$$g_m^2 > \frac{4}{R_1 R_2} \tag{1.27}$$

the input energy $\varepsilon(t)$ to the transistor is always negative for sufficiently large t because the integrand in the second term is always positive for nonzero $v_1(t)$, indicating that the device is active. If, on the other hand, $g_m^2 \leqslant 4/R_1 R_2$, then the device becomes passive and behaves like an energy sink or absorber.

Before we turn our attention to another concept, we remark that it has been suggested (and used by many) that we use average power in defining the concept of passivity instead of energy. There is nothing fundamentally wrong with this except that we have to regard the network as operating at steady state under sinusoidal excitations and responses. Also, in the case of lossless n-ports, the average power into the n-port would always be zero whether the reactive elements are positive or negative. The average power flow into a pure capacitor, for example, is always zero regardless whether the capacitance is positive or negative. It would be inconceivable to call a capacitor with negative capacitance a passive device.

1.4 CAUSALITY AND NONCAUSALITY

The term *causality* connotes the existence of a *cause-effect* relationship. Intuitively, it states that a causal n-port network cannot yield any response until after the excitation is applied. In other words, a causal n-port is not anticipative; it cannot predict the future behavior of the excitation. Of course, all the physical networks must be nonanticipative, but the networks that we deal with and analyze are models made up by the elements that are idealizations of actual physical devices such as resistors, capacitors, inductors, transformers, and generators. We cannot always assume that they are nonanticipative. Furthermore, physical networks may be noncausal.

In this section, we formally introduce the concept of causality in terms of the excitation-response properties of an n-port and then show that any linear passive n-port with the exception of a few mostly trivial cases is causal. Since causality is defined in terms of excitation and response, we must first specify the excitation and response variables of the n-port under consideration. To keep our discussion general, let $\mathbf{u}(t)$ be the vector associated with the excitation signals, and $\mathbf{y}(t)$ the response signals. For the excitation-response situation of Fig. 1.2, the excitation vector $\mathbf{u}(t)$ and the response vector $\mathbf{y}(t)$ are given by

$$\mathbf{u}(t) = [i_1(t), i_2(t), \ldots, i_k(t), v_{k+l}(t), \ldots, v_n(t)]' \qquad (1.28a)$$

$$\mathbf{y}(t) = [v_1(t), v_2(t), \ldots, v_k(t), i_{k+l}(t), \ldots, i_n(t)]' \qquad (1.28b)$$

With these preliminaries, the intuitive notion of causality can now be stated formally as follows.

Definition 1.5: Causality and noncausality An n-port network is said to be *causal* if

$$\mathbf{u}_a(t) = \mathbf{u}_b(t) \quad -\infty < t < t_1 \qquad (1.29a)$$

then

$$\mathbf{y}_a(t) = \mathbf{y}_b(t) \quad -\infty < t < t_1 \qquad (1.29b)$$

for any t_1, where $\mathbf{y}_a(t)$ and $\mathbf{y}_b(t)$ are the responses of arbitrary excitations $\mathbf{u}_a(t)$ and $\mathbf{u}_b(t)$, respectively, of the n-port. An n-port network is *noncausal* if it is not causal.

The definition states in effect that two identical excitations over any time interval must result in identical responses over the same time interval for an n-port to be causal. To test for noncausality, it is sufficient to find *some* excitations $\mathbf{u}_a(t)$ and $\mathbf{u}_b(t)$ such that for *some* time t_1, $\mathbf{u}_a(t) = \mathbf{u}_b(t)$ for $t < t_1$ gives rise to $\mathbf{y}_a(t) \neq \mathbf{y}_b(t)$ for *some* $t < t_1$. Accordingly, an n-port is noncausal because it is either anticipative or the response is not a unique function of the excitation.

Examples of the causal networks are the one-port linear resistor, capacitor, and inductor under either voltage or current excitation. The two-port networks of Figs. 1.13 and 1.14 are also examples of causal networks under all excitations. For an example of a noncausal one-port, consider the nonlinear resistor of Eq. (1.14) under voltage excitation-current response. The vi-relation of this nonlinear resistor, as repeated here, is

$$v(t) = i(t) + i^2(t) \tag{1.30}$$

Choose two input voltages $v_a(t)$ and $v_b(t)$ such that, for $t < t_1$,

$$v_a(t) = v_b(t) = e^{-t} \tag{1.31}$$

For this input, there correspond two different responses

$$i_a(t) = -\frac{1}{2} + \sqrt{\frac{1}{4} + e^{-t}} \tag{1.32a}$$

$$i_b(t) = -\frac{1}{2} - \sqrt{\frac{1}{4} + e^{-t}} \tag{1.32b}$$

for $t < t_1$. Hence, this nonlinear resistor is noncausal under voltage excitation-current response. Now suppose that the excitation is a current source, and the response is the port voltage. Then for any two input currents $i_a(t)$ and $i_b(t)$ such that

$$i_a(t) = i_b(t) \quad t < t_1 \tag{1.33}$$

the corresponding voltages $v_a(t)$ and $v_b(t)$ are equal,

$$v_a(t) = i_a(t) + i_a^2(t) = i_b(t) + i_b^2(t) = v_b(t) \quad t < t_1 \tag{1.34}$$

showing that the resistor is causal under current excitation-voltage response.

Likewise, the unbiased tunnel diode of Fig. 1.11 is a causal one-port under voltage excitation-current response, and a noncausal one-port under current excitation-voltage response (see Prob. 1.3).

Example 1.3 Consider a two-port ideal transformer of turns ratio $k{:}1$, whose port voltages and currents are related by the equation (Fig. 1.15)

$$\begin{bmatrix} v_2(t) \\ i_2(t) \end{bmatrix} = \begin{bmatrix} \dfrac{1}{k} & 0 \\ 0 & -k \end{bmatrix} \begin{bmatrix} v_1(t) \\ i_1(t) \end{bmatrix} \tag{1.35}$$

If $\mathbf{u}(t) = [v_1(t), i_2(t)]'$ is taken as the excitation, then the response $\mathbf{y}(t) = [i_1(t), v_2(t)]'$ is uniquely determined through Eq. (1.35). Consequently, for any two identical excitations we obtain identical responses, meaning that the transformer is causal under the chosen excitation and response. However, if $\mathbf{u}(t) = [v_1(t), v_2(t)]'$

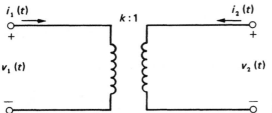

Figure 1.15 The two-port ideal transformer with turns ratio k:1.

is taken as the excitation, then we can choose two inputs $\mathbf{u}_a(t)$ and $\mathbf{u}_b(t)$ such that, for all t,

$$\mathbf{u}_a(t) = \begin{bmatrix} kv(t) \\ v(t) \end{bmatrix} = \mathbf{u}_b(t) \tag{1.36}$$

$v(t)$ being arbitrary. The corresponding responses are

$$\mathbf{y}_a(t) = \begin{bmatrix} i(t) \\ -ki(t) \end{bmatrix} \quad \text{and} \quad \mathbf{y}_b(t) = \begin{bmatrix} \hat{i}(t) \\ -k\hat{i}(t) \end{bmatrix} \tag{1.37}$$

where $i(t)$ and $\hat{i}(t)$ are arbitrary. Choosing $i(t) \neq \hat{i}(t)$ for some t shows that two identical excitations do not result in identical responses. Thus, the ideal transformer is noncausal under voltage excitation-current response. Likewise, we can show that the transformer is noncausal under current excitation-voltage response.

From the above discussion, we emphasize that in characterizing an n-port as being causal or noncausal, we must specify the excitation and response. An n-port may be causal under one set of excitation and response and noncausal under a different set. In general, most linear networks are causal, and noncausal linear networks are very rare. We shall show shortly that with the exception of a few mostly trivial cases, any linear passive n-port network is causal.

It has also been suggested in the literature that causality or nonanticipativeness of an n-port network be defined in such a way that zero excitation over a time interval implies zero response over the same time interval; that is, $\mathbf{u}(t) = \mathbf{0}$ for $t \leqslant t_1$ implies $\mathbf{y}(t) = 0$ for $t \leqslant t_1$. This criterion for causality is not satisfactory in that some obviously causal and nonanticipative n-ports may be rendered noncausal and anticipative by this approach. Consider, for example, the one-port network of Fig. 1.16, comprised of a series connection of a resistor and a battery. Shorting the two terminals gives rise to a nonzero current $i = -E/R$, indicating that a zero excitation $v(t) = 0$ does not imply a zero response $i(t) \neq 0$. It would be incorrect to say that this simple one-port is noncausal and anticipative.

Before we show that causality is a consequence of linearity and passivity, we review briefly the concept of linearity. As indicated in Sec. 1.1, an n-port is said

$i(t)$

R

$+$

$v(t)$

$+$

E

$-$

$-$

Figure 1.16 A causal and nonanticipative oneport network.

to be *linear* if the superposition principle holds, that is, if $\mathbf{y}_a(t)$ and $\mathbf{y}_b(t)$ are the responses of the excitations $\mathbf{u}_a(t)$ and $\mathbf{u}_b(t)$, respectively, of an n-port, then for any choice of real scalars α and β, the vector $\alpha\mathbf{y}_a(t) + \beta\mathbf{y}_b(t)$ represents the response of the excitation $\alpha\mathbf{u}_a(t) + \beta\mathbf{u}_b(t)$. When we speak of an n-port that can support n *linearly independent* excitations, we mean that if $\{\mathbf{u}_k(t), \mathbf{y}_k(t); \ k = 1, 2, \ldots\}$ denotes the set of all excitation-response pairs that can be supported by the n-port, there exist n linearly independent excitation vectors $\mathbf{u}_k(t)$'s in the set.

Theorem 1.1 A linear passive n-port network that can support n linearly independent excitations for all time is also causal.

PROOF. Let $\mathbf{y}(t)$ and $\mathbf{y}_0(t)$ be the responses of the excitations $\mathbf{u}(t)$ and $\mathbf{u}_0(t)$, respectively. Assume that for arbitrary finite $t_1 \geqslant t_0$,

$$\mathbf{u}(t) = 0 \quad \text{for } t < t_1 \tag{1.38}$$

Write, for any scalar α,

$$\hat{\mathbf{u}}(t) = \mathbf{u}_0(t) + \alpha\mathbf{u}(t) \tag{1.39}$$

Then $\hat{\mathbf{u}}(t)$ coincides with $\mathbf{u}_0(t)$ for $t < t_1$; that is,

$$\hat{\mathbf{u}}(t) = \mathbf{u}_0(t) \quad \text{for } t < t_1 \tag{1.40}$$

Denote by $\hat{\mathbf{y}}(t)$ the response corresponding to the excitation $\hat{\mathbf{u}}(t)$. Then, by linearity,

$$\hat{\mathbf{y}}(t) = \mathbf{y}_0(t) + \alpha\mathbf{y}(t) \tag{1.41}$$

Appealing to passivity yields

$$\varepsilon(t) = \varepsilon(t_0) + \int_{t_0}^{t} \hat{\mathbf{u}}'(x)\hat{\mathbf{y}}(x)dx \geqslant 0 \tag{1.42}$$

for all t (see Prob. 1.5). Thus, for all $t < t_1$ and for any real scalar α,

$$\varepsilon(t_0) + \int_{t_0}^{t} \mathbf{u}_0'(x)\mathbf{y}_0(x)dx + \alpha \int_{t_0}^{t} \mathbf{u}_0'(x)\mathbf{y}(x)dx \geqslant 0 \tag{1.43}$$

Since $\varepsilon(t_0)$ is finite and α is any real scalar, for the inequality (1.43) to hold it is necessary that

$$\int_{t_0}^{t} \mathbf{u}_0'(x)\mathbf{y}(x)dx = 0 \quad \text{for } t < t_1 \tag{1.44}$$

for, if not, we can choose an appropriate value of α to make the left-hand side of (1.43) negative. Equation (1.44) is possible for any $t < t_1$ only if

$$\mathbf{u}_0'(t)\mathbf{y}(t) = 0 \quad \text{for } t < t_1 \tag{1.45}$$

Let $\mathbf{u}_k(t)$, $k = 1, 2, \ldots, n$, be n linearly independent excitations that can be supported by the n-port under consideration. Since in deriving Eq. (1.45) $\mathbf{u}_0(t)$ is an arbitrary excitation, Eq. (1.45) is also applicable to $\mathbf{u}_k(t)$. Applying it to $\mathbf{u}_k(t)$ successively gives

$$\mathbf{u}_k'(t)\mathbf{y}(t) = 0 \quad k = 1, 2, \ldots, n \quad \text{and} \quad t < t_1 \tag{1.46}$$

or, more compactly,

$$[\mathbf{u}_1(t), \mathbf{u}_2(t), \ldots, \mathbf{u}_n(t)]'\mathbf{y}(t) = \mathbf{0} \quad \text{for } t < t_1 \tag{1.47}$$

Since the vectors $\mathbf{u}_k(t)$ are linearly independent, the coefficient matrix is nonsingular. Taking the inverse of this matrix yields

$$\mathbf{y}(t) = \mathbf{0} \quad \text{for } t < t_1 \tag{1.48}$$

To complete our proof, let $\mathbf{y}_a(t)$ and $\mathbf{y}_b(t)$ be the responses of the n-port resulting from the excitations $\mathbf{u}_a(t)$ and $\mathbf{u}_b(t)$, respectively, and set

$$\mathbf{u}_a(t) = \mathbf{u}_b(t) \quad \text{for } t < t_1 \tag{1.49}$$

Then we have

$$\mathbf{u}(t) = \mathbf{u}_a(t) - \mathbf{u}_b(t) \quad \text{for } t < t_1 \tag{1.50}$$

Appealing once more to linearity, we get

$$\mathbf{y}(t) = \mathbf{y}_a(t) - \mathbf{y}_b(t) \quad \text{for } t < t_1 \tag{1.51}$$

which according to (1.48) must be identically zero for $t < t_1$, giving

$$\mathbf{y}_a(t) = \mathbf{y}_b(t) \quad \text{for } t < t_1 \tag{1.52}$$

This completes the proof of the theorem.

The ideal one-port short circuit is noncausal under voltage excitation-current response because it admits an infinitude of responses for the sole excitation

$v(t) = 0$. It fails the sufficient condition of the above theorem in that the one-port does not support any independent excitation. However, it is causal under current excitation-voltage response because at each instant of time all the current excitations are related by a multiplicative constant, meaning that the ideal one-port short circuit supports one independent excitation, thus fulfilling the sufficient requirement of the theorem. As for the ideal transformer considered in Example 1.3, if $\mathbf{u}(t) = [v_1(t), i_2(t)]'$ is taken as the excitation, the transformer clearly can support two independent excitations. One could be chosen, for example, as $[v_1(t), 0]'$ and the other as $[0, i_2(t)]'$ for arbitrary but positive $v_1(t)$ and $i_2(t)$. Thus, according to the theorem, the ideal transformer is causal under the specified excitation and response. Now suppose that $\mathbf{u}(t) = [v_1(t), v_2(t)]'$ is taken as the excitation. Then the transformer can support only excitations of the type

$$\mathbf{u}(t) = \begin{bmatrix} kv(t) \\ v(t) \end{bmatrix} = v(t) \begin{bmatrix} k \\ 1 \end{bmatrix} \tag{1.53}$$

$v(t)$ being arbitrary. This shows that the ideal transformer can support only one independent excitation and hence the sufficient condition of the theorem is not satisfied, and, as demonstrated in Example 1.3, the ideal transformer is noncausal.

From the definition of causality, it is obvious that a causal n-port network must have its response as a single-valued function of its excitation. The converse is, however, not necessarily true. Even a linear passive n-port for which the response is a single-valued function of the excitation may be noncausal (see Prob. 1.9). According to Theorem 1.1, the response can be a multiple-valued function of the excitation only if the linear passive n-port does not support n linearly independent excitations for all time. The following theorem demonstrates that the instantaneous power entering a linear passive n-port is uniquely determined by its excitation.

Theorem 1.2 The instantaneous power entering a linear passive n-port network is uniquely determined by its excitation.

PROOF. Let $\mathbf{y}_a(t)$ and $\mathbf{y}_b(t)$ be the responses of two arbitrary excitations $\mathbf{u}_a(t)$ and $\mathbf{u}_b(t)$ of a linear passive n-port, respectively. It is sufficient to show that if $\mathbf{u}_a(t) = \mathbf{u}_b(t)$, then $\mathbf{u}_a'(t)\mathbf{y}_a(t) = \mathbf{u}_b'(t)\mathbf{y}_b(t)$.

Let $\mathbf{y}(t)$ and $\mathbf{y}_0(t)$ be the responses of the arbitrary excitations $\mathbf{u}(t)$ and $\mathbf{u}_0(t)$, respectively. Assume that $\mathbf{u}_a(t) = \mathbf{u}_b(t) = \mathbf{u}_0(t)$ and

$$\mathbf{u}(t) = \mathbf{u}_a(t) - \mathbf{u}_b(t) = \mathbf{0} \tag{1.54}$$

By linearity we have

$$\mathbf{y}(t) = \mathbf{y}_a(t) - \mathbf{y}_b(t) \tag{1.55}$$

Following the steps outlined in Eqs. (1.38)–(1.45) of Theorem 1.1 yields

$$\mathbf{u}_0'(t)\mathbf{y}(t) = 0 \tag{1.56}$$

Substituting (1.55) in (1.56) in conjunction with $\mathbf{u}_a(t) = \mathbf{u}_b(t) = \mathbf{u}_0(t)$ gives

$$\mathbf{u}_a'(t)\mathbf{y}_a(t) = \mathbf{u}_b'(t)\mathbf{y}_b(t) \tag{1.57}$$

This completes the proof of the theorem.

Theorem 1.2 indicates that even if the response of a linear passive n-port network is a multiple-valued function of its excitation, the instantaneous power entering the n-port is uniquely determined by its excitation, and Theorem 1.1 gives a sufficient condition under which the response can be uniquely determined by its excitation.

As an example, let us again consider the ideal transformer of Fig. 1.15 and take $\mathbf{u}(t) = [v_1(t), v_2(t)]'$ to be the excitation. Then the transformer can support only excitations of the type (1.53), which yields an arbitrary current response $\mathbf{y}(t)$ of the type

$$\mathbf{y}(t) = \begin{bmatrix} i(t) \\ -ki(t) \end{bmatrix} = i(t) \begin{bmatrix} 1 \\ -k \end{bmatrix} \tag{1.58}$$

where $i(t)$ is an arbitrary real function. The instantaneous power entering the transformer is always zero,

$$P(t) = \mathbf{u}'(t)\mathbf{y}(t) = v(t)i(t)(k - k) = 0 \tag{1.59}$$

showing that the excitation uniquely determines the instantaneous power, namely the zero power, entering the transformer.

1.5 MATRIX CHARACTERIZATIONS OF n-PORT NETWORKS

In the foregoing, we have presented a set of postulates characterizing the physical nature of a network. However, these descriptions, although very general, are not very useful because they are difficult to apply. In the present section, we discuss the frequency-domain characterizations of the class of linear time-invariant n-port networks, which are then used to derive the equivalent frequency-domain conditions of passivity. Since in the remainder of this book we deal exclusively with this class of networks, the adjectives *linear* and *time-invariant* will be omitted in the discussion unless they are used for emphasis. For our purposes, we shall deal with the Laplace-transformed variables and assume that all independent sources and initial conditions have been set to zero in the interior of the n-port network.

As indicated at the beginning of this chapter, any n independent functions of the $2n$ port variables (Fig. 1.1), taking one each of the two variables associated with each of the n ports, can be regarded as the excitation, and the remaining n variables as the response. In the present section, the restriction can be relaxed. Any n independent functions of the $2n$ port variables can be regarded as the excitation and the remaining n variables as the response. For example, both $i_1(t)$ and $v_1(t)$ can be taken as the excitation signals instead of one of them, as previously required. The reason is that in the preceding sections we require that $\mathbf{u}'(t)\mathbf{y}(t)$ represent instantaneous power entering the n-port. This is not possible if both variables such as $v_1(t)$ and $i_1(t)$ are chosen as the excitation signals.

For linear time-invariant n-port networks, their port behavior is completely characterized by giving the relationships among excitation and response signals. Let $\mathbf{y}(t)$ be the response of the excitation $\mathbf{u}(t)$. Then the square matrix $\mathbf{H}(s)$ of order n relating the Laplace transform $\tilde{\mathbf{y}}(s)$ of the response vector $\mathbf{y}(t)$ to the Laplace transform $\tilde{\mathbf{u}}(s)$ of the excitation vector $\mathbf{u}(t)$ is called the *general hybrid matrix* of the n-port, that is,[†]

$$\tilde{\mathbf{y}}(s) = \mathbf{H}(s)\tilde{\mathbf{u}}(s) \tag{1.60}$$

The elements of $\mathbf{H}(s)$ are referred to as the *general hybrid parameters*. We write

$$\tilde{\mathbf{u}}(s) = [\tilde{u}_1(s), \tilde{u}_2(s), \ldots, \tilde{u}_n(s)]' \tag{1.61a}$$

$$\tilde{\mathbf{y}}(s) = [\tilde{y}_1(s), \tilde{y}_2(s), \ldots, \tilde{y}_n(s)]' \tag{1.61b}$$

$$\mathbf{H}(s) = [h_{ij}(s)] \tag{1.61c}$$

From (1.60) it is quite clear that

$$h_{ij}(s) = \frac{\tilde{y}_i(s)}{\tilde{u}_j(s)}\bigg|_{\tilde{u}_x(s)=0,\,x\neq j} \tag{1.62}$$

This indicates that $h_{ii}(s)$ represents the input immittance looking into the ith port when the excitation signals at all other ports are set to zero, and that $h_{ij}(s)$, $i \neq j$, denotes the transfer function from port j to port i when all the excitation signals except $\tilde{u}_j(s)$ are set to zero. By setting an excitation signal to zero, we mean the open-circuiting of a current source and the short-circuiting of a voltage source.

[†]Capital letters are customarily used to denote the Laplace transforms of the corresponding lowercase time-domain functions, but $\mathbf{Y}(s)$ is the standard symbol for the admittance matrix and U the U-function. Here we use the tilde to distinguish $\mathbf{y}(t)$ and $\mathbf{u}(t)$ from their transforms $\tilde{\mathbf{y}}(s)$ and $\tilde{\mathbf{u}}(s)$.

In the special situation, when all the excitation signals are currents, Eq. (1.60) becomes

$$\mathbf{V}(s) = \mathbf{Z}(s)\mathbf{I}(s) \tag{1.63}$$

where $\mathbf{V}(s)$ and $\mathbf{I}(s)$ denote the transforms of the port-voltage and the port-current vectors $\mathbf{v}(t)$ and $\mathbf{i}(t)$, as shown in Eq. (1.1). The matrix $\mathbf{Z}(s)$ is called the *open-circuit impedance matrix* or simply the *impedance matrix*, whose elements are referred to as the *open-circuit impedance parameters* or simply the *impedance parameters*. The reason is that the elements are obtained by open-circuiting the appropriate ports. Likewise, if all the excitation signals are voltages, Eq. (1.60) becomes

$$\mathbf{I}(s) = \mathbf{Y}(s)\mathbf{V}(s) \tag{1.64}$$

The matrix $\mathbf{Y}(s)$ is called the *short-circuit admittance matrix* or simply the *admittance matrix*, whose elements are referred to as the *short-circuit admittance parameters* or *admittance parameters*.

We illustrate the above results by the following example.

Example 1.4 Consider the high-frequency equivalent network of a bipolar transistor as shown in Fig. 1.17. Let $\tilde{\mathbf{u}}(s) = [I_1(s), V_2(s)]'$ and $\tilde{\mathbf{y}}(s) = [V_1(s), I_2(s)]'$, where $V_k(s)$ and $I_k(s)$ are the transforms of $v_k(t)$ and $i_k(t)$, respectively. Then, from (1.62), we have

$$h_{11}(s) = \frac{V_1(s)}{I_1(s)}\bigg|_{V_2(s)=0} = \frac{1}{G_1 + s(C_1 + C_2)} \tag{1.65a}$$

$$h_{21}(s) = \frac{I_2(s)}{I_1(s)}\bigg|_{V_2(s)=0} = \frac{g_m - sC_2}{G_1 + s(C_1 + C_2)} \tag{1.65b}$$

Figure 1.17 A high-frequency equivalent network of a bipolar transistor.

where $G_1 = 1/R_1$. They are obtained from the network of Fig. 1.17 by short-circuiting the output port (port 2). Likewise, we compute $h_{12}(s)$ and $h_{22}(s)$. The resulting hybrid matrix is given by (see Prob. 1.8)

$$\mathbf{H}(s) = \frac{1}{G_1 + s(C_1 + C_2)} \begin{bmatrix} 1 & sC_2 \\ g_m - sC_2 & q(s) \end{bmatrix} \qquad (1.66a)$$

with $G_2 = 1/R_2$, and

$$q(s) = (G_1 + sC_1)(G_2 + sC_2) + sC_2(G_2 + g_m) \qquad (1.66b)$$

To compute the impedance matrix, which is conventionally written as

$$\mathbf{Z}(s) = \begin{bmatrix} z_{11}(s) & z_{12}(s) \\ z_{21}(s) & z_{22}(s) \end{bmatrix} \qquad (1.67)$$

we let $\tilde{\mathbf{u}}(s) = [I_1(s), I_2(s)]'$ and $\tilde{\mathbf{y}}(s) = [V_1(s), V_2(s)]'$ From Eq. (1.62) we get

$$z_{11}(s) = \left. \frac{V_1(s)}{I_1(s)} \right|_{I_2(s)=0} = \frac{G_2 + sC_2}{q(s)} \qquad (1.68a)$$

$$z_{21}(s) = \left. \frac{V_2(s)}{I_1(s)} \right|_{I_2(s)=0} = \frac{sC_2 - g_m}{q(s)} \qquad (1.68b)$$

which are obtained from the network of Fig. 1.17 by open-circuiting the output port. In a similar manner, we can compute $z_{12}(s)$ and $z_{22}(s)$. The impedance matrix is given by

$$\mathbf{Z}(s) = \frac{1}{q(s)} \begin{bmatrix} G_2 + sC_2 & sC_2 \\ sC_2 - g_m & G_1 + s(C_1 + C_2) \end{bmatrix} \qquad (1.69)$$

To determine the admittance matrix $\mathbf{Y}(s)$, conventionally written as

$$Y(s) = \begin{bmatrix} y_{11}(s) & y_{12}(s) \\ y_{21}(s) & y_{22}(s) \end{bmatrix} \qquad (1.70)$$

we let $\tilde{\mathbf{u}}(s) = [V_1(s), V_2(s)]'$ and $\tilde{\mathbf{y}}(s) = [I_1(s), I_2(s)]'$. From Eq. (1.62) we have

$$y_{11}(s) = \left. \frac{I_1(s)}{V_1(s)} \right|_{v_2(s)=0} = G_1 + s(C_1 + C_2) \qquad (1.71a)$$

$$y_{21}(s) = \left. \frac{I_2(s)}{V_1(s)} \right|_{v_2(s)=0} = g_m - sC_2 \qquad (1.71b)$$

They are obtained by short-circuiting the output port of the two-port network of Fig. 1.17. In a similar fashion, we can compute $y_{12}(s)$ and $y_{22}(s)$. The resulting admittance matrix is given by

$$\mathbf{Y}(s) = \begin{bmatrix} G_1 + s(C_1 + C_2) & -sC_2 \\ g_m - sC_2 & G_2 + sC_2 \end{bmatrix} \tag{1.72}$$

Finally, suppose that we take $\tilde{\mathbf{u}}(s) = [V_2(s), -I_2(s)]'$ as the excitation and $\tilde{\mathbf{y}}(s) = [V_1(s), I_1(s)]'$ as the response. The hybrid matrix $\mathbf{H}(s)$, conventionally written as

$$\mathbf{H}(s) = \mathbf{T}(s) = \begin{bmatrix} A(s) & B(s) \\ C(s) & D(s) \end{bmatrix} \tag{1.73}$$

is called the *transmission* or *chain matrix*, whose elements are referred to as the *transmission* or *chain parameters*. They are also known as the *ABCD parameters*. Again, from Eq. (1.62) we have

$$A(s) = \left. \frac{V_1(s)}{V_2(s)} \right|_{I_2=0} = \frac{G_2 + sC_2}{sC_2 - g_m} \tag{1.74a}$$

$$C(s) = \left. \frac{I_1(s)}{V_2(s)} \right|_{I_2(s)=0} = \frac{q(s)}{sC_2 - g_m} \tag{1.74b}$$

which are obtained from the two-port by open-circuiting the output port. Likewise, we can compute $B(s)$ and $D(s)$ by short-circuiting the output port. The resulting transmission matrix is given by

$$\mathbf{T}(s) = \frac{1}{sC_2 - g_m} \begin{bmatrix} G_2 + sC_2 & 1 \\ q(s) & G_1 + s(C_1 + C_2) \end{bmatrix} \tag{1.75}$$

The above discussion indicates that there are many matrix representations for an n-port network. Depending on the applications, an n-port network can be represented by an impedance, admittance, hybrid, or transmission matrix. They are all, of course, special cases of the general hybrid matrix defined in Eq. (1.60). It is sometimes necessary to convert one set of parameters to another in order to achieve the desired result. The hybrid parameters, for example, may be specified for a transistor by the manufacturer, and we wish to use the admittance parameters in our design. Some conversions are straightforward, and others are not. Comparing Eqs. (1.63) and (1.64), for example, we see that if $\mathbf{Z}(s)$ is nonsingular, the admittance matrix $\mathbf{Y}(s)$ is simply the inverse of the impedance matrix and vice versa. In the following, we present a procedure that permits a simple conversion from one set of parameters to another.

Express Eq. (1.60) in a slightly different form as

$$[-\mathbf{1}_n, \mathbf{H}(s)]\begin{bmatrix} \tilde{\mathbf{y}}(s) \\ \tilde{\mathbf{u}}(s) \end{bmatrix} = \mathbf{0} \qquad (1.76)$$

where $\mathbf{1}_n$ denotes the identity matrix of order n. Observe the pattern of the coefficient matrix. The vector $[\tilde{\mathbf{y}}(s), \tilde{\mathbf{u}}(s)]'$ contains all the $2n$ port variables of the n-port under consideration. Suppose that we premultiply the coefficient matrix of Eq. (1.76) by the negative of the inverse of the submatrix consisting of the columns corresponding to the variables chosen as the new response signals. Then the submatrix formed by the other columns in the resulting matrix product is the general hybrid matrix with respect to the new excitation signals. We illustrate this procedure by the following example.

Example 1.5 Consider the same equivalent network of a bipolar transistor of Fig. 1.17. Suppose that the admittance matrix (1.72) is known, and we wish to determine the hybrid matrix (1.66a). To this end, we write (1.72) as

$$\begin{bmatrix} -1 & 0 & G_1 + s(C_1 + C_2) & -sC_2 \\ 0 & -1 & g_m - sC_2 & G_2 + sC_2 \end{bmatrix} \begin{bmatrix} I_1(s) \\ I_2(s) \\ V_1(s) \\ V_2(s) \end{bmatrix} = \mathbf{0} \qquad (1.77)$$

Since the new response signals are $I_2(s)$ and $V_1(s)$, we premultiply the coefficient matrix of (1.77) by the negative of the inverse of the submatrix consisting of second and third columns, which is

$$-\begin{bmatrix} 0 & G_1 + s(C_1 + C_2) \\ -1 & g_m - sC_2 \end{bmatrix}^{-1} = \frac{1}{G_1 + s(C_1 + C_2)}$$

$$\times \begin{bmatrix} sC_2 - g_m & G_1 + s(C_1 + C_2) \\ -1 & 0 \end{bmatrix} \qquad (1.78)$$

We obtain the new coefficient matrix as

$$\frac{1}{G_1 + s(C_1 + C_2)}$$

$$\times \begin{bmatrix} g_m - sC_2 & -G_1 - s(C_1 + C_2) & 0 & q(s) \\ 1 & 0 & -G_1 - s(C_1 + C_2) & sC_2 \end{bmatrix} \qquad (1.79)$$

The submatrix formed by the first and fourth columns of (1.79) is the desired hybrid matrix relating the response vector $[I_2(s), V_1(s)]'$ to the excitation vector $[I_1(s), V_2(s)]'$. By interchanging the positions of $I_2(s)$ and $V_1(s)$ in the response vector, which corresponds to the interchange of row 1 and row 2 in the hybrid matrix, we obtain (1.66a).

To compute the transmission matrix $\mathbf{T}(s)$ of (1.75), it is convenient to rewrite (1.77) as

$$
\begin{bmatrix}
G_1 + s(C_1 + C_2) & -1 & -sC_2 & 0 \\
g_m - sC_2 & 0 & G_2 + sC_2 & 1
\end{bmatrix}
\begin{bmatrix}
V_1(s) \\
I_1(s) \\
V_2(s) \\
-I_2(s)
\end{bmatrix}
= \mathbf{0}
\qquad (1.80)
$$

Premultiplying the coefficient matrix of (1.80) by the negative of the inverse of the submatrix consisting of the first two columns, which is

$$
\frac{1}{sC_2 - g_m}
\begin{bmatrix}
0 & 1 \\
sC_2 - g_m & G_1 + s(C_1 + C_2)
\end{bmatrix}
\qquad (1.81)
$$

we obtain

$$
\frac{1}{sC_2 - g_m}
\begin{bmatrix}
g_m - sC_2 & 0 & G_2 + sC_2 & 1 \\
0 & g_m - sC_2 & q(s) & G_1 + s(C_1 + C_2)
\end{bmatrix}
\qquad (1.82)
$$

The submatrix formed by the last two columns is the desired transmission matrix.

We remark that in deriving the various representations for an n-port, we implicitly assume that the inverses of the various submatrices exist. It is clear that if any one of them is singular, the conversion is not possible, and the corresponding set of parameters does not exist. For example, the ideal transformer of Fig. 1.15 does not possess the impedance, admittance, or transmission representation except the hybrid representation of Eq. (1.35) or its inverse.

In addition to the port characterizations of a network, it is sometimes advantageous to consider a network with n accessible terminals. Terminal characterization of a network and its applications will be considered in the following chapter.

1.6   EQUIVALENT FREQUENCY-DOMAIN CONDITIONS OF PASSIVITY

The concept of passivity was introduced in Sec. 1.3 as a useful qualitative measure of the time-domain behavior of networks. As pointed out earlier, the definition, although very general, is difficult to apply. The examples we presented were either simple one-ports or two-ports. In this section, the time-domain passivity criterion is translated into frequency-domain passivity conditions for linear time-invariant n-port networks.

Let $\mathbf{y}(t)$ and $\mathbf{u}(t)$ be the response and the excitation of a linear time-invariant n-port network N. As in Secs. 1.3 and 1.4, we restrict that the elements of $\mathbf{u}(t)$ be taken one each of the two variables associated with each of the n ports, so that the product $\mathbf{u}'(t)\mathbf{y}(t)$ represents the instantaneous power entering the n-port. Specifically, we shall show that the linear time-invariant n-port network is passive if and only if its general hybrid matrix is a positive-real matrix. For this reason, we first introduce the concept of positive realness.

1.6.1 Positive-Real Matrix and Passivity

Let

$$\mathbf{A} = [a_{ij}] \tag{1.83}$$

be a matrix of order n. We write

$$\overline{\mathbf{A}} = [\overline{a}_{ij}] \tag{1.84a}$$

$$\mathbf{A}^* = \overline{\mathbf{A}}' \tag{1.84b}$$

$$\mathbf{A}_h = \frac{1}{2}(\mathbf{A} + \mathbf{A}^*) \tag{1.84c}$$

$$\mathbf{A}_s = \frac{1}{2}(\mathbf{A} + \mathbf{A}') \tag{1.84d}$$

where $\overline{a}_{ij}$ denotes the complex conjugate of a_{ij}. The matrix $\mathbf{A}_h$ is called the *hermitian part* of $\mathbf{A}$, and $\mathbf{A}_s$ the *symmetric part*, because $\mathbf{A}_h$ is hermitian and $\mathbf{A}_s$ is symmetric. The *right half* of the complex s-plane will be abbreviated as RHS. Likewise, the *left half* of the complex s-plane will be abbreviated as LHS. The open RHS is the region defined by Re $s > 0$ (Re denotes the real part of), and the closed RHS is the region defined by Re $s \geqslant 0$. When we say that $\mathbf{A}$ is analytic in a region, we mean that every element of $\mathbf{A}$ is analytic in the region. On the other hand, when we say that $\mathbf{A}$ has a pole at s_0, we mean that at least one element of $\mathbf{A}$ has a pole at s_0.

Definition 1.6: Positive-real matrix An $n \times n$ matrix function $\mathbf{A}(s)$ of the complex variable s is said to be a *positive-real matrix* if it satisfies the following three conditions:

1. $\mathbf{A}(s)$ is analytic in the open RHS.
2. $\overline{\mathbf{A}}(s) = \mathbf{A}(\overline{s})$ for all s in the open RHS.
3. Its hermitian part $\mathbf{A}_h(s)$ is nonnegative definite for all s in the open RHS.

Recall that a hermitian matrix $\mathbf{A}_h$ is said to be nonnegative definite if its associated hermitian form $\mathbf{X}^*\mathbf{A}_h\mathbf{X}$ is nonnegative; that is,

$$\mathbf{X}^*\mathbf{A}_h\mathbf{X} \geqslant 0 \tag{1.85}$$

for all complex n-vectors $\mathbf{X}$. The second condition of the definition states that each element of $\mathbf{A}(s)$ is real when s is real. If $\mathbf{A}(s)$ is a matrix of rational functions, condition 1 is redundant because it is implied by conditions 2 and 3. However, if general functions are considered, condition 1 must be included.

Definition 1.7: Positive-real function A positive-real matrix of order 1 is called a *positive-real function*.

We illustrate the above definitions by the following example.

Example 1.6 Consider the admittance matrix (1.72):

$$\mathbf{Y}(s) = \begin{bmatrix} G_1 + s(C_1 + C_2) & -sC_2 \\ g_m - sC_2 & G_2 + sC_2 \end{bmatrix} \tag{1.86}$$

Write $s = \sigma + j\omega$. Then the hermitian part of $\mathbf{Y}(s)$ is given by

$$\mathbf{Y}_h(s) = \frac{1}{2}[\mathbf{Y}(s) + \mathbf{Y}^*(s)] = \begin{bmatrix} G_1 + \sigma(C_1 + C_2) & \frac{1}{2}g_m - \sigma C_2 \\ \frac{1}{2}g_m - \sigma C_2 & G_2 + \sigma C_2 \end{bmatrix} \tag{1.87}$$

We now investigate the three conditions of Definition 1.6. Since $\mathbf{Y}(s)$ is devoid of poles in the open RHS and since $\overline{\mathbf{Y}}(s) = \mathbf{Y}(\bar{s})$, the first two conditions are satisfied. To show that $\mathbf{Y}_h(s)$ is nonnegative definite for $\sigma > 0$, we recall that a hermitian matrix is nonnegative definite if and only if all of its principal minors are nonnegative (see App. I). This is equivalent to requiring that

$$G_1 + \sigma(C_1 + C_2) \geqslant 0 \tag{1.88a}$$

$$G_2 + \sigma C_2 \geqslant 0 \tag{1.88b}$$

$$\det \mathbf{Y}_h(s) \geqslant 0 \tag{1.88c}$$

where $\sigma > 0$. The third inequality is equivalent to

$$4(G_1 + \sigma C_1)(G_2 + \sigma C_2) + 4\sigma C_2 G_2 + 4\sigma g_m C_2 - g_m^2 \geqslant 0 \quad \sigma > 0 \tag{1.89}$$

By assuming that $G_1, G_2, C_1, C_2,$ and g_m are all nonnegative and

$$4G_1 G_2 \geqslant g_m^2 \tag{1.90}$$

(1.88) and (1.89) are all satisfied, and the hermitian matrix $Y_h(s)$ is nonnegative definite for all s in the open RHS. Consequently, the admittance matrix is positive real. On the other hand, if $4G_1G_2 < g_m^2$, for a sufficiently small value of σ the inequality (1.89) can be violated, showing that $Y(s)$ is not positive real.

To use the definition to test for positive realness of a matrix is usually complicated. The first two conditions are relatively easy to check. The third constraint is somewhat difficult to verify, because the nonnegative definiteness of the hermitian part must be investigated for all s in the open RHS. In the case where a matrix is rational, each element being the ratio of two polynomials, an equivalent set of conditions can be stated that reduces the test for the points in the vast open RHS to the points on the boundary, the real-frequency axis (jw-axis). Its justification can be found in Chen (1976).

Theorem 1.3 An $n \times n$ rational matrix $A(s)$ is positive real if and only if the following four conditions are satisfied:

1. $\overline{A}(s) = A(\overline{s})$.
2. $A(s)$ has no poles in the open RHS.
3. Poles of $A(s)$ on the real-frequency axis, if they exist, are simple, and the associated residue matrix K evaluated at each of these poles is hermitian and nonnegative definite.
4. Its hermitian part $A_h(jw)$ is nonnegative definite whenever it is defined.

We remark that the associated residue matrix K evaluated at a pole is the matrix of residues of the elements of $A(s)$ evaluated at the pole. We illustrate this by the following examples.

Example 1.7 Consider the hybrid matrix (1.66a) of Fig. 1.17, as repeated below:

$$H(s) = \frac{1}{G_1 + s(C_1 + C_2)} \begin{bmatrix} 1 & sC_2 \\ g_m - sC_2 & q(s) \end{bmatrix} \qquad (1.91a)$$

where

$$q(s) = (G_1 + sC_1)(G_2 + sC_2) + sC_2(G_2 + g_m) \qquad (1.91b)$$

The matrix is analytic in the closed RHS because the only singularity is at the pole $s = -\sigma_0 = -G_1/(C_1 + C_2)$. Thus, conditions 1, 2, and 3 are all satisfied. To test condition 4, we first compute the real-frequency hermitian part

$\mathbf{H}_h(j\omega)$, yielding

$$\mathbf{H}_h(j\omega) = \frac{1}{(C_1 + C_2)(\omega^2 + \sigma_0^2)}$$

$$\times \begin{bmatrix} \sigma_0 & \frac{1}{2}g_m\sigma_0 + j\omega\left(C_2\sigma_0 + \frac{1}{2}g_m\right) \\ \frac{1}{2}g_m\sigma_0 - j\omega\left(C_2\sigma_0 + \frac{1}{2}g_m\right) & p(\omega) \end{bmatrix}$$

$$(1.92a)$$

where

$$p(\omega) = \omega^2(C_1G_2 + C_2G_2 + G_1C_2 - C_1C_2\sigma_0 + C_2g_m) + G_1G_2\sigma_0 \qquad (1.92b)$$

For $\mathbf{H}_h(j\omega)$ to be nonnegative definite, it is necessary and sufficient that all of its principal minors are nonnegative (App. I). This is equivalent to requiring that for real frequencies ω,

$$\sigma_0 \geqslant 0 \qquad (1.93a)$$

$$p(\omega) \geqslant 0 \qquad (1.93b)$$

$$\det \mathbf{H}_h(j\omega) \geqslant 0 \qquad (1.93c)$$

By assuming that G_1, G_2, C_1, C_2, and g_m are all nonnegative, Eqs. (1.93a) and (1.93b) are clearly satisfied, because $G_1C_2 - C_1C_2\sigma_0 \geqslant 0$. The third inequality is equivalent to

$$(\sigma_0^2 + \omega^2)(4G_1G_2 - g_m^2) \geqslant 0 \qquad (1.94)$$

As in Example 1.6, if $4G_1G_2 \geqslant g_m^2$, Eq. (1.94) is satisfied and $\mathbf{H}_h(j\omega)$ is nonnegative definite. Consequently, the hybrid matrix $\mathbf{H}(s)$ is positive real. If $4G_1G_2 < g_m^2$, Eq. (1.94) can be violated for any value of ω on the real-frequency axis, indicating that $\mathbf{H}_h(j\omega)$ is not nonnegative definite and $\mathbf{H}(s)$ is not positive real. To appreciate the significance of Theorem 1.3, it is suggested that the reader use the definition to verify the above results.

Example 1.8 We use Theorem 1.3 to verify the results obtained in Example 1.6. The first two conditions are clearly satisfied by the matrix $\mathbf{Y}(s)$ of Eq. (1.86). To verify condition 3, we compute the associated residue matrix. The admittance matrix $\mathbf{Y}(s)$ has a pole at the infinity, which is customarily considered to be on the real-frequency axis. The associated residue matrix evaluated at this pole is given by

$$\mathbf{K} = \begin{bmatrix} C_1 + C_2 & -C_2 \\ -C_2 & C_2 \end{bmatrix} \tag{1.95}$$

which is hermitian and nonnegative definite. Thus, condition 3 is satisfied. For condition 4, we compute the real-frequency hermitian part of $\mathbf{Y}(s)$, giving

$$\mathbf{Y}_h(j\omega) = \begin{bmatrix} G_1 & \dfrac{1}{2}g_m \\ \dfrac{1}{2}g_m & G_2 \end{bmatrix} \tag{1.96}$$

which is nonnegative definite if and only if $4G_1G_2 \geqslant g_m^2$, confirming Eq. (1.90). Thus, the admittance matrix $\mathbf{Y}(s)$ is positive real if and only if $4G_1G_2 \geqslant g_m^2$.

With these preliminaries, we are now in a position to state the equivalent frequency-domain conditions of passivity. In order to exhibit the result succinctly, we relegate the proof to the next section.

Theorem 1.4 A linear, time-invariant n-port network possessing a general hybrid matrix,[†] which is analytic in the open RHS, is passive if and only if the general hybrid matrix is positive real.

The theorem does not require any specific hybrid matrix; any characterization will do. The only requirement is that the product $\mathbf{u}'(t)\mathbf{y}(t)$ of the associated excitation $\mathbf{u}(t)$ and response $\mathbf{y}(t)$ denote the instantaneous power entering the n-port network under consideration. Thus, the impedance matrix, the admittance matrix, and the hybrid matrix discussed in Example 1.4 are all permitted, but the transmission matrix is not. Although all the permissible characterizations would eventually result in the same conclusion, the amount of computation involved may be quite different. The availability of so many different characterizations permits one to choose a particular approach or a combination of approaches, so that the problem at hand can be solved in the simplest and most satisfying manner.

Consider, again, the small-signal high-frequency equivalent network of a bipolar transistor of Fig. 1.17. Three permissible matrix characterizations were derived in Example 1.4 and are given in Eqs. (1.66a), (1.69), and (1.72). To determine the conditions under which the transistor is passive, we need only derive conditions under which the general hybrid matrix is positive real. In Examples 1.7

[†]The general hybrid matrix is restricted to the one that is defined by an excitation vector whose elements are taken one each of the two variables associated with each of the n ports.

and 1.8, we have shown that the hybrid matrix (1.66a) or the admittance matrix (1.72) is positive real if and only if $4G_1G_2 \geqslant g_m^2$. Thus, according to Theorem 1.4, the transistor is passive if and only if $4G_1G_2 \geqslant g_m^2$. Since activity is the formal negation of passivity, the transistor is active if and only if $4G_1G_2 < g_m^2$. The same conclusion can be reached from the impedance matrix (1.69), but the amount of work involved is considerable (see Prob. 1.10). Thus, to test for passivity or activity of a device, the choice of an appropriate matrix is as important as the test itself. The reader is urged to compare the amount of labor involved in the above three cases.

1.6.2 Outline of a proof of Theorem 1.4

Let $\mathbf{y}(t)$ be the response of an excitation $\mathbf{u}(t)$ of an n-port network N. We first demonstrate that the n-port is passive if and only if

$$\varepsilon = \varepsilon(t_0) + \int_{t_0}^{\infty} \mathbf{u}'(x)\mathbf{y}(x)dx \geqslant 0 \qquad (1.97)$$

for all initial time t_0 and for all excitations $\mathbf{u}(t)$.

The condition is clearly necessary, because Eq. (1.10) must be true for all time t and, in particular, for $t = \infty$, and because $\mathbf{u}'(t)\mathbf{y}(t) = \mathbf{v}'(t)\mathbf{i}(t)$. To show that Eq. (1.97) also implies Eq. (1.10), we assume that $\varepsilon(t)$ of (1.10) is negative for some $t = t_1$. We can now simulate this condition in (1.97) by setting

$$\mathbf{u}(t) = \mathbf{v}(t) \quad t < t_1 \qquad (1.98a)$$
$$= \mathbf{0} \qquad t \geqslant t_1 \qquad (1.98b)$$

Then ε of (1.97) will also be negative, contradicting our hypothesis. Thus, $\varepsilon(t)$ of (1.10) must be nonnegative for all $t \geqslant t_0$.

Without loss of generality, we assume that the n-port under consideration is initially relaxed. The passivity condition becomes

$$\varepsilon = \int_{-\infty}^{\infty} \mathbf{u}'(x)\mathbf{y}(x)dx \geqslant 0 \qquad (1.99)$$

for all excitations $\mathbf{u}(t)$.

Let $\tilde{\mathbf{y}}(s)$ and $\tilde{\mathbf{u}}(s)$ be the Laplace transforms of $\mathbf{y}(t)$ and $\mathbf{u}(t)$, respectively. Then $\tilde{\mathbf{y}}(s)$ and $\tilde{\mathbf{u}}(s)$ are related by

$$\tilde{\mathbf{y}}(s) = \mathbf{H}(s)\tilde{\mathbf{u}}(s) \qquad (1.100)$$

The time function $\mathbf{y}(t)$ may be expressed as the convolution product

$$\mathbf{y}(t) = \int_0^t \mathbf{h}(t - \lambda)\mathbf{u}(\lambda)d\lambda \qquad (1.101a)$$

where $\mathbf{h}(t)$ represents the response resulting from the unit impulse excitation; that is,

$$\mathbf{h}(t) = \mathscr{L}^{-1}\mathbf{H}(s)\mathbf{g} \qquad (1.101b)$$

where $\mathbf{g}$ is a vector comprised only of 1's. Consider the function

$$e(w) = \int_{-\infty}^{\infty} \mathbf{u}'(x - w)y(x)dx \qquad (1.102)$$

and observe that

$$e(0) = \varepsilon \qquad (1.103)$$

the total energy stored in the n-port. On the real-frequency axis, Eq. (1.100) becomes

$$\tilde{\mathbf{y}}(jw) = \mathbf{H}(jw)\tilde{\mathbf{u}}(jw) \qquad (1.104)$$

Consider $\tilde{\mathbf{y}}(jw)$, $\tilde{\mathbf{u}}(jw)$, and $\mathbf{H}(jw)$ as the Fourier transforms of $\mathbf{y}(t)$, $\mathbf{u}(t)$, and $\mathbf{h}(t)$, respectively. Then through the convolution integral, the time-domain response can be written as

$$\mathbf{y}(t) = \int_{-\infty}^{\infty} \mathbf{h}(t - \lambda)\mathbf{u}(\lambda)d\lambda \qquad (1.105)$$

Substituting (1.105) in (1.102) yields

$$e(w) = \int_{-\infty}^{\infty}\int_{-\infty}^{\infty} \mathbf{u}'(x - w)\mathbf{h}(x - \lambda)\mathbf{u}(\lambda)d\lambda\,dx \qquad (1.106)$$

which is recognized as the convolution of the functions of the forms $\mathbf{u}'(-t)$, $\mathbf{h}(t)$, and $\mathbf{u}(t)$, whose Fourier transforms are $\tilde{\mathbf{u}}'(-j\omega)$, $\mathbf{H}(j\omega)$, and $\tilde{\mathbf{u}}(j\omega)$. Thus, the Fourier transform on both sides of Eq. (1.106) is given by

$$\mathcal{F}e(w) = \tilde{\mathbf{u}}^*(j\omega)\mathbf{H}(j\omega)\tilde{\mathbf{u}}(j\omega) \qquad (1.107)$$

where $\tilde{\mathbf{u}}'(-j\omega) = \tilde{\mathbf{u}}'(\overline{j\omega}) = \overline{\tilde{\mathbf{u}}'}(j\omega) = \tilde{\mathbf{u}}^*(j\omega)$, so that

$$e(w) = \frac{1}{2\pi}\int_{-\infty}^{\infty} \tilde{\mathbf{u}}^*(j\omega)\mathbf{H}(j\omega)\tilde{\mathbf{u}}(j\omega)e^{jw\omega}d\omega \qquad (1.108)$$

Finally, from Eq. (1.103) we see that the total energy ε stored in the n-port can now be expressed in terms of the frequency-domain excitation $\tilde{\mathbf{u}}(j\omega)$ and the characterization $\mathbf{H}(j\omega)$ by setting $w = 0$ in (1.108), yielding

$$\varepsilon = \frac{1}{2\pi} \int_{-\infty}^{\infty} \tilde{\mathbf{u}}^*(j\omega)\mathbf{H}(j\omega)\tilde{\mathbf{u}}(j\omega)d\omega \qquad (1.109)$$

Necessity We show that if the n-port is passive, then the general hybrid matrix $\mathbf{H}(s)$ must be positive real. To this end, we apply a particular excitation of the form

$$\mathbf{u}(t) = [u_1(t), u_2(t), \ldots, u_n(t)]' \qquad (1.110)$$

where

$$u_k(t) = 0 \qquad t < 0 \qquad (1.111a)$$

$$= c_k\sigma_0^{1/2}e^{-\sigma_0 t}\cos(\omega_0 t - \phi_k - \phi_0) \qquad t \gg 0 \qquad (1.111b)$$

with $k = 1, 2, \ldots, n$; where $s_0 = \sigma_0 + j\omega_0$ is an arbitrary point in the open RHS, and c_k, ϕ_k, and ϕ_0 are arbitrary real constants. Note that each excitation signal has the same associated σ_0, ω_0, and ϕ_0, but different c_k and ϕ_k. The Fourier transform $\tilde{u}_k(j\omega)$ of the signal $u_k(t)$ is given by

$$\tilde{u}_k(j\omega) = c_k\sigma_0^{1/2}\frac{(\sigma_0 + j\omega)\cos(\phi_k + \phi_0) + \omega_0\sin(\phi_k + \phi_0)}{(\sigma_0 + j\omega)^2 + \omega_0^2} \qquad (1.112)$$

Consider the product $\bar{\tilde{u}}_i(j\omega)\tilde{u}_k(j\omega)$, which can be simplified to

$$\bar{\tilde{u}}_i(j\omega)\tilde{u}_k(j\omega) = \frac{c_i c_k \sigma_0 A_1}{[(\sigma_0 + j\omega)^2 + \omega_0^2][(\sigma_0 - j\omega)^2 + \omega_0^2]} \qquad (1.113a)$$

where

$$A_1 = (\omega^2 + \sigma_0^2)\cos(\phi_i + \phi_0)\cos(\phi_k + \phi_0) + \omega_0^2\sin(\phi_i + \phi_0)\sin(\phi_k + \phi_0)$$
$$+ \omega_0\sigma_0\sin(\phi_i + \phi_k + 2\phi_0) + j\omega_0\omega\sin(\phi_i - \phi_k) \qquad (1.113b)$$

For our purposes, we express Eq. (1.113a) as

$$\bar{\tilde{u}}_i(j\omega)\tilde{u}_k(j\omega) = \frac{1}{4}c_i c_k(A_2 + A_3 + A_4 + \overline{A}_4) \qquad (1.114a)$$

where

$$A_2 = \frac{\sigma_0 e^{j(\phi_i - \phi_k)}}{\sigma_0^2 + (\omega - \omega_0)^2} \qquad (1.114b)$$

$$A_3 = \frac{\sigma_0 e^{-j(\phi_i - \phi_k)}}{\sigma_0^2 + (\omega + \omega_0)^2} \qquad (1.114c)$$

$$A_4 = \frac{\sigma_0 e^{j(\phi_i + \phi_k + 2\phi_0)}}{[\sigma_0 - j(\omega - \omega_0)][\sigma_0 + j(\omega + \omega_0)]} \qquad (1.114d)$$

The validity of (1.114a) can be verified by combining the terms on the right-hand side of (1.114a) and comparing the results with (1.113a). Expressing $\mathbf{H}(j\omega)$ explicitly as in (1.61c) and substituting (1.110) in (1.109) yield

$$\varepsilon = \frac{1}{2\pi} \int_{-\infty}^{\infty} \sum_{i=1}^{n} \sum_{k=1}^{n} \bar{u}_i(j\omega) h_{ik}(j\omega) \tilde{u}_k(j\omega) d\omega$$

$$= \frac{1}{8\pi} \sum_{i=1}^{n} \sum_{k=1}^{n} c_i c_k \left[\int_{-\infty}^{\infty} h_{ik}(j\omega) A_2 d\omega + \int_{-\infty}^{\infty} h_{ik}(j\omega) A_3 d\omega \right.$$

$$\left. + \int_{-\infty}^{\infty} h_{ik}(j\omega)(A_4 + \overline{A}_4) d\omega \right] \qquad (1.115)$$

In the theory of functions of a complex variable there is a well-known integral called *Poisson's integral* [see, for example, Guillemin (1969)], which states that

$$F(s) = F(\sigma + j\omega) = \frac{1}{\pi} \int_{-\infty}^{\infty} \frac{\sigma F(jx)}{\sigma^2 + (\omega - x)^2} dx \qquad (1.116)$$

is valid for any s in the closed RHS. Applying this integral to the first two integrals on the right-hand side of (1.115) gives

$$\varepsilon = \frac{1}{4}(\varepsilon_1 + \varepsilon_2) \qquad (1.117a)$$

where

$$\varepsilon_1 = \frac{1}{2} \sum_{i=1}^{n} \sum_{k=1}^{n} c_i c_k [h_{ik}(\sigma_0 + j\omega_0) e^{j(\phi_i - \phi_k)} + h_{ik}(\sigma_0 - j\omega_0) e^{-j(\phi_i - \phi_k)}]$$

$$= \mathrm{Re} \sum_{i=1}^{n} \sum_{k=1}^{n} (c_i e^{j\phi_i}) h_{ik}(\sigma_0 + j\omega_0)(c_k e^{-j\phi_k}) = \mathrm{Re}\mathbf{X}^*\mathbf{H}(s_0)\mathbf{X} \qquad (1.117b)$$

in which, as before, $s_0 = \sigma_0 + j\omega_0$, and

$$\mathbf{X} = [c_1 e^{-j\phi_1}, c_2 e^{-j\phi_2}, \ldots, c_n e^{-j\phi_n}]' \qquad (1.117c)$$

$$\varepsilon_2 = \frac{1}{2\pi} \sum_{i=1}^{n} \sum_{k=1}^{n} c_i c_k \int_{-\infty}^{\infty} h_{ik}(j\omega)(A_4 + \overline{A}_4) d\omega \qquad (1.117d)$$

Since the n-port network is passive, ε is nonnegative, showing that

$$\varepsilon_1 + \varepsilon_2 \geqslant 0 \qquad (1.118)$$

In a similar manner, if we apply an excitation of the type (1.111) with $\phi_0 + \frac{1}{2}\pi$ replacing ϕ_0 in (1.111b), everything else being the same, we arrive at an expression similar to (1.117a) for the total energy stored in the n-port, which again must be nonnegative. The result is given by

$$\varepsilon_1 - \varepsilon_2 \geqslant 0 \qquad (1.119)$$

The reason for this is that, by replacing ϕ_0 by $\phi_0 + \frac{1}{2}\pi$, Eqs. (1.114b) and (1.114c) remain the same and (1.114d) becomes $-A_4$. Thus, (1.117b) remains the same as before but (1.117d) is changed to $-\varepsilon_2$. Adding (1.118) and (1.119) shows that, for an arbitrary point $s_0 = \sigma_0 + j\omega_0$ in the open RHS,

$$2\varepsilon_1 = 2\text{Re } \mathbf{X}^*\mathbf{H}(s_0)\mathbf{X} = \mathbf{X}^*\mathbf{H}(s_0)\mathbf{X} + \mathbf{X}^*\mathbf{H}^*(s_0)\mathbf{X}$$

$$= 2\mathbf{X}^*\mathbf{H}_h(s_0)\mathbf{X} \geqslant 0 \qquad (1.120)$$

for all complex $\mathbf{X}$, because c_k and ϕ_k of (1.117c) are arbitrary. This, together with the fact that $\mathbf{H}(\bar{s}) = \overline{\mathbf{H}}(s)$ and that $\mathbf{H}(s)$ is analytic in the open RHS, shows that $\mathbf{H}(s)$ is positive real.

Sufficiency To show that positive realness is also sufficient, we write (1.109) as

$$\varepsilon = \frac{1}{2\pi} \int_{-\infty}^{\infty} [\text{Re } \tilde{\mathbf{u}}^*(j\omega)\mathbf{H}(j\omega)\tilde{\mathbf{u}}(j\omega) + j\text{Im } \tilde{\mathbf{u}}^*(j\omega)\mathbf{H}(j\omega)\tilde{\mathbf{u}}(j\omega)]d\omega$$

$$= \frac{1}{2\pi} \int_{-\infty}^{\infty} \text{Re } \tilde{\mathbf{u}}^*(j\omega)\mathbf{H}(j\omega)\tilde{\mathbf{u}}(j\omega)d\omega \qquad (1.121)$$

where Im denotes the *imaginary part of*. The second integral of the first line is zero because Im $\tilde{\mathbf{u}}^*(j\omega)\mathbf{H}(j\omega)\tilde{\mathbf{u}}(j\omega)$ is an odd function of ω. Since $\mathbf{H}(s)$ is positive real, its hermitian part $\mathbf{H}_h(s)$ must be nonnegative definite for all s in the open RHS, or

$$\mathbf{X}^*\mathbf{H}_h(s)\mathbf{X} = \text{Re } \mathbf{X}^*\mathbf{H}(s)\mathbf{X} \geqslant 0 \qquad \sigma = \text{Re } s > 0 \qquad (1.122)$$

for all complex $\mathbf{X}$. In the limit as σ approaches to zero, (1.122) becomes

$$\text{Re } \mathbf{X}^*\mathbf{H}(j\omega)\mathbf{X} \geqslant 0 \qquad (1.123)$$

for all complex $\mathbf{X}$ and for all ω whenever $\mathbf{H}(j\omega)$ is defined. This shows that the integrand of (1.121) is nonnegative, and ε must be nonnegative. Thus, the n-port network is passive. This completes the proof of the theorem.

1.7 DISCRETE-FREQUENCY CONCEPTS OF PASSIVITY AND ACTIVITY

In the preceding section we showed that for a linear, time-invariant n-port network, the time-domain definition of passivity can be translated into the equivalent frequency-domain condition that states that the general hybrid matrix of the n-port must be positive real. In the present section we extend these results by introducing the discrete-frequency concepts of passivity and activity. These concepts are very useful for studying the behavior and limitations of active networks.

In showing that positive realness is a necessary condition for an n-port to be passive, we apply a particular signal of the type (1.111). Thus, the logical set of signals with which to connect directly the time and frequency domains are those of the form

$$u_k(t) = c_k e^{\sigma_0 t} \cos(\omega_0 t - \phi_k) = \mathrm{Re}\ (\hat{c}_k e^{s_0 t}) \tag{1.124}$$

where

$$s_0 = \sigma_0 + j\omega_0 \qquad \sigma_0 \geqslant 0 \tag{1.125a}$$

$$\hat{c}_k = c_k e^{-j\phi_k} \tag{1.125b}$$

and c_k and ϕ_k are arbitrary real constants. To avoid the use of Re, which is not distributive over the product, we express (1.124) as

$$u_k(t) = \frac{1}{2}(\hat{c}_k e^{s_0 t} + \overline{\hat{c}}_k e^{\overline{s}_0 t}) \tag{1.126}$$

with each of these time-domain complex exponential signals corresponding to a simple pole of the signal transform at the point s_0 or $\overline{s}_0$ in the complex frequency s-plane. This class of signals is the most significant in the study of linear time-invariant networks.

Let $\mathbf{y}(t)$ and $\mathbf{u}(t)$ be the response and the excitation of an n-port. Assume that each component of $\mathbf{u}(t)$ is of the form (1.124) or (1.126). Then we can write

$$\mathbf{u}(t) = \frac{1}{2}(\mathbf{u}_0 e^{s_0 t} + \overline{\mathbf{u}}_0 e^{\overline{s}_0 t}) \tag{1.127}$$

where $\mathbf{u}_0$ is an arbitrary complex but fixed n-vector. Provided that we set up the appropriate initial conditions, the response of the n-port is given by (see Prob. 1.44

for an illustration)

$$\mathbf{y}(t) = \frac{1}{2}[\mathbf{H}(s_0)\mathbf{u}_0 e^{s_0 t} + \overline{\mathbf{H}}(s_0)\overline{\mathbf{u}}_0 e^{\bar{s}_0 t}] \tag{1.128}$$

where $\mathbf{H}(s)$ is the general hybrid matrix of the n-port as defined in (1.60). By substituting (1.127) and (1.128) in (1.10), the total energy stored in the n-port at any time t can be determined, and for $\sigma \neq 0$, with $\measuredangle$ denoting *the angle of*, we have

$$
\begin{aligned}
\varepsilon(t) &= \varepsilon(t_0) + \int_{t_0}^{t} \mathbf{u}'(x)\mathbf{y}(x)\, dx \\
&= \varepsilon(t_0) + \frac{1}{2}\mathrm{Re} \int_{t_0}^{t} [\mathbf{u}_0^* \mathbf{H}(s_0)\mathbf{u}_0 e^{2\sigma_0 x} + \mathbf{u}_0' \mathbf{H}(s_0)\mathbf{u}_0 e^{2s_0 x}]\, dx \\
&= \varepsilon(t_0) + \frac{1}{4}e^{2\sigma_0 t} \left\{ \frac{1}{\sigma_0}\, \mathrm{Re}\, \mathbf{u}_0^* \mathbf{H}(s_0)\mathbf{u}_0 + \frac{1}{|s_0|}|\mathbf{u}_0' \mathbf{H}(s_0)\mathbf{u}_0| \right. \\
&\quad \times \left. \cos\left[2\omega_0 t + \measuredangle \mathbf{u}_0' \mathbf{H}(s_0)\mathbf{u}_0 - \measuredangle s_0\right] \right\} + C_0 \geqslant 0
\end{aligned}
$$

$$\tag{1.129}$$

where the constant C_0 is introduced so that for $t = t_0$, $\varepsilon(t) = \varepsilon(t_0)$. Since the terms containing $\exp(2\sigma_0 t)$ grow exponentially without bound as t is increased, ultimately dominating the other terms, for (1.129) to hold for all $t \geqslant t_0$, it is necessary that

$$\mathrm{Re}\, \mathbf{u}_0^* \mathbf{H}(s_0)\mathbf{u}_0 + \frac{\sigma_0}{|s_0|}|\mathbf{u}_0' \mathbf{H}(s_0)\mathbf{u}_0| \cos\left[2\omega_0 t + \measuredangle \mathbf{u}_0' \mathbf{H}(s_0)\mathbf{u}_0 - \measuredangle s_0\right] \geqslant 0 \tag{1.130}$$

for sufficiently large t. Even so, if $\omega_0 \neq 0$, the cosine function can assume the value -1 for some t, necessitating that

$$\mathrm{Re}\, \mathbf{u}_0^* \mathbf{H}(s_0)\mathbf{u}_0 - \frac{\sigma_0}{|s_0|}|\mathbf{u}_0' \mathbf{H}(s_0)\mathbf{u}_0| \geqslant 0 \tag{1.131}$$

For $\omega_0 = 0$, Eq. (1.130) becomes $\mathrm{Re}\, [\mathbf{u}_0^* \mathbf{H}(\sigma_0)\mathbf{u}_0 + \mathbf{u}_0' \mathbf{H}(\sigma_0)\mathbf{u}_0]$. Writing $\mathbf{u}_0 = \mathbf{u}_1 + j\mathbf{u}_2$ with $\mathbf{u}_1$ and $\mathbf{u}_2$ real, the inequality becomes $\mathbf{u}_1' \mathbf{H}(\sigma_0)\mathbf{u}_1 \geqslant 0$. But, since the imaginary part $\mathbf{u}_2$ of $\mathbf{u}_0$, being the real part of some other complex n-vector, must also satisfy this constraint, that is, $\mathbf{u}_2' \mathbf{H}(\sigma_0)\mathbf{u}_2 \geqslant 0$, we conclude that $\mathrm{Re}\, \mathbf{u}_0^* \mathbf{H}(\sigma_0)\mathbf{u}_0 \geqslant 0$, which is implied by (1.131) (see Prob. 1.14). In the situation where $\sigma = 0$, a similar inequality can be derived. The only difference is that in

(1.129) we replace $e^{2\sigma_0 t}/\sigma_0$ by t. In fact, we can show that (1.131) is valid for all $s_0 \geqslant 0, \omega_0 \neq 0$. The details are left as an exercise (see Prob. 1.15). Hence, we obtain the necessary condition (1.131) for $\omega_0 \neq 0$ and Re $\mathbf{u}_0^* \mathbf{H}(s_0)\mathbf{u}_0 \geqslant 0$ for $\omega_0 = 0$.

We remark that for $s_0 = \sigma_0$ there is no need to take $\mathbf{u}_0$ complex, because any excitation of the type (1.127) with $\mathbf{u}_0$ complex corresponds to an excitation $\hat{\mathbf{u}}\exp(\sigma_0 t)$ with $\hat{\mathbf{u}}_0$ real, where $\hat{\mathbf{u}}_0 = $ Re $\mathbf{u}_0$. Also, when we state that $\mathbf{u}(t)$ of (1.127) is the excitation and $\mathbf{y}(t)$ of (1.128) is the response, we implicitly assume that we have already set up appropriate initial conditions inside the n-port network such that when the excitation $\mathbf{u}(t)$ is applied at the time $t = t_0$, *no transient occurs* at t_0 and thereafter, and the voltages and currents appearing at the ports behave exactly as specified. The amount of energy required to set up the appropriate initial conditions is represented by $\varepsilon(t_0)$ in (1.129). These initial conditions need never be evaluated in this study; it is sufficient to know that they exist.

The condition (1.131) will now be employed in characterizing the single-frequency behavior of a linear, time-invariant n-port network.

Definition 1.8: Passivity at a complex frequency A linear time-invariant n-port network possessing a general hybrid matrix[†] $\mathbf{H}(s)$ is said to be *passive at a complex frequency* s_0 in the closed RHS if, for all finite nonzero complex n-vectors $\mathbf{u}_0$,

$$\mathbf{u}_0^* \mathbf{H}_h(s_0)\mathbf{u}_0 - \frac{\sigma_0}{|s_0|}|\mathbf{u}_0' \mathbf{H}_s(s_0)\mathbf{u}_0| \geqslant 0 \tag{1.132}$$

where $\mathbf{H}_h(s)$ and $\mathbf{H}_s(s)$ denote the hermitian part and the symmetric part of $\mathbf{H}(s)$, respectively, and $s_0 = \sigma_0 + j\omega_0$. The second term on the left-hand side of (1.132) is defined to be zero for $s_0 = 0$. For $s_0 = \sigma_0$, Eq. (1.132) is to be taken for all finite nonzero real n-vectors $\mathbf{u}_0$.

The statement of the definition of passivity at a single point s_0 is not really meaningful for real networks, because passivity must always occur at the complex conjugate pair of the complex frequencies s_0 and $\bar{s}_0$. To avoid unnecessary complications and cumbersome notation, we shall be satisfied with the above definition with the understanding that it must always occur at the complex conjugate pair.

Definition 1.9: Activity at a complex frequency A linear, time-invariant n-port network possessing a general hybrid matrix is said to be *active at a complex frequency* s_0 in the closed RHS if it is not passive at s_0.

[†] See the footnote on page 33.

The condition (1.132) is identical to (1.131) on recognition of the relations

$$\text{Re } \mathbf{u}_0^* \mathbf{H}(s_0)\mathbf{u}_0 = \mathbf{u}_0^* \mathbf{H}_h(s_0)\mathbf{u}_0 \qquad (1.133a)$$

$$\mathbf{u}_0' \mathbf{H}(s_0)\mathbf{u}_0 = \mathbf{u}_0' \mathbf{H}_s(s_0)\mathbf{u}_0 \qquad (1.133b)$$

To test activity for an n-port at s_0, it is sufficient to demonstrate that

$$\mathbf{u}_0^* \mathbf{H}_h(s_0)\mathbf{u}_0 - \frac{\sigma_0}{|s_0|}|\mathbf{u}_0' \mathbf{H}_s(s_0)\mathbf{u}_0| < 0 \qquad (1.134)$$

for *some* finite complex n-vector $\mathbf{u}_0$.

We emphasize the difference of the concepts that an n-port is passive and that an n-port is passive at a closed RHS point s_0. An n-port is passive if and only if its general hybrid matrix $\mathbf{H}(s)$ is positive real, implying that the hermitian matrix $\mathbf{H}_h(s)$ is nonnegative definite for all s in the open RHS; that is,

$$\mathbf{u}_0^* \mathbf{H}_h(s)\mathbf{u}_0 \geqslant 0 \quad \sigma = \text{Re } s > 0 \qquad (1.135)$$

for all complex n-vectors $\mathbf{u}_0$. An n-port is passive at a single complex frequency s_0 in the closed RHS if

$$\mathbf{u}_0^* \mathbf{H}_h(s_0)\mathbf{u}_0 \geqslant \frac{\sigma_0}{|s_0|}|\mathbf{u}_0' \mathbf{H}_s(s_0)\mathbf{u}_0| \qquad (1.136)$$

for all complex n-vectors $\mathbf{u}_0$. As demonstrated in Eq. (1.131), Eq. (1.136) is a consequence of positive realness of $\mathbf{H}(s)$. However, the converse is not necessarily true. An n-port may be passive at s_0 but it may not be a passive n-port in the sense of Definition 1.3. Definition 1.3 is applicable to any network and for any excitation, but Definition 1.8 is valid only for linear time-invariant n-ports and for the excitations of the type (1.127).

The definitions of passivity and activity at a single complex frequency s_0 in the closed RHS suggest that for any given n-port, the closed right half of the complex frequency s-plane can be partitioned into regions of passivity and activity. The *active region* is the set of points in the closed RHS at which the n-port is active, and the *passive region* is the set of points in closed RHS at which the n-port is passive. These regions are clearly mutually exclusive. We know a priori, then, that the device characterized by the n-port may be used as an amplifier only for signals with frequencies in the active region. In fact, if an n-port is passive at the complex frequency s_0, then the n-port may be replaced by a passive n-port that exhibits identical behavior with respect to all real signals of the form (1.127) comprised of an arbitrary combination of two complex exponential signals at the frequencies s_0 and $\bar{s}_0$. The realization generally involves a $2n$-port transformer and a nonreciprocal Darlington type-E section. We shall not discuss this general aspect of the subject

Figure 1.18 The equivalent network of a tunnel diode biased in the linear negative-resistance region.

any further, because it would take us far afield into network synthesis. As a result, the n-port may neither amplify nor oscillate at this complex frequency.

1.7.1 Single Complex Frequency Passivity and Activity of One-Port Networks

For one-port networks, the condition (1.132) for passivity at a single complex frequency can be greatly simplified. It is more usefully stated as

$$\text{Re } H(s_0) - \frac{\sigma_0}{|s_0|}|H(s_0)| \geqslant 0 \qquad (1.137)$$

for a one-port characterized by its immittance $H(s)$ to be passive at a closed RHS point s_0. On the σ-axis, Eq. (1.137) becomes

$$H(\sigma_0) \geqslant 0 \quad \text{for } \sigma_0 \geqslant 0 \qquad (1.138)$$

For the remainder of the closed RHS, we can use the equivalent condition

$$\left|\frac{\text{Im } H(s_0)}{\text{Re } H(s_0)}\right| \leqslant \left|\frac{\omega_0}{\sigma_0}\right| \qquad (1.139)$$

to determine the region of passivity.

We illustrate the above results by the following example.

Example 1.9 Figure 1.18 is the equivalent network of a tunnel diode biased in the linear negative-resistance region. The input impedance of the device is

given by

$$z(s) = R_1 + sL + \frac{R}{RCs - 1} \qquad (1.140)$$

whose real and imaginary parts are computed as

$$\text{Re } z(s) = R_1 + \sigma L + \frac{R(RC\sigma - 1)}{(1 - RC\sigma)^2 + R^2 C^2 \omega^2} \qquad (1.141a)$$

$$\text{Im } z(s) = \omega L - \frac{R^2 \omega C}{(1 - RC\sigma)^2 + R^2 C^2 \omega^2} \qquad (1.141b)$$

To test passivity on the σ-axis we use Eq. (1.138), which yields

$$\frac{RLC\sigma^2 + (R_1 RC - L)\sigma + R - R_1}{RC\sigma - 1} \geqslant 0 \quad \text{for } \sigma \geqslant 0 \qquad (1.142)$$

To satisfy (1.142), the numerator and the denominator on the left-hand side of (1.142) must have the same sign. For the tunnel diode, R is usually much larger than R_1. Assume that

$$(R_1 RC - L)^2 < 4RLC(R - R_1) \qquad (1.143)$$

meaning that the numerator polynomial has no real root and therefore is always positive. Thus, on the σ-axis the device is passive for all

$$\sigma \geqslant \sigma_m = \frac{1}{RC} \qquad (1.144)$$

On the real-frequency axis, the condition (1.137) becomes

$$\text{Re } z(j\omega) = R_1 - \frac{R}{1 + R^2 C^2 \omega^2} \geqslant 0 \qquad (1.145a)$$

requiring that

$$|\omega| \geqslant \omega_m = \frac{1}{RC} \sqrt{\frac{R}{R_1} - 1} \qquad (1.145b)$$

For the remainder of the closed RHS, we apply (1.139), yielding

$$|\omega| \left| R_1 + \sigma L + \frac{R(RC\sigma - 1)}{(1 - RC\sigma)^2 + R^2 C^2 \omega^2} \right| \geqslant \left| \sigma \omega L - \frac{R^2 \omega \sigma C}{(1 - RC\sigma)^2 + R^2 C^2 \omega^2} \right| \qquad (1.146)$$

The boundaries of the passive and active regions occur at the points when (1.139) is satisfied with the equality sign. This gives

$$\omega \operatorname{Re} z(s) = \pm \sigma \operatorname{Im} z(s) \tag{1.147}$$

Choosing first the positive sign results in the equation

$$R_1 = \frac{R - 2R^2 C \sigma}{(1 - RC\sigma)^2 + R^2 C^2 \omega^2} \tag{1.148}$$

As ω approaches zero, the curve intersects the positive σ-axis at

$$\sigma_1 = \frac{1}{RC} - \frac{1}{R_1 C} + \frac{1}{R_1 C} \sqrt{1 - \frac{R_1}{R}} \tag{1.149}$$

The partition of the closed RHS into the regions of passivity and activity is sketched in Fig. 1.19. The shaded area and the solid bar from $\sigma = 0$ to any $\sigma < 1/RC$ denote the active region, and the remainder of the closed RHS is the passive region. The boundary between active and passive regions is determined by (1.148).

In the case that we choose the negative sign in (1.147), the boundary of the active and passive regions is described by the equation

$$R_1 + 2\sigma L = \frac{R}{(1 - RC\sigma)^2 + R^2 C^2 \omega^2} \tag{1.150}$$

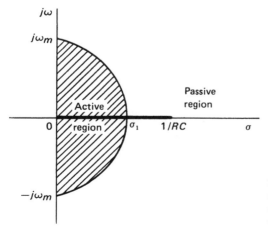

Figure 1.19 The partition of the closed RHS into the regions of passivity and activity by a properly biased tunnel diode.

As ω approaches zero, this curve intersects the positive σ-axis at a point determined by the cubic equation

$$2R^2C^2L\sigma^3 + RC(R_1RC - 4L)\sigma^2 - 2(R_1RC - L)\sigma + R_1 - R = 0 \tag{1.151}$$

Suppose that we choose a point $s_0 = j\omega_0$, $\omega_0 > \omega_m$, on the real-frequency axis; according to (1.146) the device is passive at $j\omega_0$. This device may be replaced by an equivalent passive one-port comprised of the series connection of a resistor of resistance

$$r = R_1 - \frac{R}{1 + R^2C^2\omega_0^2} \geqslant 0 \tag{1.152a}$$

and an inductor of inductance

$$l = L - \frac{R^2C}{1 + R^2C^2\omega_0^2} \geqslant 0 \tag{1.152b}$$

which exhibits identical behavior with respect to all real signals of the type (1.127). The general replacement formulas for a one-port will be discussed in the following example.

Example 1.10 Consider a one-port impedance $z(s)$ that is passive at a point s_0 in the closed RHS. The replacement of this one-port by a passive equivalent one-port that exhibits identical behavior with respect to all real signals of the type (1.127) may be effected in terms of one of the four simple passive one-ports, depending on the location of s_0 and the sign of Im $z(s_0)$.

For $s_0 = \sigma_0 > 0$, the impedance $z(\sigma_0)$ can be represented by a single resistor of resistance

$$r_1 = z(\sigma_0) \geqslant 0 \tag{1.153}$$

For $s_0 = j\omega_0$, $\omega_0 > 0$ and Im $z(j\omega_0) \geqslant 0$, or $\omega_0 < 0$ and Im $z(j\omega_0) \leqslant 0$, the impedance $z(j\omega_0)$ is the series combination of a resistor of resistance

$$r_2 = \text{Re } z(j\omega_0) \geqslant 0 \tag{1.154a}$$

and an inductor of inductance

$$l_2 = \frac{\text{Im } z(j\omega_0)}{\omega_0} \geqslant 0 \tag{1.154b}$$

For $s_0 = j\omega_0$, $\omega_0 > 0$ and $\text{Im } z(j\omega_0) < 0$, or $\omega_0 < 0$ and $\text{Im } z(j\omega_0) > 0$, $z(j\omega_0)$ is the series combination of a resistor of resistance

$$r_3 = \text{Re } z(j\omega_0) \geqslant 0 \tag{1.155a}$$

and a capacitor of capacitance

$$c_3 = \frac{1}{\omega_0 \text{ Im } z(j\omega_0)} \geqslant 0 \tag{1.155b}$$

Finally, for $s_0 = \sigma_0 + j\omega_0$, $\sigma_0 > 0$ and $\omega_0 \neq 0$, the impedance $z(s_0)$ can be represented equivalently by the series combination of an inductor of inductance l_4 and a capacitor of capacitance c_4; that is,

$$z(s_0) = l_4 s_0 + \frac{1}{s_0 c_4} \tag{1.156a}$$

Substituting $z(s_0) = \text{Re } z(s_0) + j \text{ Im } z(s_0)$ in Eq. (1.156a) and solving for l_4 and c_4 yield

$$l_4 = \frac{1}{2\sigma_0} \text{Re } z(s_0) + \frac{1}{2\omega_0} \text{Im } z(s_0) \tag{1.156b}$$

and

$$c_4 = \frac{1}{(|s_0|^2/2\sigma_0) \text{ Re } z(s_0) - (|s_0|^2/2\omega_0) \text{ Im } z(s_0)} \tag{1.156c}$$

which according to (1.137) are nonnegative because $z(s)$ is passive at the point s_0. The replacement one-port networks are presented in Fig. 1.20.

Consider, for example, the equivalent network of a tunnel diode of Fig. 1.18. Some typical values of its parameters are given below:

$$R_1 = 1\,\Omega \qquad R = 50\,\Omega$$

$$L = 0.4\,\text{nH} \qquad C = 10\,\text{pF}$$

Using these values and the results obtained in Example 1.9, we conclude that the device is passive on the nonnegative σ-axis for all $\sigma \geqslant \sigma_m = 2 \cdot 10^9$ Np/s, and on the real-frequency axis for all $|\omega| \geqslant \omega_m = 14 \cdot 10^9$ rad/s. Suppose that we choose $s_0 = (2 + j10) \cdot 10^9$. It is straightforward to demonstrate that the device is passive at this point s_0. Thus, it can be replaced by an equivalent network comprised of the series connection of an inductor of inductance l_4 and a capacitor of capacitance c_4. From (1.156) we obtain $l_4 = 0.15\,\text{nH}$ and $c_4 = 12.82\,\text{pF}$, where $z(s_0) = 1.8 - j6$. This equivalent one-port exhibits identical behavior with respect to all real signals

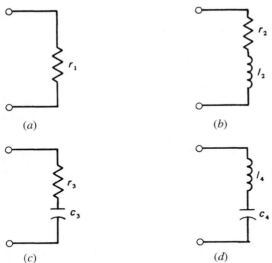

Figure 1.20 The replacement of a passive one-port impedance $z(s_0)$ by a passive equivalent one-port network with $s_0 = \sigma_0 + j\omega_0$. (a) $s_0 = \sigma_0 > 0$. (b) $s_0 = j\omega_0, \omega_0 > 0$ and Im $z(j\omega_0) \geqslant 0$, or $\omega_0 < 0$ and Im $z(j\omega_0) \leqslant 0$. (c) $s_0 = j\omega_0, \omega_0 > 0$ and Im $z(j\omega_0) < 0$, or $\omega_0 < 0$ and Im $z(j\omega_0) > 0$. (d) $s_0 = \sigma_0 + j\omega_0, \sigma_0 > 0$ and $\omega_0 \neq 0$.

comprised of two complex exponential signals at the complex conjugate pair of frequencies $s_0 = (2 + j10) \cdot 10^9$ and $\bar{s}_0 = (2 - j10) \cdot 10^9$ as shown in (1.127).

1.7.2 Active Networks

In the foregoing, we have demonstrated that by applying the definitions of passivity and activity at a single complex frequency, a given n-port can divide the closed RHS into regions of passivity and activity. The physical significance is that, in the passive region, the given n-port network can be replaced by a passive n-port that exhibits identical behavior with respect to all real signals of the type (1.127). In the active region, the n-port network can be made to achieve power gain or to oscillate by means of an appropriate passive imbedding network. In practical applications, we are concerned mainly with the behavior of the n-port on the real-frequency axis and, less frequently, on the positive σ-axis rather than the entire closed RHS. On these axes, the passivity condition (1.132) can be greatly simplified.

First, on the real-frequency axis, (1.132) becomes

$$\mathbf{u}_0^* \mathbf{H}_h(j\omega_0))\mathbf{u}_0 \geqslant 0 \qquad (1.157)$$

for all finite nonzero complex n-vectors $\mathbf{u}_0$, showing that the hermitian part $\mathbf{H}_h(s)$ of the general hybrid matrix $\mathbf{H}(s)$ of the given n-port must be nonnegative definite at the point $s = s_0 = j\omega_0$. On the positive σ-axis, $\mathbf{H}(\sigma_0)$ is real and (1.132) becomes

$$\mathbf{u}_0^* \mathbf{H}_h(\sigma_0)\mathbf{u}_0 \geqslant |\mathbf{u}_0' \mathbf{H}_s(\sigma_0)\mathbf{u}_0| = |\mathbf{u}_0' \mathbf{H}_h(\sigma_0)\mathbf{u}_0| \qquad (1.158)$$

because $\mathbf{H}_s(\sigma_0) = \mathbf{H}_h(\sigma_0)$. In computing the quadratic form $\mathbf{u}_0^* \mathbf{H}_h(\sigma_0)\mathbf{u}_0$ for real $\mathbf{H}_h(\sigma_0)$, there is no need to consider all complex $\mathbf{u}_0$; real $\mathbf{u}_0$ would be sufficient. To see this, we show that if

$$\mathbf{u}' \mathbf{H}_h(\sigma_0)\mathbf{u} \geqslant 0 \qquad (1.159)$$

for all real $\mathbf{u}$, then

$$\mathbf{u}^* \mathbf{H}_h(\sigma_0)\mathbf{u} \geqslant 0 \qquad (1.160)$$

for all complex $\mathbf{u}$, and vice versa. Clearly, (1.160) implies (1.159). To demonstrate that (1.159) also implies (1.160), we consider any complex n-vector $\mathbf{u}$ and write

$$\mathbf{u} = \mathbf{u}_1 + j\mathbf{u}_2 \qquad (1.161)$$

where $\mathbf{u}_1$ and $\mathbf{u}_2$ are real n-vectors. Substituting (1.161) in $\mathbf{u}^* \mathbf{H}_h(\sigma_0)\mathbf{u}$ gives

$$\begin{aligned} \mathbf{u}^* \mathbf{H}_h(\sigma_0)\mathbf{u} &= \mathbf{u}_1' \mathbf{H}_h(\sigma_0)\mathbf{u}_1 + j\mathbf{u}_1' \mathbf{H}_h(\sigma_0)\mathbf{u}_2 - j\mathbf{u}_2' \mathbf{H}_h(\sigma_0)\mathbf{u}_1 + \mathbf{u}_2' \mathbf{H}_h(\sigma_0)\mathbf{u}_2 \\ &= \mathbf{u}_1' \mathbf{H}_h(\sigma_0)\mathbf{u}_1 + \mathbf{u}_2' \mathbf{H}_h(\sigma_0)\mathbf{u}_2 \geqslant 0 \end{aligned} \qquad (1.162)$$

because $\mathbf{u}_1' \mathbf{H}_h(\sigma_0)\mathbf{u}_2 = \mathbf{u}_2' \mathbf{H}_h(\sigma_0)\mathbf{u}_1$. The second line follows directly from (1.159). The equivalence of (1.159) and (1.160) does not mean that their values are equal. In general, they are not:

$$\mathbf{u}^* \mathbf{H}_h(\sigma_0)\mathbf{u} \neq \mathbf{u}' \mathbf{H}_h(\sigma_0)\mathbf{u} \qquad (1.163)$$

Consider, for example, the case where $\mathbf{H}(\sigma_0) = \mathbf{1}_2$, the identity matrix of order 2. Choosing $\mathbf{u} = [j, 1]'$ shows that the left-hand side is 2 whereas the right-hand side is 0.

Theorem 1.5 On the real-frequency axis or the positive σ-axis, a linear time-invariant n-port network is passive at a point $s_0 = j\omega_0$ or $s_0 = \sigma_0$ if and only if the hermitian part of its associated general hybrid matrix is nonnegative definite at s_0.

PROOF. From Eq. (1.157) we see that the theorem is clearly true for all the points on the real-frequency axis. Thus, we consider the situation where $s_0 = \sigma_0 > 0$. If the n-port is passive at σ_0, meaning that (1.158) holds, $\mathbf{H}_h(\sigma_0)$ must be nonnegative definite. Conversely, if $\mathbf{H}_h(\sigma_0)$ is nonnegative definite, then for all real $\mathbf{u}_0$, $\mathbf{u}_0^* \mathbf{H}_h(\sigma_0)\mathbf{u}_0 = \mathbf{u}_0' \mathbf{H}_h(\sigma_0)\mathbf{u}_0$ and (1.158) follows. Note that on the σ-axis $\mathbf{H}(\sigma)$

is real, and there is no need in taking $\mathbf{u}_0$ complex, as indicated in Definition 1.8. This completes the proof of the theorem.

A problem of great practical interest is to determine the maximum frequency of sinusoidal oscillation. For this we define the following.

Definition 1.10: Maximum frequency of oscillation The largest ω on the real-frequency axis at which an n-port network is active is called the *maximum frequency of oscillation* of the n-port network.

In the piecewise-linear analysis of a regenerative oscillator, such as the blocking oscillator, a somewhat similar problem is posed by the question of designing a pulse circuit of minimum rise time. More precisely, given an active device, what is the smallest time constant for an exponentially increasing signal that the device can produce? For this we introduce the concept of fastest regenerative mode.

Definition 1.11: Fastest regenerative mode The largest σ on the nonnegative σ-axis at which an n-port network is active is called the *fastest regenerative mode* of the n-port network.

We illustrate the above results by the following examples.

Example 1.11 Consider the small-signal high-frequency equivalent network of a bipolar transistor of Fig. 1.21, whose admittance matrix was computed in Example 1.4 and is given by

$$\mathbf{Y}(s) = \begin{bmatrix} G_1 + s(C_1 + C_2) & -sC_2 \\ g_m - sC_2 & G_2 + sC_2 \end{bmatrix} \tag{1.164}$$

Figure 1.21 A small-signal high-frequency equivalent network of a bipolar transistor.

The hermitian part of $\mathbf{Y}(s)$ is obtained as

$$\mathbf{Y}_h(s) = \begin{bmatrix} G_1 + \sigma(C_1 + C_2) & \frac{1}{2}g_m - \sigma C_2 \\ \frac{1}{2}g_m - \sigma C_2 & G_2 + \sigma C_2 \end{bmatrix} \qquad (1.165)$$

On the real-frequency axis, $\mathbf{Y}_h(j\omega)$ is nonnegative definite if and only if

$$G_1 G_2 \geqslant \frac{1}{4}g_m^2 \qquad (1.166)$$

Thus, the device is passive for all real frequencies provided that (1.166) is satisfied. On the other hand, if

$$G_1 G_2 < \frac{1}{4}g_m^2 \qquad (1.167)$$

the device becomes active for all real frequencies, and the maximum frequency of oscillation is therefore infinity. The situation is evidently physically impossible. The reason for this seemingly inconsistent result is that we assume the equivalent network of the transistor to be an adequate representation of the transistor for all frequencies. That this is not the case follows directly from the observation that at higher frequencies, parasitic effects must be taken into consideration, yielding a finite maximum frequency of oscillation.

On the positive σ-axis, $\mathbf{Y}_h(\sigma)$ is nonnegative definite if and only if

$$G_2 + \sigma C_2 \geqslant 0 \qquad (1.168a)$$

$$G_1 + \sigma(C_1 + C_2) \geqslant 0 \qquad (1.168b)$$

$$C_1 C_2 \sigma^2 + (C_1 G_2 + C_2 G_2 + C_2 G_1 + C_2 g_m)\sigma + G_1 G_2 - \frac{1}{4}g_m^2 \geqslant 0 \qquad (1.168c)$$

Thus, if Eq. (1.166) is true, the device is passive for all nonnegative σ. Assuming that $G_1 G_2 < \frac{1}{4}g_m^2$ and setting the left-hand side of (1.168c) to zero yield the minimum value σ_m for which (1.168c) is satisfied:

$$\sigma_m = \frac{1}{2C_1 C_2}\left[\sqrt{b^2 + C_1 C_2(g_m^2 - 4G_1 G_2)} - b\right] \qquad (1.169a)$$

where

$$b = C_1 G_2 + C_2(G_1 + G_2 + g_m) \qquad (1.169b)$$

Figure 1.22 The small-signal T-equivalent network of a transistor.

Thus, the fastest regenerative mode of the device is given by (1.169a).[†] The device is passive for all $\sigma \geqslant \sigma_m$ and active for all $\sigma < \sigma_m$. In the piecewise-linear analysis of a regenerative oscillator, the value σ_m imposes a limitation on the period of relaxation oscillation.

Example 1.12 Figure 1.22 is the small signal T-equivalent network of a transistor. The two-port network is most conveniently characterized by its impedance matrix with

$$
\mathbf{Z}(s) = \begin{bmatrix} r_e + r_b & r_b \\ r_b + \dfrac{\alpha}{sC} & r_b + \dfrac{1}{sC} \end{bmatrix}
\tag{1.170}
$$

On the real-frequency axis, its hermitian part becomes

$$
\mathbf{Z}_h(j\omega) = \begin{bmatrix} r_e + r_b & r_b + \dfrac{j\alpha}{2\omega C} \\ r_b - \dfrac{j\alpha}{2\omega C} & r_b \end{bmatrix}
\tag{1.171}
$$

which is nonnegative definite if and only if

$$
\det \mathbf{Z}_h(j\omega) = r_e r_b - \frac{\alpha^2}{4\omega^2 C_2} \geqslant 0
\tag{1.172}
$$

[†] Strictly speaking, the fastest regenerative mode is $\sigma_m - \epsilon, \epsilon \to 0$.

Thus, the matrix remains nonnegative definite for all real frequencies

$$\omega \geqslant \omega_m = \frac{\alpha}{2C\sqrt{r_e r_b}} \tag{1.173}$$

showing that ω_m is the maximum frequency of oscillation[†] for the transistor.

On the nonnegative σ-axis, the hermitian part is obtained as

$$\mathbf{Z}_h(\sigma) = \begin{bmatrix} r_b + r_e & r_b + \dfrac{\alpha}{2\sigma C} \\ r_b + \dfrac{\alpha}{2\sigma C} & r_b + \dfrac{1}{\sigma C} \end{bmatrix} \tag{1.174}$$

which is nonnegative definite if and only if

$$\det \mathbf{Z}_h(\sigma) = 4C^2 r_e r_b \sigma^2 + 4C(r_b + r_e - \alpha r_b)\sigma - \alpha^2 \geqslant 0 \tag{1.175}$$

Choosing the equality and solving for σ yield the smallest value of σ at which $\mathbf{Z}_h(\sigma)$ is nonnegative, giving

$$\sigma_m = \frac{-(1-\alpha)r_b - r_e + \sqrt{[(1-\alpha)r_b + r_e]^2 r_e r_b \alpha^2}}{2C r_b r_e} \tag{1.176}$$

This is also the fastest regenerative mode for the transistor.

In the special situation where $\alpha = 0$, we have $\omega_m = 0$ and $\sigma_m = 0$. It can easily be shown that $\mathbf{Z}_h(s)$ is nonnegative definite for all s in the closed RHS, meaning that the device is passive at every point in the closed RHS. Thus, the impedance matrix $\mathbf{Z}(s)$ is a positive real matrix and the device is simply passive, as discussed in Sec. 1.6.1.

Example 1.13 Consider the two-port network of Fig. 1.21. We wish to partition the closed RHS into regions of passivity and activity by this two-port. To simplify the computation, let $C_2 = 0$. The admittance matrix becomes

$$\mathbf{Y}(s) = \begin{bmatrix} G_1 + sC_1 & 0 \\ g_m & G_2 \end{bmatrix} \tag{1.177}$$

As indicated in Example 1.8, the two-port is passive, its admittance matrix being positive real, if $4G_1 G_2 \geqslant g_m^2$. To exhibit activity in the closed RHS, we assume that $4G_1 G_2 < g_m^2$. This does not mean that every point in the closed RHS is active; there is still a region of passivity.

[†] Strictly speaking, the maximum frequency of oscillation is $\omega_m - \epsilon, \epsilon \to 0$.

We first investigate the behavior of the network for the positive σ-axis and the real-frequency axis points. According to Theorem 1.5, we need only test for nonnegative definiteness of the hermitian part of $\mathbf{Y}(s)$, which is given by

$$\mathbf{Y}_h(s) = \begin{bmatrix} G_1 + \sigma C_1 & \dfrac{1}{2}g_m \\[2mm] \dfrac{1}{2}g_m & G_2 \end{bmatrix} \tag{1.178}$$

Since $\mathbf{Y}_h(s)$ is independent of ω, on the real-frequency axis $\mathbf{Y}_h(j\omega)$ is not nonnegative definite, showing that the device is active for all real frequencies. On the positive σ-axis, $\mathbf{Y}_h(\sigma)$ is nonnegative definite if and only if

$$\sigma \geqslant \sigma_m = \frac{g_m^2 - 4G_1 G_2}{4G_2 C_1} > 0 \tag{1.179}$$

Thus, the device is active for all $\sigma < \sigma_m$ and passive for $\sigma \geqslant \sigma_m$.

For the remainder of the closed RHS, we must apply Eq. (1.136). Consider the points s_0 in the region defined by the vertical strip $0 < \sigma < \sigma_m$, as indicated in Fig. 1.23. At any of these points s_0, $\mathbf{Y}_h(s_0)$ is not nonnegative definite, implying that there exists a vector $\mathbf{u}_0$ such that $\mathbf{u}_0^* \mathbf{Y}_h(s_0)\mathbf{u}_0 < 0$. Clearly, for this $\mathbf{u}_0$ Eq. (1.136) cannot be satisfied, and the device is therefore active in this region. For $\sigma = \sigma_m$, $\mathbf{Y}_h(\sigma_m + j\omega)$ is singular and there exists a nontrivial solution $\mathbf{u} = [u_1, u_2]'$, $u_1 \neq 0$, such that $\mathbf{Y}_h(\sigma_m + j\omega)\mathbf{u} = \mathbf{0}$. Now we can write Eq. (1.136) as

$$|\sigma_m + j\omega| \mathbf{u}^* \mathbf{Y}_h(\sigma_m + j\omega)\mathbf{u} \geqslant \sigma_m |\mathbf{u}' \mathbf{Y}_h(\sigma_m + j\omega)\mathbf{u} + j\omega C_1 u_1^2| \tag{1.180}$$

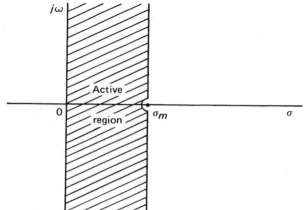

Figure 1.23 The partition of the closed RHS into regions of passivity and activity by the two-port network of Fig. 1.21.

which reduces to $|\sigma_m \omega C_1 u_1^2| \leqslant 0$. This inequality cannot be satisfied unless $\omega = 0$. Thus, the device is active along the vertical line $\sigma_m + j\omega$, $\omega \neq 0$. The results are sketched in Fig. 1.23.

1.7.3 Physical Significance of Single Complex Frequency Passivity and Activity

In introducing the concepts of passivity and activity at a single complex frequency s_0 in the closed RHS, we started from the time-energy definition (Definition 1.3) of passivity. We showed that if we apply an excitation of the type (1.127) to a given n-port network, which is known to be passive, then (1.131) or (1.132) follows. In the present section, we demonstrate that if an n-port is passive at a single complex frequency s_0, then for all excitations of the type (1.127) the energy into the n-port is always nonnegative. Furthermore, if an n-port is active at s_0, there exists an excitation source of the type (1.127) such that, given sufficient time, an arbitrarily large amount of energy can be extracted from the n-port.

Theorem 1.6 If a linear, time-invariant n-port network is passive at a single complex frequency s_0 in the closed RHS, then for all excitations of the type (1.127) the total energy delivered to the n-port network is nonnegative.

PROOF. Let $\mathbf{H}(s)$ be the general hybrid matrix of the n-port, which is passive at $s_0 = \sigma_0 + j\omega_0$, $\sigma_0 \geqslant 0$. Assume first that $\sigma_0 \neq 0$ and $\omega_0 \neq 0$. Suppose that there exists an excitation of the type (1.127) such that the total energy delivered to the n-port is negative for some t, say t_1. Then we can obtain an expression for the total energy $\varepsilon(t_1)$ similar to that given in (1.129) with t_1 replacing t. If $\varepsilon(t)$ is negative at t_1, it is negative at $t_1 + k\pi/\omega_0$ for any positive integer k, however large. Since $\varepsilon(t_0)$ and C_0 in (1.129) are constants and, for $\sigma > 0$, the terms containing $\exp\left[2\sigma_0(t_1 + k\pi/\omega_0)\right]$ in $\varepsilon(t_1 + k\pi/\omega_0)$ can be made arbitrarily large. This implies that the sum of the terms inside the curl brackets of (1.129) is negative at $t_1 + k\pi/\omega_0$. Hence, the left-hand side of (1.131) must be negative. This contradicts the hypothesis that the n-port is passive at s_0. In a similar manner, we are led to the same conclusion for $\sigma = 0$ or $\omega_0 = 0$, the details being left as an exercise (see Prob. 1.18). The only difference is that for $\sigma_0 = 0$, $e^{2\sigma_0 t/\sigma_0}$ is replaced by t in (1.129), and the same argument follows. This completes the proof of the theorem.

The following theorem establishes a relation between activity at a single complex frequency and energy-delivering capability of an n-port network.

Theorem 1.7 If a linear, time-invariant n-port network is active at a single complex frequency $s_0 = \sigma_0 + j\omega_0$ with $\sigma_0 \geqslant 0$ and $\omega_0 \neq 0$, then there exists an excitation of the type (1.127) such that, given sufficient time, an arbitrary, large amount of energy can be extracted from the n-port network.

PROOF. Since the n-port is active at s_0, there exists a fixed complex vector $\mathbf{u}_o$ such that for $\sigma_0 \neq 0$,

$$\frac{1}{\sigma_0}\mathbf{u}_0^*\mathbf{H}_h(s_0)\mathbf{u}_0 - \frac{1}{|s_0|}|\mathbf{u}_0'\mathbf{H}_s(s_0)\mathbf{u}_0| < 0 \qquad (1.181)$$

where $\mathbf{H}(s)$ is the general hybrid matrix of the n-port. Following Eqs. (1.127)–(1.129), we obtain, for $\sigma > 0$ and for large t,

$$4\varepsilon(t) \approx e^{2\sigma_0 t}\left[\frac{1}{\sigma_0}\mathrm{Re}\,\mathbf{u}_0^*\mathbf{H}(s_0)\mathbf{u}_0 + \frac{1}{|s_0|}|\mathbf{u}_0'\mathbf{H}(s_0)\mathbf{u}_0|\cos(2\omega_0 t + \theta)\right] \qquad (1.182)$$

where θ is the phase angle contributed by both the quadratic form $\mathbf{u}_0'\mathbf{H}(s_0)\mathbf{u}_0$ and s_0. Thus, if $t_k = (k + \frac{1}{2})\pi/\omega_0 - \theta/2\omega_0$, for sufficiently large integer k,

$$4\varepsilon(t_k) \approx e^{2\sigma_0 t_k}\left[\frac{1}{\sigma_0}\mathrm{Re}\,\mathbf{u}_0^*\mathbf{H}(s_0)\mathbf{u}_0 - \frac{1}{|s_0|}|\mathbf{u}_0'\mathbf{H}(s_0)\mathbf{u}_0|\right]$$

$$= e^{2\sigma_0 t_k}\left[\frac{1}{\sigma_0}\mathbf{u}_0^*\mathbf{H}_h(s_0)\mathbf{u}_0 - \frac{1}{|s_0|}|\mathbf{u}_0'\mathbf{H}_s(s_0)\mathbf{u}_0|\right] \qquad (1.183)$$

Combining Eqs. (1.181) and (1.183) yields

$$4\varepsilon(t_k) \approx -\left[\frac{1}{\sigma_0}\mathbf{u}_0^*\mathbf{H}_h(s_0)\mathbf{u}_0 - \frac{1}{|s_0|}|\mathbf{u}_0'\mathbf{H}_s(s_0)\mathbf{u}_0|\right]e^{2\sigma_0 t_k} \qquad (1.184)$$

or

$$\varepsilon(t_k) < -M \qquad (1.185)$$

where M is any prescribed positive number, however large. For the case $\sigma_0 = 0$, the above reasoning still applies provided that we replace $e^{2\sigma_0 t}/\sigma_0$ by t. Thus, for sufficient time, an arbitrary, large amount of energy can be extracted from the given n-port network. This completes the proof of the theorem.

Thus, in the passive region, an active n-port network behaves like a passive n-port network and can be replaced by a passive n-port network. If an n-port is passive at every point in the closed RHS, the n-port is simply passive in the sense of Definition 1.3. In the active region, the n-port can be made to achieve power gain or to oscillate by means of an appropriate passive imbedding network. We illustrate these by the following examples.

Figure 1.24 A small-signal high-frequency equivalent network of a field-effect transistor.

Example 1.14 Figure 1.24 is the small-signal high-frequency equivalent network of a field-effect transistor, whose admittance matrix is given by

$$\mathbf{Y}(s) = \begin{bmatrix} (C_1 + C_2)s & -sC_2 \\ g_m - sC_2 & G + sC_2 \end{bmatrix} \tag{1.186}$$

The hermitian part is found to be

$$\mathbf{Y}_h(s) = \begin{bmatrix} \sigma(C_1 + C_2) & \frac{1}{2}g_m - \sigma C_2 \\ \frac{1}{2}g_m - \sigma C_2 & G + \sigma C_2 \end{bmatrix} \tag{1.187}$$

On the nonnegative σ-axis, $\mathbf{Y}_h(\sigma)$ is nonnegative definite if and only if $\det \mathbf{Y}_h(\sigma) \geqslant 0$, which is equivalent to

$$C_1 C_2 \sigma^2 + (C_1 G + C_2 G + g_m C_2)\sigma - \frac{1}{2}g_m^2 \geqslant 0 \tag{1.188}$$

Setting $\det \mathbf{Y}_h(\sigma) = 0$, we obtain the positive root of the above polynomial in σ, giving the fastest regenerative mode

$$\sigma_m = \frac{-(C_1 + C_2)G - g_m C_2 + \sqrt{(C_1 G + C_2 G + g_m C_2)^2 + g_m^2 C_1 C_2}}{2C_1 C_2} \tag{1.189}$$

To demonstrate that this mode can be realized by an appropriate passive imbedding, we consider a gyrator with gyration resistance $2/g_m$, as shown in

Figure 1.25 A gyrator with gyration resistance $2/g_m$.

Figure 1.26 The parallel connection of a gyrator and a transistor.

Fig. 1.25. The admittance matrix of the gyrator is given by

$$
\mathbf{Y}_g(s) =
\begin{bmatrix}
0 & \dfrac{1}{2}g_m \\[2mm]
-\dfrac{1}{2}g_m & 0
\end{bmatrix}
\tag{1.190}
$$

If the transistor and the gyrator are connected in parallel as shown in Fig. 1.26, the admittance matrix of the overall two-port network is given by

$$
\mathbf{Y}(s) + \mathbf{Y}_g(s) =
\begin{bmatrix}
(C_1 + C_2)s & \dfrac{1}{2}g_m - sC_2 \\[2mm]
\dfrac{1}{2}g_m - sC_2 & G + sC_2
\end{bmatrix}
\tag{1.191}
$$

The open-circuit natural frequencies of the resulting two-port network are the zeros of det $[\mathbf{Y}(s) + \mathbf{Y}_g(s)]$. Since Eqs. (1.187) and (1.191) are identical except that σ is

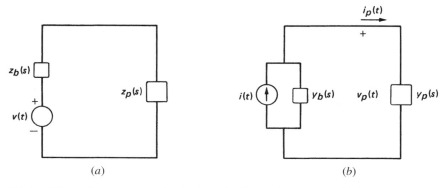

Figure 1.27 (*a*) A passive one-port network excited by a voltage source. (*b*) A passive one-port network excited by a current source.

replaced by s, the regenerative mode of the network is therefore σ_m, which is the largest σ for this transistor under arbitrary passive imbedding.

Example 1.15 In this example, we demonstrate that if a one-port network is active at a single complex frequency s_0 in the closed RHS, then it may be employed for energy gain with respect to all excitations of the type (1.127).

Let $z_p(s) = 1/y_p(s)$ be a positive real impedance characterizing a passive one-port network N_p. Assume that the one-port N_p is excited by a source combination as shown in Fig. 1.27a or 1.27b with $u(t) = v(t)$ or $i(t)$ and

$$u(t) = \frac{1}{2}(u_0 e^{s_0 t} + \bar{u}_0 e^{\bar{s}_0 t}) \tag{1.192}$$

where u_0 is an arbitrary complex but fixed constant, and $z_b(s) = 1/y_b(s)$ is another passive impedance. To be specific, we shall consider only the situation depicted in Fig. 1.27b and leave the justification for the other configuration as an exercise (see Prob. 1.20). With appropriate initial conditions, the voltage $v_p(t)$ and current $i_p(t)$ at the terminals of the one-port admittance y_p in Fig. 1.27b are obtained as

$$v_p(t) = \frac{1}{2}\left(\frac{I_0 e^{s_0 t}}{y_b(s_0) + y_p(s_0)} + \frac{\bar{I}_0 e^{\bar{s}_0 t}}{y_b(\bar{s}_0) + y_p(\bar{s}_0)}\right) \tag{1.193a}$$

$$i_p(t) = \frac{1}{2}\left(\frac{I_0 y_p(s_0) e^{s_0 t}}{y_b(s_0) + y_p(s_0)} + \frac{\bar{I}_0 y_p(\bar{s}_0) e^{\bar{s}_0 t}}{y_b(\bar{s}_0) + y_p(\bar{s}_0)}\right) \tag{1.193b}$$

Figure 1.28 An active one-port network emp-loyed for energy gain.

where $I_0 = u_0$. Then the total energy delivered to the initially relaxed one-port N_p up to any time t can be determined by (1.18) and is given by

$$\varepsilon(t) = \frac{|I_0|^2}{4|y_b(s_0) + y_p(s_0)|^2} e^{2\sigma_0 t}$$

$$\times \left[\frac{1}{\sigma_0} \operatorname{Re} y_p(s_0) + \frac{1}{|s_0|} |y_p(s_0)| \cos(2\omega_0 t + \theta) \right] \quad (1.194a)$$

where

$$\frac{I_0^2 y_p(s_0)}{s_0[y_b(s_0) + y_b(s_0)]^2} = \frac{|I_0|^2 |y_p(s_0)|}{|s_0||y_b(s_0) + y_p(s_0)|^2} e^{j\theta} \quad (1.194b)$$

To maximize $\varepsilon(t)$ for a fixed $y_p(s_0)$, we minimize $|y_b(s_0) + y_p(s_0)|$. For this we choose

$$\operatorname{Im} y_b(s_0) = -\operatorname{Im} y_p(s_0) \quad (1.195a)$$

$$B_0 \equiv \operatorname{Re} y_b(s_0) = \frac{\sigma_0 |\operatorname{Im} y_p(s_0)|}{|\omega_0|} \quad (1.195b)$$

Equation (1.195b) represents the minimum value that the real part of $y_b(s)$ can have at the point s_0, because at this point (1.139) applies. Thus, the maximum energy that can be delivered to the one-port N_p up to any time t is given by

$$\varepsilon_m(t) = \frac{|I_0|^2}{4\sigma_0[\operatorname{Re} y_p(s_0) + B_0]^2} e^{2\sigma_0 t}$$

$$\times \left[\operatorname{Re} y_p(s_0) + \frac{\sigma_0}{|s_0|} |y_p(s_0)| \cos(2\omega_0 t + \theta) \right] \quad (1.196)$$

Now suppose that we have a one-port network characterized by its admittance function $y_a(s)$, which is active at the point s_0. We show that by using this admittance, we can achieve an energy input to the one-port N_p greater than $\varepsilon_m(t)$. For this we consider the network of Fig. 1.28. The total energy delivered to the one-port N_p

up to any time t can be computed and is given by

$$\varepsilon_a(t) = \frac{|I_0|^2}{4\sigma_0|y_a(s_0) + y_p(s_0)|^2}e^{2\sigma_0 t}$$

$$\times \left[\operatorname{Re} y_p(s_0) + \frac{\sigma_0}{|s_0|}|y_p(s_0)|\cos(2\omega_0 t + \phi) \right] \quad (1.197a)$$

where

$$\frac{I_0^2 y_p(s_0)}{s_0[y_a(s_0) + y_p(s_0)]^2} = \frac{|I_0|^2|y_p(s_0)|}{|s_0||y_a(s_0) + y_p(s_0)|^2}e^{j\phi} \quad (1.197b)$$

Assume that we can choose a passive $y_p(s)$ so that $y_a(s_0) + y_p(s_0)$ is real. Under this situation, the phase angles θ and ϕ are equal, and the expression for net energy gain becomes

$$\frac{\varepsilon_a(t)}{\varepsilon_m(t)} = \frac{\{\operatorname{Re} y_p(s_0) + [\sigma_0|\operatorname{Im} y_p(s_0)|/|\omega_0|]\}^2}{|\operatorname{Re} y_p(s_0) + \operatorname{Re} y_a(s_0)|^2} \quad (1.198)$$

Since $y_a(s)$ is active at s_0, from (1.137) we have

$$\operatorname{Re} y_a(s_0) - \sigma_0\left|\frac{y_a(s_0)}{s_0}\right| < 0 \quad (1.199)$$

giving

$$|\omega_0| \operatorname{Re} y_a(s_0) < \sigma_0|\operatorname{Im} y_a(s_0)| \quad (1.200)$$

Since by choice $\operatorname{Im} y_p(s_0) = -\operatorname{Im} y_a(s_0)$, the expression (1.198) is always indicative of energy gain greater than unity. Thus, we conclude that if a one-port network is active at a point in the closed RHS, then it can be employed for energy gain with respect to signals of the type (1.127).

1.8 SUMMARY

We began the chapter by introducing the concepts of portwise linearity and time invariance. A portwise linear n-port network need not be a linear network in the usual sense. Conversely, a linear network may not be a portwise linear n-port. Similar statements can be made for the time invariance. We then defined passivity and activity in terms of the universally encountered quantities time and energy, and we showed that causality is a consequence of linearity and passivity. From the definition of causality, a causal n-port network must have its response as a single-valued function of its excitation. The converse, however, is not necessarily true. In the situation where the response of a linear passive n-port network is a multiple-valued function of its excitation, we showed that the instantaneous power entering the n-port is uniquely determined by its excitation. This was followed by a brief review of the general characterizations of linear, time-invariant n-port networks in the frequency domain. The characterizations are stated in terms of the general hybrid matrix. The translation of the time-domain passivity criteria into the equivalent frequency-domain passivity condition was taken up next. We showed that an n-port is passive if and only if its general hybrid matrix is positive real.

Finally, we introduced the discrete-frequency concepts of passivity and activity for a class of most significant and useful signals comprised of two complex exponential signals at the complex conjugate pair of frequencies. In this way, the closed RHS can be divided into regions of passivity and activity by a given n-port network. In the passive region, the net energy delivered to the n-port is always nonnegative. Thus, the n-port behaves like a passive network and can be replaced by a passive n-port that exhibits identical behavior with respect to all real signals comprised of two complex exponential signals at the complex conjugate pair of frequencies. In the active region, the n-port network can be made to achieve power gain or to oscillate by means of an appropriate passive imbedding network. In fact, we demonstrated that there exists an excitation of the above type such that, for sufficient time, an arbitrary, large amount of energy can be extracted from the n-port network. For practical applications, we are concerned mainly with the behavior of the n-port networks on the real-frequency axis and, less frequently, on the positive σ-axis. For these purposes, we introduced the notions of maximum frequency of oscillation and fastest regenerative mode. The former determines the maximum frequency of sinusoidal oscillation and the latter imposes a limitation on the period of relaxation oscillation, such as multivibrators and blocking oscillators, that an active n-port device can attain under passive imbedding.

PROBLEMS

1.1 Show that a properly biased tunnel diode of Fig. 1.12*a* with its vi-characteristic as shown in Fig. 1.12*b* is active.

1.2 Use Definition 1.3 to show that a linear time-invariant inductor is passive if its inductance is nonnegative and active if its inductance is negative.

1.3 Demonstrate that the unbiased tunnel diode of Fig. 1.11 is a causal one-port under voltage excitation-current response, and a noncausal one-port under current excitation-voltage response.

1.4 Consider a linear time-varying resistor whose resistance is characterized by the equation

$$R(t) = r_0 + r_1 \cos(\omega_0 t + \theta) \tag{1.201}$$

where $r_0 \geqslant r_1 > 0$ and $\omega_0 > 0$. Determine whether this device is passive or active.

1.5 Let $\mathbf{y}(t)$ be the response vector of an n-port network resulting from the excitation vector $\mathbf{u}(t)$. If $\mathbf{v}(t)$ and $\mathbf{i}(t)$ denote the port-voltage and port-current vectors, is it generally true that

$$\mathbf{v}'(t)\mathbf{i}(t) = \mathbf{u}'(t)\mathbf{y}(t) \tag{1.202}$$

1.6 Demonstrate that the open circuit is noncausal under current excitation-voltage response.

1.7 For the given one-port network of Fig. 1.29, determine the regions of passivity and activity in the closed RHS. Choose a point in the passive region and demonstrate that this one-port has a passive equivalent one-port at the chosen point. Choose a point in the active region and show that this one-port can be employed for energy gain at the chosen complex frequency.

1.8 For the two-port network considered in Example 1.4, compute the hybrid parameters $h_{12}(s)$ and $h_{22}(s)$, the impedance parameters $z_{12}(s)$ and $z_{22}(s)$, the admittance parameters $y_{12}(s)$ and $y_{22}(s)$, and the transmission parameters $B(s)$ and $D(s)$.

1.9 Suppose that we have a two-port network that can admit only excitation signals of the type

$$\mathbf{u}(t) = a \begin{bmatrix} 1 \\ 1 \end{bmatrix} \quad t > 0 \tag{1.203a}$$

$$= \begin{bmatrix} 0 \\ 0 \end{bmatrix} \quad t \leqslant 0 \tag{1.203b}$$

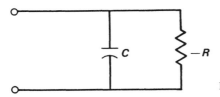

Figure 1.29 An active one-port network.

where a is real and arbitrary. Assume that the response of the two-port caused by the above excitation (1.203) is given by

$$\mathbf{y}(t) = a \begin{bmatrix} 1 \\ -1 \end{bmatrix} \tag{1.204}$$

Show that the two-port is linear and passive. The response is evidently a single-valued function of excitation. Demonstrate that this two-port is noncausal.

1.10 Determine the condition under which the impedance matrix (1.69) is positive real.

1.11 Confirm that the Fourier transform of the signal (1.111) is given by (1.112).

1.12 For the two-port network of Fig. 1.17, use the procedure outlined in Example 1.5 to compute the hybrid matrix and the transmission matrix from the known impedance matrix (1.69).

1.13 Repeat Prob. 1.12 if the impedance parameters are given as $z_{ij}(s)$, i, $j = 1, 2$.

1.14 Show that for $\omega_0 = 0$ Eq. (1.130) requires that Re $\mathbf{u}_9^* \mathbf{H}(s_0)\mathbf{u}_0 \geqslant 0$, and this inequality is equivalent to (1.131) for all real $\mathbf{u}_0$.

1.15 In deriving (1.131) from (1.129), we assume that $\sigma_0 \neq 0$. Show that for $\sigma = 0$, the passivity requirement implies Re $\mathbf{u}_0^* \mathbf{H}(j\omega_0)\mathbf{u}_0 \geqslant 0$. This indicates that the inequality (1.131) is valid for all s_0 in the closed RHS for a passive n-port network.

1.16 Consider a one-port admittance $y(s)$ that is passive at a point s_0 in the closed RHS. Obtain an equivalent passive replacement one-port network that exhibits identical behavior with respect to all real signals of the type (1.127).

1.17 A nonlinear capacitor is characterized by the equation

$$i(t) = 2v(t)\frac{dv(t)}{dt} \tag{1.205}$$

Determine whether this capacitor is passive or active.

1.18 Prove that Theorem 1.6 is valid for $\sigma_0 = 0$ or $\omega_0 = 0$, where $s_0 = \sigma_0 + j\omega_0$ is a point in the closed RHS where the given n-port is passive.

1.19 A nonlinear inductor is characterized by the equation

$$v(t) = 2i^2(t)\frac{di(t)}{dt} \tag{1.206}$$

Determine whether this inductor is passive or active.

Figure 1.30 The small-signal high-frequency hybrid-pi equivalent network of a bipolar transistor.

1.20 Repeat Example 1.15 for a one-port impedance $z_a(s)$ that is active at a closed RHS point s_0, and show that this one-port can be employed for energy gain at s_0.

1.21 Figure 1.30 shows the small-signal high-frequency hybrid-pi equivalent network of a bipolar transistor. Compute its admittance matrix.

1.22 Let $\tilde{u}(s) = [I_1(s), V_2(s)]'$ be the excitation vector of the two-port network of Fig. 1.30. Compute the associated hybrid matrix.

1.23 The matrix given below is known to be positive real. Confirm this claim.

$$H(s) = \frac{1}{2s(s^2 + 1)} \begin{bmatrix} 2s^2 + 1 & 1 \\ 1 & 2s^2 + 1 \end{bmatrix} \qquad (1.207)$$

1.24 Typical values of the hybrid-pi parameters for a type 2N2614 transistor of Fig. 1.30 are given as foilows:

$$
\begin{array}{ll}
r_x = 300\,\Omega & g_m = 0.0385\,\text{mho} \\
r_\pi = 4\,\text{k}\Omega & C_\pi = 750\,\text{pF} \\
r_0 = 167\,\text{k}\Omega & C_\mu = 9\,\text{pF}
\end{array}
\qquad (1.208)
$$

These values pertain at $I_C = 1\,\text{mA}$ and $V_{CE} = 6\,\text{V}$. Using these values, find the maximum frequency of oscillation and the fastest regenerative mode of the transistor.

1.25 For the one-port network of Fig. 1.31, determine the regions of passivity and activity.

1.26 Test the following matrix to see if it is positive real:

$$H(s) = \frac{1}{46s + 1} \begin{bmatrix} 2s + 1 & 2s - 10 \\ 2s - 10 & 2s + 1 \end{bmatrix} \qquad (1.209)$$

Figure 1.31 An active one-port network.

Figure 1.32 A linear time-varying one-port network.

1.27 Assume that the one-port network of Fig. 1.32 is initially relaxed. Determine if this linear time-varying one-port is active.

1.28 A nonlinear resistor is characterized by the equation

$$v(t) = i(t) + 2i^2(t) \qquad (1.210)$$

Is this resistor active? Is this one-port network causal under voltage excitation-current response? If the roles of $v(t)$ and $i(t)$ are interchanged in (1.210), can one draw the same conclusion?

1.29 A nonlinear resistor is characterized by the equation

$$v(t) = \tanh i(t) \qquad (1.211)$$

Is this resistor active? Is this resistor causal under voltage excitation-current response?

1.30 Show that a linear time-varying inductor of inductance $L(t)$ is passive if and only if $L(t) \geqslant 0$ and $dL(t)/dt \geqslant 0$ for all t.

1.31 Show that a linear time-varying capacitor of capacitance $C(t)$ is passive if and only if $C(t) \geqslant 0$ and $dC(t)/dt \geqslant 0$ for all t.

1.32 Show that a nonlinear time-varying resistor is passive if and only if its vi-characteristic is in the first and third quadrants of the vi-plane for all time.

1.33 Consider the one-port network of Fig. 1.33. Determine the condition under which the one-port is passive.

1.34 Let a and b denote real and nonnegative constants. Determine conditions under which each of the following elements is active:

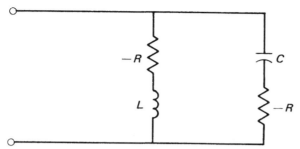

Figure 1.33 A one-port network.

1. A resistor characterized by the resistance $R(t) = a + b \sin \omega_0 t$
2. A capacitor characterized by the capacitance $C(t) = a + b \sin \omega_0 t$
3. An inductor characterized by the inductance $L(t) = a + b \sin \omega_0 t$

1.35 Under what circumstances is the emitter-follower of Fig. 1.34 passive? Use the equivalent network of Fig. 1.17 for the transistor and ignore the biasing circuit.

1.36 Determine whether the independent sources are linear one-port networks. Are controlled sources nonlinear two-ports? If so, justify your conclusion.

1.37 Figure 1.35 shows a one-port network containing two ideal diodes. Determine the condition under which the one-port is linear.

1.38 Suppose that we have a transformer whose turns ratio is time-varying. As shown in Fig. 1.36, a capacitor of positive capacitance C is connected at the output of the transformer. Determine condition under which the resulting one-port network is not lossless.

1.39 Determine whether the ideal diode is causal or noncausal. Is it lossless?

1.40 The inductance of a linear time-varying inductor is given by

$$L(t) = t + \sin t \qquad (1.212)$$

Determine whether this inductor is passive or active.

1.41 A charge-controlled capacitor is a capacitor whose terminal voltage $v(t)$ is a single-valued function of the stored electric charge $q(t)$. A charge-controlled capacitor is characterized by the equation

$$v(t) = 1 + q^2(t) \qquad (1.213)$$

Determine whether this capacitor is passive or active, linear or nonlinear. Is it causal under voltage excitation-charge response?

Figure 1.34 A transistor emitter-follower.

Figure 1.35 A one-port network containing two ideal diodes.

1.42 A linear time-varying resistor of conductance

$$G(t) = g_0 + g_1 \cos \omega_p t \qquad (1.214)$$

can be made active by appropriately choosing g_0 and g_1. This idea has been applied successfully to the design of *parametric amplifiers*. Figure 1.37 shows a simple parametric amplifier, where the signal to be amplified is represented by the current source $i_s(t)$. Show that power amplification at the source is possible only if $g_1 > g_0$.

1.43 A flux-controlled inductor is an inductor whose current $i(t)$ is a single-valued function of the magnetic flux $\phi(t)$. Consider a time-invariant flux-controlled

$k(t):1$

Figure 1.36 An ideal transformer whose turns ratio is time-varying.

$i_s(t) = I_0 \cos \omega_0 t$
L
C
$R(t) = 1/G(t)$

Figure 1.37 A simple parametric amplifier.

inductor with a characteristic like the one shown in Fig. 1.38. Determine the condition under which the inductor is passive. Express the condition in terms of the ϕi-curve.

1.44 In the two-port network of Fig. 1.17, let

$$\mathbf{u}(t) = \mathbf{i}(t) = \begin{bmatrix} i_1(t) \\ i_2(t) \end{bmatrix} = \frac{1}{2}(\mathbf{I}_0 e^{s_0 t} + \bar{\mathbf{I}}_0 e^{\bar{s}_0 t}) \qquad (1.215)$$

be the excitation, where $\mathbf{I}_0$ is an arbitrary complex 2-vector, and let

$$\mathbf{y}(t) = \mathbf{v}(t) = \begin{bmatrix} v_1(t) \\ v_2(t) \end{bmatrix} = \frac{1}{2}(\mathbf{V}_0 e^{s_0 t} + \bar{\mathbf{V}}_0 e^{\bar{s}_0 t}) \qquad (1.216)$$

be the response. Show that

$$\mathbf{V}_0 = \mathbf{Z}(s_0)\mathbf{I}_0 \qquad (1.217)$$

where $\mathbf{Z}(s)$ is the impedance matrix of the two-port given by (1.69). At the initial time t_0, the port voltages are required to be

$$v(t_0) = \frac{1}{2}[\mathbf{Z}(s_0)\mathbf{I}_0 e^{s_0 t_0} + \bar{\mathbf{Z}}(s_0)\bar{\mathbf{I}}_0 e^{\bar{s}_0 t_0}] \qquad (1.218)$$

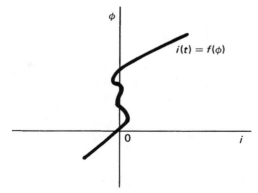

Figure 1.38 The characteristic of a time-invariant flux-controlled inductor.

Determine the initial voltages across the two capacitors at t_0. What is the amount of energy required at the initial time t_0 to set up the appropriate initial conditions so that no transient will occur?

BIBLIOGRAPHY

Anderson, B. D. O. and S. Vongpanitlerd: "Network Analysis and Synthesis: A Modern Systems Theory Approach," Englewood Cliffs, N.J.: Prentice-Hall, 1973.

Chen, W. K.: Relationships Between Scattering Matrix and Other Matrix Representations of Linear Two-Port Networks, *Int. J. Electronics,* vol. 38, no. 4, pp. 433–441, 1975.

Chen, W. K.: "Theory and Design of Broadband Matching Networks," Oxford: Pergamon, 1976.

Chen, W. K.: The Hybrid Matrix in Linear Multiple-Loop Feedback Networks, *IEEE Trans. Circuits and Systems,* vol. CAS-24, no. 9, pp. 469–474, 1977.

Desoer, C. A. and E. S. Kuh: Bounds on Natural Frequencies of Linear Active Networks, *Proc. Symp. on Active Networks and Feedback Systems*, Polytechnic Inst. of Brooklyn, New York, vol. 10, pp. 415–436, 1960.

Guillemin, E. A.: "The Mathematics of Circuit Analysis," Cambridge, Mass.: The M.I.T. Press, 1969.

Kuh, E. S.: Regenerative Modes of Active Networks, *IRE Trans. Circuit Theory,* vol. CT-7, no. 1, pp. 62–63, 1960.

Kuh, E. S. and R. A. Rohrer: "Theory of Linear Active Networks," San Francisco, Calif.: Holden-Day, 1967.

Kuo, Y. L.: A Note on the n-Port Passivity Criterion, *IEEE Trans. Circuit Theory*, vol. CT-15, no. 1, p. 74, 1968.

Lam, Y. F.: Characterization of Time-Invariant Network Elements, *J. Franklin Inst.*, vol. 298, no. 1, pp. 1–7, 1974.

Nedunuri, R.: On the Definition of Passivity, *IEEE Trans. Circuit Theory*, vol CT-19, no. 1, p. 72, 1972.

Newcomb, R. W.: Synthesis of Passive Networks for Networks Active at p_0, *IRE Int. Convention Record*, part 4, pp. 162–175, 1961.

Newcomb, R. W.: On Causality, Passivity and Single-Valuedness, *IRE Trans. Circuit Theory*, vol. CT-9, no. 1, pp. 87–89, 1962.

Newcomb, R. W.: On the Definition of a Network, *Proc. IEEE*, vol. 53, no. 5, pp. 547–548, 1965.

Newcomb, R. W.: "Linear Multiport Synthesis," New York: McGraw-Hill, 1966.

Peikari, B.: "Fundamentals of Network Analysis and Synthesis" Englewood Cliffs, N.J.: Prentice-Hall, 1974.

Raisbeck, G.: A Definition of Passive Linear Networks in Terms of Time and Energy, *J. Appl. Phys.*, vol. 25, no. 12, pp. 1510–1514, 1954.

Resh, J. A.: A Note Concerning the n-port Passivity Condition, *IEEE Trans. Circuit Theory*, vol. CT-13, no. 2, pp. 238–239, 1966.

Rohrer, R. A.: Lumped Network Passivity Criteria, *IEEE Trans. Circuit Theory*, vol. CT-15, no. 1, pp. 24–30, 1968.

Youla, D. C.: Physical Realizability Criteria, *IRE Trans. Circuit Theory*, vol. CT-7, Special Supplement, pp. 50–68, 1960.

Youla, D. C., L. J. Castriota, and H. J. Carlin: Bounded Real Scattering Matrices and the Foundations of Linear Passive Network Theory, *IRE Trans. Circuit Theory*, vol. CT-6, no. 1, pp. 102–124, 1959.

Zemanian, A. H.: The Passivity and Semipassivity of Time-Varying Systems Under the Admittance Formulism, *SIAM J. Appl. Math.*, vol. 21, no. 4, pp. 533–541, 1971.

TWO

THE INDEFINITE-ADMITTANCE MATRIX

In the preceding chapter, networks were characterized by their port behaviors. Fundamental to the concept of a port is the assumption that the instantaneous current entering one terminal of the port is always equal to the instantaneous current leaving the other terminal of the port. However, we recognize that upon the interconnection of networks, this port constraint may be violated. Thus, it is sometimes desirable and more advantageous to consider n-terminal networks, as depicted in Fig. 2.1.

In this chapter, we discuss a useful description of the external behavior of a multiterminal network in terms of the indefinite-admittance matrix and demonstrate how it can be employed effectively for the computation of network functions. Specifically, we derive formulas expressing the network functions in terms of the first-order and the second-order cofactors of the elements of the indefinite-admittance matrix. The significance of this approach is that the indefinite-admittance matrix can usually be written down directly from the network by inspection.

Since in the remainder of this book we deal exclusively with linear, lumped, and time-invariant networks, the adjectives *linear, lumped*, and *time-invariant* are omitted in the discussion unless they are used for emphasis.

2.1 THE INDEFINITE-ADMITTANCE MATRIX

Referring to Fig. 2.1, let N be an n-terminal network comprised of an arbitrary number of active and passive network elements connected in any way whatsoever. Let $V_1, V_2, \ldots, V_n$ be the Laplace-transformed potentials measured between

Reference-potential point

Figure 2.1 The general symbolic representation of an n-terminal network.

terminals $1, 2, \ldots, n$ and some arbitrary but unspecified reference point, and let $I_1, I_2, \ldots, I_n$ be the Laplace-transformed currents entering the terminals $1, 2, \ldots, n$ from outside the network.

Since the network N together with its loading is linear, the terminal currents and voltages are related by the equation[†]

$$
\begin{bmatrix} I_1 \\ I_2 \\ \vdots \\ I_n \end{bmatrix} = \begin{bmatrix} y_{11} & y_{12} & \cdots & y_{1n} \\ y_{21} & y_{22} & \cdots & y_{2n} \\ \cdots & \cdots & \cdots & \cdots \\ y_{n1} & y_{n2} & \cdots & y_{nn} \end{bmatrix} \begin{bmatrix} V_1 \\ V_2 \\ \vdots \\ V_n \end{bmatrix} + \begin{bmatrix} J_1 \\ J_2 \\ \vdots \\ J_n \end{bmatrix} \tag{2.1a}
$$

or more compactly as

$$
\mathbf{I}(s) = \mathbf{Y}(s)\mathbf{V}(s) + \mathbf{J}(s) \tag{2.1b}
$$

where J_k $(k = 1, 2, \ldots, n)$ denotes the current flowing into the kth terminal when all terminals of N are grounded to the reference point. The coefficient matrix $\mathbf{Y}(s)$ is called the *indefinite-admittance matrix* by Shekel (1952) because the reference point for the potentials is some arbitrary but unspecified point outside the network. The short-circuit currents J_k result from the independent sources and/or initial conditions in the interior of the n-terminal network. For our purposes, we shall consider all independent sources outside the network and set all initial conditions

---

[†]Instead of introducing additional symbols, $\mathbf{Y}(s)$ is used to denote either the admittance matrix or the indefinite-admittance matrix. This should not create any confusion, because the context will tell. Occasionally, we use $\mathbf{Y}_{sc}(s)$ to represent the admittance matrix in order to distinguish it from the indefinite-admittance matrix. This is similarly valid for the admittance parameters y_{ij}.

to zero. Hence, $\mathbf{J}(s)$ is considered to be zero, and Eq. (2.1) becomes

$$\mathbf{I}(s) = \mathbf{Y}(s)\mathbf{V}(s) \tag{2.2}$$

The elements of $\mathbf{Y}(s)$ are short-circuit admittances because they can be obtained by the equation

$$y_{ij} = \left.\frac{I_i}{V_j}\right|_{V_x=0,x\neq j} \tag{2.3}$$

showing that y_{ii} is the driving-point admittance looking into the terminal i and the reference point when all other terminals are grounded to the reference point, and that y_{ij}, $i \neq j$, is the transfer admittance from terminal j to terminal i when all terminals except the jth one are grounded to the reference point. In fact, this relation may be employed to compute the elements of $\mathbf{Y}(s)$. As will be shown in Sec. 2.2, the elements of $\mathbf{Y}(s)$ can usually be obtained directly from the network by inspection.

Example 2.1 Figure 2.2 is the network model of a transistor. Assume that each of its nodes is an accessible terminal. The suppression of inaccessible terminals will be considered later. Using Eq. (2.3), we compute the elements of its indefinite-admittance matrix $\mathbf{Y}(s)$, as follows.

For elements y_{11}, y_{21}, and y_{31}, we set $V_2 = V_3 = 0$ and apply a voltage source V_1 between the terminal 1 and the reference point as shown in Fig. 2.3. This gives

$$y_{11} = \left.\frac{I_1}{V_1}\right|_{V_2=V_3=0} = g_1 + sC_1 + sC_2 \tag{2.4a}$$

$$y_{21} = \left.\frac{I_2}{V_1}\right|_{V_2=V_3=0} = g_m - sC_2 \tag{2.4b}$$

Figure 2.2 An equivalent network of a transistor.

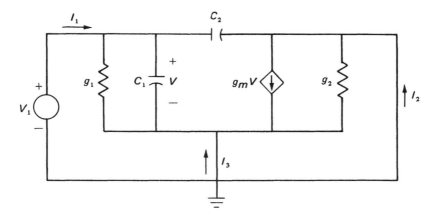

Figure 2.3 The network used to evaluate the admittance parameters y_{11}, y_{21}, and y_{31}.

$$y_{31} = \left.\frac{I_3}{V_1}\right|_{V_2=V_3=0} = -g_1 - sC_1 - g_m \qquad (2.4c)$$

In a similar fashion, to compute y_{12}, y_{22}, and y_{32} we set $V_1 = V_3 = 0$ and apply a voltage source between terminal 2 and the reference point; and to compute y_{13}, y_{23}, and y_{33} we set $V_1 = V_2 = 0$ and apply a voltage source between terminal 3 and the reference point. The details are left as an exercise (see Prob. 2.1). The indefinite-admittance matrix of the transistor is given by

$$\mathbf{Y}(s) = \begin{bmatrix} g_1 + sC_1 + sC_2 & -sC_2 & -g_1 - sC_1 \\ g_m - sC_2 & g_2 + sC_2 & -g_2 - g_m \\ -g_1 - sC_1 - g_m & -g_2 & g_1 + g_2 + g_m + sC_1 \end{bmatrix} \qquad (2.5)$$

It is significant to observe that the sum of elements of each row or column is equal to zero. That these properties are valid in general for the indefinite-admittance matrix will now be demonstrated.

First, we show that the sum of the elements in each column of $\mathbf{Y}(s)$ equals zero. To this end we add all n equations of (2.1) to yield

$$\sum_{i=1}^{n}\sum_{j=1}^{n} y_{ji} V_i = \sum_{m=1}^{n} I_m - \sum_{m=1}^{n} J_m = 0 \qquad (2.6)$$

The last equation is obtained by appealing to Kirchhoff's current law for the node corresponding to the reference point. Setting all the terminal voltages to zero except the kth one, which is nonzero, gives

$$V_k \sum_{j=1}^{n} y_{jk} = 0 \tag{2.7}$$

Since $V_k \neq 0$, it follows that the sum of the elements of each column of $\mathbf{Y}(s)$ equals zero. Thus, the indefinite-admittance matrix is always singular.

To demonstrate that each row sum of $\mathbf{Y}(s)$ is also zero, we recognize that because the point of zero potential may be chosen arbitrarily, the currents J_k and I_k remain invariant when all the terminal voltages V_k are changed by the same but arbitrary constant amount. Thus, if $\mathbf{V}_0$ is an n-vector, each of its elements being $v_0 \neq 0$, then

$$\mathbf{I}(s) - \mathbf{J}(s) = \mathbf{Y}(s)\big[\mathbf{V}(s) + \mathbf{V}_0\big] = \mathbf{Y}(s)\mathbf{V}(s) + \mathbf{Y}(s)\mathbf{V}_0 \tag{2.8}$$

which after invoking $(2.1b)$ shows that

$$\mathbf{Y}(s)\mathbf{V}_0 = \mathbf{0} \tag{2.9}$$

or

$$\sum_{j=1}^{n} y_{ij} = 0 \quad i = 1, 2, \ldots, n \tag{2.10}$$

indicating that each row sum of $\mathbf{Y}(s)$ equals zero.

As a consequence of the zero-row-sum and zero-column-sum properties, all the cofactors of the elements of the indefinite-admittance matrix are equal. Before we justify this statement, we need the following definition.

Definition 2.1: Equicofactor matrix A square matrix is said to be an *equicofactor matrix* if all the cofactors of its elements are equal.

By using this definition, a fundamental property of the indefinite-admittance matrix can now be stated.

Theorem 2.1 The indefinite-admittance matrix of a linear, lumped, and time-invariant multiterminal network is an equicofactor matrix.

PROOF. Let Y_{ij} be the cofactor of the element y_{ij} of $\mathbf{Y}(s)$. We first show that $Y_{iu} = Y_{iv}$ for all i, u, and v. Without loss of generality, assume that $u > v$. Since each row sum is zero, we may replace the elements y_{jv} ($v = 1, 2, \ldots, n$) in the submatrix $\mathbf{Y}_{iu}$ obtained from $\mathbf{Y}(s)$ by deleting the ith row and uth column by

$$-\sum_{\substack{k=1 \\ k \neq v}}^{n} y_{jk} \tag{2.11}$$

without changing the value of Y_{iu}. Let the submatrix thus obtained be denoted by $\hat{\mathbf{Y}}_{iu}$. Now, adding all the columns of $\hat{\mathbf{Y}}_{iu}$ to column v and then shifting column v to the right of column $u - 1$ if $v \neq u - 1$ result in

$$\det \hat{\mathbf{Y}}_{iu} = (-1)^{u-v-1}(-1)\det \mathbf{Y}_{iv} \tag{2.12}$$

$\mathbf{Y}_{iv}$ being the submatrix obtained from $\mathbf{Y}(s)$ by deleting row i and column v. From Eq. (2.12) we have

$$Y_{iu} = (-1)^{i+u}(-1)^{u-v-1}(-1)\det \mathbf{Y}_{iv} = Y_{iv} \tag{2.13}$$

Similarly, by considering the transpose of $\mathbf{Y}(s)$, we can show that $Y_{ui} = Y_{vi}$ for all i, u, and v. Thus, we conclude that all the cofactors of the elements of the indefinite-admittance matrix $\mathbf{Y}(s)$ are equal. This completes the proof of the theorem.

As an illustration, we consider the indefinite-admittance matrix $\mathbf{Y}(s)$ of Eq. (2.5) computed in Example 2.1. It is easy to verify that all of its nine cofactors are equal to

$$s^2 C_1 C_2 + s(C_1 g_2 + C_2 g_1 + C_2 g_2 + g_m C_2) + g_1 g_2 \tag{2.14}$$

2.2 RULES FOR WRITING DOWN THE PRIMITIVE INDEFINITE-ADMITTANCE MATRIX

Suppose that every node of a network N is an accessible terminal to which external connections are to be made. The indefinite-admittance matrix associated with such a network is referred to as the *primitive indefinite-admittance matrix*. In the present section, we show that for networks comprising multiterminal devices and voltage-controlled current sources, their primitive indefinite-admittance matrices can easily be written down by inspection. We recognize that confinement to voltage-controlled current sources should not be deemed restrictive, because other types of controlled sources can usually be converted to the above type by applying Norton's theorem if necessary. In the following section, we shall describe a procedure for *suppressing* inaccessible terminals.

Figure 2.4 A voltage-controlled current source.

Consider a voltage-controlled current source, as shown in Fig. 2.4, whose indefinite-admittance matrix is clearly given by

$$
\begin{array}{c}
\begin{array}{cccc} + & - & & \\ V_a & V_b & V_c & V_d \end{array} \\
\begin{array}{c} I_a \\ I_b \\ + I_c \\ - I_d \end{array}
\left[
\begin{array}{cccc}
0 & 0 & 0 & 0 \\
0 & 0 & 0 & 0 \\
y & -y & 0 & 0 \\
-y & y & 0 & 0
\end{array}
\right]
\end{array}
\tag{2.15}
$$

where V's and I's are the terminal voltages and currents of the controlled source. Observe that the admittance y enters the indefinite-admittance matrix in a rectangular pattern that is not necessarily centered on the main diagonal. The admittance y is actuated by the voltage V_{ab} from terminal a to terminal b that corresponds to the first two columns of Eq. (2.15) and affects the currents of the terminals c and d associated with the third and fourth rows of the matrix. Thus, the matrix Eq. (2.15) can easily be written down by inspection from the terminal labelings of the controlled source. The sign associated with the admittance y is determined according to the usual multiplication rules, as indicated in Eq. (2.15). This representation is, of course, quite general. Terminals $a, b, c,$ and d need not be distinct. For example, if $a = c$ and $b = d$, then it represents a one-port admittance y. Thus, if these elements are imbedded in a network, its indefinite-admittance matrix is simply the sum of the indefinite-admittance matrices of the component subnetworks that form the complete network when combined in parallel by connecting together their corresponding terminals. Each of the subnetworks has n terminals, some of which may consist only of isolated nodes. The reason for this is that in parallel connection, it implies equality of voltages and addition of currents at each terminal, and the addition of the indefinite-admittance matrices is a direct result. This leads directly to the following set of rules for writing down the primitive indefinite-admittance matrix by inspection:

1. Write down that part of the indefinite-admittance matrix corresponding to the subnetwork composed of the one-port admittances according to the formulas

$$y_{ii} = \Sigma \text{ admittances incident at the terminal } i$$

and, for $i \neq j$,

$$y_{ij} = -\Sigma \text{ admittances connected between terminals } i \text{ and } j$$

2. The contribution resulting from the voltage-controlled current sources of Fig. 2.4 is equivalent to adding the matrices of the type Eq. (2.15).

The primitive indefinite-admittance matrix is then obtained by adding all the component matrices discussed above after bringing them to the same order by the insertion of an appropriate number of zero rows and columns.

Example 2.2 Consider the active network N of Fig. 2.5, in which all the five nodes are assumed to be accessible terminals. The indefinite-admittance matrix corresponding to the subnetwork composed of the five one-port admittances is given by

$$
\begin{bmatrix}
g_1 & -g_1 & 0 & 0 & 0 \\
-g_1 & g_1 + g_2 + sC & -g_2 & 0 & -sC \\
0 & -g_2 & g_2 + \dfrac{1}{sL} & -\dfrac{1}{sL} & 0 \\
0 & 0 & -\dfrac{1}{sL} & g_3 + \dfrac{1}{sL} & -g_3 \\
0 & -sC & 0 & -g_3 & g_3 + sC
\end{bmatrix}
\qquad (2.16)
$$

Figure 2.5   An active network used to illustrate the computation of the primitive indefinite-admittance matrix.

The indefinite-admittance matrix of the controlled source can be obtained in accordance with Eq. (2.15). The controlled source is connected between nodes 3 and 5, whose controlling voltage is the potential from node 2 to node 3. Thus, the rectangular pattern appears in rows 3 and 5 and columns 2 and 3 with the plus sign associated with column 2 and row 3. This yields

$$
\begin{array}{c}
 \\
 \\
\begin{array}{cc} + & - \end{array} \\
\begin{array}{ccccc} 1 & 2 & 3 & 4 & 5 \end{array}
\end{array}
\begin{array}{c}
1 \\ 2 \\ +\,3 \\ 4 \\ -\,5
\end{array}
\begin{bmatrix}
0 & 0 & 0 & 0 & 0 \\
0 & 0 & 0 & 0 & 0 \\
0 & \beta g_2 & -\beta g_2 & 0 & 0 \\
0 & 0 & 0 & 0 & 0 \\
0 & -\beta g_2 & \beta g_2 & 0 & 0
\end{bmatrix}
\tag{2.17}
$$

Observe that matrix Eq. (2.16) is symmetric because the corresponding subnetwork is reciprocal, and that in Eq. (2.17) if the sign associated with one of the terms βg_2 is known, the signs of other terms can be identified cyclically by assigning the plus and minus signs alternately as shown in Eq. (2.15). The complete primitive indefinite-admittance matrix of the active network is obtained simply by adding the two matrices Eqs. (2.16) and (2.17), and the result is given by

$$
\begin{bmatrix}
g_1 & -g_1 & 0 & 0 & 0 \\
-g_1 & g_1 + g_2 + sC & -g_2 & 0 & -sC \\
0 & \beta g_2 - g_2 & g_2 - \beta g_2 + \dfrac{1}{sL} & -\dfrac{1}{sL} & 0 \\
0 & 0 & -\dfrac{1}{sL} & g_3 + \dfrac{1}{sL} & -g_3 \\
0 & -\beta g_2 - sC & \beta g_2 & -g_3 & g_3 + sC
\end{bmatrix}
\tag{2.18}
$$

In fact, a moment's thought would indicate that the above two steps can be combined to yield the desired matrix directly from the given network. It is significant to point out that the one-port admittances g_1, g_2, g_3, sC, and $1/sL$ enter the indefinite-admittance matrix Eq. (2.18) in a beautifully symmetric form. Each admittance appears in a square pattern of four element positions, as in Eq. (2.15), but centered about the main diagonal. The two rows and two columns associated with the square pattern have the same labels as the nodes to which the branch in question

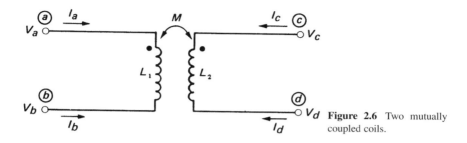

Figure 2.6 Two mutually coupled coils.

is connected in the network. The admittance enters with a positive sign on the main diagonal and with a negative sign for the off-diagonal elements. This follows directly from an earlier observation that a one-port admittance is a special type of voltage-controlled current source in which the controlling voltage is also its terminal voltage.

The rules outlined at the beginning of this section can be extended to include the situation where multiterminal devices are imbedded in the network. Like the voltage-controlled current source, all we need to derive is their associated indefinite-admittance matrices. The complete primitive indefinite-admittance matrix is then obtained by adding all the component matrices after bringing them to the same order by the insertion of an appropriate number of zero rows and columns.

An element that frequently appears in networks is the mutual inductance. Refer to Fig. 2.6; let V's and I's be the terminal voltages and currents. Then they are related by the equation

$$\begin{bmatrix} V_a - V_b \\ V_c - V_d \end{bmatrix} = \begin{bmatrix} L_1 s & M s \\ M s & L_2 s \end{bmatrix} \begin{bmatrix} I_a \\ I_c \end{bmatrix} \tag{2.19}$$

Assuming that the coupling is imperfect, that is, $L_1 L_2 - M^2 \neq 0$, Eq. (2.19) can be inverted, giving

$$\begin{bmatrix} I_a \\ I_c \end{bmatrix} = \frac{1}{s(L_1 L_2 - M^2)} \begin{bmatrix} L_2 & -M \\ -M & L_1 \end{bmatrix} \begin{bmatrix} V_a - V_b \\ V_c - V_d \end{bmatrix} \tag{2.20}$$

Since $I_b = -I_a$ and $I_d = -I_c$, Eq. (2.20) can be expanded and rewritten as

$$
\begin{bmatrix} I_a \\ I_b \\ I_c \\ I_d \end{bmatrix} = \frac{1}{s(L_1 L_2 - M^2)} \begin{bmatrix} L_2 & -L_2 & -M & M \\ -L_2 & L_2 & M & -M \\ -M & M & L_1 & -L_1 \\ M & -M & -L_1 & L_1 \end{bmatrix} \begin{bmatrix} V_a \\ V_b \\ V_c \\ V_d \end{bmatrix} \tag{2.21}
$$

the coefficient matrix being the indefinite-admittance matrix of the imperfectly coupled transformer of Fig. 2.6.

The ideal transformer by itself has infinite short-circuit admittances and hence cannot be formally characterized on the nodal basis. However, in practice a transformer usually appears in series with some other elements having finite admittance as shown in Fig. 2.7. In such a case, its indefinite-admittance matrix can be determined in a similar fashion (see Prob. 2.2):

$$
\begin{bmatrix} y_0 & -y_0 & -n y_0 & n y_0 \\ -y_0 & y_0 & n y_0 & -n y_0 \\ -n y_0 & n y_0 & n^2 y_0 & -n^2 y_0 \\ n y_0 & -n y_0 & -n^2 y_0 & n^2 y_0 \end{bmatrix} \tag{2.22}
$$

Another useful element having two pairs of terminals is the *gyrator*, the diagrammatic symbol for which is shown in Fig. 2.8. The element is characterized by the terminal relation

$$
\begin{bmatrix} I_a \\ I_c \end{bmatrix} = \begin{bmatrix} 0 & g \\ -g & 0 \end{bmatrix} \begin{bmatrix} V_a - V_b \\ V_c - V_d \end{bmatrix} \tag{2.23}
$$

Figure 2.7 An ideal transformer in series with some other element having finite admittance.

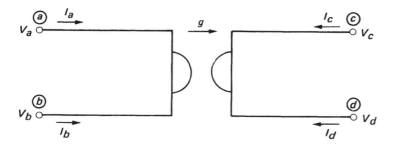

Figure 2.8   The diagrammatic symbol for a gyrator with gyration conductance g.

In the diagram, the direction of the arrow signifies the *direction of gyration*. Like the ideal transformer, the gyrator is characterized by a single parameter g, called the *gyration conductance*. The indefinite-admittance matrix of the gyrator of Fig. 2.8 is obtained as (see Prob. 2.4)

$$
\begin{bmatrix}
0 & 0 & g & -g \\
0 & 0 & -g & g \\
-g & g & 0 & 0 \\
g & -g & 0 & 0
\end{bmatrix}
\tag{2.24}
$$

Example 2.3 In the active network of Fig. 2.9, assume that all five nodes are accessible terminals. Applying rules to that part of the subnetwork composed of g_1, g_2, g_3, C, and the controlled source, we can write down the indefinite-admittance matrix directly from the network by inspection:

$$
\begin{array}{c}
\\
\\
1 \\
2 \\
+3 \\
4 \\
-5
\end{array}
\begin{array}{cc}
& + \\
1 \quad\;\; 2 \\
\begin{bmatrix}
0 & 0 \\
0 & g_2 + sC \\
0 & \beta g_2 - g_2 \\
0 & 0 \\
0 & -\beta g_2 - sC
\end{bmatrix}
\end{array}
$$

$$
\begin{array}{c}
\\
\\
1 \\
2 \\
+3 \\
4 \\
-5
\end{array}
\left[
\begin{array}{ccccc}
& & - & & \\
1 & 2 & 3 & 4 & 5 \\
0 & 0 & 0 & 0 & 0 \\
0 & g_2 + sC & -g_2 & 0 & -sC \\
0 & \beta g_2 - g_2 & g_1 + g_2 - \beta g_2 & 0 & -g_1 \\
0 & 0 & 0 & g_3 & -g_3 \\
0 & -\beta g_2 - sC & \beta g_2 - g_1 & -g_3 & g_1 + g_3 + sC
\end{array}
\right]
\tag{2.25}
$$

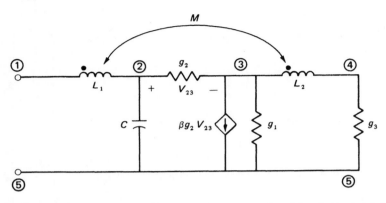

Figure 2.9 An active network used to illustrate the computation of the primitive indefinite-admittance matrix.

The matrix of the two imperfectly coupled coils can be obtained from the coefficient matrix of Eq. (2.21) by inserting a zero row at the bottom and a zero column at the right-hand side. The sum of this matrix and Eq. (2.25) yields the desired indefinite-admittance matrix of the network:

$$
\begin{bmatrix}
\dfrac{L_2}{\delta} & -\dfrac{L_2}{\delta} & -\dfrac{M}{\delta} & \dfrac{M}{\delta} & 0 \\[2.2ex]
-\dfrac{L_2}{\delta} & g_2 + sC + \dfrac{L_2}{\delta} & \dfrac{M}{\delta} - g_2 & -\dfrac{M}{\delta} & -sC \\[2.2ex]
-\dfrac{M}{\delta} & \beta g_2 - g_2 + \dfrac{M}{\delta} & g_1 + g_2 - \beta g_2 + \dfrac{L_1}{\delta} & -\dfrac{L_1}{\delta} & -g_1 \\[2.2ex]
\dfrac{M}{\delta} & -\dfrac{M}{\delta} & -\dfrac{L_1}{\delta} & g_3 + \dfrac{L_1}{\delta} & -g_3 \\[2.2ex]
0 & -\beta g_2 - sC & \beta g_2 - g_1 & -g_3 & g_1 + g_3 + sC
\end{bmatrix}
$$

$$(2.26)$$

where $\delta = s(L_1 L_2 - M^2)$

2.3 TERMINAL CONTRACTION AND SUPPRESSION

In the preceding section, we presented a procedure for writing down the primitive indefinite-admittance matrix of a network by inspection. Basic to the concept of the primitive indefinite-admittance matrix is the assumption that every node of the network is an accessible terminal. Also, not all the terminals of a multiterminal device are distinct; some of them may be combined together to form new terminals.

Thus, in order to apply the procedure we must know a way of finding the indefinite-admittance matrix of the network in terms of those discussed in the foregoing section.

2.3.1 Terminal Contraction

By the *contraction* of a multiterminal device is meant the joining together of two or more of its terminals. We recognize that a contraction involving the joining of more than two terminals can be regarded as a composition of contractions involving only two terminals.

Observe that the joining together of two or more terminals of a multiterminal device implies the equality of voltages and addition of currents at the composite terminal. As a consequence, the indefinite-admittance matrix of the resulting multiterminal device can be derived from that of the original one by adding the rows and columns corresponding to the terminals being joined together to form a new row and a new column for the composition terminal (see Prob. 2.5).

As an illustration, consider the two imperfectly coupled coils shown in Fig. 2.6. On joining together terminals b and d, we get a three-terminal device. Its indefinite-admittance matrix can be derived from the coefficient matrix of Eq. (2.21) by adding row 2 to row 4 and column 2 to column 4 and then deleting row 2 and column 2 in the resulting matrix. The result is given by

$$\frac{1}{s(L_1 L_2 - M^2)} \begin{bmatrix} L_2 & -M & M - L_2 \\ -M & L_1 & M - L_1 \\ M - L_2 & M - L_1 & L_1 + L_2 - 2M \end{bmatrix} \qquad (2.27)$$

2.3.2 Terminal Suppression

A terminal is said to be *suppressed* if it is open-circuited. In other words, the associated terminal current is constrained to be zero, and the terminal is inaccessible. As in the case of contraction, the suppression of two or more terminals can be regarded as a succession of suppressions each involving a single terminal. To obtain the indefinite-admittance matrix of a network that possesses inaccessible internal terminals from its primitive indefinite-admittance matrix, we simply suppress the internal terminals to yield only the accessible ones. In terms of network equations (2.1a), this procedure is equivalent to eliminating the unwanted variables.

Let $\mathbf{I}_b$, $\mathbf{J}_b$, and $\mathbf{V}_b$ be the vectors of currents and voltages corresponding to the terminals to be suppressed. Partition the matrix equation (2.1b) according to unsuppressed and suppressed terminals as follows[†]:

$$\begin{bmatrix} \mathbf{I}_a \\ \mathbf{I}_b \end{bmatrix} = \begin{bmatrix} \mathbf{W}_{11} & \mathbf{W}_{12} \\ \mathbf{W}_{21} & \mathbf{W}_{22} \end{bmatrix} \begin{bmatrix} \mathbf{V}_a \\ \mathbf{V}_b \end{bmatrix} + \begin{bmatrix} \mathbf{J}_a \\ \mathbf{J}_b \end{bmatrix} \tag{2.28}$$

where the elements in $\mathbf{I}_a$, $\mathbf{J}_a$, and $\mathbf{V}_a$ correspond to unsuppressed terminals. Since suppressing a terminal is equivalent to open-circuiting that terminal, this requires that we set $\mathbf{I}_b = \mathbf{0}$. Using this and assuming that $\mathbf{W}_{22}$ is nonsingular, we can eliminate $\mathbf{V}_b$ in Eq. (2.28), yielding

$$\mathbf{I}_a = (\mathbf{W}_{11} - \mathbf{W}_{12}\mathbf{W}_{22}^{-1}\mathbf{W}_{21})\mathbf{V}_a - \mathbf{W}_{12}\mathbf{W}_{22}^{-1}\mathbf{J}_b + \mathbf{J}_a \tag{2.29}$$

Thus, the new indefinite-admittance matrix is identified as

$$\mathbf{Y}'' = \mathbf{W}_{11} - \mathbf{W}_{12}\mathbf{W}_{22}^{-1}\mathbf{W}_{21} \tag{2.30}$$

In particular, if a single terminal is to be suppressed, the procedure is exceedingly simple. Under this situation, to suppress the kth terminal, the ith row and jth column element y_{ij}'' of $\mathbf{Y}''$ is given by

$$y_{ij}'' = y_{ij} - \frac{y_{ik}y_{kj}}{y_{kk}} \tag{2.31}$$

In other words, suppressing the kth terminal is equivalent to subtracting from each of the original admittances y_{ij} the amount $y_{ik}y_{kj}/y_{kk}$ to yield the element values of the new matrix. The details of this procedure are demonstrated below:

$$
\begin{array}{c}
 \quad\quad\quad\quad j \quad\quad\quad\quad\quad\quad\quad k \\
\begin{array}{c} i \\ \\ \\ k \end{array}
\left[
\begin{array}{ccc}
\cdots\cdots y_{ij} - \dfrac{y_{ik}y_{kj}}{y_{kk}} \cdots\cdots\cdots & y_{ik} & \cdots\cdots \\
& & \\
\rule{5cm}{0.4pt}\, y_{kj} \rule{2cm}{0.4pt} & y_{kk} & \rule{2cm}{0.4pt} \\
\end{array}
\right]
\end{array}
\tag{2.32}
$$

[†] As indicated in Eq. (2.2), if we consider all independent sources outside the network and set all initial conditions to zero, then $\mathbf{J}(s) = \mathbf{0}$. It is included here for completeness.

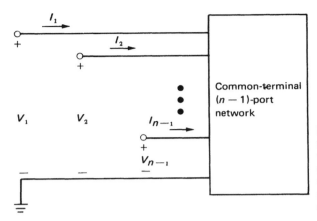

Figure 2.10 A common-terminal $(n - 1)$-port network obtained from an n-terminal network by pairing one of the terminals with $n - 1$ other terminals to form $n - 1$ ports.

We recognize that the result is identical with that yielded by the standard method of pivotal condensation. In determinantal form, the element y''_{ij} can be expressed as

$$y''_{ij} = \frac{1}{y_{kk}} \begin{vmatrix} y_{ij} & y_{ik} \\ y_{kj} & y_{kk} \end{vmatrix} \tag{2.33}$$

In general, if more than one terminal is suppressed, it can be shown from Eq. (2.30) that $(\det \mathbf{W_{22}}) y''_{ij}$ is equal to the determinant of the submatrix of $\mathbf{Y}$ formed by row i and column j and those rows and columns corresponding to the suppressed terminals.

After suppressing the unwanted terminals, an n-terminal network can be transformed into a common-terminal $(n - 1)$-port network by pairing one of the terminals, say, terminal n, with $n-1$ other terminals to form $n-1$ ports, as indicated in Fig. 2.10. In doing so, we also specify terminal n to be the reference potential point for $n - 1$ other terminals. This is equivalent to setting V_n in Eq. (2.1a) to zero. Hence the last column in $\mathbf{Y}(s)$ can be removed. Since, according to Kirchhoff's current law [see Eq. (2.6)], the last equation is redundant, the last row of $\mathbf{Y}(s)$ can also be removed. This results in the admittance matrix of the common-terminal $(n - 1)$-port network.

We shall illustrate the above results by the following examples.

Example 2.4 Consider the transistor network of Fig. 2.11, whose primitive indefinite-admittance matrix can easily be obtained by inspection according to the

Figure 2.11 The T-equivalent network of a transistor.

rules established in the preceding section, and is given by

$$
\begin{array}{c}
\quad\quad\overset{+}{1}\quad\quad 2\quad\quad 3\quad\quad\quad\overset{-}{4}\\
\begin{array}{c}1\\-\,2\\3\\+\,4\end{array}
\left[
\begin{array}{cccc}
g_e & 0 & 0 & -g_3\\
-\alpha g_3 & g_c & 0 & \alpha g_e - g_c\\
0 & 0 & g_b & -g_b\\
\alpha g_e - g_e & -g_c & -g_b & g_e + g_b + g_c - \alpha g_e
\end{array}
\right]
\end{array}
\tag{2.34}
$$

Suppose that terminal 4 is inaccessible and should be suppressed. By using Eqs. (2.30) or (2.31), the new indefinite-admittance matrix becomes

$$
\left[
\begin{array}{ccc}
g_e - \dfrac{g_e(1-\alpha)g_e}{w} & -\dfrac{g_e g_c}{w} & -\dfrac{g_e g_b}{w}\\[3mm]
-\alpha g_e - \dfrac{(g_c-\alpha g_e)(1-\alpha)g_e}{w} & g_c - \dfrac{g_c(g_c-\alpha g_e)}{w} & -\dfrac{g_b(g_c-\alpha g_e)}{w}\\[3mm]
-\dfrac{g_b(1-\alpha)g_e}{w} & -\dfrac{g_b g_c}{w} & g_b - \dfrac{g_b g_b}{w}
\end{array}
\right]
\tag{2.35}
$$

where $w = g_e + g_c + g_b - \alpha g_e$. By expressing $r_e = 1/g_e, r_b = 1/g_b,$ and $r_c = 1/g_c,$ matrix Eq. (2.35) can be simplified to

$$
\frac{1}{r_b r_c r_e w}
\left[
\begin{array}{ccc}
r_b + r_c & -r_b & -r_c\\[2mm]
-\alpha r_c - r_b & r_b + r_e & -r_e + \alpha r_c\\[2mm]
-(1-\alpha)r_c & -r_e & r_e + (1-\alpha)r_c
\end{array}
\right]
\tag{2.36}
$$

If node 3 is chosen as the common terminal, the short-circuit admittance matrix of the resulting two-port network is obtained from Eq. (2.36) by deleting row 3 and column 3, yielding

$$\mathbf{Y}_{sc} = \frac{1}{r_b r_c r_e w} \begin{bmatrix} r_b + r_c & -r_b \\ -\alpha r_c - r_b & r_b + r_e \end{bmatrix} \tag{2.37}$$

which is recognized to be the admittance matrix of the common-base transistor configuration. In fact, once the indefinite-admittance matrix Eq. (2.36) is known, the admittance matrices of the common-emitter and common-collector configurations are obtained by simply deleting row 1 and column 1, and row 2 and column 2, respectively. We shall discuss this aspect further in Sec. 2.4.

Suppose now that in Eq. (2.36) we continue our process by suppressing terminal 3. The indefinite-admittance matrix of the resulting two-terminal network becomes

$$\frac{1}{r_e + (1 - \alpha)r_c} \begin{bmatrix} 1 & -1 \\ -1 & 1 \end{bmatrix} \tag{2.38}$$

the diagonal elements being its input admittance. Alternatively, the elements of Eq. (2.38) can be computed directly from Eq. (2.34), as follows: Since we know in advance that matrix (2.38) must be an equicofactor matrix, it suffices to determine only its first row and first column element y_{11}'' by the procedure outlined in the paragraph following Eq. (2.33), yielding

$$y_{11}'' = \frac{1}{\det \mathbf{W}_{22}} \det \begin{bmatrix} g_e & 0 & -g_e \\ 0 & g_b & -g_b \\ \alpha g_e - g_e & -g_b & g_e + g_b + g_c - \alpha g_e \end{bmatrix}$$

$$= \frac{1}{r_e + (1 - \alpha)r_c} \tag{2.39}$$

where

$$\mathbf{W}_{22} = \begin{bmatrix} g_b & -g_b \\ -g_b & g_e + g_b + g_c - \alpha g_e \end{bmatrix} \tag{2.40}$$

which is the submatrix formed by the rows and columns corresponding to the suppressed terminals. The third-order matrix in Eq. (2.39) is the submatrix of Eq. (2.34) formed by rows 1, 3, and 4 and columns 1, 3, and 4.

Example 2.5 Figure 2.12 is the hybrid-pi equivalent network of a transistor. By applying the procedure outlined in Sec. 2.2, the primitive indefinite-admittance

Figure 2.12 The hybrid-pi equivalent network of a transistor.

matrix of the network can easily be obtained by inspection and is given by

$$
\begin{array}{c}
\\
+ \\
\\
-
\end{array}
\begin{bmatrix}
0.02 & 0 & -0.02 & 0 \\
0 & 5 \cdot 10^{-12}s & 0.2 - 5 \cdot 10^{-12}s & -0.2 \\
-0.02 & -5 \cdot 10^{-12}s & 0.024 + 105 \cdot 10^{-12}s & -0.004 - 10^{-10}s \\
0 & 0 & -0.204 - 10^{-10}s & 0.204 + 10^{-10}s
\end{bmatrix} \tag{2.41}
$$

Choosing node 4 as the reference potential and letting $p = s/10^9$ yield

$$
\begin{bmatrix}
0.02 & 0 & -0.02 \\
0 & 0.005p & 0.2 - 0.005p \\
-0.02 & -0.005p & 0.024 + 0.105p
\end{bmatrix} \tag{2.42}
$$

which is derived from Eq. (2.41) by deleting the fourth row and fourth column. The admittance matrix of the two-port network is obtained from Eq. (2.42) by suppressing the terminal corresponding to the third row or third column. Using Eqs. (2.30) or (2.31) gives

$$
\frac{1}{240 + 1050p}
\begin{bmatrix}
0.8 + 21p & -p \\
40 - p & 11.2p + 5p^2
\end{bmatrix} \tag{2.43}
$$

It is significant to observe that the order of the operations of deletion and suppressing is immaterial. Of course, the suppressed terminals cannot be deleted before the operations are completed.

2.4 INTERRELATIONSHIPS OF TRANSISTOR MODELS

The transistor is a three-terminal device as shown in Fig. 2.13. If any one of the terminals is chosen as the reference-potential point, the device can be transformed into the common-terminal two-port network of Fig. 2.14, which can then be characterized by the short-circuit admittance parameters, the open-circuit impedance parameters, or the hybrid parameters. In order to distinguish the parameters associated with various terminals and configurations, the following subscripts are commonly used:

i *i*nput
o *o*utput
f *f*orward transfer
r *reverse* transfer
b common-*b*ase
e common-*e*mitter
c common-*c*ollector

The first subscript (i, o, f, or r) identifies the element in the two-port parameter matrix, and the second (b, e, or c) determines the common terminal used. In Fig. 2.15, the base terminal is chosen as the input, the collector terminal as the output, and the emitter terminal as the common terminal. Then the transistor equations of the common-emitter configuration in terms of the short-circuit

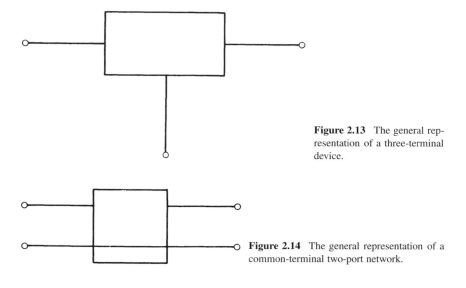

Figure 2.13 The general representation of a three-terminal device.

Figure 2.14 The general representation of a common-terminal two-port network.

Figure 2.15 The common-emitter representation of a transistor.

admittance parameters and the hybrid parameters can be written as

$$
\begin{bmatrix} I_b \\ I_c \end{bmatrix} = \begin{bmatrix} y_{ie} & y_{re} \\ y_{fe} & y_{oe} \end{bmatrix} \begin{bmatrix} V_{be} \\ V_{ce} \end{bmatrix}
\tag{2.44}
$$

$$
\begin{bmatrix} V_{be} \\ I_c \end{bmatrix} = \begin{bmatrix} h_{ie} & h_{re} \\ h_{fe} & h_{oe} \end{bmatrix} \begin{bmatrix} I_b \\ V_{ce} \end{bmatrix}
\tag{2.45}
$$

Thus, y_{ie} denotes the input admittance and h_{fe} the forward current-transfer ratio of the common-emitter configuration when the output port is short-circuited. In general, the admittance matrix and the hybrid matrix of a common-terminal two-port device can be expressed by

$$
\mathbf{Y}_x = \begin{bmatrix} y_{ix} & y_{rx} \\ y_{fx} & y_{ox} \end{bmatrix}
\tag{2.46a}
$$

$$
\mathbf{H}_x = \begin{bmatrix} h_{ix} & h_{rx} \\ h_{fx} & h_{ox} \end{bmatrix}
\tag{2.46b}
$$

where $x = b, e,$ or c. The elements of $\mathbf{Y}_x$ and $\mathbf{H}_x$ are related, and using the procedure outlined in Sec. 1.5, we can express the elements of $\mathbf{Y}_x$ in terms of those of $\mathbf{H}_x$ and vice versa. The results are given by

$$
\mathbf{Y}_x = \begin{bmatrix} y_{ix} & y_{rx} \\ y_{fx} & y_{ox} \end{bmatrix} = \frac{1}{h_{ix}} \begin{bmatrix} 1 & -h_{rx} \\ h_{fx} & \Delta_{hx} \end{bmatrix}
\tag{2.47a}
$$

$$
\mathbf{H}_x = \begin{bmatrix} h_{ix} & h_{rx} \\ h_{fx} & h_{ox} \end{bmatrix} = \frac{1}{y_{ix}} \begin{bmatrix} 1 & -y_{rx} \\ y_{fx} & \Delta_{yx} \end{bmatrix}
\tag{2.47b}
$$

where

$$
\Delta_{hx} = \det \mathbf{H}_x = h_{ix}h_{ox} - h_{rx}h_{fx}
\tag{2.48a}
$$

$$
\Delta_{yx} = \det \mathbf{Y}_x = y_{ix}y_{ox} - y_{rx}y_{fx}
\tag{2.48b}
$$

The most important mode of operation for transistors in practical applications is the common-emitter configuration, and the transistor parameters are usually specified by the manufacturer in this mode. In the following, we shall derive formulas expressing the parameters of other configurations in terms of those of the common-emitter connection.

2.4.1 From the Common-Emitter Admittance Model to Other Models

The common-emitter configuration of a transistor is shown in Fig. 2.16a and its admittance model is presented in Fig. 2.16b. From Fig. 2.16b, the indefinite-admittance matrix is given by

$$
\mathbf{Y} = \begin{array}{c} \\ b \\ e \\ c \end{array}
\begin{array}{ccc} b & e & c \end{array}
\begin{bmatrix}
y_{ie} & -y_{ie} - y_{re} & y_{re} \\
-y_{fe} - y_{ie} & y_{fe} + y_{ie} + y_{oe} + y_{re} & -y_{oe} - y_{re} \\
y_{fe} & -y_{fe} - y_{oe} & y_{oe}
\end{bmatrix} \qquad (2.49)
$$

To obtain the admittance matrix of the common-base configuration, we delete row 1 and column 1 from $\mathbf{Y}$, resulting in

$$
\mathbf{Y}_b = \begin{bmatrix} y_{ib} & y_{rb} \\ y_{fb} & y_{ob} \end{bmatrix} = \begin{bmatrix} y_{fe} + y_{ie} + y_{oe} + y_{re} & -y_{oe} - y_{re} \\ -y_{fe} - y_{oe} & y_{oe} \end{bmatrix} \qquad (2.50)
$$

whose network model is as shown in Fig. 2.17. Applying Eq. (2.47b), we obtain the hybrid parameters of the common-base configuration of Fig. 2.18:

$$
\mathbf{H}_b = \begin{bmatrix} h_{ib} & h_{rb} \\ h_{fb} & h_{ob} \end{bmatrix} = \frac{1}{y_{fe} + y_{ie} + y_{oe} + y_{re}} \begin{bmatrix} 1 & y_{oe} + y_{re} \\ -y_{fe} - y_{oe} & y_{ie}y_{oe} - y_{fe}y_{re} \end{bmatrix}
$$

$$(2.51)$$

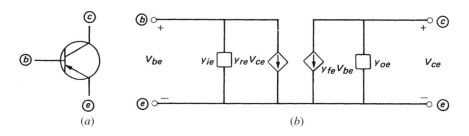

Figure 2.16 (a) The common-emitter configuration of a transistor and (b) its admittance model.

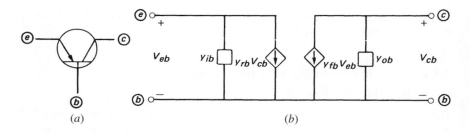

Figure 2.17 (*a*) The common-base configuration of a transistor and (*b*) its admittance model.

Figure 2.18 (*a*) The common-base configuration of a transistor and (*b*) its hybrid model.

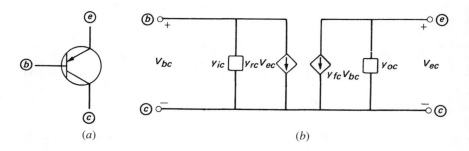

Figure 2.19 (*a*) The common-collector configuration of a transistor and (*b*) its admittance model.

For the admittance matrix of the common-collector configuration of Fig. 2.19, we delete row 3 and column 3 from $\mathbf{Y}$, yielding

$$\mathbf{Y}_c = \begin{bmatrix} y_{ic} & y_{rc} \\ y_{fc} & y_{oc} \end{bmatrix} = \begin{bmatrix} y_{ie} & -y_{ie} - y_{re} \\ -y_{fe} - y_{ie} & y_{fe} + y_{ie} + y_{oe} + y_{re} \end{bmatrix} \qquad (2.52)$$

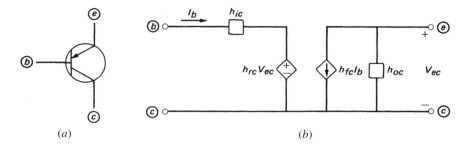

Figure 2.20 (*a*) The common-collector configuration of a transistor and (*b*) its hybrid model.

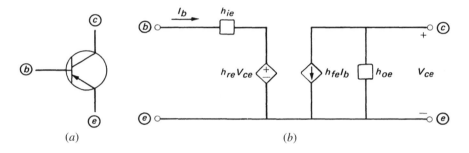

Figure 2.21 (*a*) The common-emitter configuration of a transistor and (*b*) its hybrid model.

and the corresponding hybrid matrix of Fig. 2.20 is found to be

$$\mathbf{H}_c = \begin{bmatrix} h_{ic} & h_{rc} \\ h_{fc} & h_{oc} \end{bmatrix} = \frac{1}{y_{ie}} \begin{bmatrix} 1 & y_{ie} + y_{re} \\ -y_{fe} - y_{ie} & y_{ie}y_{oe} - y_{fe}y_{re} \end{bmatrix} \tag{2.53}$$

We remark that the hybrid matrix of the common-emitter configuration of Fig. 2.21 is that of Eq. (2.47*b*) with *e* replacing *x* in all the second subscripts.

The transistor parameters vary with changes in the temperature as well as with bias-point variations. In normal operation, some of the parameters can be ignored in comparison with others. The values of the parameters listed below are typical for a variety of types of transistors for the small-signal, low-frequency, two-port model of Fig. 2.16*b*:

$$y_{ie} = 7 \cdot 10^{-4} \, \text{mho} \qquad y_{fe} = 0.07 \, \text{mho}$$
$$y_{re} = -7 \cdot 10^{-8} \, \text{mho} \qquad y_{oe} = 8 \, \mu\text{mho} \tag{2.54}$$

Hence, y_{re} and y_{oe} can usually be ignored in actual computations of various parameters. This approximation provides considerable simplification in design. However, at higher frequencies the admittances may be complex, and care must be taken in the simplification. For example, considering the transistor 2N697 operating at 30 MHz with $V_{CE} = 20$ V and $I_C = 20$ mA, we have[†]

$$
\begin{aligned}
y_{ie} &= (22.5 + j14.7) \cdot 10^{-3} \text{ mho} \\
y_{re} &= (-0.8 - j0.38) \cdot 10^{-3} \text{ mho} \\
y_{fe} &= (36.6 - j91.6) \cdot 10^{-3} \text{ mho} \\
y_{oe} &= (1.7 + j5.7) \cdot 10^{-3} \text{ mho}
\end{aligned}
\tag{2.55}
$$

2.4.2 From the Common-Emitter Hybrid Model to Other Models

In this section, we derive formulas expressing parameters of various models in terms of those of the common-emitter hybrid configuration.

Figure 2.21b is the common-emitter hybrid model of a transistor, whose admittance matrix is given by Eq. (2.47a) as

$$
\mathbf{Y}_e = \begin{bmatrix} y_{ie} & y_{re} \\ y_{fe} & y_{oe} \end{bmatrix} = \frac{1}{h_{ie}} \begin{bmatrix} 1 & -h_{re} \\ h_{fe} & \Delta_{he} \end{bmatrix}
\tag{2.56}
$$

where $\Delta_{he} = h_{ie}h_{oe} - h_{re}h_{fe}$. The indefinite-admittance matrix may now be formed by adding the row and column corresponding to the emitter terminal. The elements of the new row and the new column can easily be determined from Eq. (2.56) in that each row sum and each column sum in the resulting matrix must be identically zero. For our purposes, the added row and column will occupy the second row and second column positions. We obtain

$$
\mathbf{Y} = \frac{1}{h_{ie}} \begin{array}{c} \begin{array}{ccc} \quad b & \qquad\quad e & \qquad\quad c \end{array} \\ \begin{bmatrix} 1 & h_{re} - 1 & -h_{re} \\ -1 - h_{fe} & 1 + h_{fe} - h_{re} + \Delta_{he} & h_{re} - \Delta_{he} \\ h_{fe} & -h_{fe} - \Delta_{he} & \Delta_{he} \end{bmatrix} \begin{array}{c} b \\ e \\ c \end{array} \end{array}
\tag{2.57}
$$

[†]Uppercase E, C, or B is used to represent the quiescent value of voltage or current, as is commonly done in electronics literature.

To obtain the admittance matrix of the common-base configuration of Fig. 2.17, we delete row 1 and column 1 from $\mathbf{Y}$, giving

$$\mathbf{Y}_b = \frac{1}{h_{ie}} \begin{bmatrix} 1 + h_{fe} - h_{re} + \Delta_{he} & h_{re} - \Delta_{he} \\ -h_{fe} - \Delta_{he} & \Delta_{he} \end{bmatrix} \tag{2.58}$$

Applying Eq. (2.47b), we obtain the hybrid matrix of the common-base configuration of Fig. 2.18:

$$\mathbf{H}_c = \frac{1}{1 + h_{fe} - h_{re} + \Delta_{he}} \begin{bmatrix} h_{ie} & \Delta_{he} - h_{re} \\ -h_{fe} - \Delta_{he} & h_{oe} \end{bmatrix} \tag{2.59}$$

In a similar way, if we delete row 3 and column 3 from $\mathbf{Y}$, we obtain the admittance matrix of the common-collector configuration of Fig. 2.19 as

$$\mathbf{Y}_c = \frac{1}{h_{ie}} \begin{bmatrix} 1 & h_{re} - 1 \\ -1 - h_{fe} & 1 + h_{fe} - h_{re} + \Delta_{he} \end{bmatrix} \tag{2.60}$$

and appealing once more to Eq. (2.47b), we obtain the hybrid matrix of Fig. 2.20:

$$\mathbf{H}_c = \begin{bmatrix} h_{ie} & 1 - h_{re} \\ -1 - h_{fe} & h_{oe} \end{bmatrix} \tag{2.61}$$

Typical values of the hybrid parameters of a transistor operating at an emitter current $I_E = 1.3\,\text{mA}$ for the small-signal, low-frequency, two-port model of Fig. 2.21 are given by

$$\begin{aligned} h_{ie} &= 1.1\,\text{k}\Omega & h_{fe} &= 50 \\ h_{re} &= 2.5 \cdot 10^{-4} & h_{oe} &= 25\,\mu\text{mho} \end{aligned} \tag{2.62}$$

Thus, normally, we can assume that $h_{re} \ll 1$ and $\Delta_{he} \ll h_{fe}$, which leads to considerable simplification of the formulas Eqs. (2.58)–(2.61).

2.4.3 From the Common-Base T-Model to Other Models

Another common way of characterizing a transistor is by means of its equivalent T-model as shown in Fig. 2.22. This model is valid for small-signal, low-frequency operations. At higher frequencies, the model must be modified like that shown in Fig. 2.23, in which two capacitances are required. One is the junction capacitance C_c of the reverse-biased collector junction, and the other is the capacitance C_e comprised of the junction capacitance of the forward-biased emitter junction and

Figure 2.22 The equivalent T-model of a transistor.

the diffusion capacitance, which is the domain component. A typical range of values of these parameters is listed below:

$$r_e = 5\,\Omega\,(\text{at } 25^\circ\text{C and } 5\,\text{mA}) \qquad r_c = 1\,\text{M}\Omega$$

$$\alpha_0 = 0.95{\sim}0.99 \qquad C_c = 1{\sim}20\,\text{pF}$$

$$r_b = 20{\sim}100\,\Omega \quad C_e = 20{\sim}5000\,\text{pF} \tag{2.63}$$

$$\alpha = \frac{\alpha_0}{1 + j\omega/\omega_\alpha} \qquad \omega_\alpha = 4\cdot 10^9\,\text{rad/s}$$

In this section, we derive formulas expressing parameters of various models in terms of those of Fig. 2.22.

As shown in Example 2.4, the indefinite-admittance matrix of the T-model of Fig. 2.22, after an interchange of rows and columns, can be written as

$$
\mathbf{Y} = \frac{1}{r_e(r_b + r_c) + (1-\alpha)r_b r_c}
\begin{array}{c}
\begin{array}{ccc} b & e & c \end{array} \\
\begin{bmatrix}
r_e + (1-\alpha)r_c & -(1-\alpha)r_c & -r_e \\
-r_c & r_b + r_c & -r_b \\
\alpha r_c - r_e & -\alpha r_c - r_b & r_b + r_e
\end{bmatrix}
\begin{array}{c} b \\ e \\ c \end{array}
\end{array}
\tag{2.64}
$$

For the common-base configuration of Fig. 2.17, we delete row 1 and column 1 from $\mathbf{Y}$, resulting in

$$
\mathbf{Y}_b = \frac{1}{r_e(r_b + r_c) + (1-\alpha)r_b r_c}
\begin{bmatrix}
r_b + r_c & -r_b \\
-\alpha r_c - r_b & r_b + r_e
\end{bmatrix}
\tag{2.65}
$$

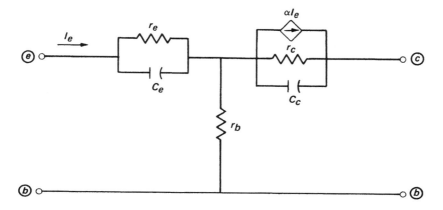

Figure 2.23 The high-frequency equivalent T-model of a transistor.

and from Eq. (2.47b) we obtain the hybrid matrix of Fig. 2.18 as

$$\mathbf{H}_b = \frac{1}{r_b + r_c} \begin{bmatrix} r_e(r_b + r_c) + (1 - \alpha)r_b r_c & r_b \\ -\alpha r_c - r_b & 1 \end{bmatrix} \qquad (2.66a)$$

$$\mathbf{H}_b \approx \begin{bmatrix} r_e + (1 - \alpha)r_b & \frac{r_b}{r_c} \\ -\alpha & \frac{1}{r_c} \end{bmatrix} \qquad (2.66b)$$

The second equation Eq. (2.66b) is an approximation and is obtained by invoking $r_b \ll r_c$ and $\alpha \approx 1$. In fact, from Eq. (2.66b) we can identify the elements of the T-model in terms of those of $\mathbf{H}_b$:

$$\alpha \approx -h_{fb} \quad r_c \approx \frac{1}{h_{ob}}$$
$$r_b \approx \frac{h_{rb}}{h_{ob}} \quad r_e \approx h_{ib} - \frac{(1 + h_{fb})h_{rb}}{h_{ob}} \qquad (2.67)$$

To obtain the admittance matrix of the common-emitter configuration of Fig. 2.16, we delete row 2 and column 2 from $\mathbf{Y}$, yielding

$$\mathbf{Y}_e = \frac{1}{r_e(r_b + r_c) + (1 - \alpha)r_b r_c} \begin{bmatrix} r_e + (1 - \alpha)r_c & -r_e \\ \alpha r_c - r_e & r_b + r_e \end{bmatrix} \qquad (2.68)$$

and by appealing to Eq. (2.47b) we get the hybrid matrix of Fig. 2.21 as

$$\mathbf{H}_e = \frac{1}{r_e + (1-\alpha)r_c} \begin{bmatrix} r_e(r_b + r_c) + (1-\alpha)r_b r_c & r_e \\ \alpha r_c - r_e & 1 \end{bmatrix} \quad (2.69a)$$

$$\approx \frac{1}{(1-\alpha)r_c} \begin{bmatrix} r_e r_c + (1-\alpha)r_b r_c & r_e \\ \alpha r_c & 1 \end{bmatrix} \quad (2.69b)$$

As in Eq. (2.67), the elements of Fig. 2.22 can be expressed in terms of those of Fig. 2.21 as follows:

$$\alpha = \frac{h_{fe}}{1 + h_{fe}} \quad r_c \approx \frac{1 + h_{fe}}{h_{oe}}$$

$$r_b \approx h_{ie} - \frac{(1 + h_{fe})h_{re}}{h_{oe}} \quad r_e \approx \frac{h_{re}}{h_{oe}} \quad (2.70)$$

Consider the inverse $\mathbf{Z}_e$ of $\mathbf{Y}_e$ of Eq. (2.68):

$$\mathbf{Z}_e = \begin{bmatrix} r_b + r_e & r_e \\ r_e - \alpha r_c & r_e + (1-\alpha)r_c \end{bmatrix} \quad (2.71)$$

This matrix is identified as the impedance matrix of the two-port network of Fig. 2.24. Thus, Fig. 2.24 is another T-model for the transistor operating in the common-emitter mode.

Finally, the admittance matrix and the hybrid matrix of the common-collector configurations of Figs. 2.19 and 2.20 are obtained from $\mathbf{Y}$ as follows:

$$\mathbf{Y}_c = \frac{1}{r_e(r_b + r_c) + (1-\alpha)r_b r_c} \begin{bmatrix} r_e + (1-\alpha)r_c & -(1-\alpha)r_c \\ -r_c & r_b + r_c \end{bmatrix} \quad (2.72)$$

Figure 2.24 The equivalent T-model of a transistor operating in the common-emitter mode.

$$\mathbf{H}_c = \frac{1}{r_e + (1-\alpha)r_c} \begin{bmatrix} r_e(r_b + r_c) + (1-\alpha)r_b r_c & (1-\alpha)r_c \\ -r_c & 1 \end{bmatrix} \quad (2.73a)$$

$$\approx \begin{bmatrix} r_b + \dfrac{r_e}{1-\alpha} & 1 \\ -\dfrac{1}{1-\alpha} & \dfrac{1}{(1-\alpha)r_c} \end{bmatrix} \quad (2.73b)$$

From Eq. (2.73*b*) we can identify

$$\alpha \approx \frac{1 + h_{fc}}{h_{fc}} \qquad r_c \approx -\frac{h_{fc}}{h_{oc}}$$

$$r_b \approx h_{ic} + \frac{(1 - h_{rc})h_{fc}}{h_{oc}} \qquad r_e \approx \frac{1 - h_{rc}}{h_{oc}} \quad (2.73c)$$

2.4.4 Conversion Formulas among the Hybrid Parameters

In addition to the conversion formulas derived in Sec. 2.4.2 from the common-emitter hybrid configuration to other configurations, similar relationships among the hybrid parameters of different configurations are summarized below with the approximate formulas being given in Probs. 2.31, 2.32, and 2.40:

$$\begin{bmatrix} h_{ie} & h_{re} \\ h_{fe} & h_{oe} \end{bmatrix} = \frac{1}{D_b} \begin{bmatrix} h_{ib} & \Delta_{hb} - h_{rb} \\ -\Delta_{hb} - h_{fb} & h_{ob} \end{bmatrix} = \begin{bmatrix} h_{ic} & 1 - h_{rc} \\ -(1 + h_{fc}) & h_{oc} \end{bmatrix}$$

$$(2.74a)$$

$$\begin{bmatrix} h_{ib} & h_{rb} \\ h_{fb} & h_{ob} \end{bmatrix} = \frac{1}{D_e} \begin{bmatrix} h_{ie} & \Delta_{he} - h_{re} \\ -\Delta_{he} - h_{fe} & h_{oe} \end{bmatrix} = \frac{1}{\Delta_{hc}} \begin{bmatrix} h_{ic} & h_{fc} + \Delta_{hc} \\ h_{rc} - \Delta_{rc} & h_{oc} \end{bmatrix}$$

$$(2.74b)$$

$$\begin{bmatrix} h_{ic} & h_{rc} \\ h_{fc} & h_{oc} \end{bmatrix} = \frac{1}{D_b} \begin{bmatrix} h_{ib} & 1 + h_{fb} \\ h_{rb} - 1 & h_{ob} \end{bmatrix} = \begin{bmatrix} h_{ie} & 1 - h_{re} \\ -(1 + h_{fe}) & h_{oe} \end{bmatrix}$$

$$(2.74c)$$

where

$$D_x = (1 + h_{fx})(1 - h_{rx}) + h_{ox}h_{ix} \quad (2.74d)$$

$$\Delta_{hx} = h_{ix}h_{ox} - h_{rx}h_{fx} \quad (2.74e)$$

2.5 THE FIRST- AND SECOND-ORDER COFACTORS

For an indefinite-admittance matrix $\mathbf{Y}(s)$ or simply $\mathbf{Y}$, denote by $\mathbf{Y}_{ij}$ the submatrix obtained from $\mathbf{Y}$ by deleting the ith row and jth column. Write $\mathbf{Y}$ explicitly as in Eq. (2.1a). Following the notation adopted in the proof of Theorem 2.1, the (*first-order*) *cofactor*, denoted by the symbol Y_{ij} of the element y_{ij} of $\mathbf{Y}$, as is well known, is defined by

$$Y_{ij} = (-1)^{i+j} \det \mathbf{Y}_{ij} \tag{2.75}$$

Denote by $\mathbf{Y}_{pq,rs}$ the submatrix derived from $\mathbf{Y}$ by striking out rows p and r and columns q and s. Then the *second-order cofactor*, denoted by the symbol $Y_{pq,rs}$, of the elements y_{pq} and y_{rs} of $\mathbf{Y}$ is a scalar quantity defined by the relation

$$Y_{pq,rs} = \text{sgn}\,(p-r)\text{sgn}\,(q-s)(-1)^{p+q+r+s}\,\det \mathbf{Y}_{pq,rs} \tag{2.76}$$

where $p \neq r$ and $q \neq s$; and $\text{sgn}\,u = +1$ if $u > 0$ and $\text{sgn}\,u = -1$ if $u < 0$.

The symbols $\mathbf{Y}_{ij}$ and Y_{ij} or $\mathbf{Y}_{pq,\mathrm{YS}}$ and $Y_{pq,rs}$ should create no confusion, because one is in boldface whereas the other is not. Also, for our purposes, it is convenient to define $Y_{pq,rs} = 0$ for $p = r$ or $q = s$, or $\text{sgn}\,0 = 0$. This convention will be followed throughout the remainder of this book.

As an illustration, consider the matrix $\mathbf{Y}$ given in (2.41). The second-order cofactors $Y_{31,42}$ and $Y_{11,34}$ of the elements of $\mathbf{Y}$ are given by

$$Y_{31,42} = \text{sgn}\,(3-4)\text{sgn}(1-2)(-1)^{3+1+4+2}\det \begin{bmatrix} -0.02 & 0 \\ 0.2 - 5 \cdot 10^{-12}s & -0.2 \end{bmatrix}$$

$$= 0.004 \tag{2.77}$$

$$Y_{11,34} = \text{sgn}(1-3)\text{sgn}(1-4)(-1)^{1+1+3+4}\det \begin{bmatrix} 5 \cdot 10^{-12}s & 0.2 - 5 \cdot 10^{-12}s \\ 0 & -0.204 - 10^{-10}s \end{bmatrix}$$

$$= 5 \cdot 10^{-12}s(0.204 + 10^{-10}s) \tag{2.78}$$

As indicated in Theorem 2.1, the indefinite-admittance matrix $\mathbf{Y}$ is an equicofactor matrix, meaning that all the first-order cofactors of its elements are equal. In the following, we shall demonstrate that not all of its second-order cofactors are independent; some may be obtained from the others by linear combinations. In fact, we shall show that there are only $(n-1)^2$ independent second-order cofactors for $\mathbf{Y}(s)$.

Theorem 2.2 The second-order cofactors of the elements of the indefinite-admittance matrix are related by

$$Y_{pq,rs} = Y_{pq,uv} + Y_{rs,uv} - Y_{ps,uv} - Y_{rq,uv} \qquad (2.79)$$

for arbitrary u and v.

PROOF. Assume that $p \neq u \neq r$ and $q \neq v \neq s$. Then in $\mathbf{Y}_{pq,rs}$, replace the column corresponding to the vth column of $\mathbf{Y}$ by the sum of all the columns of $\mathbf{Y}_{pq,rs}$. By virtue of Eq. (2.10), the element of this column in the ith row is now $-(y_{iq} + y_{is})$. If we remove the -1 from this column and let the resulting matrix be denoted by $\hat{\mathbf{Y}}_{pq,rs}$, then $\det \mathbf{Y}_{pq,rs} = -\det \hat{\mathbf{Y}}_{pq,rs}$. In $\hat{\mathbf{Y}}_{pq,rs}$, let $\hat{\mathbf{Y}}_1$ and $\hat{\mathbf{Y}}_2$ be the matrices obtained from $\hat{\mathbf{Y}}_{pq,rs}$ by setting $y_{is} = 0$ and $y_{iq} = 0$ $(i = 1, 2, \ldots, n)$, respectively. It is obvious that we have $\det \hat{\mathbf{Y}}_{pq,rs} = \det \hat{\mathbf{Y}}_1 + \det \hat{\mathbf{Y}}_2$.

At this stage, the numerical order in which the columns q, s, and v occur is very relevant. Apart from a sign, $\det \hat{\mathbf{Y}}_1$ and $\det \hat{\mathbf{Y}}_2$ are the second-order cofactors $Y_{ps,rv}$, and $Y_{pq,rv}$, respectively.

In $\hat{\mathbf{Y}}_1$, let us shift the column containing the elements y_{iq} to the position such that the columns in the resulting matrix appear in the same relative order as those in $\mathbf{Y}$. Doing so requires $v - q - 1$ adjacent transpositions for the cases $s < q < v$ and $q < v < s$; $q - v - 1$ for $s < v < q$ and $v < q < s$; $v - q - 2$ for $q < s < v$; and $q - v - 2$ for $v < s < q$. It is not difficult to check that all the cases are clearly contained in the single equation

$$\det \hat{\mathbf{Y}}_1 = \text{sgn } (q - s)\text{sgn}(v - s)(-1)^{v-q-1} \det \mathbf{Y}_{ps,rv} \qquad (2.80)$$

Thus, we have

$$Y_{ps,rv} = \text{sgn}(p - r)\text{sgn } (s - v)(-1)^{p+s+r+v} \det \mathbf{Y}_{ps,rv}$$
$$= \text{sgn } (p - r)\text{sgn } (q - s)(-1)^{p+q+r+s} \det \hat{\mathbf{Y}}_1 \qquad (2.81)$$

In a similar way, we can show that

$$Y_{pq,rv} = \text{sgn}(p - r)\text{sgn}(s - q)(-1)^{p+s+r+q} \det \hat{\mathbf{Y}}_2 \qquad (2.82)$$

Combining these results yields

$$Y_{pq,rs} = \text{sgn } (p - r) \text{ sgn } (q - s)(-1)^{p+q+r+s} \det \mathbf{Y}_{pq,rs}$$
$$= \text{sgn } (p - r) \text{ sgn } (q - s)(-1)^{p+q+r+s-1}(\det \hat{\mathbf{Y}}_1 + \det \hat{\mathbf{Y}}_2)$$
$$= Y_{pq,rv} - Y_{ps,rv} \qquad (2.83)$$

Observe that in deriving Eq. (2.83) we have invoked only the fact that each row sum of **Y** equals zero. Now consider the transpose **Y**′ of **Y** and appeal to Eq. (2.7), which states that each column sum of **Y** or each row sum of **Y**′ equals zero. By virtue of Eq. (2.83), we have

$$Y'_{pq,rs} = Y'_{pq,rv} - Y'_{ps,rv} \tag{2.84}$$

or, equivalently,

$$Y_{qp,sr} = Y_{qp,vr} - Y_{sp,vr} \tag{2.85}$$

By using Eq. (2.85), the two terms on the right-hand side of Eq. (2.83) can be expressed as

$$Y_{pq,rv} = Y_{pq,uv} - Y_{rq,uv} \tag{2.86a}$$

$$Y_{ps,rv} = Y_{ps,uv} - Y_{rs,uv} \tag{2.86b}$$

Substituting (2.86) in Eq. (2.83) yields the desired formula Eq. (2.79).

Suppose that the conditions $p \neq u \neq r$ and $q \neq v \neq s$ are not fulfilled. It is easy to check that either Eq. (2.79) is trivially satisfied or it reduces to Eqs. (2.83) or (2.85), which in turn is either trivially satisfied or can be established without invoking the assumption (see Prob. 2.27). This completes the proof of the theorem.

As an example, consider the indefinite-admittance matrix **Y** given in Eq. (2.34). The following second-order cofactors are computed:

$$Y_{12,44} = \alpha g_e g_b \tag{2.87a}$$

$$Y_{33,44} = g_e g_c \tag{2.87b}$$

$$Y_{13,44} = 0 \tag{2.87c}$$

$$Y_{32,44} = 0 \tag{2.87d}$$

$$Y_{12,33} = \alpha g_e g_b + g_e g_c \tag{2.87e}$$

Thus, we have

$$Y_{12,33} = Y_{12,44} + Y_{33,44} - Y_{13,44} - Y_{32,44} \tag{2.88}$$

confirming Eq. (2.79).

It is significant to note that, by virtue of Eq. (2.79), all the second-order cofactors of the elements of **Y** can be obtained linearly in terms of the first-order cofactors of the elements of $\mathbf{Y}_{nn}$. Hence there are only $(n-1)^2$ linearly independent second-order cofactors for the nth-order indefinite-admittance matrix.

2.6 COMPUTATION OF NETWORK FUNCTIONS

The usefulness of the indefinite-admittance matrix lies in the fact that it facilitates the formulation of the driving-point or transfer functions between any pair of nodes or from any pair of nodes to any other pair of nodes in the network. In the present section, we derive explicit formulas that express the network functions in terms of the ratios of the first- and/or second-order cofactors of the elements of the indefinite-admitíance matrix in an elegant and compact form.

2.6.1 Transfer Impedance and Voltage Gain

Assume that a current source is connected between any two nodes r and s so that a current I_{sr} is injected into the rth node and at the same time is extracted from the sth node. Suppose also that an ideal voltmeter is connected from node p to node q so that it indicates the potential rise from node q to node p, as depicted symbolically in Fig. 2.25. Then the *transfer impedance*, denoted by the symbol $z_{rp,sq}$, between the node pairs rs and pq of the network of Fig. 2.25 is defined by the relation

$$z_{rp,sq} = \frac{V_{pq}}{I_{sr}} \tag{2.89}$$

with all initial conditions and independent sources in the network N being set to zero. The representation is, of course, quite general. When $r = p$ and $s = q$, the transfer impedance $z_{rp,sq}$ becomes the *driving-point impedance $z_{rr,ss}$* between the terminal pair rs of the network.

With reference to Fig. 2.25, set all initial conditions and independent sources in N to zero and choose terminal or node q to be the reference-potential point for all other terminals. In terms of the equations of (2.1), the operations are equivalent to setting $\mathbf{J} = \mathbf{0}$, $V_q = 0$, $I_x = 0$ for $x \neq r, s$ and $I_r = -I_s = I_{sr}$. Since $\mathbf{Y}$ is an equicofactor matrix, the equations of (2.1) are not linearly independent and so

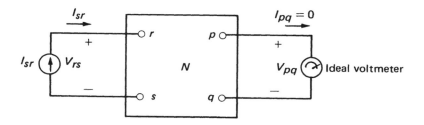

Figure 2.25 The symbolic representation for the measurement of the transfer impedance.

one of them is superfluous. Let us suppress the sth equation from (2.1), which then reduces to

$$\mathbf{I}_{-s} = \mathbf{Y}_{sq}\mathbf{V}_{-q} \tag{2.90}$$

where $\mathbf{I}_{-s}$ and $\mathbf{V}_{-q}$ denote the subvectors obtained from $\mathbf{I}$ and $\mathbf{V}$ of Eq. (2.2) by deleting the sth row and qth row, respectively. Applying Cramer's rule to solve for V_p yields

$$V_P = \frac{\det \tilde{\mathbf{Y}}_{sq}}{\det \mathbf{Y}_{sq}} \tag{2.91}$$

where $\tilde{\mathbf{Y}}_{sq}$ is the matrix derived from $\mathbf{Y}_{sq}$ by replacing the column corresponding to V_p by $\mathbf{I}_{-s}$. We now expand $\det \mathbf{Y}_{sq}$ along the column $\mathbf{I}_{-s}$ that has $n - 2$ zeros and I_{sr}. We recognize that $\mathbf{I}_{-s}$ is in the pth column if $p < q$ but in the $(p - 1)$th column if $p > q$. Furthermore, the row in which I_{sr} appears is the rth row if $r < s$ but is the $(r - 1)$th row if $r > s$. With these in mind, it is not difficult to confirm that

$$(-1)^{s+q} \det \tilde{\mathbf{Y}}_{sq} = I_{sr}Y_{rp,sq} \tag{2.92}$$

In addition, we have

$$\det \mathbf{Y}_{sq} = (-1)^{s+q}Y_{sq} \tag{2.93}$$

Substituting Eqs. (2.92) and (2.93) in (2.91) in conjunction with Eq. (2.89), we obtain

$$z_{rp,sq} = \frac{Y_{rp,sq}}{Y_{uv}} \tag{2.94}$$

$$z_{rr,ss} = \frac{Y_{rr,ss}}{Y_{uv}} \tag{2.95}$$

in which we have invoked the fact that $Y_{sq} = Y_{uv}$.

The *voltage gain*, denoted by the symbol $g_{rp,sq}$, between the node pairs rs and pq of the network of Fig. 2.25 is defined as

$$g_{rp,sp} = \frac{V_{pq}}{V_{rs}} \tag{2.96}$$

again with all initial conditions and independent sources in N being set to zero. Thus, from Eqs. (2.94) and (2.95) we get

$$g_{rp,sq} = \frac{z_{rp,sq}}{z_{rr,ss}} = \frac{Y_{rp,sq}}{Y_{rr,ss}} \tag{2.97}$$

The symbols have been chosen to help us to remember. In the numerators of Eqs. (2.94) and (2.95), the order of the subscripts is as follows: r, the current injecting node; p, the voltage measurement node; s, the current extracting node; and q, the voltage reference node. Nodes r and p designate the input and output transfer measurement, and nodes s and q form a sort of double datum. This is similarly valid for the numerator of $g_{rp,sq}$. In fact, formula Eq. (2.94) for the transfer impedance $z_{rp,sq}$ is all that we need to remember. The other two expressions Eqs. (2.95) and (2.97) can always be deduced easily from the first. The formula Eq. (2.95) is a special case of the first in which $r = p$ and $s = q$, and Eq. (2.97) is simply the quotient of the other two. The results presented in these formulas are, of course, quite general, because the letters r, p, s, and q may refer to the labels of any four nodes or terminals in the network.

We illustrate these results by the following examples.

Example 2.6 Consider the transistor amplifier of Fig. 2.26. By using the procedure outlined in Sec. 2.2, the primitive indefinite-admittance matrix of the amplifier is found to be

$$
\mathbf{Y} =
\begin{bmatrix}
& 1 & 2 & 3 & 4 \\
g_e + sC_e & -g_e - sC_e & 0 & 0 \\
(\alpha - 1)(g_e + sC_e) & g_b + g_c + (1 - \alpha)(g_e + sC_e) + sC_c & -g_c - sC_c & -g_b \\
-\alpha(g_e + sC_e) & \alpha(g_e + sC_e) - g_c - sC_c & g_c + sC_c + y_l & -y_l \\
0 & -g_b & -y_l & g_b + y_l
\end{bmatrix}
$$

$$(2.98)$$

To compute the transfer impedance $z_{13,44}$, we use Eq. (2.94), yielding

$$
z_{13,44} = \frac{V_{34}}{I_e} = \frac{Y_{13,44}}{Y_{uv}} = \frac{Y_{13,44}}{Y_{22}} = \frac{g_c + sC_c + \alpha g_b}{(g_c + sC_c)(g_b + y_l) + g_b y_l} \tag{2.99}
$$

The voltage gain $g_{13,44}$ of the amplifier is determined by

$$
g_{13,44} = \frac{V_{34}}{V_{14}} = \frac{Y_{13,44}}{Y_{11,44}}
$$

$$
= \frac{(g_e + sC_e)(g_c + sC_c + \alpha g_b)}{(g_b + g_e + sC_e + y_l)(g_c + sC_c) + (1 - \alpha)(g_e + sC_e)y_l + g_b y_l} \tag{2.100}
$$

Figure 2.26 A transistor amplifier used to illustrate the computation of the transfer and driving- point impedances.

From Eq. (2.95), we can compute the amplifier input impedance, which is given by

$$z_{11,44} = \frac{V_{14}}{I_e} = \frac{Y_{11,44}}{Y_{uv}} = \frac{Y_{11,44}}{Y_{22}}$$

$$= \frac{(g_b + g_e + sC_e + Y_l)(g_c + sC_c) + (1 - \alpha)(g_e + sC_e)y_l + g_b y_l}{[(g_c + sC_c)(g_b + y_l) + g_b y_l](g_e + sC_e)}$$

(2.101)

For the amplifier output impedance with the input open-circuited, we have

$$z_{33,44} = \frac{Y_{33,44}}{Y_{uv}} = \frac{Y_{33,44}}{Y_{22}} = \frac{g_b + g_c + sC_c}{(g_c + sC_c)(g_b + y_l) + g_b y_l}$$

(2.102)

Suppose that we wish now to compute the open-circuit impedance matrix $\mathbf{Z}(s)$ of the transistor with respect to the ports formed by the terminal pairs 1, 4 and 3, 4 after the load admittance y_l has been removed. We write

$$z_{rp,sq}^0 = z_{rp,sq}|_{y_l=0}$$

(2.103)

Then according to our convention, $\mathbf{Z}(s)$ can be expressed as

$$\mathbf{Z}(s) = \begin{bmatrix} z^0_{11,44} & z^0_{31,44} \\ z^0_{13,44} & z^0_{33,44} \end{bmatrix} = \frac{1}{Y^0_{22}} \begin{bmatrix} Y^0_{11,44} & Y^0_{31,44} \\ Y^0_{13,44} & Y^0_{33,44} \end{bmatrix}$$

$$= \begin{bmatrix} r_b + \dfrac{r_e}{1 + sr_eC_e} & r_b \\ r_b + \dfrac{\alpha r_c}{1 + sr_cC_c} & r_b + \dfrac{r_c}{1 + sr_cC_c} \end{bmatrix} \tag{2.104}$$

where $r_b = 1/g_b$, $r_e = 1/g_e$, $r_c = 1/g_c$, and

$$Y^0_{uv} = Y_{uv}\big|_{y_l=0} \quad \text{and} \quad Y^0_{rp,sq} = Y_{rp,sq}\big|_{y_l=0} \tag{2.105}$$

Example 2.7 In the hybrid-pi equivalent network of a transistor as shown in Fig. 2.12, suppose that we connect a 100-Ω load between nodes 2 and 4 as shown in Fig. 2.27. This is equivalent to adding 0.01 to y_{22} and y_{44} and -0.01 to y_{24} and y_{42} in (2.41). As before, let $p = s/10^9$. The indefinite-admittance matrix of the resulting amplifier is given by

$$\mathbf{Y} = \begin{bmatrix} 0.02 & 0 & -0.02 & 0 \\ 0 & 0.01 + 0.005p & 0.2 - 0.005p & -0.21 \\ -0.02 & -0.005p & 0.024 + 0.105p & -0.004 - 0.1p \\ 0 & -0.01 & -0.204 - 0.1p & 0.214 + 0.1p \end{bmatrix} \tag{2.106}$$

Figure 2.27 A transistor amplifier used to illustrate the computation of the voltage gain.

To compute the voltage gain $g_{12,44}$, we appeal to Eq. (2.97), which yields

$$g_{12,44} = \frac{Y_{12,44}}{Y_{11,44}} = \frac{p-40}{5p^2 + 21.7p + 2.4} \tag{2.107}$$

The input impedance looking into terminals 1 and 4 is determined by

$$z_{11,44} = \frac{Y_{11,44}}{Y_{uv}} = \frac{Y_{11,44}}{Y_{44}} = \frac{50p^2 + 217p + 24}{p^2 + 4.14p + 0.08} \tag{2.108}$$

Finally, to compute the current gain of the amplifier, which is defined as the ratio of the current I_{24} in the 100-Ω load to the current I_{41} at terminals 1 and 4 as indicated in Fig. 2.27, we apply Eq. (2.94), which shows that

$$\frac{I_{24}}{I_{41}} = 0.01z_{12,44} = 0.01\frac{Y_{12,44}}{Y_{44}} = \frac{0.1p-4}{p^2 + 4.14p + 0.08} \tag{2.109}$$

The transfer admittance, which is defined as the ratio of the current I_{24} to the voltage V_{14} of Fig. 2.27, is related to the voltage gain $g_{12,44}$ by

$$\frac{I_{24}}{V_{41}} = 0.01g_{12,44} = 0.01\frac{Y_{12,44}}{Y_{11,44}} = \frac{p-40}{500p^2 + 2170p + 240} \tag{2.110}$$

Thus, formulas Eqs. (2.94), (2.95), and (2.97) can be employed to compute nearly all the driving-point and transfer functions except those to be discussed in the following section where the short-circuit current is involved.

2.6.2 Short-Circuit Transfer Admittance and Current Gain

Assume that a voltage source is connected between any two nodes r and s so that a current I_{sr} is injected into the rth node and at the same time is extracted from the sth node. Suppose also that an ideal ammeter is connected between nodes p and q so that it indicates a short-circuit current flow from p to q, as depicted symbolically in Fig. 2.28. Then the *short-circuit current gain*, denoted by the symbol $\alpha_{rp,sq}$, between the node pairs rs and pq of the network of Fig. 2.28 is defined by

$$\alpha_{rp,sq} = \frac{I_{pq}}{I_{sr}} \tag{2.111}$$

with all initial conditions and independent sources in the network N being set to zero. For $r = p$ and $s = q$, $\alpha_{rp,sq}$ is defined to be unity.

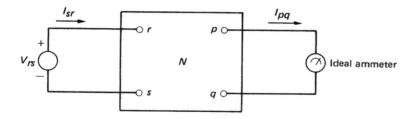

Figure 2.28 The symbolic representation for the measurement of the short-circuit current gain.

Likewise, the *short-circuit transfer admittance*, denoted by the symbol $y_{rp,sq}$, between the node pairs rs and pq of the network of Fig. 2.28 is defined as

$$y_{rp,sq} = \frac{I_{pq}}{V_{rs}} \tag{2.112}$$

again with all initial conditions and independent sources in N being set to zero. Note that for $r = p$ and $s = q$, $y_{rp,sq}$ cannot be meaningfully defined.

Refer to Fig. 2.28. Let $\hat{N}$ be the network obtained from N by connecting a branch of admittance y between nodes p and q. For simplicity, let us use the cap to indicate the similar quantities defined for $\hat{N}$. For example, $\hat{\mathbf{Y}}$, $\hat{z}_{rp,sq}$, and $\hat{V}_{pq}$ denote the indefinite-admittance matrix, the transfer impedance, and the voltage rise from q to p in $\hat{N}$, respectively. Thus, from Eq. (2.94) we have

$$y\hat{z}_{rp,sq} = y\frac{\hat{V}_{pq}}{\hat{I}_{sr}} = y\frac{\hat{Y}_{rp,sq}}{\hat{Y}_{qq}} \tag{2.113}$$

Now consider the limit by letting y approach infinity. The left-hand side of Eq. (2.113) becomes the short-circuit current gain $\alpha_{rp,sq}$ in N and the denominator $\hat{Y}_{qq}$ approaches $yY_{pp,qq}$. Since $\hat{Y}_{rp,sq} = Y_{rp,sq}$ Eq. (2.113) becomes

$$\alpha_{rp,sq} = \frac{Y_{rp,sq}}{Y_{pp,qq}} \tag{2.114}$$

Likewise, using Eq. (2.114) and the limiting process, we can show that

$$y_{rp,sq} = \frac{Y_{rp,sq}}{Y_{rr,ss,pp} + Y_{rr,ss,qq} - Y_{rr,ss,pq} - Y_{rr,ss,qp}} \tag{2.115}$$

in which the *third-order cofactor* $Y_{rr,ss,pq}$ of the elements y_{rr}, y_{ss} and y_{pq} of $\mathbf{Y}$ is defined by the relation

$$Y_{rr,ss,pq} = \text{sgn}(r - p)\,\text{sgn}(r - q)\,\text{sgn}(s - p)\,\text{sgn}(s - q)(-1)^{p+q}\det \mathbf{Y}_{rr,ss,pq} \tag{2.116}$$

$\mathbf{Y}_{rr,ss,pq}$ being the submatrix obtained from $\mathbf{Y}$ by deleting the rows r, s, p and columns r, s, q. $\mathbf{Y}_{rr,ss,pq}$ is defined to be zero if any two of the row or column indices are identical, or alternatively we can define sgn $0 = 0$. The details of the derivation of Eq. (2.115) are left as an exercise (see Prob. 2.9).

As a consequence of Eqs. (2.94), (2.97), and (2.114), we conclude that for a reciprocal network, whose indefinite-admittance matrix is symmetric, the transfer impedances in opposite directions are equal, the transfer admittances in opposite directions are equal, and the open-circuit voltage gain in one direction is equal to the short-circuit current gain in the opposite direction, all being measured from one node pair to another. These are three manifestations of reciprocity, and if one of these three is known, the other two must follow. More specifically, for reciprocal networks we have $z_{rp,sq} = z_{pr,qs}$ and $g_{rp,sq} = \alpha_{pr,qs}$ (see Probs. 2.28 and 2.29).

Example 2.8 Suppose that we wish to use the above formulas to compute the short-circuit admittance parameters of the two-port network of Fig. 2.11. Then according to our convention, the short-circuit admittance matrix $\mathbf{Y}_{sc}(s)$ can be expressed as

$$\mathbf{Y}_{sc}(s) = \begin{bmatrix} \dfrac{1}{\hat{z}_{11,33}} & -y_{21,33} \\ -y_{12,33} & \dfrac{1}{\tilde{z}_{22,33}} \end{bmatrix} \qquad (2.117)$$

The elements $\hat{z}_{11,33}$ and $\tilde{z}_{22,33}$ will be defined shortly.

The indefinite-admittance matrix of the network of Fig. 2.11 was computed earlier and is given by Eq. (2.34). Thus, from Eq. (2.115) we have

$$y_{12,33} = \frac{Y_{12,33}}{Y_{11,33,22}} = \frac{g_e(g_c + \alpha g_b)}{w} = \frac{\alpha r_c + r_b}{r_b r_c r_e w} \qquad (2.118a)$$

$$y_{21,33} = \frac{Y_{21,33}}{Y_{22,33,11}} = \frac{g_c g_e}{w} = \frac{r_b}{r_b r_c r_e w} \qquad (2.118b)$$

The input impedance $\hat{z}_{11,33}$ looking into terminals 1 and 3 when terminals 2 and 3 are joined together can be computed by Eq. (2.95) with the indefinite-admittance matrix $\hat{\mathbf{Y}}$ being derived from Eq. (2.34) by adding row 3 to row 2, column 3 to column 2, and then deleting row 3 and column 3 in the resulting matrix. The result is given by

$$
\hat{\mathbf{Y}} = \begin{array}{c} \\ 1 \\ 2,3 \\ 4 \end{array}
\begin{array}{ccc} 1 & 2,3 & 4 \\ \end{array}
\left[\begin{array}{ccc}
g_e & 0 & -g_e \\
-\alpha g_e & g_b + g_c & \alpha g_e - g_b - g_c \\
\alpha g_e - g_e & -g_b - g_c & w
\end{array} \right] \tag{2.119}
$$

yielding

$$
\hat{z}_{11,33} = \frac{\hat{Y}_{11,22}}{\hat{Y}_{33}} = \frac{w}{g_e(g_b + g_c)} = \frac{r_b r_c r_e w}{r_b + r_c} \tag{2.120}
$$

Finally, $\tilde{z}_{22,33}$ is the impedance looking into terminals 2 and 3 when terminals 1 and 3 are joined together. The corresponding indefinite-admittance matrix is obtained as

$$
\tilde{\mathbf{Y}} = \begin{array}{c} \\ 1,3 \\ 2 \\ 4 \end{array}
\begin{array}{ccc} 1,3 & 2 & 4 \\ \end{array}
\left[\begin{array}{ccc}
g_b + g_e & 0 & -g_e - g_b \\
-\alpha g_e & g_c & \alpha g_e - g_c \\
\alpha g_e - g_e - g_b & -g_c & w
\end{array} \right] \tag{2.121}
$$

giving

$$
\tilde{z}_{22,33} = \frac{\tilde{Y}_{22,11}}{\tilde{Y}_{33}} = \frac{w}{g_c(g_b + g_e)} = \frac{r_b r_c r_e w}{r_b + r_e} \tag{2.122}
$$

The short-circuit admittance matrix of the two-port network becomes

$$
\mathbf{Y}_{sc}(s) = \frac{1}{r_b r_c r_e w} \left[\begin{array}{cc}
r_b + r_c & -r_b \\
-\alpha r_c - r_b & r_b + r_e
\end{array} \right] \tag{2.123}
$$

confirming Eq. (2.37). Alternatively, the elements of $\mathbf{Y}_{sc}(s)$ can be deduced directly from Eq. (2.36) instead of Eq. (2.34), the details being left as an exercise (see Prob. 2.11).

Example 2.9 We wish to compute the short-circuit admittance matrix $\mathbf{Y}_{sc}(s)$ of the active network of Fig. 2.5 with respect to the ports formed by the terminal pairs 1, 5 and 4, 5. As before, we can write

$$
\mathbf{Y}_{sc}(s) = \left[\begin{array}{cc}
\dfrac{1}{\hat{z}_{11,55}} & -y_{41,55} \\[3mm]
-y_{14,55} & \dfrac{1}{\tilde{z}_{44,55}}
\end{array} \right] \tag{2.124}
$$

Appealing to Eq. (2.115) and using the indefinite-admittance matrix $\mathbf{Y}$ given in Eq. (2.18) yield

$$y_{41,55} = \frac{Y_{41,55}}{Y_{44,55,11}}$$

$$= \frac{g_1 g_2}{(1-\beta)g_2 L C s^2 + [(1-\beta)g_1 g_2 L + C]s + g_1 + g_2} \quad (2.125a)$$

$$y_{14,55} = \frac{Y_{14,55}}{Y_{11,55,44}}$$

$$= \frac{(1-\beta)g_1 g_2}{(1-\beta)g_2 L C s^2 + [(1-\beta)g_1 g_2 L + C]s + g_1 + g_2} \quad (2.125b)$$

The input impedance $\hat{z}_{11,55}$ looking into terminals 1 and 5 with terminals 4 and 5 joined together can be determined from Eq. (2.95) by using the indefinite-admittance matrix

$$\hat{\mathbf{Y}} = \begin{bmatrix} g_1 & -g_1 & 0 & 0 \\ -g_1 & g_1 + g_2 + sC & -g_2 & -sC \\ 0 & \beta g_2 - g_2 & g_2 - \beta g_2 + \dfrac{1}{sL} & -\dfrac{1}{sL} \\ 0 & -\beta g_2 - sC & \beta g_2 - \dfrac{1}{sL} & sC + \dfrac{1}{sL} \end{bmatrix} \quad (2.126)$$

which is derived from Eq. (2.18) by adding row 5 to row 4, column 5 to column 4, and then deleting row 5 and column 5 in the resulting matrix. The result is given by

$$\hat{z}_{11,55} = \frac{\hat{Y}_{11,44}}{\hat{Y}_{44}} = \frac{(g_1 + sC)[(1-\beta)g_2 Ls + 1] + g_2}{g_1[(1-\beta)g_2 L C s^2 + sC + g_2]} \quad (2.127)$$

Finally, the input impedance $\tilde{z}_{44,55}$ is computed from the matrix

$$\tilde{\mathbf{Y}} = \begin{bmatrix} g_1 + g_3 + sC & -g_1 - \beta g_2 - sC & \beta g_2 & -g_3 \\ -g_1 - sC & g_1 + g_2 + sC & -g_2 & 0 \\ 0 & \beta g_2 - g_2 & g_2 - \beta g_2 + \dfrac{1}{sL} & -\dfrac{1}{sL} \\ -g_3 & 0 & -\dfrac{1}{sL} & g_3 + \dfrac{1}{sL} \end{bmatrix} \quad (2.128)$$

which is obtained from Eq. (2.18) by adding row 5 to row 1, column 5 to column 1, and then deleting row 5 and column 5 in the resulting matrix, giving

$$\tilde{z}_{44,55} = \frac{\tilde{Y}_{44,11}}{\tilde{Y}_{11}} = \frac{(g_1 + sC)[(1 - \beta)g_2Ls + 1] + g_2}{g_3(g_1 + g_2 + sC) + (g_1 + sC)(1 - \beta)g_2(g_3Ls + 1)} \tag{2.129}$$

The short-circuit admittance matrix of the two-port network becomes

$$\mathbf{Y}_{sc}(s)$$

$$= \frac{1}{q(s)} \begin{bmatrix} g_1[(1 - \beta)g_2LCs^2 \\ + Cs + g_2] & -g_1g_2 \\ (\beta - 1)g_1g_2 & (g_1 + sC)[g_3 + (1 - \beta)g_2(g_3Ls + 1)] \\ & + g_2g_3 \end{bmatrix} \tag{2.130a}$$

where

$$q(s) = (g_1 + sC)\big[(1 - \beta)g_2Ls + 1\big] + g_2 \tag{2.130b}$$

2.7 ANALYSIS OF CONSTRAINED ACTIVE NETWORKS

An important network element that is used extensively to perform a wide variety of functions is the operational amplifier, abbreviated as *op-amp*. It is a direct-coupled high-gain amplifier to which feedback is added to control its overall response characteristic. The schematic diagram of an op-amp is shown in Fig. 2.29a, and its equivalent network in Fig. 2.29b. V^+ and V^- denote the voltages from terminals + and – to the ground, respectively. The + terminal is referred to as the *noninverting terminal* and the — terminal as the *inverting terminal*. Usually, one of the input terminals is grounded, and nearly all op-amps have only one output terminal. The values of the parameters of a typical op-amp are listed below:

$$A \geqslant 10,000 \quad \text{at frequencies } f \leqslant 10\,\text{kHz}$$
$$r_i = 500\,\text{k}\Omega, \quad r_o = 300\,\Omega \tag{2.131}$$

The *ideal op-amp* is the idealization of the real-world op-amp with the following characteristics:

1. infinite input impedance, $r_i = \infty$.
2. Zero output impedance, $r_o = 0$.

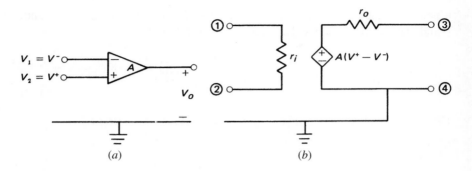

Figure 2.29 (*a*) The schematic diagram of an op-amp and (*b*) its equivalent network.

3. Infinite voltage gain, $A = \infty$.
4. Infinite bandwidth.
5. Zero output signal when the input voltage is zero: $V_o = 0$ for $V^+ = V^-$. This is known as the *zero offset*.
6. Characteristics do not drift with environmental factors such as temperature.

In practice, the ideal op-amp is a good approximation to the physical op-amp of Fig. 2.29, and it can be used with considerable simplification in the design and analysis of various networks where op-amps are employed. Since the output of the ideal op-amp must be finite, $V_o = A(V^+ - V^-)$ implies that the potential difference between the input terminals of the amplifier must be zero. Moreover, because the input impedance is infinite, the input current must be identically zero, Thus, for an ideal op-amp we should apply the following rules:

Rule 1 The potential difference between the input terminals is zero.

Rule 2 The current at each of the input terminals is zero.

It is clear from the above analysis that the indefinite-admittance matrix does not exist for the ideal op-amp. For the indefinite-admittance matrix to exist, we must assume finite parasitic input and output impedances and the finite gain as shown in Fig. 2.29*b*, whose indefinite-admittance matrix is defined by the equation

$$
\begin{bmatrix} I_1 \\ I_2 \\ I_3 \\ I_4 \end{bmatrix}
\begin{bmatrix} g_i & -g_i & 0 & 0 \\ -g_i & g_i & 0 & 0 \\ g_o A & -g_o A & g_o & -g_o \\ -g_o A & g_o A & -g_o & g_o \end{bmatrix}
\begin{bmatrix} V_1 \\ V_2 \\ V_3 \\ V_4 \end{bmatrix}
\tag{2.132}
$$

where $g_i = 1/r_i$ and $g_o = 1/r_o$. To compute network functions when ideal op-amp conditions are assumed, we can let $A \to \infty$, $g_i \to 0$, and $g_o \to \infty$ after the calculations involving the indefinite-admittance matrix have been completed. This approach, although general, complicates the computation. In the following, we present a simplified procedure for computing network functions when ideal op-amps are involved. However, before we do this, we illustrate the above procedure by the following example.

Example 2.10 Consider the inverting amplifier of Fig. 2.30a together with its equivalent network of Fig. 2.30b. The indefinite-admittance matrix of the amplifier is found to be

$$
\mathbf{Y} = \begin{bmatrix} y_1 + y_2 + g_i & -y_2 & -y_1 - g_i \\ -y_2 + Ag_o & g_o + y_2 & -Ag_o - g_o \\ -Ag_o - y_1 - g_i & -g_o & y_1 + g_i + (1+A)g_o \end{bmatrix} \tag{2.133}
$$

Appealing to Eq. (2.94) yields

$$
z_{12,33} = \frac{V_o}{y_1 V_{\text{in}}} = \frac{Y_{12,33}}{Y_{uv}} = \frac{Y_{12,33}}{Y_{33}} \tag{2.134}
$$

or

$$
\frac{V_o}{V_{\text{in}}} = y_1 \frac{Y_{12,33}}{Y_{33}} = \frac{y_1(y_2 - Ag_o)}{(y_1 + g_i)(y_2 + g_o) + (1+A)y_2 g_o} \tag{2.135}
$$

As $g_i \to 0$ and $g_o \to \infty$ Eq. (2.135) becomes

$$
\frac{V_0}{V_{\text{in}}} = -\frac{Ay_1}{y_1 + (1+A)y_2} \tag{2.136}
$$

For the ideal op-amp, we let $A \to \infty$ giving

$$
\frac{V_0}{V_{\text{in}}} = -\frac{y_1}{y_2} \tag{2.137}
$$

For constant y_1 and y_2, Eq. (2.137) represents an ideal voltage-controlled voltage source. Thus, the inverting amplifier of Fig. 2.30a is a good approximation to the ideal voltage-controlled voltage source.

2.7.1 Voltage-Constrained Terminals

Suppose that an ideal voltage-controlled voltage source, abbreviated as VCVS, with a finite gain A is connected between terminals c and d of an n-terminal network, as depicted in Fig. 2.31. The VCVS introduces a constraint on the network by forcing

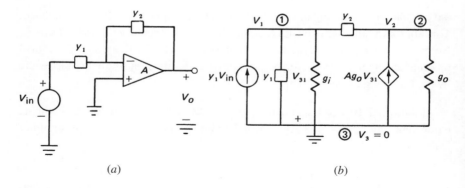

Figure 2.30 (a) An inverting amplifier and (b) its equivalent network.

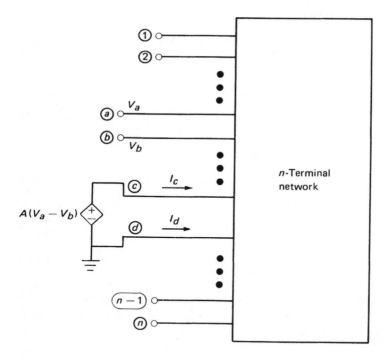

Figure 2.31 The general representation of two voltage-constrained terminals of an n-terminal network.

the voltage between terminals c and d to follow that between terminals a and b, so that the relation

$$V_{cd} = V_c - V_d = A V_{ab} = A(V_a - V_b) \qquad (2.138)$$

holds, where as before V_x denotes the voltage of terminal x to the reference point and V_{ij} is the potential drop from terminal i to terminal j.

Consider the network equations of (2.1) with $\mathbf{J} = \mathbf{0}$. To exhibit the rows and columns corresponding to terminals $a, b, c,$ and d, we write

$$
\begin{array}{c}
\begin{array}{cccc}
 a & b & c & d
\end{array} \\
\begin{array}{c}
\\
\\
a \\
b \\
c \\
d \\
\\
\\
\end{array}
\begin{bmatrix}
\cdots & y_{1a} & y_{1b} & y_{1c} & y_{1d} & \cdots \\
\cdots & \cdots & \cdots & \cdots & \cdots & \cdots \\
\cdots & y_{aa} & y_{ab} & y_{ac} & y_{ad} & \cdots \\
\cdots & y_{ba} & y_{bb} & y_{bc} & y_{bd} & \cdots \\
\cdots & y_{ca} & y_{cb} & y_{cc} & y_{cd} & \cdots \\
\cdots & y_{da} & y_{db} & y_{dc} & y_{dd} & \cdots \\
\cdots & \cdots & \cdots & \cdots & \cdots & \cdots \\
\cdots & y_{na} & y_{nb} & y_{nc} & y_{nd} & \cdots
\end{bmatrix}
\begin{bmatrix}
V_1 \\ \vdots \\ V_a \\ V_b \\ V_c \\ V_d \\ \vdots \\ V_n
\end{bmatrix}
\begin{bmatrix}
I_1 \\ \vdots \\ I_a \\ I_b \\ I_c \\ I_d \\ \vdots \\ I_n
\end{bmatrix}
\end{array}
\qquad (2.139)
$$

From Eq. (2.138) we have

$$V_c = A V_a - A V_b + V_d \qquad (2.140)$$

Substituting this in Eq. (2.139), the variable V_c can be eliminated. This operation is equivalent to adding column c multiplied by A to column a, adding column c multiplied by $-A$ to column b, and adding column c to column d. We then delete column c and variable V_c. This results in a system of n equations in $n - 1$ voltage variables. Since each column sum of $\mathbf{Y}$ and the sum of elements of $\mathbf{I}$ are zero, one of the equations is redundant and can be deleted. The current I_c resulting from the VCVS is not known. For our purposes, we delete the equation corresponding to I_c. Moreover, if terminal d is chosen as the reference-potential point, we set $V_d = 0$. This operation is equivalent to deleting column d and variable V_d in Eq. (2.139), yielding a system of $n - 1$ equations in $n - 1$ unknowns, $n - 2$ unknown terminal voltages, and the unknown terminal current $I_d = -I_c$. To compute the $n - 2$ unknown terminal voltages $V_x, x \neq c, d$, there is no need to include the equation corresponding to I_d. Thus, it can be deleted from Eqs. (2.139), and (2.139) becomes a system of $n - 2$ equations in $n - 2$ unknowns V_x $(x = 1, 2, \ldots, n$ and $x \neq c, d)$:

$$
\begin{array}{c}
\quad\quad\quad\quad a \quad\quad\quad b \\
\begin{array}{c} \\ \\ a \\ b \\ \\ \\ \end{array}
\left[
\begin{array}{ccccc}
\cdots & y_{1a} + Ay_{1c} & y_{1b} - Ay_{1c} & \cdots \\
\cdots\cdots & \cdots\cdots & \cdots\cdots & \cdots \\
\cdots & y_{aa} + Ay_{ac} & y_{ab} - Ay_{ac} & \cdots \\
\cdots & y_{ba} + Ay_{bc} & y_{bb} - Ay_{bc} & \cdots \\
\cdots\cdots & \cdots\cdots & \cdots\cdots & \cdots \\
\cdots\cdots & y_{na} + Ay_{nc} & y_{nb} - Ay_{nc} & \cdots \\
\end{array}
\right]
\left[
\begin{array}{c}
V_1 \\ \vdots \\ V_a \\ V_b \\ \vdots \\ V_n
\end{array}
\right]
\left[
\begin{array}{c}
I_1 \\ \vdots \\ I_a \\ I_b \\ \vdots \\ I_n
\end{array}
\right]
\end{array}
\tag{2.141}
$$

The coefficient matrix of Eq. (2.141) is the admittance matrix of the constrained common-terminal $(n-2)$-port network. The rule for obtaining this matrix for finite gain A can now be stated as follows:

Rule In the unconstrained indefinite-admittance matrix $\mathbf{Y}$ of Fig. 2.31, add column c multiplied by A to column a and column c multiplied by $-A$ to column b. Then delete rows and columns c and d from $\mathbf{Y}$. The resulting matrix is the coefficient matrix of Eq. (2.141).

We illustrate the above results by analyzing several active networks in which an op-amp is used as a VCVS.

Example 2.11 Consider the inverting amplifier of Fig. 2.30a whose equivalent network is presented in Fig. 2.32. The indefinite-admittance matrix of the unconstrained network can be obtained directly from Fig. 2.32 by inspection and

Figure 2.32 An equivalent network of the inverting amplifier of Fig. 2.30a.

is given by

$$
\begin{bmatrix}
y_1 + y_2 + g_i & -y_2 & 0 & -y_1 - g_i \\
-y_2 & y_2 + g_o & -g_o & 0 \\
0 & -g_o & g_o & 0 \\
-y_1 - g_i & 0 & 0 & y_1 + g_i
\end{bmatrix}
\tag{2.142}
$$

Since $a = 4$, $b = 1$ $c = 3$ and $d = 4$, according to the rule we add column 3 multiplied by A to column 4 and column 3 multiplied by $-A$ to column 1. Then we delete rows and columns 3 and 4 in the resulting matrix. This yields a desired system of nodal equations for the constrained network as

$$
\begin{bmatrix}
y_1 + y_2 + g_i & -y_2 \\
Ag_o - y_2 & y_2 + g_o
\end{bmatrix}
\begin{bmatrix}
V_1 \\
V_2
\end{bmatrix}
=
\begin{bmatrix}
y_1 V_{in} \\
0
\end{bmatrix}
\tag{2.143}
$$

Solving for V_2 yields the voltage gain

$$
\frac{V_2}{V_{in}} = \frac{V_o}{V_{in}} = \frac{y_1(y_2 - Ag_o)}{(y_1 + g_i)(y_2 + g_o) + (1 + A)y_2 g_o}
\tag{2.144}
$$

confirming Eq. (2.135)

Example 2.12 Figure 2.33a is a noninverting amplifier containing an op-amp in the noninverting mode to provide the VCVS characteristic. The equivalent network of the amplifier is shown in Fig. 2.33b. The indefinite admittance matrix of the unconstrained network can be written down directly from Fig. 2.33b by

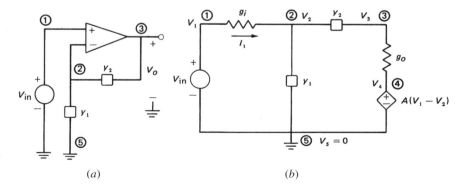

(a)   (b)

Figure 2.33 (a) A noninverting amplifier and (b) its equivalent network.

inspection and is given by

$$
\begin{bmatrix}
g_i & -g_i & 0 & 0 & 0 \\
-g_i & y_1 + y_2 + g_i & -y_2 & 0 & -y_1 \\
0 & -y_2 & y_2 + g_o & -g_o & 0 \\
0 & 0 & -g_o & g_o & 0 \\
0 & -y_1 & 0 & 0 & y_1
\end{bmatrix}
\tag{2.145}
$$

To obtain the constrained admittance matrix, we add column 4 multiplied by A to column 1 and column 4 multiplied by $-A$ to column 2. We then delete rows and columns 4 and 5. This yields a system of nodal equations for the constrained network as

$$
\begin{bmatrix}
g_i & -g_i & 0 \\
-g_i & y_1 + y_2 + g_i & -y_2 \\
-Ag_o & Ag_o - y_2 & y_2 + g_o
\end{bmatrix}
\begin{bmatrix}
V_1 \\
V_2 \\
V_3
\end{bmatrix}
=
\begin{bmatrix}
I_1 \\
0 \\
0
\end{bmatrix}
\tag{2.146}
$$

Since $V_1 = V_{\text{in}}$ and I_1 is unknown, the first equation need not be considered and the last two become

$$
\begin{bmatrix}
y_1 + y_2 + g_i & -y_2 \\
Ag_o - y_2 & y_2 + g_o
\end{bmatrix}
\begin{bmatrix}
V_2 \\
V_3
\end{bmatrix}
=
\begin{bmatrix}
g_i V_{\text{in}} \\
Ag_o V_{\text{in}}
\end{bmatrix}
\tag{2.147}
$$

Solving for V_3 gives the voltage gain

$$
\frac{V_3}{V_{\text{in}}} = \frac{V_o}{V_{\text{in}}} = \frac{Ag_o(y_1 + y_2) + y_2 g_i}{(y_1 + y_2 + g_i)g_o + (y_1 + g_i)y_2 + Ay_2 g_o}
\tag{2.148}
$$

In the limit, as $g_o \to \infty$, $g_i \to 0$, and $A \to \infty$, we have

$$
\frac{V_o}{V_{\text{in}}} = 1 + \frac{y_1}{y_2}
\tag{2.149}
$$

Thus, for the ideal op-amp and constant y_1/y_2, Fig. 2.33a represents an ideal VCVS.

Example 2.13 A general configuration of a second-order active filter is shown in Fig. 2.34a. The equivalent network of the filter is presented in Fig. 2.34b. The unconstrained indefinite-admittance matrix is obtained by inspection directly from

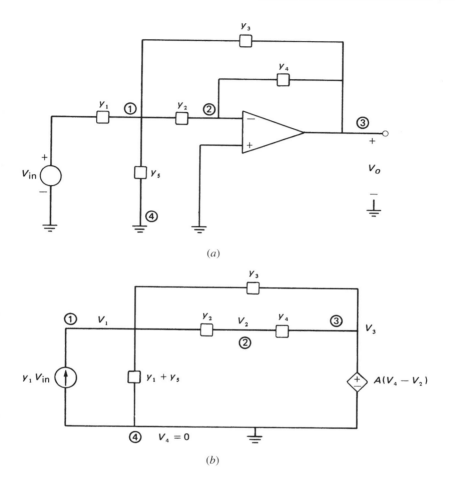

Figure 2.34 (a) A general configuration of a second-order active filter and (b) its equivalent network.

Fig. 2.34a after a simple source transformation or from Fig. 2.34b:

$$\begin{bmatrix} y_1 + y_2 + y_3 + y_5 & -y_2 & -y_3 & -y_1 - y_5 \\ -y_2 & y_2 + y_4 & -y_4 & 0 \\ -y_3 & -y_4 & y_3 + y_4 & 0 \\ -y_1 - y_5 & 0 & 0 & y_1 + y_5 \end{bmatrix} \qquad (2.150)$$

in which the op-amp is assumed to have infinite input impedance, zero output impedance, and a finite gain A. To obtain the admittance matrix of the constrained

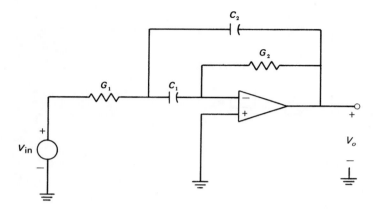

Figure 2.35 A band-pass active filter.

network, according to the rule we add column 3 multiplied by $-A$ to column 2 and then delete rows and columns 3 and 4. The resulting equation becomes

$$\begin{bmatrix} y_1 + y_2 + y_3 + y_5 & Ay_3 - y_2 \\ -y_2 & y_2 + (1 + A)y_4 \end{bmatrix} \begin{bmatrix} V_1 \\ V_2 \end{bmatrix} = \begin{bmatrix} y_1 V_{in} \\ 0 \end{bmatrix} \qquad (2.151)$$

Solving for V_2 and substituting it in $V_3 = -AV_2$ yield the voltage gain

$$\frac{V_3}{V_{in}} = \frac{V_o}{V_{in}} = \frac{Ay_1y_2}{(y_1 + y_2 + y_3 + y_5)(y_2 + y_4 + Ay_4) - y_2(y_2 - Ay_3)} \qquad (2.152)$$

In the limit, as $A \to \infty$, the gain reduces to

$$\frac{V_o}{V_{in}} = \frac{y_1y_2}{y_4(y_1 + y_2 + y_3 + y_5) + y_2y_3} \qquad (2.153)$$

As an application, consider the band-pass filter of Fig. 2.35. Comparing this with Fig. 2.34a, we can identify that $y_1 = G_1$, $y_2 = sC_1$, $y_3 = sC_2$, $y_4 = G_2$, and $y_5 = 0$. Substituting these in Eq. (2.152) yields

$$\frac{V_o}{V_{in}} = -\frac{C_1G_1s}{\begin{array}{c} C_1C_2s^2 + (C_1 + C_2)G_2s + G_1G_2 \\ +[C_1C_2s^2 + (C_1G_1 + C_2G_2 + C_1G_2)s + G_1G_2]/A \end{array}} \qquad (2.154)$$

For the ideal op-amp, the gain is simplified to

$$\frac{V_o}{V_{in}} = -\frac{C_1G_1s}{C_1C_2s^2 + (C_1 + C_2)G_2s + G_1G_2} \qquad (2.155)$$

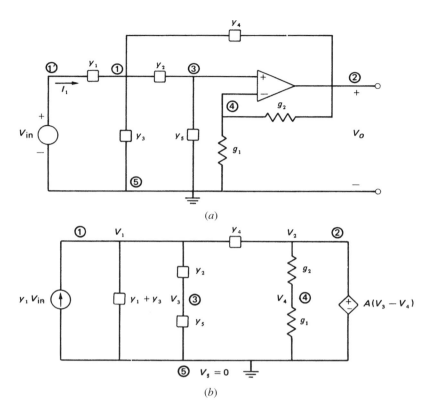

Figure 2.36 (a) A general biquadratic active filter and (b) its equivalent network.

Example 2.14 Consider the negative feedback network of Fig. 2.36a, which can realize the general biquadratic function of an active filter. The equivalent network of the filter is presented in Fig. 2.36b, assuming infinite input impedance and zero output impedance for the op-amp. The indefinite-admittance matrix of the unconstrained network is obtained from Fig. 2.36a after replacing the source combination V_{in} and y_1 by its Norton equivalent or from Fig. 2.36b:

$$
\begin{bmatrix}
y_1 + y_2 + y_3 + y_4 & -y_4 & -y_2 & 0 & -y_1 - y_3 \\
-y_4 & g_2 + y_4 & 0 & -g_2 & 0 \\
-y_2 & 0 & y_2 + y_5 & 0 & -y_5 \\
0 & -g_2 & 0 & g_1 + g_2 & -g_1 \\
-y_1 - y_3 & 0 & -y_5 & -g_1 & g_1 + y_1 + y_3 + y_5
\end{bmatrix}
$$

$$(2.156)$$

To obtain the admittance matrix of the constrained network, according to the rule we add column 2 multiplied by A to column 3 and column 2 multiplied by $-A$ to column 4. We then delete rows and columns 2 and 5. The resulting nodal equation is given by

$$
\begin{bmatrix}
y_1 + y_2 + y_3 + y_4 & -y_2 - Ay_4 & Ay_4 \\
-y_2 & y_2 + y_5 & 0 \\
0 & -Ag_2 & g_1 + (1+A)g_2
\end{bmatrix}
\begin{bmatrix}
V_1 \\
V_3 \\
V_4
\end{bmatrix}
\begin{bmatrix}
y_1 V_{\text{in}} \\
0 \\
0
\end{bmatrix}
$$

(2.157)

Solving for V_3 and V_4 gives

$$
V_3 = \frac{y_1 y_2 (g_1 + Ag_2 + g_2) V_{\text{in}}}{q}
$$

(2.158a)

$$
V_4 = \frac{y_1 y_2 Ag_2 V_{\text{in}}}{q}
$$

(2.158b)

where

$$
q = (g_1 + g_2 + Ag_2)[y_5(y_1 + y_2 + y_3 + y_4) + y_2(y_1 + y_3 + y_4) - Ay_2 y_4]
$$
$$
+ A^2 g_2 y_2 y_4
$$

(2.158c)

The voltage gain of the filter is obtained as

$$
\frac{V_o}{V_{\text{in}}} = \frac{V_2}{V_{\text{in}}} = \frac{A(V_3 - V_4)}{V_{\text{in}}} = \frac{Ay_1 y_2 (g_1 + g_2)}{q}
$$
$$
= \frac{ky_1 y_2}{[(y_2 + y_5)(y_1 + y_3 + y_4) + y_2 y_5](1 + k/A) - ky_2 y_4}
$$

(2.159)

where $k = 1 + g_1/g_2$.

As an application, consider the low-pass filter of Fig. 2.37, from which we can identify $y_1 = G_1$, $y_2 = G_2$, $y_3 = 0$, $y_4 = sC_1$, and $y_5 = sC_2$. Substituting these in Eq. (2.159), we get

$$
\frac{V_o}{V_{\text{in}}} = \frac{kG_1 G_2}{[(G_1 + G_2 + sC_1)(G_2 + sC_2) - G_2^2](1 + k/A) - kC_1 G_2 s}
$$

(2.160)

which is recognized as the gain function of a low-pass characteristic.

Figure 2.37 A low-pass active filter.

2.7.2 The Ideal Operational Amplifier Constraint

In the preceding section, we considered the situation where the terminal voltages
are constrained by an ideal VCVS with a finite gain A, as depicted in Fig. 2.31.
In this section, we present rules for writing down the admittance matrix of the
constrained network when ideal op-amps are employed.

Refer again to the general configuration of Fig. 2.31, where the VCVS
represents the ideal op-amp with $A \to \infty$. The noninverting terminal is terminal
a, and the inverting terminal is terminal b. Since the output of the ideal op-amp
must be finite, $V_{cd} = A V_{ab} = A(V_a - V_b)$ implies that the potential difference
between the input terminals a and b must be zero, requiring that

$$V_a = V_b \qquad\qquad (2.161)$$

By substituting Eqs. (2.161) in (2.139), the variable V_b can be eliminated. This
operation is equivalent to adding column b to column a and then deleting column
b and V_b. This results in a system of n equations in $n - 1$ voltage variables. As
before, one of the equations is redundant and can be deleted. Since the current I_c is
not known, we delete the equation corresponding to I_c. Moreover, since terminal
d is chosen as the reference-potential point, we can delete column d and V_d and
the equation corresponding to $I_d = -I_c$. The above procedure can be summarized
by the following simple rule:

Rule (Ideal op-amp constraint) In the unconstrained indefinite-admittance
matrix **Y** of Fig. 2.31, add column b to column a (the controlling terminals),

and delete rows c and d (the controlled terminals) and columns b and d ($b \neq d$). The resulting matrix is the admittance matrix of the constrained network. For grounded terminal a or b ($a = d$ or $b = d$), the constrained admittance matrix can be obtained from $\mathbf{Y}$ simply by deleting columns a and b and rows c and d.

We illustrate the above rule by the following examples, using the same networks as in the previous examples.

Example 2.15 In Fig. 2.33a, assume that the op-amp is ideal. The unconstrained indefinite-admittance matrix can be obtained by inspection directly from Fig. 2.33a, and is given by

$$
\begin{array}{c} \\ 1 \\ 2 \\ 3 \\ 5 \end{array}
\begin{array}{cccc}
1 & 2 & 3 & 5 \\
\end{array}
\begin{bmatrix}
0 & 0 & 0 & 0 \\
0 & y_1 + y_2 & -y_2 & -y_1 \\
0 & -y_2 & y_2 & 0 \\
0 & -y_1 & 0 & y_1
\end{bmatrix}
\tag{2.162}
$$

According to the rule, to obtain the constrained admittance matrix we add column 2 to column 1 and then delete rows 3 and 4 corresponding to the constrained terminals 3 and 5 and columns 2 and 4 corresponding to terminals 2 and 5. The equation corresponding to the second row of the resulting matrix becomes

$$
(y_1 + y_2)V_1 - y_2 V_3 = 0 \tag{2.163}
$$

which in conjunction with the facts that $V_1 = V_{\text{in}}$ and $V_3 = V_o$ gives

$$
\frac{V_o}{V_{\text{in}}} = 1 + \frac{y_1}{y_2} \tag{2.164}
$$

confirming Eq. (2.149).

Example 2.16 In Fig. 2.34a, replace the source combination of V_{in} and y_1 by its Norton equivalent network. For the ideal op-amp, the unconstrained indefinite-admittance matrix can easily be obtained by inspection and is given by Eq. (2.150). Since one of the input terminals of the op-amp is grounded, to obtain the constrained admittance matrix we delete rows 3 and 4 corresponding to the controlled terminals 3 and 4, and columns 2 and 4 corresponding to the controlling terminals 2 and 4.

The resulting nodal equation is obtained as

$$
\begin{array}{c}
\quad\quad 1 \quad\quad\quad\quad 3 \\
\begin{matrix} 1 \\ 2 \end{matrix}
\begin{bmatrix} y_1 + y_2 + y_3 + y_5 & -y_3 \\ -y_2 & -y_4 \end{bmatrix}
\begin{bmatrix} V_1 \\ V_3 \end{bmatrix}
=
\begin{bmatrix} y_1 V_{\text{in}} \\ 0 \end{bmatrix}
\end{array}
\tag{2.165}
$$

Solving for V_3 yields

$$
\frac{V_3}{V_{\text{in}}} = \frac{V_0}{V_{\text{in}}} = -\frac{y_1 y_2}{y_4(y_1 + y_2 + y_3 + y_5) + y_2 y_3}
\tag{2.166}
$$

confirming Eq. (2.153).

Example 2.17 In Fig. 2.36a, replace the source combination of V_{in} and y_1 by its Norton equivalent network and assume that the op-amp is ideal. The unconstrained indefinite-admittance matrix can be obtained by inspection, and is given by Eq. (2.156). To obtain the constrained admittance matrix, we add column 4 to column 3 and then delete rows 2 and 5 (controlled terminals) and columns 4 and 5. The resulting matrix is the constrained admittance matrix of the network. The corresponding nodal system of equations is found to be

$$
\begin{array}{c}
\quad\quad 1 \quad\quad\quad\quad\quad 2 \quad\quad 3+4 \\
\begin{matrix} 1 \\ 3 \\ 4 \end{matrix}
\begin{bmatrix} y_1 + y_2 + y_3 + y_4 & -y_4 & -y_2 \\ -y_2 & 0 & y_2 + y_5 \\ 0 & -g_2 & g_1 + g_2 \end{bmatrix}
\begin{bmatrix} V_1 \\ V_2 \\ V_3 \end{bmatrix}
=
\begin{bmatrix} y_1 V_{\text{in}} \\ 0 \\ 0 \end{bmatrix}
\end{array}
\tag{2.167}
$$

Solving for V_2 yields

$$
\frac{V_2}{V_{\text{in}}} \frac{V_o}{V_{\text{in}}} = \frac{k y_1 y_2}{(y_2 + y_5)(y_1 + y_3 + y_4) + y_2 y_5 - k y_2 y_4}
\tag{2.168}
$$

where $k = 1 + g_1/g_2$, confirming Eq. (2.159) for $A = \infty$.

We remark that in writing down the unconstrained indefinite-admittance matrix, the source combination of V_{in} and y_1 need not be replaced by its Norton equivalent network. The only difference is that we have to add a row and a column corresponding to the input terminal $1'$. In so doing, the constrained nodal system of equations becomes

$$
\begin{array}{c}
\begin{array}{cccc} 1' & 1 & 2 & 3+4 \end{array} \\
\begin{array}{c} 1' \\ 1 \\ 3 \\ 4 \end{array}
\begin{bmatrix}
y_1 & -y_1 & 0 & 0 \\
-y_1 & y_1+y_2+y_3+y_4 & -y_4 & -y_2 \\
0 & -y_2 & 0 & y_2+y_5 \\
0 & 0 & -g_2 & g_1+g_2
\end{bmatrix}
\begin{bmatrix} V_{in} \\ V_1 \\ V_2 \\ V_3 \end{bmatrix}
=
\begin{bmatrix} I_1 \\ 0 \\ 0 \\ 0 \end{bmatrix}
\end{array}
\qquad (2.169)
$$

Since the current I_1 due to the source V_{in} is unknown and since V_{in} is known, the last three equations of Eq. (2.169) can be rewritten as in Eq. (2.167), yielding the same set of equations. The details are left as an exercise (see Prob. 2.30).

In the above discussion, we have demonstrated how to obtain the constrained nodal system of equations. The procedure requires that we first write down the unconstrained indefinite-admittance matrix and then apply the rule to derive the desired constrained admittance matrix. A moment's thought would indicate that we could obtain the constrained admittance matrix directly from the network without the necessity of first writing down the unconstrained indefinite-admittance matrix. For example, the coefficient matrix of Eq. (2.165) can be written down directly from Fig. 2.34a as follows: The (1, 1)-element denotes the sum of admittances of all the branches incident at the terminal or node 1. The (1,2)- and (2, 1)-elements are the negative of the admittances of the branches connecting the terminals 1 and 3, and 2 and 1, respectively. Finally, the (2, 2)-element represents the negative of the admittance of the branch connecting the terminals 2 and 3. In general, if $y(i, j)$ denotes the element of the constrained admittance matrix in the row corresponding to terminal i and in the column corresponding to terminal j, then it can be written down directly from the network by the formulas [Chen (1978)]

$$ y(i, i) = \Sigma \text{ admittances incident at terminal } i \qquad (2.170a) $$

and for $i \neq j$,

$$ y(i, j) = -\Sigma \text{ admittances connected between terminals } i \text{ and } j \qquad (2.170b) $$

We emphasize that $y(i, j)$ does not necessarily represent the ith row and jth column element. In Eq. (2.167), for example, the diagonal elements are $y(1, 1) = y_1+y_2+y_3+y_4$, $y(3, 2) = 0$, and $y(4, 3+4) = y(4, 3)+y(4, 4) = 0+g_1+g_2 = g_1+g_2$. The (2,3)-element is $y(3, 3+4) = y(3, 3)+y(3, 4) = y_2+y_5$ and the (1,3)-element is $y(1, 3+4) = y(1, 3)+y(1, 4) = -y_2-0 = -y_2$. These elements are obtained directly from the network of Fig. 2.36a. In a similar fashion, we can obtain the other elements.

Example 2.18 One of the first three-amplifier biquads, proposed by Kerwin, Huelsman, and Newcomb (1967), is shown in Fig. 2.38. Assume that the three op-amps are ideal. To obtain the constrained admittance matrix from the unconstrained indefinite-admittance matrix, according to the rule we add column 1 to column 7 and then delete columns 1, 3, 5, and 8 and rows 2, 4, 6, and 8. The resulting nodal system of equations becomes

$$
\begin{array}{c}
\begin{array}{cccc} 2 & 4 & 6 & 1+7 \end{array} \\
\begin{array}{c} 1 \\ 3 \\ 5 \\ 7 \end{array}
\begin{bmatrix}
-G_3 & 0 & -G_2 & G_2 + G_3 \\
-G_4 & -sC_1 & 0 & 0 \\
0 & -G_5 & -sC_2 & 0 \\
0 & -G_6 & 0 & G_1 + G_6
\end{bmatrix}
\begin{bmatrix}
V_2 \\ V_4 \\ V_6 \\ V_7
\end{bmatrix}
=
\begin{bmatrix}
0 \\ 0 \\ 0 \\ G_1 V_{in}
\end{bmatrix}
\end{array}
\tag{2.171}
$$

The $(1, 1)$-element, for example, is $y(1, 2) = -G_3$, and the $(1,4)$-element is $y(1, 1 + 7) = y(1, 1) + y(1, 7) = G_2 + G_3 + 0 = G_2 + G_3$, where $G_i = 1/R_i (i = 1, 2, 3, 4, 5, 6)$. Solving for V_2, V_4, and V_6 yields

$$
\frac{V_2}{V_{in}} = \frac{G_1 C_1 C_2 (G_2 + G_3) s^2}{C_1 C_2 G_3 (G_1 + G_6) s^2 + C_2 G_4 G_6 (G_2 + G_3) s + G_2 G_4 G_5 (G_1 + G_6)}
\tag{2.172a}
$$

$$
\frac{V_4}{V_{in}} = \frac{G_1 C_2 G_4 (G_2 + G_3) s}{C_1 C_2 G_3 (G_1 + G_6) s^2 + C_2 G_4 G_6 (G_2 + G_3) s + G_2 G_4 G_5 (G_1 + G_6)}
\tag{2.172b}
$$

$$
\frac{V_6}{V_{in}} = \frac{G_1 G_4 G_5 (G_2 + G_3)}{C_1 C_2 G_3 (G_1 + G_6) s^2 + C_2 G_4 G_6 (G_2 + G_3) s + G_2 G_4 G_5 (G_1 + G_6)}
\tag{2.172c}
$$

which are recognized as the functions possessing the high-pass, band-pass, and low-pass characteristics, respectively.

2.8 GENERALIZED NORTON'S THEOREM

In this section, we establish a generalized version of Norton's theorem for multiterminal active networks. Specifically, we show that any active multiterminal network can be represented as a passive multiterminal network in parallel with equivalent current sources.

Let N be an active n-terminal network connected in parallel with an arbitrary passive n-terminal network M that is initially relaxed, as shown in Fig. 2.39a.

Figure 2.38 A three-amplifier biquad.

Suppose that additional current sources J_k are inserted between the common terminals and the reference-potential point as depicted in Fig. 2.39b. Assume that it is possible to choose the currents J_k, whose values will be determined shortly, so that all the terminal voltages $\hat{V}_k$ in Fig. 2.39b are zero. Since M is passive, the vanishing of its terminal voltages implies the vanishing of its terminal currents. Thus, M may be disconnected from the other part without affecting conditions to the right. This shows that J_k is equal to the kth terminal current of N when all of its terminal voltages vanish. In other words, J_k is the current flowing in the kth terminal of N when all of its terminals are short-circuited to the reference point. By the principle of superposition, each $\hat{V}_k$ in Fig. 2.39b can be regarded as the sum of two components $\hat{V}_k = \hat{V}'_k + \hat{V}''_k$, where $\hat{V}'_k$ is the kth terminal voltage when all the current sources J_k are removed, and $\hat{V}''_k$ is the kth terminal voltage when all independent current sources in N are open-circuited, all independent voltage sources are short-circuited, and all initial conditions are set to zero. Since $\hat{V}_k = 0$, we have $-\hat{V}''_k = \hat{V}'_k$, which is also equal to the kth terminal voltage V_k in Fig. 2.39a, or $\hat{V}''_k = -\hat{V}'_k = -V_k$. Thus, we have established that if we place the current sources J_k, being equal to the short-circuit terminal currents of N, in the positions and with reference directions as indicated in Fig. 2.39b, and open-circuit all independent current sources, short-circuit all independent voltage sources, and set all initial conditions to zero in N, then we shall obtain terminal voltages $-V_k$ in Fig. 2.39b. Suppose that we reverse the reference directions of J_k; then we must get V_k at the kth common terminal in Fig. 2.39b. Consequently, we can state the following:

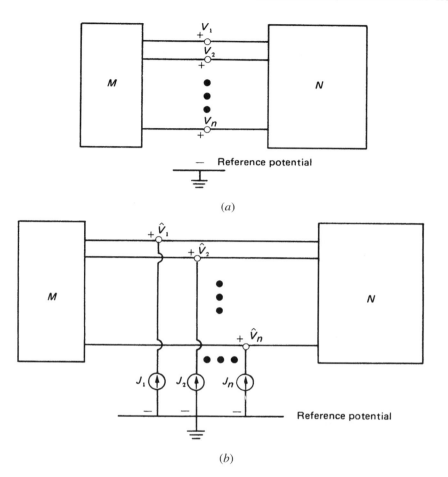

Figure 2.39 (*a*) An active *n*-terminal network *N* connected in parallel with an arbitrary passive initially relaxed *n*-terminal network *M* and (*b*) the insertion of additional current sources between the common terminals and the reference-potential point.

An active multiterminal network N can be represented as a parallel connection of a passive multiterminal network $\tilde{N}$ and current sources J_k as shown in Fig. 2.40. The passive network $\tilde{N}$ is derived from N by open-circuiting all of its independent current sources, short-circuiting all of its independent voltage sources, and reducing all of its initial conditions to zero. The value of the current source J_k is equal to the current flowing in the kth terminal of N when all of its terminals are short-circuited to the reference-potential point.

Figure 2.40 The generalized Norton's equivalent network of an n-terminal network.

We remark that in deriving the result we assumed that the multiterminal network M was passive and initially relaxed. As a matter of fact, this assumption is not necessary, and the result applies equally well if M is active. The only restriction is that M and N not be magnetically coupled. The justification of this is too long to warrant its inclusion here.

Example 2.19 Consider the three-terminal active network N of Fig. 2.41. We wish to determine its Norton equivalent network. For convenience, we choose terminal 3 as the reference-potential point.

To obtain the values of the current sources J_1 and J_2, we short-circuit terminals 1 and 2 to terminal 3 as shown in Fig. 2.42. Referring to Fig. 2.5, as discussed in Example 2.9, we have

$$J_1 = \frac{V_g}{\hat{z}_{11,55}} = \frac{g_1[(1-\beta)g_2LCs^2 + sC + g_2]V_g}{q(s)} \qquad (2.173a)$$

$$J_2 = -y_{14,55}V_g = \frac{(\beta - 1)g_1g_2V_g}{q(s)} \qquad (2.173b)$$

$q(s)$ being given in Eq. (2.130b). The corresponding passive network $\tilde{N}$ is derived from N by short-circuiting the voltage source V_g. Thus, N can be represented equivalently by a three-terminal network as indicated in Fig. 2.43.

Figure 2.41 A three-terminal active network used to illustrate the determination of its Norton equivalent network.

Figure 2.42 The network used to compute the equivalent current sources J_1 and J_2.

Figure 2.43 The Norton equivalent network of the three-terminal network of Fig. 2.41.

2.9 SUMMARY

In this chapter, we described a useful characterization of the external behavior of a multiterminal network in terms of its indefinite-admittance matrix, and we presented rules for its derivation. The rules included a procedure for writing down the primitive indefinite-admittance matrix from the network by inspection and ways to contract and suppress terminals and their effects on the matrix. One of the fundamental properties of the indefinite-admittance matrix is that it is an equicofactor matrix, the sum of elements of each row or column being zero. This is a direct consequence of the fact that in solving an electrical network problem, it is immaterial which terminal or node is chosen as the reference-potential point.

One of the most common and useful three-terminal devices is the transistor. If any one of its terminals is chosen as the reference-potential point, the device can be transformed into a common-terminal two-port network, yielding different configurations of the transistor model. We have demonstrated how the various configurations are interrelated.

The usefulness of the indefinite-admittance-matrix formulation of network equations lies in the fact that it facilitates the computations of the driving-point or transfer functions between any pair of terminals or nodes or from any pair of nodes to any other pair of nodes in the network. For this we introduced the concept of the second-order cofactors and derived explicit formulas for computing the network functions. We showed that these formulas can be expressed compactly and elegantly in terms of the ratios of the first- and/or second-order cofactors of the elements of the indefinite-admittance matrix.

An important network element that is used extensively to perform a wide variety of functions is the operational amplifier. In practice, the ideal op-amp is a good approximation to the physical op-amp and can be used with considerable simplification in the design and analysis of various networks where op-amps are employed. For this we presented rules for obtaining the admittance matrix of the constrained network from the indefinite-admittance matrix of the corresponding unconstrained network.

Finally, we established a generalized version of Norton's theorem by showing that any multiterminal network can be represented equivalently as a passive network in parallel with current sources that are short-circuit currents of the original multiterminal network. Clearly, an analogous extension of Thévenin's theorem can be obtained in a similar fashion by making use of the series connection of multiterminal networks. However, if we demand that series connection should correspond to an addition of the impedance parameters of the component multiterminal networks, then the desired result cannot be achieved in general without the use of ideal transformers. Thus, the result does not seem to be satisfactory. For completeness, we state the theorem as follows:

> An active multiterminal network N can be represented as a series connection of a passive network $\tilde{N}$, which has the significance as in Norton's theorem, and voltage sources whose values are equal to the open-circuit terminal potentials when all the terminals of N are open-circuited. The voltage sources are connected in series with the corresponding terminals of the passive network $\tilde{N}$.

PROBLEMS

2.1 With Eq. (2.3), confirm that the elements y_{k2} and y_{k3} ($k = 1, 2, 3$) of the indefinite-admittance matrix of the three-terminal network of Fig. 2.2 are those given in Eq. (2.5).

2.2 Show that the indefinite-admittance matrix of the four-terminal network of Fig. 2.7 is given by Eq. (2.22).

2.3 Compute the indefinite-admittance matrix of the network of Fig. 2.9 with respect to the terminals or nodes 1, 4, and 5.

2.4 Show that the indefinite-admittance matrix of the ideal gyrator of Fig. 2.8 is that given in Eq. (2.24).

2.5 Let M be a multiterminal network derived from another one N by joining together terminals i and j. Show that the indefinite-admittance matrix of M can be deduced from that of N by adding row j to row i, column j to column i, and then deleting row j and column j in the resulting matrix.
Hint: Make use of Eq. (2.1).

2.6 Figure 2.24 is an equivalent network of the common-emitter transistor configuration.

(*a*) Write down by inspection its primitive indefinite-admittance matrix.

(*b*) Suppress the inaccessible terminal corresponding to node d. Compare your result with Eq. (2.68).

2.7 The network of a difference amplifier is presented in Fig. 2.44. Write down its primitive indefinite-admittance matrix. By using this matrix, compute the short-circuit admittance matrix of the amplifier.

2.8 Repeat Prob. 2.7 by computing the open-circuit impedance matrix.

2.9 By applying the limiting process used in the derivation of Eq. (2.114), derive the formula Eq. (2.115).

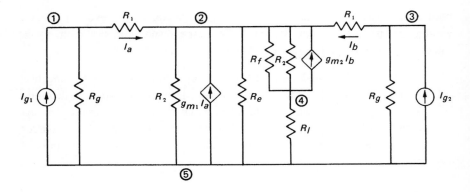

Figure 2.44 An equivalent network of a difference amplifier.

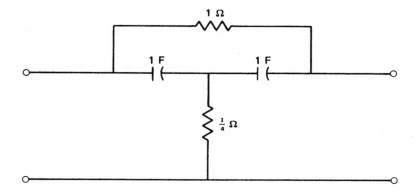

Figure 2.45 A bridged-T RC two-port network.

2.10 By using the indefinite-admittance matrix, compute the short-circuit admittance parameters of the two-port network of Fig. 2.45.

2.11 By using Eq. (2.36), compute $\hat{z}_{11,33}$ and $\hat{z}_{22,33}$ as defined in Example 2.8. Compare your results with Eqs. (2.120) and (2.122).

2.12 Compute the hybrid parameters and the open-circuit impedance parameters of the two-port network of Fig. 2.45.

2.13 Determine the short-circuit admittance parameters of the two-port network of Fig. 2.46.

2.14 Determine the open-circuit impedance parameters of the two-port network of Fig. 2.46.

Figure 2.46 A two-port network.

2.15 Refer to Fig. 2.27. A three-terminal network can be formed with respect to the nodes 1, 2, and 3. Determine its Norton equivalent three-terminal network.

2.16 Connect a voltage source V_g between nodes 1 and 5 in Fig. 2.9, so that there is a voltage rise from node 5 to node 1. Determine a three-terminal Norton equivalent network with respect to the terminals connected at nodes 2, 3, and 5.

2.17 By using the indefinite-admittance-matrix technique, confirm that the short-circuit admittance matrix of the two-port network of Fig. 2.47 is given by

$$\mathbf{Y}_{sc} = \frac{1}{p(s)} \begin{bmatrix} (s+1)(4s^4 + 8s^3 & \\ \quad +6s^2 + 6s + 1) & -(s+1)(2s+1) \\ & (2s+1)(2s^4 + 2s^3 \\ -(s+1)(2s+1) & \quad +4s^2 + s + 1) \end{bmatrix} \qquad (2.174)$$

where

$$p(s) = 4s^5 + 12s^4 + 18s^3 + 18s^2 + 7s + 2 \qquad (2.175)$$

2.18 By using the indefmite-admittance-matrix technique, confirm that the open-circuit impedance matrix of the two-port network of Fig. 2.47 is given by

$$\mathbf{Z}_{oc} = \frac{1}{2s(s^2+1)} \begin{bmatrix} 2s^2+1 & 1 \\ 1 & 2s^2+1 \end{bmatrix} + \begin{bmatrix} \dfrac{s}{s+1} & 0 \\ 0 & \dfrac{2(s+1)}{2s+1} \end{bmatrix} \qquad (2.176)$$

Figure 2.47 A passive two-port network.

Figure 2.48 A given two-port network.

2.19 Determine the indefinite-admittance matrix of the network of Fig. 2.48 with respect to the three terminals connected at nodes 1, 2, and 3. Obtain the short-circuit admittance matrix of the two-port network formed by the terminal pairs 1, 3 and 2, 3.

2.20 Refer to the network of Fig. 2.48. Compute the short-circuit current gain $\alpha_{12,33}$ and short-circuit transfer admittance $y_{12,33}$.

2.21 Compute the short-circuit current gain $\alpha_{14,55}$ and the short-circuit transfer admittance $y_{14,55}$ in the network of Fig. 2.47.

2.22 For Fig. 2.47, determine the impedance looking into terminals 1 and 5 when the output terminal 4 is shorted to terminal 5. Under this situation, also determine the transfer impedance $z_{13,55}$.

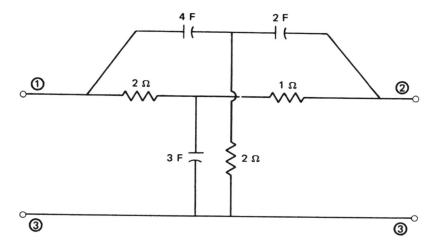

Figure 2.49 An *RC* twin-T two-port network used in the design of equalizers.

2.23 Figure 2.49 is an *RC* twin-T used in the design of equalizers. Compute the transfer impedance between the terminal pairs 1, 3 and 2, 3 when the output pair 2, 3 is open-circuited. Repeat the problem if the output pair 2, 3 is loaded by a 2-Ω resistor.

2.24 By using the indefinite-admittance-matrix technique, compute the short-circuit admittance parameters of the twin-T network of Fig. 2.49 with respect to the terminal pairs 1, 3 and 2, 3.

2.25 For Fig. 2.49, compute $g_{12,33}$ and $y_{12,33}$.

2.26 For the network of Fig. 2.50, find the indefinite-admittance matrix with respect to terminals 1, 2, 3, and 4. What is the resulting matrix when terminals 2 and 4 are joined together?

2.27 Show that

$$Y_{pq,rs} = -Y_{ps,rq} \qquad (2.177a)$$

$$Y_{pq,rs} = -Y_{rs,pq} \qquad (2.177b)$$

Use these relations to confirm that Eqs. (2.83) and (2.85) are valid for all p, q, r, s, and v.

2.28 Use Eq. (2.94) to show that the transfer impedances or admittances in opposite directions in a reciprocal network are equal.

2.29 Apply Eqs. (2.97) and (2.114) to show that, in a reciprocal network, the voltage gain in one direction is equal to the short-circuit current gain in the opposite direction.

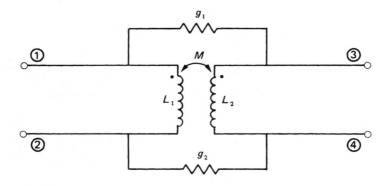

Figure 2.50 A two-port network with two mutually coupled coils.

2.30 By applying the rule derived in Sec. 2.7.2, obtain the constrained nodal system of equations Eq. (2.169).

2.31 Typical values of the hybrid parameters of a common-base transistor for small-signal, low-frequency operation are given by

$$h_{ib} = 20\,\Omega \quad h_{fb} = -0.98$$
$$h_{rb} = 3 \cdot 10^{-4} \quad h_{ob} = 0.5\,\mu\text{mho} \tag{2.178}$$

With these relative values, derive the following approximate formulas:

$$h_{ie} \approx \frac{h_{ib}}{1 + h_{fb}} \quad h_{fe} \approx -\frac{h_{fb}}{1 + h_{fb}}$$
$$h_{re} \approx \frac{h_{ib}h_{ob} - h_{rb}h_{fb} - h_{rb}}{1 + h_{fb}} \quad h_{oe} \approx \frac{h_{ob}}{1 + h_{fb}} \tag{2.179}$$

2.32 With the relative values given in Eq. (2.178), derive the following approximate formulas:

$$h_{ic} \approx \frac{h_{ib}}{1 + h_{fb}} \quad h_{fc} \approx \frac{1}{1 + h_{fb}}$$
$$h_{rc} \approx 1 \quad h_{oc} \approx \frac{h_{ob}}{1 + h_{fb}} \tag{2.180}$$

2.33 Figure 2.51 is the hybrid-pi equivalent network of a transistor. Express the parameters of the T-model of Fig. 2.22 in terms of those of the hybrid-pi

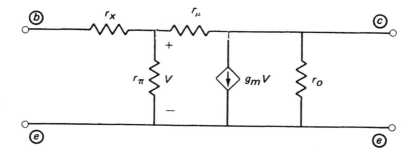

Figure 2.51 The low-frequency hybrid-pi equivalent network of a transistor.

model, as follows:

$$\alpha = \frac{g_m r_\pi}{1 + g_m r_\pi} \qquad r_b = r_x + \frac{r_\pi r_\mu}{r_\mu + (1 + g_m r_\pi) r_o}$$

$$\frac{1}{r_c} = \frac{1}{r_\mu} + \frac{1}{r_o(1 + g_m r_\pi)} \qquad r_e = \frac{r_\pi r_o}{r_o + g_m r_\pi r_o + r_\mu} \qquad (2.181)$$

2.34 With the relative values given in Eq. (2.62), obtain the following approximate formulas for the hybrid-pi parameters of Fig. 2.51:

$$r_\pi \approx h_{ie} - r_x \qquad r_\mu \approx \frac{h_{ie} - r_x}{h_{re}}$$

$$g_m \approx \frac{h_{fe}}{h_{ie} - r_x} \qquad \frac{1}{r_o} \approx h_{oe} - \frac{h_{fe} h_{re}}{h_{ie} - r_x} \qquad (2.182)$$

2.35 Compute the voltage gain of the active filter of Fig. 2.52 under the following situations:

(a) The op-amp has finite input and output impedances with finite gain.
(b) The op-amp has infinite input impedance and zero output impedance with finite gain.
(c) The op-amp is ideal.
(d) Compare the results obtained in (b) and (c) by limiting process ($r_i \rightarrow \infty$, $r_o \rightarrow 0$, and $A \rightarrow \infty$ and by applying the rules outlined in Sec. 2.7.2.

2.36 Repeat Prob. 2.35 for the active filter configuration of Fig. 2.53.
2.37 Repeat Prob. 2.35 for the active filter configuration of Fig. 2.54.
2.38 Derive Eqs. (2.72)–(2.74) for the common-collector configurations of Figs. 2.19 and 2.20.

Figure 2.52 An active filter.

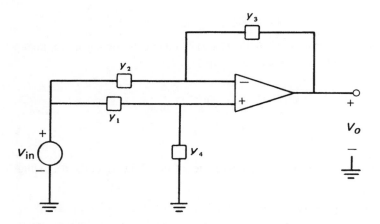

Figure 2.53 An active filter configuration.

2.39 Compute the output impedance looking into terminals 3 and 4 of the network of Fig. 2.26 with input terminals 1 and 4 joined together.

2.40 Typical values of the hybrid parameters of a common-collector transistor for small-signal, low-frequency operation are given by

$$h_{ic} = 1\,\text{k}\Omega \quad h_{fc} = -50$$

$$h_{rc} = 1 \quad h_{oc} = 25\,\mu\text{mho}$$

(2.183)

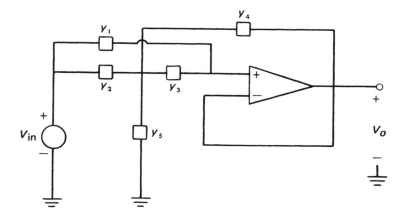

Figure 2.54   An active filter configuration.

Using these relative values, derive the following approximate formulas:

$$h_{ib} \approx \frac{h_{ic}}{h_{fc}} \quad h_{fb} \approx -1 - \frac{1}{h_{fc}}$$

$$h_{rb} \approx h_{rc} - 1 - \frac{h_{ic}h_{oc}}{h_{fc}} \quad h_{ob} \approx \frac{h_{oc}}{h_{fc}} \tag{2.184}$$

BIBLIOGRAPHY

Alderson, G. E. and P. M. Lin: Computer Generation of Symbolic Network Functions — A New Theory and Implementation, *IEEE Trans. Circuit Theory,* vol. CT-20, no. 1, pp. 48–56, 1973.

Barabaschi, S. and E. Gatti: Moderni Metodi di Analisi delle Reti Elettriche Lineari Attive con Particolare Riguardo ai Sistemi Controreazionati, *Energia Nucleare,* vol. 2, no. 12, pp. 105–119, 1954.

Chen, W. K.: A Generalization of the Equicofactor Matrix, *IEEE Trans. Circuit Theory,* vol. CT-13, no. 4, pp. 440–442, 1966.

Chen, W. K.: On Equicofactor and Indefinite-Admittance Matrices, *Matrix and Tensor Quart.,* vol. 23, no. 1, pp. 26–28, 1972.

Chen, W. K.: "Applied Graph Theory: Graphs and Electrical Networks," 2d rev. ed., chap. 4, New York: American Elsevier, and Amsterdam: North-Holland, 1976a.

Chen, W. K.: Indefinite-Admittance Matrix Formulation of Feedback Amplifier Theory, *IEEE Trans. Circuits and Systems,* vol. CAS-23, no. 8, pp. 498–505, 1976b.

Chen, W. K.: Analysis of Constrained Active Networks, *Proc. IEEE,* vol. 66, no. 12, pp. 1655–1657, 1978.

Kerwin, W. J., L. P. Huelsman, and R. W. Newcomb: State-Variable Synthesis for Insensitive Integrated Circuit Transfer Functions, *IEEE J. Solid-State Circuits,* vol. SC-2, no. 3, pp. 87–92, 1967.

Moschytz, G. S.: "Linear Integrated Networks: Fundamentals," New York: Van Nostrand, 1974.

Nathan, A.: Matrix Analysis of Constrained Networks, *Proc. IEE (London)*, vol. 108, part C, pp. 98–106, 1954.

Nathan, A.: Matrix Analysis of Networks Having Infinite-Gain Operational Amplifiers, *Proc. IEEE*, vol. 49, no. 10, pp. 1577–1578, 1961.

Puckett, T. H.: A Note on the Admittance and Impedance Matrices of an n-Terminal Network, *IRE Trans. Circuit Theory*, vol. CT-3, no. 1, pp. 70–75, 1956.

Sharpe, G. E. and B. Spain: On the Solution of Networks by Means of the Equicofactor Matrix, *IRE Trans. Circuit Theory*, vol. CT-7, no. 3, pp. 230–239, 1960.

Sharpe, G. E. and G. P. H. Styan: Circuit Duality and the General Network Inverse, *IEEE Trans. Circuit Theory*, vol. CT-12, no. 1, pp. 22–27, 1965.

Shekel, J.: Matrix Representation of Transistor Circuits, *Proc. IRE*, vol. 40, no. 11, pp. 1493–1497, 1952.

Shekel, J.: Indefinite Admittance Representation of Linear Network Elements, *Bull. Res. Council Israel*, vol. 3, pp. 390–394, 1954a.

Shekel, J.: Matrix Analysis of Multi-terminal Transducers, *Proc. IRE*, vol. 42, no. 5, pp. 840–847, 1954b.

Skelboe, S.: A Universal Formula for Network Functions, *IEEE Trans. Circuits and Systems*, vol. CAS-22, no. 1, pp. 58–60, 1975.

Yokomoto, C. F.: An Efficient Method of Obtaining Port Parameters from the Indefinite Admittance Matrix, *IEEE Trans. Circuit Theory*, vol. CT-19, no. 5, pp. 521–524, 1972.

Zadeh, L. A.: On Passive and Active Networks and Generalized Norton's and Thevenin's Theorems, *Proc. IRE*, vol. 44, no. 3, p. 378, 1956.

Zadeh, L. A.: Multipole Analysis of Active Networks, *IRE Trans. Circuit Theory,* vol. CT-4, no. 3, pp. 97–105, 1957.

THREE

ACTIVE TWO-PORT NETWORKS

In Chap. 1 we introduced many fundamental concepts related to linear, time-invariant n-port networks. Some of the results, although very general, are difficult to apply. In Chap. 2 we discussed a useful description of the external behavior of a multiterminal network in terms of the indefinite-admittance matrix, and demonstrated how it can be employed effectively for the computation of network functions.

In practical applications, the most useful class of n-port or n-terminal networks is that of two-port or three-terminal networks. Many active devices of practical importance such as transistors are naturally subsumed in this class. In this chapter, we consider the specialization of the general passivity condition for n-port networks in terms of the more immediately useful two-port parameters. We introduce various types of power gains, sensitivity, and the notion of absolute stability as opposed to potential instability. Llewellyn's conditions for absolute stability and the optimum terminations for absolutely stable two-port networks at a single frequency will be derived.

3.1 TWO-PORT PARAMETERS

In Sec. 1.5 we observed that the port behavior of an n-port network is completely characterized by giving the relationships among excitation and response signal transforms. Any n independent functions of the $2n$ port variables can be regarded as the excitation and the remaining n variables as the response. The characterization is given in the form of the general hybrid matrix whose elements are called the general hybrid parameters. For our purposes, we consider four different sets of two-port parameters, as follows:

$$\begin{bmatrix} V_1 \\ V_2 \end{bmatrix} = \begin{bmatrix} z_{11} & z_{12} \\ z_{21} & z_{22} \end{bmatrix} \begin{bmatrix} I_1 \\ I_2 \end{bmatrix} \tag{3.1a}$$

$$\begin{bmatrix} I_1 \\ I_2 \end{bmatrix} = \begin{bmatrix} y_{11} & y_{12} \\ y_{21} & y_{22} \end{bmatrix} \begin{bmatrix} V_1 \\ V_2 \end{bmatrix} \tag{3.1b}$$

$$\begin{bmatrix} V_1 \\ I_2 \end{bmatrix} = \begin{bmatrix} h_{11} & h_{12} \\ h_{21} & h_{22} \end{bmatrix} \begin{bmatrix} I_1 \\ V_2 \end{bmatrix} \tag{3.1c}$$

$$\begin{bmatrix} I_1 \\ V_2 \end{bmatrix} = \begin{bmatrix} g_{11} & g_{12} \\ g_{21} & g_{22} \end{bmatrix} \begin{bmatrix} V_1 \\ I_2 \end{bmatrix} \tag{3.1d}$$

The coefficient matrices are referred to as the *impedance matrix* (z-*matrix*), the *admittance matrix* (y-*matrix*), the *hybrid matrix* (h-*matrix*), and the *inverse hybrid matrix* (g-*matrix*), respectively, whose elements are called the z-*parameters*, y-*parameters*, h-*parameters*, and g-*parameters*. They are all special situations of the general representation of (1.60). Since h's are customarily used for (3.1c), the general two-port representation of (1.60) is written as

$$\begin{bmatrix} y_1 \\ y_2 \end{bmatrix} = \begin{bmatrix} k_{11} & k_{12} \\ k_{21} & k_{22} \end{bmatrix} \begin{bmatrix} u_1 \\ u_2 \end{bmatrix} \tag{3.2}$$

where $u_1, y_1 = V_1, I_1$ and $u_2, y_2 = V_2, I_2$. As indicated in Sec. 1.5, the coefficient matrix of (3.2) is referred to as the general hybrid matrix, whose elements are called the general hybrid parameters, which are different from the hybrid parameters h_{ij} of (3.1c).

For each parameter set of (3.1), there is a corresponding representation of the source and load terminations, as shown in Fig. 3.1. We shall demonstrate that formulas derived for any model are equally valid for the other three representations provided that we use the corresponding variables and parameter set. Thus, we can obtain the general formulas for network quantities that may be used with any consistent parameter set by simply substituting the appropriate quantities. Table 3.1 gives the corresponding quantities in the four parameter representations.

To illustrate the above procedure, we consider the impedance representation of Fig. 3.1a. Substituting $V_2 = -I_2 Z_2$ in Eq. (3.1a) and solving for I_1 yield

$$Z_{11} = \frac{V_1}{I_1} = z_{11} - \frac{z_{12} z_{21}}{z_{22} + Z_2} \tag{3.3}$$

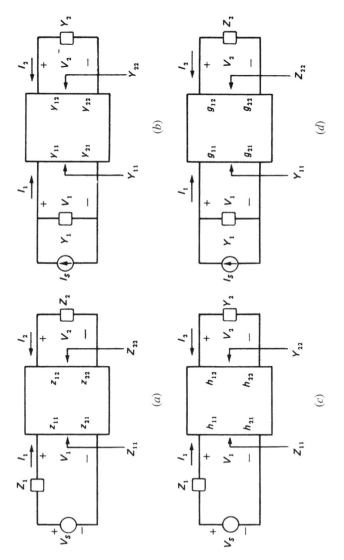

Figure 3.1 The four parameter representations of a two-port network. (*a*) The *z*-parameter representation. (*b*) The *y*-parameter representation. (*c*) The *h*-parameter representation. (*d*) The *g*-parameter representation.

Table 3.1. Corresponding quantities
in four parameter representations.

z's	y's	h's	g's	k's
z_{11}	y_{11}	h_{11}	g_{11}	k_{11}
z_{22}	y_{22}	h_{22}	g_{22}	k_{22}
z_{12}	y_{12}	h_{12}	g_{12}	k_{12}
z_{21}	y_{21}	h_{21}	g_{21}	k_{21}
V_s	I_s	V_s	I_s	u_s
I_1	V_1	I_1	V_1	u_1
I_2	V_2	V_2	I_2	u_2
Z_1	Y_1	Z_1	Y_1	M_1
Z_2	Y_2	Y_2	Z_2	M_2
Z_{11}	Y_{11}	Z_{11}	Y_{11}	M_{11}
Z_{22}	Y_{22}	Y_{22}	Z_{22}	M_{22}
V_1	I_1	V_1	I_1	y_1
V_2	I_2	I_2	V_2	y_2

Likewise, if we substitute $I_2 = -Y_2 V_2$ in Eq. (3.1b) and solve for V_1, we obtain

$$Y_{11} = \frac{I_1}{V_1} = y_{11} - \frac{y_{12} y_{21}}{y_{22} + Y_2} \tag{3.4}$$

A similar derivation using the h-parameters and g-parameters gives

$$Z_{11} = h_{11} - \frac{h_{12} h_{21}}{h_{22} + Y_2} \tag{3.5}$$

$$Y_{11} = g_{11} - \frac{g_{12} g_{21}}{g_{22} + Z_2} \tag{3.6}$$

Thus, the general formula for the input immittance can be written as

$$M_{11} = k_{11} - \frac{k_{12} k_{21}}{k_{22} + M_2} \tag{3.7}$$

The formulas (3.3)–(3.6) are the special cases of (3.7) that are obtained when appropriate substitutions are made from Table 3.1.

In a similar way, the output immittance can be expressed as

$$M_{22} = k_{22} - \frac{k_{12} k_{21}}{k_{11} + M_1} \tag{3.8}$$

3.2 POWER GAINS

Refer to the general representation of a two-port network N of Fig. 3.1. The simplest measure of power flow in N is the *power gain*. Power gain is denoted by the symbol $\mathcal{G}_p$ and defined by

$$\mathcal{G}_p = \frac{\text{average power delivered to the load}}{\text{average power entering the input port}} = \frac{P_2}{P_1} \tag{3.9}$$

Clearly, it is a function of the two-port parameters and the load impedance, and does not depend on the source impedance. For a passive lossless two-port network, $\mathcal{G}_p = 1$.

The second measure of power flow is called the *available power gain*, denoted by the symbol $\mathcal{G}_a$. The available power gain is defined by

$$\mathcal{G}_a = \frac{\text{maximum available average power at the load}}{\text{maximum available average power at the source}} = \frac{P_{2a}}{P_{1a}} \tag{3.10}$$

The quantity $\mathcal{G}_a$ is a function of the two-port parameters and the source impedance, being independent of the load impedance.

The third and most useful measure of power flow is called the *transducer power gain*, denoted by the symbol $\mathcal{G}$. The transducer power gain is defined by

$$\mathcal{G} = \frac{\text{average power delivered to the load}}{\text{maximum available average power at the source}} = \frac{P_2}{P_{1a}} \tag{3.11}$$

It is a function of the two-port parameters and the source and load impedances. Its importance arises from the fact that it compares the power delivered to the load with the power that the source is capable of supplying under the optimum conditions, and therefore it measures the efficacy of using the active device and provides the most meaningful description of the power transfer capabilities of the two-port network.

To illustrate these definitions, we shall derive expressions for them in terms of the impedance parameters of Fig. 3.1a. Substituting $V_2 = -I_2 Z_2$ in Eq. (3.1a) and solving for I_1 and I_2 yield the current gain

$$\frac{I_2}{I_1} = -\frac{z_{21}}{z_{22} + Z_2} \tag{3.12}$$

The average power P_1 entering the input port and the average power P_2 delivered to the load Z_2 are given by

$$P_1 = |I_1|^2 \operatorname{Re} Z_{11} \tag{3.13a}$$

$$P_2 = |I_2|^2 \operatorname{Re} Z_2 \tag{3.13b}$$

Figure 3.2 The Thévenin equivalent network looking into the output port of the network of Fig. 3.1a.

The maximum available average power P_{1a} from the source is

$$P_{1a} = \frac{|V_s|^2}{4 \operatorname{Re} Z_1} \tag{3.14}$$

which represents the average power delivered by the given source to a conjugately matched load $Z_{11} = \overline{Z}_1$. Combining (3.3) and (3.12)–(3.14) gives

$$\mathcal{G}_p = \frac{P_2}{P_1} = \frac{|z_{21}|^2 \operatorname{Re} Z_2}{|z_{22} + Z_2|^2 \operatorname{Re} Z_{11}} \tag{3.15a}$$

$$\mathcal{G} = \frac{P_2}{P_{1a}} = \frac{4|z_{21}|^2 \operatorname{Re} Z_1 \operatorname{Re} Z_2}{|(z_{11} + Z_1)(z_{22} + Z_2) - z_{12}z_{21}|^2} \tag{3.15b}$$

The maximum available average power P_{2a} at the output port can easily be appreciated by means of the Thévenin equivalent network of Fig. 3.2 when the input is terminated in V_s in series with Z_1 giving

$$Z_{\mathrm{eq}} = z_{22} - \frac{z_{12}z_{21}}{z_{11} + Z_1} \tag{3.16a}$$

$$V_{\mathrm{eq}} = \frac{z_{21}V_s}{z_{11} + Z_1} \tag{3.16b}$$

The maximum available average power at the output port is obtained when $Z_2 = \overline{Z}_{\mathrm{eq}}$, the complex conjugate of Z_{eq}, yielding

$$P_{2a} = \frac{|z_{21}|^2|V_s|^2}{4|z_{11} + Z_1|^2 \operatorname{Re} Z_{\mathrm{eq}}} \tag{3.17}$$

Thus, the available power gain is given by

$$\mathcal{G}_a = \frac{P_{2a}}{P_{1a}} = \frac{|z_{21}|^2 \operatorname{Re} Z_1}{|z_{11} + Z_1|^2 \operatorname{Re} Z_{\mathrm{eq}}} \tag{3.18}$$

In a similar manner, we can evaluate the three power gains in terms of other two-port parameters of Fig. 3.1, $b-d$. The general formulas are obtained as (see Prob. 3.1)

$$G_p = \frac{|k_{21}|^2 \operatorname{Re} M_2}{|k_{22} + M_2|^2 \operatorname{Re} [k_{11} - k_{12}k_{21}/(k_{22} + M_2)]} \tag{3.19a}$$

$$G = \frac{4|k_{21}|^2 \operatorname{Re} M_1 \operatorname{Re} M_2}{|(k_{11} + M_1)(k_{22} + M_2) - k_{12}k_{21}|^2} \tag{3.19b}$$

$$G_a = \frac{|k_{21}|^2 \operatorname{Re} M_1}{|k_{11} + M_1|^2 \operatorname{Re} [k_{22} - k_{12}k_{21}/(k_{11} + M_1)]} \tag{3.19c}$$

When appropriate substitutions are made from Table 3.1, we get the various gain formulas in terms of the chosen two-port parameters and the source and load immittances. We remark that the above three power gains are defined at a single real frequency, which may be at any point on the $j\omega$-axis. The variable $s = j\omega$ was dropped in all the expressions, for simplicity.

3.3 SENSITIVITY

Active networks are designed to perform certain functions such as signal amplification or processing. Given perfect components, there would be little difference among the many possible realizations. In practice, however, real components deviate from their nominal values because of changes in environmental conditions or aging of the components. As a result, the performance of the built networks differ from the nominal design. A practical solution to this problem is to design a circuit that has a low sensitivity to these changes. In the following, we consider the sensitivity of the network functions to the changes in the parameters. Techniques of reducing sensitivity by means of feedback will be discussed in the following chapter.

Sensitivity is a measure of the change of the overall transfer function to the change of a particular parameter in the network. It is formally defined as follows.

Definition 3.1: Sensitivity function The *sensitivity function* is defined as the ratio of the fractional change in a transfer function to the fractional change in an element for the situation when all changes concerned are differentially small.

Thus, if w is the transfer function and x is the element of interest, the sensitivity function, written as $S(x)$, is defined by the equation

$$S(x) = \lim_{\Delta x \to 0} \frac{\Delta w/w}{\Delta x/x} = \frac{x}{w} \frac{\partial w}{\partial x} \tag{3.20}$$

In terms of logarithms, Eq. (3.20) can be simplified to

$$S(x) = \frac{\partial \ln w}{\partial \ln x} = x \frac{\partial \ln w}{\partial x} \tag{3.21}$$

To compute the sensitivity function of a transfer function of Fig. 3.1 to the change of a two-port parameter, we must first express the transfer function in terms of the two-port parameters and the terminations. To this end, we shall work out the case of h-parameters in detail, leaving the other possibilities as an exercise for the reader.

For Fig. 3.1c, we have $V_1 = V_s - Z_1 I_1$ and $I_2 = -Y_2 V_2$. Substituting these in Eq. (3.1c) and solving for V_2 yield the ratio of the output voltage to source voltage as

$$\frac{V_2}{V_s} = -\frac{h_{21}}{(h_{11} + Z_1)(h_{22} + Y_2) - h_{12}h_{21}} \tag{3.22}$$

If one is interested in the ratio of output current to source voltage, (3.22) is changed to

$$\frac{I_2}{V_s} = \frac{h_{21}Y_2}{(h_{11} + Z_1)(h_{22} + Y_2) - h_{12}h_{21}} \tag{3.23}$$

For the transfer impedance V_2/I_1, we have

$$\frac{V_2}{I_1} = -\frac{h_{21}}{h_{22} + Y_2} \tag{3.24}$$

Similar results can be derived for other configurations of Fig. 3.1. The general formulas for the transfer functions become (see Prob. 3.2)

$$\frac{y_2}{u_s} = \frac{k_{21}M_2}{(k_{11} + M_1)(k_{22} + M_2) - k_{12}k_{21}} \tag{3.25a}$$

$$\frac{u_2}{u_s} = \frac{k_{21}}{(k_{11} + M_1)(k_{22} + M_2) - k_{12}k_{21}} \tag{3.25b}$$

$$\frac{u_2}{u_1} = \frac{k_{21}}{k_{22} + M_2} \tag{3.25c}$$

Since (3.22) and (3.23) are related only by $-Y_2$, their sensitivity functions with respect to the two-port parameters h_{ij} will be the same. To be more specific, we compute the sensitivity function of (3.22) with respect to the forward-current transfer ratio h_{21}:

$$S(h_{21}) = \frac{h_{21}}{V_2/V_s} \frac{\partial(V_2/V_s)}{\partial h_{21}} = \frac{(h_{11} + Z_1)(h_{22} + Y_2)}{(h_{11} + Z_1)(h_{22} + Y_2) - h_{12}h_{21}} \quad (3.26)$$

It is significant to observe that if $h_{12}h_{21}$ is small in comparison with $(h_{11} + Z_1)(h_{22} + Y_2)$, the sensitivity to h_{21} is nearly unity. This means that a 5% change in h_{21} gives a 5% change in V_2/V_S. On the other hand, if $h_{12}h_{21}$ is large in comparison with the other term in the denominator, the sensitivity decreases about inversely proportional to $h_{12}h_{21}$. This is precisely the effect that is obtained when negative feedback is employed.

In a similar way, the sensitivity functions of V_2/V_s to changes of other two-port parameters are obtained as (see Prob. 3.4)

$$S(h_{12}) = \frac{h_{12}h_{21}}{D_h} \quad (3.27a)$$

$$S(h_{11}) = -\frac{h_{11}(h_{22} + Y_2)}{D_h} \quad (3.27b)$$

$$S(h_{22}) = -\frac{h_{22}(h_{11} + Z_1)}{D_h} \quad (3.27c)$$

where

$$D_h = (h_{11} + Z_1)(h_{22} + Y_2) - h_{12}h_{21} \quad (3.28)$$

The general expressions of the sensitivity functions of the transfer functions y_2/u_s or u_2/u_s to changes of the general hybrid parameters $k's$ are given by (see Prob. 3.5)

$$S(k_{11}) = -\frac{k_{11}(k_{22} + M_2)}{D_k} \quad (3.29a)$$

$$S(k_{12}) = \frac{k_{12}k_{21}}{D_k} \quad (3.29b)$$

$$S(k_{21}) = \frac{(k_{11} + M_1)(k_{22} + M_2)}{D_k} \quad (3.29c)$$

$$S(k_{22}) = \frac{k_{22}(k_{11} + M_1)}{D_k} \quad (3.29d)$$

where

$$D_k = (k_{11} + M_1)(k_{22} + M_2) - k_{12}k_{21} \quad (3.30)$$

Figure 3.3 The hybrid-pi equivalent network of a transistor.

Example 3.1 Figure 3.3 shows the hybrid-pi equivalent network of a transistor. Its admittance matrix was computed earlier, as Eq. (2.43), and is repeated here:

$$\mathbf{Y}(p) = \frac{1}{240 + 1050p} \begin{bmatrix} 0.8 + 21p & -p \\ 40 - p & 11.2p + 5p^2 \end{bmatrix} \quad (3.31)$$

where $p = s/10^9$. Assume that the transistor is terminated at the load with $2\,\text{k}\Omega$ and the source with $100\,\Omega$. Then $Y_1 = 0.01$ mho and $Y_2 = 5 \cdot 10^{-4}$ mho. From (3.29), the sensitivity functions of the transfer impedance V_2/I_s or the current gain I_2/I_s to the changes of the y-parameters are computed as follows:

$$S(y_{11}) = -\frac{y_{11}(y_{22} + Y_2)}{D_y} = -\frac{105p^3 + 250.225p^2 + 11.9p + 0.096}{157.5p^3 + 384.338p^2 + 81.3p + 0.384}$$

$$(3.32a)$$

$$S(y_{12}) = \frac{y_{12}y_{21}}{D_y} = \frac{p^2 - 40p}{157.5p^3 + 384.338p^2 + 81.3p + 0.384}$$

$$(3.32b)$$

$$S(y_{21}) = \frac{(y_{11} + Y_1)(y_{22} + Y_2)}{D_y}$$

$$= \frac{157.5p^3 + 385.338p^2 + 41.3p + 0.384}{157.5p^3 + 384.338p^2 + 81.3p + 0.384}$$

$$(3.32c)$$

$$S(y_{22}) = -\frac{y_{22}(y_{11} + Y_1)}{D_y} = \frac{157.5p^3 + 368.8p^2 + 35.84p}{157.5p^3 + 384.338p^2 + 81.3p + 0.384}$$

$$(3.32d)$$

where

$$D_y = (y_{11} + Y_1)(y_{22} + Y_2) - y_{12}y_{21}$$

$$= \frac{157.5p^3 + 384.338p^2 + 81.3p + 0.384}{(1050p + 240)^2} \tag{3.33}$$

3.4 PASSIVITY AND ACTIVITY

In Chap. 1 we introduced the time-energy definition of passivity and activity for an n-port network. We showed that for a linear, time-invariant n-port network to be passive, it is necessary and sufficient that its general hybrid matrix be positive real. We extended this result by introducing the discrete-frequency concepts of passivity and activity. We demonstrated that by applying the definitions of passivity and activity at a single complex frequency, a given n-port can divide the closed RHS (right half of the complex frequency s-plane) into regions of passivity and activity. In the passive region, the n-port network can be replaced by an equivalent passive n-port network. In the active region, the n-port network can be made to achieve power gain or to oscillate by means of an appropriate passive imbedding network. However, in practical applications, we are concerned mainly with the behavior of the two-port networks on the real-frequency axis and, less frequently, on the positive σ-axis rather than the n-port networks and the entire closed RHS. In this section, we consider the specialization of the general passivity condition for n-port networks in terms of the more immediately useful two-port parameters.

As indicated in Theorem 1.5, a two-port network is passive at a point s_0 on the real-frequency axis or the positive σ-axis if and only if the hermitian part of its general hybrid matrix is nonnegative definite at s_0. The general hybrid matrix referred to in the theorem is the coefficient matrix of (3.2), which is rewritten as[†]

$$\mathbf{H}(s) = \begin{bmatrix} k_{11} & k_{12} \\ k_{21} & k_{22} \end{bmatrix} = \begin{bmatrix} m_{11} & m_{12} \\ m_{21} & m_{22} \end{bmatrix} + j \begin{bmatrix} n_{11} & n_{12} \\ n_{21} & n_{22} \end{bmatrix} \tag{3.34}$$

[†]The letters m and n are chosen to denote the real and imaginary parts of k because M is even with respect to its midpoint and N is odd with respect to its midpoint. This helps us to remember the symbols.

where the m's and n's are real. The hermitian part of $\mathbf{H}(s)$ is found to be

$$\mathbf{H}_h(s) = \frac{1}{2}[\mathbf{H}(s) + \mathbf{H}^*(s)]$$

$$= \begin{bmatrix} m_{11} & \frac{1}{2}(m_{12} + m_{21} + jn_{12} - jn_{21}) \\ \frac{1}{2}(m_{21} + m_{12} + jn_{21} - jn_{12}) & m_{22} \end{bmatrix}$$

$$(3.35)$$

The hermitian matrix $\mathbf{H}_h(s)$ is nonnegative define if and only if

$$m_{11} \geqslant 0 \tag{3.36a}$$

$$m_{22} \geqslant 0 \tag{3.36b}$$

$$\det \mathbf{H}_h(s) \geqslant 0 \tag{3.36c}$$

The third condition (3.36c) is equivalent to

$$4m_{11}m_{22} - |k_{12}|^2 - |k_{21}|^2 - 2\operatorname{Re}(k_{12}k_{21}) \geqslant 0 \tag{3.37a}$$

or

$$4m_{11}m_{22} \geqslant |k_{12} + \bar{k}_{21}|^2 \tag{3.37b}$$

In the case of the reciprocal two-port network, $z_{12} = z_{21}$, $y_{12} = y_{21}$, $h_{12} = -h_{21}$, or $g_{12} = -g_{21}$, and (3.37b) is reduced to

$$m_{11}m_{22} \geqslant m_{12}^2 \tag{3.38a}$$

for z- or y-parameters, and

$$m_{11}m_{22} \geqslant n_{12}^2 \tag{3.38b}$$

for h- or g-parameters. Equation (3.38) together with Eqs. (3.36a) and (3.36b) is known as the *Gewertz condition for passivity* of a reciprocal two-port network, and was first given by Gewertz (1933). We summarize the above results by stating the following theorem.

Theorem 3.1 A two-port network is passive at a point s_0 on the real-frequency axis or the positive σ-axis if and only if

$$m_{11} \geqslant 0 \quad m_{22} \geqslant 0 \tag{3.39a}$$

$$4m_{11}m_{22} \geqslant |k_{12} + \bar{k}_{21}|^2 \tag{3.39b}$$

It is strictly passive at s_0 if and only if the three conditions of (3.39) are satisfied with the strict inequality.

A two-port network is active at s_0 if it is not passive at s_0. Thus, if any one of the three conditions (3.39) is violated at s_0, the two-port is active at s_0. As stated in Sec. 1.7.2, the largest ω on the real-frequency axis at which a two-port network is active is called the maximum frequency of oscillation, and the largest σ on the nonnegative σ-axis at which a two-port is active is called the fastest regenerative mode.

Example 3.2 Consider again the hybrid-pi equivalent network of a transistor of Fig. 3.3. The admittance matrix $\mathbf{Y}(p)$ of the two-port network is given by (3.31). On the real-frequency axis, the hermitian part of $\mathbf{Y}(p)$ is obtained as

$$
\mathbf{Y}_h(j\omega) = \frac{1}{576 + 11{,}025\hat{\omega}^2}
\begin{bmatrix}
1.92 + 220.5\hat{\omega}^2 & 48 - 10.5\hat{\omega}^2 + j210\hat{\omega} \\
48 - 10.5\hat{\omega}^2 - j210\hat{\omega} & 105.6\hat{\omega}^2
\end{bmatrix}
$$

$$(3.40)$$

where $p = j\hat{\omega} = j\omega/10^9$. Thus, we have

$$
m_{11} = \operatorname{Re} y_{11}(j\hat{\omega}) = \frac{1.92 + 220.5\hat{\omega}^2}{576 + 11{,}025\hat{\omega}^2} \geqslant 0 \tag{3.41a}
$$

$$
m_{22} = \operatorname{Re} y_{22}(j\hat{\omega}) = \frac{105.6\hat{\omega}^2}{576 + 11{,}025\hat{\omega}^2} \geqslant 0 \tag{3.41b}
$$

$$
\begin{aligned}
4m_{11}m_{22} - |k_{12} + \bar{k}_{21}|^2 &= 4(\operatorname{Re} y_{11})(\operatorname{Re} y_{22}) - |y_{12} + \bar{y}_{21}|^2 \\
&= \frac{4(23{,}174.55\hat{\omega}^4 - 42{,}889.25\hat{\omega}^2 - 2304)}{(576 + 11{,}025\hat{\omega}^2)^2}
\end{aligned} \tag{3.41c}
$$

By setting the numerator of (3.41c) to zero, we obtain the minimum value of $|\hat{\omega}|$ for which (3.41c) is nonnegative:

$$
23.175\hat{\omega}^4 - 42.889\hat{\omega}^2 - 2.304 = 0 \tag{3.42}
$$

giving $\hat{\omega}_m^2 = 1.9$ or $\hat{\omega}_m = 1.38$. Thus, the device is passive for all $|\hat{\omega}| \geqslant \hat{\omega}_m$, and active for all $|\hat{\omega}| < \hat{\omega}_m$. The maximum frequency of oscillation of the device occurs at (see footnote on page 53)

$$
\omega_m = \hat{\omega}_m \cdot 10^9 = 1.38 \cdot 10^9 \text{ rad/s} \tag{3.43}
$$

which corresponds to 219.6 MHz.

On the σ-axis, conditions (3.39) become

$$m_{11} = y_{11}(\sigma) = \frac{0.8 + 21\hat{\sigma}}{240 + 1050\hat{\sigma}} \geqslant 0 \tag{3.44a}$$

$$m_{22} = y_{22}(\sigma) = \frac{11.2\hat{\sigma} + 5\hat{\sigma}^2}{240 + 1050\hat{\sigma}} \geqslant 0 \tag{3.44b}$$

for nonnegative $\hat{\sigma}$, where $\hat{\sigma} = \sigma/10^9$, and

$$4m_{11}m_{22} - |k_{12} + \bar{k}_{21}|^2 = 4y_{11}(\sigma)y_{22}(\sigma) - [y_{12}(\sigma) + y_{21}(\sigma)]^2$$

$$= \frac{4(105\hat{\sigma}^3 + 238.2\hat{\sigma}^2 + 48.96\hat{\sigma} - 400)}{(240 + 1050\hat{\sigma})^2}$$

$$\tag{3.44c}$$

Setting the numerator of (3.44c) to zero yields the fastest regenerative mode of the device as (see footnote on page 52)

$$\sigma_m = \hat{\sigma}_m \cdot 10^9 = 1.01 \cdot 10^9 \, \text{Np/s} \tag{3.45}$$

Thus, on the nonnegative σ-axis, the device is passive for all $\sigma \geqslant \sigma_m$ and active for all $\sigma < \sigma_m$.

Example 3.3 Consider the high-frequency, small-signal equivalent network of a bipolar transistor as shown in Fig. 3.4. Its hybrid matrix was computed earlier in Example 1.4 and is given by

$$\mathbf{H}(s) = \frac{1}{G_1 + s(C_1 + C_2)} \begin{bmatrix} 1 & sC_2 \\ g_m - sC_2 & q(s) \end{bmatrix} \tag{3.46a}$$

where $G_1 = 1/R_1$, $G_2 = 1/R_2$, and

$$q(s) = (G_1 + sC_1)(G_2 + sC_2) + sC_2(G_2 + g_m) \tag{3.46b}$$

Figure 3.4 A high-frequency, small-signal equivalent network of a bipolar transistor.

To test the passivity conditions (3.39) on the real-frequency axis, we compute the following: Let $a = G_1/(C_1 + C_2)$. Then we have

$$m_{11} = \operatorname{Re} h_{11}(j\omega) = \frac{a}{(C_1 + C_2)(\omega^2 + a^2)} \geqslant 0 \qquad (3.47a)$$

$$m_{22} = \operatorname{Re} h_{22}(j\omega)$$

$$= \frac{(C_1 G_2 + C_2 G_2 + G_1 C_2 - C_1 C_2 a + C_2 g_m)\omega^2 + G_1 G_2 a}{(C_1 + C_2)(\omega^2 + a^2)} \geqslant 0$$

$$(3.47b)$$

$$4m_{11}m_{22} - |k_{12} + \bar{k}_{21}|^2 = 4(\operatorname{Re} h_{11})(\operatorname{Re} h_{22}) - |h_{12} + \bar{h}_{21}|^2$$

$$= \frac{4G_1 G_2 - g_m^2}{(C_1 + C_2)(\omega^2 + a^2)} \qquad (3.47c)$$

The third equation is nonnegative if and only if $4G_1 G_2 \geqslant g_m^2$. Thus, the device is passive for all real frequencies if $4G_1 G_2 \geqslant g_m^2$ and active for all real frequencies if $4G_1 G_2 < g_m^2$. This situation is evidently physically impossible. The reason is that we assume the model of Fig. 3.4 to be an adequate representation of the transistor for all frequencies. At higher frequencies, parasitic effects must be taken into consideration, yielding a finite maximum frequency of oscillation.

Now suppose that we consider the admittance matrix of the two-port network of Fig. 3.4, which is given by

$$\mathbf{Y}(s) = \begin{bmatrix} G_1 + s(C_1 + C_2) & -sC_2 \\ g_m - sC_2 & G_2 + sC_2 \end{bmatrix} \qquad (3.48)$$

To test passivity on the two axes, we again appeal to (3.39). On the real-frequency axis, we have

$$m_{11} = \operatorname{Re} y_{11}(j\omega) = G_1 \geqslant 0 \qquad (3.49a)$$

$$m_{22} = \operatorname{Re} y_{22}(j\omega) = G_2 \geqslant 0 \qquad (3.49b)$$

$$4m_{11}m_{22} - |k_{12} + \bar{k}_{21}|^2 = 4(\operatorname{Re} y_{11})(\operatorname{Re} y_{22}) - |y_{12} + \bar{y}_{21}|^2$$
$$= 4G_1 G_2 - g_m^2 \qquad (3.49c)$$

Thus, the device is passive for all real frequencies if $4G_1 G_2 \geqslant g_m^2$ and active for all real frequencies if $4G_1 G_2 < g_m^2$, confirming an earlier assertion. On the positive

σ-axis, the device is passive if and only if

$$m_{11} = y_{11}(\sigma) = G_1 + \sigma(C_1 + C_2) \geqslant 0 \qquad (3.50a)$$

$$m_{22} = y_{22}(\sigma) = G_2 + \sigma C_2 \geqslant 0 \qquad (3.50b)$$

$$4m_{11}m_{22} - |k_{12} + \bar{k}_{21}|^2 = 4y_{11}(\sigma)y_{22}(\sigma) - [y_{12}(\sigma) + y_{21}(\sigma)]^2$$

$$= 4 \left[C_1 C_2 \sigma^2 + (C_1 G_2 + C_2 G_2 + G_1 C_2 + g_m C_2)\sigma \right.$$

$$\left. + G_1 G_2 - \frac{1}{4} g_m^2 \right] \geqslant 0 \qquad (3.50c)$$

resulting in the same set of inequalities as in Eq. (1.168).

The example demonstrates that even though the same conclusion can be reached from any one of the four matrix characterizations, the amount of computation involved is quite different. Thus, to test for passivity or activity of a device, the choice of an appropriate matrix description is as important as the test itself. The reader is urged to compare the amount of labor involved in the four cases.

3.5 THE U-FUNCTIONS

In this section, we introduce a useful parameter associated with a two-port network N and study some of its properties first given by Mason (1954). Specifically, we show that the parameter has the physical meaning of the maximum unilateral power gain of a two-port network under a lossless reciprocal imbedding, and it is invariant under all lossless reciprocal imbedding. Before we proceed, we comment on the difference between the words *imbedding* and *terminating*. The general connection of imbedding allows external connections to be made between the ports, whereas the specialized connection of terminating does not.

Depending on the choice of the two-port parameters, the *U-function* is defined by the expression

$$U = \frac{|k_{21} - k_{12}|^2}{4(\text{Re } k_{11} \text{ Re } k_{22} - \text{Re } k_{12} \text{ Re } k_{21})}$$

$$= \frac{|k_{21} - k_{12}|^2}{4(m_{11}m_{22} - m_{12}m_{21})} \qquad (3.51a)$$

for z- or y-parameters, and

$$U = \frac{|k_{21} + k_{12}|^2}{4(\operatorname{Re} k_{11} \operatorname{Re} k_{22} - \operatorname{Im} k_{12} \operatorname{Im} k_{21})}$$

$$= \frac{|k_{21} + k_{12}|^2}{4(m_{11}m_{22} + n_{12}n_{21})} \tag{3.51b}$$

for h- or g-parameters. The reason that the U-function has different expressions for the immittance and hybrid parameters is that, as will be shown shortly, it is dependent on reciprocity, and reciprocity is expressed differently for the immittance and hybrid parameters. For a reciprocal two-port network, we have

$$z_{12} = z_{21} \quad y_{12} = y_{21} \tag{3.52a}$$

or

$$h_{12} = -h_{21} \quad g_{12} = -g_{21} \tag{3.52b}$$

To motivate our discussion, we first demonstrate that the U-function is closely related to the determinant of the hermitian part of the general hybrid matrix $\mathbf{H}(s)$, which is not to be confused with (3.1c) although they share the same symbol. The hermitian part of $\mathbf{H}(s)$ is given by (3.35), whose determinant is found to be

$$4 \det \mathbf{H}_h(s) = 4m_{11}m_{22} - m_{12}^2 - m_{21}^2 - n_{12}^2 - n_{21}^2 - 2m_{12}m_{21} + 2n_{12}n_{21}$$

$$= 4m_{11}m_{22} - 4m_{12}m_{21} - |k_{21} - k_{12}|^2$$

$$= 4(m_{11}m_{22} - m_{12}m_{21})(1 - U) \tag{3.53a}$$

for z- or y-parameters, provided that $m_{11}m_{22} - m_{12}m_{21} \neq 0$, and

$$4 \det \mathbf{H}_h(s) = 4m_{11}m_{22} + 4n_{12}n_{21} - |k_{21} + k_{12}|^2$$

$$= 4(m_{11}m_{22} + n_{12}n_{21})(1 - U) \tag{3.53b}$$

for h- or g-parameters, provided that $m_{11}m_{22} + n_{12}n_{21} \neq 0$.

According to Theorem 3.1, the two-port network is passive at a point $s_0 = \sigma_0$ or $j\omega_0$ if and only if the three conditions of (3.39) are satisfied. The third condition (3.39b) is equivalent to $\det \mathbf{H}_h(s_0) \geqslant 0$. In the case where N is reciprocal, $U = 0$

because of (3.52), and (3.53) is nonnegative if and only if

$$m_{11}m_{22} \geqslant m_{12}^2 \tag{3.54a}$$

for z- or y-parameters, and

$$m_{11}m_{22} \geqslant n_{12}^2 \tag{3.54b}$$

for h- or g-parameters. On the other hand, if N is nonreciprocal, then (3.53) is nonnegative if and only if

$$0 \leqslant U \leqslant 1 \tag{3.55}$$

This follows directly from the observation that $m_{11}m_{22} - m_{12}m_{21}$ or $m_{11}m_{22} + n_{12}n_{21}$ and U must have the same sign, as dictated by (3.51). For $U = 1$, (3.53) is identically zero or (3.39b) is satisfied with the equality sign. Thus, the condition $U = 1$ serves as a convenient value for the computation of the maximum frequency of oscillation and the fastest regenerative mode of N.

Example 3.4 Consider the two-port network of Fig. 3.3, whose admittance matrix is given by (3.31). On the real-frequency axis, the admittance matrix becomes

$$\mathbf{Y}(j\omega) = \frac{1}{576 + 11,025\hat{\omega}^2}$$
$$\times \begin{bmatrix} 1.92 + 220.5\hat{\omega}^2 + j42\hat{\omega} & -10.5\hat{\omega} - j2.4\hat{\omega} \\ 96 - 10.5\hat{\omega}^2 - j422.4\hat{\omega} & 105.6\hat{\omega}^2 + j(26.88\hat{\omega} + 52.5\hat{\omega}^3) \end{bmatrix} \tag{3.56}$$

where $\hat{\omega} = \omega/10^9$, and the associated U-function is obtained as

$$U(j\omega) = \frac{|y_{21} - y_{12}|^2}{4(\text{Re } y_{11} \text{ Re } y_{22} - \text{Re } y_{12} \text{ Re } y_{21})}$$
$$= \frac{44,100\hat{\omega}^2 + 2304}{23,174.55\hat{\omega}^4 + 1210.752\hat{\omega}^2} \tag{3.57}$$

Setting $U = 1$ yields (3.42), which determines the maximum frequency of oscillation.

On the σ-axis, the admittance matrix becomes

$$\mathbf{Y}(\sigma) = \frac{1}{240 + 1050\hat{\sigma}} \begin{bmatrix} 0.8 + 21\hat{\sigma} & -\hat{\sigma} \\ 40 - \hat{\sigma} & 11.2\hat{\sigma} + 5\hat{\sigma}^2 \end{bmatrix} \tag{3.58}$$

where $\hat{\sigma} = \sigma/10^9$, and the U-function is found to be

$$
\begin{aligned}
U(\sigma) &= \frac{1600}{4[(0.8 + 21\hat{\sigma})(11.2\hat{\sigma} + 5\hat{\sigma}^2) - (40 - \hat{\sigma})(-\hat{\sigma})]} \\
&= \frac{400}{105\hat{\sigma}^3 + 238.2\hat{\sigma}^2 + 48.96\hat{\sigma}}
\end{aligned}
\tag{3.59}
$$

Setting $U = 1$ and solving for $\hat{\sigma}$, we get the fastest regenerative mode $\sigma_m = 1.01 \cdot 10^9$ Np/s, as in (3.45).

Before we turn our attention to other properties of the U-function, we remark that the significance of passivity is that it is the formal negation of activity. Thus, if $U(j\omega)$ is larger than unity, the two-port network is active. With appropriate passive reciprocal imbedding, power gain can be achieved for the device. In fact, we shall demonstrate in the succeeding section that $U(j\omega)$ has the physical meaning of the maximum unilateral power gain of a device under a lossless reciprocal imbedding, thereby making it a good measure of the inherent power-amplifying ability of the device.

3.5.1 The Invariance of the U-Function

In this section, we show that the U-function is invariant under the lossless reciprocal imbedding. This property was first obtained by Mason (1954) when he studied some invariant characteristics of two-port networks. For our purposes, it is sufficient to consider only the admittance parameters, leaving the other representations as being obvious.

The physical significance of this result is that if $U(j\omega)$ is used to characterize the power-amplifying capability of a device, then it is unique in that, among other reasons, it is independent of the measuring circuit, provided only that the circuit uses lossless reciprocal elements. Consequently, it measures an inherent characteristic of the device, not the device used in a particular way.

Consider a two-port network N that is imbedded in a lossless reciprocal four-port network N_0, as shown in Fig. 3.5. Let $\mathbf{Y}(s)$ and $\mathbf{Y}_0(s)$ be the admittance matrices characterizing the two-port network N and the four-port network N_0, respectively. For the time being, assume that $\mathbf{Y}_0(s)$ exists. Refer to Fig. 3.5. The matrix $\mathbf{Y}_0(s)$ is defined by the relation

$$
\begin{bmatrix} \mathbf{I}_a \\ -\mathbf{I} \end{bmatrix} = \begin{bmatrix} \mathbf{Y}_{11}^0 & \mathbf{Y}_{12}^0 \\ \mathbf{Y}_{21}^0 & \mathbf{Y}_{22}^0 \end{bmatrix} \begin{bmatrix} \mathbf{V}_a \\ \mathbf{V} \end{bmatrix}
\tag{3.60}
$$

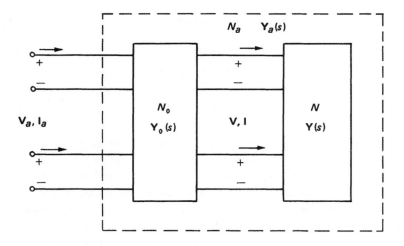

Figure 3.5 The lossless reciprocal four-port imbedding of a two-port network N.

where the coefficient matrix is the matrix $\mathbf{Y}_0(s)$ in partitioned form. Substituting

$$\mathbf{I} = \mathbf{Y}(s)\mathbf{V} \tag{3.61}$$

in the second equation of (3.60) and solving for $\mathbf{V}$ yield

$$\mathbf{V} = -(\mathbf{Y} + \mathbf{Y}_{22}^0)^{-1}\mathbf{Y}_{21}^0\mathbf{V}_a \tag{3.62}$$

where $(\mathbf{Y} + \mathbf{Y}_{22}^0)$ is assumed to be nonsingular, to be justified shortly. Substituting (3.62) in the first equation of (3.60) gives

$$\mathbf{I}_a = [\mathbf{Y}_{11}^0 - \mathbf{Y}_{12}^0(\mathbf{Y} + \mathbf{Y}_{22}^0)^{-1}\mathbf{Y}_{21}^0]\mathbf{V}_a \tag{3.63}$$

showing that the admittance matrix $\mathbf{Y}_a(s)$ of the composite two-port network N_a of Fig. 3.5 is given by

$$\mathbf{Y}_a(s) = \mathbf{Y}_{11}^0 - \mathbf{Y}_{12}^0(\mathbf{Y} + \mathbf{Y}_{22}^0)^{-1}\mathbf{Y}_{21}^0 \tag{3.64}$$

Our objective is to show that, on the real-frequency axis, the U-functions defined by the elements of $\mathbf{Y}(s)$ and those of $\mathbf{Y}_a(s)$ are the same. To this end, let U and U_a be the U-functions associated with $\mathbf{Y}(s)$ and $\mathbf{Y}_a(s)$, respectively. Then, from the definition of the U-function, U and U_a can be manipulated into the forms

$$U = \frac{|\det[\mathbf{Y}(s) - \mathbf{Y}'(s)]|}{\det[\mathbf{Y}(s) + \overline{\mathbf{Y}}(s)]} \tag{3.65a}$$

$$U_a = \frac{|\det[\mathbf{Y}_a(s) - \mathbf{Y}'_a(s)]|}{\det[\mathbf{Y}_a(s) + \overline{\mathbf{Y}}_a(s)]} \tag{3.65b}$$

where the prime, as before, denotes the matrix transpose. In the following, we show that $U(j\omega) = U_a(j\omega)$.

Since N_0 is assumed to be lossless and reciprocal, on the $j\omega$-axis we have

$$\mathbf{Y}_{ik}^0(j\omega) = j\mathbf{B}_{ik} \quad i, k = 1, 2 \tag{3.66a}$$

$$\mathbf{B}_{ik} = \mathbf{B}'_{ki} \quad i, k = 1, 2 \tag{3.66b}$$

$\mathbf{B}_{ik}$ being matrices of real elements. Using (3.66) in (3.64) yields

$$\mathbf{Y}_a(j\omega) + \overline{\mathbf{Y}}_a(j\omega) = \mathbf{B}_{12}(\mathbf{W}^{-1} + \overline{\mathbf{W}}^{-1})B'_{12}$$

$$= \mathbf{B}_{12}\mathbf{W}^{-1}[\mathbf{Y}(j\omega) + \overline{\mathbf{Y}}(j\omega)]\overline{\mathbf{W}}^{-1}\mathbf{B}'_{12} \tag{3.67}$$

where

$$\mathbf{W} = \mathbf{Y}(j\omega) + \mathbf{Y}_{22}^0(j\omega) \tag{3.68}$$

The determinant of (3.67) is found to be

$$\det[\mathbf{Y}_a(j\omega) + \overline{\mathbf{Y}}_a(j\omega)] = \frac{(\det \mathbf{B}_{12})^2 \det[\mathbf{Y}(j\omega) + \overline{\mathbf{Y}}(j\omega)]}{|\det\mathbf{W}|^2} \tag{3.69}$$

In a similar manner, we can show that

$$\mathbf{Y}_a(j\omega) - \mathbf{Y}'_a(j\omega) = \mathbf{B}_{12}\mathbf{W}^{-1}\left[\mathbf{Y}'(j\omega) - \mathbf{Y}(j\omega)\right]\mathbf{W}'^{-1}\mathbf{B}'_{12} \tag{3.70}$$

whose determinant is given by

$$\det[\mathbf{Y}_a(j\omega) - \mathbf{Y}'_a(j\omega)] = \frac{(\det \mathbf{B}_{12})^2 \det[\mathbf{Y}'(j\omega) - \mathbf{Y}(j\omega)]}{(\det \mathbf{W})^2} \tag{3.71}$$

Substituting (3.69) and (3.71) in (3.65b) in conjunction with (3.65a) yields

$$U_a(j\omega) = U(j\omega) \tag{3.72}$$

In the case where the admittance matrix $\mathbf{Y}_0(s)$ does not exist or $\mathbf{W}$ is identically singular, we first insert some lossless reciprocal elements of finite nonzero values

Figure 3.6 An active two-port network with a feedback admittance y_f connected in shunt from output to input.

in N_0 so that in the resulting four-port network, $\mathbf{Y}_0(s)$ exists and $\mathbf{W}$ is nonsingular. We then take the limit in the resulting $U_a(j\omega)$ of (3.65b) as the values of these added elements approach zero or infinity so that the resulting network becomes N_0. Since $U_a(j\omega)$ is independent of the added elements, we arrive at the same conclusion as in (3.72).

For future reference, we summarize the above result as follows.

Theorem 3.2 On the real-frequency axis, the U-function of a linear, time-invariant two-port network is invariant under the lossless reciprocal imbedding.

We illustrate this theorem by the following examples.

Example 3.5 Figure 3.6 shows an active two-port network N_a in which a feedback admittance y_f is in shunt from output to input. Assume that the one-port admittances y_a, y_f, and y_b are lossless and reciprocal. Then N_a can be viewed as the lossless reciprocal four-port imbedding of the active two-port network N, as shown in Fig. 3.7. To compute the admittance matrix $\mathbf{Y}_a(s)$ of N_a, it is convenient to consider N_a as the parallel combination of two two-port networks, so that $\mathbf{Y}_a(s)$ can be expressed as the sum of the admittance matrices of the component two-ports, as follows:

$$\mathbf{Y}_a(s) = \begin{bmatrix} y_{11} & y_{12} \\ y_{21} & y_{22} \end{bmatrix} + \begin{bmatrix} y_a + y_f & -y_f \\ -y_f & y_b + y_f \end{bmatrix} \tag{3.73}$$

where y_{ij} are the y-parameters of N. Since y_a, y_f, and y_b are lossless, on the real-frequency axis their real parts are zero, and the U-function of N_a is

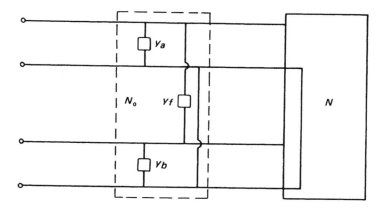

Figure 3.7 The lossless reciprocal four-port imbedding of an active two-port network N.

Figure 3.8 A three-terminal device N with a series feedback provided by the admittances y_a and y_b.

found to be

$$U_a(j\omega) = \frac{|y_{a21} - y_{a12}|^2}{4(\mathrm{Re}\ y_{a11}\ \mathrm{Re}\ y_{a22} - \mathrm{Re}\ y_{a12}\ \mathrm{Re}\ y_{a21})}$$

$$= \frac{|y_{21} - y_{12}|^2}{4(\mathrm{Re}\ y_{11}\ \mathrm{Re}\ y_{22} - \mathrm{Re}\ y_{12}\ \mathrm{Re}\ y_{21})} = U(jw) \qquad (3.74)$$

the U-function of N, where y_{aij} are the y-parameters of N_a. This confirms that the U-function is invariant under the lossless reciprocal imbedding.

Example 3.6 A three-terminal device N with a series feedback is shown in Fig. 3.8. Assume that the feedback admittances y_a and y_b are lossless and reciprocal. Then the complete network N_a of Fig. 3.8 can be viewed as the lossless

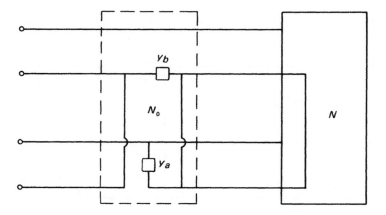

Figure 3.9 The lossless reciprocal four-port imbedding of an active device N.

reciprocal four-port imbedding of the active device N, as shown in Fig. 3.9. Since in the foregoing we have concentrated on the admittance basis, in this example we shall pick the impedance matrix for a change.

Let y_{ij} and z_{aij} be the y-parameters and the z-parameters of N and N_a, respectively. To compute the impedance matrix $\mathbf{Z}_a(s)$ of N_a, it is convenient to represent N_a as the series combination of two two-ports as shown in Fig. 3.10, so that $\mathbf{Z}_a(s)$ can be written as the sum of the impedance matrices of the two component two-ports, as follows:

$$
Z_a(s) = \begin{bmatrix} y_{11} & y_{12} \\ y_{21} & y_{22} + y_a \end{bmatrix}^{-1} + \begin{bmatrix} \dfrac{1}{y_b} & \dfrac{1}{y_b} \\ \dfrac{1}{y_b} & \dfrac{1}{y_b} \end{bmatrix}
$$

$$
= \frac{1}{\Delta_y + y_{11} y_a} \begin{bmatrix} y_{22} + y_a & -y_{12} \\ -y_{21} & y_{11} \end{bmatrix} + \begin{bmatrix} \dfrac{1}{y_b} & \dfrac{1}{y_b} \\ \dfrac{1}{y_b} & \dfrac{1}{y_b} \end{bmatrix} \tag{3.75}
$$

where $\Delta_y = y_{11} y_{22} - y_{12} y_{21}$. To compute the U-function $U_a(j\omega)$ of N_a, we first calculate $|z_{a21} - z_{a12}|^2$. From (3.75) we find that

$$
|z_{a21} - z_{a12}|^2 = \frac{|y_{21} - y_{12}|^2}{|\Delta_y + y_{11} y_a|^2} \tag{3.76}
$$

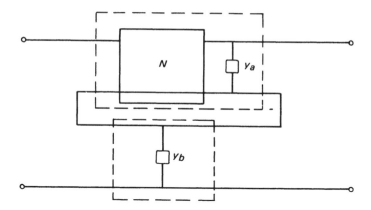

Figure 3.10 The series combination of two two-port networks.

To compute the denominator of (3.51a), we use

$$
\begin{aligned}
4(\operatorname{Re} z_{a11} \operatorname{Re} z_{a22} &- \operatorname{Re} z_{a12} \operatorname{Re} z_{a21}) \\
&= \det[\mathbf{Z}_a(j\omega) + \overline{\mathbf{Z}}_a(j\omega)] \\
&= \det[\mathbf{Z}_b(j\omega) + \overline{\mathbf{Z}}_b(j\omega)] \\
&= \frac{2 \operatorname{Re} (y_{11}\bar{y}_{22} + y_{11}\bar{y}_a - y_{12}\bar{y}_{21} + \Delta_y + y_{11}y_a)}{|\Delta_y + y_{11}y_a|^2} \\
&= \frac{4(\operatorname{Re} y_{11} \operatorname{Re} y_{22} - \operatorname{Re} y_{12}\operatorname{Re} y_{21})}{|\Delta_y + y_{11}y_a|^2}
\end{aligned}
\tag{3.77}
$$

in which we have invoked the relation $\bar{y}_a(j\omega) = -y_a(j\omega)$, where

$$
\mathbf{Z}_b(j\omega) = \frac{1}{\Delta_y + y_{11}y_a}
\begin{bmatrix} y_{22} + y_a & -y_{12} \\ -y_{21} & y_{11} \end{bmatrix}
\tag{3.78}
$$

Substituting (3.76) and (3.77) in (3.51a) yields

$$
U_a(j\omega) = \frac{|y_{21} - y_{12}|^2}{4(\operatorname{Re} y_{11} \operatorname{Re} y_{22} - \operatorname{Re} y_{12}\operatorname{Re} y_{21})}
\tag{3.79}
$$

Next we compute the U-function of N on the impedance basis. The impedance matrix of N, being the inverse of its admittance matrix, is given by (3.78) with

$y_a = 0$. Thus, if z_{ij} denote the z-parameters of N, we have

$$|z_{21} - z_{12}|^2 = \frac{|y_{21} - y_{12}|^2}{|\Delta_y|^2} \qquad (3.80a)$$

$$4(\text{Re } z_{11} \text{ Re } z_{22} - \text{Re } z_{12} \text{ Re } z_{21}) = \det[Z(j\omega) + \overline{Z}(j\omega)]$$

$$= \frac{2 \text{ Re } (\Delta_y + y_{22}\bar{y}_{11} + \bar{y}_{12}y_{21})}{|\Delta_y|^2}$$

$$= \frac{4(\text{Re } y_{11} \text{ Re } y_{22} - \text{Re } y_{12}\text{Re } y_{21})}{|\Delta_y|^2}$$

$$\qquad (3.80b)$$

where

$$\mathbf{Z}(j\omega) = \frac{1}{\Delta_y} \begin{bmatrix} y_{22} & -y_{12} \\ -y_{21} & y_{11} \end{bmatrix} \qquad (3.81)$$

Substituting (3.80) in (3.51a) gives

$$U(j\omega) = \frac{|y_{21} - y_{12}|^2}{4(\text{Re } y_{11} \text{ Re } y_{22} - \text{Re } y_{12}\text{Re } y_{21})} \qquad (3.82)$$

which when compared with (3.79) shows that

$$U_a(j\omega) = U(j\omega) \qquad (3.83)$$

again confirming that the U-function is invariant under the lossless reciprocal imbedding (see Prob. 3.32). In fact, (3.79) or (3.82) is also the U-function of N defined on the admittance basis. Thus, a two-port network will possess the same U-function regardless of the matrix employed in computation.

3.5.2 The Maximum Unilateral Power Gain

In the preceding section, we showed that the U-function of a two-port network is invariant under the lossless reciprocal imbedding. In the present section, we demonstrate that for a given active two-port network, its U-function is identifiable as the maximum unilateral power gain under a lossless reciprocal imbedding, the resulting structure being unilateral.

Consider a three-terminal device N that has two lossless reciprocal admittances y_α and y_f, one in series with the output and the other in shunt from output to input, as shown in Fig. 3.11. The network can be viewed as the lossless reciprocal four-port imbedding of the active device N as shown in Fig. 3.12. To compute the admittance matrix $\mathbf{Y}_a(s)$ of the network N_a of Fig. 3.11, it is advantageous to

Figure 3.11 A three-terminal device N with two lossless reciprocal admittances y_α and y_f.

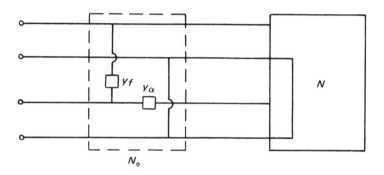

Figure 3.12 The lossless reciprocal four-port imbedding of an active device N.

consider N_a as being composed of two two-ports connected in cascade and then in parallel with another, as indicated in Fig. 3.13. The transmission matrix of the two two-ports N and N_α connected in cascade is obtained as

$$\frac{1}{y_a y_{21}} \begin{bmatrix} y_{22} & 1 \\ \Delta_y & y_{11} \end{bmatrix} \begin{bmatrix} -y_\alpha & -1 \\ 0 & -y_\alpha \end{bmatrix} \tag{3.84}$$

where y_{ij} are the y-parameters of N and $\Delta_y = y_{11}y_{22} - y_{12}y_{21}$. The corresponding admittance matrix becomes

$$\frac{y_\alpha}{y_\alpha + y_{22}} \begin{bmatrix} y_{11} + \dfrac{\Delta_y}{y_\alpha} & y_{12} \\ y_{21} & y_{22} \end{bmatrix} \tag{3.85}$$

Figure 3.13 The decomposition of the two-port network of Fig. 3.11 into two two-ports connected in cascade and then in parallel with another.

The admittance matrix $\mathbf{Y}_a(s)$ of the overall network N_a of Fig. 3.11 is found to be

$$\mathbf{Y}_a(s) = \frac{y_\alpha}{y_\alpha + y_{22}} \begin{bmatrix} y_{11} + \dfrac{\Delta_y}{y_\alpha} & y_{12} \\ y_{21} & y_{22} \end{bmatrix} + \begin{bmatrix} y_f & -y_f \\ -y_f & y_f \end{bmatrix} \tag{3.86}$$

Our objective is to determine the admittances y_α and y_f so that the resulting network N_a is unilateralized at a specified frequency on the real-frequency axis. To this end, we set the reverse transfer admittance, the (1,2)-element of $\mathbf{Y}_a(j\omega)$, to zero, yielding

$$y_f = \frac{y_{12} y_\alpha}{y_{22} + y_\alpha} \tag{3.87}$$

Since y_α and y_f are lossless, we can write

$$y_\alpha(j\omega) = jb_\alpha \quad y_f(j\omega) = jb_f \tag{3.88}$$

b_α and b_f being real susceptances. Substituting (3.88) in (3.87) results in

$$b_\alpha = b_f \frac{\text{Re } y_{22}}{\text{Re } y_{12}} \tag{3.89a}$$

$$b_f = \text{Im } y_{12} - \frac{\text{Re } y_{12}}{\text{Re } y_{22}} \text{Im } y_{22} \tag{3.89b}$$

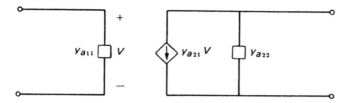

Figure 3.14 An equivalent network of the admittance matrix (3.90).

provided that Re $y_{22} \neq 0$. By choosing these values for b_α and b_f, the two-port network N_a is unilateralized, whose admittance matrix is found to be

$$
\mathbf{Y}_a(s) = \frac{j\,\mathrm{Im}(\bar{y}_{22}y_{12})}{y_{12}\,\mathrm{Re}\,y_{22}}
\begin{bmatrix}
y_{11} + y_{12} - j\dfrac{\Delta_y\,\mathrm{Re}\,y_{12}}{\mathrm{Im}(\bar{y}_{22}y_{12})} & 0 \\[2ex]
y_{21} - y_{12} & y_{22} + y_{12}
\end{bmatrix}
\tag{3.90}
$$

The U-function associated with the unilateralized two-port network N_a becomes

$$
U_a(j\omega) = \frac{|y_{a21}|^2}{4\mathrm{Re}\,y_{a11}\,\mathrm{Re}\,y_{a22}}
\tag{3.91}
$$

where y_{aij} are the elements of $\mathbf{Y}_a(s)$. The equivalent network of (3.90) is presented in Fig. 3.14. Thus, any active three-terminal device can be unilateralized by the above procedure provided that its output short-circuit conductance Re $y_{22} \neq 0$. In the case where Re $y_{22} = 0$, as will be discussed in the succeeding section, the original device is potentially unstable, and an infinite power gain can be achieved, rendering the concept of maximum unilateral power gain meaningless.

Now we demonstrate that the maximum transducer power gain or the available power gain that can be achieved for the unilateralized two-port network N_a is precisely $U(j\omega)$. To this end, let us conjugate match the input and output ports as shown in Fig. 3.15. Under this situation, the maximum transducer power gain becomes equal to the available power gain and is given by

$$
\mathcal{G}_a = \frac{|V_2|^2\,\mathrm{Re}\,y_{a22}}{|V_1|^2\,\mathrm{Re}\,y_{a11}} = \frac{|y_{a21}|^2}{4\,\mathrm{Re}\,y_{a11}\,\mathrm{Re}\,y_{a22}} = U_a(j\omega)
\tag{3.92}
$$

all being evaluated on the $j\omega$-axis. The last equality follows from (3.91). Since the U-function is invariant under all lossless reciprocal imbedding, the U-function of

Figure 3.15 A unilateral two-port network with conjugate-matched input and output ports.

N must be the same as $U_a(j\omega)$ (see Prob. 3.6),

$$\mathcal{G}_a = U_a(j\omega) = U(j\omega) \qquad (3.93)$$

In fact, any lossless reciprocal imbedding that unilateralizes a three-terminal device will yield the same maximum transducer power gain of (3.93). Thus, the U-function of an active three-terminal device has the physical meaning of the maximum unilateral transducer power gain. Furthermore, from (3.86) we have

$$\text{Re}\, y_{a22}(j\omega) = \text{Re}\, \frac{jb_\alpha y_{22}(j\omega)}{y_{22}(j\omega) + jb_\alpha} = \frac{b_\alpha^2\,\text{Re}\, y_{22}}{(\text{Re}\, y_{22})^2 + (b_\alpha + \text{Im}\, y_{22})^2} \qquad (3.94)$$

showing that $\text{Re}\, y_{a22}$ and $\text{Re}\, y_{22}$ have the same sign. For the unilateralized two-port network N_a, the short-circuit input and output conductances can be calculated directly from (3.90), giving

$$\text{Re}\, y_{a11} = \text{Re}\left\{ \frac{j\,\text{Im}(\bar{y}_{22}y_{12})}{y_{12}\,\text{Re}\, y_{22}} \left[y_{11} + y_{12} - j\frac{\Delta_y\,\text{Re}\, y_{12}}{\text{Im}(\bar{y}_{22}y_{12})} \right] \right\}$$

$$= \frac{\text{Re}\{ y_{11}\bar{y}_{12}\,[\,j\,\text{Im}(\bar{y}_{22}y_{12}) + y_{22}\,\text{Re}\, y_{12}] - y_{21}|y_{12}|^2\,\text{Re}\, y_{12}\}}{|y_{12}|^2\,\text{Re}\, y_{22}}$$

$$= \frac{\text{Re}\, y_{11}\,\text{Re}\, y_{22} - \text{Re}\, y_{12}\,\text{Re}\, y_{21}}{\text{Re}\, y_{22}} = \frac{|y_{21} - y_{12}|^2}{4U\,\text{Re}\, y_{22}} \qquad (3.95a)$$

$$\text{Re}\, y_{a22} = \text{Re}\left[\frac{j(y_{22} + y_{12})\,\text{Im}(\bar{y}_{22}y_{12})}{y_{12}\,\text{Re}\, y_{22}} \right] = \frac{|\text{Im}(\bar{y}_{22}y_{12})|^2}{|y_{12}|^2\,\text{Re}\, y_{22}} \qquad (3.95b)$$

Observe from (3.95) that $\text{Re}\, y_{a11}$ is positive if U and $\text{Re}\, y_{22}$ have the same sign, and that $\text{Re}\, y_{a22}$ is positive if $\text{Re}\, y_{22}$ is positive. Thus, the input and output conductances are both positive provided that U and Re y_{22} are both positive. Our conclusion is, therefore, that for positive U a unilateral network can always be chosen to give Re y_{a11} and Re y_{a22} the same sign as Re y_{22}. By moving the admittance y_α from the output port to the input port, the same conclusion can be reached with respect to Re y_{11} (see Prob. 3.33). We summarize the above results by the following theorem.

Theorem 3.3 At a specified frequency on the real-frequency axis, a three-terminal device whose U-function and at least one of whose short-circuit driving-point conductances are positive can be transformed by means of a lossless reciprocal imbedding into a unilateral, common-terminal two-port network having positive short-circuit input and output conductances, its available power gain always being equal to its U-function.

We remark that, in the above theorem, the U is not the greatest attainable power gain under all passive imbedding; it is the maximum obtainable power gain under all lossless reciprocal imbedding (see Probs. 3.36 and 3.37).

The basic performance such as the power-amplifying ability of an active device can be measured meaningfully by its U-function. The measurement is unique in that it is independent of the measuring circuit, provided only that the circuit is unilateral and uses lossless reciprocal elements. As demonstrated above, such a measurement is usually physically possible and can be achieved by the network proposed above. For the transistor, for example, it is a three-terminal device. Its power-amplifying ability is usually measured in terms of the maximum power gain available in a specified network. Such a specification depends critically on the details of the particular network used. It is meaningless to compare different measurements unless the same measuring network is used. Thus, it is difficult to relate the results to basic transistor properties. Furthermore, the orientation of the transistor terminals such as the common-base, common-emitter, or common-collector configuration results in a critical difference in the measurements. The U-function, on the other hand, provides a unique measure of the degree of inherent transfer activity exhibited by a device. Since the U-function is invariant under the lossless reciprocal imbedding, it is invariant under all orientations of the device. Thus, it measures a general characteristic of the device, not merely a particular way the device is used.

As will be shown in the following section, a unilateral two-port network is "absolutely stable" at a specified frequency if and only if the real parts of its short-circuit input and output admittances are positive. It follows from Theorem 3.3 that the network of Fig. 3.11 is absolutely stable and can always be used to measure the unilateral power gain, provided that U is positive and at least one of the short-circuit driving-point conductances of the device is positive. However, if both conductances are negative and U exceeds unity, the network of Fig. 3.11 cannot be used to measure U, and a different lossless reciprocal unilateralizing network should be used (see, for example, Prob. 3.11). The situation seems unlikely to be encountered in practice. Transistors in particular obey the conditions stipulated in Theorem 3.3 in their normal operating range. The technique has been confirmed experimentally for a few transistors in the frequency range, from 100 to 900 MHz by Rollett (1965).

Example 3.7 At 30 MHz, $V_{CE} = 20$ V, and $I_c = 20$ mA, the following values of the y-parameters apply to a sample type 2N697 transistor:

$$
\begin{aligned}
y_{11} &= y_{ie} = (22.5 + j14.7) \cdot 10^{-3} \text{ mho} \\
y_{12} &= y_{re} = (-0.8 - j0.38) \cdot 10^{-3} \text{ mho} \\
y_{21} &= y_{fe} = (36.6 - j91.6) \cdot 10^{-3} \text{ mho} \\
y_{22} &= y_{oe} = (1.7 + j5.7) \cdot 10^{-3} \text{ mho}
\end{aligned}
\tag{3.96}
$$

We wish to design a unilateral two-port network to measure the U-value of this transistor at the frequency of 30 MHz.

By using the network of Fig. 3.11, the susceptances b_α and b_f are computed from Eq. (3.89):

$$b_f = 2.30 \cdot 10^{-3} \text{ mho} \tag{3.97a}$$

$$b_\alpha = -4.89 \cdot 10^{-3} \text{ mho} \tag{3.97b}$$

Let $y_f = j\omega C = jb_f$ and $y_\alpha = 1/j\omega L = jb_\alpha$, where $\omega = 60\pi \cdot 10^6$ rad/s. Then we have

$$C = 12.2 \text{ pF} \tag{3.98a}$$

$$L = 1.08 \, \mu\text{H} \tag{3.98b}$$

The desired unilateral network is shown in Fig. 3.16. The U-value associated with the transistor at 30 MHz is found to be

$$U = \frac{|37.4 - j91.22|^2}{4\,[(22.5)(1.7) - (-0.8)(36.6)]} = 35.98 \tag{3.99}$$

which is the maximum transducer power gain or the available power gain of the transistor operating at 30 MHz under all lossless reciprocal imbedding.

Figure 3.16 A unilateral two-port network used to measure the U-value of a transistor.

From (3.90), the admittance matrix of the unilateral two-port network of Fig. 3.16 evaluated at $\omega = 60\pi \cdot 10^6$ rad/s is given by

$$\mathbf{Y}_a(j60\pi \cdot 10^6) = 10^{-3} \begin{bmatrix} 39.72 - j26.10 & 0 \\ -255.91 + j13.92 & 11.49 - j8.05 \end{bmatrix} \quad (3.100)$$

Appealing to (3.92) yields $U_a = 35.98$, as expected.

3.6 POTENTIAL INSTABILITY AND ABSOLUTE STABILITY

A network is said to be *stable* if all of its natural frequencies are restricted to the open LHS (*l*eft *h*alf of the complex frequency s-plane). The $j\omega$-axis poles of the transfer functions are not permitted for stable networks because such poles will yield an unbounded time response. If the transfer function, say, has a simple pole $j\omega_0$ on the real-frequency axis, an excitation of the type $\exp(j\omega_0 t)$ will result in an unbounded time response. For a two-port network, its stability cannot be determined by the two-port itself; it also depends on the terminations. Thus, a two-port network is said to be *potentially unstable* if it is possible to select passive one-port immittances that, when terminated at the ports, produce an unstable network. If no such passive immittances can be found, the two-port network is called *absolutely stable*. Like the discrete-frequency concepts of passivity discussed in Sec. 3.4, it is convenient to introduce single frequency concepts of stability for two-port networks.

Definition 3.2: Potential instability at $j\omega_0$. A two-port network is said to be *potentially unstable at $j\omega_0$* on the real-frequency axis if there exist two passive one-port immittances that, when terminated at the ports, produce a natural frequency at $j\omega_0$ in the overall network.

Definition 3.3: Absolute stability at $j\omega_0$. A two-port network is said to be *absolutely stable at $j\omega_0$* on the real-frequency axis if it is not potentially unstable at $j\omega_0$.

It is possible to extend the above definitions to include all the points in the closed RHS. Then a two-port network is potentially unstable if it is potentially unstable at a point in the closed RHS and is absolutely stable if it is absolutely stable at every point in the closed RHS. The restriction to the real-frequency axis is sufficient for all practical purposes.

Using the above definitions to test the stability of a two-port network is clearly difficult and tedious, because we have to examine all passive terminations, and they are not intended for this purpose. In the following, we develop equivalent criteria for these concepts.

3.6.1 Preliminary Considerations

The natural frequencies of a linear system are the frequencies for which signals can be supported by the system without external excitations. Thus, for a two-port network N, characterized by the general hybrid equation

$$\begin{bmatrix} y_1 \\ y_2 \end{bmatrix} = \begin{bmatrix} k_{11} & k_{12} \\ k_{21} & k_{22} \end{bmatrix} \begin{bmatrix} u_1 \\ u_2 \end{bmatrix} \tag{3.101}$$

and terminated in the immittances M_1 and M_2 as shown in Fig. 3.17, the natural frequencies are the values of s for which the above equation together with the

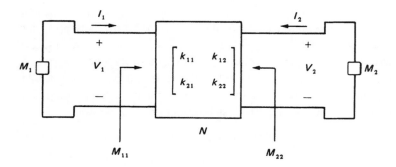

Figure 3.17 The general representation of a terminated two-port network.

boundary conditions

$$y_1 = -M_1 u_1 \tag{3.102a}$$

$$y_2 = -M_2 u_2 \tag{3.102b}$$

are satisfied simultaneously. Table 3.1 and Eq. (3.1) define the variables for various parameter sets. Combining (3.101) and (3.102) gives

$$\begin{bmatrix} k_{11} + M_1 & k_{12} \\ k_{21} & k_{22} + M_2 \end{bmatrix} \begin{bmatrix} u_1 \\ u_2 \end{bmatrix} = \mathbf{0} \tag{3.103}$$

For the above equation to possess a nontrivial solution, it is necessary and sufficient that the coefficient matrix be singular, requiring that

$$\det \begin{bmatrix} k_{11} + M_1 & k_{12} \\ k_{21} & k_{22} + M_2 \end{bmatrix} = 0 \tag{3.104a}$$

or

$$(k_{11} + M_1)(k_{22} + M_2) - k_{12}k_{21} = 0 \tag{3.104b}$$

which is referred to as the *characteristic equation* of the network. By using Eqs. (3.7) and (3.8), Eq. (3.104b) becomes

$$M_1(s) + M_{11}(s) = 0 \tag{3.105a}$$

$$M_2(s) + M_{22}(s) = 0 \tag{3.105b}$$

The left-hand sides of (3.105) are recognized as the input immittances at the input and output ports of Fig. 3.17. These provide alternative means of obtaining the characteristic equation. One may equate the impedance of the input or output loop of Fig. 3.17 to zero, or the admittance appearing across any port to zero. The zeros of these functions are the natural frequencies of the network. For example, if (3.101) is defined on the admittance basis, the left-hand sides of (3.105) denote the total admittances appearing across the input and output ports of Fig. 3.1b, and (3.105) can be written as

$$Y_1(s) + Y_{11}(s) = 0 \tag{3.106a}$$

$$Y_2(s) + Y_{22}(s) = 0 \tag{3.106b}$$

If (3.101) represents (3.1c), the corresponding equations are given by

$$Z_1(s) + Z_{11}(s) = 0 \qquad (3.107a)$$

$$Y_2(s) + Y_{22}(s) = 0 \qquad (3.107b)$$

The left-hand side of (3.107a) is the impedance looking into the voltage source V_s of Fig. 3.1c. The various means described above for arriving at the characteristic equation all lead to equivalent results in that the same characteristic equation is obtained.

A two-port network is therefore absolutely stable at $j\omega_0$ if and only if, for all passive terminations, its characteristic equation (3.104) does not possess a root at $j\omega_0$. In other words, at $j\omega_0$ the equation (3.103) has only a trivial solution. Thus, we can state the following.

Theorem 3.4 A linear time-invariant two-port network is absolutely stable at $j\omega_0$ if and only if the port voltages and currents are zero at $j\omega_0$ under all passive one-port terminations.

The stability of a two-port network can also be determined by its input and output conductances under all passive one-port terminations.

Theorem 3.5 A linear time-invariant two-port network is absolutely stable at $j\omega_0$ if and only if the driving-point immittance at each of its two ports has a positive-real part for all passive one-port terminations at the other port at the frequency $j\omega_0$.

PROOF. It is sufficient to consider only the admittances. Referring to Fig. 3.17 and from Eq. (3.106), we conclude that to create a natural frequency at $j\omega_0$, it is necessary and sufficient that

$$\text{Re } Y_1(j\omega_0) + \text{Re } Y_{11}(j\omega_0) = 0 \qquad (3.108a)$$

$$\text{Re } Y_2(j\omega_0) + \text{Re } Y_{22}(j\omega_0) = 0 \qquad (3.108b)$$

Since by hypothesis $Y_1(s)$ and $Y_2(s)$ are passive, being positive-real functions, their real parts at $j\omega_0$ must be nonnegative; that is,

$$\text{Re } Y_1(j\omega_0) \geqslant 0 \qquad (3.109a)$$

$$\text{Re } Y_2(j\omega_0) \geqslant 0 \qquad (3.109b)$$

With the constraints (3.109), (3.108) possesses a solution if and only if Re $Y_{11}(j\omega_0) \leqslant 0$ and Re $Y_{22}(j\omega_0) \leqslant 0$ because $Y_1(s)$ and $Y_2(s)$ are arbitrary passive

Figure 3.18 A network possessing a natural frequency at $s = j\frac{1}{4}$.

admittances. In other words, the two-port network is absolutely stable at $j\omega_0$ if and only if Re $Y_{11}(j\omega_0) > 0$ and Re $Y_{22}(j\omega_0) > 0$. This completes the proof of the theorem.

We illustrate the above results by the following example.

Example 3.8 Consider the two-port network N shown in Fig. 3.18, whose impedance matrix is given by

$$\mathbf{Z} = \begin{bmatrix} -1 & -2 \\ -2 & -2 \end{bmatrix} \tag{3.110}$$

In particular, if we choose the open-circuit termination, the input impedances of N become $Z_{11} = z_{11} = -1\ \Omega$ and $Z_{22} = z_{22} = -2\ \Omega$. According to Theorem 3.5, N is potentially unstable over the entire real-frequency axis. Assume that the output of N is terminated in a 1-F capacitor. By applying (3.7), the input impedance is found to be

$$Z_{11}(s) = \frac{2s+1}{2s-1} \tag{3.111}$$

Choose

$$s = j\omega_0 = j\frac{1}{4} \tag{3.112}$$

We demonstrate that a passive input termination $Z_1(s)$ can always be found, so that the resulting network possesses a natural frequency at $j\frac{1}{4}$. From (3.111) we have

$$Z_{11}\left(j\frac{1}{4}\right) = -\frac{3}{5} - j\frac{4}{5} \tag{3.113}$$

To create a natural frequency at $j\frac{1}{4}$, we set

$$Z_1\left(j\frac{1}{4}\right) = -Z_{11}\left(j\frac{1}{4}\right) = \frac{3}{5} + j\frac{4}{5} \tag{3.114}$$

showing that $Z_1(s)$ can be realized as the series combination of a resistor of $\frac{3}{5}$ Ω and an inductor of $\frac{16}{5}$ H. The complete network is presented in Fig. 3.18. It is straightforward to confirm that the network possesses a natural frequency at $j\frac{1}{4}$.

At $s = j\frac{1}{4}$, the two equations of (3.103) are linearly dependent, and the general solution can be written as

$$I_1 = -(1 + j2)x \tag{3.115a}$$

$$I_2 = x \tag{3.115b}$$

where x is arbitrary. Thus, for $x \neq 0$ the network can support nonzero port currents I_1 and I_2, as expected from Theorem 3.4.

3.6.2 Llewellyn's Stability Criteria

In this section, we establish Llewellyn's conditions for absolute stability of a two-port network at a specified frequency on the real-frequency axis. These conditions were first obtained by Llewellyn (1952) and are relatively simple to test. To keep our discussion general, we again consider the general hybrid matrix (3.34), whose elements are, as before, written as

$$k_{ij} = \operatorname{Re} k_{ij} + j \operatorname{Im} k_{ij} = m_{ij} + jn_{ij} \tag{3.116}$$

Definition 3.4: Stability parameter The dimensionless quantity

$$\eta = \frac{2 \operatorname{Re} k_{11} \operatorname{Re} k_{22} - \operatorname{Re} (k_{12}k_{21})}{|k_{12}k_{21}|} \tag{3.117}$$

associated with a two-port network is called the *stability parameter* of the two-port network.

An important property of this stability parameter is that it is invariant when any one set of impedance, admittance, hybrid, or inverse hybrid parameters is replaced by any other set. The operation is termed as the *immittance substitution*. The y- and

h-parameters, for example, are related by the equations (see App. II)

$$y_{11} = \frac{1}{h_{11}} \qquad y_{12} = -\frac{h_{12}}{h_{11}} \qquad (3.118a)$$

$$y_{21} + \frac{h_{21}}{h_{11}} \qquad y_{22} = \frac{h_{11}h_{22} - h_{12}h_{21}}{h_{11}} \qquad (3.118b)$$

By substituting these in (3.117), it is straightforward to confirm that

$$\begin{aligned}
\eta &= \frac{2 \, \mathrm{Re} \; y_{11} \, \mathrm{Re} \; y_{22} - \mathrm{Re} \; (y_{12}y_{21})}{|y_{12}y_{21}|} \\
&= \frac{2 \, \mathrm{Re} \; h_{11} \, \mathrm{Re} \; h_{22} - \mathrm{Re} \; (h_{12}h_{21})}{|h_{12}h_{21}|}
\end{aligned} \qquad (3.119)$$

The verification for other sets of parameters is left as an exercise (see Prob. 3.14). Since η is invariant under immittance substitution, it is immaterial as to which set is chosen in defining (3.117), and they all lead to the same result.

With these preliminaries, we now state Llewellyn's criteria for absolute stability.

Theorem 3.6 A linear time-invariant two-port network is absolutely stable at $j\omega_0$ if and only if

$$\mathrm{Re} \; k_{11} = m_{11} > 0 \qquad (3.120a)$$

$$\mathrm{Re} \; k_{22} = m_{22} > 0 \qquad (3.120b)$$

$$\eta > 1 \qquad (3.120c)$$

all being evaluated at $s = j\omega_0$.

PROOF. *Necessity*: Assume that a given two-port network N is absolutely stable at $j\omega_0$. Then according to Theorem 3.5 the input immittances $M_{11}(j\omega_0)$ and $M_{22}(j\omega_0)$, as defined in Fig. 3.17, of the terminated two-port must have positive-real parts for all passive one-port terminations $M_2(j\omega_0)$ and $M_1(j\omega_0)$. In particular, if we choose the open-circuit or short-circuit termination, $M_1 = \infty$ and $M_2 = \infty$, and we have

$$\mathrm{Re} \; M_{11}(j\omega_0) = m_{11}(\omega_0) > 0 \qquad (3.121a)$$

$$\mathrm{Re} \; M_{22}(j\omega_0) = m_{22}(\omega_0) > 0 \qquad (3.121b)$$

To show that (3.120c) is necessary, two cases are distinguished, the second case being based on the first.

Case 1: N is reciprocal Since the stability parameter is invariant under immittance substitution, it is sufficient to consider only the admittance parameters. Thus, we write $k_{ij} = y_{ij}$, $M_{ii} = Y_{ii}$ and $M_i = Y_i$.

For a reciprocal two-port network N, $y_{12} = y_{21}$ and (3.120c) is reduced to

$$\text{Re } y_{11} \text{ Re } y_{22} > (\text{Re } y_{12})^2 \tag{3.122}$$

(see Prob. 3.16). By appealing to (3.7), the input admittance $Y_{11}(j\omega_0)$ is related to the y-parameters $y_{ij}(j\omega_0)$ and the terminating admittance $Y_2(j\omega_0)$ by

$$
\begin{aligned}
Y_{11}(j\omega_0) &= y_{11} - \frac{y_{12}^2}{y_{22} + Y_2} \\
&= y_{11} + y_{12} + \frac{1}{1/(y_{22} + y_{12} + Y_2) + 1/(-y_{12})}
\end{aligned}
\tag{3.123}
$$

all being evaluated at $s = j\omega_0$. Choose $Y_2(j\omega_0) = jB_2$ such that

$$\frac{\text{Im } y_{22} + \text{Im } y_{12} + B_2}{\text{Re } y_{22} + \text{Re } y_{12}} = \frac{\text{Im } y_{12}}{\text{Re } y_{12}} \tag{3.124}$$

Under this situation, the admittances $(y_{22} + y_{12} + Y_2)$ and $-y_{12}$ will have the same phase at $j\omega_0$. Consequently, the real part of these two admittances in series, which corresponds to the last term on the right-hand side of (3.123), equals the series combination of their real parts (see Prob. 3.18), giving

$$
\begin{aligned}
\text{Re } Y_{11}(j\omega_0) &= \text{Re } y_{11} + \text{Re } y_{12} - \frac{\text{Re } y_{12} \text{ Re } (y_{22} + y_{12})}{\text{Re } (y_{22} + y_{12} - y_{12})} \\
&= \text{Re } y_{11} - \frac{(\text{Re } y_{12})^2}{\text{Re } y_{22}}
\end{aligned}
\tag{3.125}
$$

which, according to Theorem 3.5, must be positive. Hence we have (3.122).

Case 2: N is nonreciprocal Consider the associated reciprocal two-port network N_a with admittance matrix

$$\mathbf{Y}_a(s) = \begin{bmatrix} y_{11} & \sqrt{y_{12}y_{21}} \\ \sqrt{y_{12}y_{21}} & y_{22} \end{bmatrix} \tag{3.126}$$

By applying (3.7) and (3.8), it is easy to show that the admittances looking into the corresponding ports of N and N_a are identical for identical terminations. Then, by Theorem 3.5, N is absolutely stable at $j\omega_0$ if and only if N_a is. Since N_a is reciprocal, from Case 1 it is necessary that

$$\text{Re } y_{11} \text{ Re } y_{22} > (\text{Re } \sqrt{y_{12}y_{21}})^2 \qquad (3.127)$$

Now we demonstrate that (3.127) is equivalent to (3.120c). To this end, let

$$\sqrt{y_{12}y_{21}} = Le^{j\theta} \qquad (3.128)$$

Substituting it in (3.127) yields

$$2 \text{ Re } y_{11} \text{ Re } y_{22} > 2L^2 \cos^2 \theta = L^2(1 + \cos 2\theta)$$
$$= |y_{12}y_{21}| + \text{Re } y_{12}y_{21} \qquad (3.129)$$

Since the stability parameter η is invariant under immittance substitution, (3.120c) follows.

Sufficiency: Assume that the conditions (3.120) hold at $s = j\omega_0$. We show that N is absolutely stable at $j\omega_0$ by demonstrating that the real parts of the driving-point immittances $M_{11}(j\omega_0)$ and $M_{22}(j\omega_0)$ are positive for all passive one-port terminations M_1 and M_2.

The real part of the input immittance M_{11}, as given in (3.7), can be manipulated into the form

$$\frac{|k_{22} + M_2|^2 \text{ Re } M_{11}}{\text{Re } k_{11}} = |k_{22} + M_2|^2 - \frac{\text{Re } \left[k_{12}k_{21}(\bar{k}_{22} + \overline{M}_2)\right]}{\text{Re } k_{11}}$$
$$= (A_2 + \alpha)^2 + (B_2 + \beta)^2 - \frac{|k_{12}k_{21}|^2}{(2 \text{ Re } k_{11})^2} \qquad (3.130)$$

where $M_2 = A_2 + jB_2$ and

$$\alpha = \text{Re } k_{22} - \frac{\text{Re } k_{12}k_{21}}{2 \text{ Re } k_{11}} \qquad (3.131a)$$

$$\beta = \text{Im } k_{22} - \frac{\text{Im } k_{12}k_{21}}{2 \text{ Re } k_{11}} \qquad (3.131b)$$

Since $\eta > 1$ and Re $k_{11} > 0$, we have

$$\alpha = \frac{\eta|k_{12}k_{21}|}{2 \text{ Re } k_{11}} \geqslant 0 \qquad (3.131c)$$

This, together with the fact that $A_2 \geqslant 0$, indicates that the minimum of the quantity on the right-hand side of (3.130) is attained at

$$A_2 = 0 \quad B_2 = -\beta \qquad (3.132)$$

and (3.130) becomes

$$|k_{22} + M_2|^2 \, \text{Re} \, M_{11} = \alpha^2 \, \text{Re} \, k_{11} - \frac{|k_{12}k_{21}|^2}{4 \, \text{Re} \, k_{11}} = \frac{|k_{12}k_{21}|^2(\eta^2 - 1)}{4 \, \text{Re} \, k_{11}} \qquad (3.133)$$

Since by hypothesis $\eta > 1$ and $\text{Re} \, k_{11} > 0$, $\text{Re} \, M_{11}$ is positive at $j\omega_0$. In a similar fashion, we can show that $\text{Re} \, M_{22}(j\omega_0) > 0$. Thus, by Theorem 3.5, the two-port network N is absolutely stable at $j\omega_0$. This completes the proof of the theorem.

A number of important observations can be made from the above discussions. First, as stated in Theorem 3.5, because the stability of a two-port network depends on the positiveness of the real part of the immittance looking in at either port, it must be invariant under interchange of input and output. The inequalities (3.120) are clearly invariant under the interchange of subscripts 1 and 2. Second, if the stability test is performed for one set of parameters, the same conclusions will be reached for other sets of parameters. However, the quantities m_{11} and m_{22} may be different for different sets of parameters. Finally, we mention that the stability parameter η is a measure of the degree of two-port stability. For positive m_{11} and m_{22}, the value of η lies between -1 and ∞ (see Prob. 3.19). When η is positive and large compared with unity, the degree of absolute stability is high; when η is near unity, the two-port network is close to the boundary between absolute stability and potential instability, which is defined by $\eta = 1$. When $1 \geqslant \eta \geqslant -1$, the two-port network is in the region of potential instability, meaning that we can always choose passive one-port terminations that will result in oscillations.

Two special cases are worthy of mentioning. First, for a unilateral two-port network, $k_{12} = 0$ and Eq. (3.120c) is reduced to $2m_{11}m_{22} > 0$, showing that $m_{11} > 0$ and $m_{22} > 0$ are enough to guarantee absolute stability. In other words, a unilateral two-port network is absolutely stable at $j\omega_0$ if and only if the short-circuit or open-circuit input and output immittances possess positive-real parts at $j\omega_0$. Next, for a reciprocal two-port network, (3.52) applies and (3.120c) is reduced to (see Prob. 3.16)

$$4m_{11}m_{22} > |k_{12} + \bar{k}_{21}|^2 \qquad (3.134)$$

which together with $m_{11} > 0$ and $m_{22} > 0$ is precisely (3.39) with strict inequality. Thus, a reciprocal two-port network is absolutely stable at $j\omega_0$ if and only if it is strictly passive at $j\omega_0$. Mere passivity, however, will not suffice, because the passivity conditions (3.39) allow the equality to hold whereas the absolute stability conditions require the strict inequality. A reciprocal LC two-port is passive but is potentially unstable. On the other hand, an absolutely stable nonreciprocal two-port network may be active. In fact, we can show that any strictly passive two-port network must be absolutely stable. However, not all absolutely stable two-port

Figure 3.19 A two-port network used to illustrate absolute stability.

networks are strictly passive. To see this, we consider (3.39b) with strict inequality, which can be written in the equivalent form

$$4m_{11}m_{22} - 2\operatorname{Re} k_{12}k_{21} > |k_{12}|^2 + |k_{21}|^2 \qquad (3.135a)$$

or

$$\eta > \frac{|k_{12}|^2 + |k_{21}|^2}{2|k_{12}k_{21}|} = 1 + \frac{(|k_{12}| - |k_{21}|)^2}{2|k_{12}k_{21}|} \qquad (3.135b)$$

showing that $\eta > 1$. This together with (3.39a) with strict inequality guarantees the absolute stability of the two-port network.

We illustrate the above results by the following examples.

Example 3.9 Consider the two-port network N of Fig. 3.19, whose hybrid matrix is given by

$$\mathbf{H}(s) = \frac{1}{s-1} \begin{bmatrix} s & 1 \\ -1 & s-2 \end{bmatrix} \qquad (3.136)$$

On the real-frequency axis, its value is found to be

$$\mathbf{H}(j\omega) = \frac{1}{\omega^2 + 1} \begin{bmatrix} \omega^2 - j\omega & -1 - j\omega \\ 1 + j\omega & \omega^2 + 2 + j\omega \end{bmatrix} \qquad (3.137)$$

Thus, according to Theorem 3.6, the device is absolutely stable if and only if

$$\operatorname{Re} h_{11} = \frac{\omega^2}{\omega^2 + 1} > 0 \qquad (3.138a)$$

$$\operatorname{Re} h_{22} = \frac{\omega^2 + 2}{\omega^2 + 1} > 0 \qquad (3.138b)$$

$$\eta = \frac{2\operatorname{Re} h_{11}\operatorname{Re} h_{22} - \operatorname{Re} h_{12}h_{21}}{|h_{12}h_{21}|} = 2\omega^2 + 1 > 1 \qquad (3.138c)$$

Thus, the device is absolutely stable for all nonzero frequencies.

To see if the device is also passive, we compute the hermitian part of $\mathbf{H}(j\omega)$ giving

$$\mathbf{H}_h(j\omega) = \frac{1}{\omega^2 + 1}\begin{bmatrix} \omega^2 & -j\omega \\ j\omega & \omega^2 + 2 \end{bmatrix} \tag{3.139}$$

The matrix $\mathbf{H}_h(j\omega)$ is clearly nonnegative definite for all ω, indicating that the device is passive for all frequencies, zero included. As a matter of fact, $\mathbf{H}_h(j\omega)$ is positive definite for all nonzero ω. The two-port network is therefore strictly passive for all nonzero ω. This in conjunction with the fact that N is reciprocal, $\mathbf{H}(s)$ being skew symmetric, $h_{12} = -h_{21}$, implies that the device is absolutely stable and strictly passive for all nonzero real frequencies.

To demonstrate the invariance of the stability parameter under immittance substitution, we compute the impedance matrix of Fig. 3.19:

$$\mathbf{Z}(s) = \frac{1}{s-2}\begin{bmatrix} s-1 & 1 \\ 1 & s-1 \end{bmatrix} \tag{3.140}$$

which, when evaluated on the $j\omega$-axis, yields

$$\mathbf{Z}(j\omega) = \frac{1}{\omega^2 + 4}\begin{bmatrix} \omega^2 + 2 - j\omega & -2 - j\omega \\ -2 - j\omega & \omega^2 + 2 - j\omega \end{bmatrix} \tag{3.141}$$

The corresponding stability parameter is found to be

$$\eta = \frac{2\,\mathrm{Re}\,z_{11}\,\mathrm{Re}\,z_{22} - \mathrm{Re}\,z_{12}z_{21}}{|z_{12}z_{21}|} = 2\omega^2 + 1 \tag{3.142}$$

which is the same as that given in (3.138c). Using (3.142) in conjunction with the fact that $\mathrm{Re}\,z_{11} = \mathrm{Re}\,z_{22} > 0$, we arrive at the same conclusion as before. Note that the invariance of the stability parameter under immittance substitution is valid for all s, not merely for $s = j\omega$, the latter being demonstrated above.

Example 3.10 Figure 3.20 shows the equivalent T-model of a transistor. On the real-frequency axis, its impedance matrix is found to be

$$\mathbf{Z}(j\omega) = \begin{bmatrix} r_e + r_b & r_b \\ r_b - \dfrac{j\alpha}{\omega C} & r_b - \dfrac{j}{\omega C} \end{bmatrix} \tag{3.143}$$

Figure 3.20 The equivalent T-model of a transistor.

Thus, according to Theorem 3.6, the transistor is absolutely stable at $j\omega_0$ if and only if

$$r_e + r_b > 0 \qquad (3.144a)$$

$$r_b > 0 \qquad (3.144b)$$

$$2(r_e + r_b)r_b - \text{Re}\left[r_b\left(r_b - \frac{j\alpha}{\omega_0 C}\right)\right] > \left|r_b\left(r_b - \frac{j\alpha}{\omega_0 C}\right)\right| \qquad (3.144c)$$

The last inequality (3.144c) is simplified to

$$4\omega_0^2 C^2 r_e(r_e + r_b) > \alpha^2 \qquad (3.145)$$

In Example 1.12, we showed that the transistor is active for all real frequencies

$$\omega < \frac{\alpha}{2C\sqrt{r_e r_b}} \qquad (3.146)$$

Using this in conjunction with (3.145), we conclude that the transistor is both active and absolutely stable at $j\omega_0$ only if

$$\frac{\alpha^2}{4\omega_0^2 C^2} < r_e(r_e + r_b) < \frac{\alpha^2}{4\omega_0^2 C^2} + r_e^2 \qquad (3.147)$$

To put it differently, for the device to be both active and absolutely stable, we must restrict the frequencies in the range

$$\frac{\alpha^2}{4C^2 r_e(r_e + r_b)} < \omega^2 < \frac{\alpha^2}{4C^2 r_b r_e} \qquad (3.148)$$

Figure 3.21 A high-frequency, small-signal equivalent network of a transistor.

Example 3.11 Figure 3.21 shows the high-frequency, small-signal equivalent network of a transistor. We wish to determine the frequency range for which the device is potentially unstable.

First, we compute the admittance matrix, which, when evaluated on the real-frequency axis, is given by

$$\mathbf{Y}(j\omega) = \begin{bmatrix} 0.004 + j105 \cdot 10^{-12}\omega & -j5 \cdot 10^{-12}\omega \\ 0.2 - j5 \cdot 10^{-12}\omega & 0.01 + j5 \cdot 10^{-12}\omega \end{bmatrix} \quad (3.149)$$

To simplify the notation, let $\hat{\omega} = \omega/10^9$, giving

$$\mathbf{Y}(j\omega) = 10^{-3} \begin{bmatrix} 4 + j105\hat{\omega} & -j5\hat{\omega} \\ 200 - j5\hat{\omega} & 10 + j5\hat{\omega} \end{bmatrix} \quad (3.150)$$

Since

$$\operatorname{Re} y_{11} = 0.004 > 0 \quad (3.151a)$$

$$\operatorname{Re} y_{22} = 0.01 > 0 \quad (3.151b)$$

for the device to be potentially unstable, it is necessary and sufficient that $\eta \leqslant 1$ or

$$80 \leqslant -25\hat{\omega}^2 + (625\hat{\omega}^4 + 10^6\hat{\omega}^2)^{1/2} \quad (3.152)$$

yielding

$$\hat{\omega} \geqslant 80.16 \cdot 10^{-3} \quad (3.153a)$$

or

$$\omega \geqslant 80.16 \cdot 10^6 \text{ rad/s} \quad (3.153b)$$

which corresponds to 12.76 MHz. Thus, the transistor is potentially unstable for all real frequencies not less than 12.76 MHz.

3.7 OPTIMUM TERMINATIONS FOR ABSOLUTELY STABLE TWO-PORT NETWORKS

In Sec. 3.2 we introduced three types of power gains for a terminated two-port network and derived expressions for these power gains in terms of the two-port parameters and the source and load immittances. Our task here is to select the source and load terminations so that the power gain of the two-port network is maximized. We limit our discussion to absolutely stable two-port networks, because in a potentially unstable two-port network oscillation will occur for certain passive terminations. In such situations, the maximum power gain is infinite and the concept of optimum power gain has no significance. For absolutely stable two-port networks, there exists a set of *optimum* source and load immittances in relation to obtaining maximum power gain. We recognize that other performance characteristics such as the sensitivity to variations in parameters or availability of practical elements may be equally important in the selection of the terminations.

The optimum terminations to be presented in the following are optimum at a single frequency, which may be at any point on the real-frequency axis. Thus, the results are useful only for narrow-band applications.

According to the definitions of various power gains (3.9)–(3.11), the transducer power gain $\mathcal{G}$ is clearly less than or equal to the power gain $\mathcal{G}_p$:

$$\frac{P_2}{P_{1a}} = \mathcal{G} \leqslant \mathcal{G}_p = \frac{P_2}{P_1} \tag{3.154}$$

The input power P_1 to the two-port can at most equal the available power P_{1a} from the source, and this upper bound is reached when the input port is conjugate-matched. As indicated in (3.19a), $\mathcal{G}_p$ is a function only of the two-port parameters and the load immittance M_2. When the load $M_2 = M_{2,\text{opt}}$ is chosen to give the maximum value of the power gain $\mathcal{G}_{p,\text{max}}$ and when the input port is conjugate-matched, the maximum value of the transducer power gain $\mathcal{G}_{\text{max}}$ becomes equal to the maximum value of the power gain $\mathcal{G}_{p,\text{max}}$ and we can write

$$\mathcal{G}_{\text{max}} = \mathcal{G}_{p,\text{max}} \tag{3.155}$$

The transducer power gain also can never exceed the available power gain $\mathcal{G}_a$, as indicated by the equation

$$\frac{P_2}{P_{1a}} = \mathcal{G} \leqslant \mathcal{G}_a = \frac{P_{2a}}{P_{1a}} \tag{3.156}$$

because the output power can at most equal the available power at the output. Equation (3.19c) shows that $\mathcal{G}_a$ is a function only of the two-port parameters and the source immittance M_1. When the source immittance $M_1 = M_{1,\text{opt}}$ is chosen to

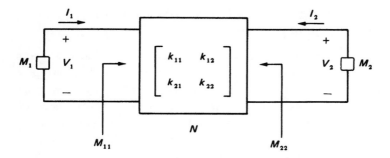

Figure 3.22 The general representation of a terminated two-port network.

give the maximum value of the available power gain $\mathcal{G}_{a,\max}$ and when the output port is conjugate-matched, we can write

$$\mathcal{G}_{\max} = \mathcal{G}_{a,\max} \qquad (3.157)$$

Thus, the three power gains attain a common maximum value with

$$\mathcal{G}_{\max} = \mathcal{G}_{p,\max} = \mathcal{G}_{a,\max} \qquad (3.158)$$

There are many ways of determining the optimum source and load terminations of a two-port network. One way is to differentiate the transducer power gain expression (3.19b) with respect to the real and imaginary parts of the source and load immittances M_1 and M_2 as shown in Fig. 3.22 and to solve for the optimum M_1 and M_2 after the partial derivatives are set equal to zero. This process is rather complicated.

A second possibility is to determine the optimum terminations by simultaneous conjugate match at the ports. Refer to Fig. 3.22. The maximum transducer power gain will result if the input immittance M_{11} is the conjugate of the source immittance M_1, whereas the output immittance M_{22} is the conjugate of the load immittance M_2. Under this situation and from Eqs. (3.7) and (3.8), we have

$$M_{11} = k_{11} - \frac{k_{12}k_{21}}{k_{22} + M_2} = \overline{M}_1 \qquad (3.159a)$$

$$M_{22} = k_{22} - \frac{k_{12}k_{21}}{k_{11} + M_1} = \overline{M}_2 \qquad (3.159b)$$

Solving for M_1 and M_2 will yield the optimum terminations of the two-port network, but the algebra involved is rather tedious.

A simpler approach is as follows. We first determine the load termination $M_{2,\text{opt}}$ that will give the maximum power gain $\mathcal{G}_{p,\max}$ and then determine the source immittance by means of (3.159a). Alternatively, we can find the source immittance $M_{1,\text{opt}}$ that will give the maximum available power gain $\mathcal{G}_{a,\text{max}}$ and then determine the load immittance by means of (3.139b). In the first case, the input of the two-port is conjugate-matched by construction. We assert that the output must also be conjugate-matched, for, if not, we could get more power out by adjusting the load immittance, thus increasing $\mathcal{G}_{\text{max}}$ above $\mathcal{G}_{p,\text{max}}$. This is impossible because it violates Eq. (3.154). In the second case, the output port is conjugate-matched by construction. Hence, $\mathcal{G}$ is equal to $\mathcal{G}_a$. But $\mathcal{G}_a$ is already at its maximum $\mathcal{G}_{a,\text{max}}$. Therefore, $\mathcal{G}$ is also $\mathcal{G}_{\text{max}}$. In the following, we use this procedure to compute the optimum terminations.

From (3.19a), the power gain can be expressed as

$$\mathcal{G}_p = \frac{|k_{21}|^2 \, \text{Re} \, M_2}{|k_{22} + M_2|^2 \, \text{Re} \, [k_{11} - k_{12}k_{21}/(k_{22} + M_2)]} \tag{3.160}$$

To simplify our notation, as in (3.34) we write

$$k_{ik} = m_{ik} + j n_{ik} \quad i, k = 1, 2 \tag{3.161a}$$

$$M_i = A_i + j B_i \quad i = 1, 2 \tag{3.161b}$$

$$k_{12}k_{21} = P + j Q = L e^{j\phi} \tag{3.161c}$$

all being evaluated at the real frequency ω_0. The argument $j\omega_0$ will be dropped, as above, in all the equations, for simplicity.

By using the symbols introduced in Eq. (3.161), Eq. (3.160) becomes

$$\mathcal{G}_p = \frac{|k_{21}|^2 A_2}{m_{11}|k_{22} + M_2|^2 - P(m_{22} + A_2) - Q(n_{22} + B_2)} \tag{3.162}$$

Taking the partial derivative of $\mathcal{G}_p$ with respect to B_2 and setting it to zero yield

$$n_{22} + B_2 = \frac{Q}{2m_{11}} \tag{3.163}$$

giving the optimum load susceptance or reactance

$$\text{Im} \, M_{2,\text{opt}} = B_2 = \frac{Q}{2m_{11}} - n_{22} \tag{3.164a}$$

Similarly, taking the partial derivative of (3.162) with respect to A_2 and setting it to zero, we obtain

$$\text{Re} \, M_{2,\text{opt}} = A_2 = \frac{1}{2m_{11}}\sqrt{(2m_{11}m_{22} - P)^2 - L^2} \tag{3.164b}$$

Substituting $M_{2,\text{opt}}$ from (3.164) in (3.162) yields the maximum power gain of the two-port network as

$$\mathcal{G}_{p,\text{max}} = \frac{|k_{21}|^2}{2m_{11}m_{22} - P + \sqrt{(2m_{11}m_{22} - P)^2 - L^2}} \tag{3.165}$$

Once the optimum load immittance $M_{2,\text{opt}}$ is known, the optimum source immittance $M_{1,\text{opt}}$ can easily be found from (3.159a). However, from the symmetry of the formulas for power gain (3.160) or (3.162) and available power gain (3.19c), we can write down the value of $M_{1,\text{opt}}$ directly from (3.164) as

$$\text{Re } M_{1,\text{opt}} = \frac{1}{2m_{22}}\sqrt{(2m_{11}m_{22} - P)^2 - L^2} \tag{3.166a}$$

$$\text{Im } M_{1,\text{opt}} = \frac{Q}{2m_{22}} - n_{11} \tag{3.166b}$$

A slight digression at this point will allow us to shine some light on our assumption that the two-port network must be absolutely stable at $j\omega_0$. This assumption is equivalent to requiring that the input immittances M_{11} and M_{22} possess positive-real parts at $j\omega_0$. From the conjugate-match conditions of (3.159), Eqs. (3.164b) and (3.166a) must be positive. This is possible if and only if the term inside the radical is positive or

$$2m_{11}m_{22} - P > L \tag{3.167}$$

showing that the stability parameter

$$\eta = \frac{2m_{11}m_{22} - P}{L} > 1 \tag{3.168}$$

Both the optimum terminations (3.164) and (3.166) and the maximum power gain (3.165) can be expressed in terms of the stability parameter as follows:

$$M_{1,\text{opt}} = \text{Re } M_{1,\text{opt}} + j\text{Im } M_{1,\text{opt}} = \frac{k_{12}k_{21} + |k_{12}k_{21}|(\eta + \sqrt{\eta^2 - 1})}{2\,\text{Re } k_{22}} - k_{11}$$
$$\tag{3.169a}$$

$$M_{2,\text{opt}} = \text{Re } M_{2,\text{opt}} + j\text{Im } M_{2,\text{opt}} = \frac{k_{12}k_{21} + |k_{12}k_{21}|(\eta + \sqrt{\eta^2 - 1})}{2\,\text{Re } k_{11}} - k_{22}$$
$$\tag{3.169b}$$

$$\mathcal{G}_{p,\text{max}} = \mathcal{G}_{\text{max}} = \mathcal{G}_{a,\text{max}} = \left|\frac{k_{21}}{k_{12}}\right| \frac{1}{\eta + \sqrt{\eta^2 - 1}} \tag{3.170}$$

Since the two-port network is assumed to be absolutely stable at $j\omega_0$, $\eta > 1$, and the expression (3.170) is therefore meaningful.

It is significant to observe that the expression for maximum gain is made up of two factors. The one involving the stability parameter is reciprocal in that it is invariant with respect to the interchange of forward and reverse transfer parameters k_{21} and k_{12} (and also the self-immittances k_{11} and k_{22}), whereas the other is nonreciprocal. Thus, we can write

$$\mathcal{G}_{\text{max}} = \left| \frac{k_{21}}{k_{12}} \right| \gamma \tag{3.171}$$

where

$$\gamma = \left(\eta + \sqrt{\eta^2 - 1} \right)^{-1} = \eta - \sqrt{\eta^2 - 1} \tag{3.172a}$$

$$= 2\left(\sqrt{\eta + 1} + \sqrt{\eta - 1} \right)^{-2} = \frac{1}{2}\left(\sqrt{n + 1} - \sqrt{\eta - 1} \right)^2 \tag{3.172b}$$

The first factor $|k_{21}/k_{12}|$ is a measure of the nonreciprocity of the two-port and is called the *nonreciprocal gain* of the two-port network. The second factor γ is usually known as the *efficiency of the reciprocal part* of the two-port network. Since $\eta > 1$ for absolutely stable two-port networks, γ is allowed to vary between 0 and 1. This implies that for an absolutely stable reciprocal two-port network, its maximum gain, given by $\mathcal{G}_{\text{max}} = \gamma$, is bounded by unity, and therefore the two-port network must be passive, a fact that was pointed out earlier.

For a unilateral two-port network, $k_{12} = 0$ and both the nonreciprocal gain and the stability parameter are infinite. The maximum unilateral power gain can be obtained directly from (3.165) and is given by

$$\mathcal{G}_{\text{max}} = \frac{|k_{21}|^2}{4m_{11}m_{22}} \tag{3.173}$$

which agrees with (3.92).

Example 3.12 Consider the two-port network of Fig. 3.23, whose admittance matrix is given by

$$\mathbf{Y}(s) = \begin{bmatrix} g_1 + sC & -sC \\ g_m - sC & g_2 + sC \end{bmatrix} \tag{3.174}$$

It is straightforward to show that the device is both active and absolutely stable for all real frequencies ω satisfying the inequalities

$$\frac{1}{2}g_m^2 > 2g_1 g_2 > \omega C[(g_m^2 + \omega^2 C^2)^{1/2} - \omega C] \tag{3.175}$$

Figure 3.23 An active and absolutely stable two-port device.

The details are left as an exercise (see Prob. 3.22). To determine the optimum terminating admittances

$$Y_{1,\text{opt}} = G_{1,\text{opt}} + jB_{1,\text{opt}} \tag{3.176a}$$

$$Y_{2,\text{opt}} = G_{2,\text{opt}} + jB_{2,\text{opt}} \tag{3.176b}$$

at the frequencies where the two-port is absolutely stable, we first compute

$$y_{12}y_{21} = -j\omega C(g_m - j\omega C) = -\omega^2 C^2 - j\omega C g_m \tag{3.177}$$

From (3.166) we have

$$
\begin{aligned}
G_{1,\text{opt}} &= \frac{1}{2\,\text{Re}\,y_{22}}\sqrt{(2\,\text{Re}\,y_{11}\,\text{Re}\,y_{22} - \text{Re}\,y_{12}y_{21})^2 - |y_{12}y_{21}|^2} \\
&= \frac{1}{2g_2}\sqrt{(2g_1g_2 + \omega^2 C^2)^2 - \omega^4 C^4 - \omega^2 C^2 g_m^2} \\
&= \frac{1}{2g_2}\sqrt{4g_1^2 g_2^2 + \omega^2 C^2(4g_1g_2 - g_m^2)} > 0 \tag{3.178a}
\end{aligned}
$$

the last inequality following from (3.175),

$$B_{1,\text{opt}} = \frac{\text{Im}\,y_{12}y_{21}}{2\text{Re}\,y_{22}} - \text{Im}\,y_{11} = -\frac{\omega C g_m}{2g_2} - \omega C = -\omega C\left(1 + \frac{g_m}{2g_2}\right) \tag{3.178b}$$

and from (3.164) we get

$$
\begin{aligned}
G_{2,\text{opt}} &= \frac{1}{2\,\text{Re}\,y_{11}}\sqrt{(2\,\text{Re}\,y_{11}\,\text{Re}\,y_{22} - \text{Re}\,y_{12}y_{21})^2 - |y_{12}y_{21}|^2} \\
&= \frac{1}{2g_1}\sqrt{4g_1^2 g_2^2 + \omega^2 C^2(4g_1g_2 - g_m^2)} > 0 \tag{3.179a}
\end{aligned}
$$

$$B_{2,\text{opt}} = \frac{\text{Im } y_{12}y_{21}}{2 \text{ Re } y_{11}} - \text{Im } y_{22} = -\frac{\omega C g_m}{2g_1} - \omega C = -\omega C \left(1 + \frac{g_m}{2g_1}\right) \quad (3.179b)$$

Under these optimum port terminations, the three power gains assume a common maximum value determined by (3.165):

$$\mathcal{G}_{\max} = \mathcal{G}_{p,\max} = \mathcal{G}_{a,\max}$$

$$= \frac{|y_{21}|^2}{2 \text{ Re } y_{11} \text{ Re } y_{22} - \text{Re } y_{12}y_{21}}$$
$$\qquad + \sqrt{(2 \text{ Re } y_{11} \text{ Re } y_{22} - \text{Re } y_{12}y_{21})^2 - |y_{12}y_{21}|^2}$$

$$= \frac{2g_1g_2 + \omega^2 C^2 - \sqrt{4g_1^2 g_2^2 + \omega^2 C^2 (4g_1 g_2 - g_m^2)}}{\omega^2 C^2} \quad (3.180)$$

The nonreciprocal gain of the two-port network is found to be

$$\left|\frac{y_{21}}{y_{12}}\right| = \left|\frac{g_m - j\omega C}{-j\omega C}\right| = \sqrt{1 + \frac{g_m^2}{\omega^2 C^2}} \quad (3.181)$$

and the efficiency of the reciprocal part of the two-port network is calculated from (3.172a) and is given by

$$\gamma = \eta - \sqrt{\eta^2 - 1} = \frac{2g_1g_2 + \omega^2 C^2 - \sqrt{4g_1^2 g_2^2 + \omega^2 C^2 (4g_1 g_2 - g_m^2)}}{\omega C \sqrt{\omega^2 C^2 + g_m^2}} \quad (3.182)$$

where the stability parameter is given by

$$\eta = \frac{2g_1g_2 + \omega^2 C^2}{\omega C \sqrt{\omega^2 C^2 + g_m^2}} \quad (3.183)$$

3.7.1 Representation of Nonreciprocal Two-Port Networks

As indicated in Eq. (3.171), the maximum power gain $\mathcal{G}_{\max}$ can be expressed as the product of the nonreciprocal gain and the efficiency of the reciprocal part of the two-port network. In this section, we demonstrate that this interpretation may be employed to decompose a general two-port network into two two-ports connected in cascade, one reciprocal and the other nonreciprocal.

Consider a general two-port network N characterized by its general hybrid equation

$$\begin{bmatrix} y_1 \\ y_2 \end{bmatrix} = \begin{bmatrix} k_{11} & k_{12} \\ k_{21} & k_{22} \end{bmatrix} \begin{bmatrix} u_1 \\ u_2 \end{bmatrix} \quad (3.184)$$

where the variables u and y for different immittance representations are tabulated in Table 3.1. Let

$$k_s = \sqrt{k_{12}k_{21}} \qquad (3.185a)$$

$$\lambda = \sqrt{\frac{k_{21}}{k_{12}}} \qquad (3.185b)$$

With these, Eq. (3.184) can be written as

$$\begin{bmatrix} y_1 \\ y_2/\lambda \end{bmatrix} = \begin{bmatrix} k_{11} & k_s \\ k_s & k_{22} \end{bmatrix} \begin{bmatrix} u_1 \\ u_2/\lambda \end{bmatrix} \qquad (3.186a)$$

or

$$\begin{bmatrix} \lambda y_1 \\ y_2 \end{bmatrix} = \begin{bmatrix} k_{11} & k_s \\ k_s & k_{22} \end{bmatrix} \begin{bmatrix} \lambda u_1 \\ u_2 \end{bmatrix} \qquad (3.186b)$$

Equation (3.186) can be represented by the cascade connection of a reciprocal two-port, whose general hybrid matrix is its coefficient matrix, and a nonreciprocal two-port on either its input or output side, as shown in Fig. 3.24. The reciprocal two-port network is the reciprocal part of the original two-port network, and the nonreciprocal two-port is characterized by the equation

$$\frac{V_\beta}{V_\alpha} = -\frac{I_\beta}{I_\alpha} = \lambda \qquad (3.187)$$

where V_α and V_β are its port voltages and I_α and I_β are its port currents. Evidently, the maximum power gain for the reciprocal two-port is γ, the efficiency of the reciprocal part of the original two-port, and the power gain for the nonreciprocal two-port is $|k_{21}/k_{12}|$, the nonreciprocal gain of the original two-port network. It is significant to note that $|k_{21}/k_{12}|$ is invariant under immittance substitution, as it must be with the above interpretation.

3.7.2 Physical Significance of the Efficiency of the Reciprocal Part

In the foregoing, we have shown that the maximum gain $\mathcal{G}_{\max}$ of a general two-port network is made up of two factors. One is a measure of the nonreciprocity and the other the efficiency of the reciprocal part of the two-port network. The latter

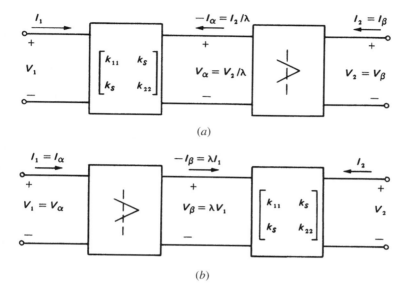

Figure 3.24 The decomposition of a general two-port network into two two-port networks connected in cascade (a) A reciprocal two-port followed by a nonreciprocal one. (b) A nonreciprocal two-port followed by a reciprocal one.

involves the stability parameter and is given by

$$\gamma = \eta - \sqrt{\eta^2 - 1} \tag{3.188}$$

In this section, we present a physical interpretation of γ.

Let $w(s)$ denote the transfer function of either (3.25a) or (3.25b), Then $w(s)$ can be expressed as

$$w(s) = \frac{\tilde{w}(s)}{1 - f(s)} \tag{3.189}$$

where

$$\tilde{w}(s) = w(s)|_{k_{12}=0} \tag{3.190a}$$

$$f(s) = \frac{k_{12}k_{21}}{(k_{11} + M_1)(k_{22} + M_2)} \tag{3.190b}$$

The term $f(s)$ is called the *loop transmission*. The reason is that, as will be shown in the following chapter, the closed-loop transfer function of a feedback network can also be expressed as shown in (3.189). Then $f(s)$ is identified as the loop transmission in the terminated two-port network when the forward transmission is interrupted. Observe that $f(s)$ is a function of the two-port parameters and terminations.

Assume that the two-port network is conjugate-matched at the ports. The corresponding loop transmission, written as $f_m(s)$, can be expressed explicitly in terms of the two-port parameters k_{ij} and the stability parameter η, as follows. From (3.169) we have

$$f_m(s) = \frac{k_{12}k_{21}}{(k_{11} + M_{1,\text{opt}})(k_{22} + M_{2,\text{opt}})} = \frac{4k_{12}k_{21} \operatorname{Re} k_{11} \operatorname{Re} k_{22}}{[k_{12}k_{21} + |k_{12}k_{21}|(\eta + \sqrt{\eta^2 - 1})]^2}$$

$$(3.191)$$

whose magnitude, by using the symbols of (3.161), is found to be

$$|f_m(s)| = \frac{1}{\eta + \sqrt{\eta^2 - 1}} = \eta - \sqrt{\eta^2 - 1} \qquad (3.192a)$$

Thus, on the real-frequency axis, we get

$$|f_m(j\omega)| = \eta - \sqrt{\eta^2 - 1} = \gamma \qquad (3.192b)$$

showing that the efficiency of the reciprocal part of the two-port network is equal to the real-frequency axis magnitude of its loop transmission under the situation where the two-port is conjugate-matched at its ports.

The factors $|k_{21}/k_{12}|$ and γ can also be determined experimentally. For this, let $\mathcal{G}_{\text{max},r}$ be the maximum power gain of the two-port network in the reverse direction, which is found by interchanging the input and the output ports:

$$\mathcal{G}_{\text{max,r}} = \left| \frac{k_{12}}{k_{21}} \right| \gamma \qquad (3.193)$$

which is obtained from (3.171) by interchanging the subscripts 1 and 2. Combining (3.171) and (3.193) yields

$$\frac{\mathcal{G}_{\text{max}}}{\mathcal{G}_{\text{max},r}} = \left| \frac{k_{21}}{k_{12}} \right|^2 \qquad (3.194a)$$

$$\sqrt{\mathcal{G}_{\text{max}}\mathcal{G}_{\text{max},r}} = \eta - \sqrt{\eta^2 - 1} = \gamma \qquad (3.194b)$$

which can then be used to determine the nonreciprocal gain $|k_{21}/k_{12}|$ and the efficiency γ of the reciprocal part of the two-port network or its stability parameter η.

3.7.3 Maximum Stable Power Gain and Nonreciprocal Gain

If a two-port network N is potentially unstable, then Re $k_{11} < 0$ or Re $k_{22} < 0$, or $\eta < 1$. Under these situations, lossy padding immittances $M_{1,\text{pad}}$ and $M_{2,\text{pad}}$ can always be placed at the two ports so as to make the real parts of $M_1 + M_{1,\text{pad}}$ and $M_2 + M_{2,\text{pad}}$ positive and the overall stability parameter $\tilde{\eta}$ approaching unity or, for real-positive $M_{1,\text{pad}}$ and $M_{2,\text{pad}}$,

$$M_{1,\text{pad}} + \text{Re } M_1 > 0 \tag{3.195a}$$

$$M_{2,\text{pad}} + \text{Re } M_2 > 0 \tag{3.195b}$$

$$\tilde{\eta} = \frac{2(M_{1,\text{pad}} + \text{Re } M_1)(M_{2,\text{pad}} + \text{Re } M_2) - \text{Re } k_{12}k_{21}}{|k_{12}k_{21}|} \to 1 \tag{3.195c}$$

The maximum power gain then tends toward its maximum stable value,

$$\mathcal{G}_{\text{max}} \to \mathcal{G}_0 = \left| \frac{k_{21}}{k_{12}} \right| \tag{3.196}$$

$\mathcal{G}_0$ is called the *maximum stable power gain* of the two-port network.

The above concept is defined for a potentially unstable two-port network. This definition can also be extended to include absolutely stable two-port networks by the use of negative padding resistances. The only difference is that $M_{1,\text{pad}}$ and $M_{2,\text{pad}}$ are now real and nonpositive and Eq. (3.195c) may be employed to determine the values of the padding resistances, so that the stability boundary $\tilde{\eta} = 1$ can again be approached. Thus, $\mathcal{G}_0$ can be defined as the maximum stable power gain for all devices.

For an absolutely stable two-port network, the maximum stable power gain is now identified as the nonreciprocal gain of the two-port, and (3.171) is reduced to

$$\mathcal{G}_{\text{max}} = \mathcal{G}_0 \gamma \tag{3.197}$$

In words, it states that the maximum power gain of an absolutely stable two-port network is equal to the maximum stable power gain times the efficiency of its reciprocal part.

Example 3.13 In the two-port network N of Fig. 3.23 considered in Example 3.12, let

$$g_1 = 4 \cdot 10^{-3} \text{ mho} \qquad g_2 = 0.01 \text{ mho}$$
$$g_m = 0.2 \text{ mho} \qquad C = 5 \text{ pF} \tag{3.198}$$

Figure 3.25 The stabilization of a two-port network by introducing at the input port a conductance g.

Suppose that the two-port is operating at 30 MHz. Then the second inequality of (3.175) is violated, meaning that the two-port network is active but potentially unstable at 30 MHz. One way to stabilize the two-port is to introduce sufficient loss by connecting at the input port a conductance g as shown in Fig. 3.25. To determine the value of this padding conductance so that the stability boundary $\tilde{\eta} = 1$ can be approached, we set (3.195c) to unity. From (3.183), this is equivalent to

$$\frac{2(g_1 + g)g_2 + \omega^2 C^2}{\omega C \sqrt{\omega^2 C^2 + g_m^2}} = 1 \tag{3.199}$$

giving

$$g = 5.382 \cdot 10^{-3} \, \text{mho} \tag{3.200}$$

or 185.81 Ω. The maximum stable power gain is determined from (3.196) as

$$\mathcal{G}_0 = \left| \frac{y_{21}}{y_{12}} \right| = \sqrt{1 + \frac{g_m^2}{\omega^2 C^2}} = 212.21 \tag{3.201}$$

Now we lower the operating frequency from 30 MHz to 5 MHz. The inequalities (3.175) are satisfied, and the two-port network is both active and absolutely stable at 5 MHz. From (3.183) we calculate the stability parameter

$$\eta = \frac{2g_1 g_2 + \omega^2 C^2}{\omega C \sqrt{\omega^2 C^2 + g_m^2}} = 2.547 \tag{3.202}$$

The efficiency of the reciprocal part of the two-port network is found to be

$$\gamma = \eta - \sqrt{\eta^2 - 1} = 0.2045 \tag{3.203}$$

yielding the maximum power gain

$$\mathcal{G}_{\max} = \mathcal{G}_0 \gamma = 260.38 \tag{3.204}$$

where $\mathcal{G}_0 = 1273.24$.

Suppose that we wish to determine the value of the padding conductance g connected across the input port at 5 MHz so that the stability boundary $\tilde{\eta} = 1$ is approached. Again, from (3.199) with $\omega = \pi \cdot 10^7$ rad/s, we get

$$g = -2.431 \cdot 10^{-3} \text{ mho} \tag{3.205}$$

or $-411.35 \,\Omega$.

3.7.4 Maximum Power Gain and the U-Function

In Sec. 3.5.2 we demonstrated that for a given active two-port network, its U-function is identifiable as the maximum unilateral power gain under a lossless reciprocal imbedding. In the present section, we express this maximum power gain $\mathcal{G}_{\max}$ in terms of the U-function and the complex nonreciprocal gain. Let

$$\pm\delta = \frac{k_{21}}{k_{12}} = \mathcal{G}_0 e^{j\xi} \tag{3.206}$$

The plus sign is chosen for the z- and y-parameters, and the minus sign for the h- and g-parameters. As discussed previously, the magnitude of δ is called the nonreciprocal gain of the two-port network and can be interpreted as the maximum stable power gain. Thus, it is convenient to call δ the *complex nonreciprocal gain* of the two-port network, because it is a complex measure of nonreciprocity.

From Eqs. (3.51), (3.117), and (3.171), we have

$$\begin{aligned}
|k_{12}|^2|\delta - \mathcal{G}_{\max}|^2 &= |k_{21}|^2 + |k_{12}|^2\mathcal{G}_{\max}^2 \mp 2\mathcal{G}_{\max}\text{Re } k_{12}\bar{k}_{21} \\
&= 2\mathcal{G}_{\max}|k_{12}k_{21}|\eta \mp 2\mathcal{G}_{\max}\text{Re } k_{12}\bar{k}_{21} \\
&= \mathcal{G}_{\max}\left[4\text{Re } k_{11}\text{Re } k_{22} - 2\text{Re}(k_{12}k_{21} \pm k_{12}\bar{k}_{21})\right] \\
&= \frac{\mathcal{G}_{\max}|k_{21} \mp k_{12}|^2}{U} \tag{3.207}
\end{aligned}$$

which can be rewritten as

$$\frac{\mathcal{G}_{\max}}{U} = \left|\frac{\delta - \mathcal{G}_{\max}}{\delta - 1}\right|^2 \tag{3.208}$$

showing that $\mathcal{G}_{max}$ is a function of only two basic parameters: the U-function and the complex nonreciprocal gain. Since U is invariant under all lossless reciprocal imbedding, and, in particular, under all device orientations such as the common-emitter, common-base, and common-collector configurations of the transistor, a simple way to control $\mathcal{G}_{max}$ is by modifying δ with a lossless reciprocal imbedding or by changing the device orientation without changing U. Alternatively, we can alter U by lossy padding at the input and output ports, as suggested in the preceding section, without changing δ.

3.8 SUMMARY

We began this chapter by reviewing briefly the general representation of a two-port network and by defining several measures of power flow. We found that the transducer power gain is the most meaningful description of power transfer capabilities of a two-port network as it compares the power delivered to the load with the power that the source is capable of supplying under optimum conditions. We then introduced the concepts of sensitivity and derived the general expressions for various power gains, transfer functions, and the sensitivity functions in terms of the general two-port parameters and the source and load immittances. The specialization of the general passivity condition for n-port networks in terms of the more immediately useful two-port parameters was taken up next. A useful parameter associated with a two-port network called the U-function was introduced. We showed that this function is invariant under all lossless reciprocal imbedding. The physical significance of this result is that the U-function can be used to characterize the power-amplifying capability of a device. The characterization is unique in that, among other reasons, it is independent of the measuring circuit, provided only that the circuit uses lossless reciprocal elements. Furthermore, we demonstrated that for a given active two-port network, its U-function is identifiable as the maximum unilateral power gain under a lossless reciprocal imbedding, the resulting structure being unilateral.

Whenever we consider active networks, we must discuss stability. For this we introduced the single frequency concepts of stability for a two-port network: potential instability and absolute stability. Necessary and sufficient conditions for absolute stability were derived and are known as *Llewellyn's criteria*. Specifically, we showed that a two-port network is absolutely stable if and only if its open-circuit or short-circuit input and output conductances are positive and the stability parameter is greater than unity. We indicated that the stability parameter is invariant under immittance substitution, and its value remains unaltered regardless of the two-port matrix employed in computation.

For a potentially unstable two-port network, there exists a set of terminations so that the overall network possesses a natural frequency on the real-frequency axis. In such situations, the maximum power gain is infinite and the concept of optimum power gain has no significance. For an absolutely stable two-port network, we can find a set of optimum source and load immittances so that the power gain is maximized. It turns out that under this optimum matching situation, the three power gains assume a common maximum value. This maximum power gain can be written as the product of two factors. The first factor is a measure of the nonreciprocity of the two-port network and is identifiable as the maximum stable power gain. The second factor is a measure of the efficiency of the reciprocal part of the two-port network, and is bounded between 0 and 1. This product expression for the maximum power gain was employed to decompose a general two-port network into two two-ports connected in cascade, one reciprocal and the other nonreciprocal. Finally, we derived a useful relation among the maximum power gain, the U-function, and the complex nonreciprocal gain, a complex measure of nonreciprocity, of a two-port network.

PROBLEMS

3.1 Confirm that the three power gain expressions (3.19) are valid for (a) the h-parameters, (b) the y-parameters, and (c) the g-parameters.

3.2 Confirm that the general expressions (3.25) for the transfer functions are valid for (a) the z-parameters, (b) the y-parameters, and (c) the g-parameters.

3.3 Figure 3.26 shows the equivalent network of a field-effect transistor (FET). Determine if the device can be both active and absolutely stable.

3.4 Derive the sensitivity expressions (3.27).

3.5 Confirm that the sensitivity expressions (3.29) are valid for (a) the z-parameters, (b) the y-parameters, and (c) the g-parameters.

Figure 3.26 An equivalent network of an FET.

Figure 3.27 A given two-port network.

3.6 Figure 3.11 can be viewed as the lossless reciprocal four-port imbedding of the active device N as shown in Fig. 3.12. Show that the U-function is invariant under this imbedding.

3.7 Consider the two-port network of Fig. 3.27. Determine the maximum frequency of oscillation and the fastest regenerative mode of the device from its (*a*) y-parameters, (*b*) z-parameters, (*c*) h-parameters, and (*d*) g-parameters. Compare the amount of effort involved in arriving at the same conclusion.

3.8 A transistor has the following parameter values at 3 MHz:

$$h_{11} = 50 + j1.2\,\Omega \quad h_{12} = (1.5 + j1.6) \cdot 10^{-3}$$
$$h_{21} = -0.9 + j0.06 \quad h_{22} = 7 + j25\,\mu\text{mho}$$

(3.209)

Find the optimum source and load terminations, along with the maximum power gain.

3.9 A three-terminal device N with a shunt and series feedback is shown in Fig. 3.28. Assume that the two feedback admittances are lossless and reciprocal. Show that the U-function of N is invariant under this lossless reciprocal imbedding.

3.10 Refer to the network of Fig. 3.28. Can appropriate feedback admittances y_f and y_α be chosen so that, at a particular frequency, the overall two-port network is unilateral?

3.11 Let N be a three-terminal device characterized by its impedance matrix. By using the configuration shown in Fig. 3.29, show that a lossless reciprocal four-port network can always be chosen so that, at a particular frequency, the overall two-port network is unilateral.

3.12 The stability parameter η and the efficiency of the reciprocal part γ of a two-port network are related by (3.172). If η lies between -1 and 1, γ is complex.

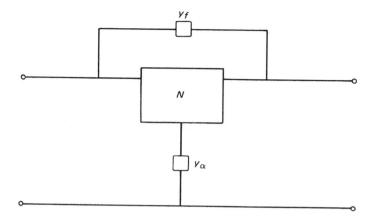

Figure 3.28 A three-terminal device N with a shunt and series feedback.

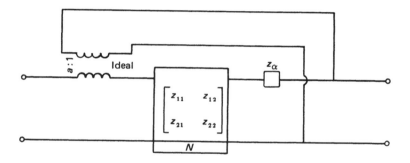

Figure 3.29 A three-terminal device N together with its feedback circuit.

Show that the locus of γ on the complex plane for values of η from -1 to 1 is a circle with the origin as center and unit radius.

3.13 Show that the efficiency of the reciprocal part γ of a two-port network can be expressed in the form

$$\gamma^2 = \frac{(1 - \theta_r)^2 + \theta_x^2}{(1 + \theta_r)^2 + \theta_x^2} \tag{3.210}$$

where

$$\theta_r = \frac{\operatorname{Re} M_{1,\mathrm{opt}}}{\operatorname{Re} k_{11}} = \frac{\operatorname{Re} M_{2,\mathrm{opt}}}{\operatorname{Re} k_{22}} \tag{3.211a}$$

$$\theta_x = \frac{\text{Im}\,(k_{11} + M_{1,\text{opt}})}{\text{Re}\,k_{11}} = \frac{\text{Im}\,(k_{22} + M_{2,\text{opt}})}{\text{Re}\,k_{22}} \qquad (3.211b)$$

3.14 By substituting one set of the impedance, admittance, hybrid, or inverse hybrid parameters for any other, show that the stability parameter is invariant under this substitution.

3.15 The values of the y-parameters of a transistor operating at 30 MHz are given by (3.96). Find the optimum source and load immittances, along with the maximum power gain at this frequency.

3.16 For a reciprocal two-port network, show that (3.120c) and (3.134) are equivalent.

3.17 Use (3.206) and show that the U-function of a two-port network can be expressed as

$$U = \frac{\frac{1}{2}(\mathcal{G}_0 + 1/\mathcal{G}_0) \pm \cos\xi}{\eta \pm \cos\xi} \qquad (3.212)$$

where the plus sign applies to h- and g-parameters and the minus sign to z- and y-parameters.

3.18 Justify the statement that if two admittances have the same phase, then the real part of these two admittances in series is equal to the series combination of their real parts.

3.19 Assume that $\text{Re}\,k_{11}$ and $\text{Re}\,k_{22}$ are positive. Show that the value of the stability parameter lies between -1 and ∞.

3.20 Let

$$\epsilon = \frac{2}{\eta + 1} \qquad (3.213)$$

Show that the efficiency of the reciprocal part γ of a two-port network can be written in the form

$$\gamma = \frac{1 - \sqrt{1 - \epsilon}}{1 + \sqrt{1 - \epsilon}} \qquad (3.214)$$

3.21 Refer to Prob. 3.8. Determine the values of the padding resistances so that the stability boundary of the device can be approached.

3.22 Show that the two-port network of Fig. 3.23 is absolutely stable for all real frequencies satisfying the inequality

$$\omega C[(g_m^2 + \omega^2 C^2)^{1/2} - \omega C] < 2g_1 g_2 \qquad (3.215)$$

Figure 3.30 A given two-port network.

3.23 The absolute stability constraint $\eta > 1$ can be expressed equivalently as

$$\zeta = \frac{2 \operatorname{Re} k_{11} \operatorname{Re} k_{22}}{\operatorname{Re} k_{12}k_{21} + |k_{12}k_{21}|} > 1 \tag{3.216}$$

Show that this quantity is *not* invariant under immittance substitution.

3.24 Consider the two-port network of Fig. 3.30. Determine the range of frequencies in which the device is both active and absolutely stable. Also, find the maximum frequency of oscillation and the fastest regenerative mode of the device.

3.25 An absolutely stable two-port network is characterized by its general hybrid parameters k_{ij}. Show that the optimum source and load immittances can be expressed as

$$\overline{M}_{1,\text{opt}} = k_{11} - \frac{k_{12}k_{21}(1 - \gamma e^{j\phi})}{2 \operatorname{Re} k_{22}} \tag{3.217a}$$

$$\overline{M}_{2,\text{opt}} = -k_{22} + \frac{2 \operatorname{Re} k_{22}}{1 - \gamma e^{j\phi}} \tag{3.217b}$$

where γ is the efficiency of the reciprocal part of the two-port, and

$$-\bar{k}_{12}\bar{k}_{21} = |k_{12}k_{21}|e^{j\phi} \tag{3.217c}$$

3.26 From (3.7), the input immittance M_{11} of a two-port network when its output is terminated in M_2 can be expressed as

$$M_{11} = \frac{aM_2 + b}{M_2 + c} \tag{3.218}$$

where $a = k_{11}$, $b = k_{11}k_{22} - k_{12}k_{21}$, and $c = k_{22}$. The above relation is a simple bilinear transformation of the terminating immittance M_2. Show that for all passive M_2, that is, the closed right half of the M_2-plane, M_{11} is transformed into a circle centered at

$$p_1 = k_{11} - \frac{k_{12}k_{21}}{2\,\mathrm{Re}\,k_{22}} \qquad (3.219a)$$

with radius

$$r_1 = \frac{|k_{12}k_{21}|}{2\,\mathrm{Re}\,k_{22}} \qquad (3.219b)$$

If the two-port network is absolutely stable, then

$$\mathrm{Re}\,p_1 - r_1 > 0 \qquad (3.220)$$

Show that (3.220) is equivalent to the familiar constraint $\eta > 1$. Instead of starting from M_{11}, demonstrate that the same conclusion can be reached if we start from the output immittance of the two-port network, as given in (3.8).

3.27 Consider the two-port network N_a of Fig. 3.28. Let y_{aij} and y_{ij} be the y-parameters of N_a and N, respectively. Show that

$$y_{a11} = y_f + \frac{y_{11}y_\alpha + y_{11}y_{22} - y_{12}y_{21}}{y_\alpha + y_{11} + y_{12} + y_{21} + y_{22}} \qquad (3.221a)$$

$$y_{a22} = y_f + \frac{y_{22}y_\alpha + y_{11}y_{22} - y_{12}y_{21}}{y_\alpha + y_{11} + y_{12} + y_{21} + y_{22}} \qquad (3.221b)$$

3.28 By using the parameter values of a transistor given in (3.209), determine the optimum terminations at 3 MHz that lead to the maximum power gain. If the transistor is terminated at the load with a 20-kΩ resistor and at the source with a 100-Ω resistor in shunt with a 10-pF capacitor, what is the power gain at 3 MHz? What is the transducer power gain at this frequency? Is the amplifier stable?

3.29 Repeat the second part of Prob. 3.28 if the source impedance is represented by the series connection of a 100-Ω resistor and a 10-μH inductor, everything else being the same.

3.30 For the hybrid-pi equivalent network of a transistor shown in Fig. 3.3, determine the frequency range in which the transistor is both active and absolutely stable.

3.31 Show that a lossless two-port network that is conjugate-matched at the input port also provides a conjugate match at the output port. Can the same conclusion be drawn for the lossy two-port network? If not, construct a counterexample.

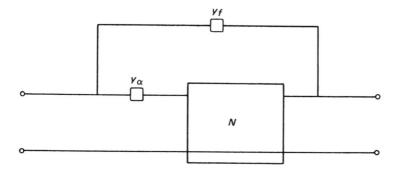

Figure 3.31 A three-terminal device with two lossless reciprocal admittances y_α and y_f.

3.32 Show that the U-function, as defined in (3.51), for a two-port network will possess the same value regardless of the matrix employed in computation. More specifically, demonstrate that the U-function is invariant when any one set of z-, y-, h-, or g-parameters is replaced by any other set.

3.33 Refer to Fig. 3.11. Suppose that we move the admittance y_α from the output port to the input port, as shown in Fig. 3.31. Verify that if y_{ij} are the y-parameters of N, then the admittance matrix of the two-port with y_α connected in series at the input port is

$$\frac{y_\alpha}{y_\alpha + y_{11}} \begin{bmatrix} y_{11} & y_{12} \\ y_{21} & y_{22} + \dfrac{\Delta_y}{y_\alpha} \end{bmatrix} \tag{3.222}$$

Determine the admittances and y_α and y_f so that the resulting network is unilateralized at a specified frequency on the real-frequency axis. Derive expressions similar to those given by (3.95).

3.34 Assume that the two-port parameters k_{ij} are real. Show that the optimum terminations $M_{1,\text{opt}}$ and $M_{2,\text{opt}}$ and the maximum power gain $G_{\max}$ can be expressed as

$$M_{1,\text{opt}} = \sqrt{\frac{\Delta_k k_{11}}{k_{22}}} \tag{3.223a}$$

$$M_{2,\text{opt}} = \sqrt{\frac{\Delta_k k_{22}}{k_{11}}} \tag{3.223b}$$

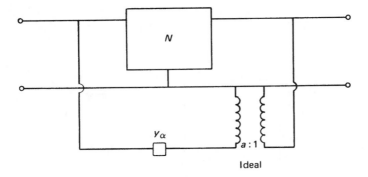

Figure 3.32 An active device N connected in parallel with an ideal transformer and an admittance y_α.

$$\mathcal{G}_{\max} = \frac{k_{21}^2}{k_{11}k_{22}[1 + (1 - k_{12}k_{21}/k_{11}k_{22})^{1/2}]^2} \qquad (3.223c)$$

where $\Delta_k = k_{11}k_{22} - k_{12}k_{21}$.

3.35 An active device N is connected in parallel with an ideal transformer and an admittance y_α, as shown in Fig. 3.32. Demonstrate that a passive but not necessarily lossless y_α can always be chosen so that, at a particular frequency, the overall two-port network is unilateral, where a can be positive or negative. By using (3.92), the maximum unilateral transducer power gain $\mathcal{G}_a$ is obtained. Show that by optimizing $\mathcal{G}_a$ versus a, the gain expression becomes

$$\mathcal{G}_{a,\max} = \frac{|y_{21} - y_{12}|^2}{4(\sqrt{\text{Re } y_{11} \text{ Re } y_{22}} + |\text{Re } y_{12}|)^2} \qquad (3.224)$$

where y_{ij} are the y-parameters of N.

3.36 Consider the unilateral two-port network N_a of Fig. 3.15. For simplicity, let $y_{a11} = y_{a22} = 1$ mho and $y_{a21} = -g_m$. Suppose that N_a is connected in parallel with an ideal gyrator N_g with the admittance matrix

$$\mathbf{Y}_g = \begin{bmatrix} 0 & -g \\ g & 0 \end{bmatrix} \qquad (3.225)$$

The composite two-port network N_c can be unilateralized by a lossless reciprocal imbedding. Demonstrate that the maximum transducer power gain that can be achieved for the unilateralized N_c is precisely $U_c(j\omega)$, the U-function of N_c:

$$U_c(j\omega) = \frac{|2g - g_m|^2}{4(g^2 - g_m g + 1)} \qquad (3.226)$$

This indicates that for $|g_m| > 2$ there exists a real g for which $U_c(j\omega)$ is infinite. In other words, if N_a is active, then there exists a lossless *nonreciprocal* imbedding under which the maximum unilateral power gain can be made as large as one desires whereas the lossless reciprocal imbedding limits the gain to the value U, provided that the unilateralized two-port network is absolutely stable.

3.37 Consider the unilateral two-port network N_a of Fig. 3.15. Let $y_{a11} = y_{a22} = 1$ mho and $y_{a21} = -g_m$. By connecting transformers at the ports, the new admittance matrix $\mathbf{Y}_t$ can be written as

$$\mathbf{Y}_t = \begin{bmatrix} 1 & n \\ n & 1 \end{bmatrix} \begin{bmatrix} 1 & 0 \\ -g_m & 1 \end{bmatrix} \begin{bmatrix} 1 & n \\ n & 1 \end{bmatrix}$$

$$= \begin{bmatrix} 1 + n(n - g_m) & n(2 - ng_m) \\ 2n - g_m & 1 + n(n - g_m) \end{bmatrix} \qquad (3.227)$$

Show that by carrying out an optimum unilateralization including losses as described in Prob. 3.35, the gain expression (3.224) can be made infinite if $g_m > 2$. This result indicates that if N_a is active, then there exists a *lossy* reciprocal imbedding under which the maximum unilateral power gain can be made as large as one desires whereas the lossless reciprocal imbedding limits

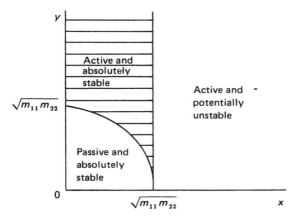

Figure 3.33 The regions of activity and stability of a two-port network in terms of its general hybrid parameters.

the gain to the value U, provided that the unilateralized two-port network is absolutely stable.

3.38 Choose $x = \text{Re}\ \sqrt{|k_{12}k_{21}|}$ and

$$ y = \frac{1}{2}||k_{12}| - |k_{21}|| = \left| \sqrt{\frac{1}{4}|k_{12} + \bar{k}_{21}|^2 - x^2} \right| \tag{3.228} $$

as the coordinates of a two-dimensional diagram. Show that the regions of activity and stability can be represented as in Fig. 3.33.

BIBLIOGRAPHY

Aurell, C. G.: Representation of the General Linear Four-Terminal Network and Some of Its Properties, *Ericsson Technics*, vol. 11, no. 1, pp. 155–179, 1955.

Aurell, C. G.: Some Tools for the Analysis and Representation of Linear Two-Port Networks, *IEEE Trans. Circuit Theory*, vol. CT-12, no. 1, pp. 18–21, 1965.

Bolinder, E. F.: Survey of Some Properties of Linear Networks, *IRE Trans. Circuit Theory*, vol. CT4, no. 3, pp. 70–78, 1957 (correction in vol. CT-5, no. 2, p. 139, 1958).

Brodie, J. H.: The Stability of Linear Two-Ports, *IEEE Trans. Circuit Theory*, vol. CT-12, no. 4, pp. 608–610, 1965.

Chen, W. K.: The Scattering Matrix and the Passivity Condition, *Matrix and Tensor Quart.*, vol. 24, nos. 1 and 2, pp. 30–32 and 74–75, 1973.

Chen, W. K.: Relationships between Scattering Matrix and Other Matrix Representations of Linear Two-Port Networks, *Int. J. Electronics*, vol. 38, no. 4, pp. 433–441, 1975.

Dietl, A.: Der optimale Wirkungsgrad von Vierpolen im Dezimeter- und Zentimeterwellen-bereich, *Hochfrequenz. u. Elak.*, vol. 66, pp. 25–29, 1944.

Fjallbrant, T.: Activity and Stability of Linear Networks, *IEEE Trans. Circuit Theory*, vol. CT-12, no. 1, pp. 12–17, 1965.

Gartner, W. W.: Maximum Available Power Gain of Linear Fourpoles, *IRE Trans. Circuit Theory*, vol. CT-5, no. 4, pp. 375–376, 1958.

Gewertz, C. M.: Synthesis of a Finite, Four-Terminal Network from Its Prescribed Driving-Point and Transfer Functions, *J. Math. Phys.*, vol. 12, nos. 1 and 2, pp. 1–257, 1933.

Jørsboe, H.: Criterion of Absolute Stability of Two-Ports, *IEEE Trans. Circuit Theory*, vol. CT-17, no. 4, pp. 639–640, 1970.

Ku, W. H.: Stability of Linear Active Nonreciprocal n-Ports, *J. Franklin Inst.*, vol. 276, no. 3, pp. 207–224, 1963.

Ku, W. H.: A Simple Derivation for the Stability Criterion of Linear Active Two-Ports, *Proc. IEEE*, vol. 53, no. 3, pp. 310–311, 1965.

Ku, W. H.: Extension of the Stability Criterion of Linear Active Two-Ports to the Entire Complex Frequency Plane, *Proc. IEEE*, vol. 58, no. 4, pp. 591–592, 1970.

Kuh, E. S. and R. A. Rohrer: "Theory of Linear Active Networks," San Francisco, Calif.: Holden-Day, 1967.

Leine, P. O.: On the Power Gain of Unilaterized Active Networks, *IRE Trans. Circuit Theory*, vol. CT-8, no. 3, pp. 357–358, 1961.

Linvill, J. G. and L. G. Schimpf: The Design of Tetrode Transistor Amplifiers, *Bell Syst. Tech. J.*, vol. 35, no. 4, pp. 813–840, 1956.

Llewellyn, F. B.: Some Fundamental Properties of Transmission Systems, *Proc. IRE*, vol. 40, no. 3, pp. 271–283, 1952.

Mason, S. J.: Power Gain in Feedback Amplifiers, *IRE Trans. Circuit Theory*, vol. CT-1, no. 2, pp. 20–25, 1954.

Mason, S. J.: Some Properties of Three-Terminal Devices, *IRE Trans. Circuit Theory*, vol. CT-4, no. 4, pp. 330–332, 1957.

Mathis, H. F.: Experimental Procedures for Determining the Efficiency of Four-Terminal Networks, *J. Appl. Phys.*, vol. 25, no. 8, pp. 982–986, 1954.

Page, D. F. and A. R. Boothroyd: Instability in Two-Port Active Networks, *IRE Trans. Circuit Theory*, vol. CT-5, no. 2, pp. 133–139, 1958.

Rollett, J. M.: Stability and Power-Gain Invariants of Linear Twoports, *IRE Trans. Circuit Theory*, vol. CT-9, no. 1, pp. 29–32, 1962 (correction in vol. CT-10, no. 1, p. 107, 1963).

Rollett, J. M.: The Measurement of Transistor Unilateral Gain, *IEEE Trans. Circuit Theory*, vol. CT-12, no. 1,pp. 91–97, 1965.

Sathe, S. T.: A Stability Test for a Class of Linear, Active Two-Ports with Uncertain Parameters, *IEEE Trans. Commun.*, vol. COM-23, no. 11, pp. 1357–1361, 1975.

Scanlan, J. O. and J. S. Singleton: The Gain and Stability of Linear Two-Port Amplifiers, *IRE Trans. Circuit Theory*, vol. CT-9, no. 3, pp. 240–246, 1962a.

Scanlan, J. O. and J. S. Singleton: Two-Ports-Maximum Gain for a Given Stability Factor, *IRE Trans. Circuit Theory*, vol. CT-9, no. 4, pp. 428–429, 1962b.

Sharpe, G. E., J. L. Smith, and J. R. W. Smith: A Power Theorem on Absolutely Stable Two-Ports, *IRE Trans. Circuit Theory*, vol. CT-6, no. 2, pp. 159–163, 1959.

Singhakowinta, A.: On Analysis and Representation of Two-Port, *IEEE Trans. Circuit Theory*, vol. CT-13, no. 1, pp. 102–103, 1966.

Stern, A. P.: Considerations on the Stability of Active Elements and Applications to Transistors, *IRE Natl. Conv. Rec.*, part 2, pp. 46–52, 1956.

Van der Puije, P. D.: On Negative Conductance Activity in 2-Port Networks, *IEEE Trans. Circuits and Systems*, vol. CAS-21, no. 2, pp. 194–196, 1974.

Van Heuven, J. H. C. and T. E. Rozzi: The Invariance Properties of a Multivalue n-Port in a Linear Embedding, *IEEE Trans. Circuit Theory*, vol. CT-19, no. 2, pp. 176–183, 1972.

Venkateswaran, S.: An Invariant Stability Factor and Its Physical Significance, *Proc. IEE (London)*, part C, Mono. 468E, pp. 98–102, 1961.

Venkateswaran, S. and A. R. Boothroyd: Power Gain and Bandwidth of Tuned Transistor Amplifier Stages, *Proc. IEE (London)*, vol. 106B, suppl. 15, pp. 518–529, 1960.

Woods, D.: Reappraisal of the Unconditional Stability Criteria for Active 2-Port Networks in Terms of s Parameters, *IEEE Trans. Circuits and Systems*, vol. CAS-23, no. 2, pp. 73–81, 1976.

Youla, D.: A Stability Characterization of the Reciprocal Linear Passive n Port, *Proc. IRE*, vol. 47, no. 6, pp. 1150–1151, 1959.

Youla, D.: A Note on the Stability of Linear, Nonreciprocal n-Ports, *Proc. IRE*, vol. 48, no. 1, pp. 121–122, 1960.

Zawels, J.: Gain-Stability Relationship, *IEEE Trans. Circuit Theory*, vol. CT-10, no. 1, pp. 109–110, 1963.

FOUR
THEORY OF FEEDBACK AMPLIFIERS I

In the preceding chapter, we demonstrated that by introducing physical feedback loops externally to an active device, we can produce a particular change in the performance of the network. Specifically, we showed that a three-terminal device can be unilateralized by a lossless reciprocal imbedding. In this and following chapters, we shall study the subject of feedback in detail and demonstrate that feedback may be employed to make the gain of an amplifier less sensitive to variations in the parameters of the active components, to control its transmission and driving-point properties, to reduce the effects of noise and nonlinear distortion, and to affect the stability or instability of the network.

We first discuss the conventional treatment of feedback amplifiers, which is based on the ideal feedback model, and analyze several simple feedback networks. We then present Bode's feedback theory in detail. Bode's theory is based on the concepts of return difference and null return difference and is applicable to both simple and complicated feedback amplifiers, where the analysis by conventional method for the latter breaks down. We show that return difference is a generalization of the concept of the feedback factor of the ideal feedback model and can be interpreted physically as the returned voltage. The relationships between the network functions and return difference and null return difference are derived and are employed to simplify the calculation of driving-point impedance of an active network.

4.1 IDEAL FEEDBACK MODEL

A *feedback amplifier* is a network in which some variable, either the output variable or any other controllable one, is fed back as the input variable to some part of the network in such a way that it is able to affect its own value. The simplest form of a feedback amplifier can be represented by the ideal block diagram of Fig. 4.1, in which $\mu(s)$ represents the transfer function of the *unilateral forward path* and is conventionally called the *open-loop transfer function* or *forward amplifier gain*, defined by the relation

$$\mu(s) = \left.\frac{y(s)}{u(s)}\right|_{u_f(s)=0} = \frac{y(s)}{u_i(s)} \tag{4.1}$$

where $y(s)$ and the u's denote either the current and/or voltage variables. $\beta(s)$ is the transfer function of a *unilateral feedback path* and is defined as

$$\beta(s) = \frac{u_f(s)}{y(s)} \tag{4.2}$$

Thus, in this highly idealized model, none of the input signal is transmitted through the feedback path and none of the output signal is transmitted in the reverse direction through the forward path. Also, the presence of the ideal adder in the model indicates that there is no loading effect at the input. In other words, the transmission of signal is permitted only in the direction of the arrows as shown in the block diagram of Fig. 4.1. Under these assumptions, it is straightforward to show that the overall transfer function of the amplifier can be expressed as

$$w(s) = \frac{y(s)}{u(s)} = \frac{\mu(s)}{1 - \mu(s)\beta(s)} \tag{4.3}$$

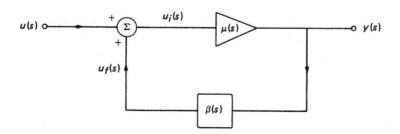

Figure 4.1 The block diagram of the ideal feedback model.

which is referred to as the *closed-loop transfer function*. In terms of logarithms, we have

$$\ln w(s) = \ln \mu(s) - \ln [1 - \mu(s)\beta(s)] \tag{4.4}$$

whose $j\omega$-axis real parts are related by

$$\ln |w(j\omega)| = \ln |\mu(j\omega)| - \ln |1 - \mu(j\omega)\beta(j\omega)| \tag{4.5}$$

Observe that in the application of feedback, the open-loop transfer function is modified by a factor $1 - \mu(s)\beta(s)$. The factor $1 - \mu(s)\beta(s)$ is referred to as the *feedback factor*. At a specified frequency on the real-frequency axis, if its magnitude is greater than unity, we have *negative* or *degenerative feedback,* and if it is smaller than unity, we have *positive or regenerative feedback*. From Eq. (4.5) we see that negative feedback reduces the forward amplifier gain by the amount of feedback $\ln |1 - \mu(j\omega)\beta(j\omega)|$ measured in nepers, or $20 \log |1 - \mu(j\omega)\beta(j\omega)|$ measured in decibels. In exchange, as we shall see, the gain becomes less sensitive to the variations of the parameters, the effects of noise and nonlinear distortion are usually reduced, and the transmission and driving-point characteristics are modified, all by the feedback factor $1 - \mu(j\omega)\beta(j\omega)$. On the other hand, the positive feedback increases the forward amplifier gain, the sensitivity, and the effects of noise and nonlinear distortion all by the factor $1 - \mu(j\omega)\beta(j\omega)$, and is therefore seldom used in amplifier design. Nevertheless, it is useful in the design of oscillator circuits. Finally, under the condition

$$1 - \mu(s)\beta(s) = 0 \tag{4.6}$$

a feedback amplifier will function as an oscillator. This condition is known as the *Barkhausen criterion,* which states that the frequency of sinusoidal oscillator is determined by the condition that the phase shift of $\mu(j\omega)\beta(j\omega)$ is zero provided that the magnitude of $\mu(j\omega)\beta(j\omega)$ equals unity. The factor $\mu(s)\beta(s)$ is often called the *loop transmission* and can be interpreted physically as follows. Suppose that no input excitation is applied to the amplifier and that the input side of the unilateral forward path is broken, as shown in Fig. 4.2, with a unit input applied to the right of the break. Then the signal appearing at the left of the break is precisely $\mu(s)\beta(s)$.

To employ the ideal feedback model to analyze a practical feedback amplifier, it becomes necessary to separate the feedback amplifier into two blocks: the basic unilateral amplifier $\mu(s)$ and the feedback network $\beta(s)$. The procedure is difficult and sometimes virtually impossible, because the forward path may not be strictly unilateral, the feedback path is usually bilateral, and the input and output coupling networks are often complicated. Thus, the ideal feedback model of Fig. 4.1 is

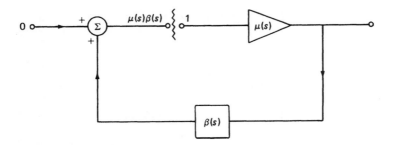

Figure 4.2 The physical interpretation of the loop transmission.

not an adequate representation of a practical feedback amplifier. In the remainder of this chapter, we shall develop Bode's feedback theory, which is applicable to general network configuration and avoids the necessity of identifying $\mu(s)$ and $\beta(s)$. However, before we do this, we demonstrate a method for identifying $\mu(s)$ and $\beta(s)$ from given simple feedback amplifier configurations.

4.2 FEEDBACK AMPLIFIER CONFIGURATIONS

In this section, we present four simple feedback amplifier configurations based on the two-port representation, together with a general configuration.

Let an active two-port device N_a be characterized by its general hybrid matrix

$$\begin{bmatrix} y_{a1} \\ y_{a2} \end{bmatrix} = \begin{bmatrix} k_{a11} & k_{a12} \\ k_{a21} & k_{a22} \end{bmatrix} \begin{bmatrix} u_{a1} \\ u_{a2} \end{bmatrix} \tag{4.7}$$

We can consider another two-port feedback network N_f described by the general hybrid matrix

$$\begin{bmatrix} y_{f1} \\ y_{f2} \end{bmatrix} = \begin{bmatrix} k_{f11} & k_{f12} \\ k_{f21} & k_{f22} \end{bmatrix} \begin{bmatrix} u_{f1} \\ u_{f2} \end{bmatrix} \tag{4.8}$$

Upon interconnection of these two two-ports in such a way that forces u_{a1} and u_{a2} to be equal to u_{f1} and u_{f2}, respectively, the composite two-port network N is characterized by the equation

$$\begin{bmatrix} y_1 \\ y_2 \end{bmatrix} = \begin{bmatrix} k_{11} & k_{12} \\ k_{21} & k_{22} \end{bmatrix} \begin{bmatrix} u_1 \\ u_2 \end{bmatrix} \tag{4.9}$$

where

$$k_{ij} = k_{aij} + k_{fij} \quad i, j = 1, 2 \tag{4.10}$$

More compactly, the coefficient matrix of (4.9) can be expressed as

$$\mathbf{H}(s) = \mathbf{H}_a(s) + \mathbf{H}_f(s) \tag{4.11}$$

$\mathbf{H}_a(s)$ and $\mathbf{H}_f(s)$ being the coefficient matrices of (4.7) and (4.8), respectively.

As discussed in Sec. 3.1, Eq. (4.9) can be applied to all four representations (3.1) and results in four basic feedback configurations, as shown schematically in Fig. 4.3. It is important to remember that in applying these configurations, care must be taken to ensure the validity of (4.10), which requires that the instantaneous current entering one terminal of a port must equal to the instantaneous current leaving the other terminal of the port. In Fig. 4.3a, for example, we must have

$$u_1 = u_{a1} = u_{f1} = I_1 \tag{4.12a}$$

$$u_2 = u_{a2} = u_{f2} = I_2 \tag{4.12b}$$

after interconnection. To this end, Brune proposes a simple procedure known as the *Brune tests.* His tests are exceedingly simple and are both necessary and sufficient. Depending on the configurations of Fig. 4.3, the tests for each case are presented in Fig. 4.4. An interconnection is permissible if the voltage marked V is zero. If this condition is not satisfied, the matrix addition, as given in (4.11), will give incorrect answers for the parameters of the composite network, unless isolating ideal transformers are introduced at one of the two ends. However, the introduction of ideal transformers is often undesirable because of network performance or cost. Therefore, the most practical networks will be those that pass the Brune tests.

If the feedback parameter k_{f12} of the feedback network N_f is such that it satisfies the equation

$$k_{f12} = -k_{a12} \tag{4.13}$$

then the composite two-port network N becomes *unilateral*, and we can say that the internal feedback of the active device N_a is *neutralized*.

As indicated in Sec. 3.1, for each parameter set there is a corresponding representation of the source and load terminations, as shown in Fig. 4.3. The formulas that we shall derive are general and can be applied to all four two-port representations provided that we use the corresponding variables and parameters as listed in Table 3.1.

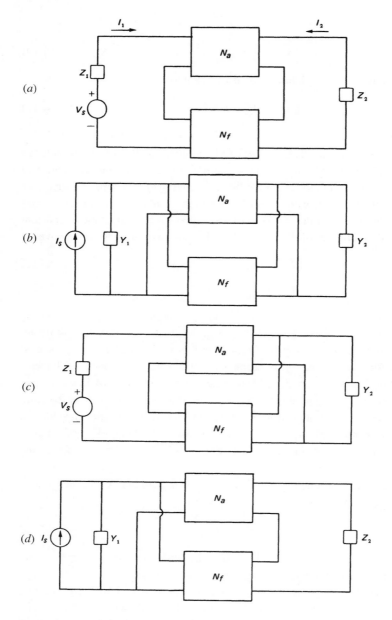

Figure 4.3 Simple feedback amplifier configurations, (*a*) The series-series or current-series feedback. (*b*) The parallel-parallel or voltage-shunt feedback, (*c*) The series-parallel or voltage-series feedback. (*d*) The parallel-series or current-shunt feedback.

Refer to Fig. 4.3 and Table 3.1. The overall transfer function y_2/u_s, as given in (3.25a), of the composite two-port network N is

$$w(s) = \frac{y_2}{u_s} = \frac{k_{21}M_2}{(k_{11} + M_1)(k_{22} + M_2) - k_{12}k_{21}} \qquad (4.14)$$

To identify the forward amplifier gain $\mu(s)$ of Fig. 4.1, as defined in Eq. (4.1), we set $k_{12} = 0$ in Eq. (4.14), yielding

$$\mu(s) = \frac{y_2}{u_s}\bigg|_{k_{12}=0} = \frac{k_{21}M_2}{(k_{11} + M_1)(k_{22} + M_2)} \qquad (4.15)$$

Next, we write (4.14) in the form of (4.3) by dividing the numerator and denominator of (4.14) by $(k_{11} + M_1)(k_{22} + M_2)$, thus identifying the transfer function of the feedback path as

$$\beta(s) = \frac{k_{12}}{M_2} = \frac{k_{12}u_2}{M_2u_2} = \frac{\tau_f}{y_2} \qquad (4.16a)$$

where

$$\tau_f = -k_{12}u_2 \qquad (4.16b)$$

is defined as the *feedback signal*, which can either be voltage or current. Combining (4.15) and (4.16a), we obtain the loop transmission as

$$\mu(s)\beta(s) = \frac{k_{12}k_{21}}{(k_{11} + M_1)(k_{22} + M_2)} \qquad (4.17)$$

as previously defined in Eq. (3.190b).

With $\mu(s)$ and $\beta(s)$ identified, the ideal feedback model can now be employed to determine the effects of feedback on network performance. One of the main disadvantages of this approach is that the quantities $\mu(s)$ and $\beta(s)$ cannot be measured experimentally, because in computing $\mu(s)$ the forward transmission of the feedback network is lumped with the active device, whereas the backward transmission of the active device is lumped with the feedback network. Nevertheless, for most practical amplifiers, the forward transmission of the active device is much larger than that of the feedback network, whereas the backward transmission of feedback network is much larger than that of the active device. Thus, we can often make the following approximation:

$$k_{21} \approx k_{a21} \qquad (4.18a)$$

$$k_{12} \approx k_{f12} \qquad (4.18b)$$

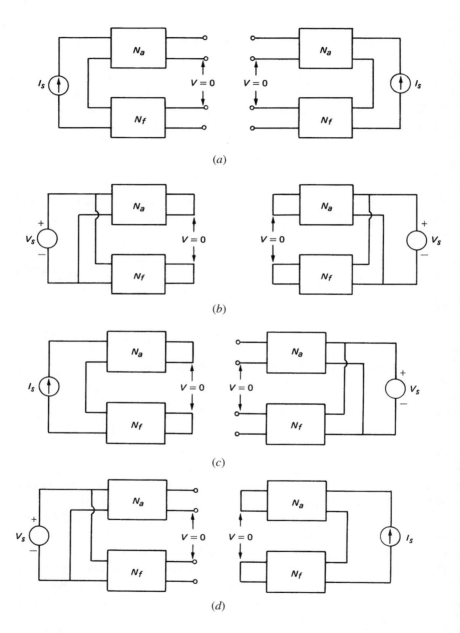

Figure 4.4 The Brune tests for (*a*) the series-series feedback, (*b*) the parallel-parallel feedback, (*c*) the series-parallel feedback, and (*d*) the parallel-series feedback.

Because they are interconnected, the four feedback topologies of Fig. 4.3, $a-d$ are usually referred to as the *series-series feedback, parallel-parallel feedback, series-parallel feedback,* and *parallel-series feedback,* respectively. The word *shunt* is sometimes used in lieu of *parallel.* In electronics, the four feedback configurations of Fig. 4.3, $a-d$ are frequently called the *current-series feedback, voltage-shunt feedback, voltage-series feedback,* and *current-shunt feedback,* respectively. The word *voltage* refers to connecting the output voltage as input to the feedback network, *current* to tapping off some output current through the feedback network, *series* to connecting the feedback signal in series with the input voltage, and *shunt* to connecting the feedback signal in shunt or parallel with an input current source.

In addition to the reduction of gain by negative feedback, we shall now discuss the effects of feedback on the input and output immittances of the four feedback configurations of Fig. 4.3. From Eq. (3.7), the input immittance $M_{11}(s) + M_1(s)$ looking into the source of the composite two-port network N is found to be

$$M_{11}(s) + M_1(s) = k_{11} + M_1 - \frac{k_{12}k_{21}}{k_{22} + M_2}$$

$$= (k_{11} + M_1)\left[1 - \frac{k_{12}k_{21}}{(k_{11} + M_1)(k_{22} + M_2)}\right]$$

$$= (k_{11} + M_1)\left[1 - \mu(s)\beta(s)\right] \tag{4.19a}$$

Likewise, from Eq. (3.8) we can show that the output immittance $M_{22}(s) + M_2(s)$ looking into the output port including the termination can be written as

$$M_{22}(s) + M_2(s) = (k_{22} + M_2)\left[1 - \mu(s)\beta(s)\right] \tag{4.19b}$$

Thus, we conclude that feedback affects the short-circuit or open-circuit input immittance $k_{11} + M_1$ and output immittance $k_{22} + M_2$ by the feedback factor $1 - \mu(s)\beta(s)$. Assume that $1 - \mu(0)\beta(0) > 1$. Then the effects of feedback on the low-frequency input and output immittances for the four feedback configurations are illustrated in Table 4.1.

We remark that *not* every feedback amplifier can be classified as being in one of the above four categories. A general feedback configuration may contain an input coupling network and an output coupling network as shown in Fig. 4.5. A typical example is the bridge feedback configuration of Fig. 4.6, which does not belong to any one of the topologies of Fig. 4.3.

We illustrate the above results by considering a few typical feedback amplifiers.

Table 4.1. Effect of feedback on impedances.

Feedback configuration	Input impedance	Output impedance
Series-series	Increases	Increases
Parallel-parallel	Decreases	Decreases
Series-parallel	Increases	Decreases
Parallel-series	Decreases	Increases

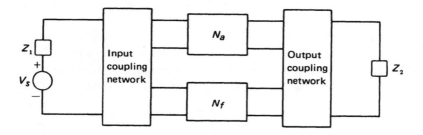

Figure 4.5 A general feedback configuration containing an input coupling network and an output coupling network.

Figure 4.6 A typical example of a bridge feedback configuration.

4.2.1 Series-Series Feedback

An amplifier with an unbypassed emitter resistor, as shown in Fig. 4.7, is an example of series-series or current-series feedback. Assume that the transistor is characterized by its hybrid matrix

$$\mathbf{H}_e = \begin{bmatrix} h_{ie} & h_{re} \\ h_{fe} & h_{oe} \end{bmatrix} \tag{4.20}$$

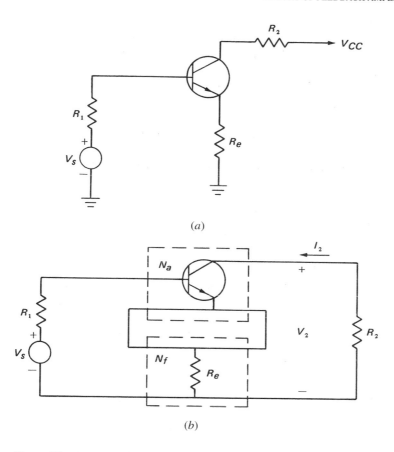

(a)

(b)

Figure 4.7 (a) An amplifier with an unbypassed emitter resistor and (b) its series-series or current-series feedback representation.

The impedance matrix of the transistor N_a becomes (see App. II)

$$\mathbf{Z}_a = \frac{1}{h_{oe}} \begin{bmatrix} h_{ie}h_{oe} - h_{fe}h_{re} & h_{re} \\ -h_{fe} & 1 \end{bmatrix} \tag{4.21}$$

The impedance matrix of the feedback network N_f is given by

$$\mathbf{Z}_f = \begin{bmatrix} R_e & R_e \\ R_e & R_e \end{bmatrix} \tag{4.22}$$

Since the Brune tests of Fig. 4.4a are satisfied when applied to the network configuration of Fig. 4.7b, the connection is permissible and the impedance matrix

$\mathbf{Z}$ of the composite two-port N is simply the sum of the impedance matrices $\mathbf{Z}_a$ and $\mathbf{Z}_f$, giving

$$
\mathbf{Z} = \begin{bmatrix} R_e + h_{ie} - \dfrac{h_{fe}h_{re}}{h_{oe}} & R_e + \dfrac{h_{re}}{h_{oe}} \\[3mm] R_e - \dfrac{h_{fe}}{h_{oe}} & R_e + \dfrac{1}{h_{oe}} \end{bmatrix} \tag{4.23}
$$

From (4.15), the open-loop voltage gain μ can be expressed in terms of the elements z_{ij} of $\mathbf{Z}$ and the terminating resistances R_1 and R_2:

$$
\begin{aligned}
\mu = \frac{V_2}{V_s}\bigg|_{z_{12}=0} &= \frac{z_{21}R_2}{(z_{11}+R_1)(z_{22}+R_2)} \\[2mm]
&= \frac{(R_e - h_{fe}/h_{oe})R_2}{(R_e + h_{ie} - h_{fe}h_{re}/h_{oe} + R_1)(R_e + 1/h_{oe} + R_2)}
\end{aligned} \tag{4.24}
$$

The transfer function of the unilateral feedback path is calculated from (4.16) and is given by

$$
\beta = \frac{V_f}{V_2} = \frac{z_{12}}{R_2} = \frac{R_e + h_{re}/h_{oe}}{R_2} \tag{4.25a}
$$

where the feedback voltage V_f is defined by

$$
V_f = -z_{12}I_2 = -\left(R_e + \frac{h_{re}}{h_{oe}}\right)I_2 \tag{4.25b}
$$

As indicated in (4.18), for practical amplifiers, we can usually assume that

$$
\frac{h_{fe}}{h_{oe}} \gg R_e \tag{4.26a}
$$

$$
R_e \gg \frac{h_{re}}{h_{oe}} \tag{4.26b}
$$

Also, if $1/h_{oe} \gg R_e$, then Eqs. (4.24) and (4.25a) can be simplified to

$$
\mu \approx \frac{h_{fe}R_2}{(h_{ie} + R_e + R_1 - h_{fe}h_{re}/h_{oe})(1 + R_2h_{oe})} \tag{4.27a}
$$

$$
\beta \approx \frac{R_e}{R_2} \tag{4.27b}
$$

From (4.3), the closed-loop voltage gain is obtained as

$$w \approx -\frac{h_{fe}R_2}{(h_{ie} + R_e + R_1 - h_{fe}h_{re}/h_{oe})(1 + R_2 h_{oe}) + h_{fe}R_e} \tag{4.28}$$

The above analysis can be explained by the schematic diagram of Fig. 4.8a. The unilateral forward path is represented by the controlled voltage source $z_{21}I_1$, whereas the unilateral feedback path is represented by the controlled voltage source $z_{12}I_2$. The calculation of μ is shown in Fig. 4.8b. Thus, μ can be calculated either from (4.24) or directly from the network of Fig. 4.8b.

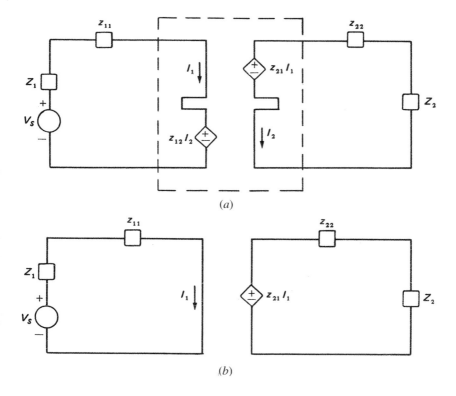

(a)

(b)

Figure 4.8   (a) The schematic diagram of the analysis of the series-series feedback configuration, and (b) the network used to calculate the voltage gain of the unilateral forward path.

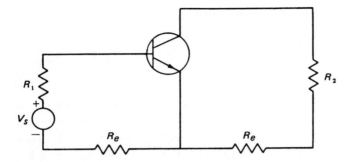

Figure 4.9 The transistor network used to calculate the open-loop voltage gain.

Example 4.1 In the feedback amplifier of Fig. 4.7, let

$$h_{fe} = 50 \quad h_{ie} = 1\,\mathrm{k}\Omega$$

$$h_{re} = 2.5 \cdot 10^{-4} \quad h_{oe} = 25\,\mu\mathrm{mho} \tag{4.29}$$

$$R_1 = R_e = 1\,\mathrm{k}\Omega \quad R_2 = 10\,\mathrm{k}\Omega$$

Substituting these in (4.24) yields the open-loop voltage gain

$$\mu = -156.78 \tag{4.30}$$

which can also be computed directly from the network of Fig. 4.9. From (4.28) the closed-loop voltage gain is found to be

$$w = \frac{V_2}{V_s} = \frac{\mu}{1 - \mu\beta} = -9.41 \tag{4.31}$$

where $\beta = 0.1$, a reduction from the gain without feedback by a factor of 16.68.

The above analysis is restricted to low-frequency applications. When transistors are used to amplify high-frequency signals, the h-parameters of (4.20) are generally complex. The most useful high-frequency model of the bipolar transistor is shown in Fig. 4.10. This model is very widely used in the literature and is called the *hybrid-pi equivalent network*. The typical element values of the model are given below:

$$g_m \approx 40 I_e \approx 0.4\,\mathrm{mho} \quad \text{for } I_e = 1\,\mathrm{mA}$$

$$r_x = 20{\sim}100\,\Omega \quad r_\pi = 75{\sim}250\,\Omega$$

$$C_\mu = 1{\sim}20\,\mathrm{pF} \quad C_\pi = 10{\sim}1000\,\mathrm{pF} \tag{4.32}$$

$$r_o = 10{\sim}100\,\mathrm{k}\Omega$$

Figure 4.10 The hybrid-pi equivalent network of a transistor.

Figure 4.11 A two-port network terminated at its output by the admittance Y_2.

We may wonder at this time if we can make some simplifications of the hybrid-pi model at the outset. If the approximation can be made, we must be aware, of course, of the conditions under which it is made.

Consider the two-port network of Fig. 4.11, with the admittance matrix

$$\mathbf{Y}(s) = \begin{bmatrix} (C_1 + C_2)s & -sC_2 \\ g_m - sC_2 & sC_2 \end{bmatrix} \tag{4.33}$$

By appealing to Eq. (3.7), the input admittance is determined as

$$Y_{\text{in}}(s) = s(C_1 + C_2) + \frac{sC_2(g_m - sC_2)}{Y_2 + sC_2} \tag{4.34}$$

If we assume that in the range of frequencies of interest

$$\omega C_2 \ll \begin{cases} g_m \\ |Y_2| \end{cases} \tag{4.35}$$

then Eq. (4.34) can be approximated by

$$Y_{in}(s) \approx s \left[C_1 + C_2 \left(1 + \frac{g_m}{Y_2} \right) \right] = sC_1 + Y_{eq} \tag{4.36}$$

where

$$Y_{eq}(s) = sC_2 \left[1 + \frac{g_m}{Y_2(s)} \right] \tag{4.37}$$

showing that the network of Fig. 4.11 can be approximated by the unilateralized network of Fig. 4.12 as far as the input admittance is concerned. The voltage gain of Fig. 4.11 is found to be

$$\frac{V_2}{V_1} = -\frac{g_m - sC_2}{Y_2 + sC_2} \tag{4.38}$$

Under the assumption of (4.35), the above expression simplifies to

$$\frac{V_2}{V_1} \approx -\frac{g_m}{Y_2} \tag{4.39}$$

which can also be computed directly from the network of Fig. 4.12. Hence, we conclude that if the approximation (4.35) is valid, the unilateralized network of Fig. 4.12 gives fairly correct results as far as the input admittance and the forward gain are concerned. However, it *does not* give the correct *output admittance* and *reverse transmission*. In situations where these two quantities are important, the simplified model should not be used.

Now we consider an important special case where the load is purely resistive with $Y_2 = 1/R_2$. Under this situation, the network of Fig. 4.12 reduces to that shown in Fig. 4.13. The input capacitance is now $C_1 + C_2(1 + g_m R_2)$. This increase results from the voltage gain of the amplifying device and is known as the *Miller effect*. The Miller effect will now be employed to unilateralize the hybrid-pi

Figure 4.12 A unilateralized approximation of the network of Fig. 4.11.

Figure 4.13 A unilateralized approximation of the network of Fig. 4.11 with resistive termination.

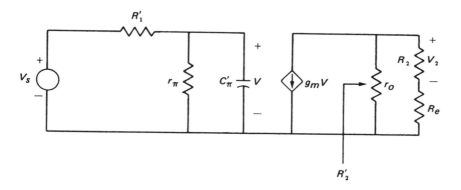

Figure 4.14 A unilateralized approximation of the network of Fig. 4.9.

model of Fig. 4.10 when it is used for the transistor in the feedback amplifier of Fig. 4.7.

The open-loop voltage gain $\mu(s)$ of the amplifier of Fig. 4.7 can be calculated directly from the network of Fig. 4.9. By using the high-frequency model of Fig. 4.10 and applying the Miller effect, the network of Fig. 4.9 can be simplified to that of Fig. 4.14, with

$$R'_1 = R_1 + R_e + r_x \qquad (4.40a)$$

$$R'_2 = \frac{r_0(R_2 + R_e)}{r_0 + R_2 + R_e} \qquad (4.40b)$$

$$C'_\pi = C_\pi + C_\mu(1 + g_m R'_2) \qquad (4.40c)$$

From Fig. 4.14 we obtain the open-loop voltage gain

$$\mu(s) = -\frac{A_0}{s + s_0} \qquad (4.41a)$$

where

$$A_0 = \frac{g_m R_2 r_o}{R_1' C_\pi' (r_o + R_2 + R_e)} \qquad (4.41b)$$

$$s_0 = \frac{1/r_\pi + 1/R_1'}{C_\pi'} \qquad (4.41c)$$

Using (4.3) in conjunction with (4.27b) yields the closed-loop voltage gain

$$w(s) = -\frac{A_0}{s + s_0 + A_0 R_e / R_2} \qquad (4.42)$$

Example 4.2 As a numerical example, let

$$g_m = 0.4 \, \text{mho} \quad r_x = 50 \, \Omega$$
$$r_\pi = 250 \, \Omega \quad C_\pi = 195 \, \text{pF}$$
$$C_\mu = 5 \, \text{pF} \quad r_o = 50 \, \text{k}\Omega \qquad (4.43)$$
$$R_2 = R_e = 1 \, \text{k}\Omega \quad R_1 = 100 \, \Omega$$

From (4.40) we have

$$R_1' = 1.15 \, \text{k}\Omega \quad R_2' = 1.923 \, \text{k}\Omega \quad C_\pi' = 4.05 \, \text{nF} \qquad (4.44)$$

giving

$$A_0 = 82.66 \cdot 10^6 \quad s_0 = 1.20 \cdot 10^6 \, \text{rad/s} \qquad (4.45)$$

Thus, the closed-loop voltage gain is obtained from (4.42) as

$$w(s) = -\frac{82.66 \cdot 10^6}{s + 83.86 \cdot 10^6} \qquad (4.46)$$

showing a midband gain $w(0) = -0.99$. The midband feedback factor is

$$1 - \mu(0)\beta(0) = 1 + \frac{A_0 R_e}{s_0 R_2} = 69.88 \qquad (4.47)$$

where the midband open-loop gain is –68.88. From (4.19) the midband open-circuit input and output impedances are increased by a factor of 69.88 from their values $R_1' + r_\pi = 1.4 \, \text{k}\Omega$ and $R_2 + R_e + r_o = 52 \, \text{k}\Omega$, respectively.

4.2.2 Parallel-Parallel Feedback

The network of Fig. 4.15 shows a common-emitter stage with a resistor R_f connected from the output to the input. It is clear that this configuration conforms to the parallel-parallel or voltage-shunt topology of Fig. 4.3*b*. Assume that the transistor is described by (4.20). Then, from (2.47*a*), the admittance matrix of the transistor N_a is given by

$$\mathbf{Y}_a = \frac{1}{h_{ie}} \begin{bmatrix} 1 & -h_{re} \\ h_{fe} & \Delta_{he} \end{bmatrix} \tag{4.48}$$

(a)

(b)

Figure 4.15 (*a*) A common-emitter stage with a resistor R_f providing the feedback, and (*b*) its parallel-parallel or voltage-shunt feedback representation.

where $\Delta_{he} = h_{ie}h_{oe} - h_{fe}h_{re}$. The admittance matrix of the feedback network N_f with $G_f = 1/R_f$ is obtained by inspection as

$$\mathbf{Y}_f = \begin{bmatrix} G_f & -G_f \\ -G_f & G_f \end{bmatrix} \tag{4.49}$$

Since by the Brune tests of Fig. 4.4b the connection of N_a and N_f is permissible, the admittance matrix of the composite two-port N can be written as

$$\mathbf{Y} = \mathbf{Y}_a + \mathbf{Y}_f = \begin{bmatrix} G_f + \dfrac{1}{h_{ie}} & -G_f - \dfrac{h_{re}}{h_{ie}} \\ -G_f + \dfrac{h_{fe}}{h_{ie}} & G_f + \dfrac{\Delta_{he}}{h_{ie}} \end{bmatrix} \tag{4.50}$$

From (4.15) and (4.16), the open-loop current gain μ and the transfer function of the unilateral feedback path β, being expressible in terms of the elements y_{ij} of $\mathbf{Y}$ and the terminating conductances $G_1 = 1/R_1$ and $G_2 = 1/R_2$, are calculated as

$$\mu = \frac{I_2}{I_s}\bigg|_{y_{12}=0} = -\frac{(G_f - h_{fe}/h_{ie})G_2}{(G_1 + G_f + 1/h_{ie})(G_2 + G_f + \Delta_{he}/h_{ie})} \tag{4.51}$$

$$\beta = \frac{I_f}{I_2} = \frac{y_{12}}{G_2} = -\frac{G_f + h_{re}/h_{ie}}{G_2} \tag{4.52a}$$

where the feedback current I_f is defined by

$$I_f = -y_{12}V_2 = \left(G_f + \frac{h_{re}}{h_{ie}}\right)V_2 \tag{4.52b}$$

As indicated in (4.18), for practical amplifiers, we can make the following approximations:

$$\frac{h_{fe}}{h_{ie}} \gg G_f \tag{4.53a}$$

$$G_f \gg \frac{h_{re}}{h_{ie}} \tag{4.53b}$$

If, in addition, $G_2 + G_f \gg \Delta_{he}/h_{ie}$, then we have

$$\mu \approx \frac{h_{fe}G_2}{(G_2 + G_f)[1 + (G_1 + G_f)h_{ie}]} \tag{4.54a}$$

$$\beta \approx -\frac{G_f}{G_2} = -\frac{R_2}{R_f} \tag{4.54b}$$

and the closed-loop current ratio or gain is found to be

$$w = \frac{I_2}{I_s} = \frac{\mu}{1 - \mu\beta} \approx \frac{h_{fe}G_2}{(G_2 + G_f)[1 + h_{ie}(G_1 + G_f)] + h_{fe}G_f} \tag{4.55}$$

Example 4.3 In the feedback amplifier of Fig. 4.15, let

$$h_{fe} = 50 \quad h_{ie} = 1\,\text{k}\Omega$$

$$h_{re} = 2.5 \cdot 10^{-4} \quad h_{oe} = 25\,\mu\text{mho}$$

$$R_1 = 10\,\text{k}\Omega \quad R_2 = 4\,\text{k}\Omega \tag{4.56}$$

$$R_f = 40\,\text{k}\Omega$$

Then from (4.54a) we compute the current gain of the amplifier without feedback as

$$\mu = 40.40 \tag{4.57}$$

The amplifier current gain with feedback is found to be

$$w = \frac{I_2}{I_s} = \frac{\mu}{1 - \mu\beta} = 8.01 \tag{4.58}$$

where $\beta = -0.1$, indicating a gain reduction by a factor of 5.04. If the exact formulas (4.51) and (4.52a) are used, the corresponding values are found to be $\mu = 38.63$, $\beta = -0.101$, and $w = 7.88$.

At high frequencies, we use the hybrid-pi equivalent network of Fig. 4.10 for the transistor in the feedback amplifier of Fig. 4.15.

As before, let y_{aij}, y_{fij}, and y_{ij}, where $y_{ij} = y_{aij} + y_{fij}$, be the y-parameters of the transistor N_a, the feedback network N_f, and the composite two-port N of

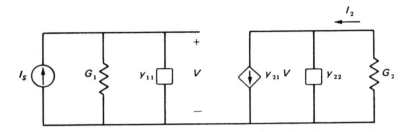

Figure 4.16 The network used to calculate the open-loop current gain of the amplifier of Fig. 4.15

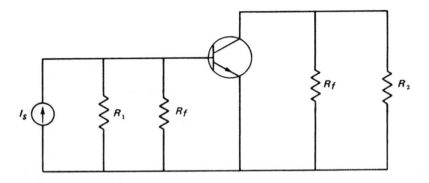

Figure 4.17 The amplifier approximating the network of Fig. 4.16.

Fig. 4.15, respectively. Then, according to Eq. (4.15), the forward path is specified by the current gain

$$\mu(s) = \left.\frac{I_2}{I_s}\right|_{y_{12}=0} \tag{4.59}$$

This situation can be depicted as in Fig. 4.16. Since $y_{21} \approx y_{a21}$ and $y_{a12} \approx 0$, Fig. 4.16 can be approximated by the network of Fig. 4.17. By using the hybrid-pi model of Fig. 4.10 and applying the Miller effect, the network of Fig. 4.17 can be unilateralized as shown in Fig. 4.18. The open-loop current gain or the forward amplifier gain is found to be

$$\mu(s) = \frac{A_0}{1 + s/s_0} \tag{4.60}$$

Figure 4.18 A unilateralized approximation of the network of Fig. 4.17.

where

$$A_0 = \frac{g_m r_\pi R'_1 R'_2}{(R'_1 + r_\pi + r_x) R_2} \tag{4.61a}$$

$$s_0 = \frac{1}{C} \left(\frac{1}{r_\pi} + \frac{1}{r_x + R'_1} \right) \tag{4.61b}$$

$$C = C_\pi + C_\mu (1 + g_m R'_2) \tag{4.61c}$$

$$R'_1 = \frac{R_1 R_f}{R_1 + R_f} \tag{4.61d}$$

$$R'_2 = \frac{r_o R_2 R_f}{r_o R_2 + r_o R_f + R_2 R_f} \tag{4.61e}$$

Since the backward transmission of a feedback network is much larger than that of the transistor, (4.53b) applies. The closed-loop current gain of the amplifier becomes

$$w(s) = \frac{I_2}{I_s} = \frac{A_0 s_0}{s + s_0 (1 + A_0 R_2 / R_f)} \tag{4.62}$$

Example 4.4 As a numerical example, let

$$g_m = 0.4 \,\text{mho} \quad r_x = 50 \,\Omega$$

$$r_\pi = 250 \,\Omega \quad C_\pi = 195 \,\text{pF}$$

$$C_\mu = 5 \,\text{pF} \quad r_o = 50 \,\text{k}\Omega \tag{4.63}$$

$$R_1 = 10 \,\text{k}\Omega \quad R_f = 40 \,\text{k}\Omega$$

$$R_2 = 4 \,\text{k}\Omega$$

From (4.61) we have

$$R_1' = 8\,\text{k}\Omega \quad R_2' = 3.39\,\text{k}\Omega$$

$$C = 6.98\,\text{nF} \quad s_0 = 5.9 \cdot 10^5\,\text{rad/s} \tag{4.64}$$

$$A_0 = 81.69$$

The closed-loop current gain from (4.62) is

$$w(s) = \frac{81.69}{p + 9.169} \tag{4.65}$$

where $p = s/s_0$ and $\beta = -R_2/R_f = -0.1$. The midband feedback factor is 9.169, and it reduces the midband current gain without feedback by a factor of 9.169.

From (4.19), the approximate midband input admittance facing the current source is given by

$$Y_{\text{in}} = (y_{11} + G_1)(1 - \mu\beta) = \left(\frac{1}{r_x + r_\pi} + G_f + G_1 \right)(1 - \mu\beta)$$

$$= 31.71 \cdot 10^{-3}\,\text{mho or } 32\,\Omega \tag{4.66a}$$

and the approximate output admittance looking into the two terminals of R_2, including R_2, is found to be

$$Y_{\text{out}} = (y_{22} + G_2)(1 - \mu\beta) = \left(\frac{1}{r_0} + G_f + G_2 \right)(1 - \mu\beta)$$

$$= 2.71 \cdot 10^{-3}\,\text{mho or } 370\,\Omega \tag{4.66b}$$

Thus, for the parallel-parallel feedback, the midband input and output impedances are decreased by a factor of 9.169 from their values without feedback. We remark that the unilateralized hybrid-pi equivalent model is not valid for the calculations of the output impedance at high frequencies, and we must use the original model of Fig. 4.10 for such calculations.

In the situation where there are several amplifier stages in cascade, as shown in Fig. 4.19, the same technique can be applied to identify $\mu(s)$ and $\beta(s)$. The open-loop current gain $\mu(s)$, for example, can be determined by the network of Fig. 4.20. However, in invoking Miller effect to decouple each amplifier stage, as shown in Fig. 4.18, only the last stage is directly applicable, because the equivalent terminations for the first two stages are not purely resistive. However, we can usually assume that Fig. 4.18 is a valid approximation for each amplifier stage.

Figure 4.19 A three-transistor voltage-shunt feedback amplifier.

Figure 4.20 The equivalent network used to calculate the open-loop current gain of the amplifier of Fig. 4.19.

Under this assumption, it is straightforward to demonstrate that the open-loop current gain can be expressed as (see Prob. 4.2)

$$\mu(s) = \frac{I_2}{I_s} = \frac{A_1 A_2 A_3 R_f}{(1 + s/s_1)(1 + s/s_2)(1 + s/s_3)(R_2 + R_f)} \tag{4.67}$$

where A_k and s_k ($k = 1, 2, 3$) have the same significance as A_0 and s_0 in (4.61) for a single stage and are given by

$$A_k = \frac{g_{mk} \hat{R}_k r_{\pi k}}{r_{xk} + \hat{R}_k + r_{\pi k}} \tag{4.68a}$$

$$s_k = \frac{r_{xk} + \hat{R}_k + r_{\pi k}}{C_k r_{\pi k}(r_{xk} + \hat{R}_k)} \tag{4.68b}$$

in which $\hat{R}_1 = R_1 R_f/(R_1+R_f)$, $\hat{R}_2 = R_3$, and $\hat{R}_3 = R_4$. In the above expressions, the output impedance r_o of a common-emitter stage has been ignored, because it is usually large. Otherwise, it should be included in R_3, R_4, and R_f at the output port. C_k denotes the total capacitance including the capacitance due to the Miller effect.

4.2.3 Series-Parallel Feedback

The amplifier of Fig. 4.21 makes use of series-parallel or voltage-series feedback by connecting the second collector to the first emitter through the voltage divider

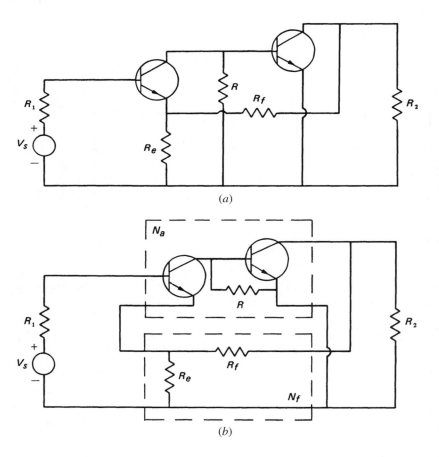

(a)

(b)

Figure 4.21 (a) A series-parallel or voltage-series feedback amplifier, and (b) its series-parallel or voltage-series feedback representation.

$R_f R_e$. As shown in Fig. 4.21b, the basic amplifier is represented by N_a and the feedback network by N_f, whereas the composite two-port is denoted by N. Let h_{aij}, h_{fij}, and h_{ij} be the h-parameters of N_a, N_f, and N, respectively. Applying the Brune tests of Fig. 4.4c to the network of Fig. 4.21b shows that the conditions are *not* satisfied, meaning that $h_{ij} \neq h_{aij} + h_{fij}$. To confirm this assertion, we compute the short-circuit input impedances h_{11}, h_{a11}, and h_{f11} by assuming identical transistors with $h_{re} = 0$ and $h_{oe} = 0$. The results are given by

$$h_{11} = h_{ie} + \frac{(1 + h_{fe})R_e R_f}{(R_e + R_f)} \tag{4.69a}$$

$$h_{a11} = h_{ie} \tag{4.69b}$$

$$h_{f11} = \frac{R_e R_f}{R_e + R_f} \tag{4.69c}$$

In fact, the hybrid matrix of the composite two-port network N is found to be (see Prob. 4.3)

$$\mathbf{H} = \begin{bmatrix} h_{ie} + (1 + h_{fe})R_e' & \dfrac{R_e}{R_e + R_f} \\ -\dfrac{h_{fe}^2 R}{R + h_{ie}} - \dfrac{(1 + h_{fe})R_e'}{R_f} & \dfrac{1}{R_e + R_f} \end{bmatrix} \tag{4.70a}$$

$$\approx \begin{bmatrix} h_{ie} + (1 + h_{fe})R_e' & \dfrac{R_e}{R_e + R_f} \\ -\dfrac{h_{fe}^2 R}{R + h_{ie}} & \dfrac{1}{R_e + R_f} \end{bmatrix} \tag{4.70b}$$

where

$$R_e' = \frac{R_e R_f}{R_e + R_f} \tag{4.70c}$$

The approximation is obtained under the assumption that $h_{fe}^2 R/(R + h_{ie}) \gg (1 + h_{fe})R_e'/R_f$, which is usually valid for practical amplifiers.

Even though the connection failed the validity tests, formulas (4.15) and (4.16) are still valid. Thus, with $G_2 = 1/R_2$, the open-loop transfer admittance is calculated as

$$
\mu = \frac{I_2}{V_s}\bigg|_{h_{12}=0} = \frac{h_{21}G_2}{(h_{11}+R_1)(h_{22}+G_2)}
$$

$$
= -\frac{G_2[h_{fe}^2 R(R_e+R_f)+R_e(1+h_{fe})(R+h_{ie})]}{(R+h_{ie})[h_{ie}+(1+h_{fe})R'_e+R_1][1+G_2(R_e+R_f)]} \tag{4.71a}
$$

$$
\approx -\frac{h_{fe}^2 RG_2(R_e+R_f)}{(R+h_{ie})[h_{ie}+(1+h_{fe})R'_e+R_1][1+G_2(R_e+R_f)]} \tag{4.71b}
$$

the approximate formula (4.71b) being computed from (4.70b), and the transfer impedance of the feedback network N_f is found to be

$$
\beta = \frac{V_f}{I_2} = \frac{h_{12}}{G_2} = \frac{R_e R_2}{R_e+R_f} \tag{4.72a}
$$

where the feedback voltage V_f is defined by

$$
V_f = -h_{12}V_2 = -\frac{R_e V_2}{R_e+R_f} \tag{4.72b}
$$

Finally, the closed-loop transfer admittance is determined from (4.3) and is given by (see Prob. 4.23)

$$
w = \frac{I_2}{V_s}
$$

$$
= \frac{-h_{fe}^2 R(R_e+R_f)+R_e(1+h_{fe})(R+h_{ie})}{(R+h_{ie})\{(R_2+R_f)[R_1+h_{ie}+(1+h_{fe})R_e]+(R_1+h_{ie})R_e\}+h_{fe}^2 RR_e R_2}
$$

$$
\approx -\frac{h_{fe}^2 R(R_e+R_f)}{(R+h_{ie})[h_{ie}+(1+h_{fe})R'_e+R_1](R_2+R_e+R_f)+h_{fe}^2 RR_e R_2} \tag{4.73}
$$

As in (4.71b), the approximate formula was obtained by using (4.70b).

Like the series-series feedback, the above analysis can also be explained as follows: The unilateral forward path is represented by the controlled current source $h_{21}I_1$, whereas the unilateral feedback path is represented by the controlled voltage source $h_{12}V_2$. Therefore, the open-loop transfer admittance μ can be computed directly from the network of Fig. 4.22 with reasonable accuracy. With $h_{re}=0$

Figure 4.22 The network used to calculate the open-loop transfer admittance of the feedback amplifier of Fig. 4.21.

and $h_{oe} = 0$, the equivalent network of Fig. 4.22 is presented in Fig. 4.23, whose indefinite-admittance matrix is given by

$$\mathbf{Y} = \begin{bmatrix} G_1 + \dfrac{1}{h_{ie}} & -\dfrac{1}{h_{ie}} & 0 & 0 & -G_1 \\[2mm] -\alpha - \dfrac{1}{h_{ie}} & G'_e + \dfrac{1}{h_{ie}} + \alpha & 0 & 0 & -G'_e \\[2mm] \alpha & -\alpha & G' & 0 & -G' \\[2mm] 0 & 0 & \alpha & G'_2 & -G'_2 - \alpha \\[2mm] -G_1 & -G'_e & -G' - \alpha & -G'_2 & G_1 + G'_e + G'_2 + G' + \alpha \end{bmatrix}$$

(4.74)

where $G_1 = 1/R_1$, $G'_e = 1/R'_e$, $G'_2 = G_2 + 1/(R_e + R_f)$, $G' = 1/R + 1/h_{ie}$, and $\alpha = h_{fe}/h_{ie}$. By applying (2.94), the open-loop transfer admittance can be written as

$$\mu = \frac{I_2}{V_s} = -\frac{V_2 G_2}{I_s R_1} = -G_1 G_2 z_{14,55} = -G_1 G_2 \frac{Y_{14,55}}{Y_{55}}$$

$$= -\frac{G_1 G_2 G'_e \alpha^2}{G'_2 (1/R + 1/h_{ie})[G'_e(G_1 + 1/h_{ie}) + G_1(\alpha + 1/h_{ie})]}$$

$$= -\frac{h_{fe}^2 R G_2}{[h_{ie} + (1 + h_{fe})R'_e + R_1][1/(R_f + R_e) + G_2](h_{ie} + R)} \qquad (4.75)$$

confirming (4.71).

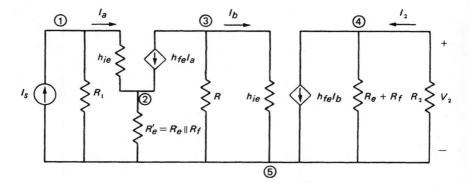

Figure 4.23 An equivalent network of Fig. 4.22.

Example 4.5 In Fig. 4.24, assume that the two transistors are identical, with

$$h_{ie} = 1.1\,\text{k}\Omega \quad h_{fe} = 50 \quad h_{re} = h_{oe} = 0 \qquad (4.76a)$$

After the biasing and coupling circuitry have been removed, the network is reduced to that of Fig. 4.21, with R denoting the parallel combination of the 10-kΩ, 47-kΩ, and 33-kΩ resistors. The corresponding element values are

$$R_f = 4.7\,\text{k}\Omega \quad R_e = 100\,\Omega$$
$$R_1 = 0 \quad R_2 = 4.7\,\text{k}\Omega \qquad (4.76b)$$
$$R'_e = 98\,\Omega \quad R = 6.6\,\text{k}\Omega$$

Substituting these in (4.71) and (4.72) yields

$$\mu = -177.55\,\text{mA/V} \qquad (4.77a)$$
$$\beta = 98\,\Omega \qquad (4.77b)$$

The closed-loop transfer admittance becomes

$$w = \frac{\mu}{1 - \mu\beta} = -9.65\,\text{mA/V} \qquad (4.78)$$

The corresponding open-loop voltage gain is related to μ by

$$\mu_v = -\frac{I_2 R_2}{V_s} = -\mu R_2 = 834.5 \qquad (4.79)$$

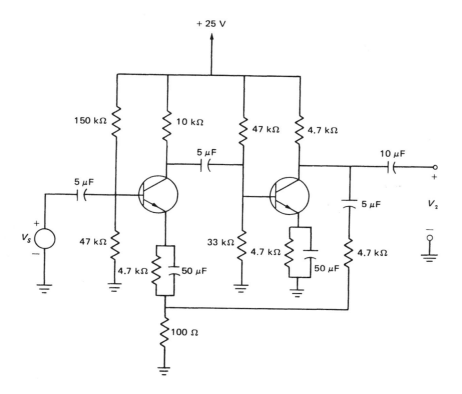

Figure 4.24 A voltage-series feedback amplifier together with its biasing and coupling circuitry.

giving the closed-loop voltage gain

$$w_V = \frac{\mu_V}{1 - \mu_V \beta_V} = 45.38 \tag{4.80}$$

where $\beta_V = -R_e/(R_e + R_f) = \frac{1}{48}$. This gain value is to be compared with the approximate solution $w_V = 48$ obtained by letting $\mu_V \to \infty$.

At high frequencies, we use the hybrid-pi equivalent model for the two transistors of Fig. 4.22. We assume that the resistance r_π of the second stage is small, so that the first-stage load is approximately resistive, as shown in Fig. 4.25a, with

$$R' = \frac{R(r_{x2} + r_{\pi 2})}{R + r_{x2} + r_{\pi 2}} \tag{4.81}$$

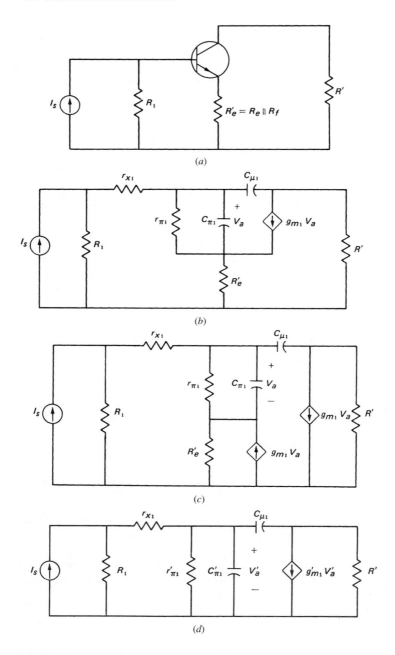

Figure 4.25 (*a*) The network approximating the first-stage of the amplifier of Fig. 4.22. (*b*) The equivalent network of (*a*). (*c*) An equivalent representation of (*b*). (*d*) An approximate representation of (*c*).

where the subscript 2 denotes the transistor parameters of the second stage. The equivalent network of Fig. 4.25a is presented in Fig. 4.25b, which can be transformed equivalently to the network of Fig. 4.25c with the subscript 1 denoting the transistor parameters of the first stage. The input impedance of the one-port formed by the mid-four branches $r_{\pi 1}$, $C_{\pi 1}$, R'_e, and $g_{m1}V_a$ is given by

$$Z' = \frac{r_{\pi 1}(1 + g_{m1}R'_e)}{C_{\pi 1}r_{\pi 1}s + 1} + R'_e \approx \frac{r_{\pi 1}(1 + g_{m1}R'_e)}{C_{\pi 1}r_{\pi 1}s + 1} \qquad (4.82)$$

which can be realized by the parallel combination of

$$r'_{\pi 1} = r_{\pi 1}(1 + g_{m1}R'_e) \qquad (4.83a)$$

$$C'_{\pi 1} = \frac{C_{\pi 1}}{1 + g_{m1}R'_e} \qquad (4.83b)$$

as indicated in Fig. 4.25d, where the effective transconductance becomes

$$g'_{m1} = \frac{g_{m1}}{1 + g_{m1}R'_e} \qquad (4.83c)$$

By using this model for the first stage in Fig. 4.22 and applying the Miller effect, the amplifier of Fig. 4.22 can be represented by the network of Fig. 4.26 with $r_{o1} = r_{o2} = \infty$, $R'_2 = R_2(R_f + R_e)/(R_2 + R_f + R_e)$, and

$$C''_{\pi 1} = C'_{\pi 1} + C_{\mu 1}(1 + g'_{m1}R') \qquad (4.84a)$$

$$C''_{\pi 2} = C_{\pi 2} + C_{\mu 2}(1 + g_{m2}R'_2) \qquad (4.84b)$$

From Fig. 4.26, the open-loop transfer admittance $\mu(s)$ is determined as follows:

$$\begin{aligned}
\mu(s) &= \frac{I_2}{V_s} = \frac{I_2}{I'_2} \cdot \frac{I'_2}{V} \cdot \frac{V}{I_s} \cdot \frac{I_s}{V_s} \\[2mm]
&= \frac{R_f + R_e}{R_2 + R_f + R_e} \cdot \frac{-g'_{m1}A_2R}{1 + s/s_2} \cdot \frac{R_1A_1/g'_{m1}}{1 + s/s_1} \cdot \frac{1}{R_1} \\[2mm]
&= \frac{A_1A_2(R_e + R_f)R}{(1 + s/s_1)(1 + s/s_2)(R_f + R_2 + R_e)}
\end{aligned} \qquad (4.85)$$

Figure 4.26 A unilateralized approximation of the network of Fig. 4.22.

where

$$A_1 = \frac{g'_{m1} r'_{\pi 1}}{R_1 + r_{x1} + r'_{\pi 1}} \qquad (4.86a)$$

$$A_2 = \frac{g_{m2} r_{\pi 2}}{R + r_{x2} + r_{\pi 2}} \qquad (4.86b)$$

$$s_1 = \frac{r'_{\pi 1} + r_{x1} + R_1}{C''_{\pi 1} r'_{\pi 1} (r_{x1} + R_1)} \qquad (4.86c)$$

$$s_2 = \frac{r_{\pi 2} + r_{x2} + R}{C''_{\pi 2} r_{\pi 2} (r_{x2} + R)} \qquad (4.86d)$$

The transfer impedance of the unilateral feedback path is given by (4.72a). Thus, the closed-loop transfer admittance is obtained as

$$w(s) = \frac{I_2}{V_s} = \frac{\mu(s)}{1 - \mu(s)\beta(s)} = -\frac{(R_e + R_f)A}{s^2 + (s_1 + s_2)s + s_1 s_2 + R_e R_2 A} \qquad (4.87a)$$

where

$$A = \frac{A_1 A_2 R s_1 s_2}{R_f + R_e + R_2} \qquad (4.87b)$$

As indicated in (4.19), the input impedance facing the voltage source of Fig. 4.21 is increased by a factor of $1 - \mu(s)\beta(s)$ from its value when the output port is short-circuited, whereas the output impedance looking into the two terminals of the load R_2, including R_2, is decreased by the same factor from its value when the input port is open-circuited.

Example 4.6 In Fig. 4.24, assume that the two transistors are identical, with

$$g_m = 0.4 \, \text{mho} \quad r_x = 50 \, \Omega$$
$$r_\pi = 250 \, \Omega \quad C_\pi = 195 \, \text{pF} \tag{4.88}$$
$$C_\mu = 5 \, \text{pF} \quad r_o = \infty$$

Then from (4.81), (4.83), and (4.84), we have

$$R' = 287 \, \Omega \quad r'_{\pi 1} = 10.05 \, \text{k}\Omega$$
$$C'_{\pi 1} = 4.85 \, \text{pF} \quad g'_{m1} = 9.95 \cdot 10^{-3} \, \text{mho} \tag{4.89a}$$
$$C''_{\pi 1} = 24.13 \, \text{pF} \quad C''_{\pi 2} = 4.95 \, \text{nF}$$

yielding, from (4.86),

$$A_1 = 9.9 \cdot 10^{-3} \quad A_2 = 14.49 \cdot 10^{-3}$$
$$s_1 = 83.3 \cdot 10^7 \, \text{rad/s} \quad s_2 = 8.39 \cdot 10^5 \, \text{rad/s} \tag{4.89b}$$

Substituting these in (4.85) gives the open-loop transfer admittance

$$\mu(s) = -\frac{474.95}{(p + 992.85)(p + 1)} \tag{4.90}$$

where $p = s/s_2$. The closed-loop transfer admittance is computed from (4.87) as

$$w(s) = \frac{474.95}{p^2 + 993.85p + 47,499} \tag{4.91}$$

where $\beta = 98 \, \Omega$. The closed-loop voltage gain is related to $w(s)$ by

$$w_V(s) = -R_2 w(s) = \frac{22.33 \cdot 10^5}{p^2 + 993.85p + 47,499} \tag{4.92}$$

with $\beta_V = -\frac{1}{48}$, giving the midband voltage gain with feedback as $w_V(0) = 47$. This corresponds to a reduction from its midband value without feedback $\mu_V(0) = 2248$, by a factor of 47.84, the feedback factor.

As shown in (4.19), the midband input impedance with feedback is increased from its value without feedback by the feedback factor,

$$R_{\text{in}} \approx (R_1 + R_e + r_{x1} + r_{\pi 1}) [1 - \mu(0)\beta(0)] = 19.14 \, \text{k}\Omega \qquad (4.93a)$$

whereas the output impedance with feedback is decreased from its value without feedback by the same factor:

$$R_{\text{out}} \approx \frac{R_2 R_f / (R_2 + R_f)}{1 - \mu(0)\beta(0)} = 49 \, \Omega \qquad (4.93b)$$

Thus, the series-parallel configuration can be used to realize approximately a voltage-controlled voltage source.

4.2.4 Parallel-Series Feedback

The amplifier of Fig. 4.27 shows two transistors in cascade with feedback from the second emitter to the first base through the feedback resistor R_f. As demonstrated in Fig. 4.27b, it conforms with the parallel-series or current-shunt feedback. Let g_{aij}, g_{fij}, and g_{ij} be the g-parameters of the basic amplifier N_a, the feedback network N_f, and the composite two-port network N of Fig. 4.27b, respectively. Like the series-parallel topology of Fig. 4.21b, the Brune tests of Fig. 4.4d, when applied to the network of Fig. 4.27b, are *not* satisfied, showing that $g_{ij} \neq g_{aij} + g_{fij}$. However, formulas (4.15) and (4.16) are still valid. Thus, with $G_1 = 1/R_1$, we have

$$\mu = \left. \frac{V_2}{I_s} \right|_{g_{12}=0} = \frac{g_{21} R_2}{(g_{11} + G_1)(g_{22} + R_2)} \qquad (4.94)$$

$$\beta = \frac{I_f}{V_2} = \frac{g_{12}}{R_2} \qquad (4.95a)$$

where the feedback current I_f is defined by

$$I_f = -g_{12} I_2 \qquad (4.95b)$$

Assuming that the two transistors are identical, with $h_{re} = 0$ and

$$\frac{1}{h_{oe}} \gg R \quad R_f \gg R_e \qquad (4.96)$$

the g-parameters g_{ij} of the composite two-port network N are found to be (see Prob. 4.5)

$$
\mathbf{G} = \begin{bmatrix} \dfrac{1}{h_{ie}} + \dfrac{h_{ie} + R_e + R'(1 + h_{fe}R_e/h_{ie})}{(R' + h_{ie})(R_e + R_f) + R_e R_f} & -\dfrac{R_e(R' + h_{ie})}{\begin{array}{c}(R' + h_{ie})(R_e + R_f)\\ + R_e R_f\end{array}} \\[2em] \dfrac{h_{fe}\big[h_{fe}R'(R_e + R_f)/h_{ie} + R_e\big]}{h_{oe}\big[(R' + h_{ie})(R_e + R_f) + R_e R_f\big]} & \dfrac{1}{h_{oe}}\left[1 + \dfrac{R_e R_f h_{fe}}{\begin{array}{c}(R' + h_{ie})(R_e + R_f)\\ + R_e R_f\end{array}}\right] \end{bmatrix}
$$

$$
\approx \begin{bmatrix} \dfrac{1}{h_{ie}} + \dfrac{h_{ie} + R_e + R(1 + h_{fe}R_e/h_{ie})}{R_f(R + h_{ie} + R_e)} & -\dfrac{R_e(R + h_{ie})}{R_f(R + h_{ie} + R_e)} \\[2em] \dfrac{h_{fe}(h_{fe}RR_f/h_{ie} + R_e)}{R_f(R + h_{ie} + R_e)h_{oe}} & \dfrac{1}{h_{oe}}\left[1 + \dfrac{R_e R_f h_{fe}}{R_f(R + h_{ie} + R_e)}\right] \end{bmatrix}
$$

$$(4.97)$$

where

$$
R' = \frac{R}{1 + Rh_{oe}} \approx R \tag{4.98a}
$$

$$
R'_e = \frac{R_e R_f}{R_e + R_f} \approx R_e \tag{4.98b}
$$

In the limit, as h_{oe} approaches zero, the open-loop transfer impedance becomes

$$
\mu = \frac{h_{fe}[h_{fe}R(R_e + R_f) + R_e h_{ie}]R_2 q}{[(1 + G_1 h_{ie})q + h_{ie}(h_{ie} + R_e + R) + h_{fe}RR_e](q + R_e R_f h_{fe})} \tag{4.99a}
$$

where $q = (R_e + R_f)(R + h_{ie}) + R_e R_f \approx R_f(R + R_e + h_{ie})$. From $(4.95a)$ the transfer admittance of the feedback network N_f is obtained as

$$
\beta = -\frac{R_e G_2(R + h_{ie})}{q} \tag{4.99b}
$$

Finally, the closed-loop transfer impedance is determined from (4.3) and is given by (see Prob. 4.24)

$$w = \frac{h_{fe}[h_{fe}R(R_e + R_f) + h_{ie}R_e]R_2}{(R + h_{ie})[(R_e + R_f)(1 + G_1 h_{ie}) + h_{ie}] + R_e(1 + h_{fe})(R_f + h_{ie} + h_{fe}R + G_1 R_f h_{ie})} \tag{4.100}$$

We remark that in deriving (4.100) there was a cancellation of the common factor q in the denominator of (4.3) or (4.14).

Example 4.7 In Fig. 4.27, assume that the two transistors are identical, with

$$h_{ie} = 1.1\,\text{k}\Omega \quad h_{fe} = 50$$

$$h_{re} = h_{oe} = 0 \quad R = 3\,\text{k}\Omega$$

$$R_e = 50\,\Omega \quad R_f = 1.2\,\text{k}\Omega \tag{4.101}$$

$$R_1 = 1.2\,\text{k}\Omega \quad R_2 = 500\,\Omega$$

For illustrative purposes, we first compute the inverse hybrid matrix $\mathbf{G}$ by assuming that $h_{oe} \neq 0$, since for $h_{oe} = 0$, $\mathbf{G}$ does not exist. Substituting (4.101) in (4.97) yields

$$\mathbf{G} = \begin{bmatrix} 3.03 \cdot 10^{-3} & -0.04 \\ \dfrac{1.644}{h_{oe}} & \dfrac{1.58}{h_{oe}} \end{bmatrix} \tag{4.102}$$

Thus, from (4.94) and (4.95a) we have

$$\mu = 135\,\text{k}\Omega \tag{4.103a}$$

$$\beta = -80\,\mu\text{mho} \tag{4.103b}$$

giving the closed-loop transfer impedance

$$w = 11.56\,\text{k}\Omega \tag{4.103c}$$

The closed-loop current gain w_I is related to w by

$$w_I = -\frac{w}{R_2} = -23.13 \tag{4.103d}$$

a reduction from its open-loop current gain $\mu_I = -270$ by a factor of 11.67, the feedback factor. Finally, the closed-loop voltage gain w_v is related

Figure 4.27 (*a*) A transistor feedback amplifier and (*b*) its parallel-series or current-shunt feedback representation.

to w by

$$w_V = \frac{w}{R_1} = 9.64 \qquad (4.103e)$$

with the open-loop voltage gain $\mu_V = 112.50$.

As before, at high frequencies, we use the hybrid-pi equivalent model for the two transistors of Fig. 4.27. The open-loop transfer function $\mu(s)$ can be computed by the network of Fig. 4.28. Using the subscripts 1 and 2 to distinguish the parameters of the two transistors, the second transistor, as shown in (4.83)

Figure 4.28 The transistor network used to calculate the open-loop transfer impedance of the amplifier of Fig. 4.27.

Figure 4.29 A unilateralized approximation of the second transistor of the amplifier of Fig. 4.28.

and (4.84), can be represented by the unilateralized model of Fig. 4.29 with $r_{o1} = r_{o2} = \infty$ and

$$C''_{\pi2} = C'_{\pi2} + C_{\mu2}(1 + g'_{m2}R_2) \qquad (4.104a)$$

$$g'_{m2} = \frac{g_{m2}}{1 + g_{m2}R'_e} \qquad (4.104b)$$

$$r'_{\pi2} = r_{\pi2}(1 + g_{m2}R'_e) \qquad (4.104c)$$

$$C'_{\pi2} = \frac{C_{\pi2}}{1 + g_{m2}R'_e} \qquad (4.104d)$$

where R'_e is defined in (4.98b). The first stage sees approximately a parallel RC load, as indicated in Fig. 4.30a, because the effect of r_{x2} is negligible for R,

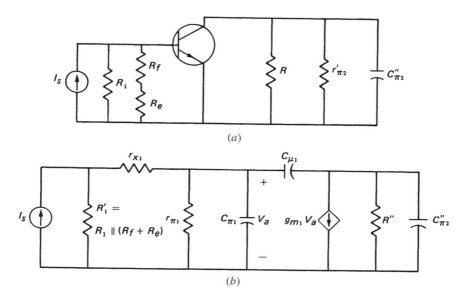

Figure 4.30 (*a*) An approximation of the first stage of the amplifier of Fig. 4.28, and (*b*) its equivalent network.

$r'_{\pi 2} \gg r_{x2}$. The equivalent network of Fig. 4.30*a* is presented in Fig. 4.30*b*. By using this RC load for $Y_2(s)$ in (4.37), the admittance facing the capacitor $C_{\pi 1}$ can be written as

$$Y_{\text{eq}}(s) = sC_{\mu 1}\left(1 + \frac{g_{m1}}{Y_2}\right) = sC_{\mu 1} + \frac{1}{1/(R'' g_{m_1} C_{\mu 1} s) + C''_{\pi 2}/g_{m1}C_{\mu 1}}\right)$$

(4.105)

which can be realized by the network as shown in Fig. 4.31 with

$$R_\alpha = \frac{C''_{\pi 2}}{g_{m1}C_{\mu 1}}$$

(4.106*a*)

$$C_\alpha = R'' g_{m1} C_{\mu 1}$$

(4.106*b*)

where

$$R'' = \frac{Rr'_{\pi 2}}{R + r'_{\pi 2}}$$

(4.106*c*)

Figure 4.31 A unilateralized approximation of the network of Fig. 4.30.

Figure 4.32 A unilateralized approximation of the amplifier of Fig. 4.28.

The network of Fig. 4.28 can now be represented approximately by that of Fig. 4.32 for the computation of the open-loop transfer impedance $\mu(s)$ with

$$C_1 = C_{\pi 1} + C_{\mu 1} \tag{4.107a}$$

$$R_1' = \frac{R_1(R_e + R_f)}{R_1 + R_e + R_f} \tag{4.107b}$$

The open-loop transfer impedance is found to be

$$\mu(s) = \frac{V_2'}{I_s} = \frac{V_2'}{V_b} \frac{V_b}{V} \frac{V}{I_s} = \frac{A_0 R_2}{a_2 s^2 + a_1 s + a_0} \tag{4.108}$$

where

$$A_0 = g_{m1} g_{m2}' R_1' R'' \tag{4.109a}$$

$$a_2 = C_1 C_{\pi 2}'' R'' (R_1' + r_{x1}) \tag{4.109b}$$

$$a_1 = C_{\pi 2}'' R'' + (R_1' + r_{x1}) \left(C_1 + R'' g_{m1} C_{\mu 1} + \frac{R'' C_{\pi 2}''}{r_{\pi 1}} \right) \quad (4.109c)$$

$$a_0 = 1 + \frac{R_1' + r_{x1}}{r_{\pi 1}} \quad (4.109d)$$

The transfer admittance of the feedback network N_f is the same as that given in (4.95) except that h_{ie} in g_{12}, as given in (4.97), is replaced by $r_{x2} + r_{\pi 2}$, yielding

$$\beta = -\frac{R_e/R_2}{R_e + R_f \left[1 + R_e/(R + r_{x2} + r_{\pi 2}) \right]} \quad (4.110a)$$

For $R \gg R_e$, β is reduced to

$$\beta \approx -\frac{R_e}{R_2(R_e + R_f)} \quad (4.110b)$$

The closed-loop transfer impedance is found to be

$$w(s) = \frac{A_0 R_2}{a_2 s^2 + a_1 s + a_0 + A_0 R_e'/R_f} \quad (4.111)$$

We illustrate the above results by considering the following numerical example.

Example 4.8 In Fig. 4.27a, assume that the two transistors are identical, with

$$g_m = 0.4 \, \text{mho} \qquad r_x = 50 \, \Omega$$
$$r_\pi = 250 \, \Omega \qquad C_\pi = 195 \, \text{pF} \qquad (4.112a)$$
$$C_\mu = 5 \, \text{pF} \qquad r_0 = \infty$$

The other parameters of the amplifier are given by

$$R_1 = 1.2 \, \text{k}\Omega \quad R_2 = 500 \, \Omega$$
$$R_f = 1.2 \, \text{k}\Omega \quad R_e = 50 \, \Omega \qquad (4.112b)$$
$$R = 3 \, \text{k}\Omega$$

From (4.98b), (4.104), (4.106c), and (4.107), we have

$$R'_e = 48\,\Omega \qquad C'_{\pi 2} = 9.65\,\text{pF}$$
$$g'_{m2} = 0.0198\,\text{mho} \qquad r'_{\pi 2} = 5.05\,\text{k}\Omega$$
$$C''_{\pi 2} = 64.15\,\text{pF} \qquad R'' = 1.88\,\text{k}\Omega \tag{4.113}$$
$$C_1 = 200\,\text{pF} \qquad R'_1 = 612.2\,\Omega$$

Substituting these in (4.109) yields

$$A_0 = 9115.41 \qquad a_2 = 1.597 \cdot 10^{-14}$$
$$a_1 = 30.62 \cdot 10^{-7} \qquad a_0 = 3.65 \tag{4.114}$$

The open-loop transfer impedance becomes

$$\mu(p) = \frac{5707.8\,R_2}{(p + 0.1199)(p + 19.056)} \tag{4.115}$$

where $p = s/10^7$. The transfer admittance of the feedback network from (4.110b) is obtained as

$$\beta \approx -\frac{R_e}{R_2(R_e + R_f)} = -80\,\mu\text{mho} \tag{4.116}$$

giving the closed-loop transfer impedance

$$w(s) = \frac{5707.8\,R_2}{p^2 + 19.18p + 230.61} = -w_I(s)\,R_2 \tag{4.117}$$

where $w_I(s)$ denotes the closed-loop current gain I_2/I_s (see Fig. 4.27a). Thus, the midband current gain with feedback is

$$w_I(0) = -24.75 \tag{4.118}$$

a reduction from the midband value without feedback $\mu_I(0) = -2498.1$ by the feedback factor $1 - \mu_I(0)\beta_I(0) = 100.9$, where $\beta_I(0) = 0.04$.

The closed-loop voltage gain $w_V(s)$ is related to $w(s)$ by $w(s)/R_1$. The midband voltage gain with feedback is

$$w_V(0) = 10.31 \tag{4.119}$$

This corresponds to a reduction from the midband value without feedback, $\mu_V(0) = 1040.9$, by the same factor $1 - \mu_V(0)\beta_V(0) = 100.9$ with $\beta_V = 0.096$.

As indicated in (4.19), the midband input admittance facing the current source of Fig. 4.27a is increased by the factor $1 - \mu(0)\beta(0)$ from its value without feedback, which corresponds to the open-circuiting of the output port,

$$G_{\text{in}} = \left(\frac{1}{R_1} + \frac{1}{r_{x1} + r_{\pi 1}} + \frac{1}{R_e + R_f} \right) [1 - \mu(0)\beta(0)] \approx \frac{1}{2} \text{ mho}$$

$$(4.120a)$$

whereas the output impedance is increased from its value without feedback, which corresponds to the short-circuiting of the input port, by the same factor:

$$R_{\text{out}} = (r_0 + R_2 + R'_e)[1 - \mu(0)\beta(0)] \qquad (4.120b)$$

where r_0 is the common-emitter transistor output resistance. Since r_0 is on the order of $10{\sim}100\,\text{k}\Omega$, the output resistance R_{out} is very high. The parallel-series configuration is therefore attractive for situations where low input and high output impedances are required, such as for cascading and stagger tuning to obtain multistage amplifiers with a large gain-bandwidth product.

4.3 GENERAL FEEDBACK THEORY

Up to now, we have used the ideal feedback model of Fig. 4.1 to study the properties of feedback amplifiers. The model is useful only if we can separate a feedback amplifier into two blocks: the basic amplifier $\mu(s)$ and the feedback network $\beta(s)$. For this we presented four useful feedback configurations and demonstrated in each case by a practical example how to identify $\mu(s)$ and $\beta(s)$. The four configurations are essentially the series-series, parallel-parallel, series-parallel, and parallel-series connections of two-port networks. In practice, we often run into feedback amplifiers that cannot be classified as being in one of the above four categories. As indicated earlier, a general feedback amplifier may contain an input coupling network and an output coupling network, as shown in Fig. 4.5. A typical example is the bridge feedback configuration of Fig. 4.6.

The calculation of $\mu(s)$ and $\beta(s)$, as demonstrated in the foregoing, is based on the assumption that the complete equivalent networks of the active devices and the passive two-port networks are known. Therefore they cannot be measured experimentally, because the forward transmission of the passive network is included in the basic amplifier, whereas the backward transmission of the active two-port network is lumped with the feedback network. Since the zeros of the feedback factor $1 - \mu(s)\beta(s)$ are poles of the closed-loop transfer function, they are also the natural frequencies of the feedback amplifier. If $\mu(s)$ and $\beta(s)$ are known, the roots of $1 - \mu(s)\beta(s)$ can readily be computed explicitly with the aid of a computer if necessary, and the stability problem can then be settled directly. However, for a physical amplifier, there remains the difficulty of getting an accurate formulation and representation of the network itself, because every equivalent network is, to a greater or lesser extent, an idealization of the physical reality. What is really needed is an equivalent measurement of the feedback factor that provides some kind of experimental verification that the system is stable and will remain so under certain prescribed conditions. Thus, the ideal feedback model is not an adequate representation of a practical feedback amplifier. In the following, we shall develop Bode's feedback theory, which is applicable to general network configuration and avoids the necessity of identifying $\mu(s)$ and $\beta(s)$.

Bode's feedback theory is based on the concept of return difference, which is defined in terms of network determinants. We shall show that the return difference is a generalization of the concept of the feedback factor of the ideal feedback model, and can be measured physically from the amplifier itself. We shall then introduce the notion of null return difference and discuss its physical significance. Among the many important properties, we shall find in the next chapter that return difference and null return difference are closely related to the sensitivity function and that they are basic to the study of the stability of the feedback amplifier. As indicated in the preceding section, feedback affects the input and output impedances. Here, we show that return difference possesses the same significance, which together with null return difference can be employed to simplify the calculation of driving-point impedance of an active network, thus observing the effects of feedback on amplifier impedance and gain.

4.3.1 The Return Difference

In the study of a feedback amplifier, we usually single out an element for particular attention. The element is generally one that is either crucial in terms of its effect on the entire system or of primary concern to the designer. The element may be the transfer function of an active device, the gain of an amplifier, or the immittance of a one-port network. For our purposes, we assume that the element x of interest is the controlling parameter of a voltage-controlled current source defined by the relation

$$I = xV \tag{4.121}$$

To focus our attention on the element x, Fig. 4.33 is the general configuration of a feedback amplifier in which the controlled source is brought out as a two-port network connected to a general four-port network, along with the input source combination of I_s and Y_1 and the load admittance Y_2.

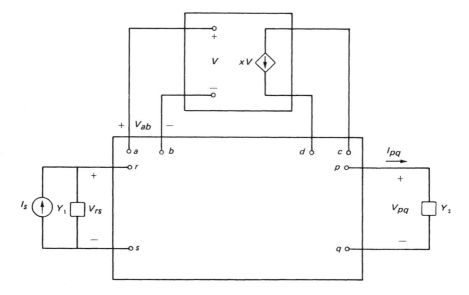

Figure 4.33 The general configuration of a feedback amplifier.

We remark that the two-port network representation of a controlled source (4.121) is quite general. It includes as a special case a one-port element characterized by its immittance. In (4.121) if the controlling voltage V is the terminal voltage of the controlled current source I, then x represents the one-port admittance. Also, the consideration of voltage-controlled current source should not be deemed to be restrictive, because the other three types of controlled sources (current-controlled current source, voltage-controlled voltage source, and current-controlled voltage source) can usually be converted to the above type by a simple transformation and by appealing to Norton's theorem if necessary. The element x in (4.121) can either be the transconductance or more generally a transfer admittance function, depending on the type of equivalent circuit being used.

We first give a mathematical definition of return difference and then discuss its physical significance.

Definition 4.1: Return difference The *return difference* $F(x)$ of a feedback amplifier with respect to an element x is the ratio of the two functional values assumed by the first-order cofactor of an element of its indefinite-admittance matrix under the condition that the element x assumes its nominal value and the condition that the element x assumes the value zero.

Let $\mathbf{Y}$ be the indefinite-admittance matrix of a feedback amplifier. As shown in Theorem 2.1, $\mathbf{Y}$ is an equicofactor matrix, all of its cofactors being equal. Thus, it is immaterial which cofactor is chosen to formulate the return difference. To emphasize the importance of the feedback element x, we express $\mathbf{Y}$ as a function of x alone for the present discussion, even though it is also a function of the complex-frequency variable s, and write $\mathbf{Y} = \mathbf{Y}(x)$. Following the convention adopted in Chap. 2 for the cofactors, $Y_{ij}(x)$ denotes the cofactor of the ith row and jth column element of $\mathbf{Y}(x)$. Then the return difference $F(x)$ with respect to the element x can be expressed as

$$F(x) \equiv \frac{Y_{ij}(x)}{Y_{ij}(0)} \tag{4.122}$$

where $Y_{ij}(0) = Y_{ij}(x)|_{x=0}$

The physical significance of the return difference will now be considered. In the network of Fig. 4.33, we label the terminals of the input, the output, the controlling branch, and the controlled source as indicated. With this designation, the element x enters the indefinite-admittance matrix $\mathbf{Y}$ in a rectangular pattern as shown below:

$$
\mathbf{Y} = \begin{array}{c}
\\ a \\ b \\ c \\ d
\end{array}
\begin{array}{cccc}
a & b & c & d \\
\left[\begin{array}{cccc}
 & & & \\
 & & & \\
x & -x & & \\
-x & x & &
\end{array}\right]
\end{array}
\tag{4.123}
$$

If in Fig. 4.33 we replace the controlled current source xV by an independent current source of x amperes and set the excitation I_s to zero, the corresponding indefinite-admittance matrix of the resulting network is simply $\mathbf{Y}(0)$. By appealing to formula (2.94), the new voltage V'_{ab} appearing at terminals a and b of the controlling branch is given by

$$V'_{ab} = x \frac{Y_{da,cb}(0)}{Y_{uv}(0)} = -x \frac{Y_{ca,db}(0)}{y_{uv}(0)} \tag{4.124}$$

Note that the current injecting point is terminal d, not c. Alternatively, we can reverse the direction of the current source by replacing x by $-x$ and apply the formula (2.94) directly.

The above manipulation of replacing the controlled current source by an independent current source and setting the excitation I_s to zero can be represented

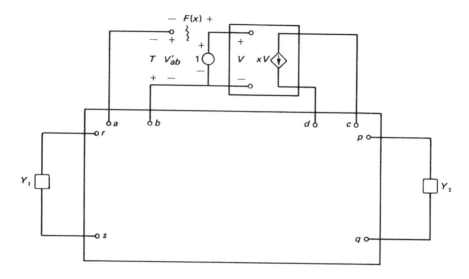

Figure 4.34 The physical interpretation of the return difference with respect to the controlling parameter of a voltage-controlled current source.

symbolically as shown in Fig. 4.34. The controlling branch is broken off as marked and a voltage source of 1 V is applied to the right of the breaking mark. This 1-V sinusoidal voltage of a fixed angular frequency produces a current of x amperes at the controlled current source. The voltage appearing at the left of the breaking mark caused by the 1-V excitation is then V'_{ab}, as indicated in Fig. 4.34. This returned voltage V'_{ab} has the same physical significance as the loop transmission $\mu\beta$ defined for the ideal feedback model of Fig. 4.1. To see this, we set the input excitation to the ideal feedback model to zero, break the forward path, and apply a unit input to the right of the break, as shown in Fig. 4.35. The signal appearing at the left of the break is precisely the loop transmission $\mu\beta$. For this reason, we introduce the concept of return ratio.

Definition 4.2: Return ratio The *return ratio* T with respect to a voltage-controlled current source $I = xV$ is the negative of the voltage appearing at the controlling branch when the controlled current source is replaced by an independent current source of x amperes and the input excitation is set to zero.

Thus, the return ratio T is simply the negative of the returned voltage V'_{ab}, that is, $T = -V'_{ab}$. With this in mind, we next compute the difference between the 1-V excitation and the returned voltage V'_{ab}. As before, we denote by $Y_{ab,cd}(x)$ the

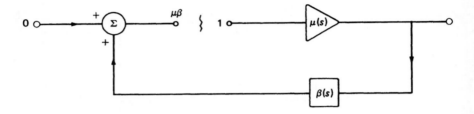

Figure 4.35 The physical interpretation of the loop transmission.

second-order cofactor of the elements of $\mathbf{Y}(x)$. Then we have

$$
1 - V'_{ab} = 1 + x \frac{Y_{ca,db}}{Y_{uv}(0)} = \frac{Y_{uv}(0) + x Y_{ca,db}}{Y_{uv}(0)}
$$

$$
= \frac{Y_{db}(0) + x Y_{ca,db}}{Y_{db}(0)} = \frac{Y_{db}(x)}{Y_{db}(0)} = \frac{Y_{ij}(x)}{Y_{ij}(0)} = F(x) \qquad (4.125)
$$

in which we have invoked the identities $Y_{uv} = Y_{ij}$ and

$$
Y_{db}(x) = Y_{db}(0) + x Y_{ca,db} \qquad (4.126)
$$

$Y_{ca,db}$ being independent of x. In other words, the return difference $F(x)$ is simply the difference of the 1-V excitation and the returned voltage V'_{ab} as illustrated in Fig. 4.34, and hence its name. Since $F(x) = 1 + T = 1 - \mu\beta$, we conclude that the return difference has the same physical significance as the feedback factor of the ideal feedback model of Fig. 4.1.

The significance of the above physical interpretations is that it permits us to determine the return ratio T or $-\mu\beta$ by measurement. Once the return ratio is measured, the other quantities such as return difference and loop transmission are known. A discussion of the measurement techniques will be presented in the following chapter.

Example 4.9 Figure 4.36a is a common-emitter amplifier with an admittance G_f purposely added to produce an external physical feedback loop to affect specific change in the performance of the amplifier. The equivalent network of the feedback amplifier is presented in Fig. 4.36b, using the T-model of the transistor for a change. The amplifier is of the parallel-parallel configuration and was discussed in Sec. 4.2.2. The indefinite-admittance matrix of the amplifier can be written down

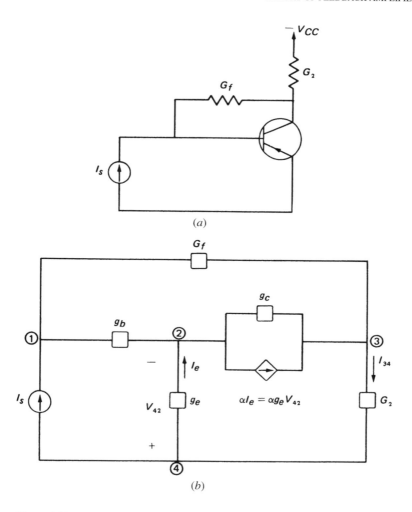

Figure 4.36 (*a*) A common-emitter feedback amplifier, and (*b*) its equivalent network.

by inspection as

$$
\mathbf{Y} =
\begin{bmatrix}
G_f + g_b & -g_b & -G_f & 0 \\
-g_b & g_b + g_e + g_c - \alpha g_e & -g_c & -g_e + \alpha g_e \\
-G_f & -g_c + \alpha g_e & G_f + g_c + G_2 & -G_2 - \alpha g_e \\
0 & -g_e & -G_2 & G_2 + g_e
\end{bmatrix}
$$

(4.127)

Suppose that the controlling parameter αg_e is the element of interest. Then, according to (4.122), the return difference with respect to αg_e is given by

$$
\begin{aligned}
F(\alpha g_e) &= \frac{Y_{ij}(\alpha g_e)}{Y_{ij}(0)} = \frac{Y_{22}(\alpha g_e)}{Y_{22}(0)} \\
&= \frac{(G_2 + g_e)(g_c G_f + g_b g_c + g_b G_f) + (1 - \alpha)G_2 g_e(g_b + G_f)}{(G_2 + g_e)(g_c G_f + g_b g_c + g_b G_f) + G_2 g_e(g_b + G_f)} \\
&= 1 + T
\end{aligned}
\tag{4.128}
$$

where the return ratio

$$
T = \frac{\alpha g_e G_2(g_b + G_f)}{(G_2 + g_e)(g_c G_f + g_b g_c + g_b G_f) + G_2 g_e(g_b + G_f)}
\tag{4.129}
$$

corresponds to the negative of the loop transmission $\mu\beta$ of the ideal feedback model.

For illustrative purposes, we compute the return ratio T by its definition. In the network of Fig. 4.36b, we remove the current source I_s and replace the controlled current source by an independent current source of αg_e amperes. The resulting network is shown in Fig. 4.37, in which the negative of the voltage V_{42}' is the return ratio T. The indefinite-admittance matrix of the network of Fig. 4.37 can

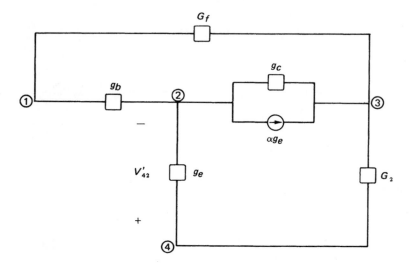

Figure 4.37 The network used to calculate the return ratio with respect to the controlling parameter αg_e of the controlled source in Fig. 4.36b.

be obtained from (4.127) by setting $\alpha g_e = 0$, yielding $\mathbf{Y}(0)$. Appealing again to formula (2.94) gives

$$
\begin{aligned}
T = -V'_{42} &= -\alpha g_e \frac{Y_{34,22}(0)}{Y_{uv}(0)} \\
&= \frac{\alpha g_e G_2 (g_b + G_f)}{(G_2 + g_e)(g_c G_f + g_b g_c + g_b G_f) + G_2 g_e (g_b + G_f)}
\end{aligned}
\tag{4.130}
$$

confirming (4.129).

As illustrated in this example, terminals a, b, c, and d of the controlled current source need not be distinct. In fact, Eqs. (4.124)–(4.126) remain valid even for the situation where $a = b$ or $c = d$ provided that we follow the earlier convention by defining sgn $0 = 0$.

Let us now examine the physical significance of return difference with respect to a one-port admittance x. The result can easily be deduced from (4.125) by letting $a = c$ and $b = d$ as follows:

$$
\begin{aligned}
F(x) &= \frac{Y_{ij}(x)}{Y_{ij}(0)} = \frac{Y_{dd}(x)}{Y_{dd}(0)} = \frac{Y_{dd}(0) + x Y_{cc,dd}(0)}{Y_{dd}(0)} \\
&= 1 + x \frac{Y_{cc,dd}(0)}{Y_{dd}(0)} = 1 + \frac{x}{y}
\end{aligned}
\tag{4.131}
$$

where y is the admittance that x faces, as depicted in Fig. 4.38. The last equation in (4.131) follows directly from formula (2.95). The return difference can therefore be written as

$$
F(x) = 1 + T = 1 - \mu\beta = 1 + \frac{x}{y} = \frac{x + y}{y}
\tag{4.132}
$$

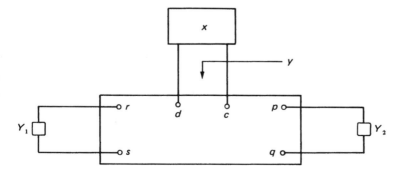

Figure 4.38 The physical interpretation of the return difference with respect to a one-port admittance.

In other words, the return ratio T for the one-port admittance x is equal to the ratio of admittance x to the admittance that x faces, and the return difference $F(x)$ with respect to the one-port admittance x is equal to the ratio of total admittance looking into the node pair where x is connected to the admittance that x faces.

Example 4.10 Consider the network of Fig. 4.36b. Suppose that we wish to compute the return difference with respect to the one-port admittance G_2. From (4.122) in conjunction with (4.127), we obtain

$$F(G_2) = \frac{Y_{ij}(G_2)}{Y_{ij}(0)} = \frac{Y_{22}(G_2)}{Y_{22}(0)}$$

$$= \frac{(G_2 + g_e)(g_c G_f + g_b g_c + g_b G_f) + (1 - \alpha)G_2 g_e(g_b + G_f)}{g_e(g_c G_f + g_b g_c + g_b G_f)} = 1 + T$$

$$(4.133)$$

where

$$T = G_2 \left[\frac{1}{g_e} + \frac{(1 - \alpha)(g_b + G_f)}{g_c G_f + g_b g_c + g_b G_f} \right] \qquad (4.134)$$

To compute the admittance that G_2 faces, we appeal to formula (2.95) and use the matrix $\mathbf{Y}(0)$, giving

$$\frac{1}{y} = \frac{Y_{33,44}(0)}{Y_{22}(0)} = \frac{G_f(g_b + g_e + g_c - \alpha g_e) + g_b(g_e + g_c - \alpha g_e)}{g_e(g_c G_f + g_b g_c + g_b G_f)}$$

$$= \frac{1}{g_e} + \frac{(1 - \alpha)(g_b + G_f)}{g_c G_f + g_b g_c + g_b G_f} \qquad (4.135)$$

This shows that

$$F(G_2) = 1 + T = 1 + \frac{G_2}{y} = \frac{G_2 + y}{y} \qquad (4.136)$$

Example 4.11 Consider the series-parallel feedback amplifier of Fig. 4.24. As in Example 4.5, assume that the transistors are identical, with $h_{ie} = 1.1\,\text{k}\Omega$, $h_{fe} = 50$, and $h_{re} = h_{oe} = 0$. After the biasing and coupling circuitry have been removed, its equivalent network is presented in Fig. 4.39a. The effective load R' of the first transistor is composed of the parallel combination of 10-kΩ, 33-kΩ, 47-kΩ and 1.1-kΩ resistors; or the parallel combination of 6.6-kΩ and 1.1-kΩ resistors, as shown in Fig. 4.39a. The effect of the 150-kΩ and 47-kΩ

Figure 4.39 (a) An equivalent network of the series-parallel feedback amplifier of Fig. 4.24, and (b) an equivalent representation of (a).

biasing resistors can be ignored; they are included in the equivalent network to show their insignificance in the computation. We remark that in some cases the effect of biasing resistors may not be entirely negligible and should therefore be included. With all the conductances denoted in mho, the network of Fig. 4.39a is redrawn as in Fig. 4.39b with

$$\tilde{\alpha}_k = \alpha_k \cdot 10^{-4} = \frac{h_{fe}}{h_{ie}} = 455 \cdot 10^{-4} \qquad (4.137)$$

where $k = 1, 2$ is used to distinguish the transconductances of the first and second transistors. The indefinite-admittance matrix can now be written down by inspection and is given by

$$\mathbf{Y} = 10^{-4}$$

$$\times \begin{bmatrix} 9.37 & 0 & -9.09 & 0 & -0.28 \\ 0 & 4.256 & -2.128 & \alpha_2 & -2.128 - \alpha_2 \\ -9.09 - \alpha_1 & -2.128 & 111.218 + \alpha_1 & 0 & -100 \\ \alpha_1 & 0 & -\alpha_1 & 10.61 & -10.61 \\ -0.28 & -2.128 & -100 & -10.61 - \alpha_2 & 113.02 + \alpha_2 \end{bmatrix}$$

$$(4.138)$$

By applying formula (2.97), the voltage gain of the amplifier is found to be

$$w = \frac{V_{25}}{V_s} = \frac{Y_{12,55}}{Y_{11,55}} = \frac{211.54 \cdot 10^{-7}}{4.66 \cdot 10^{-7}} = 45.39 \qquad (4.139)$$

confirming (4.80), where

$$Y_{12,55} = (102.13\alpha_1\alpha_2 + 22.58\alpha_1 + 205.24) \cdot 10^{-12} = 211.535 \cdot 10^{-7}$$

$$(4.140a)$$

$$Y_{11,55} = (2.128\alpha_1\alpha_2 + 45.156\alpha_1 + 4974.132\alpha_2) \cdot 10^{-12} = 4.661 \cdot 10^{-7}$$

$$(4.140b)$$

To compute the return difference with respect to the transconductances $\tilde{\alpha}_k$ of the transistors, we short-circuit the voltage source V_s. The corresponding indefinite-admittance matrix is obtained by adding the first row of (4.138) to the fifth row and the first column to the fifth column and then deleting the first row and column. The first-order cofactor of the resulting matrix is simply $Y_{11,55}$. Thus, the return differences with respect to $\tilde{\alpha}_1$ and $\tilde{\alpha}_2$ are given by

$$F(\tilde{\alpha}_1) = \frac{Y_{11,55}(\tilde{\alpha}_1)}{Y_{11,55}(0)} = \frac{466.1 \cdot 10^{-9}}{4.97 \cdot 10^{-9}} = 93.70 \qquad (4.141a)$$

$$F(\tilde{\alpha}_2) = \frac{Y_{11,55}(\tilde{\alpha}_2)}{Y_{11,55}(0)} = \frac{466.1 \cdot 10^{-9}}{25.52 \cdot 10^{-9}} = 18.26 \qquad (4.141b)$$

For illustrative purposes, we compute the voltage gains of the first and second stages. Appealing once again to formula (2.97), we obtain

$$\frac{V_{45}}{V_{15}} = \frac{Y_{14,55}}{Y_{11,55}} = -\frac{195.7 \cdot 10^{-9}}{466.1 \cdot 10^{-9}} = -0.42 \tag{4.142a}$$

$$\frac{V_{25}}{V_{45}} = \frac{Y_{42,55}}{Y_{44,55}} = -\frac{494.53 \cdot 10^{-9}}{4.58 \cdot 10^{-9}} = -107.98 \tag{4.142b}$$

where, as in Chap. 2, V_{ij} denotes the voltage drop from terminal i to terminal j. This gives the overall amplifier voltage gain

$$w = \frac{V_{25}}{V_{15}} = \frac{V_{25}}{V_{45}} \frac{V_{45}}{V_{15}} = 45.35 \tag{4.143}$$

confirming (4.139) within computational accuracy.

4.3.2 A Useful Identity in Feedback Amplifier Theory

In this section, we establish a useful identity that is needed in interpreting physically the algebraic definition of null return difference to be presented in the following section.

Let $\mathbf{Y}$ be the indefinite-admittance matrix of the feedback amplifier of Fig. 4.33. The element x of interest enters $\mathbf{Y}$ in a rectangular pattern as shown in (4.123). Our objective is to establish the useful relation

$$Y_{uv}\dot{Y}_{rp,sq} = Y_{ra,sb}Y_{dp,cq} - Y_{da,cb}Y_{rp,sq} \tag{4.144}$$

where $\dot{Y}_{rp,sq}$ denotes the derivative of the function $Y_{rp,sq}$ with respect to x. This relation is very similar to the well-known identity [see, for example, Bocher (1938)]

$$\Delta\Delta_{ab,cd} = \Delta_{ab}\Delta_{cd} - \Delta_{ad}\Delta_{cb} \tag{4.145}$$

where $\Delta_{ab,cd}$ is the abth cofactor of Δ_{cd}, which in turn is the cdth cofactor of the determinant Δ. For our purposes, we define the *third-order cofactor* of the elements of $\mathbf{Y}$ as

$$Y_{ab,cd,nn} = \text{sgn}\,(a-c)\,\text{sgn}\,(b-d)(-1)^{a+b+c+d}\,\det\mathbf{Y}_{ab,cd,nn} \tag{4.146}$$

where $\mathbf{Y}_{ab,cd,nn}$ denotes the submatrix obtained from $\mathbf{Y}$, which is of order $n \geqslant 4$, by deleting rows a, c, n and columns b, d, n for $a, c \neq n$ or $b, d \neq n$; and $Y_{ab,cd,nn} =$

0 for a, b, c, or $d = n$. Also, it is convenient to define that $Y_{11,22} = 1$ for $n = 2$ and $Y_{11,22,33} = 1$ for $n = 3$. By using this, Eq. (4.145) can be written as

$$Y_{uv}Y_{ab,cd,nn} = Y_{ab,nn}Y_{cd,nn} - Y_{ad,nn}Y_{cb,nn} \tag{4.147}$$

Also, as is known, not all the second-order cofactors are independent, and they are related by the equation [see, for example, Chen (1976a)]

$$Y_{rp,sq} = Y_{rp,uv} + Y_{sq,uv} - Y_{rq,uv} - Y_{sp,uv} \tag{4.148}$$

The other relations needed in establishing (4.144) are (see Prob. 4.18)

$$Y_{rp,sq} = -Y_{sp,rq} = -Y_{rq,sp} \tag{4.149a}$$

$$Y_{rp,sq}(x) = Y_{rp,sq}(0) + x\dot{Y}_{rp,sq} \tag{4.149b}$$

Note that $\dot{Y}_{rp,sq}$ is independent of x. With these preliminaries, we are now in a position to derive the identity (4.144).

With (4.148), the first term on the right-hand side of (4.144) can be expanded to 16 terms, as follows:

$$\begin{aligned}
Y_{ra,sb}Y_{dp,cq} &= (Y_{ra,nn} + Y_{sb,nn} - Y_{rb,nn} - Y_{sa,nn}) \\
&\quad \times (Y_{dp,nn} + Y_{cq,nn} - Y_{dq,nn} - Y_{cp,nn}) \\
&= Y_{ra,nn}Y_{dp,nn} + Y_{ra,nn}Y_{cq,nn} + Y_{sb,nn}Y_{dp,nn} \\
&\quad + Y_{sb,nn}Y_{cq,nn} + Y_{rb,nn}Y_{dq,nn} + Y_{rb,nn}Y_{cp,nn} + Y_{sa,nn}Y_{dq,nn} \\
&\quad + Y_{sa,nn}Y_{cp,nn} - Y_{ra,nn}Y_{cp,nn} - Y_{sa,nn}Y_{cq,nn} - Y_{rb,nn}Y_{dp,nn} \\
&\quad - Y_{sb,nn}Y_{dq,nn} - Y_{rb,nn}Y_{cq,nn} - Y_{sb,nn}Y_{cp,nn} - Y_{ra,nn}Y_{dq,nn} \\
&\quad - Y_{sa,nn}Y_{dp,nn} \tag{4.150}
\end{aligned}$$

Likewise, the second term on the right-hand side of (4.144) can be expanded:

$$\begin{aligned}
-Y_{da,cb}Y_{rp,sq} &= Y_{ca,nn}Y_{rp,nn} + Y_{ca,nn}Y_{sq,nn} + Y_{db,nn}Y_{rp,nn} + Y_{db,nn}Y_{sq,nn} \\
&\quad + Y_{cb,nn}Y_{rq,nn} + Y_{cb,nn}Y_{sp,nn} + Y_{da,nn}Y_{rq,nn} + Y_{da,nn}Y_{sp,nn} \\
&\quad - Y_{da,nn}Y_{rp,nn} - Y_{ca,nn}Y_{rq,nn} - Y_{db,nn}Y_{sp,nn} - Y_{cb,nn}Y_{sq,nn} \\
&\quad - Y_{db,nn}Y_{rq,nn} - Y_{cb,nn}Y_{rp,nn} - Y_{da,nn}Y_{sq,nn} - Y_{ca,nn}Y_{sp,nn} \tag{4.151}
\end{aligned}$$

By appealing to (4.147), the first eight terms on the right-hand side of (4.150) can be combined with the corresponding last eight terms on the right-hand side of (4.151), and the first eight terms of (4.151) with the last eight terms of (4.150). The result is given by

$$
Y_{ra,sb}Y_{dp,cq} - Y_{da,cb}Y_{rp,sq} = Y_{uv}(Y_{ra,dp,nn} + Y_{ra,cq,nn} + Y_{sb,dp,nn} + Y_{sb,cq,nn}
$$

$$
+ Y_{rb,dq,nn} + Y_{rb,cp,nn} + Y_{sa,dq,nn} + Y_{sa,cp,nn}
$$

$$
+ Y_{ca,rp,nn} + Y_{ca,sq,nn} + Y_{db,rp,nn} + Y_{db,sq,nn}
$$

$$
+ Y_{cb,rq,nn} + Y_{cb,sp,nn} + Y_{da,rq,nn} + Y_{da,sp,nn})
$$

$$(4.152)$$

Referring to (4.123), we have the following expansions:

$$
Y_{rp,nn}(x) = Y_{rp,nn}(0) + x(Y_{ca,rp,nn} + Y_{ra,dp,nn} + Y_{rb,cp,nn} + Y_{db,rp,nn})
$$

$$(4.153a)$$

$$
Y_{sq,nn}(x) = Y_{sq,nn}(0) + x(Y_{ca,sq,nn} + Y_{sa,dq,nn} + Y_{sb,cq,nn} + Y_{db,sq,nn})
$$

$$(4.153b)$$

$$
Y_{rq,nn}(x) = Y_{rq,nn}(0) + x(Y_{ca,rq,nn} + Y_{db,rq,nn} - Y_{da,rq,nn} - Y_{cb,rq,nn})
$$

$$
= Y_{rq,nn}(0) - x(Y_{ra,cq,nn} + Y_{rb,dq,nn} + Y_{da,rq,nn} + Y_{cb,rq,nn})
$$

$$(4.153c)$$

$$
Y_{sp,nn}(x) = Y_{sp,nn}(0) + x(Y_{ca,sp,nn} + Y_{db,sp,nn} - Y_{da,sp,nn} - Y_{cb,sp,nn})
$$

$$
= Y_{sp,nn}(0) - x(Y_{sa,cp,nn} + Y_{sb,dp,nn} + Y_{da,sp,nn} + Y_{cb,sp,nn})
$$

$$(4.153d)$$

Appealing to (4.148) in conjunction with (4.152) and (4.153) yields

$$
Y_{rp,sq}(x) = Y_{rp,nn}(x) + Y_{sq,nn}(x) - Y_{rq,nn}(x) - Y_{sp,nn}(x)
$$

$$
= Y_{rp,nn}(0) + Y_{sq,nn}(0) - Y_{rq,nn}(0) - Y_{sp,nn}(0)
$$

$$
+ \frac{x[Y_{ra,sb}(x)Y_{db,cq}(x) - Y_{da,cb}(x)Y_{rp,sq}(x)]}{Y_{uv}(x)} \qquad (4.154)
$$

giving

$$
x(Y_{ra,sb}Y_{dp,cq} - Y_{da,cb}Y_{rp,sq}) = Y_{uv}[Y_{rp,sq}(x) - Y_{rp,sq}(0)] \qquad (4.155)
$$

Since from (4.149b) we have $\left[Y_{rp,sq}(x) - Y_{rp,sq}(0)\right] = x\dot{Y}_{rp,sq}$ Eq. (4.155) is simplified to (4.144), and we obtain the desired identity.

Example 4.12 Consider the network of Fig. 4.39b, whose indefinite-admittance matrix is given by Eq. (4.138). Suppose that $x = \tilde{\alpha}_2 = \alpha_2/10^4$ is the element of interest. According to the labels of Fig. 4.33, we have

$$a = 4 \quad b = 5 \quad c = 2 \quad d = 5$$
$$r = 1 \quad p = 2 \quad s = 5 \quad q = 5 \tag{4.156}$$

We remark that r, p, s, and q are rather arbitrary. Any two terminals can be the input terminals, and any other two the output terminals. In the present situation, we choose 1 and 5 as the input terminals and 2 and 5 as the output terminals.

With the above designations, (4.144) becomes

$$Y_{uv}\dot{Y}_{12,55} = Y_{14,55}Y_{52,25} - Y_{54,25}Y_{12,55} \tag{4.157}$$

where $\dot{Y}_{12,55}$ is the derivative of $Y_{12,55}$ with respect to $\tilde{\alpha}_2$. From (4.138), the desired quantities are calculated, as follows:

$$Y_{uv} = Y_{55} = (271.11\alpha_2 + 48629.46) \cdot 10^{-16} = 171.983 \cdot 10^{-13}$$
$$Y_{12,55} = (46,468.2\alpha_2 + 10,478.23) \cdot 10^{-12} = 211.53 \cdot 10^{-7}$$
$$\dot{Y}_{12,55} = 464.682 \cdot 10^{-6}$$
$$Y_{14,55} = -195.71 \cdot 10^{-9} \tag{4.158}$$
$$Y_{52,25} = -Y_{22,55} = -115.32 \cdot 10^{-10}$$
$$Y_{54,25} = -Y_{24,55} = -271.11 \cdot 10^{-12}$$

where $\alpha_2 = 455$, giving

$$Y_{uv}\dot{Y}_{12,55} = 799.174 \cdot 10^{-17} \tag{4.159a}$$
$$Y_{14,55}Y_{52,25} - Y_{54,25}Y_{12,55} = 799.172 \cdot 10^{-17} \tag{4.159b}$$

confirming (4.157) within computational accuracy.

4.3.3 The Null Return Difference

The null return difference is found to be very useful in measurement situations and in the computation of the sensitivity for the feedback amplifiers. In this section, we introduce this concept by first giving a mathematical definition and then discussing its physical significance.

Definition 4.3: Null return difference The *null return difference* $\hat{F}(x)$ of a feedback amplifier with respect to an element x is the ratio of the two functional values assumed by the second-order cofactor $Y_{rp,sq}$ of the elements of its indefinite-admittance matrix **Y** under the condition that the element x assumes its nominal value and the condition that the element x assumes the value zero where r and s are input terminals and p and q are output terminals of the amplifier.

Thus, in computing the null return difference, we must first identify the input and output ports. Referring to the general configuration of Fig. 4.33, we have

$$\hat{F}(x) \equiv \frac{Y_{rp,sq}(x)}{Y_{rp,sq}(0)} \tag{4.160}$$

As in the case of return difference, the null return difference is simply the return difference in the network under the constraint that the input excitation I_s has been adjusted so that the output is identically zero. For this we introduce the concept of null return ratio.

Definition 4.4: Null return ratio The *null return ratio* $\hat{T}$ with respect to a voltage-controlled current source $I = xV$ is the negative of the voltage appearing at the controlling branch when the controlled current source is replaced by an independent current source of x amperes and when the input excitation is adjusted so that the output of the amplifier is identically zero.

Consider the network of Fig. 4.33. Suppose that we replace the controlled current source by an independent current source of x amperes. Then, by applying formula (2.94) and the principle of superposition, the output current I_{pq} at the load can be calculated and is found to be

$$I_{pq} = Y_2 \left[I_s \frac{Y_{rp,sq}(0)}{Y_{uv}(0)} + x \frac{Y_{dp,cq}(0)}{Y_{uv}(0)} \right] \tag{4.161}$$

Setting $V_{pq} = 0$ or $I_{pq} = 0$ yields

$$I_s \equiv I_0 = -x \frac{Y_{dp,cq}(0)}{Y_{rp,sq}(0)} \tag{4.162}$$

$Y_{dp,cq}$ being independent of x. This adjustment is possible only if there is a direct transmission from the input to the output when x is set to zero. Thus, in the network of Fig. 4.34, if we connect an independent current source of this strength at its input port, the voltage V'_{ab} is the negative of the null return ratio $\hat{T}$. Using (2.94) we obtain

$$
\hat{T} = V'_{ab} = -x\frac{Y_{da,cb}(0)}{Y_{uv}(0)} - I_0\frac{Y_{ra,sb}(0)}{Y_{uv}(0)}
$$

$$
= -\frac{x[Y_{da,cb}(0)Y_{rp,sq}(0) - Y_{ra,sb}(0)Y_{dp,cq}(0)]}{Y_{uv}(0)Y_{rp,sq}(0)} \tag{4.163a}
$$

$$
= \frac{x\dot{Y}_{rp,sq}}{Y_{rp,sq}(0)} = \frac{Y_{rp,sq}(x)}{Y_{rp,sq}(0)} - 1 \tag{4.163b}
$$

Equation (4.163b) follows directly from Eqs. (4.144) and (4.149b). This leads to

$$
\hat{F}(x) = 1 + \hat{T} = 1 - V'_{ab} \tag{4.164}
$$

Thus, like return difference, the null return difference $\hat{F}(x)$ is simply the difference of the 1-V excitation applied to the right of the breaking mark of the broken controlling branch of the controlled source and the returned voltage V'_{ab} appearing at the left of the breaking mark under the situation that the input signal I_s is adjusted so that the output is identically zero. Since, from (4.149a),

$$
Y_{rp,sq} = -Y_{sp,rq} = -Y_{rq,sp} = Y_{sq,rp} \tag{4.165}
$$

we have

$$
\hat{F}(x) = \frac{Y_{rp,sq}(x)}{Y_{rp,sq}(0)} = \frac{Y_{sp,rq}(x)}{Y_{sp,rq}(0)} = \frac{Y_{rq,sp}(x)}{Y_{rq,sp}(0)} = \frac{Y_{sq,rp}(x)}{Y_{sq,rp}(0)} \tag{4.166}
$$

showing that the null return difference is independent of the choice of the references for the voltages and currents at the input and output ports. In other words, in computing the null return difference $\hat{F}(x)$ it is sufficient to identify the input and the output ports.

We illustrate the above result by the following examples.

Example 4.13 In the feedback network of Fig. 4.36b, suppose that we wish to compute the null return difference with respect to the element αg_e. The terminals of the input port are 1 and 4 and those of the output port are 3 and 4. In terms of the labels of Fig. 4.33, we have $a = 4, b = 2, c = 2, d = 3, r = 1, p = 3, s = 4$, and $q = 4$. Thus, from (4.160) in conjunction with (4.127), we obtain

$$\hat{F}(\alpha g_e) = \frac{Y_{13,44}(\alpha g_e)}{Y_{13,44}(0)}$$

$$= \frac{g_b(g_c - \alpha g_e) + G_f(g_b + g_e + g_c - \alpha g_e)}{G_f(g_b + g_e + g_c) + g_b g_c}$$

$$= 1 + \hat{T} \qquad (4.167)$$

where

$$\hat{T} = -\frac{\alpha g_e(g_b + G_f)}{G_f(g_b + g_e + g_c) + g_b g_c} \qquad (4.168)$$

is the null return ratio.

For illustrative purposes, we compute $\hat{T}$ by (4.163a), yielding

$$\hat{T} = -\alpha g_e \frac{Y_{34,22}(0)Y_{13,44}(0) - Y_{14,42}(0)Y_{33,24}(0)}{Y_{22}(0)Y_{13,44}(0)}$$

$$= \frac{\alpha g_e}{Y_{22}(0)} \left\{ -G_2(G_f + g_b) - \frac{g_e(G_f + g_b)[g_b(G_f + g_c + G_2) + g_c G_f]}{G_f(g_b + g_e + g_c) + g_b g_c} \right\}$$

$$= -\frac{\alpha g_e(g_b + G_f)}{G_f(g_b + g_e + g_c) + g_b g_c} \qquad (4.169)$$

confirming (4.168). Alternatively, as can be seen from Fig. 4.36b, if the output current I_{34} is zero, the returned voltage V_{42} is simply $-I_s/g_e$, which in conjunction with (4.162) gives

$$\hat{T} = -V_{42} = -\alpha \frac{Y_{33,24}(0)}{Y_{13,44}(0)} = -\frac{\alpha g_e(g_b + G_f)}{G_f(g_b + g_e + g_c) + g_b g_c} \qquad (4.170)$$

Example 4.14 Consider the voltage-series feedback amplifier of Fig. 4.24, whose equivalent network is shown in Fig. 4.39b. The indefinite-admittance matrix was computed earlier in Example 4.11 and is given by Eq. (4.138). Suppose that we wish to compute the null return differences with respect to the transconductances

All conductances in mho

Figure 4.40 The network used to compute the null return difference $\hat{F}(\tilde{\alpha}_1)$ by its physical interpretation.

$\tilde{\alpha}_k = \alpha_k/10^4$, as defined in (4.137). Applying definition (4.160) gives

$$\hat{F}(\tilde{\alpha}_l) = \frac{Y_{12,55}(\tilde{\alpha}_1)}{Y_{12,55}(0)} = \frac{211.54 \cdot 10^{-7}}{205.24 \cdot 10^{-12}} = 103.07 \cdot 10^3 \qquad (4.171)$$

$$\hat{F}(\tilde{\alpha}_2) = \frac{Y_{12,55}(\tilde{\alpha}_2)}{Y_{12,55}(0)} = \frac{211.54 \cdot 10^{-7}}{104.79 \cdot 10^{-10}} = 2018.70 \qquad (4.172)$$

where $Y_{12,55}$ is given by (4.140a).

Alternatively, $\hat{F}(\tilde{\alpha}_1)$ can be obtained as follows: Replace the controlled current source $\tilde{\alpha}_1 V_{13}$ in Fig. 4.39b by an independent current source of $\tilde{\alpha}_1$ amperes. We then adjust the voltage source V_s so that the output current I_{25} is identically zero. Let I_0 be the corresponding input current resulting from this source. The resulting network is depicted in Fig. 4.40. According to (4.164), we have in Fig. 4.40,

$$\hat{F}(\tilde{\alpha}_1) = 1 - V'_{13} = 1 - \frac{100V'_{35} + \alpha_2 V'_{45} - \alpha_1}{9.09} = 103.07 \cdot 10^3 \qquad (4.173)$$

confirming (4.171), where $V'_{35} = \alpha_2 V'_{45}/2.128$ and $V'_{45} = -\alpha_1/10.61$. V'_{ij}, as before, denotes the voltage drop from terminal i to terminal j. In a similar manner we can compute $\hat{F}(\tilde{\alpha}_2)$, the details being left as an exercise (see Prob. 4.7).

4.4 THE NETWORK FUNCTIONS AND FEEDBACK

In this section, we apply the formulas derived in the foregoing directly to calculate the network functions. We show that the computation of driving-point impedance of an active network can be simplified. As a consequence, we can observe the effects of feedback on amplifier impedance and gain and obtain some useful relations among the return difference, the null return difference, and impedance functions in general.

Refer to the general feedback configuration of Fig. 4.33. Let w be a transfer function. As before, to emphasize the importance of the feedback element x, we write $w = w(x)$ even though it is also a function of the complex-frequency variable s. To be definitive, let $w(x)$ for the time being be the current gain between the input and output ports. Then from (2.94) we obtain immediately

$$w(x) = \frac{I_{pq}}{I_s} = \frac{Y_2 V_{pq}}{I_s} = \frac{Y_{rp,sq}(x)}{Y_{uv}(x)} Y_2 \qquad (4.174)$$

yielding

$$\frac{w(x)}{w(0)} = \frac{Y_{rp,sq}(x)}{Y_{uv}(x)} \frac{Y_{uv}(0)}{Y_{rp,sq}(0)} = \frac{\hat{F}(x)}{F(x)} \qquad (4.175)$$

provided that $w(0) \neq 0$. In fact, (4.175) remains valid if $w(x)$ represents the transfer impedance $z_{rp,sq} = V_{pq}/I_s$ (see Prob. 4.8). This gives a very useful formula for computing the current gain I_{pq}/I_s or the transfer impedance $z_{rp,sq} = V_{pq}/I_s$:

$$w(x) = w(0) \frac{\hat{F}(x)}{F(x)} \qquad (4.176)$$

In the particular situation where $r = p$ and $s = q$, $w(x)$ represents the driving-point impedance $z_{rr,ss}(x)$ looking into the terminals r and s, and we have a somewhat different physical interpretation. In this case, $F(x)$ is the return difference with respect to the element x under the condition $I_s = 0$. Thus, $F(x)$ can be considered to be the return difference for the situation when the port where the input impedance is defined is left open without a source and we write $F(x) = F$ (input open-circuited). Likewise, $\hat{F}(x)$ is the return difference with respect to the element x for the input excitation I_s and output response V_{rs} (Fig. 4.33) under the condition that I_s is adjusted so that V_{rs} is identically zero. This means that $\hat{F}(x)$ can be considered to be the return difference for the situation when the port where the input impedance is defined is short-circuited and we write $\hat{F}(x) = F$ (input short-circuited). Consequently, the input impedance $Z(x)$ looking into a terminal pair can be conveniently expressed as

$$Z(x) = Z(0) \frac{F(\text{input short-circuited})}{F(\text{input open-circuited})} \qquad (4.177)$$

which is the well-known *Blackman's formula* for an active impedance, which was first derived by Blackman (1943). The formula is extremely useful in computing the impedance of an active network, because the right-hand side of Eq. (4.177) can usually be determined rather easily. If x represents the controlling parameter of a controlled source in a single-loop feedback amplifier, then setting $x = 0$ opens the feedback loop and $Z(0)$ is simply a passive impedance. The return difference for x when the input port is short-circuited or open-circuited is relatively simple to compute because shorting out or opening a terminal pair frequently breaks the feedback loop. On the other hand, Blackman's formula may be used to determine the return difference by measurements. However, because it involves two return differences, only one of them can be identified; the other must be known in advance. In the case of single-loop feedback amplifiers, it is usually possible to choose a terminal pair so that either the numerator or the denominator on the right-hand side of (4.177) is unity. If F(input short-circuited) $= 1$ and F(input open-circuited) $= F(x)$, $F(x)$ being the return difference under normal operating conditions, then we have

$$F(x) = \frac{Z(0)}{Z(x)} \tag{4.178}$$

On the other hand, if F(input open-circuited) $= 1$ and F(input short-circuited) $= F(x)$, then

$$F(x) = \frac{Z(x)}{Z(0)} \tag{4.179}$$

We illustrate the above results by the following examples.

Example 4.15 Consider the network shown in Fig. 4.41, which is the equivalent network of a simplified transistor feedback amplifier. The indefinite-admittance matrix of the network is given by

$$\mathbf{Y} = \begin{bmatrix} G_1 + g_3 + sC_1 + sC_2 & -sC_2 & -G_1 - g_3 - sC_1 \\ \alpha - sC_2 & G_2 + sC_2 & -\alpha - G_2 \\ -\alpha - G_1 - g_3 - sC_1 & -G_2 & G_1 + G_2 + g_3 + sC_1 + \alpha \end{bmatrix} \tag{4.180}$$

Suppose that the controlling parameter α is the element of interest, I_s forms the input port, and G_2 is the output load. By using (2.94), the current gain I_{23}/I_s

Figure 4.41 An equivalent network of a simplified transistor feedback amplifier.

when α is set to zero, is found to be

$$w(0) = \left.\frac{I_{23}}{I_s}\right|_{\alpha=0} = \frac{G_2 Y_{12,33}(0)}{Y_{11}(0)} = \frac{sC_2 G_2}{(G_2 + sC_2)(G_1 + g_3 + sC_1) + sC_2 G_2}$$

(4.181)

The return difference with respect to α is obtained as

$$F(\alpha) = \frac{Y_{11}(\alpha)}{Y_{11}(0)} = \frac{(G_2 + sC_2)(G_1 + g_3 + sC_1) + sC_2(\alpha + G_2)}{(G_2 + sC_2)(G_1 + g_3 + sC_1) + sC_2 G_2}$$

(4.182)

and the null return difference $\hat{F}(\alpha)$ is computed by

$$\hat{F}(\alpha) = \frac{Y_{12,33}(\alpha)}{Y_{12,33}(0)} = \frac{sC_2 - \alpha}{sC_2} = 1 - \frac{\alpha}{sC_2}$$

(4.183)

Substituting (4.181)–(4.183) in (4.176) yields the current gain

$$w(a) = \frac{I_{23}}{I_s} = \frac{G_2(sC_2 - \alpha)}{(G_2 + sC_2)(G_1 + g_3 + sC_1) + sC_2(\alpha + G_2)}$$

(4.184)

which can also be computed directly by the formula $w(x) = G_2 Y_{12,33}(\alpha)/Y_{11}(\alpha)$, giving, of course, the same result.

Assume that we wish to compute the impedance $Z(\alpha)$ facing the input current I_s by Blackman's formula. F(input open-circuited) is the return difference with respect to α when the input excitation I_s is removed, yielding the return difference as shown in (4.182). F(input short-circuited) is the return difference with respect to α in the network obtained from that of Fig. 4.41 by shorting the terminals 1 and 3, resulting in F(input short-circuited) $= 1$. Finally, the input impedance $Z(0)$

when α is set to zero is given by

$$Z(0) = \frac{Y_{11,33}(0)}{Y_{11}(0)} = \frac{G_2 + sC_2}{(G_2 + sC_2)(G_1 + g_3 + sC_1) + sC_2G_2} \qquad (4.185)$$

Substituting these in (4.177) yields the active impedance

$$Z(\alpha) = \frac{G_2 + sC_2}{(G_2 + sC_2)(G_1 + g_3 + sC_1) + sC_2(\alpha + G_2)} \qquad (4.186)$$

which can also be computed directly by the formula $Z(\alpha) = Y_{11,33}(\alpha)/Y_{11}(\alpha)$.

As another example, let us compute the impedance facing the capacitor C_2 in the network of Fig. 4.41. The desired quantities are given by

$$F(\text{input short-circuited}) = 1 + \frac{\alpha}{G_1 + G_2 + g_3 + sC_1} \qquad (4.187a)$$

$$F(\text{input open-circuited}) = 1 \qquad (4.187b)$$

$$Z(0) = \frac{1}{G_2} + \frac{1}{G_1 + g_3 + sC_1} \qquad (4.188)$$

in which the input port is formed by removing the capacitor C_2 and the element of interest is α. The details are left as an exercise (see Prob. 4.10). Thus, using Blackman's formula, the active impedance facing the capacitor C_2 is given by

$$Z(\alpha) = Z(0)\frac{F(\text{input short-circuited})}{F(\text{input open-circuited})} = \frac{G_1 + G_2 + g_3 + sC_1 + \alpha}{G_2(G_1 + g_3 + sC_1)} \qquad (4.189)$$

Example 4.16 The network of Fig. 4.42 is the equivalent network of a feedback amplifier. We use Blackman's formula to compute the driving-point impedance facing the admittance y_3.

First, we write down the indefinite-admittance matrix by the rules outlined in Sec. 2.2:

$$\mathbf{Y} = \begin{bmatrix} y_1 + y_2 & \alpha_2 - y_2 & -y_1 - \alpha_2 \\ \alpha_1 - y_2 & y_2 + y_3 & -\alpha_1 - y_3 \\ -\alpha_1 - y_1 & -y_3 - \alpha_2 & y_1 + y_3 + \alpha_1 + \alpha_2 \end{bmatrix} \qquad (4.190)$$

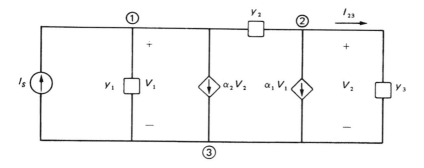

Figure 4.42 An equivalent network of a feedback amplifier.

Let α_1 be the element of interest. Then F(input open-circuited) is the return difference with respect to α_1 with the admittance y_3 removed:

$$F(\text{input open-circuited}) = F(\alpha_1)|_{y_3=0} = \left.\frac{Y_{33}(\alpha_1)}{Y_{33}(0)}\right|_{y_3=0}$$

$$= \frac{y_2(y_1 + \alpha_1 + \alpha_2) - \alpha_1\alpha_2}{y_2(y_1 + \alpha_2)} = 1 + \frac{\alpha_1(y_2 - \alpha_2)}{y_2(y_1 + \alpha_2)} \tag{4.191}$$

the last term on the right-hand side being the return ratio when $y_3 = 0$ (see Prob. 4.11). F(input short-circuited) is the return difference with respect to α_1 when the terminals of y_3 are short-circuited. Since shorting y_3 breaks the feedback loop, no voltage is returned and we have F(input short-circuited) $= 1$. Finally, the impedance facing y_3 when α_1 is set to zero is obtained as

$$Z(0) = \left.\frac{Y_{22,33}(0)}{Y_{33}(0)}\right|_{y_3=0} = \frac{y_1 + y_2}{y_2(y_1 + \alpha_2)} \tag{4.192}$$

Substituting these in (4.177) yields the impedance facing y_3 as

$$Z(\alpha_1) = \frac{y_1 + y_2}{y_2(y_1 + \alpha_1 + \alpha_2) - \alpha_1\alpha_2} \tag{4.193}$$

which can also be obtained directly by formula (2.95) (see Prob. 4.14).

Before we turn our attention to other transfer functions, we apply Blackman's formula to one of the four common types of feedback amplifier configurations of Fig. 4.3, as discussed in Sec. 4.2.

To determine the effect of feedback on the input and output impedances, we choose the series-parallel feedback configuration of Fig. 4.3c. Using Blackman's formula, we shall determine the impedance looking into the two terminals of the load admittance Y_2. Since by shorting the terminals of Y_2 we interrupt the feedback loop, formula (4.178) applies and the output impedance across the load admittance Y_2 becomes

$$Z_{\text{out}}(x) = \frac{Z_{\text{out}}(0)}{F(x)} \tag{4.194}$$

meaning that the impedance measured across the path of the feedback is reduced by the factor that is the normal value of the return difference with respect to the element x, where x is an arbitrary element of interest. For the input impedance of the amplifier looking into the voltage source V_s of Fig. 4.3c, we recognize that by open-circuiting or removing the voltage source V_s, we break the feedback loop, and formula (4.179) applies. Therefore, the input impedance becomes

$$Z_{\text{in}}(x) = F(x)Z_{\text{in}}(0) \tag{4.195}$$

showing that impedance measured in series lines is increased by the same factor $F(x)$. The analyses of the other three configurations are left as exercises (see Probs. 4.15–4.17).

Referring to the network of Fig. 4.33, assume that $w(x)$ represents the voltage gain V_{pq}/V_{rs} or transfer admittance I_{pq}/V_{rs}. From (2.97) we can write

$$\frac{w(x)}{w(0)} = \frac{Y_{rp,sq}(x)}{Y_{rp,sq}(0)} \frac{Y_{rr,ss}(0)}{Y_{rr,ss}(x)} \tag{4.196}$$

The first term in the product on the right-hand side is the null return difference $\hat{F}(x)$ with respect to x for the input terminals r and s and output terminals p and q. The second term is the reciprocal of the null return difference with respect to x for the same input and output port at terminals r and s. As in (4.177), this reciprocal can then be interpreted as the return difference with respect to x when the input port of the amplifier is short-circuited. Thus, the voltage gain or transfer admittance can be expressed as

$$w(x) = w(0) \frac{\hat{F}(x)}{F(\text{input short-circuited})} \tag{4.197}$$

Likewise, if $w(x)$ denotes the short-circuit current gain I_{pq}/I_s as Y_2 approaches infinity, from (2.114) we have

$$\frac{w(x)}{w(0)} = \frac{Y_{rp,sq}(x)}{Y_{rp,sq}(0)} \frac{Y_{pp,qq}(0)}{Y_{pp,qq}(x)} \tag{4.198}$$

As before, the second term in the product is the reciprocal of the return difference with respect to x when the output port of the amplifier is short-circuited, giving a formula for the short-circuit current gain

$$w(x) = w(0) \frac{\hat{F}(x)}{F(\text{output short-circuited})} \tag{4.199}$$

Example 4.17 For the network of Fig. 4.42, we compute the voltage gain V_2/V_1 and the short-circuit current gain I_{23}/I_s as $y_3 \to \infty$. For a change, we choose α_2 to be the element of interest. The null return difference $\hat{F}(\alpha_2)$ can easily be determined from the network of Fig. 4.42. First, we replace the controlled source $\alpha_2 V_2$ by an independent current source of α_2 amperes and then adjust I_s so that V_2 is identically zero, giving* $I_s = \alpha_2$ for $\alpha_1 \neq y_2$. Under this situation, the returned voltage V_2 is clearly zero and $\hat{F}(\alpha_2) = 1$. Alternatively, from the definition we have $\hat{F}(\alpha_2) = Y_{12,33}(\alpha_2)/Y_{12,33}(0) = 1$.

To compute $F(\text{input short-circuited})$, we short-circuit I_s and compute the return difference with respect to α_2 in the resulting network. Since short-circuiting I_s breaks the feedback loop, we have $F(\text{input short-circuited}) = 1$. Substituting the above in (4.197), we obtain the voltage gain

$$w(\alpha_2) = \frac{V_2}{V_1} = w(0) \frac{\hat{F}(\alpha_2)}{F(\text{input short-circuited})} = \frac{V_2}{V_1}\bigg|_{\alpha_2=0} = \frac{y_2 - \alpha_1}{y_2 + y_3} \tag{4.200}$$

which can also be derived directly by formula (2.97) in conjunction with (4.190).

Finally, to determine $F(\text{output short-circuited})$, we short-circuit y_3 and compute the return difference with respect to α_2 in the resulting network. Again, the return difference is unity. Thus, from (4.199) the short-circuit current gain is obtained as

$$w(\alpha_2) = \frac{I_{23}}{I_s}\bigg|_{y_3 \to \infty} = w(0) \frac{\hat{F}(\alpha_2)}{F(\text{output short-circuited})}$$

$$= \frac{I_{23}}{I_s}\bigg|_{\substack{y_3 \to \infty \\ \alpha_2=0}} = \frac{y_2 - \alpha_1}{y_1 + y_2} \tag{4.201}$$

This gain can, of course, be computed directly by means of formula (2.114), giving

$$w(\alpha_2) = \frac{Y_{12,33}(\alpha_2)}{Y_{22,33}(\alpha_2)} = \frac{y_2 - \alpha_1}{y_1 + y_2} \tag{4.202}$$

---

*For $\alpha_1 = y_2$, V_2 is identically zero for any value of I_s.

Observe that in (4.200) and (4.201) we have $w(\alpha_2) = w(0)$, meaning that they are independent of the controlled current source $\alpha_2 V_2$.

Example 4.18 Consider again the voltage-series feedback amplifier of Fig. 4.24. In Examples 4.11 and 4.14, we obtained the following:

$$F(\tilde{\alpha}_1) = 93.70 \quad F(\tilde{\alpha}_2) = 18.26 \tag{4.203a}$$

$$\hat{F}(\tilde{\alpha}_1) = 103.07 \cdot 10^3 \quad \hat{F}(\tilde{\alpha}_2) = 2018.70 \tag{4.203b}$$

$$w = \frac{V_2}{V_s} = w(\tilde{\alpha}_1) = w(\tilde{\alpha}_2) = 45.39 \tag{4.203c}$$

Appealing to (4.197) yields

$$w(\tilde{\alpha}_1) = w(0) \frac{\hat{F}(\tilde{\alpha}_1)}{F(\text{input short-circuited})} = 0.04126 \frac{103.07 \cdot 10^3}{93.699} = 45.39 \tag{4.204a}$$

where, from (4.138),

$$F(\text{input short-circuited}) = \frac{Y_{11,55}(\tilde{\alpha}_1)}{Y_{11,55}(0)} = \frac{466.07 \cdot 10^{-9}}{4.9741 \cdot 10^{-9}} = 93.699 \tag{4.204b}$$

$$w(0) = \left. \frac{Y_{12,55}(\tilde{\alpha}_1)}{Y_{11,55}(\tilde{\alpha}_1)} \right|_{\tilde{\alpha}_1 = 0} = \frac{205.24 \cdot 10^{-12}}{497.41 \cdot 10^{-11}} = 0.04126 \tag{4.204c}$$

and

$$w(\tilde{\alpha}_2) = w(0) \frac{\hat{F}(\tilde{\alpha}_2)}{F(\text{input short-circuited})} = 0.41058 \frac{2018.70}{18.26} = 45.39 \tag{4.205a}$$

where

$$F(\text{input short-circuited}) = \frac{Y_{11,55}(\tilde{\alpha}_2)}{Y_{11,55}(0)} = \frac{466.07 \cdot 10^{-9}}{25.52 \cdot 10^{-9}} = 18.26 \tag{4.205b}$$

$$w(0) = \left. \frac{Y_{12,55}(\tilde{\alpha}_2)}{Y_{11,55}(\tilde{\alpha}_2)} \right|_{\tilde{\alpha}_2 = 0} = \frac{104.79 \cdot 10^{-10}}{255.22 \cdot 10^{-10}} = 0.41058 \tag{4.205c}$$

4.5 SUMMARY

We began this chapter by introducing the ideal feedback model. To employ the ideal feedback model to analyze a practical feedback amplifier, it becomes necessary to separate the feedback amplifier into two blocks: the basic unilateral amplifier and the feedback network. For this we considered four common types of feedback amplifier configurations. They are referred to as the series-series feedback, parallel-parallel feedback, series-parallel feedback, and parallel-series feedback, and they can be reduced to the ideal single-loop feedback model by employing z-parameters, y-parameters, h-parameters, and g-parameters, respectively. We demonstrated in each case by a practical example how to identify the transfer functions of the basic unilateral amplifier and the feedback network. We found that the gain, input, and output immittances are all affected by the same factor called the feedback factor. More specifically, for negative feedback the gain is reduced by the amount of feedback measured either in nepers or in decibels. The input and output immittances, including the terminating immittances, are increased by the feedback factor from their values when the input and output ports are either open-circuited or short-circuited. Since open-circuiting or short-circuiting the ports clearly breaks the feedback loop, these values can be considered as the input and output immittances of the amplifier without feedback.

In practice, we often run into feedback amplifiers that cannot be classified as being in one of the above four categories. A general feedback configuration may contain an input coupling network and an output coupling network in addition to the basic amplifier and the feedback network. Thus, the ideal feedback model is not an adequate representation of a practical feedback amplifier. What is needed is a more general theory that avoids the necessity of identifying the forward and backward transmission paths. To this end, we presented Bode's theory on feedback amplifiers.

Bode's feedback theory is based on the concepts of return difference and null return difference. We showed that the return difference is a generalization of the concept of the feedback factor of the ideal feedback model and can be measured physically from the amplifier itself. Also, we introduced the notion of null return difference and discussed its physical significance. We demonstrated that the ratio of the two functional values assumed by a transfer function under the condition that the element x of interest assumes its nominal value and the condition that x assumes the value zero is equal to the ratio of the return differences with respect to the element x under the condition that the input excitation I_s has been adjusted so that the output is identically zero and the condition that the input excitation I_s is removed. This result degenerates into the well-known Blackman's formula for an active impedance when the transfer impedance becomes the driving-point impedance. This relation can be employed to simplify the calculation of driving-point impedance of an active network, making it possible to observe the effects of feedback on amplifier impedance and gain.

The effect of feedback on sensitivity, stability, and other related topics will be discussed in the following chapters.

PROBLEMS

4.1 In the feedback amplifier of Fig. 4.43, assume that the two transistors are identical, with $h_{re} = h_{oe} = 0$. Compute the return differences and null return differences with respect to h_{fe}/h_{ie} of the two transistors.

4.2 Show that A_k and s_k in Eq. (4.67) are given by Eq. (4.68).

4.3 Consider the feedback amplifier of Fig. 4.21, which is of the series-parallel or voltage-series type. Assume that the two transistors are identical, with $h_{re} = h_{oe} = 0$. Confirm that the hybrid matrix of the composite two-port network is given by (4.70).

4.4 Use the equivalent network of Fig. 4.23 to compute the voltage gains of the first and second stages. Compare the results with (4.75).

4.5 The feedback amplifier of Fig. 4.27 conforms with the parallel-series or current-shunt feedback configuration. Assume that $h_{re} = 0$ and conditions

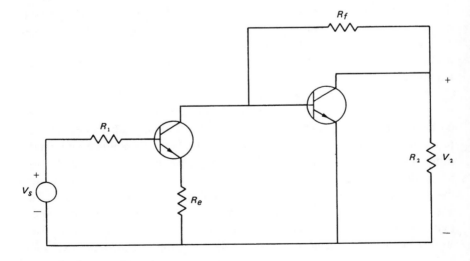

Figure 4.43 A given feedback amplifier.

(4.96) are satisfied. Confirm that the inverse hybrid matrix of the composite two-port network is given by (4.97).

4.6 Use Blackman's formula to compute the impedance facing the conductance $G_2 = 1/R_2$ in the network of Fig. 4.43.

4.7 Compute the null return difference $\hat{F}(\tilde{\alpha}_2)$ of (4.172) by its physical interpretation.

4.8 Demonstrate that (4.176) remains valid if $w(x)$ represents the transfer impedance $z_{rp,sq} = V_{pq}/I_s$ of the generai feedback configuration of Fig. 4.33. Is it still true if $w(x)$ represents the transfer admittance I_{pq}/V_{rs}? If not, how should one modify the formula?

4.9 In the feedback network of Fig. 4.36, compute the return difference and the null return difference with respect to the one-port conductance G_f. Repeat the problem by appealing to their physical significance in terms of the returned voltages.

4.10 Derive (4.187) and (4.188).

4.11 Use physical interpretation of the return difference to compute (4.191).

4.12 Use Blackman's formula to show that the impedance facing the independent voltage source V_s in the network of Fig. 4.44 is given by

$$Z(s) = \frac{y_2 y_f + (y_1 + y_g)(y_2 + y_f) + y_f g_m}{y_1(y_2 y_f + y_g y_f + y_2 y_g + y_f g_m)} \tag{4.206}$$

4.13 By applying Blackman's formula, confirm that the admittance facing y_2 in the feedback network of Fig. 4.44 is given by

$$Y(s) = \frac{y_f(y_1 + y_g + g_m)}{y_1 + y_f + y_g} \tag{4.207}$$

4.14 Apply formula (2.95) to confirm (4.193).

Figure 4.44 A given feedback network.

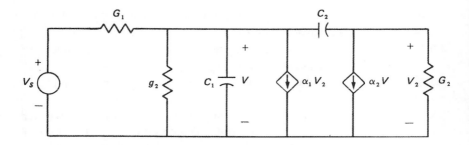

Figure 4.45 A given feedback network with two controlled sources.

4.15 Apply Blackman's formula and determine the effect of feedback on the input and output impedances of the series-series feedback configuration of Fig. 4.3a.

4.16 Repeat Prob. 4.15 for the parallel-parallel feedback configuration of Fig. 4.3b.

4.17 Repeat Prob. 4.15 for the parallel-series feedback configuration of Fig. 4.3d.

4.18 Prove the identities (4.149).

Hint: Use (4.123).

4.19 For the feedback network of Fig. 4.45, compute the following:

(a) The return difference with respect to the element α_2

(b) The impedance facing the capacitor C_2

(c) The return difference with respect to the one-port admittance sC_2

(d) The null return difference with respect to the element α_1

(e) The return difference with respect to the element α_1

4.20 In the compound-feedback amplifier of Fig. 4.46, compute the following:

(a) The return difference with respect to the conductance $G_1 = 1/R_1$

(b) The null return difference with respect to the controlling parameter of the forward transmission of the transistor

(c) The voltage gain of the amplifier

(d) The impedance facing the resistor R_f

4.21 In the two-stage feedback amplifier of Fig. 4.47, assume that the two transistors are identical, with $h_{ie} = 1\,\text{k}\Omega, h_{fe} = 45$, and $h_{re} = h_{oe} = 0$. Determine the following:

(a) The return difference with respect to the controlling parameter h_{fe}/h_{ie} of the second transistor

(b) The null return difference with respect to the controlling parameter h_{fe}/h_{ie} of the second transistor

Figure 4.46 A compound-feedback amplifier.

    (c) The input impedance of the feedback amplifier

    (d) The output impedance of the feedback amplifier

4.22 Repeat (a) and (b) of Prob. 4.21 for the controlling parameter h_{fe}/h_{ie} of the first transistor. Also compute the current gain of the feedback amplifier.

4.23 To distinguish the transistor parameters in the voltage-series feedback amplifier of Fig. 4.21, we use subscripts 1 and 2. Assume that $R_1 = 0$, $h_{re1} = h_{re2} = 0$, and $h_{oe1} = h_{oe2} = 0$. Determine the h-parameters of the composite two-port network N. Use these parameters to show that the closed-loop voltage gain is given by

$$w_V = \frac{h_{fe1}h_{fe2}RR_2(R_e + R_f) + R_e R_2(1 + h_{fe1})(R + h_{ie2})}{(R + h_{ie2})\{h_{ie1}R_e + (R_f + R_2)\left[h_{ie1} + (1 + h_{fe1})R_e\right]\}} \quad (4.208)$$
$$+ h_{fe1}h_{fe2}RR_e R_2$$

Figure 4.47 A two-stage feedback amplifier.

For large h_{fe1} and h_{fe2}, demonstrate that the voltage gain w_V approaches the limiting value $1 + R_f/R_e$, thereby making it practically independent of the transistor parameters.

4.24 Use the subscripts 1 and 2 to distinguish the transistor parameters of Fig. 4.27. Let $h_{oe1} = h_{oe2} = 0, h_{re1} = h_{re2} = 0$, and $R_1 = \infty$. Determine the g-parameters of the composite two-port network N. Use these parameters to show that the closed-loop current gain I_2/I_s is given by

$$w_I = \frac{-h_{fe1}h_{fe2}R(R_e + R_f) - h_{fe2}h_{ie1}R_e}{(R + h_{ie2})(R_e + R_f + h_{ie1})}$$
$$+ R_e(1 + h_{fe2})(Rh_{fe1} + h_{ie1} + R_f) \qquad (4.209)$$

For large h_{fe1} and h_{fe2}, demonstrate that the current gain w_I approaches the limiting value $-(1 + R_f/R_e)$, thereby making it practically independent of the transistor parameters.

4.25 The emitter-follower as shown in Fig. 4.48 is an example of voltage-series feedback amplifier. Compute the following:

(a) The return difference with respect to the element $\alpha = h_{fe}/h_{ie}$ both by its definition and by its physical interpretation

(b) The impedance facing the resistor R_2

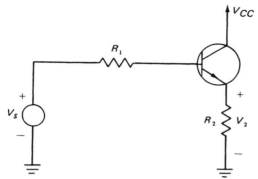

Figure 4.48   A voltage-series feedback amplifier.

(c) The null return difference with respect to the element α both by its definition and by its physical interpretation

(d) The input and output impedances of the amplifier

4.26 In the series-parallel feedback amplifier of Fig. 4.21, assume that the two transistors are identical, with

$$h_{fe} = 60 \qquad h_{ie} = 1.8\,\mathrm{k}\Omega$$
$$h_{re} = h_{oe} = 0 \qquad R_1 = 0$$
$$R = 8\,\mathrm{k}\Omega \qquad R_f = 2\,\mathrm{k}\Omega \tag{4.210}$$
$$R_e = 200\,\Omega \qquad R_2 = 6.8\,\mathrm{k}\Omega$$

Determine the voltage gains of the amplifier with and without feedback. Also compute the input and output immittances of the amplifier.

4.27 Consider the feedback amplifier of Fig. 4.49. To simplify the computation, assume that the transistors are identical, with $h_{re} = h_{oe} = 0$. Determine the open-loop and closed-loop voltage gains and the transfer function of the feedback network.

4.28 At high frequencies, we use the hybrid-pi equivalent model of Fig. 4.10 for the two transistors of Fig. 4.49. Determine the open-loop and closed-loop voltage gains of the amplifier.

4.29 In Fig. 4.15, replace the feedback resistor R_f by a capacitor of capacitance C_f. Show that the voltage gain V_2/V_s of the amplifier with $G_1 = 1/R_1$ and $G_2 = 1/R_2$ is given by

$$w = \frac{-h_{fe} + sC_f h_{ie}}{\begin{aligned} &(G_2 + h_{oe})(R_1 + h_{ie}) - h_{re}h_{fe} \\ &\quad + sC_f R_1[1 + h_{ie}(h_{oe} + G_2 + G_1) + h_{fe}] \end{aligned}} \tag{4.211}$$

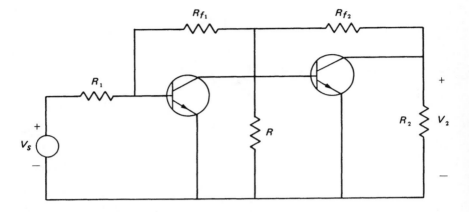

Figure 4.49 A transistor feedback amplifier.

BIBLIOGRAPHY

Black, H. S.: Inventing the Negative Feedback Amplifier, *IEEE Spectrum*, vol. 14, no. 12, pp. 55–60, 1977.

Blackman, R. B.: Effect of Feedback on Impedance, *Bell Syst. Tech. J.*, vol. 22, no. 4, pp. 268–277, 1943.

Bocher, M.: "Introduction to Higher Algebra," New York: Macmillan, 1938.

Bode, H. W.: "Network Analysis and Feedback Amplifier Design," Princeton, N.J.: Van Nostrand, 1945.

Chen, W. K.: Graph-Theoretic Considerations on the Invariance of Return Difference, *J. Franklin Inst.*, vol. 298, no. 2, pp. 81–100, 1974a.

Chen, W. K.: Invariance and Mutual Relations of the General Null-Return-Difference Functions, *Proc. 1974 Eur. Conf Circuit Theory and Design,* IEE Conf. Pubi. No. 116, Inst. of Electrical Engineers, London, pp. 371–376, 1974b.

Chen, W. K.: Graph Theory and Feedback Systems, *Proc. Ninth Asilo mar Conf. on Circuits, Systems, and Computers,* Pacific Grove, Calif., pp. 26–30, 1975.

Chen, W. K.: "Applied Graph Theory: Graphs and Electrical Networks," 2d rev. ed., chap. 4, New York: American Elsevier, and Amsterdam: North-Holland, 1976a.

Chen, W. K.: Indefinite-Admittance Matrix Formulation of Feedback Amplifier Theory, *IEEE Trans. Circuits and Systems,* vol. CAS-23, no. 8, pp. 498–505, 1976b.

Chen, W. K.: Network Functions and Feedback, *Int. J. Electronics,* vol. 42, no. 6, pp. 617–618, 1977.

Chen, W. K.: On Second-Order Cofactors and Null Return Difference in Feedback Amplifier Theory, *Int. J. Circuit Theory and Applications,* vol. 6, no. 3, pp. 305–312, 1978.

Cheng, C. C.: Neutralization and Unilateralization, *IRE Trans. Circuit Theory,* vol. CT-2, no. 2, pp. 138–145, 1955.

Chu, G. Y.: Unilateralization of Junction-Transistor Amplifiers at High Frequencies, *Proc. IRE*, vol. 43, no. 8, pp. 1001–1006, 1955.

Ghausi, M. S.: Optimum Design of the Shunt-Series Feedback Pair with a Maximally Flat Magnitude Response, *IRE Trans. Circuit Theory*, vol. CT-8, no. 4, pp. 448–453, 1961.

Ghausi, M. S.: "Principles and Design of Linear Active Circuits," New York: McGraw-Hill, 1965.

Ghausi, M. S.: "Electronic Circuits," New York: Van Nostrand, 1971.

Hakim, S. S.: Aspects of Return-Difference Evaluation in Transistor Feedback Amplifiers, *Proc. IEE (London)*, vol. 112, no. 9, pp. 1700–1704, 1965.

Haykin, S. S.: "Active Network Theory," Reading, Mass.: Addison-Wesley, 1970.

Hoskins, R. F.: Definition of Loop Gain and Return Difference in Transistor Feedback Amplifiers, *Proc. IEE (London)*, vol. 112, no. 11, pp. 1995–2001, 1965.

Kuh, E. S. and R. A. Rohrer: "Theory of Linear Active Networks," San Francisco, Calif.: Hoiden-Day, 1967.

Millman, J. and C. C. Halkias: "Electronic Fundamentals and Applications: For Engineers and Scientists," New York: McGraw-Hill, 1976.

Mulligan, J. H., Jr.: Signal Transmission in Nonreciprocal Systems, *Proc. Symp. on Active Networks and Feedback Systems*, Polytechnic Inst. of Brooklyn, New York, vol. 10, pp. 125–153, 1960.

Truxal, J. G.: "Automatic Feedback Control System Synthesis," New York: McGraw-Hill, 1955.

THEORY OF FEEDBACK AMPLIFIERS II

In the preceding chapter, we studied the ideal feedback model and demonstrated by several practical examples how to calculate the transfer functions $\mu(s)$ and $\beta(s)$ of the basic amplifier and the feedback network of a given feedback configuration. We introduced Bode's feedback theory, which is based on the concepts of return difference and null return difference. We showed that the return difference is a generalization of the concept of the feedback factor of the ideal feedback model and can be interpreted physically as the difference between the 1-V excitation and the returned voltage. We demonstrated that return difference and null return difference are closely related to network functions and can therefore be employed to simplify the calculation of driving-point impedance of an active network, thereby observing the effects of feedback on amplifier impedance and gain.

In the present chapter, we continue our study of feedback amplifier theory. We show that feedback may be employed to make the gain of an amplifier less sensitive to variations in the parameters of the active components and to reduce the effects of noise and nonlinear distortion, and to affect the stability of the network. The concepts of return difference, null return difference, and sensitivity function will be generalized by introducing the general reference value, which is very useful in measurement situations. Since the zeros of the return difference are also the natural frequencies of the network, they are essential for the stability study. To this end, we present three procedures for the physical measurements of return difference. This is especially important in view of the fact that it is difficult to get an accurate formulation of the equivalent network, which, to a greater or lesser extent, is an idealization of the physical reality. The measurement of return difference provides an experimental verification that the system is stable and will remain so under certain prescribed conditions. Finally, we discuss the invariance of return difference and null return difference under different formulations of network equations.

5.1 SENSITIVITY FUNCTION AND FEEDBACK

One of the most important effects of feedback is its ability to make an amplifier less sensitive to the variations of its parameters because of aging, temperature change, or other environmental changes. In this section, we study this effect.

A useful quantitative measure for the degree of dependence of an amplifier on a particular parameter is sensitivity. Sensitivity is a measure of the change of the overall transfer function to the change of a particular parameter in the network. As indicated in Sec. 3.3, the sensitivity function, written as $S(x)$, with respect to an element x is defined as the ratio of the fractional change in a transfer function to the fractional change in x for the situation when all changes concerned are differentially small (see Definition 3.1). Thus, if $w(x)$ is the transfer function, the sensitivity function can be written as

$$S(x) = \frac{x}{w}\frac{\partial w}{\partial x} = \frac{\partial \ln w}{\partial \ln x} = x\frac{\partial \ln w}{\partial x} \tag{5.1}$$

Consider the general feedback configuration of Fig. 5.1, whose indefinite-admittance matrix is denoted by $\mathbf{Y}$. As before, to emphasize the importance of the element x, we express w and $\mathbf{Y}$ as functions of x alone, even though they are also functions of the complex-frequency variable s, and we write $w = w(x)$ and $\mathbf{Y} = \mathbf{Y}(x)$.

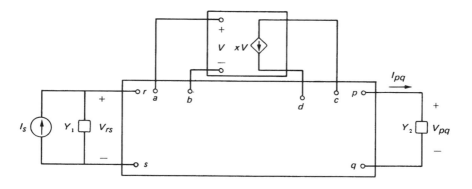

Figure 5.1 The configuration of a general feedback amplifier.

To be definitive, let $w(x)$ represent either the current gain I_{pq}/I_s or the transfer impedance V_{pq}/I_s (Fig. 5.1) for the time being. Then, from Eq. (4.174) we have

$$w(x) = Y_2 \frac{Y_{rp,sq}(x)}{Y_{uv}(x)} \quad \text{or} \quad \frac{Y_{rp,sq}(x)}{Y_{uv}(x)} \tag{5.2}$$

as the case may be. For simplicity, we write

$$\dot{Y}_{uv}(x) = \frac{\partial Y_{uv}(x)}{\partial x} \tag{5.3a}$$

$$\dot{Y}_{rp,sq}(x) = \frac{\partial Y_{rp,sq}(x)}{\partial x} \tag{5.3b}$$

Our objective is to establish a relation among the sensitivity function, the return difference, and the transfer functions. To this end, we need the following identities:

$$Y_{uv}(x) = Y_{uv}(0) + x\dot{Y}_{uv}(x) \tag{5.4a}$$

$$Y_{rp,sq}(x) = Y_{rp,sq}(0) + x\dot{Y}_{rp,sq}(x) \tag{5.4b}$$

as given in Eq. (4.149b). Substituting (5.2) in (5.1) in conjunction with (5.4), we get

$$
\begin{aligned}
S(x) &= x\frac{\dot{Y}_{rp,sq}(x)}{Y_{rp,sq}(x)} - x\frac{\dot{Y}_{uv}(x)}{Y_{uv}(x)} \\
&= \frac{Y_{rp,sq}(x) - Y_{rp,sq}(0)}{Y_{rp,sq}(x)} - \frac{Y_{uv}(x) - Y_{uv}(0)}{Y_{uv}(x)} \\
&= \frac{Y_{uv}(0)}{Y_{uv}(x)} - \frac{Y_{rp,sq}(0)}{Y_{rp,sq}(x)} = \frac{1}{F(x)} - \frac{1}{\hat{F}(x)}
\end{aligned} \tag{5.5}
$$

Combining Eqs. (4.176) and (5.5), we obtain

$$S(x) = \frac{1}{F(x)}\left[1 - \frac{w(0)}{w(x)}\right] \tag{5.6}$$

an equation relating the sensitivity function, the return difference, and the transfer functions. As pointed out in Chap. 4, the concept of null return difference is invalid if $Y_{rp,sq}(0)$ is zero. However, (5.6) remains valid, because under the stipulated condition $w(0) = 0$, showing that (5.5) and (5.6) are again equivalent. It is significant to observe that if $w(0) = 0$, then (5.6) becomes $S(x) = 1/F(x)$, meaning that the sensitivity is equal to the reciprocal of the return difference.

For the ideal feedback model of Fig. 4.1, the feedback path is unilateral. Hence $w(0) = 0$ and $S = 1/F = 1/(1 + T) = 1/(1 - \mu\beta)$. For a practical amplifier, $w(0)$ is usually very much smaller than $w(x)$ in the passband, and $F \approx 1/S$ may be used as a good estimate of the reciprocal of the sensitivity in the same frequency band. A single-loop feedback amplifier composed of a cascade of common-emitter stages with a passive network providing the desired feedback fulfills this requirement. If in such a structure any one of the transistors fails, the forward transmission is nearly zero and $w(0)$ is practically zero. Thus, we conclude that the sensitivity and reciprocal of the return difference are equal for any element whose failure would interrupt the transmission through the amplifier as a whole to zero.

In the particular situation where $r = p$ and $s = q$, the transfer impedance becomes the driving-point impedance, and formula (5.5) or (5.6) is the sensitivity function of a one-port impedance with respect to an element x of interest. We recognize that in this case $w(0)$ is not usually smaller than $w(x)$.

Refer again to Fig. 5.1. Assume that $w(x)$ represents the voltage gain V_{pq}/V_{rs}. From (2.97) we have $w(x) = Y_{rp,sq}(x)/Y_{rr,ss}(x)$. Using this in (5.1) gives

$$
\begin{aligned}
S(x) &= x\frac{\dot{Y}_{rp,sq}(x)}{Y_{rp,sq}(x)} - x\frac{\dot{Y}_{rr,ss}(x)}{Y_{rr,ss}(x)} \\[2mm]
&= \frac{Y_{rp,sq}(x) - Y_{rp,sq}(0)}{Y_{rp,sq}(x)} - \frac{Y_{rr,ss}(x) - Y_{rr,ss}(0)}{Y_{rr,ss}(x)} \\[2mm]
&= \frac{Y_{rr,ss}(0)}{Y_{rr,ss}(x)} - \frac{Y_{rp,sq}(0)}{Y_{rp,sq}(x)} \\[2mm]
&= \frac{1}{F(\text{input short-circuited})} - \frac{1}{\hat{F}(x)}
\end{aligned}
\tag{5.7}
$$

By using (4.197), the sensitivity function can be expressed as

$$
S(x) = \frac{1}{F(\text{input short-circuited})}\left[1 - \frac{w(0)}{w(x)}\right]
\tag{5.8}
$$

Likewise, if $w(x)$ represents the short-circuit current gain I_{pq}/I_s as Y_2 approaches infinity, the sensitivity function can be written as

$$
S(x) = \frac{Y_{pp,qq}(0)}{Y_{pp,qq}(x)} - \frac{Y_{rp,sq}(0)}{Y_{rp,sq}(x)} = \frac{1}{F(\text{output short-circuited})} - \frac{1}{\hat{F}(x)}
\tag{5.9}
$$

which when combined with (4.199) gives

$$S(x) = \frac{1}{F(\text{output short-circuited})} \left[1 - \frac{w(0)}{w(x)} \right] \qquad (5.10)$$

The derivation of (5.9) is left as an exercise (see Prob. 5.1). Observe that formulas (5.6), (5.8), and (5.10) are quite similar. If the return difference $F(x)$ is interpreted properly, they can all be represented by the single relation (5.6). As before, if $w(0) = 0$, the sensitivity for the voltage gain function is equal to the reciprocal of the return difference under the situation that the input port of the amplifier is short-circuited, and the sensitivity for the short-circuit current gain is equal to the reciprocal of the return difference when the output port is short-circuited.

We illustrate these by the following examples.

Example 5.1 In the feedback network of Fig. 5.2, let α_2 be the element of interest. As demonstrated in Example 4.17, the voltage gain or short-circuit current gain $w(\alpha_2)$ is independent of the controlling parameter α_2, giving $w(x) = w(0)$. From (5.8) and (5.10), we have $S(\alpha_2) = 0$, meaning that $w(\alpha_2)$ is invariant to the change of the value of α_2, as expected.

On the other hand, suppose that we pick α_1 to be the element of interest. The voltage gain was computed earlier in Example 4.17 and is given by

$$w(\alpha_1) = \frac{y_2 - \alpha_1}{y_2 + y_3} \qquad (5.11)$$

$F(\text{input short-circuited})$ is the return difference with respect to α_1 when terminals 1 and 3 are short-circuited. Since shorting terminals 1 and 3 breaks the feedback

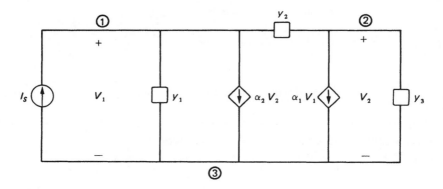

Figure 5.2 A feedback network used to illustrate the computation of the sensitivity function.

loop, F(input short-circuited) $= 1$. Substituting these in (5.8) yields the sensitivity function for the voltage gain with respect to the controlling parameter α_1:

$$S(\alpha_1) = 1 - \frac{w(0)}{w(\alpha_1)} = \frac{\alpha_1}{\alpha_1 - y_2} \tag{5.12}$$

Likewise, applying (4.201) and the fact that F(output short-circuited) $= 1$, we can show that the sensitivity function for the short-circuit current gain with respect to α_1 is again given by Eq. (5.12) (see Prob. 5.3).

Example 5.2 Consider the common-emitter transistor amplifier of Fig. 5.3a. After removing the biasing circuit, its equivalent network is presented in

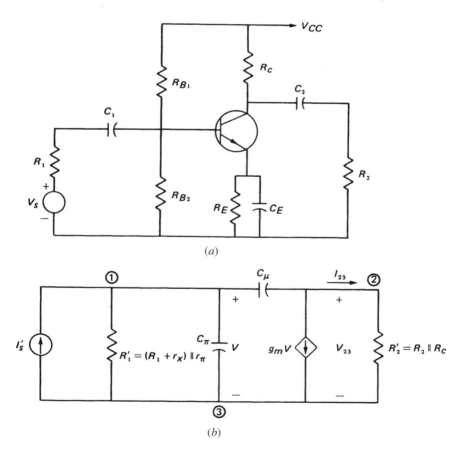

(a)

(b)

Figure 5.3 (a) A common-emitter transistor feedback amplifier and (b) its equivalent network.

Fig. 5.3*b* with

$$I'_s = \frac{V_s}{R_1 + r_x} \tag{5.13a}$$

$$G'_1 = \frac{1}{R'_1} = \frac{1}{R_1 + r_x} + \frac{1}{r_\pi} \tag{5.13b}$$

$$G'_2 = \frac{1}{R'_2} = \frac{1}{R_2} + \frac{1}{R_c} \tag{5.13c}$$

The indefinite-admittance matrix of the network of Fig. 5.3*b* can be written down directly by inspection as

$$\mathbf{Y} = \begin{bmatrix} G'_1 + sC_\pi + sC_\mu & -sC_\mu & -G'_1 - sC_\pi \\ -sC_\mu + g_m & G'_2 + sC_\mu & -G'_2 - g_m \\ -G'_1 - sC_\pi - g_m & -G'_2 & G'_1 + G'_2 + sC_\pi + g_m \end{bmatrix} \tag{5.14}$$

Suppose that the controlling parameter g_m is the element of interest. The return difference and the null return difference with respect to g_m in Fig. 5.3*b*, with I'_s as the input port and R'_2 as the output port, are found to be

$$F(g_m) = \frac{Y_{33}(g_m)}{Y_{33}(0)} = \frac{(G'_1 + sC_\pi)(G'_2 + sC_\mu) + sC_\mu(G'_2 + g_m)}{(G'_1 + sC_\pi)(G'_2 + sC_\mu) + sC_\mu G'_2} \tag{5.15}$$

$$\hat{F}(g_m) = \frac{Y_{12,33}(g_m)}{Y_{12,33}(0)} = \frac{sC_\mu - g_m}{sC_\mu} = 1 - \frac{g_m}{sC_\mu} \tag{5.16}$$

The current gain I_{23}/I'_s, as defined in Fig. 5.3*b*, is determined to be

$$w(g_m) = \frac{Y_{12,33}(g_m)}{R'_2 Y_{33}(g_m)} = \frac{sC_\mu - g_m}{R'_2[(G'_1 + sC_\pi)(G'_2 + sC_\mu) + sC_\mu(G'_2 + g_m)]} \tag{5.17}$$

Substituting (5.15) and (5.16) in (5.5) or (5.15) and (5.17) in (5.6), we obtain the sensitivity function for the current gain I_{23}/I'_s or the transfer impedance V_{23}/I'_s with respect to the transconductance g_m:

$$S(g_m) = \frac{g_m(G'_2 + sC_\mu)(G'_1 + sC_\pi + sC_\mu)}{(sC_\mu - g_m)[(G'_1 + sC_\pi)(G'_2 + sC_\mu) + sC_\mu(G'_2 + g_m)]} \tag{5.18}$$

Finally, we compute the sensitivity for the driving-point impedance facing the current source I'_s. From (5.6) we have

$$S(g_m) = \frac{1}{F(g_m)}\left[1 - \frac{Z(0)}{Z(g_m)}\right] = -\frac{sC_\mu g_m}{(G'_1 + sC_\pi)(G'_2 + sC_\mu) + sC_\mu(G'_2 + g_m)} \tag{5.19}$$

where

$$Z(g_m) = \frac{Y_{11,33}(g_m)}{Y_{33}(g_m)} = \frac{G'_2 + sC_\mu}{(G'_1 + sC_\pi)(G'_2 + sC_\mu) + sC_\mu(G'_2 + g_m)} \tag{5.20}$$

Example 5.3 Figure 5.4a represents a common-emitter stage with a resistor R_f connected from the output to the input to provide an external physical feedback. By using the common-emitter hybrid model of a transistor at low frequencies, an equivalent network of the feedback amplifier is presented in Fig. 5.4b, whose indefinite-admittance matrix is given by

$$\mathbf{Y} = \begin{bmatrix} G_1 + G_f + \dfrac{1}{h_{ie}} & -G_f & -G_1 - \dfrac{1}{h_{ie}} \\ \alpha - G_f & G_2 + G_f & -G_2 - \alpha \\ -G_1 - \dfrac{1}{h_{ie}} - \alpha & -G_2 & G_1 + G_2 + \dfrac{1}{h_{ie}} + \alpha \end{bmatrix} \tag{5.21}$$

where $\alpha \equiv h_{fe}/h_{ie}$ and h_{re} and h_{oe} are assumed to be zero. The return difference with respect to the controlling parameter α is found to be

$$F(\alpha) = \frac{Y_{33}(\alpha)}{Y_{33}(0)} = \frac{(G_2 + G_f)(G_1 + 1/h_{ie}) + G_f(\alpha + G_2)}{(G_2 + G_f)(G_1 + 1/h_{ie}) + G_2 G_f}$$

$$= 1 + \frac{h_{fe} R'_2 R}{R_f(h_{ie} + R)} \tag{5.22}$$

where $R'_2 = R_2 R_f/(R_2 + R_f)$ and $R = R_1(R_2 + R_f)/(R_1 + R_2 + R_f)$. By applying (2.97), the voltage gain V_{23}/V_{13} is obtained as

$$w(\alpha) = \frac{V_{23}}{V_{13}} = \frac{Y_{12,33}(\alpha)}{Y_{11,33}(\alpha)} = \frac{G_f - \alpha}{G_2 + G_f} = \frac{R_2(1 - \alpha R_f)}{R_2 + R_f} \tag{5.23}$$

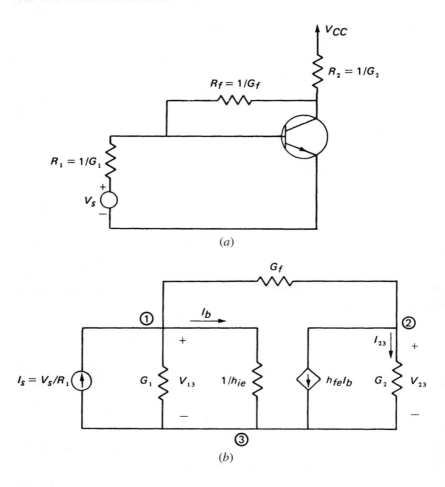

Figure 5.4 (a) A common-emitter feedback amplifier and (b) its low-frequency equivalent network.

giving

$$w(\alpha) = (1 - \alpha R_f)w(0) \qquad (5.24)$$

Since for a practical amplifier $|1 - \alpha R_f| \gg 1$, the ratio $w(0)/w(\alpha)$ is practically zero, and from (5.8) the sensitivity of the voltage gain with respect to α approximately equals the reciprocal of F(input short-circuited), which is unity. Thus, the sensitivity is about unity, being independent of α.

Suppose that we are interested in the sensitivity of the voltage gain V_{23}/V_s of the feedback amplifier with respect to α. We consider the equivalent network of

Figure 5.5 An equivalent network of the feedback amplifier of Fig. 5.4.

Fig. 5.5, whose indefinite-admittance matrix is given by

$$
\mathbf{Y} =
\begin{bmatrix}
G_1 & -G_1 & 0 & 0 \\
-G_1 & G_f + G_1 + \dfrac{1}{h_{ie}} & -G_f & -\dfrac{1}{h_{ie}} \\
0 & \alpha - G_f & G_f + G_2 & -\alpha - G_2 \\
0 & -\alpha - \dfrac{1}{h_{ie}} & -G_2 & \alpha + G_2 + \dfrac{1}{h_{ie}}
\end{bmatrix}
\tag{5.25}
$$

By appealing once more to (2.97) and using (5.25), the amplifier gain is given by

$$
w(\alpha) = \frac{V_{34}}{V_s} = \frac{Y_{13,44}(\alpha)}{Y_{11,44}(\alpha)} = \frac{G_1(G_f - \alpha)}{(G_2 + G_f)(G_1 + 1/h_{ie}) + G_f(G_2 + \alpha)}
\tag{5.26}
$$

which can be expressed in the form of (4.3) as

$$
w(\alpha) = \frac{\mu(\alpha)}{1 - \mu(\alpha)\beta(\alpha)} = \frac{\mu(\alpha)}{F(\alpha)}
\tag{5.27}
$$

where $F(\alpha)$ is given in (5.22) and (see Prob. 5.4)

$$
\mu(\alpha) = -\frac{(\alpha R_f - 1)h_{ie} R_2' R}{R_1 R_f (h_{ie} + R)}
\tag{5.28a}
$$

$$
\beta(\alpha) = \frac{R_1 \alpha}{\alpha R_f - 1}
\tag{5.28b}
$$

$\mu(\alpha)$ may be considered as the amplifier gain without feedback. Thus, the amplifier gain with feedback is reduced by a factor equal to the return difference under normal operating conditions. Consider the ratio

$$\frac{w(0)}{w(\alpha)} = \frac{F(\alpha)\mu(0)}{F(0)\mu(\alpha)} = \frac{F(\alpha)}{F(0)} \frac{1}{1 - \alpha R_f} \tag{5.29}$$

which is usually very small compared with unity because $|1 - \alpha R_f| \gg 1$. Thus, from (5.8) the sensitivity of the voltage gain of the feedback amplifier with respect to the controlling parameter α is approximately equal to the reciprocal of the return difference $F(\alpha)$ when the voltage source V_s is short-circuited. This is precisely the same return difference obtained in (5.22) for the network of Fig. 5.4b under normal operating conditions. This leads to

$$S(\alpha) \approx \frac{R_f(h_{ie} + R)}{R_f(h_{ie} + R) + h_{fe}R_2'R} \tag{5.30}$$

We note that the matrix (5.21) can be derived from (5.25) by first adding row 1 to row 4, then column 1 to column 4, and finally deleting row 1 and column 1. The operations are equivalent to shorting terminals 1 and 4.

As an illustration, we use the following set of parameters:

$$R_f = 40\,\text{k}\Omega \quad R_1 = 10\,\text{k}\Omega \quad R_2 = 4\,\text{k}\Omega$$

$$h_{fe} = 50 \quad h_{ie} = 1.1\,\text{k}\Omega \quad h_{re} = h_{oe} = 0 \tag{5.31}$$

Substituting these in (5.22) and (5.26)–(5.30), we obtain

$$F(\alpha) = 5.01$$

$$\mu(\alpha) = -16.01$$

$$w(\alpha) = -3.20 \tag{5.32}$$

$$\frac{w(0)}{w(\alpha)} = -0.0028$$

$$S(\alpha) = 0.20$$

It is instructive to compare the above results with those obtained by approximating the feedback amplifier to the ideal feedback model of Fig. 4.1. From Prob. 5.6 and using the same set of parameters as given in (5.31), we obtain the gain without feedback to be -16, the feedback factor to be 5, and the gain with feedback to be -3.2. The results are nearly the same as those given above.

5.2   THE RETURN DIFFERENCE AND TWO-PORT FUNCTIONS

In Secs. 3.1–3.3, we derived the general expressions of the network functions and power gains in terms of the general hybrid parameters. In the present section, we show the general relations between these expressions and feedback, and we demonstrate by a specific example how they can be employed to design amplifiers that achieve maximum gain with a prescribed sensitivity.

Consider an active two-port device N_a characterized by its y-parameters y_{ij}. Suppose that the two-port is terminated at its input and output ports by the admittances Y_1 and Y_2, as shown in Fig. 3.1b. The indefinite-admittance matrix is found to be

$$
\mathbf{Y} = \begin{bmatrix} y_{11} + Y_1 & y_{12} & -y_{11} - y_{12} - Y_1 \\ y_{21} & y_{22} + Y_2 & -y_{21} - y_{22} - Y_2 \\ -y_{11} - y_{21} - Y_1 & -y_{12} - y_{22} - Y_2 & y_{11} + y_{12} + y_{21} + y_{22} + Y_1 + Y_2 \end{bmatrix}
$$
(5.33)

The return difference with respect to the forward-transfer parameter y_{21} is given by

$$
F(y_{21}) = \frac{Y_{33}(y_{21})}{Y_{33}(0)} = \frac{(y_{11} + Y_1)(y_{22} + Y_2) - y_{12}y_{21}}{(y_{11} + Y_1)(y_{22} + Y_2)} = 1 + T_y
$$
(5.34)

where

$$
T_y = -\frac{y_{12}y_{21}}{(y_{11} + Y_1)(y_{22} + Y_2)}
$$
(5.35)

is the return ratio with respect to y_{21}.

From App. II, we can express the y-parameters y_{ij} in terms of the h-parameters h_{ij} of N_a, giving

$$
Y_{33} = \left(\frac{1}{h_{11}} + Y_1\right)\left(h_{22} - \frac{h_{12}h_{21}}{h_{11}} + Y_2\right) + \frac{h_{12}h_{21}}{h_{11}^2}
$$
(5.36)

The return difference with respect to h_{21} becomes

$$
F(h_{21}) = \frac{Y_{33}(h_{21})}{Y_{33}(0)} = \frac{(h_{11} + Z_1)(h_{22} + Y_2) - h_{12}h_{21}}{(h_{11} + Z_1)(h_{22} + Y_2)} = 1 + T_h
$$
(5.37)

where $Z_1 = 1/Y_1$ and

$$T_h = -\frac{h_{12}h_{21}}{(h_{11} + Z_1)(h_{22} + Y_2)} \qquad (5.38)$$

is the return ratio with respect to h_{21}. Likewise, in terms of the g-parameters g_{ij} of N_a, we have

$$Y_{33} = \left(g_{11} - \frac{g_{12}g_{21}}{g_{22}} + Y_1 \right) \left(\frac{1}{g_{22}} + Y_2 \right) + \frac{g_{12}g_{21}}{g_{22}^2} \qquad (5.39)$$

yielding the return difference with respect to g_{21} as

$$F(g_{21}) = \frac{Y_{33}(g_{21})}{Y_{33}(0)} = \frac{(g_{11} + Y_1)(g_{22} + Z_2) - g_{12}g_{21}}{(g_{11} + Y_1)(g_{22} + Z_2)} = 1 + T_g \qquad (5.40)$$

where $Z_2 = 1/Y_2$ and

$$T_g = -\frac{g_{12}g_{21}}{(g_{11} + Y_1)(g_{22} + Z_2)} \qquad (5.41)$$

is the return ratio with respect to g_{21}.

From the above discussion, we recognize that return difference and return ratio with respect to the forward-transfer parameter are invariant under immittance substitution for y-, h-, and g-parameters, and all can be written in terms of the general hybrid parameters k_{ij} by the single expression

$$F(k_{21}) = \frac{Y_{33}(k_{21})}{Y_{33}(0)} = \frac{(k_{11} + M_1)(k_{22} + M_2) - k_{12}k_{21}}{(k_{11} + M_1)(k_{22} + M_2)} = 1 + T_k \qquad (5.42)$$

where, as before, M_1 and M_2 denote the terminating immittances, and

$$T_k = -\frac{k_{12}k_{21}}{(k_{11} + M_1)(k_{22} + M_2)} \qquad (5.43)$$

We emphasize that the above expressions are valid for y-, h-, or g-parameters. For the z-parameters z_{ij}, the situation is somewhat different and will be elaborated in Sec. 5.9.2. In this case, the first-order cofactor Y_{33} becomes

$$Y_{33} = \left(\frac{z_{22}}{\Delta_z} + Y_1\right)\left(\frac{z_{11}}{\Delta_z} + Y_2\right) - \frac{z_{12}z_{21}}{\Delta_z^2} \tag{5.44}$$

where $\Delta_z = z_{11}z_{22} - z_{12}z_{21}$, yielding

$$F(z_{21}) = \frac{z_{11}z_{22}\left[(z_{11} + Z_1)(z_{22} + Z_2) - z_{12}z_{21}\right]}{\Delta_z(z_{11} + Z_1)(z_{22} + Z_2)} = \frac{z_{11}z_{22}}{\Delta_z}F^m(z_{21}) \tag{5.45a}$$

where $F^m(z_{21}) = 1 + T_z^m$ and

$$T_z^m = -\frac{z_{12}z_{21}}{(z_{11} + Z_1)(z_{22} + Z_2)} \tag{5.45b}$$

Even though the return difference $F(k_{21})$ and return ratio T_k are invariant under immittance substitution for the y-, h-, and g-parameters, it does not mean that their values are the same. For a given active two-port device, the return difference is not unique and depends on the matrix representation used to characterize its external port behavior. We illustrate this by the following example.

Example 5.4 The transistor amplifier of Fig. 5.6 is in the common-emitter connection. Suppose that the parameters of the amplifier are specified by

$$h_{ie} = 1.1\,\text{k}\Omega \quad h_{re} = 2.5 \cdot 10^{-4}$$
$$h_{fe} = 50 \quad h_{oe} = 25\,\mu\text{mho} \tag{5.46}$$
$$R_1 = 10\,\text{k}\Omega \quad R_2 = 4\,\text{k}\Omega$$

Figure 5.6   A common-emitter transistor amplifier.

From App. II, the corresponding y-, g-, and z-parameters are found to be

$$y_{11} = 0.91 \cdot 10^{-3} \, \text{mho} \quad y_{12} = -0.23 \, \mu\text{mho}$$

$$y_{21} = 45.46 \cdot 10^{-3} \, \text{mho} \quad y_{22} = 13.64 \, \mu\text{mho}$$

$$g_{11} = 1.67 \cdot 10^{-3} \, \text{mho} \quad g_{12} = -16.67 \cdot 10^{-3}$$

$$g_{21} = -3.33 \cdot 10^3 \quad g_{22} = 73.33 \, \text{k}\Omega \tag{5.47}$$

$$z_{11} = 600 \, \Omega \quad z_{12} = 10 \, \Omega$$

$$z_{21} = -2 \, \text{M}\Omega \quad z_{22} = 40 \, \text{k}\Omega$$

From these we obtain

$$T_y = 39.3 \cdot 10^{-3} \quad F(y_{21}) = 1.039$$

$$T_h = -4.10 \cdot 10^{-3} \quad F(h_{21}) = 0.996 \tag{5.48}$$

$$T_g = -0.41 \quad F(g_{21}) = 0.59$$

and $F^m(z_{21}) = 1.043$, indicating that the internal feedback for the common-emitter stage is negative for the y- and z-representations and is positive for the h- and g-representations of the transistor.

In terms of the return difference, the general formulas for the transfer functions, as defined by Table 3.1 and given by Eq. (3.25), can be written as

$$\frac{y_2}{u_s} = \frac{k_{21}M_2}{(k_{11} + M_1)(k_{22} + M_2) - k_{12}k_{21}} = -\frac{M_2 T_k/k_{12}}{F(k_{21})} \tag{5.49a}$$

$$\frac{u_2}{u_s} = -\frac{k_{21}}{(k_{11} + M_1)(k_{22} + M_2) - k_{12}k_{21}} = \frac{T_k/k_{12}}{F(k_{21})} \tag{5.49b}$$

From (3.29), the sensitivity functions of (5.49) to changes of the general hybrid parameters are given by

$$S(k_{11}) = -\frac{k_{11}/(k_{11} + M_1)}{1 + T_k} = -\frac{k_{11}/(k_{11} + M_1)}{F(k_{21})} \tag{5.50a}$$

$$S(k_{12}) = -\frac{T_k}{1 + T_k} = -\frac{T_k}{F(k_{21})} \tag{5.50b}$$

$$S(k_{21}) = -\frac{1}{1 + T_k} = \frac{1}{F(k_{21})} \tag{5.50c}$$

$$S(k_{22}) = \frac{k_{22}/(k_{22} + M_2)}{1 + T_k} = -\frac{k_{22}/(k_{22} + M_2)}{F(k_{21})} \tag{5.50d}$$

From (5.50) we observe that in making T_k or $F(k_{21})$ large and positive, the sensitivities to changes of k_{11}, k_{21}, and k_{22} are reduced, whereas the sensitivity to changes of k_{12} is close to unity. However, too much feedback may bring a system to the condition of oscillation, which is not what we want. Also, from (5.50) we have

$$S(k_{21}) - S(k_{12}) = 1 \tag{5.51}$$

In amplifier performance, the transfer functions (5.49) are less significant than the transducer power gain. Therefore, we express the transducer power gain, as given in (3.19b), in terms of the return difference, which is important in determining the sensitivities.

$$\begin{aligned}
\mathcal{G} &= \frac{4|k_{21}|^2 \operatorname{Re} M_1 \operatorname{Re} M_2}{|(k_{11} + M_1)(k_{22} + M_2) - k_{12}k_{21}|^2} = \frac{4|k_{21}|^2 \operatorname{Re} M_1 \operatorname{Re} M_2}{|k_{11} + M_1|^2|k_{22} + M_2|^2|1 + T_k|^2} \\
&= \frac{4|k_{21}|^2 \operatorname{Re} M_1 \operatorname{Re} M_2}{|k_{11} + M_1|^2|k_{22} + M_2|^2|F(k_{21})|^2} = \frac{4|k_{21}|^2 \operatorname{Re} M_1 \operatorname{Re} M_2|S(k_{21})|^2}{|k_{11} + M_1|^2|k_{22} + M_2|^2}
\end{aligned} \tag{5.52}$$

As a result, to increase feedback to reduce the sensitivity, the magnitudes of the transfer functions and transducer power gain are also reduced. In general, an improvement in sensitivity is accompanied by a reduction in gain. However, the gain reduction resulting from feedback is far outweighed by the improvement in sensitivity, because the loss of gain can easily be made up by another feedback amplifier connected in cascade if necessary. In the following, we demonstrate by a specific example the design of a feedback amplifier that attains the maximum transducer power gain with prescribed sensitivity.

Example 5.5 A transistor in common-emitter mode is characterized by its hybrid parameters:

$$h_{a11} = h_{ie} = 1.1 \,\mathrm{k\Omega} \quad h_{a12} = h_{re} = 2.5 \cdot 10^{-4}$$

$$h_{a21} = h_{fe} = 50 \quad h_{a22} = h_{oe} = 25 \,\mu\mathrm{mho} \tag{5.53}$$

The most significant parameter that will affect amplifier performance is the forward-transfer parameter h_{fe}, which is subject to variations caused by drift. Thus, we assume that the sensitivity $S(h_{fe})$ is an appropriate measure of amplifier performance. Suppose that we wish to design a feedback amplifier whose sensitivity is 10% of that of the amplifier without feedback in which the transistor is terminated for maximum transducer power gain. To this end, we first determine the terminations that should be used with the transistor to yield maximum transducer power gain. From (3.164)–(3.166), we get

$$Z_{1,\text{opt}} = 812.41 \ \Omega \tag{5.54a}$$

$$Y_{2,\text{opt}} = 18.46 \ \mu\text{mho} \tag{5.54b}$$

$$G_{\text{max}} = 30077 \tag{5.54c}$$

As indicated in Sec. 4.2, there are four basic feedback configurations, as shown schematically in Fig. 4.3. For our purposes, we choose the series-parallel topology of Fig. 4.3c to desensitize the amplifier performance to variations of h_{fe}. To simplify our computation, the passive feedback network N_f is an ideal transformer of turns ratio $b : 1$, as shown in Fig. 5.7. Our objective is to determine the turns ratio and the terminating immittances that will reduce the effect of the variations of h_{fe} on amplification by a factor of 10.

Applying the Brune tests of Fig. 4.4c to the network of Fig. 5.7 shows that the conditions are satisfied. The h-parameters h_{ij} of the composite two-port network

Figure 5.7 A transistor amplifier with the ideal transformer providing the desired external feedback.

N are therefore given by

$$h_{11} = h_{a11} + h_{f11} = h_{a11} \tag{5.55a}$$

$$h_{12} = h_{a12} + h_{f12} = h_{a12} + b \tag{5.55b}$$

$$h_{21} = h_{a21} + h_{f21} = h_{a21} - b \tag{5.55c}$$

$$h_{22} = h_{a22} + h_{f22} = h_{a22} \tag{5.55d}$$

Under the optimum terminations of (5.54), the return difference of the transistor N_a with respect to h_{fe} is found from (5.37) to be

$$F(h_{fe}) = F(h_{a21}) = 0.85 \tag{5.56}$$

showing a positive internal feedback. From (5.50c) we get

$$S(h_{fe}) = S(h_{a21}) = 1.18 \tag{5.57}$$

As indicated in (4.18a), the difference between h_{21} and h_{a21} is usually very small, and our requirement can be stated as one of reducing $S(h_{21})$ to 10% of $S(h_{a21}) = 1.18$ with minimum loss in gain.

From (5.52) we have the transducer power gain

$$\mathcal{G} = \frac{4h_{21}^2 R_1 S^2(h_{21})/R_2}{(h_{11} + R_1)^2(h_{22} + 1/R_2)^2} \tag{5.58}$$

since all the quantities are real. To maximize $\mathcal{G}$ with prescribed $S(h_{21})$, we must minimize

$$\frac{(h_{11} + R_1)^2}{R_1} \quad \text{and} \quad \left(h_{22} + \frac{1}{R_2}\right)^2 R_2 \tag{5.59}$$

by choosing appropriate values of R_1 and R_2. The minimum values of (5.59) occur at

$$R_1 = h_{11} = h_{a11} = 1.1\,\text{k}\Omega \tag{5.60a}$$

$$R_2 = \frac{1}{h_{22}} = \frac{1}{h_{a22}} = 40\,\text{k}\Omega \tag{5.60b}$$

From (5.50c), the desired amount of feedback is found to be

$$F(h_{21}) = \frac{1}{S(h_{21})} = \frac{1}{0.1 \cdot 1.18} = 8.475 \tag{5.61}$$

or $T_h = 7.475$, which when substituted in (5.38) yields

$$-h_{12}h_{21} = T_h(h_{11} + R_1)\left(h_{22} + \frac{1}{R_2}\right) = 0.822 \qquad (5.62)$$

Appealing to (5.55b) and (5.55c), we get

$$b^2 - 50b - 0.8345 = 0 \qquad (5.63)$$

giving $b = -0.01669$ and $b = 50.0164$. Since $|h_{a21}|$ should be much larger than $|h_{f21}|$, we choose

$$b = -0.01669 \qquad (5.64)$$

The minus sign indicates the polarity of the transformer. The h-parameters of the composite two-port network N become

$$h_{11} = 1.1\,\mathrm{k}\Omega \quad h_{12} = -16.44 \cdot 10^{-3}$$
$$h_{21} \approx 50 \quad h_{22} = 25\,\mu\mathrm{mho} \qquad (5.65)$$

Thus, the major effect of adding the feedback network N_f is to change the reverse-transfer parameter, and the other parameters are essentially the same as the transistor N_a. From (3.7), (3.8), and (5.58), the desensitized amplifier input and output immittances and transducer power gain are given by

$$Z_{\mathrm{in}} = Z_{11} = 17.54\,\mathrm{k}\Omega \qquad (5.66a)$$

$$Y_{\mathrm{out}} = Y_{22} = 3.99 \cdot 10^{-4}\,\mathrm{mho} \qquad (5.66b)$$

$$\mathcal{G} = 316.46 \qquad (5.66c)$$

These results are to be compared with the situation where N is optimally terminated. Again, from (3.164)–(3.166), (5.42), and (5.50c), we obtain

$$Z_{1,\mathrm{opt}} = 6.11\,\mathrm{k}\Omega \quad Y_{2,\mathrm{opt}} = 139\,\mu\mathrm{mho}$$
$$\mathcal{G}_{\mathrm{max}} = 2114 \quad S(h_{21}) = 0.59 \qquad (5.67)$$

where $F(h_{21}) = 1.695$, showing that the feedback is now negative.

Before we turn our attention to another subject, we mention the effect of feedback on input and output immittances. As shown in (4.19a), the input immittance $M_{11} + M_1$ looking into the source of the composite two-port network of Fig. 4.3 can be written as

$$M_{11} + M_1 = (k_{11} + M_1)(1 + T_k) = (k_{11} + M_1)F(k_{21}) \qquad (5.68)$$

In words, this states that the input immittance is increased by a factor that is the normal value of the return difference with respect to the forward-transfer parameter k_{21} from its value when the output port is open-circuited or short-circuited, depending on the two-port parameters used. Since by open-circuiting or short-circuiting the output port we break the feedback loop, we can say that feedback increases the input immittance by the factor $F(k_{21})$ from its value without feedback.

As an example, consider the series-parallel configuration of Fig. 4.3c. It is convenient to use the h-parameters h_{ij}. Then $k_{11} + M_1 = h_{11} + Z_1$ denotes the input impedance looking into the voltage source when the output port is short-circuited. The short-circuiting of the output port breaks the feedback loop. Consequently, the input impedance with feedback is increased by the factor $F(h_{21})$ from its value without feedback.

Likewise, from Eq. (4.19b), the output immittance $M_{22} + M_2$ looking into the output port, including the load M_2, can be written as

$$M_{22} + M_2 = (k_{22} + M_2)(1 + T_k) = (k_{22} + M_2)F(k_{21}) \qquad (5.69)$$

Thus, the output immittance with feedback is also increased from its value without feedback by the factor $F(k_{21})$.

In Example 5.5, we use (5.61), (5.65), (5.66a), and (5.66b). It is straightforward to confirm the above assertion for the h-parameters.

5.3 RETURN DIFFERENCE AND NULL RETURN DIFFERENCE WITH RESPECT TO TWO ELEMENTS

In this section, we introduce return difference and null return difference with respect to two elements, which are found helpful in feedback calculations.

Suppose that x_1 and x_2 are two elements of interest in a feedback network. To express the fact that the first- and second-order cofactors are functions of both x_1 and x_2, we write them as $Y_{uv}(x_1, x_2)$ and $Y_{rp,sq}(x_1, x_2)$ so that the elements can be exhibited explicitly. The return differences with respect to x_1 and x_2 individually can be written as

$$F(x_1) = \frac{Y_{uv}(x_1, x_2)}{Y_{uv}(0, x_2)} \qquad (5.70)$$

for the element x_1, and

$$F(x_2) = \frac{Y_{uv}(x_1, x_2)}{Y_{uv}(x_1, 0)} \qquad (5.71)$$

for the element x_2. Their ratio is therefore

$$\frac{F(x_1)}{F(x_2)} = \frac{Y_{uv}(x_1, 0)}{Y_{uv}(0, x_2)} = \frac{Y_{uv}(x_1, 0)/Y_{uv}(0, 0)}{Y_{uv}(0, x_2)/Y_{uv}(0, 0)} \qquad (5.72)$$

provided that $Y_{uv}(0, 0) \neq 0$. This is equivalent to

$$\frac{F(x_1)}{F(x_2)} = \frac{F(x_1)|_{x_2=0}}{F(x_2)|_{x_1=0}} \qquad (5.73)$$

In words, this states that the ratio of the return differences with respect to the elements x_1 and x_2 individually is equal to the ratio of the return difference with respect to x_1 when x_2 is set to zero and that with respect to x_2 when x_1 is set to zero. Thus, it may be employed to compute the return difference with respect to a given element from the known return difference with respect to another element.

Likewise, the null return differences with respect to the elements x_1 and x_2 individually can be expressed as

$$\hat{F}(x_1) = \frac{Y_{rp,sq}(x_1, x_2)}{Y_{rp,sq}(0, x_2)} \qquad (5.74)$$

$$\hat{F}(x_2) = \frac{Y_{rp,sq}(x_1, x_2)}{Y_{rp,sq}(x_1, 0)} \qquad (5.75)$$

Their ratio becomes

$$\frac{\hat{F}(x_1)}{\hat{F}(x_2)} = \frac{Y_{rp,sq}(x_1, 0)/Y_{rp,sq}(0, 0)}{Y_{rp,sq}(0, x_2)/Y_{rp,sq}(0, 0)} \qquad (5.76)$$

provided that $Y_{rp,sq}(0, 0) \neq 0$, yielding

$$\frac{\hat{F}(x_1)}{\hat{F}(x_2)} = \frac{\hat{F}(x_1)|_{x_2=0}}{\hat{F}(x_2)|_{x_1=0}} \qquad (5.77)$$

Example 5.6 In the feedback network of Fig. 5.4b, we compute the return difference with respect to the conductance G_1 by means of (5.73). Let $x_1 = G_1$ and $x_2 = \alpha \equiv h_{fe}/h_{ie}$. Then $F(\alpha)|_{G_1=0}$ can easily be deduced from (5.22) and is

given by

$$F(\alpha)|_{G_1=0} = 1 + \frac{\alpha G_f}{(G_2 + G_f)/h_{ie} + G_2 G_f} \tag{5.78}$$

To compute $F(G_1)|_{\alpha=0}$, we use the concept that the return difference with respect to a one-port admittance is equal to the ratio of the total admittance at the terminal pair where the admittance is connected to the admittance that it faces. Thus, by setting $\alpha = 0$ in Fig. 5.4b, we obtain

$$F(G_1)|_{\alpha=0} = 1 + \frac{G_1(G_2 + G_f)}{G_f G_2 + (G_2 + G_f)/h_{ie}} \tag{5.79}$$

Substituting (5.22), (5.78), and (5.79) in (5.73) yields

$$F(G_1) = F(\alpha) \frac{F(G_1)|_{\alpha=0}}{F(\alpha)|_{G_1=0}} = 1 + \frac{G_1(G_2 + G_f)}{(G_2 + G_f)/h_{ie} + G_f(G_2 + \alpha)} \tag{5.80}$$

As a check, we compute the active admittance that G_1 faces. Invoking (2.95) in conjunction with (5.21) gives the active admittance

$$y = \frac{Y_{33}}{Y_{11,33}}\bigg|_{G_1=0} = \frac{(G_2 + G_f)/h_{ie} + G_f(\alpha + G_2)}{G_2 + G_f} \tag{5.81}$$

showing that

$$F(G_1) = 1 + T = 1 + \frac{G_1}{y} = \frac{y + G_1}{y} \tag{5.82}$$

5.4 EXTENSIONS TO FEEDBACK CONCEPTS

In this section, the concepts of return difference and null return difference are generalized. Their properties, physical significance, and interrelations are discussed.

5.4.1 The General Return Difference

As a direct extension to the concept of return difference as given in Definition 4.1, we define the return difference with respect to an element for a general reference value.

Definition 5.1: General return difference The *general return difference* $F_k(x)$ of a feedback amplifier with respect to an element x for a general reference value k is the ratio of the two functional values assumed by the first-order cofactor of an element of its indefinite-admittance matrix under the condition that the element

x assumes its nominal value and the condition that the element x assumes the value k. For $k = 0$ we write $F_0(x) = F(x)$.

Thus, it reduces to ordinary return difference when the reference value k is zero. This gives

$$F_k(x) = \frac{Y_{ij}(x)}{Y_{ij}(k)} \qquad (5.83)$$

where $Y_{ij}(k) = Y_{ij}(x)|_{x=k}$. Its relation to return differences for the zero reference value can easily be derived by a simple manipulation:

$$F_k(x) = \frac{Y_{ij}(x)}{Y_{ij}(k)} = \frac{Y_{ij}(x)/Y_{ij}(0)}{Y_{ij}(k)/Y_{ij}(0)} \qquad (5.84)$$

giving

$$F_k(x) = \frac{F(x)}{F(k)} \qquad (5.85)$$

In words, this states that the return difference with respect to an element x for a general reference value k is equal to the ratio of the return difference with respect to x to that with respect to k, both for the zero reference value.

Following Definition 4.2, we define the general return ratio.

Definition 5.2: General return ratio The *general return ratio* T_k with respect to a voltage-controlled current source $I = xV$ for a general reference value k is the negative of the voltage appearing at the controlling branch when the controlled current source is replaced by a parallel combination of an independent current source of $x - k$ amperes and a controlled current source $I = kV$ and when the input excitation is set to zero. For $k = 0$ we write $T_0 = T$.

Refer to the general feedback configuration of Fig. 5.8. The general return ratio T_k is the negative of the voltage V'_{ab} appearing at terminals a and b of Fig. 5.9. We show that the general return difference $F_k(x)$ and general return ratio T_k are related by the equation $F_k(x) = 1 + T_k$.

As discussed in Chap. 4, the element x enters the indefinite-admittance matrix $\mathbf{Y}$ in a rectangular pattern as shown in (4.123). The cofactor $Y_{db}(x)$ can be expanded as

$$Y_{db}(x) = Y_{db}(0) + xY_{ca,db} \qquad (5.86)$$

where $Y_{ca,db}$ is independent of x. Replacing x by k yields

$$Y_{db}(k) = Y_{db}(0) + kY_{ca,db} \qquad (5.87)$$

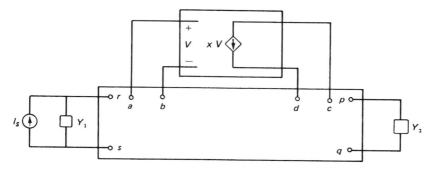

Figure 5.8 The general feedback amplifier configuration.

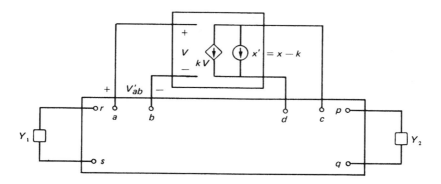

Figure 5.9 The physical interpretation of the general return ratio with respect to the controlling parameter x of a voltage-controlled current source for a general reference value k.

which, when combined with (5.86), gives

$$Y_{db}(x) = Y_{db}(k) + x'Y_{ca,db} \tag{5.88}$$

where

$$x' = x - k \tag{5.89}$$

Observe that the indefinite-admittance matrix of the network of Fig. 5.9 can be obtained from $\mathbf{Y}(x)$ by replacing the element x by k; the resulting matrix is simply $\mathbf{Y}(k)$. By appealing to formula (2.94), the voltage V'_{ab} appearing at terminals a and b of Fig. 5.9 is found to be

$$V'_{ab} = -x' \frac{Y_{ca,db}(k)}{Y_{uv}(k)} = -x' \frac{Y_{ca,db}}{Y_{uv}(k)} \tag{5.90}$$

Now we are in a position to show that

$$F_k(x) = \frac{Y_{ij}(x)}{Y_{ij}(k)} = \frac{Y_{db}(x)}{Y_{uv}(k)} = \frac{Y_{db}(k) + x'Y_{ca,db}}{Y_{db}(k)}$$

$$= 1 + x'\frac{Y_{ca,db}}{Y_{uv}(k)} = 1 - V'_{ab} = 1 + T_k \tag{5.91}$$

Thus, physically the general return difference $F_k(x)$ can be interpreted as follows: Replace the voltage-controlled current source $I = xV$ of Fig. 5.8 by two controlled current sources $I_1 = kV_1$ and $I_2 = x'V_2$ connected in parallel as shown in Fig. 5.10. The controlling branch of the controlled current source $I_2 = x'V_2$ is broken off as marked and a voltage source of 1 V is applied to the right of the breaking mark. The voltage appearing at the left of the breaking mark resulting from the 1-V excitation is then V'_{ab}, as indicated in Fig. 5.10. The negative of V'_{ab} is the general return ratio T_k. The general return difference $F_k(x)$ is simply the difference of the 1-V excitation and the returned voltage V'_{ab} as illustrated in the figure. The significance of this interpretation is that it is useful in the calculation of the sensitivity function and in the measurement of the return ratio for practical feedback amplifiers, because the interpretation corresponds to the situation where the feedback amplifier under study is made partially active rather than completely dead, as in the original interpretation of the return ratio and return difference for the zero reference value.

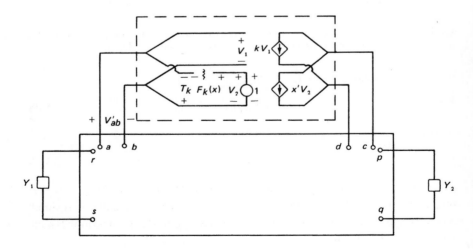

Figure 5.10 The physical interpretation of the general return difference with respect to the controlling parameter x of a voltage-controlled current source for a general reference value k.

Figure 5.11 An equivalent network of a common-emitter feedback amplifier.

Figure 5.12 The network used to compute the general return difference $F_k(\alpha)$.

Example 5.7 The network of Fig. 5.11 was considered in Example 5.3. We wish to compute the general return difference with respect to the controlling parameter $\alpha \equiv h_{fe}/h_{ie}$ for a general reference value k. For this we replace the controlled current source $h_{fe}I_b$ by a parallel combination of an independent current source of α' amperes and a voltage-controlled current source $I = kV$ as illustrated in Fig. 5.12. The voltage across the conductance $1/h_{ie}$ is the negative of the general return ratio T_k. To compute T_k, we first write down the indefinite-admittance matrix of the network as

$$
\mathbf{Y} =
\begin{bmatrix}
G_f + G_1 + \dfrac{1}{h_{ie}} & -G_f & -G_1 - \dfrac{1}{h_{ie}} \\[2ex]
k - G_f & G_2 + G_f & -k - G_2 \\[2ex]
-k - G_1 - \dfrac{1}{h_{ie}} & -G_2 & k + G_1 + G_2 + \dfrac{1}{h_{ie}}
\end{bmatrix}
\tag{5.92}
$$

By using formula (2.94), the voltage V across the conductance $1/h_{ie}$ is given by

$$V = \alpha' \frac{Y_{31,23}}{Y_{33}} = -\frac{G_f \alpha'}{(G_1 + 1/h_{ie})(G_2 + G_f) + G_f(G_2 + k)} = -T_k \quad (5.93)$$

The general return difference is obtained as

$$F_k(\alpha) = 1 + T_k = \frac{(G_1 + 1/h_{ie})(G_2 + G_f) + G_f(G_2 + \alpha)}{(G_1 + 1/h_{ie})(G_2 + G_f) + G_f(G_2 + k)} \quad (5.94)$$

which is equal to the ratio of the two functional values assumed by the cofactor of an element of the matrix (5.21) under the condition that α assumes its nominal value and the condition that α assumes the value k.

5.4.2 The General Null Return Difference

The extension is straightforward, and we first state the definition.

Definition 5.3: General null return difference The *general null return difference* $\hat{F}_k(x)$ of a feedback amplifier with respect to an element x for a general reference value k is the ratio of the two functional values assumed by the second-order cofactor $Y_{rp,sq}$ of the elements of its indefinite-admittance matrix $\mathbf{Y}$ under the condition that the element x assumes its nominal value and the condition that the element x assumes the value k, where r and s are input terminals and p and q are output terminals of the amplifier. For $k = 0$ we write $\hat{F}_0(x) = \hat{F}(x)$.

Referring to the general configuration of Fig. 5.8, we have

$$\hat{F}_k(x) = \frac{Y_{rp,sq}(x)}{Y_{rp,sq}(k)} \quad (5.95)$$

provided, of course, that $Y_{rp,sq}(k) \neq 0$. Likewise, we can define the general null return ratio.

Definition 5.4: General null return ratio The *general null return ratio* $\hat{T}_k$ with respect to a voltage-controlled current source $I = xV$ for a general reference value k is the negative of the voltage appearing at the controlling branch when the controlled current source is replaced by a parallel combination of an independent current source of $x - k$ amperes and a controlled current source $I = kV$ and when the input excitation is adjusted so that the output of the amplifier is identically zero. For $k = 0$ we write $\hat{T}_0 = \hat{T}$.

As in (5.85), we can derive a relation between the general null return difference and the null return differences. Equation (5.95) can be rewritten as

$$\hat{F}_k(x) = \frac{Y_{rp,sq}(x)}{Y_{rp,sq}(k)} = \frac{Y_{rp,sq}(x)/Y_{rp,sq}(0)}{Y_{rp,sq}(k)/Y_{rp,sq}(0)} \qquad (5.96)$$

provided that $Y_{rp,sq}(0) \neq 0$, giving

$$\hat{F}_k(x) = \frac{\hat{F}(x)}{\hat{F}(k)} \qquad (5.97)$$

Thus, the general null return difference with respect to an element x for a general reference value k is equal to the ratio of the null return difference with respect to x to that with respect to k, both for the zero reference value.

5.5 THE NETWORK FUNCTIONS AND GENERAL RETURN DIFFERENCE AND GENERAL NULL RETURN DIFFERENCE

We demonstrate that formulas (4.176), (4.197), and (4.199) can be expressed in terms of the general return difference, the general null return difference, and the transfer function $w(k)$. We also give a simple physical interpretation of the general return difference for a one-port admittance.

Combining (4.176) with (5.85) and (5.97) yields a relation among the transfer impedance or current gain and return difference and null return difference:

$$\frac{w(x)}{w(k)} = \frac{\hat{F}(x)}{F(x)} \frac{F(k)}{\hat{F}(k)} \qquad (5.98)$$

giving

$$w(x) = w(k) \frac{\hat{F}_k(x)}{F_k(x)} \qquad (5.99)$$

If $w(x)$ represents the voltage gain, then as in (4.196) we can write

$$\frac{w(x)}{w(k)} = \frac{Y_{rp,sq}(x)}{Y_{rp,sq}(k)} \frac{Y_{rr,ss}(k)}{Y_{rr,ss}(x)} \qquad (5.100)$$

The first term in the product on the right-hand side is the general null return difference $\hat{F}_k(x)$. The second term can be interpreted as the reciprocal of the general return difference with respect to x for a general reference value k when the input port of the amplifier is short-circuited. Thus, the voltage gain V_{pq}/V_{rs} (see Fig. 5.8) can be expressed as

$$w(x) = w(k)\frac{\hat{F}_k(x)}{F_k(\text{input short-circuited})} \tag{5.101}$$

Likewise, if $w(x)$ denotes the short-circuit current gain, then as in (4.198) we can write

$$\frac{w(x)}{w(k)} = \frac{Y_{rp,sq}(x)}{Y_{rp,sq}(k)}\frac{Y_{pp,qq}(k)}{Y_{pp,qq}(x)} \tag{5.102}$$

As before, the second term in the product can be interpreted as the reciprocal of the general return difference with respect to x for a general reference value k when the output port of the amplifier is short-circuited, giving a formula for the short-circuit current gain as

$$w(x) = w(k)\frac{\hat{F}_k(x)}{F_k(\text{output short-circuited})} \tag{5.103}$$

Finally, we give a physical interpretation of the general return difference $F_k(x)$ with respect to a one-port admittance x for a general reference value k. The result can easily be deduced from (5.91) by letting $a = c$ and $b = d$ as follows:

$$F_k(x) = \frac{Y_{dd}(x)}{Y_{dd}(k)} = \frac{Y_{dd}(k) + x'Y_{cc,dd}}{Y_{dd}(x)} = 1 + x'\frac{Y_{cc,dd}}{Y_{dd}(k)}$$

$$= 1 + \frac{x'}{y + k} = \frac{y + x}{y + k} \tag{5.104}$$

where y is the admittance that x faces. We remark that $\mathbf{Y}(k)$ is the indefinite-admittance matrix of the original feedback network when x is replaced by k. Thus, according to (2.94), $Y_{cc,dd}/Y_{dd}(k)$ is the impedance looking into terminals c and d when x is replaced by k, whose reciprocal is $y + k$. In other words, the general return difference $F_k(x)$ with respect to the one-port admittance x is equal to the ratio of total admittance looking into the node pair where x is connected to that when x is replaced by its reference value k.

Example 5.8 In the feedback network of Fig. 5.11, suppose that we wish to compute the transfer impedance V_{23}/I_s by means of formula (5.99). Let $\alpha \equiv h_{fe}/h_{ie}$ be the element of interest. To compute $\hat{F}_k(\alpha)$, we insert a current source I_s between terminals 1 and 3 in the network of Fig. 5.12. The resulting network is presented in Fig. 5.13. We then adjust I_s so that the output voltage across the

Figure 5.13   The network used to compute the general null return difference $\hat{F}_k(\alpha)$.

conductance G_2 is zero. This yields

$$I_s = I_0 \equiv \frac{(\alpha - k)(G_1 + G_f + 1/h_{ie})}{G_f - k} \qquad (5.105)$$

Under this situation, the returned voltage V is $I_0/(G_1 + G_f + 1/h_{ie})$, which is the negative of the general null return ratio $\hat{T}_k$. Thus, we obtain

$$\hat{F}_k(\alpha) = \frac{G_f - \alpha}{G_f - k} \qquad (5.106)$$

The transfer impedance when α is replaced by k is given by

$$w(k) = \frac{G_f - k}{(G_1 + 1/h_{ie})(G_2 + G_f) + G_f(G_2 + k)} \qquad (5.107)$$

Substituting (5.94), (5.106), and (5.107) in (5.99) gives the desired transfer impedance

$$w(\alpha) = w(k)\frac{\hat{F}_k(\alpha)}{F_k(\alpha)} = \frac{G_f - \alpha}{(G_1 + 1/h_{ie})(G_2 + G_f) + G_f(G_2 + \alpha)} \qquad (5.108)$$

Now suppose that the element of interest is the one-port admittance G_1. The admittance that G_1 faces is given by

$$y = \frac{(G_2 + G_f)/h_{ie} + G_f(G_2 + \alpha)}{G_2 + G_f} \qquad (5.109)$$

Thus, from (5.104) the general return difference $F_k(G_1)$ with respect to G_1 for a general reference value k is obtained as

$$F_k(G_1) = \frac{y + G_1}{y + k} = 1 + \frac{(G_1 - k)(G_2 + G_f)}{(G_2 + G_f)(k + 1/h_{ie}) + G_f(G_2 + \alpha)} \qquad (5.110)$$

the second term on the right-hand side being the general return ratio T_k.

Figure 5.14 A parallel-series or current-shunt feedback amplifier with its biasing and coupling circuitry.

Example 5.9 Consider the parallel-series or current-shunt feedback amplifier of Fig. 5.14, which was studied in Sec. 4.2.4. After the biasing and coupling circuitry have been removed, the network reduces to that of Fig. 5.15a. Assume that the two transistors are identical, with the following specifications:

$$h_{ie} = 1.1\,\text{k}\Omega \quad h_{fe} = 50$$

$$h_{re} = h_{oe} = 0 \quad R = 3\,\text{k}\Omega$$

$$R_e = 50\,\Omega \quad R_f = 1.2\,\text{k}\Omega \tag{5.111}$$

$$R_1 = 1.2\,\text{k}\Omega \quad R_2 = 500\,\Omega$$

Figure 5.15 (*a*) The amplifier of Fig. 5.14 after the removal of its biasing and coupling circuitry and (*b*) its equivalent network.

An equivalent network of Fig. 5.15*a* is presented in Fig. 5.15*b* with all the conductances denoted in mho and

$$\tilde{\alpha}_j \equiv \alpha_j \cdot 10^{-4} \equiv \frac{h_{fe}}{h_{ie}} = 455 \cdot 10^{-4} \tag{5.112}$$

where $j = 1, 2$ is used to distinguish the transconductances of the first and second transistors. Note that we have converted the two current-controlled current sources into the voltage-controlled current sources. All the biasing resistors have been ignored. We remark that in some situations, the effect of biasing resistors may not be entirely negligible and should therefore be included.

The indefinite-admittance matrix of the network of Fig. 5.15b can now be written down by inspection and is given by

$$Y = 10^{-4}$$
$$\times \begin{bmatrix} 25.755 & 0 & 0 & -8.333 & -17.422 \\ 0 & 20 & \alpha_2 & -\alpha_2 & -20 \\ \alpha_1 & 0 & 12.423 & -9.09 & -3.333 - \alpha_1 \\ -8.333 & 0 & -9.09 - \alpha_2 & 217.423 + \alpha_2 & -200 \\ -17.422 - \alpha_1 & -20 & -3.333 & -200 & 240.755 + \alpha_1 \end{bmatrix}$$

$$(5.113)$$

By applying formula (2.94), the closed-loop transfer impedance is found to be

$$w = \frac{V_{25}}{I_s} = \frac{Y_{12,55}}{Y_{55}} = \frac{431.42 \cdot 10^{-7}}{373.05 \cdot 10^{-11}} = 11.564 \, \text{k}\Omega \qquad (5.114)$$

confirming (4.103c), where

$$Y_{12,55} = (208.33\alpha_1\alpha_2 + 27.77\alpha_2) \cdot 10^{-12} = 431.42 \cdot 10^{-7} \qquad (5.115a)$$

$$Y_{uv} = Y_{55} = (166.66\alpha_1\alpha_2 + 1514.94\alpha_1 + 1716.83\alpha_2 + 1,331,494) \cdot 10^{-16}$$
$$= 373.05 \cdot 10^{-11} \qquad (5.115b)$$

The closed-loop voltage and current gains are obtained as

$$w_V = \frac{V_{25}}{V_s} = \frac{w}{R_1} = \frac{11.56 \cdot 10^3}{R_1} = 9.64 \qquad (5.116a)$$

$$w_I = \frac{I_2}{I_s} = -\frac{V_{25}}{I_s R_2} = -\frac{w}{R_2} = -23.13 \qquad (5.116b)$$

The return differences with respect to the transconductances $\tilde{\alpha}_j$ for a general reference value k are given by

$$F_k(\tilde{\alpha}_1) = \frac{Y_{55}(\tilde{\alpha}_1)}{Y_{55}(k)} = \frac{773.45\tilde{\alpha}_1 + 2.1127}{773.45k + 2.1127} = \frac{37.31}{773.45k + 2.1127} \qquad (5.117a)$$

$$F_k(\tilde{\alpha}_2) = \frac{Y_{55}(\tilde{\alpha}_2)}{Y_{55}(k)} = \frac{775.47\tilde{\alpha}_2 + 2.0208}{775.47k + 2.0208} = \frac{37.31}{775.47k + 2.0208} \qquad (5.117b)$$

yielding

$$F(\tilde{\alpha}_1) = F_0(\tilde{\alpha}_1) = 17.66 \qquad (5.118a)$$

$$F(\tilde{\alpha}_2) = F_0(\tilde{\alpha}_2) = 18.46 \qquad (5.118b)$$

The null return differences with respect to $\tilde{\alpha}_j$ for a general reference value k are determined from (5.95) with terminals 1 and 5 as the input and terminals 2 and 5 as the output:

$$\hat{F}_k(\tilde{\alpha}_1) = \frac{Y_{12,55}(\tilde{\alpha}_1)}{Y_{12,55}(k)} = \frac{9479\tilde{\alpha}_1 + 0.1263}{9479k + 0.1263} = \frac{431.42}{9479k + 0.1263} \qquad (5.119a)$$

$$\hat{F}_k(\tilde{\alpha}_2) = \frac{Y_{12,55}(\tilde{\alpha}_2)}{Y_{12,55}(k)} = \frac{948.18\tilde{\alpha}_2}{948.18k} = \frac{\tilde{\alpha}_2}{k} = \frac{0.0455}{k} \qquad (5.119b)$$

giving

$$\hat{F}(\tilde{\alpha}_1) = \hat{F}_0(\tilde{\alpha}_1) = 3415.84 \qquad (5.120a)$$

$$\hat{F}(\tilde{\alpha}_2) = \hat{F}_0(\tilde{\alpha}_2) = \infty \qquad (5.120b)$$

From (5.115), the transfer impedance w can be written as

$$w = w(\tilde{\alpha}_1, \tilde{\alpha}_2)$$

$$= \frac{208.33\tilde{\alpha}_1\tilde{\alpha}_2 \cdot 10^4 + 27.76\tilde{\alpha}_2}{166.66\tilde{\alpha}_1\tilde{\alpha}_2 + (1514.94\tilde{\alpha}_1 + 1716.83\tilde{\alpha}_2 + 133.15) \cdot 10^{-4}} \qquad (5.121)$$

which leads to

$$w(k, \tilde{\alpha}_2) = \frac{94{,}790k + 1.2631}{7.735k + 211.27 \cdot 10^{-4}} \qquad (5.122a)$$

$$w(\tilde{\alpha}_1, k) = \frac{94{,}818k}{7.755k + 202.08 \cdot 10^{-4}} \qquad (5.122b)$$

Substituting (5.117), (5.119), and (5.122) in (5.99), we get

$$w(\tilde{\alpha}_1, \tilde{\alpha}_2) = w(k, \tilde{\alpha}_2)\frac{\hat{F}_k(\tilde{\alpha}_1)}{F_k(\tilde{\alpha}_1)} = 11.563\,\text{k}\Omega \qquad (5.123a)$$

$$w(\tilde{\alpha}_1, \tilde{\alpha}_2) = w(\tilde{\alpha}_1, k)\frac{\hat{F}_k(\tilde{\alpha}_2)}{F_k(\tilde{\alpha}_2)} = 11.563\,\text{k}\Omega \qquad (5.123b)$$

confirming (5.114).

Assume now that $w = g_{12,55}$ denotes the voltage gain V_{25}/V_{15}, as defined in Fig. 5.15b. Appealing to formula (2.97) in conjunction with (5.115a), we obtain

$$g_{12,55} = \frac{V_{25}}{V_{15}} = \frac{Y_{12,55}}{Y_{11,55}} = \frac{431.42 \cdot 10^{-7}}{826.98 \cdot 10^{-10}} = 521.68 \qquad (5.124)$$

where

$$Y_{11,55} = (66.66\alpha_2 + 52,368.36) \cdot 10^{-12} = 826.98 \cdot 10^{-10} \qquad (5.125)$$

As in (5.122), $g_{12,55}$ can also be written as

$$g_{12,55} = g_{12,55}(\tilde{\alpha}_1, \tilde{\alpha}_2) = \frac{208.33\tilde{\alpha}_1\tilde{\alpha}_2 \cdot 10^4 + 27.77\tilde{\alpha}_2}{66.66\tilde{\alpha}_2 + 5.2368} \qquad (5.126)$$

which gives

$$g_{12,55}(k, \tilde{\alpha}_2) = \frac{94,790k + 1.2635}{8.2698} \qquad (5.127a)$$

$$g_{12,55}(\tilde{\alpha}_1, k) = \frac{94,818k}{66.66k + 5.2368} \qquad (5.127b)$$

To compute the return difference F_k(input short-circuited), we short-circuit the input terminals 1 and 5. The corresponding indefinite-admittance matrix is obtained by adding the first row of (5.113) to the fifth row and the first column to the fifth column and then deleting the first row and column. The first-order cofactor of an element of the resulting matrix is simply $Y_{11,55}$. Thus, we have

$$F_k(\text{input short-circuited}) = \frac{Y_{11,55}(\tilde{\alpha}_1)}{Y_{11,55}(k)} = 1 \qquad (5.128a)$$

with respect to $\tilde{\alpha}_1$, and

$$F_k(\text{input short-circuited}) = \frac{Y_{11,55}(\tilde{\alpha}_2)}{Y_{11,55}(k)} = \frac{8.2698}{66.66k + 5.2368} \qquad (5.128b)$$

with respect to $\tilde{\alpha}_2$.

Substituting (5.119), (5.127), and (5.128) in (5.101) results in

$$g_{12,55} = g_{12,55}(k, \tilde{\alpha}_2) \frac{\hat{F}_k(\tilde{\alpha}_1)}{F_k(\text{input short-circuited})} = 521.68 \qquad (5.129a)$$

$$g_{12,55} = g_{12,55}(\tilde{\alpha}_1, k) \frac{\hat{F}_k(\tilde{\alpha}_2)}{F_k(\text{input short-circuited})} = 521.68 \qquad (5.129b)$$

confirming (5.124).

As in (4.177), (5.99) is also valid for the driving-point impedance. This is the particular situation where the input and output are taken to be the same port. In this case, $F_k(x)$ denotes the general return difference for the situation when the port where the driving-point impedance is defined is left open without a source, and we write $F_k(x) = F_k$(input open circuited). Likewise, $\hat{F}_k(x)$ is the general return difference for the situation when the port where the driving-point impedance is measured is short-circuited, and we write $\hat{F}_k(x) = F_k$(input short-circuited). Consequently, the driving-point impedance $Z(x)$ looking into a terminal pair can be conveniently expressed as

$$Z(x) = Z(k)\frac{F_k(\text{input short-circuited})}{F_k(\text{input open-circuited})} \tag{5.130}$$

We emphasize that the word *input* means the terminal pair where the driving-point impedance is measured, not necessarily the input port of the feedback amplifier.

Example 5.10 Consider the same current-shunt feedback amplifier of Fig. 5.14. We compute the driving-point impedances facing the current source I_s and looking into the output terminals 2 and 5 in the network of Fig. 5.15b. By appealing to (2.95) in conjunction with (5.113), the amplifier input and output impedances are found to be

$$z_{11,55} = \frac{Y_{11,55}}{Y_{55}} = \frac{82{,}698}{3731} = 22.17\,\Omega \tag{5.131a}$$

$$z_{22,55} = \frac{Y_{22,55}}{Y_{55}} = \frac{1{,}864{,}617}{3731} = 499.76\,\Omega \tag{5.131b}$$

where

$$Y_{22,55} = (8.33\alpha_1\alpha_2 + 75.75\alpha_1 + 85.84\alpha_2 + 66{,}575) \cdot 10^{-12}$$
$$= 1{,}864{,}617 \cdot 10^{-12} \tag{5.132}$$

Suppose that $\tilde{\alpha}_1$ is the element of interest. Then we can write

$$z_{11,55}(k) = \frac{826.98}{773.45k + 2.1127} \tag{5.133a}$$

$$z_{22,55}(k) = \frac{386{,}590k + 1056.32}{773.45k + 2.1127} \tag{5.133b}$$

To compute F_k(input short-circuited) for $z_{11,55}$, we short-circuit terminals 1 and 5 in the network of Fig. 5.15b, giving F_k(input short-circuited) $= 1$, as in Eq. (5.128a).

For F_k(input open-circuited), we remove the current source I_s and compute $F_k(\tilde{\alpha}_1)$, which is given by (5.117a). Substituting these in (5.130) yields

$$z_{11,55} = z_{11,55}(k) \frac{F_k(\text{input short-circuited})}{F_k(\text{input open-circuited})} = 22.17\,\Omega \tag{5.134}$$

confirming (5.131a).

To compute F_k(input short-circuited) for $z_{22,55}$, we short-circuit terminals 2 and 5 in the network of Fig. 5.15b. The first-order cofactor of an element of the resulting indefinite-admittance matrix is simply $Y_{22,55}$. Thus, we have

$$F_k(\text{input short-circuited}) = \frac{Y_{22,55}(\tilde{\alpha}_1)}{Y_{22,55}(k)} = \frac{18,646.17}{386,590k + 1056.32} \tag{5.135}$$

For F_k(input open-circuited), we leave terminals 2 and 5 open as shown in Fig. 5.15b and compute $F_k(\tilde{\alpha}_1)$, which is again given by Eq. (5.117a). Substituting (5.117a), (5.133b), and (5.135) in (5.130), we obtain

$$z_{22,55} = z_{22,55}(k) \frac{F_k(\text{input short-circuited})}{F_k(\text{input open-circuited})} = 499.76\,\Omega \tag{5.136}$$

confirming (5.131b).

5.6 THE RELATIVE SENSITIVITY FUNCTION AND FEEDBACK

As a generalization of the concept of sensitivity function discussed in Sec. 5.1, we introduce the *relative sensitivity function,* which is defined according to the equation

$$S'(x') = \frac{x'}{w} \frac{\partial w}{\partial x'} \tag{5.137}$$

where, as before, $x' = x - k$. Comparing this with (5.1) for the sensitivity function, we see that the relative sensitivity function is defined with respect to the altered element x'. From their definitions, we find that the sensitivity function and the relative sensitivity function are related by the equation

$$x'S(x) = xS'(x') \tag{5.138}$$

Since the element x enters the indefinite-admittance matrix $\mathbf{Y}$ in a rectangular pattern as shown in (4.123), we have the following expansions:

$$Y_{uv}(x) = Y_{uv}(0) + x\dot{Y}_{uv} \tag{5.139a}$$

$$Y_{rp,sq}(x) = Y_{rp,sq}(0) + x\dot{Y}_{rp,sq} \tag{5.139b}$$

as in (5.4). Note that $\dot{Y}_{uv}$ and $\dot{Y}_{rp,sq}$ are independent of x. Following (5.87) we establish the identities

$$Y_{uv}(x) = Y_{uv}(k) + x'\dot{Y}_{uv} \qquad (5.140a)$$

$$Y_{rp,sq}(x) = Y_{rp,sq}(k) + x'\dot{Y}_{rp,sq} \qquad (5.140b)$$

We now proceed to derive the general relative sensitivity functions for various network functions. To start, let $w(x)$ represent either the current gain I_{pq}/I_s or the transfer impedance V_{pq}/I_s, as indicated in Fig. 5.8. From Eq. (5.138) in conjunction with Eqs. (5.2) and (5.140) we obtain

$$
\begin{aligned}
S'(x') &= x'\frac{\dot{Y}_{rp,sq}(x)}{Y_{rp,sq}(x)} - x'\frac{\dot{Y}_{uv}(x)}{Y_{uv}(x)} = \frac{Y_{rp,sq}(x) - Y_{rp,sq}(k)}{Y_{rp,sq}(x)} - \frac{Y_{uv}(x) - Y_{uv}(k)}{Y_{uv}(x)} \\
&= \frac{Y_{uv}(k)}{Y_{uv}(x)} - \frac{Y_{rp,sq}(k)}{Y_{rp,sq}(x)}
\end{aligned}
\qquad (5.141)
$$

giving

$$S'(x') = \frac{1}{F_k(x)} - \frac{1}{\hat{F}_k(x)} \qquad (5.142)$$

Using (5.99) in (5.142) yields

$$S'(x') = \frac{1}{F_k(x)}\left[1 - \frac{w(k)}{w(x)}\right] \qquad (5.143)$$

In the special situation where the reference value k is zero, the relative sensitivity function $S'(x')$ becomes the familiar sensitivity function $S(x)$, and (5.143) reduces to (5.6). Also, we see that if $w(k) = 0$, then $S'(x') = 1/F_k(x)$, the relative sensitivity being equal to the reciprocal of the general return difference $F_k(x)$. In fact, since the reference value k is arbitrary, we can choose the particular value for which $w(k) = 0$. This yields a direct generalization of the simple relation between the sensitivity function and the return difference for the ideal feedback model. The desired reference value k_0 is obtained by setting $w(k_0)$ of (5.2) to zero, giving

$$Y_{rp,sq}(k_0) = 0 \qquad (5.144)$$

Likewise, following the derivations of (5.8) and (5.10) in conjunction with the interpretations of (5.101) and (5.103) for F_k(input short-circuited) and F_k(output short-circuited), we can show that the relative sensitivity function for the voltage gain V_{pq}/V_{rs} can be expressed as

$$S'(x)' = \frac{1}{F_k(\text{input short-circuited})} \left[1 - \frac{w(k)}{w(x)}\right] \tag{5.145}$$

and the relative sensitivity function for the short-circuit current gain can be expressed as

$$S'(x)' = \frac{1}{F_k(\text{output short-circuited})} \left[1 - \frac{w(k)}{w(k)}\right] \tag{5.146}$$

Their derivations are straightforward and are left as exercises (see Probs. 5.10 and 5.11).

Example 5.11 Again consider the feedback network of Fig. 5.11. We compute the relative sensitivity function for the transfer impedance V_{23}/I_s or the current gain I_{23}/I_s with respect to the controlling parameter α. This can easily be accomplished by simply substituting (5.94) and (5.106) in (5.142). However, if we choose the reference value $k = k_0$ satisfying the equation $Y_{12,33}(k_0) = 0$, where $Y_{12,33}(\alpha)$ is a second-order cofactor of the elements of the matrix (5.21), the reciprocal of the general null return difference $\hat{F}_k(\alpha)$ will be zero or, equivalently, $w(k_0) = 0$. The desired reference value k_0 for which $w(k_0) = 0$ occurs at $k_0 = G_f$. Under this condition, the relative sensitivity function is simply equal to the reciprocal of the general return difference with respect to α for the reference value G_f, and is given by

$$S'(\alpha') = \frac{1}{F_{G_f}(\alpha)} = \frac{(G_1 + 1/h_{ie})(G_2 + G_f) + G_f(G_2 + G_f)}{(G_1 + 1/h_{ie})(G_2 + G_f) + G_f(G_2 + \alpha)} \tag{5.147}$$

where $\alpha' = \alpha - G_f$.

Example 5.12 We compute the relative sensitivities of the network functions considered in Examples 5.10 and 5.11. To compute the relative sensitivity of the transfer impedance $w(\tilde{\alpha}_1)$ with respect to $\tilde{\alpha}_1$, we have, from (5.114), (5.117a), and (5.122a),

$$w = w(\tilde{\alpha}_1) = 11,564 \,\Omega \tag{5.148a}$$

$$F_k(\tilde{\alpha}_1) = \frac{37.31}{773.45k + 2.1127} \tag{5.148b}$$

$$w(k) = \frac{94,790k + 1.2631}{7.735k + 211.27 \cdot 10^{-4}} \tag{5.148c}$$

Substituting these in (5.143) yields

$$S'(\tilde{\alpha}'_1) = -1.2383k + 0.0563 \qquad (5.149)$$

where $\tilde{\alpha}'_1 = \tilde{\alpha}_1 - k$. Likewise, from (5.114), (5.117b), and (5.123b), we have

$$S'(\tilde{\alpha}'_2) = -1.1911k + 0.0542 \qquad (5.150)$$

where $\tilde{\alpha}'_2 = \tilde{\alpha}_2 - k$. For $k = 0$, we obtain the sensitivities of the transfer function with respect to the transconductances $\tilde{\alpha}_1$ and $\tilde{\alpha}_2$:

$$S(\tilde{\alpha}_1) = 0.0563 \approx \frac{1}{F(\tilde{\alpha}_1)} \qquad (5.151a)$$

$$S(\tilde{\alpha}_2) = 0.0542 \approx \frac{1}{F(\tilde{\alpha}_2)} \qquad (5.151b)$$

since $w = 11{,}564 \gg w(0) = 59.786$ for $\tilde{\alpha}_1$ and $w(0) = 0$ for $\tilde{\alpha}_2$.

For the relative sensitivity of the voltage gain $g_{12.55}$, we use (5.124), (5.127a), and (5.128a) in (5.145). The result is obtained as

$$S'(\tilde{\alpha}'_1) = 0.9997 - 21.971k \qquad (5.152)$$

giving the sensitivity of $g_{12.55}$ with respect to $\tilde{\alpha}_1$ as

$$S(\tilde{\alpha}_1) = 0.9997 \approx \frac{1}{F(\text{input short-circuited})} \qquad (5.153)$$

Likewise, the relative sensitivity of $g_{12.55}$ with respect to $\tilde{\alpha}_2$ is found to be

$$S'(\tilde{\alpha}'_2) = -13.9175k + 0.6332 \qquad (5.154)$$

giving the sensitivity of $g_{12.55}$ with respect to $\tilde{\alpha}_2$ as

$$S(\tilde{\alpha}_2) = 0.6332 \frac{1}{F(\text{input short-circuited})} \qquad (5.155)$$

since $g_{12.55}(0) = 0$.

The above results confirm an earlier assertion that for practical amplifiers, the transfer function $|w(0)|$ is usually very much smaller than $|w(x)|$ in the passband, and $F \approx 1/S$ may be used as a good estimate of the reciprocal of the sensitivity in the same frequency band. However, for the one-port impedance, the conclusion is not generally valid, because $|w(0)|$ is not usually smaller than $|w(x)|$. We illustrate this by calculating the relative sensitivities of the input impedance $z_{11.55}$ and the output impedance $z_{22.55}$ of the feedback network of Fig. 5.15b.

From (5.117a), (5.131a), and (5.133a), we have

$$F_k(\tilde{\alpha}_1) = \frac{37.31}{773.45k + 2.1127} \qquad (5.156a)$$

$$z_{11,55}(\tilde{\alpha}_1) = 22.17\,\Omega \qquad (5.156b)$$

$$z_{11,55}(k) = \frac{826.98}{773.45k + 2.1127} \qquad (5.156c)$$

Substituting these in (5.143) gives the relative sensitivity of $z_{11,55}$ with respect to the transconductance $\tilde{\alpha}_1$ as

$$S'(\tilde{\alpha}_1') = 20.7304k - 0.9432 \qquad (5.157)$$

giving the sensitivity of $z_{11,55}$ to $\tilde{\alpha}_1$ as

$$S(\tilde{\alpha}_1) = -0.9432 \neq \frac{1}{F(\tilde{\alpha}_1)} = 0.0566 \qquad (5.158)$$

where $z_{11,55}(0) = 391.43\,\Omega$, which, in fact, is larger than $z_{11,55}(\tilde{\alpha}_1) = 22.17\,\Omega$.

For the relative sensitivity of $z_{22,55}$ with respect to $\tilde{\alpha}_1$, we use (5.117a), (5.131b), and (5.133b). The result is given by

$$S'(\tilde{\alpha}_1') = -0.002715k - 0.000026 \qquad (5.159)$$

giving the sensitivity of $z_{22,55}$ to $\tilde{\alpha}_1$ as

$$S(\tilde{\alpha}_1) = -0.000026 \neq \frac{1}{F(\tilde{\alpha}_1)} = 0.0566 \qquad (5.160)$$

where $z_{22,55}(0) = 499.99\,\Omega$, which is about the same as $z_{22,55}(\tilde{\alpha}_1) = 499.76\,\Omega$.

Finally, we determine the reference values $k = k_0$ so that the relative sensitivities of the network functions become equal to the reciprocal of the general return difference,

$$S'(x') = \frac{1}{F_{k_0}(x)} \qquad (5.161)$$

This is equivalent to setting

$$w(k_0) = 0 \qquad (5.162)$$

For the transfer impedance $w(\tilde{\alpha}_1)$, we set (5.122a) to zero, giving

$$k_0 = -1.33 \cdot 10^{-5} \qquad (5.163)$$

and for $w(\tilde{\alpha}_2)$ we set 5.122b to zero, giving $k_0 = 0$, meaning that $F_{k_0}(\tilde{\alpha}_2) = F(\tilde{\alpha}_2)$. For the voltage gain, we set (5.127) to zero and obtain the reference values

$-1.33 \cdot 10^{-5}$ and 0 for $F_k(\tilde{\alpha}_1)$ and $F_k(\tilde{\alpha}_2)$, respectively, the results being the same as those for the transfer impedance.

For the output impedance $z_{22,55}(\tilde{\alpha}_1)$, we set (5.133b) to zero, yielding

$$k_0 = -2.73 \cdot 10^{-3} \tag{5.164}$$

The reference value for the input impedance $z_{11,55}(\tilde{\alpha}_1)$ is infinite and therefore cannot be meaningfully defined.

The above results are a direct generalization of the simple relation between the sensitivity function and the feedback factor for the ideal feedback model. From (4.3) we calculate the sensitivity function of the closed-loop transfer function $w(\mu)$ with respect to the forward amplifier gain, giving

$$S(\mu) = \frac{1}{1 - \mu\beta} = \frac{1}{F(\mu)} \tag{5.165}$$

5.7 SIGNAL-FLOW GRAPH FORMULATION OF FEEDBACK AMPLIFIER THEORY

So far the feedback amplifier theory has been formulated algebraically in terms of the return difference and the null return difference, which are the ratios of the two functional values assumed by the first- and second-order cofactors of the elements of the indefinite-admittance matrix. In the present section, we demonstrate that it can be equivalently formulated in terms of a signal-flow graph representing the signal transmissions among the various ports. One of the advantages of this formulation is that it can easily be generalized to multiple-loop feedback networks to be discussed in Chap. 7.

Consider the four-port network in the general feedback configuration of Fig. 5.8, which is redrawn in Fig. 5.16. Since the network is linear, the port current I_2' and voltage V can be expressed in terms of the port currents I_s and I, as follows:

$$I_2' = t_{11}I_s + t_{12}I \tag{5.166a}$$

$$V = t_{21}I_s + t_{22}I \tag{5.166b}$$

where the transmittances $t_{ij}\,(i, j = 1, 2)$ are all independent of the controlling parameter x. However, the port current I and voltage V are not independent, and as indicated in Fig. 5.8, they are related by the equation

$$I = xV \tag{5.167}$$

Equations (5.166) and (5.167) may be represented by the signal-flow graph of Fig. 5.17, which is known as the *fundamental feedback-flow graph*.

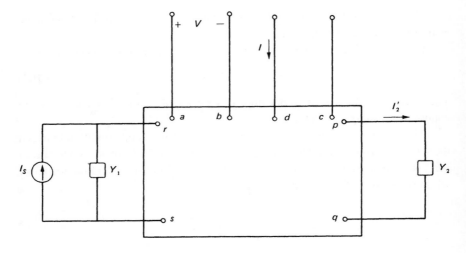

Figure 5.16 The four-port network in the general feedback configuration of Fig. 5.8.

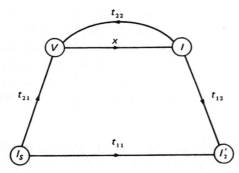

Figure 5.17 The fundamental feedback-flow graph.

Suppose that in Fig. 5.17 we break the branch with transmittance x, as illustrated in Fig. 5.18. Then the return ratio T for the element x is simply the negative of the signal returned to the left of the breaking mark when a unit signal is applied to the right of the breaking mark and when the input excitation is set to zero. This gives

$$T = -xt_{22} \tag{5.168}$$

If in (5.166b) we set $I_s = 0$ and $I = x$, we find that $V = xt_{22}$. This means that the return ratio T is equal to the negative of the voltage V appearing across the controlling branch of the controlled current source $I = xV$ of the feedback amplifier when the controlled source is replaced by an independent current source

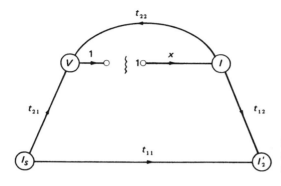

Figure 5.18 The signal-flow graph interpretation of return ratio and return difference functions.

of x amperes and when the input excitation I_s is set to zero. The return difference is then the difference between the unit signal input and the returned signal xt_{22} under the situation that the input excitation to the amplifier is set to zero:

$$F(x) = 1 - xt_{22} = 1 + T \qquad (5.169)$$

Refer again to Fig. 5.18. A unit signal is applied to the right of the breaking mark. Now we adjust the current source I_s so that the output current I'_2 is reduced to zero. This requires that $xt_{12} + I_s t_{11} = 0$, giving the required $I_s = -xt_{12}/t_{11}$. The corresponding value of the returned signal to the left of the breaking mark becomes

$$V = xt_{22} - \frac{xt_{12}t_{21}}{t_{11}} \qquad (5.170)$$

which is equal to the negative of the null return ratio $\hat{T}$ for the element x. Finally, the null return difference is the difference of the unit signal input and the returned signal (5.170), so that

$$\hat{F}(x) = 1 - V = 1 + \hat{T} = 1 - xt_{22} + \frac{xt_{12}t_{21}}{t_{11}} \qquad (5.171)$$

Combining (5.170) and (5.171) yields

$$\hat{F}(x) = F(x) + \frac{xt_{12}t_{21}}{t_{11}} \qquad (5.172)$$

From Fig. 5.17, the current gain I'_2/I_s can easily be obtained by inspection and is given by

$$w(x) = \frac{I'_2}{I_s} = t_{11} + \frac{t_{21}xt_{12}}{1 - xt_{22}} \qquad (5.173)$$

Figure 5.19 The network used to compute the transmittances t_{ij}.

By applying (5.169) and (5.172) in (5.173), the current gain can be written in the familiar form:

$$w(x) = w(0)\frac{\hat{F}(x)}{F(x)} \tag{5.174}$$

Example 5.13 We wish to compute the feedback quantities of the amplifier of Fig. 5.11 by means of the signal-flow graph technique discussed above.

The transmittances t_{ij} must first be determined. For this we replace the controlled current source $h_{fe}I_b$ in the network of Fig. 5.11 by an independent current source I as shown in Fig. 5.19. The indefinite-admittance matrix of network of Fig. 5.19 can easily be obtained by inspection and is given by

$$\mathbf{Y} = \begin{bmatrix} G_f + G_1 + \dfrac{1}{h_{ie}} & -G_f & -G_1 - \dfrac{1}{h_{ie}} \\[2ex] -G_f & G_2 + G_f & -G_2 \\[2ex] -G_1 - \dfrac{1}{h_{ie}} & -G_2 & G_1 + G_2 + \dfrac{1}{h_{ie}} \end{bmatrix} \tag{5.175}$$

From (5.166), we see that the transmittances t_{ij} can be computed directly by formula (2.94), as follows:

$$t_{11} = \left.\frac{I_{23}}{I_s}\right|_{I=0} = G_2\frac{Y_{12,33}}{Y_{33}} = \frac{G_2G_f}{(G_2 + G_f)(G_1 + 1/h_{ie}) + G_2G_f} \tag{5.176a}$$

$$t_{21} = \left.\frac{V}{I_s}\right|_{I=0} = \frac{Y_{11,33}}{Y_{33}} = \frac{G_2 + G_f}{(G_2 + G_f)(G_1 + 1/h_{ie}) + G_2G_f} \tag{5.176b}$$

$$t_{12} = \frac{I_{23}}{I}\bigg|_{I_s=0} = -G_2\frac{Y_{22,33}}{Y_{33}} = -\frac{G_2(G_f + G_1 + 1/h_{ie})}{(G_2 + G_f)(G_1 + 1/h_{ie}) + G_2 G_f}$$

$$(5.176c)$$

$$t_{22} = \frac{V}{I}\bigg|_{I_s=0} = \frac{Y_{31,23}}{Y_{33}} = -\frac{G_f}{(G_2 + G_f)(G_1 + 1/h_{ie}) + G_2 G_f}$$

$$(5.176d)$$

Substituting (5.176d) in (5.169) with $x = \alpha \equiv h_{fe}/h_{ie}$ yields the return difference with respect to α:

$$F(\alpha) = 1 - \alpha t_{22} = 1 + \frac{\alpha G_f}{(G_2 + G_f)(G_1 + 1/h_{ie}) + G_2 G_f}$$

$$(5.177)$$

confirming (5.22). The second term on the right-hand side of (5.177) is the return ratio T. Likewise, substituting (5.176) in (5.171) results in the null return difference

$$\hat{F}(\alpha) = 1 - \alpha t_{22} + \frac{\alpha t_{12} t_{21}}{t_{11}} = 1 - \frac{\alpha}{G_f}$$

$$(5.178)$$

confirming (5.106) for $k = 0$. The second term on the right-hand side of (5.178) is the null return ratio $\hat{T}$.

In the foregoing, we have demonstrated that feedback can be used to control the driving-point and transmission characteristics of an amplifier and can be employed to make an amplifier less sensitive to the variation of its parameters caused by aging, temperature change, or other environmental changes. In the following, we indicate that, in some cases, negative feedback can be used to reduce the effect of noise and nonlinear distortion generated within an amplifier.

To demonstrate this effect, consider an amplifier composed of three internal amplifying stages without feedback represented by the signal-flow graph of Fig. 5.20, in which the extraneous signals x_1 and x_2 represent noise or nonlinear distortion introduced at some arbitrary points within the amplifier. The output signal x_l in terms of the input signal x_s and the disturbing signals x_1 and x_2 is

$$x_l = \mu_1\mu_2\mu_3 x_s + \mu_2\mu_3 x_1 + \mu_3 x_2$$

$$(5.179)$$

The output *signal-to-noise ratio* in the absence of feedback is therefore given by $\mu_1\mu_2\mu_3 x_s/(\mu_2\mu_3 x_1 + \mu_3 x_2)$.

Figure 5.20 The signal-flow graph representing an amplifier composed of three internal amplifying stages without feedback.

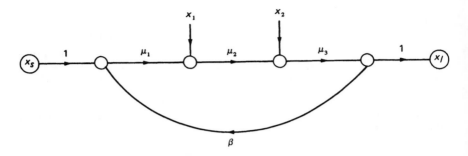

Figure 5.21 The signal-flow graph representing an amplifier composed of three internal amplifying stages with feedback.

Now, if a simple feedback loop is added as shown in Fig. 5.21, the output signal becomes

$$x_l = \frac{\mu_1\mu_2\mu_3 x_s + \mu_2\mu_3 x_1 + \mu_3 x_2}{1 - \mu_1\mu_2\mu_3\beta} \tag{5.180}$$

It is easy to check that the output signal-to-noise ratio with feedback is the same as that without feedback. Thus, feedback has no direct effect in improving the output signal-to-noise ratio. However, it does help indirectly. To see this, we observe from (5.180) that the effect of negative feedback on the output signal is the reduction of both signal and noise by a factor of $(1 - \mu_1\mu_2\mu_3\beta)$. If the input signal x_s is increased so that the output signal remains at the same level as it had before the application of feedback, we see that the net effect at the output will be the reduction of the contribution of the extraneous signals x_1 and x_2 by the same factor $(1 - \mu_1\mu_2\mu_3\beta)$. In other words, if the output of the amplifier is maintained constant with and without negative feedback, the effect of negative feedback increases the signal-to-noise ratio by the factor $(1 - \mu_1\mu_2\mu_3\beta)$. Clearly, this improvement is possible only if the extraneous signals are generated within the amplifier. The increase of input signal x_s to the amplifier without feedback is impractical, because such an increase may cause excessive distortion due to nonlinearity.

For $|\mu_1\mu_2\mu_3\beta| \gg 1$, Eq. (5.180) reduces to

$$x_l = -\frac{x_s}{\beta} - \frac{x_1}{\mu_1\beta} - \frac{x_2}{\mu_1\mu_2\beta} \tag{5.181}$$

showing that the contribution of the extraneous signal x_2 at the output has been reduced by a factor of $\mu_1\mu_2\beta$, whereas that of x_1 is reduced by $\mu_1\beta$ and that of x_s by β. Therefore, the magnitude of the reduction depends largely on the origin of the extraneous signals. It is reasonable to assume that the extraneous signals are generated in the final stage, because nonlinear distortion increases with the level of the output signal.

5.8   MEASUREMENT OF RETURN DIFFERENCE

The zeros of the determinant of the loop-impedance or node-admittance matrix of an active network are referred to as the *natural frequencies*. Their locations in the complex-frequency plane are extremely important in determining the stability of the network. A network is said to be *stable* if all of its natural frequencies are restricted to the open left half of the complex-frequency plane. If the network determinant is known, its roots can readily be computed explicitly with the aid of a computer if necessary, and the stability problem can then be settled directly. However, for a physical network there remains the difficulty of getting an accurate formulation of the network determinant itself, because every equivalent network is, to a greater or lesser extent, an idealization of the physical reality. What is really needed is some kind of experimental verification that the system is stable and will remain so under certain prescribed conditions. The measurement of return difference provides a solution to this problem.

In a feedback amplifier, the return difference with respect to a controlling parameter x is defined by $F(x) = Y_{ij}(x)/Y_{ij}(0)$. Since $Y_{ij}(x)$ also denotes the nodal determinant, we see that the zeros of the return difference are exactly the same as the zeros of the nodal determinant provided that there is no cancellation of common factors between $Y_{ij}(x)$ and $Y_{ij}(0)$. Thus, if $Y_{ij}(0)$ is known to have no zeros in the closed right half of the complex-frequency plane, which is usually the case in single-loop feedback systems, $F(x)$ gives precisely the same information about the stability of a feedback amplifier as does the nodal determinant itself. Since $F(x)$, as will be shown shortly, can be measured quite readily in practice, it can be used as the criterion of stability study for the feedback amplifiers.

The difficulty inherent in the measurement of the return difference with respect to the controlling parameter of a controlled source is that, in a physical system, the controlling branch and the controlled source both form part of a single device such as a transistor, and cannot be physically separated. Thus, we must devise a scheme that does not require the physical decomposition of a device.

In a feedback amplifier, let a device of interest be brought out as a two-port network connected to a general four-port network as shown in Fig. 5.22 along with the input and the load of the amplifier. For our purposes, assume that this device is represented by its y-parameter equivalent two-port network as indicated in Fig. 5.23, in which the parameter y_{21} controls signal transmission in the forward

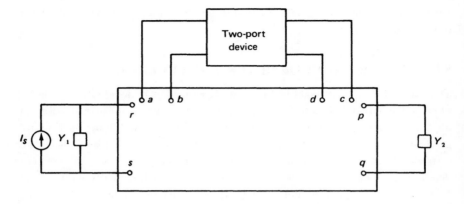

Figure 5.22 The general configuration of a feedback amplifier with a two-port device exhibited explicitly.

Figure 5.23 The representation of the two-port device in Fig. 5.22 by its y-parameters.

direction through the device whereas y_{12} gives the reverse transmission, accounting for the internal feedback within the device. Our objective is to measure the return difference with respect to the forward short-circuit transfer admittance y_{21}. In the following, we describe three procedures for measuring the return ratio and hence the return difference.

5.8.1 Blecher's Procedure [Blecher (1957) and Hakim (1965b)]

Let the device be a transistor operated in the common-emitter configuration with terminals a, $b = d$, and c representing, respectively, the base, emitter, and collector terminals. For convenience, let $a = 1$, $b = d = 3$, and $c = 2$ as indicated in Fig. 5.24. To measure $F(y_{21})$, we break the base terminal of the transistor and apply a 1-V excitation at its input as shown in Fig. 5.24. To ensure that the controlled current source $y_{21}V_{13}$ drives a replica of what it sees during normal operation, we connect an active one-port network composed of a parallel combination of the admittance y_{11} and a controlled current source $y_{12}V_{23}$ at terminals 1 and 3. The returned voltage V_{13} is precisely the negative of the return ratio with respect to the element y_{21}. If, in the frequency band of interest, the externally applied feedback is large compared with the internal feedback of the transistor, the controlled source $y_{12}V_{23}$ can be ignored. If, however, we find that this internal feedback cannot be ignored, we can simulate it by using an additional transistor, connected as shown in Fig. 5.25. This additional transistor must be matched as closely as possible to the one in question. The one-port admittance y_o denotes the admittance presented to the output port of the transistor under consideration as indicated in Figs. 5.24 and 5.25. For a common-emitter stage, it is perfectly reasonable to assume that $|y_o| \gg |y_{12}|$ and $|y_{11}| \gg |y_{12}|$. Under these assumptions, it is straightforward to show that the Norton equivalent network looking into the two-port network at terminals 1 and 3 of Fig. 5.25 can be approximated by the parallel combination of y_{11} and $y_{12}V_{23}$ as shown in Fig. 5.24 (see Prob. 5.15). Finally, we demonstrate that

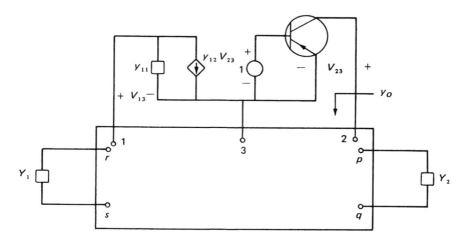

Figure 5.24 The physical interpretation of return difference $F(y_{21})$ for a transistor operated in the common-emitter configuration and represented by its y-parameters y_{ij}.

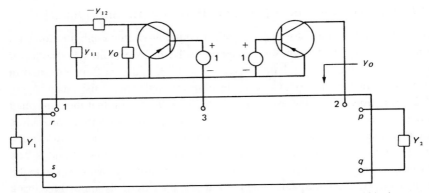

Figure 5.25 The measurement of return difference $F(y_{21})$ for a transistor operated in the common-emitter configuration and represented by its y-parameters y_{ij}.

Figure 5.26 The hybrid-pi equivalent network of a common-emitter transistor.

the admittances y_{11} and $-y_{12}$ can be realized as the input admittances of one-port RC networks.

Consider the hybrid-pi equivalent network of a common-emitter transistor of Fig. 5.26, whose indefinite-admittance matrix can easily be written down by inspection. After suppressing terminal $B' = 4$ and deleting the row and column corresponding to terminal $E = 3$, the short-circuit admittance matrix of the common-emitter transistor is given by (see Prob. 5.16)

$$
\mathbf{Y}_{sc} = \frac{1}{g_x + g_\pi + sC_\pi + sC_\mu}
$$

$$
\times \begin{bmatrix} g_x(g_\pi + sC_\pi + sC_\mu) & -g_x sC_\mu \\ g_x(g_m - sC_\mu) & sC_\mu(g_x + g_\pi + sC_\pi + g_m) \end{bmatrix} \quad (5.182)
$$

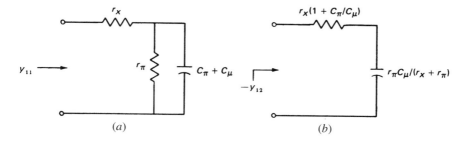

Figure 5.27 (a) The realization of y_{11} and (b) the realization of $-y_{12}$.

It is easy to confirm that y_{11} and $-y_{12}$ can be realized by the one-port networks of Fig. 5.27.

We remark that in Fig. 5.25 if the voltage sources have very low internal impedances, we can join together the two base terminals of the transistors and feed them both from a single voltage source of very low internal impedance. In this way, we avoid the need of using two separate sources.

5.8.2 Hakim's Procedure [Hakim (1965a)]

In this section, we discuss a procedure that avoids the need to use a matched transistor to simulate the controlled source responsible for the internal feedback.

Refer to the network of Fig. 5.28. The controlling parameters y_{12} and y_{21} of the two controlled sources enter the indefinite-admittance matrix of the amplifier in the following pattern:

$$
\mathbf{Y} = \begin{array}{c} \\ 1 \\ 2 \\ 3 \end{array}
\begin{array}{c} \begin{array}{ccc} 1 & 2 & 3 \end{array} \\
\left[\begin{array}{ccc}
 & y_{12} & -y_{12} \\
y_{21} & & -y_{21} \\
-y_{21} & -y_{12} & y_{12} + y_{21}
\end{array} \right]
\end{array}
\tag{5.183}
$$

As before, to indicate the dependence of $\mathbf{Y}$ on y_{12} and y_{21}, we write

$$
\mathbf{Y} = \mathbf{Y}(y_{12}, y_{21})
\tag{5.184}
$$

in Fig. 5.28, suppose that we set $y_{12} = 0$, remove the independent current source I_s, and replace the controlled current source $y_{21}V_{13}$ by an independent current source of y_{21} amperes. The resulting indefinite-admittance matrix is simply $\mathbf{Y}(0, 0)$. Under this situation, the voltages across terminals 1 and 3 and terminals 2 and 3 can be

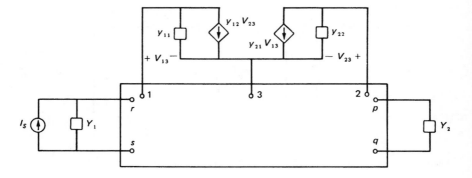

Figure 5.28 The representation of the two-port device in Fig. 5.22 by its y-parameters.

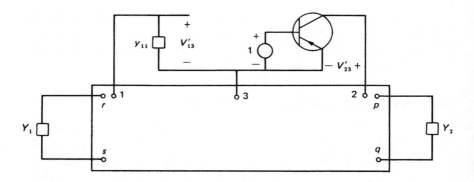

Figure 5.29 The measurement of the voltages V'_{13} and V'_{23}.

obtained directly by appealing to formula (2.94) and are given by

$$V_{13}\big|_{\substack{I_s=0 \\ y_{12}=0}} = V'_{13} = -y_{21}\frac{Y_{21,33}(0,0)}{Y_{33}(0,0)} \tag{5.185a}$$

$$V_{23}\big|_{\substack{I_s=0 \\ y_{12}=0}} = V'_{23} = -y_{21}\frac{Y_{22,33}(0,0)}{Y_{33}(0,0)} \tag{5.185b}$$

If the device is a transistor operated in the common-emitter configuration, the voltages can be measured using the arrangement of Fig. 5.29, in which the feedback loop has been broken at the base of the transistor. The desired voltages V'_{13} and V'_{23} are as indicated in the figure.

Likewise, if we set $y_{21} = 0$, remove the independent current source I_s, and replace the controlled current source $y_{12}V_{23}$ by an independent current source of

Figure 5.30 The measurement of the voltages V_{13}'' and V_{23}''.

y_{12} amperes in the network of Fig. 5.28, the voltages across terminals 1 and 3 and terminals 2 and 3 are obtained as

$$V_{13}\big|_{\substack{I_s=0 \\ y_{21}=0}} = V_{13}'' = -y_{12}\frac{Y_{11,33}(0,0)}{Y_{33}(0,0)} \tag{5.186a}$$

$$V_{23}\big|_{\substack{I_s=0 \\ y_{21}=0}} = V_{23}'' = -y_{12}\frac{Y_{12,33}(0,0)}{Y_{33}(0,0)} \tag{5.186b}$$

If the device is a transistor operated in the common-emitter configuration, these voltages can be measured using the arrangement of Fig. 5.30, in which the feedback loop has been broken at the collector of the transistor. The desired voltages V_{13}'' and V_{23}'' are as indicated in the figure. In fact, since the voltage V_{13}'' developed at the base is practically independent of termination y_{22}, and since V_{23}'' is indistinguishable from zero for all practical purposes, we could have removed y_{22} in making the measurement.

Now we show that these four measurements are sufficient to determine the return difference $F(y_{21})$ with respect to y_{21} in the original amplifier.

$$F(y_{21}) = \frac{Y_{33}(y_{12}, y_{21})}{Y_{33}(y_{12}, 0)} = \frac{Y_{33}(y_{12}, 0) + y_{21}Y_{21,33}(y_{12}, 0)}{Y_{33}(y_{12}, 0)}$$

$$= 1 + y_{21}\frac{Y_{21,33}(0,0) - y_{12}Y_{11,22,33}}{Y_{33}(y_{12}, 0)}$$

$$= 1 + y_{21}\frac{Y_{21,33}(0,0)Y_{33}(y_{12}, 0) - y_{12}Y_{11,33}(0,0)Y_{22,33}(0,0)}{Y_{33}(0,0)Y_{33}(y_{12}, 0)}$$

$$= 1 + y_{21} \frac{Y_{21,33}(0,0)}{Y_{33}(0,0)} - \frac{y_{12}y_{21}Y_{11,33}(0,0)Y_{22,33}(0,0)}{Y_{33}(0,0)[Y_{33}(0,0) + y_{12}Y_{12,33}(0,0)]}$$

$$= 1 - V_{13}' - \frac{V_{13}'' V_{23}'}{1 - V_{23}''} \tag{5.187}$$

The third line follows from (4.147) and (5.86), which can be written as

$$Y_{33}(y_{12}, 0) = Y_{33}(0,0) + y_{12}Y_{12,33}(0,0) \tag{5.188a}$$

$$Y_{11,12,33}Y_{33}(0,0) = Y_{11,33}(0,0)Y_{22,33}(0,0) - Y_{12,33}(0,0)Y_{21,33}(0,0) \tag{5.188b}$$

where $Y_{11,22,33}$ denotes the determinant of the submatrix obtained from $\mathbf{Y}$ by deleting rows 1, 2, and 3 and columns 1, 2, and 3; and the fourth line follows from (5.188a). Note that $Y_{11,22,33}$ is independent of y_{12} and y_{21}. Since for all practical purposes V_{23}'' is indistinguishable from zero, (5.187) can be simplified to

$$F(y_{21}) = 1 + T \approx 1 - V_{13}' - V_{13}'' V_{23}' \tag{5.189}$$

where T is the return ratio with respect to the element y_{21} in the original feedback amplifier.

5.8.3 Impedance Measurements

In this section, we demonstrate that the return difference can be evaluated by measuring two driving-point impedances at a convenient port in the feedback amplifier.

Refer again to the general configuration of a feedback amplifier of Fig. 5.23. Suppose that we wish to evaluate the return difference with respect to the forward short-circuit transfer admittance y_{21}. The controlling parameters y_{12} and y_{21} enter the indefinite-admittance matrix $\mathbf{Y}$ in the rectangular patterns as shown below:

$$
\mathbf{Y} = \begin{array}{c} \\ a \\ b \\ c \\ d \end{array}
\begin{array}{c} a \quad\quad b \quad\quad c \quad\quad d \end{array}
\left[\begin{array}{cccc}
 & & y_{12} & -y_{12} \\
 & & -y_{12} & y_{12} \\
y_{21} & -y_{21} & & \\
-y_{21} & y_{21} & &
\end{array} \right] \tag{5.190}
$$

By appealing to formula (2.95), the impedance looking into terminals a and b of Fig. 5.23 is given by

$$z_{aa,bb}(y_{12}, y_{21}) = \frac{Y_{aa,bb}(y_{12}, y_{21})}{Y_{dd}(y_{12}, y_{21})} \tag{5.191}$$

The return difference with respect to y_{21} is defined as

$$F(y_{21}) = \frac{Y_{dd}(y_{12}, y_{21})}{Y_{dd}(y_{12}, 0)} \tag{5.192}$$

Combining (5.191) and (5.192) yields

$$F(y_{21})z_{aa,bb}(y_{12}, y_{21}) = \frac{Y_{aa,bb}(y_{12}, y_{21})}{Y_{dd}(y_{12}, 0)} = \frac{Y_{aa,bb}(0, 0)}{Y_{dd}(y_{12}, 0)}$$

$$= \frac{Y_{aa,bb}(0, 0)}{Y_{dd}(0, 0)} \frac{Y_{dd}(0, 0)}{Y_{dd}(y_{12}, 0)} = \frac{z_{aa,bb}(0, 0)}{F(y_{12})|_{y_{21}=0}} \tag{5.193}$$

This gives a relation

$$F(y_{12})|_{y_{21}=0}F(y_{21}) = \frac{z_{aa,bb}(0, 0)}{z_{aa,bb}(y_{12}, y_{21})} \tag{5.194}$$

among the return differences and the driving-point impedances. $F(y_{12})|_{y_{21}=0}$ is the return difference with respect to y_{12} when y_{21} is set to zero, and it can be measured by the arrangement of Fig. 5.31. $z_{aa,bb}(y_{12}, y_{21})$ is the driving-point impedance looking into terminals a and b of the network of Fig. 5.23. $z_{aa,bb}(0, 0)$ is the impedance to which $z_{aa,bb}(y_{12}, y_{21})$ reduces when the controlling parameters y_{12} and y_{21} are both set to zero, and it can be measured by the arrangement of Fig. 5.32. Note that in all the three measurements, the independent current source I_s is removed.

As an illustration, suppose that the device is a transistor operated in the common-emitter configuration as shown in Fig. 5.28. Then, as indicated in Sec. 5.8.2, the return difference $F(y_{12})$ when y_{21} is set to zero is, for all practical

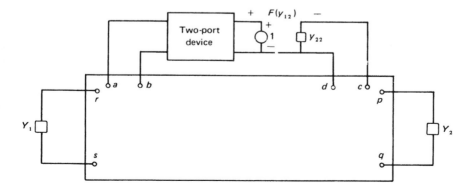

Figure 5.31 The measurement of the return difference $F(y_{12})$ with y_{21} being set to zero.

Figure 5.32 The measurement of the driving-point impedance $z_{aa,bb}(0, 0)$.

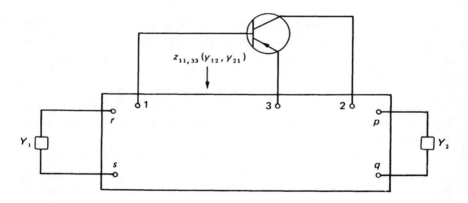

Figure 5.33 The measurement of the driving-point impedance $z_{11,33}(y_{12}, y_{21})$.

purposes, indistinguishable from unity. Consequently, it is perfectly justifiable to reduce (5.194) to the following simpler form:

$$F(y_{21}) \approx \frac{z_{11,33}(0, 0)}{z_{11,33}(y_{12}, y_{21})} \qquad (5.195)$$

showing that the return difference $F(y_{21})$ is effectively equal to the ratio of the two functional values assumed by the driving-point impedance looking into terminals 1 and 3 of Fig. 5.28 under the condition that the controlling parameters y_{12} and y_{21} are both set to zero and the condition that they assume their nominal values. These two impedances can be measured by the network arrangements of Figs. 5.33 and 5.34.

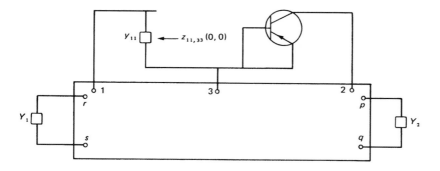

Figure 5.34 The measurement of the driving-point impedance $z_{11,33}(0,0)$.

5.8.4 Measuring the Reference Value

As mentioned in reference to Eq. (5.144), if the reference value k is chosen so that $w(k) = 0$, then the relative sensitivity function is simply the reciprocal of the general return difference for this reference value. In this section, we present a technique of Bode's (1945) that permits experimental determination of this reference value k_0.

Consider the general feedback amplifier configuration of Fig. 5.8. By using the convention adopted in Chap. 2, the various transfer impedances measured under the condition $x = 0$ can be expressed as

$$z_{rp,sq}(0) = \frac{Y_{rp,sq}(0)}{Y_{uv}(0)} \quad z_{ra,sb}(0) = \frac{Y_{ra,sb}(0)}{Y_{uv}(0)} \tag{5.196a}$$

$$z_{cp,dq}(0) = \frac{Y_{cp,dq}(0)}{Y_{uv}(0)} \quad z_{ca,db}(0) = \frac{Y_{ca,db}(0)}{Y_{uv}(0)} \tag{5.196b}$$

From (5.144), the desired reference value k_0 is obtained by setting $Y_{rp,sq}(k_0)$ equal to zero:

$$Y_{rp,sq}(k_0) = Y_{rp,sq}(0) + k_0 \dot{Y}_{rp,sq} = 0 \tag{5.197}$$

giving

$$k_0 = -\frac{Y_{rp,sq}(0)}{\dot{Y}_{rp,sq}} \tag{5.198}$$

The null return difference $\hat{F}(x)$ and the null return ratio for the zero reference value are related by

$$\hat{F}(x) = 1 + \hat{T} = \frac{Y_{rp,sq}(x)}{Y_{rp,sq}(0)} = \frac{Y_{rp,sq}(0) + x\dot{Y}_{rp,sq}}{Y_{rp,sq}(0)} = 1 - \frac{x}{k_0} \tag{5.199}$$

This shows that $\hat{T} = -x/k_0$. Using (4.163a) in conjunction with (5.196) gives

$$\frac{z_{rp,sq}(0)}{z_{ra,sb}(0)z_{cp,dq}(0) - z_{ca,db}(0)z_{rp,sq}(0)} = -\frac{x}{\hat{T}} = k_0 \qquad (5.200)$$

Since the transfer impedances on the left-hand side of (5.200) can easily be measured in the laboratory for most active devices, the desired reference value k_0 is determined by the relation (5.200).

5.9 CONSIDERATIONS ON THE INVARIANCE OF RETURN DIFFERENCE

The feedback amplifier theory developed so far is based on the concepts of return difference and null return difference, which are formulated compactly and elegantly in terms of the first- and the second-order cofactors of the elements of the indefinite-admittance matrix **Y**. The first-order cofactors of the elements of **Y** are, in fact, the determinant of the node-admittance matrix of the amplifier. In the present section, we demonstrate that the feedback amplifier theory can be dually and equally formulated in terms of the determinant of the loop-impedance matrix, and we discuss its relations to the admittance formulation.

5.9.1 Network Determinants and Their Interrelations

The term *network determinant* is usually referred to as the determinant of the loop-impedance matrix, the node-admittance matrix or, more generally, the cutset-admittance matrix. To simplify our notation, denote

Δ_m = the determinant of the loop-impedance matrix
Δ_n = the determinant of the node-admittance matrix
Δ_c = the determinant of the cutset-admittance matrix

They are called the *loop determinant,* the *nodal determinant,* and the *cutset determinant,* respectively. The value of the network determinant is dependent on the formulation of the network equations, even within the same framework of impedance or admittance consideration. If Δ_{m1} and Δ_{m2}, for example, denote the loop determinants corresponding to two different sets of loop equations, they are *not* generally equal. However, it can be shown that they are related by a real constant λ_m, depending only on the choices of the two sets of basis loops. Thus, we can write

$$\Delta_{m1} = \lambda_m \Delta_{m2} \tag{5.201}$$

Likewise, if Δ_{c1} and Δ_{c2} denote the cutset determinants corresponding to two different sets of cutset equations, we have

$$\Delta_{c1} = \lambda_c \Delta_{c2} \tag{5.202}$$

where λ_c is a real constant depending only on the choices of the two sets of basis cutsets. In particular, if Δ_{n1} and Δ_{n2} are the nodal determinants corresponding to two choices of the reference node, then (5.202) reduces to

$$\Delta_{n1} = \Delta_{n2} \tag{5.203}$$

showing that the nodal determinant is independent of the choice of the reference-potential point. We recognize that the nodal determinant is a special type of cutset determinant when basis cutsets around the nodes of the network are used.

With appropriate identifications of the basis loops and cutsets, the loop determinant and the cutset determinant of a network are related by the equation [Chen (1970)]

$$\Delta_m = \lambda_{mc} \Delta_b \Delta_c \tag{5.204}$$

where Δ_b denotes the determinant of the branch-impedance matrix of the network, and λ_{mc} is a real constant depending only on the choices of basis loops and cutsets.

We shall not discuss this aspect of the theory any further, since it is beyond the scope of this book. The proofs of the above assertions cannot be given here because they would take us far afield into the theory of graphs. For a detailed account of this subject and all of its variations and ramifications, the reader is referred to Chen (1976a).

We illustrate the above results by the following example.

Example 5.14 Figure 5.35 is an equivalent network of a general potential feedback amplifier. For our purposes, we consider the series combination of V_s and Z_1 as a single branch, and the parallel combination of $g_m V$ and Y_2 as another. For loop analysis, we use the equivalent network N_m of Fig. 5.36a, and for nodal or cutset analysis the equivalent network N_c of Fig. 5.36b. It is straightforward to calculate the loop-impedance matrix and the node-admittance matrix. Their

Figure 5.35 An equivalent network of a potential feedback amplifier.

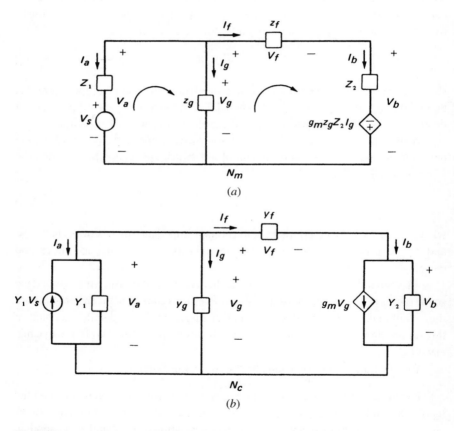

Figure 5.36 Two equivalent networks of Fig. 5.35: (a) N_m and (b) N_c.

determinants are given by

$$\Delta_m = \det \begin{bmatrix} Z_1 + z_g & -z_g \\ -z_g - g_m z_g Z_2 & Z_2 + z_f + z_g + g_m z_g Z_2 \end{bmatrix}$$

$$= Z_1 (Z_2 + z_f + z_g + g_m z_g Z_2) + z_g (Z_2 + z_f) \tag{5.205}$$

$$\Delta_n = \det \begin{bmatrix} Y_1 + y_g + y_f & -y_f \\ g_m - y_f & y_f + Y_2 \end{bmatrix}$$

$$= (Y_1 + y_g)(y_f + Y_2) + y_f (Y_2 + g_m) \tag{5.206}$$

Our objective is to confirm (5.204). For this we express (5.206) in terms of the branch impedances:

$$\Delta_n = \frac{Z_1 (Z_2 + z_f + z_g + g_m z_g Z_2) + z_g (Z_2 + z_f)}{z_g z_f Z_1 Z_2} \tag{5.207}$$

The branch voltage-current relations are described by the equation

$$\begin{bmatrix} V_a \\ V_g \\ V_f \\ V_b \end{bmatrix} \begin{bmatrix} Z_1 & 0 & 0 & 0 \\ 0 & z_g & 0 & 0 \\ 0 & 0 & z_f & 0 \\ 0 & -g_m z_g Z_2 & 0 & Z_2 \end{bmatrix} \begin{bmatrix} I_a \\ I_g \\ I_f \\ I_b \end{bmatrix} + \begin{bmatrix} V_s \\ 0 \\ 0 \\ 0 \end{bmatrix} \tag{5.208}$$

The coefficient matrix is the branch-impedance matrix, whose determinant is given by

$$\Delta_b = z_g z_f Z_1 Z_2 \tag{5.209}$$

From (5.205), (5.207), and (5.209), we obtain

$$\Delta_m = \Delta_b \Delta_n \tag{5.210}$$

confirming (5.204) with $\lambda_{mc} = 1$.

Example 5.15 Figure 5.37 is a bridged-T network terminated in the resistors R_3 and R_2. Assume that $R_i = 1 \, \Omega$ for $i = 1, 2, 3, 4, 5,$ and 6. Two sets of basis loops are chosen as follows:

Set one: $R_1 R_5 R_4$, $R_2 R_6 R_5$, $R_3 R_4 R_6$
Set two: $R_1 R_2 R_6 R_4$, $R_5 R_2 R_3 R_4$, $R_3 R_1 R_5 R_6$

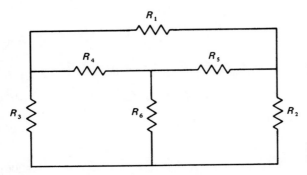

Figure 5.37 A terminated bridged-T network.

Their corresponding loop-impedance matrices are found to be

$$\mathbf{Z}_{m1} = \begin{bmatrix} 3 & -1 & -1 \\ -1 & 3 & -1 \\ -1 & -1 & 3 \end{bmatrix} \tag{5.211}$$

with

$$\Delta_{m1} = \det \mathbf{Z}_{m1} = 16 \tag{5.212}$$

and

$$\mathbf{Z}_{m2} = \begin{bmatrix} 4 & 0 & 0 \\ 0 & 4 & 0 \\ 0 & 0 & 4 \end{bmatrix} \tag{5.213}$$

with

$$\Delta_{m2} = \det \mathbf{Z}_{m2} = 64 \tag{5.214}$$

This shows that

$$\Delta_{m1} = \lambda_m \Delta_{m2} \tag{5.215}$$

with $\lambda_m = 4$.

5.9.2 The Cutset and Loop Formulations

As stated in Definition 5.1, the general return difference $F_k(x)$ is defined as the ratio of the two functional values assumed by the first-order cofactor of an element of the indefinite-admittance matrix $\mathbf{Y}$ under the condition that x assumes its nominal value and the condition that x assumes the value k:

$$F_k(x) = \frac{Y_{uv}(x)}{Y_{uv}(k)} \tag{5.216}$$

Since, from (5.203),

$$Y_{uv}(x) = \Delta_n(x) \tag{5.217}$$

and since Δ_n is a special type of cutset determinant, we obtain, from (5.202),

$$F_k(x) = \frac{Y_{uv}(x)}{Y_{uv}(k)} = \frac{\Delta_n(x)}{\Delta_n(k)} = \frac{\Delta_c(x)}{\Delta_c(k)} \tag{5.218}$$

This shows that in the cutset and hence nodal formulation of the network determinant of a feedback amplifier, the general return difference is invariant under the transformations from one such system to another. Thus, it is equally valid if $F_k(x)$ is defined in terms of the cutset determinant. A natural question that arises at this point is whether or not the general return difference can also be formulated in terms of the loop determinant. This question will be answered below.

Define the general return difference based on the loop formulation by

$$F_k^m(x) = \frac{\Delta_m(x)}{\Delta_m(k)} \tag{5.219}$$

as opposed to the general return difference $F_k(x)$ based on the cutset or nodal formulation. Like $F_k(x)$, one immediate consequence of (5.201) is that $F_k^m(x)$ is invariant under the transformations from one set of basis loops to another. Now we establish a relation between $F_k^m(x)$ and $F_k(x)$. By appealing to (5.204), (5.219) can be written as

$$F_k^m(x) = \frac{\Delta_m(x)}{\Delta_m(k)} = \frac{\Delta_b(x)\Delta_c(x)}{\Delta_b(k)\Delta_c(k)} = \frac{\Delta_b(x)F_k(x)}{\Delta_b(k)} \tag{5.220}$$

giving

$$\Delta_b(k)F_k^m(x) = \Delta_b(k)F_k(x) \tag{5.221}$$

In words, this states that if the determinant of the branch-impedance matrix is nonzero under the condition that the element x assumes its nominal value and the condition that the element x assumes the value k, then the ratio of the general return difference with respect to an element x for a reference value k and based on the loop formulation to that of the same element based on the cutset or nodal formulation is equal to the ratio of the determinant of the branch-impedance matrix when the element x assumes its nominal value to that when it assumes the value k. We summarize the above results by the following theorem.

Theorem 5.1 In the cutset or nodal (loop) formulation of the network determinant of a feedback amplifier, the general return difference is invariant with respect to the transformations from one such system to another. It is invariant under the general transformations between a system of basis loops and a system of basis cutsets if, and only if, the determinant of the branch-immittance matrix remains unaltered when the element of interest is replaced by the chosen reference value.

In the theorem we implicitly assumed that $\Delta_b(x) \neq 0$ and $\Delta_b(k) \neq 0$. As a consequence, if a feedback amplifier contains only a single controlled source, then the general return difference with respect to the controlling parameter of the controlled source is invariant under the general transformations between a system of basis loops and a system of basis cutsets. Its justification is left as an exercise (see Prob. 5.18).

Consider an active two-port device characterized by its general hybrid parameters k_{ij}. Assume that the two-port device N_a is terminated at its input and output ports by the immittances M_1 and M_2, as shown in Fig. 3.1. As indicated in Eq. (5.42), the return difference

$$F(k_{21}) = 1 + T_k = 1 - \frac{k_{12}k_{21}}{(k_{11} + M_1)(k_{22} + M_2)} \tag{5.222}$$

is invariant under immittance substitution for y-, h-, and g-parameters.

For the z-parameters of Fig. 5.38, the loop-impedance matrix is given by

$$\mathbf{Z}_m = \begin{bmatrix} z_{11} + Z_1 & z_{12} \\ z_{21} & z_{22} + Z_2 \end{bmatrix} \tag{5.223}$$

whose determinant is found to be

$$\Delta_m = \det \mathbf{Z}_m = (z_{11} + Z_1)(z_{22} + Z_2) - z_{12}z_{21} \tag{5.224}$$

From (5.219) we get

$$F^m(z_{21}) = F_0^m(z_{21}) = 1 + T_z^m = 1 - \frac{z_{12}z_{21}}{(z_{11} + Z_1)(z_{22} + Z_2)} \tag{5.225}$$

where T_z^m, called the *loop-based return ratio*, is identified as

$$T_z^m = -\frac{z_{12}z_{21}}{(z_{11} + Z_1)(z_{22} + Z_2)} \tag{5.226}$$

Comparing (5.45) with (5.225) yields

$$(z_{11}z_{22} - z_{12}z_{21})F(z_{21}) = z_{11}z_{22}F^m(z_{21}) \tag{5.227}$$

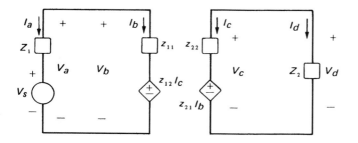

Figure 5.38 A terminated two-port network represented by its z-parameters.

From Fig. 5.38, the branch voltage-current relations are described by the equation

$$
\begin{bmatrix} V_a \\ V_b \\ V_c \\ V_d \end{bmatrix}
\begin{bmatrix} Z_1 & 0 & 0 & 0 \\ 0 & z_{11} & z_{12} & 0 \\ 0 & z_{21} & z_{22} & 0 \\ 0 & 0 & 0 & Z_2 \end{bmatrix}
\begin{bmatrix} I_a \\ I_b \\ I_c \\ I_d \end{bmatrix}
+
\begin{bmatrix} V_s \\ 0 \\ 0 \\ 0 \end{bmatrix}
\tag{5.228}
$$

The coefficient matrix in (5.228) is the branch-impedance matrix, whose determinant is found to be

$$
\Delta_b = Z_1 Z_2 (z_{11} z_{22} - z_{12} z_{21})
\tag{5.229}
$$

In terms of Δ_b, (5.227) can be written as

$$
\Delta_b(z_{21}) F(z_{21}) = \Delta_b(0) F^m(z_{21})
\tag{5.230}
$$

confirming (5.221).

As in (5.42), by substituting the corresponding parameters of z_{ij} we can show that

$$
F^m(k_{21}) = \frac{\Delta_m(k_{21})}{\Delta_m(0)} = \frac{(k_{11} + M_1)(k_{22} + M_2) - k_{12}k_{21}}{(k_{11} + M_1)(k_{22} + M_2)} = 1 + T_k^m
\tag{5.231}
$$

where

$$
T_k^m = -\frac{k_{12}k_{21}}{(k_{11} + M_1)(k_{22} + M_2)}
\tag{5.232}
$$

is invariant under immittance substitution for the z-, h-, and g-parameters. Its justification is left as an exercise (see Prob. 5.20). We remark that (5.42) and

(5.231) do not imply that $F(k_{21}) = F^m(k_{21})$. The return differences $F(k_{21})$ and $F^m(k_{21})$ are related by (5.221).

Example 5.16 Consider the feedback network of Fig. 5.39a. Suppose that g_{m2} is the element of interest and that the reference value k is zero. Two equivalent networks of Fig. 5.39a are presented in Fig. 5.39, b and c with

$$\alpha_1 = -\frac{g_{m2}g_{m3}Z_2}{q} \quad \alpha_2 = \frac{g_{m3}}{q} \tag{5.233a}$$

$$\beta_1 = \frac{g_{m2}}{q} \quad \beta_2 = -\frac{g_{m2}g_{m3}Z_3}{q} \tag{5.233b}$$

where

$$q = Y_2 Y_3 - g_{m2}g_{m3} \tag{5.233c}$$

The branch-admittance matrix of Fig. 5.39b can easily be obtained by inspection and is given by

$$\mathbf{Y}_b = \begin{bmatrix} Y_1 & 0 & 0 & 0 \\ 0 & Y_3 & g_{m2} & 0 \\ 0 & g_{m3} & Y_2 & 0 \\ 0 & 0 & 0 & Y_4 \end{bmatrix} \tag{5.234}$$

whose determinant is found to be

$$\det \mathbf{Y}_b = Y_1 Y_4 (Y_2 Y_3 - g_{m2}g_{m3}) = \frac{1}{\Delta_b} \tag{5.235}$$

Since Δ_b does not remain unaltered when g_{m2} is replaced by 0, according to Theorem 5.1, the return differences $F(g_{m2})$ and $F^m(g_{m2})$ will be different.

Consider the node-admittance matrix of Fig. 5.39b, given as

$$\mathbf{Y}_n = \begin{bmatrix} Y_1 + Y_3 + Y_4 & g_{m2} - Y_4 \\ g_{m3} - Y_4 & Y_2 + Y_4 \end{bmatrix} \tag{5.236}$$

whose determinant is also equal to the first-order cofactor of the indefinite-admittance matrix with

$$Y_{uv} = \det \mathbf{Y}_n = (Y_1 + Y_3)(Y_2 + Y_4) + Y_4(Y_2 + g_{m2} + g_{m3}) - g_{m2}g_{m3} \tag{5.237}$$

This yields

$$F(g_{m2}) = \frac{(Y_1 + Y_3)(Y_2 + Y_4) + Y_4(Y_2 + g_{m2} + g_{m3}) - g_{m2}g_{m3}}{(Y_1 + Y_3)(Y_2 + Y_4) + Y_4(Y_2 + g_{m3})} \tag{5.238}$$

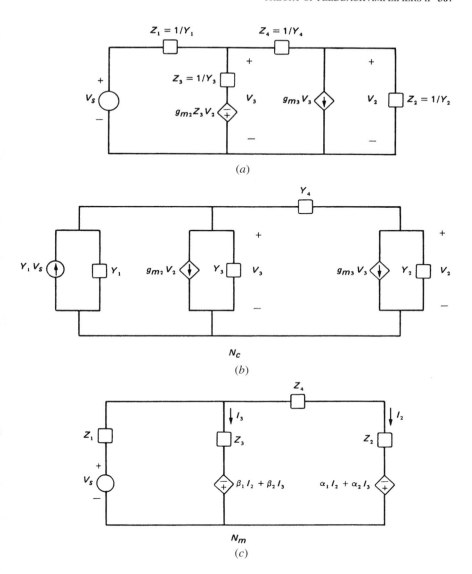

Figure 5.39 (*a*) A feedback network, (*b*) its equivalent network N_C, and (*c*) its equivalent network N_m.

By using the equivalent network of Fig. 5.39c, the branch-impedance matrix and the loop-impedance matrix are found to be

$$
\mathbf{Z}_b = \begin{bmatrix} Z_1 & 0 & 0 & 0 \\ 0 & \dfrac{Y_2}{q} & -\dfrac{g_{m2}}{q} & 0 \\ 0 & -\dfrac{g_{m3}}{q} & \dfrac{Y_3}{q} & 0 \\ 0 & 0 & 0 & Z_4 \end{bmatrix} \tag{5.239}
$$

$$
\mathbf{Z}_m = \begin{bmatrix} Z_1 + \dfrac{Y_2}{q} & -\dfrac{Y_2 + g_{m2}}{q} \\ -\dfrac{Y_2 + g_{m3}}{q} & Z_4 + \dfrac{Y_2 + Y_3 + g_{m2} + g_{m3}}{q} \end{bmatrix} \tag{5.240}
$$

giving

$$
\Delta_b = \frac{Z_1 Z_4 (Y_2 Y_3 - g_{m2} g_{m3})}{q^2} = \frac{Z_1 Z_4}{q} = \frac{1}{\det \mathbf{Y}_b} \tag{5.241}
$$

$$
\Delta_m = \det \mathbf{Z}_m = \frac{Z_1 (Z_4\, q + Y_2 + Y_3 + g_{m2} + g_{m3}) + Y_2 Z_4 + 1}{q} \tag{5.242}
$$

From (5.219), we obtain

$$
\begin{aligned}
F^m (g_{m2}) &= \frac{\Delta_m (g_{m2})}{\Delta_m (0)} \\[2mm]
&= \frac{Y_1 Y_2 Y_3 Y_4 \left[Z_1 (Z_4\, q + Y_2 + Y_3 + g_{m2} + g_{m3}) + Y_2 Z_4 + 1 \right]}{q \left[Y_2 Y_3 + Y_4 (Y_2 + Y_3 + g_{m3}) + Y_1 Y_2 + Y_1 Y_4 \right]} \\[2mm]
&= \frac{Y_2 Y_3 \left[Y_2 (Y_1 + Y_3 + Y_4) + Y_4 (Y_1 + Y_3 + g_{m2} + g_{m3}) - g_{m2} g_{m3} \right]}{(Y_2 Y_3 - g_{m2} g_{m3}) \left[Y_2 (Y_1 + Y_3 + Y_4) + Y_4 (Y_1 + Y_3 + g_{m3}) \right]}
\end{aligned} \tag{5.243}
$$

Comparing (5.238) with (5.243) gives

$$
\Delta_b (0) F^m (g_{m2}) = \Delta_b (g_{m2}) F (g_{m2}) \tag{5.244}
$$

confirming (5.221).

In the above analysis, we have used two equivalent networks N_c and N_m of a feedback amplifier N, such as the ones shown in Fig. 5.39, b and c. N_c for the cutset formulation and N_m for the loop formulation. In the equivalent network N_c, all the sources have been converted to equivalent current sources and/or voltage-controlled current sources. In N_m all the sources have been converted to equivalent voltage sources and/or current-controlled voltage sources. The n voltage-controlled current sources, characterized by the defining equations

$$I_{k_j} = x_{k_j} V_{u_j} \quad j = 1, 2, \ldots, n \tag{5.245}$$

are said to be *cyclically coupled* if $k_{v+1} = u_v$ for $v = 1, 2, \ldots, n$, where $k_{n+1} = k_1$ and I_k and V_k denote the current and voltage of the kth branch. For example, the following three voltage-controlled current sources,

$$I_2 = x_2 V_5 \quad I_5 = x_5 V_4 \quad I_4 = x_4 V_2 \tag{5.246}$$

are cyclically coupled. By interchanging the roles of I and V in (5.245), we can similarly define a set of *cyclically coupled* current-controlled voltage sources in N_m. Clearly, if every device is replaced by its equivalent network composed of the controlled sources, we can always renumber the controlling and the controlled branches in such a way that the index of any controlled branch is not higher than the index of its corresponding controlling branch. This results in a branch-immittance matrix with all the controlling parameters appearing above the main diagonal. This means that its determinant will be independent of the controlling parameter x. Hence, from Theorem 5.1, the general return difference in such a system will be invariant under the general transformations between a system of basis loops and a system of basis cutsets.

In Example 5.16, the two voltage-controlled current sources in Fig. 5.39b are cyclically coupled. If we set $g_{m3} = 0$, the resulting network has only one controlled source, so it cannot be cyclically coupled. Thus, the general return difference with respect to g_{m2} is invariant with $F(g_{m2}) = F^m(g_{m2})$, which can be confirmed from (5.244) by noting that, for $g_{m3} = 0$, $\Delta_b(0) = \Delta_b(g_{m2})$. In the network of Fig. 5.38, the two current-controlled voltage sources are cyclically coupled. To decouple them, we set $z_{12} = 0$, resulting in the invariance of the general return difference under the general transformations of the reference frame, which can again be confirmed from (5.230) by observing that for $z_{12} = 0$, $\Delta_b(z_{21}) = \Delta_b(0)$.

As shown in (5.104), the general return difference $F_k(x)$ with respect to a one-port admittance x for a general reference value k is equal to the ratio of total admittance looking into the node pair where x is connected to that when x is replaced by its reference value k:

$$F_k(x) = \frac{y + x}{y + k} \tag{5.247}$$

A similar result can be obtained based on the loop formulation. In this case, it is more convenient to consider the one-port impedance $\alpha = 1/x$.

Since from Theorem 5.1 the return difference $F_k^m(\alpha)$ is invariant with respect to the transformations from one system of basis loops to another, for simplicity and without loss of generality, we can choose a set of loop currents such that the impedance α is contained only in one of these loops, say, loop 1. With this choice of loop currents, the impedance α will appear only in the (1,1)-position of the corresponding loop-impedance matrix $\mathbf{Z}_m$. As before, we write $\Delta_m = \Delta_m(\alpha) = \det \mathbf{Z}_m$. Then Δ_m can be expanded as

$$\Delta_m(\alpha) = \Delta_m(k) + (\alpha - k)\Delta_{m11} \qquad (5.248)$$

where Δ_{m11} denotes the cofactor of the (1, 1)-element of $\mathbf{Z}_m$ in Δ_m. Combining (5.219) with (5.248) yields

$$F_k^m(\alpha) = \frac{\Delta_m(\alpha)}{\Delta_m(k)} = 1 + (\alpha - k)\frac{\Delta_{m11}}{\Delta_m(k)}$$

$$= 1 + \frac{\alpha - k}{k + z} = \frac{z + \alpha}{z + k} \qquad (5.249)$$

where z is the impedance that α faces. The second line follows from the fact that $\Delta_m(k)$ is the loop determinant of the equivalent network N_m when the element α assumes the value k, that Δ_{m11} is independent of the element α, and that if, in N_m with α assuming the value k, we insert a voltage source in series with α, the ratio $\Delta_m(k)/\Delta_{m11}$ is precisely the impedance looking into the voltage source, which is $z + k$. Thus, we conclude that $F_k^m(\alpha)$ is equal to the ratio of total impedance looking into the branch where α is situated, including α, to that when α is replaced by its reference value k. Comparing (5.247) with (5.249) yields

$$k F_k(x) = x F_\zeta^m(\alpha) \qquad (5.250)$$

where

$$\alpha = \frac{1}{x} \qquad (5.251a)$$

$$\zeta = \frac{1}{k} \qquad (5.251b)$$

Similarly, we can show that (see Prob. 5.21)

$$F_k(\alpha) = \frac{y + x}{y + \zeta} \qquad (5.252a)$$

$$F_k^m(x) = \frac{z + \alpha}{z + \zeta} \qquad (5.252b)$$

giving

$$\alpha F_k(\alpha) = k F_k^m(\alpha) \tag{5.253a}$$

$$k F_k(x) = x F_k^m(x) \tag{5.253b}$$

which also confirm (5.221) with $\Delta_b(\alpha)/\Delta_b(k) = \alpha/k$ and $\Delta_b(x)/\Delta_b(k) = k/x$, respectively.

Thus, we conclude that the general return difference of a feedback amplifier with respect to a one-port immittance x for a general reference value k is invariant under the general transformations between a system of basis loops and a system of basis cutsets if and only if $x = k$.

Example 5.17 Consider the same feedback network as in Example 5.16. Now we compute the return difference with respect to the one-port impedance Z_1 as shown in Fig. 5.39c. Applying Eq. (5.219) in conjunction with Eq. (5.242), we get

$$F_k^m(Z_1) = \frac{Z_1(Z_4\, q + Y_2 + Y_3 + g_{m2} + g_{m3}) + Y_2 Z_4 + 1}{k(Z_4\, q + Y_2 + Y_3 + g_{m2} + g_{m3}) + Y_2 Z_4 + 1} = \frac{z + Z_1}{z + k} \tag{5.254}$$

where z is the input impedance facing Z_1 and is given by

$$z = \frac{Y_2 Z_4 + 1}{Z_4\, q + Y_2 + Y_3 + g_{m2} + g_{m3}} \tag{5.255}$$

Likewise, from (5.237) we obtain

$$F_k(Y_1) = \frac{(Y_1 + Y_3)(Y_2 + Y_4) + Y_4(Y_2 + g_{m2} + g_{m3}) - g_{m2}g_{m3}}{(k + Y_3)(Y_2 + Y_4) + Y_4(Y_2 + g_{m2} + g_{m3}) - g_{m2}g_{m3}}$$

$$= \frac{y + Y_1}{y + k} \tag{5.256}$$

where $y = 1/z$ is the admittance facing Y_1, and

$$F_k(Z_1) = \frac{(Y_1 + Y_3)(Y_2 + Y_4) + Y_4(Y_2 + g_{m2} + g_{m3}) - g_{m2}g_{m3}}{(1/k + Y_3)(Y_2 + Y_4) + Y_4(Y_2 + g_{m2} + g_{m3}) - g_{m2}g_{m3}}$$

$$= \frac{y + Y_1}{y + 1/k} \tag{5.257}$$

showing that with $\zeta = l/k$,

$$F_k(Y_1) = F_\zeta(Z_1) \tag{5.258}$$

In a similar manner, we can demonstrate that

$$F_k^m(Z_1) = F_\zeta^m(Y_1) \tag{5.259}$$

where

$$F_k^m(Y_1) = \frac{Z_1(Z_4 q + Y_2 + Y_3 + g_{m2} + g_{m3}) + Y_2 Z_4 + 1}{\zeta(Z_4 q + Y_2 + Y_3 + g_{m2} + g_{m3}) + Y_2 Z_4 + 1} = \frac{z + Z_1}{z + \zeta} \tag{5.260}$$

Finally, comparing (5.254) with (5.257) and (5.256) with (5.260), we confirm (5.253).

5.10 SUMMARY

In this chapter, we continued our study of feedback amplifier theory begun in Chap. 4. We showed that feedback may be employed to make the amplifier gain less sensitive to variations in the parameters of the active components. If the magnitude of the ratio of the two functional values assumed by the amplifier forward transmission under the condition that an active device is deactivated and the condition that the active device is operating normally is small, then the sensitivity is approximately equal to the reciprocal of the return difference with respect to the forward-transfer parameter of the active device. In Chap. 3, we derived the general expressions of the transfer functions, driving-point functions, power gains, and the sensitivity functions of a two-port network in terms of the general two-port parameters. In this chapter, we exhibited the general relations between these expressions and feedback and demonstrated by a specific example how they can be employed to design amplifiers that achieve maximum gain with a prescribed sensitivity.

The concepts of return difference and null return difference introduced in the preceding chapter were extended and generalized by considering the general reference value. They are very useful in measurement situations in that they correspond to the situation where a feedback amplifier under study is made partially active rather than completely dead, as in the original interpretations of the return difference and null return difference for the zero reference value. Relations between network functions and the general return difference and null return difference were derived. They are generalizations of Blackman's formula introduced in Chap. 4, and they can be employed effectively to simplify the calculations of active impedances. We found that the network functions can be expressed in terms of the general return difference and the general null return difference in a manner similar to those discussed before. As an extension of the concept of sensitivity function, we defined the relative sensitivity function, which is expressed in terms of the altered parameters. The significance is that by choosing an appropriate reference value for the general return difference, we can interrupt the forward transmission of the amplifier at this reference value. Under this condition, the relative sensitivity becomes equal to the reciprocal of the general return difference. This yields a direct generalization of the simple relation between the sensitivity function and the return difference for the ideal feedback model. In general, the relative sensitivity function can be written as the difference of the reciprocals of the general return difference and the general null return difference.

In addition to the algebraic formulation of feedback amplifier theory, we indicated that the theory can also be formulated in terms of a signal-flow graph representing the signal transmissions among the various ports. One of the advantages of this formulation is that it can easily be generalized to multiple-loop feedback networks to be discussed later. One simple consequence is that, in many cases, negative feedback may be used to reduce the effect of noise and nonlinear distortion generated within an amplifier.

One useful relation found helpful in feedback calculations is the return difference and the null return difference with respect to two elements. We found that the ratio of the return differences with respect to two elements individually is equal to the ratio of the return differences with respect to one when the other one is set to zero.

Since the zeros of the return difference are the natural frequencies of the network, they are essential for the stability study. For this we presented three procedures for the physical measurements of the return difference, emphasizing the applications to transistors. The first technique is a direct one, but requires the use of a matched transistor and the simulation of three impedances. The second method overcomes the need for simulating the effect of internal feedback, but it essentially requires three measurements and the simulation of one driving-point impedance. The third method requires only two driving-point impedance measurements at a convenient port of the feedback amplifier, and is particularly attractive if the two impedances can be measured directly in their polar forms. As a consequence of these measurements, we can verify experimentally and study the stability problems of the amplifier under certain prescribed conditions. In addition to these, we discussed a procedure for experimental determination of the general reference value under which the amplifier becomes unilateral. The procedure requires four impedance measurements and can easily be performed in the laboratory for most active devices.

The feedback amplifier theory developed in this and the preceding chapters was formulated in terms of the first- and the second-order cofactors of the elements of the indefinite-admittance matrix, which is readily interpretable physically. This mathematical formulation is consistent, compact, and elegant, and it is easily applicable to most practical feedback amplifiers. It is also possible, although much more complicated, to formulate the theory in terms of the loop determinant or the cutset determinant. To this end, we outlined and derived relationships among the return differences based on various formulations. Specifically, we showed that in the cutset or nodal (loop) formulation of the network determinant, the general return difference is invariant with respect to the transformations from one such system to another. It is invariant under the general transformations of the reference frame if and only if the determinant of its branch-immittance matrix remains unaltered when the element of interest is replaced by the chosen reference value.

Unlike the indefinite-admittance-matrix formulation, the null return difference based on the cutset or loop determinant cannot generally be defined as the ratio of the two functional values assumed by one of its cofactors unless some specific details of the formulation of the network equations are also stated. Furthermore, this ratio is *variant* under the general transformations from a system of basis cutsets (loops) to another. To avoid this difficulty, we may introduce the concepts of "generalized cofactors," which are rather complicated. For a detailed account of these results and all the variations and ramifications, the reader is referred to Chen (1974a, 1974b).

PROBLEMS

5.1 Derive the identity (5.9).

5.2 Consider the series-parallel or voltage-series feedback amplifier of Fig. 4.24. Assume that the two transistors are identical, with $h_{re} = h_{oe} = 0$, $h_{ie} = 1.1\,k\Omega$, and $h_{fe} = 60$. Write $\tilde{\alpha}_j = \alpha_j \cdot 10^{-4} = h_{fe}/h_{ie}$ where $j = 1, 2$ is used to distinguish the transconductances of the first and second transistors. Determine the following:

(a) The general return differences $F_k(\tilde{\alpha}_1)$ and $F_k(\tilde{\alpha}_2)$
(b) The general null return differences $\hat{F}_k(\tilde{\alpha}_1)$ and $\hat{F}_k(\tilde{\alpha}_2)$
(c) The voltage gain in terms of $\tilde{\alpha}_1$ and $\tilde{\alpha}_2$
(d) The current gain in terms of $\tilde{\alpha}_1$ and $\tilde{\alpha}_2$
(e) The short-circuit current gain in terms of $\tilde{\alpha}_1$ and $\tilde{\alpha}_2$
(f) The input impedance in terms of $\tilde{\alpha}_1$ and $\tilde{\alpha}_2$
(g) The output impedance in terms of $\tilde{\alpha}_1$ and $\tilde{\alpha}_2$

By using the results obtained above, verify the identities (5.99), (5.101), and (5.103).

5.3 In the network of Fig. 5.2, confirm that the sensitivity function for the short-circuit current gain with respect to α_1 is given by Eq. (5.12).

5.4 The open-loop current gain of the voltage-shunt feedback amplifier of Fig. 5.4 is given by Eq. (4.51). Compare this with the open-loop voltage gain of Eq. (5.28a). With approximation, is one deducible from the other?

5.5 In the feedback network considered in Prob. 5.2, compute the sensitivities $S(\tilde{\alpha}_1)$ and $S(\tilde{\alpha}_2)$ and the relative sensitivities $S'(\tilde{\alpha}_1')$ and $S'(\tilde{\alpha}_2')$ where $\tilde{\alpha}_1' = \tilde{\alpha}_1 - k$ and $\tilde{\alpha}_2' = \tilde{\alpha}_2 - k$.

5.6 Using the ideal feedback model, show that the open-loop voltage gain of the feedback amplifier of Fig. 5.4 with the parameters as given in Eq. (5.31) is -16 and that the closed-loop voltage gain is -3.2.

5.7 The return ratios T_h and T_g are given by Eqs. (5.38) and (5.41). Show that for $M_1 = 0$ and $M_2 = 0$, $T_h = T_g$.

5.8 Repeat Example 5.10 if the element of interest is $\tilde{\alpha}_2$.

5.9 Repeat Example 5.5 by using the parallel-parallel topology of Fig. 4.3b.

5.10 Derive the identity (5.145).

5.11 Derive the identity (5.146).

5.12 Repeat Example 5.5 by using the series-series topology of Fig. 4.3a.

5.13 Repeat Example 5.5 by using the parallel-series topology of Fig. 4.3d

5.14 Consider the series-series or current-series feedback amplifier of Fig. 4.7. Determine the reference value for the general return difference with respect to the amplifier forward-transfer parameter under which the amplifier becomes unilateral.

5.15 Refer to Fig. 5.25. Assume that

$$|y_o| \gg |y_{12}| \quad |y_{11}| \gg |y_{12}| \qquad (5.261)$$

Show that the Norton equivalent network looking into the two-port network at terminals 1 and 3 can be approximated by the parallel combination of y_{11} and $y_{12}V_{23}$ as indicated in Fig. 5.24.

5.16 The hybrid-pi equivalent network of a common-emitter transistor is shown in Fig. 5.26, whose indefinite-admittance matrix can easily be written down by inspection. Show that after suppressing terminal $B' = 4$ and deleting the row and column corresponding to terminal $E = 3$, the admittance matrix is given by (5.182).

5.17 Consider the series-parallel feedback amplifier of Fig. 4.21. Determine the reference value for the general return difference with respect to the amplifier forward-transfer parameter under which the amplifier becomes unilateral.

5.18 Justify the statement that if a feedback amplifier contains only a single controlled source, then the general return difference with respect to the controlling parameter of the controlled source is invariant under the general transformations between a system of basis loops and a system of basis cutsets.

5.19 Repeat Prob. 5.14 for the parallel-series configuration of Fig. 4.27 using the hybrid parameters for the transistors.

5.20 Demonstrate that (5.232) is invariant under immittance substitution for the z-, h-, and g-parameters.

5.21 Derive the identities (5.252).

5.22 In the feedback amplifier of Fig. 4.27, apply Blecher's procedure to determine the return difference with respect to the forward-transfer parameter of the second transistor.

5.23 Repeat Prob. 5.22 using Hakim's procedure.

5.24 Repeat Prob. 5.22 by making only two driving-point impedance measurements at a convenient port.

5.25 Refer to the compound-feedback amplifier of Fig. 4.46. Compute the sensitivity and the relative sensitivity functions with respect to the amplifier forward-transfer parameter. Also determine the reference value of the general return difference with respect to this parameter under which the amplifier becomes unilateral.

Figure 5.40 The equivalent network of the feedback amplifier of Fig. 5.4a with the transistor being represented by its T-model.

5.26 In the feedback network of Fig. 5.39b, let $Y_1 V_s$ be the input port and Y_4 the output port. Show that

$$\hat{F}_k(g_{m2}) = 1 \tag{5.262a}$$

$$\hat{F}_k(g_{m3}) = \frac{Y_2 + g_{m3}}{Y_2 + k} \tag{5.262b}$$

Also calculate Eqs. (5.262a) and (5.262b) by their physical interpretations in terms of the returned voltages.

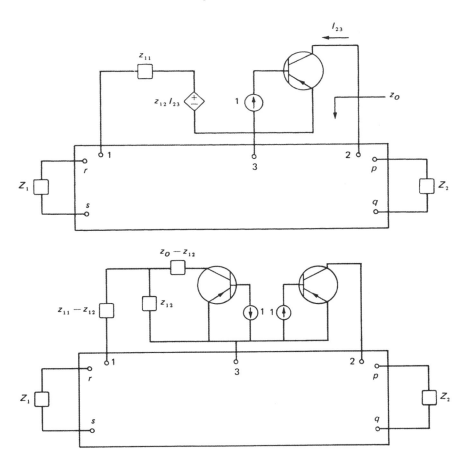

Figure 5.41 (a) The measurement of return difference $F^m(z_{21})$ for a transistor operated in the common-emitter configuration and represented by its z-parameters z_{ij}. (b) The measurement of $F^m(z_{21})$ with the simulation of the controlled source $z_{12}I_{23}$.

5.27 The transistor in the feedback amplifier of Fig. 5.4a can be represented by its equivalent T-model, as shown in Fig. 5.40. Show that the general null return difference with respect to the transconductance αg_e for a general reference value k is given by

$$\hat{F}_k(\alpha g_e) = \frac{g_b(g_c + G_f) + g_c G_f - \alpha g_e(g_b + G_f) + g_e G_f}{g_b(g_c + G_f) + g_c G_f - k(g_b + G_f) + g_e G_f} \qquad (5.263)$$

Also show that for

$$k = k_0 = \frac{g_b(g_c + G_f) + G_f(g_c + g_e)}{g_b + G_f} \qquad (5.264)$$

the relative sensitivity $S'(x')$ becomes equal to the reciprocal of the return difference $F_{k_0}(\alpha g_e)$, where $x' = \alpha g_e - k_0$.

5.28 By using Eq. (5.130), compute the impedance facing $R_f = 1/G_f$ in the network of Fig. 5.40. Also, compute the amplifier input and output impedances and determine the sensitivity function of the current gain to R_f.

5.29 Assume that the two-port network of Fig. 5.22 is a transistor and is represented by its z-parameters z_{ij}. Demonstrate that the return difference $F^m(z_{21})$ can be measured by the network of Fig. 5.41a. Also show that the controlled source $z_{12}I_{23}$ can be simulated using the arrangement of Fig. 5.41b, where z_o denotes the total impedance presented to the output port of the transistor, as shown in Fig. 5.41a.

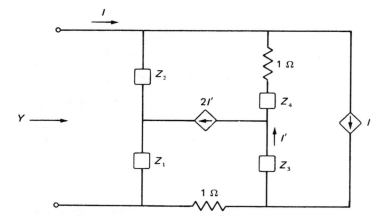

Figure 5.42 A one-port network with two current-controlled current source.

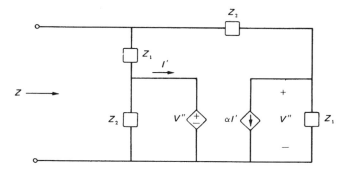

Figure 5.43 A one-port network with two controlled sources.

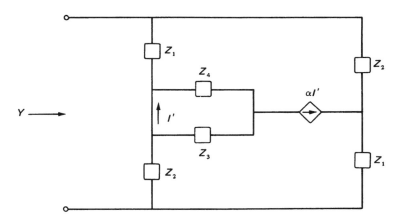

Figure 5.44 A one-port network containing a current-controlled current source.

5.30 By using Blackman's formula, show that the input admittance Y of the network of Fig. 5.42 is given by

$$Y(s) = 1 + \frac{Z_3 - Z_4}{Z_1 - Z_2} \tag{5.265}$$

5.31 By applying Blackman's formula, show that the input impedance Z of the network of Fig. 5.43 with $\alpha = Z_3/Z_4$ can be expressed as

$$Z(s) = \frac{Z_1 Z_2 (Z_4 - Z_3)}{Z_2 Z_4 - Z_1 Z_3} \tag{5.266}$$

5.32 By using Blackman's formula, demonstrate that the input admittance Y of the network of Fig. 5.44 can be written as

$$Y(s) = \frac{(\alpha + 2)Z_4 - (\alpha - 2)Z_3}{Z_4[(\alpha + 1)Z_2 + Z_1] - Z_3[(\alpha - 1)Z_1 - Z_2]} \qquad (5.267)$$

5.33 In the feedback network of Fig. 5.4a, let $h_{fe} = 60$, $h_{ie} = 1\,\text{k}\Omega$, $h_{re} = 3 \cdot 10^{-4}$, $h_{oe} = 25\,\mu\text{mho}$, $R_f = 40$ kΩ, $R_1 = 10$ kΩ, and $R_2 = 4$ kΩ. Compute the return differences with respect to the forward-transfer parameters $h_{fe} = h_{21}$, y_{21}, z_{21}, and g_{21}, respectively, of the transistor.

BIBLIOGRAPHY

Blecher, F. H.: Design Principles for Single Loop Transistor Feedback Amplifiers, *IRE Trans. Circuit Theory,* vol. CT-4, no. 3, pp. 145–156, 1957.

Bode, H. W.: "Network Analysis and Feedback Amplifier Design," Princeton, N.J.: Van Nostrand, 1945.

Chen, W. K.: Graph-Theoretic Considerations on the Invariance and Mutual Relations of the Determinants of the Generalized Network Matrices and Their Generalized Cofactors, *Quart J. Math. (Oxford),* 2d series, vol. 21, no. 84, pp. 459–479, 1970.

Chen, W. K.: Graph-Theoretic Considerations on the Invariance of Return Difference, *J. Franklin Inst.,* vol. 298, no. 2, pp. 81–100, 1974a.

Chen, W. K.: Invariance and Mutual Relations of the General Null-Return-Difference Functions, *Proc. 1974 Eur. Conf. Circuit Theory and Design,* IEE Conf. Pubi. No. 116, pp. 371–376, 1974b.

Chen, W. K.: "Applied Graph Theory : Graphs and Electrical Networks," 2d rev. ed., chaps. 2 and 4, New York: American Elsevier, and Amsterdam: North-Holland, 1976a.

Chen, W. K.: Indefinite-Admittance Matrix Formulation of Feedback Amplifier Theory, *IEEE Trans. Circuits and Systems,* vol. CAS-23, no. 8, pp. 498–505, 1976b.

Chen, W. K.: Network Functions and Feedback, *Int. J. Electronics,* vol. 42, no. 6, pp. 617–618, 1977.

Chen, W. K.: On Second-Order Cofactors and Null Return Difference in Feedback Amplifier Theory, *Int. J. Circuit Theory and Applications,* vol. 6, no. 3, pp. 305–312, 1978.

Hakim, S. S.: Return-Difference Measurement in Transistor Feedback Amplifiers, *Proc. IEE (London),* vol. 112, no. 5, pp. 914–915, 1965a.

Hakim, S. S.: Aspects of Return-Difference Evaluation in Transistor Feedback Amplifiers, *Proc. IEE (London),* vol. 112, no. 9, pp. 1700–1704, 1965b.

Haykin, S. S.: "Active Network Theory," Reading, Mass.: Addison-Wesley, 1970.

Kuh, E. S., and R. A. Rohrer: "Theory of Linear Active Networks," San Francisco, Calif.: Holden-Day, 1967.

Mason, S. J.: Feedback Theory–Some Properties of Signal Flow Graphs, *Proc. IRE,* vol. 41, no. 9, pp. 1144–1156, 1953.

Mason, S. J.: Feedback Theory–Further Properties of Signal Flow Graphs, *Proc. IRE,* vol. 44, no. 7, pp. 920–926, 1956.

Truxal, J. G.: "Automatic Feedback Control System Synthesis," New York: McGraw-Hill, 1955.

Waldhauer, F. D.: Wide-Band Feedback Amplifiers, *IRE Trans. Circuit Theory,* vol. CT-4, no. 3, pp. 178–190, 1957.

Wilts, C. H.: "Principles of Feedback Control," Reading, Mass.: Addison-Wesley, 1960.

STABILITY OF FEEDBACK AMPLIFIERS

In the preceding two chapters, we demonstrated that the application of negative feedback in amplifiers tends to make the overall gain less sensitive to variations in parameters, reduce noise and nonlinear distortion, and control the input and output impedances. These improvements are all affected by the same factor, which is the normal value of the return difference. However, the price that we paid in achieving these is the net reduction of the overall gain. In addition, we are faced with the stability problem in that, for sufficient amount of feedback, at some frequency the amplifier tends to oscillate and becomes unstable. The objective of this chapter is to discuss various stability criteria and to investigate several approaches to the stabilization of feedback amplifiers.

As indicated in Sec. 5.8, the zeros of the network determinant are called the natural frequencies. A network is stable if all of its natural frequencies are restricted to the open left half of the complex frequency s-plane (LHS). If the network determinant is known, its roots can be readily computed explicitly with the aid of a computer if necessary, and the stability problem can then be settled directly. However, for a physical network, there remains the difficulty of getting an accurate formulation of the network determinant itself. Even if we have the network determinant, the roots alone do not tell us the degree of stability when the feedback amplifier is stable, nor do they provide us with any information as to how to stabilize an unstable amplifier. These limitations are overcome by applying the Nyquist criterion to the return difference, which gives precisely the same information about the stability of a feedback amplifier as does the network determinant itself. Furthermore, the return difference can be measured physically, meaning that we can include all the parasitic effects in the stability study. The discussion of this chapter is confined to single-loop feedback amplifiers. Multiple-loop feedback amplifiers are presented in the following chapter.

We first introduce the concepts of a single-loop feedback amplifier and its stability, and then review briefly the Routh-Hurwitz criterion. This is followed by a discussion of the Nyquist stability criterion and the Bode plot. The root-locus technique and the notion of root sensitivity are taken up next. The relationship between gain and phase shift are elaborated. Finally, we discuss means of stabilizing a feedback amplifier and present Bode's design theory.

6.1 THE SINGLE-LOOP FEEDBACK AMPLIFIERS

For the sake of later discussion, in this section we define several important terms.

Definition 6.1: Single-loop feedback amplifier A *single-loop feedback amplifier* is one in which the return difference with respect to the controlling parameter of any active device is equal to unity if the controlling parameter of any other active device in the network vanishes.

This definition has two important implications, The first is that the controlling parameters of the active devices can enter the network determinant only in a simple product form. Thus, for a single-loop feedback amplifier containing n active devices, the first-order cofactor of an element of its indefinite-admittance matrix can be expressed in the form

$$Y_{uv} = g_{m1}g_{m2}\cdots g_{mn}A + B \tag{6.1}$$

where $g_{mj}(j = 1, 2, \ldots, n)$ are the controlling parameters of the active devices, and A and B are functions of s and are independent of the controlling parameters g_{mj}. As a result, the second implication is that the return differences with respect to g_{mj} of all the active devices under normal operating conditions are the same. Thus, the voltage-shunt configuration of Fig. 4.19 with its equivalent network as shown in Fig. 4.20 is a single-loop feedback amplifier. In fact, Eq. (6.1) is completely equivalent to Definition 6.1 and can be used interchangeably.

Strictly speaking, the above definition excludes amplifiers in which there is local feedback or internal feedback in one or more of the active devices. However, many of these amplifiers can be approximated by a single-loop amplifier. On the other hand, the definition includes as single-loop feedback amplifiers structures having a number of distinct feedback paths. For example, we can put another passive two-port network such as a transformer in parallel with the network of Fig. 4.19, and the resulting structure is still a single-loop feedback amplifier.

As an example of a multiple-loop feedback amplifier, we consider the parallel-series configuration of Fig. 5.14, which is represented by the equivalent network of Fig. 5.15b. From Example 5.9, the first-order cofactor or nodal determinant is found to be

$$Y_{uv} = (166.66\alpha_1\alpha_2 + 1514.94\alpha_1 + 1716.83\alpha_2 + 1,331,494) \cdot 10^{-16} \quad (6.2)$$

where α_1 and α_1 are the transconductances of the transistors. Since Eq. (6.2) cannot be represented in the form of Eq. (6.1), the feedback amplifier is not a single-loop one. Likewise, the series-parallel feedback amplifier of Fig. 4.24, as considered in Example 4.11, is another that is not a single-loop feedback amplifier.

The stability of a feedback amplifier is sometimes much affected by the relative rates at which the gains of the active devices decay with age, or the relative rates at which they increase as the devices warm up when power is first applied to the network. Therefore, a feedback amplifier that is stable under the designed operating conditions may become unstable as a result of variations in parameters. Since the performance of a feedback amplifier is most sensitive to the changes of the controlling parameters of the active devices, we introduce the following terms.

Definition 6.2: Absolutely stable single-loop feedback amplifier A single-loop feedback amplifier is said to be *absolutely stable* if the network is stable when the values of the controlling parameters of the active devices vary from zero to their nominal values.

Definition 6.3: Conditionally stable single-loop feedback amplifier A single-loop feedback amplifier is said to be *conditionally stable* if the network is stable when the controlling parameters of the active devices assume their nominal values, and it becomes unstable when they assume some values between zero and their nominal values.

These concepts will be elaborated below when we discuss various stability criteria.

6.2 THE ROUTH CRITERION, THE HURWITZ CRITERION, AND THE LIÉNARD-CHIPART CRITERION

In this section, we review briefly the Routh-Hurwitz criterion. For completeness, we also introduce the less known Liénard-Chipart criterion. As indicated at the beginning of this chapter, the criteria are of limited practical value because the network determinant or its equivalent is not usually given analytically.

During the early development of control theory, there were two sources of methods for determining the nature of the zeros of a polynomial: the Routh criterion and the Hurwitz criterion. The Routh criterion was found by E. J. Routh in 1877. Eighteen years later, in 1895, A. Hurwitz, who at the time was unaware of Routh's work, developed essentially the same result. Despite the different appearance, they have been shown to be closely related and are usually referred to as the *Routh-Hurwitz criterion*.

Consider the characteristic equation

$$P(s) = a_0 s^n + a_1 s^{n-1} + \cdots + a_{n-1}s + a_n = 0 \quad a_0 > 0 \qquad (6.3)$$

with real coefficients, which results when the network determinant is set to zero. From the coefficients of $P(s)$, we form the following array:

$$
\begin{array}{c|cccc}
s^n & a_0 & a_2 & a_4 & a_6 & \cdots \\
s^{n-1} & a_1 & a_3 & a_5 & a_7 & \cdots \\
s^{n-2} & b_1 & b_2 & b_3 & \cdots \\
s^{n-3} & c_1 & c_2 & \cdots \\
s^{n-4} & d_1 & \cdots \\
\vdots & \vdots
\end{array}
\qquad (6.4)
$$

where

$$b_i = \frac{a_1 a_{2i} - a_0 a_{2i+1}}{a_1} \qquad (6.5a)$$

$$c_i = \frac{b_1 a_{2i+1} - a_1 b_{i+1}}{b_1} \qquad (6.5b)$$

$$d_i = \frac{c_1 b_{i+1} - b_1 c_{i+1}}{c_1} \qquad (6.5c)$$

and so on. The process is continued to the right until only zeros are obtained, filling in zeros after coefficients are exhausted, and down until there are $n + 1$ rows. This array forms the basis for the next theorem, known as the *Routh criterion* [Routh (1877)].

Theorem 6.1 The number of sign changes in the left-hand column of array (6.4) is equal to the number of zeros of $P(s)$ with positive-real parts.

Occasionally, it may occur that the first element in one of the rows vanishes but not all terms vanish. In such a case, it is immediately known that the system is unstable. However, if additional information is required regarding the characteristic roots of the system, the Routh array can be constructed by multiplying $P(s)$ by $s + \alpha$, where α is almost any positive number, and then applying the criterion to the new equation. Alternatively, we simply replace the zero with an arbitrary small quantity ϵ and continue the construction process as if no zero had been obtained. The incremental ϵ is usually taken to be positive, but a negative value is also satisfactory. It may also occur that one of the rows is entirely zero, say, the jth row. In such case, we replace the jth row by the row

$$(n + 2 - j)\alpha_0, \ (n - j)\alpha_1, \ldots, (n - 2i + 2 - j)\alpha_i, \ldots \qquad (6.6)$$

where $\alpha_0, \alpha_1, \ldots$ are the elements of the $(j - 1)$th row. In fact, the elements of (6.6) correspond to the coefficients of the derivative of the *auxiliary polynomial* of degree $n + 2 - j$ defined by the $(j - 1)$th row of (6.4) such that the exponent of s in each term after the first, which is $n + 2 - j$, is the exponent of s in the preceding term diminished by 2.

The vanishing of a row of elements before the $(n + 2)$th row is caused by the existence of a pair of roots of equal magnitude but opposite in sign. In fact, all roots of this type of the characteristic equation are roots of the auxiliary polynomial of the row preceding the first vanishing row of the array. Thus, if the characteristic equation has roots on the real-frequency axis, they will be found among the roots of the auxiliary polynomial. The total number of vanishing rows is equal to the multiplicity of the pair of roots of the above type of the greatest multiplicity.

For the Hurwitz criterion, we define the nth-order determinant

$$\Delta_n = \det \begin{bmatrix} a_1 & a_3 & a_5 & a_7 & \cdots & & 0 & 0 \\ a_0 & a_2 & a_4 & a_6 & \cdots & & 0 & 0 \\ 0 & a_1 & a_3 & a_5 & a_7 & \cdots & 0 & 0 \\ 0 & a_0 & a_2 & a_4 & a_6 & \cdots & 0 & 0 \\ 0 & 0 & a_1 & a_3 & a_5 & \cdots & 0 & 0 \\ 0 & 0 & a_0 & a_2 & a_4 & \cdots & 0 & 0 \\ \cdots & \cdots & \cdots & \cdots & \cdots & \cdots & \cdots & \cdots \\ 0 & 0 & \cdots & \cdots & \cdots & \cdots & a_{n-1} & 0 \\ 0 & 0 & \cdots & \cdots & \cdots & \cdots & a_{n-2} & a_n \end{bmatrix} \qquad (6.7)$$

Denote by Δ_{n-k} the determinant of the submatrix obtained from the matrix of (6.7) by deleting the last k rows and columns. In this way, we construct a total of n such determinants, known as the *Hurwitz determinants*. The following theorem, known as the *Hurwitz criterion,* is stated in terms of the Hurwitz determinants [Hurwitz (1895)].

Theorem 6.2 A necessary and sufficient condition that the polynomial $P(s)$ has zeros with only negative-real parts is that the values of the Hurwitz determinants Δ_j $(j = 1, 2, \ldots, n)$ be all positive in sign.

A case of frequent occurrence is that in which the coefficients of $P(s)$ are functions of some parameters. In such an event, the Hurwitz criterion is reasonably easy to apply. For numerical coefficients, the Routh-Hurwitz criterion can be put into another form, as suggested by Guillemin (1949), which will now be described.

Let $P_1(s)$ and $P_2(s)$ be the polynomials derived from $P(s)$ by taking alternate terms, starting with $a_0 s^n$ and $a_1 s^{n-1}$, respectively, as follows:

$$P_1(s) = a_0 s^n + a_2 s^{n-2} + a_4 s^{n-4} + \cdots \tag{6.8a}$$

$$P_2(s) = a_1 s^{n-1} + a_3 s^{n-3} + a_5 s^{n-5} + \cdots \tag{6.8b}$$

Then form the ratio $P_1(s)/P_2(s)$ and expand it as a continued fraction, known as the *Stieltjes continued fraction*:

$$\frac{P_1(s)}{P_2(s)} = \alpha_1 s + \cfrac{1}{\alpha_2 s + \cfrac{1}{\alpha_3 s + \cfrac{1}{\ddots + \frac{1}{\alpha_m s}}}} \tag{6.9}$$

In general, there are n coefficients α's in the expansion with $m = n$. In the situation where $m < n$, we say that the expansion is terminated *prematurely*. With these we state the Routh-Hurwitz criterion in Guillemin form.

Theorem 6.3 A necessary and sufficient condition that the polynomial $P(s)$ has zeros with only negative-real parts is that the Stieltjes continued fraction expansion not terminate prematurely and yield only positive α_i for $i = 1, 2, \ldots, n$.

In applying Hurwitz criterion, we are required to evaluate n determinants. The number of these determinants can be reduced, and the result is known as the *Liénard-Chipart criterion* [Liénard and Chipart (1914)].

Theorem 6.4 A necessary and sufficient condition that the polynomial $P(s)$ has zeros with only negative-real parts is that all elements in any one of the following four sets are positive:

1. $a_n, a_{n-2}, a_{n-4}, \ldots,$ and $\Delta_n, \Delta_{n-2}, \Delta_{n-4}, \ldots$
2. $a_n, a_{n-2}, a_{n-4}, \ldots,$ and $\Delta_{n-1}, \Delta_{n-3}, \Delta_{n-5}, \ldots$
3. $a_n, a_{n-1}, a_{n-3}, \ldots,$ and $\Delta_n, \Delta_{n-2}, \Delta_{n-4}, \ldots$
4. $a_n, a_{n-1}, a_{n-3}, \ldots,$ and $\Delta_{n-1}, \Delta_{n-3}, \Delta_{n-5}, \ldots$

Unlike the Hurwitz criterion, the Liénard-Chipart criterion requires only that every other Hurwitz determinant be evaluated. Since the amount of work increases considerably with the size of the determinant, it is advantageous to choose either condition (2) or condition (4) of the theorem, since neither includes the largest Hurwitz determinant Δ_n.

We illustrate the above results by considering the stability of a feedback network.

Example 6.1 A practical version of a field-effect transistor (FET) Colpitts oscillator is shown in Fig. 6.1a. After the biasing circuitry and the radio frequency (RF) choke have been removed, the equivalent network is presented in Fig. 6.1b. Typical parameter values for the FETs are as follows:

$$g_m = 100 \sim 10,000\,\mu\text{mho} \quad C_{gs} = 2 \sim 10\,\text{pF}$$

$$C_{gd}, C_{ds} = 0.1 \sim 2\,\text{pF} \quad r_d = 0.05 \sim 1\,\text{M}\Omega \tag{6.10}$$

Thus, in computing the frequency of oscillation, the capacitances of the FET can usually be ignored. The indefinite-admittance matrix of the network of Fig. 6.1b can be written down by inspection and is given by

$$\mathbf{Y} = \begin{bmatrix} sC_1' + sC_{gd} + \dfrac{1}{sL} & -sC_{gd} - \dfrac{1}{sL} & -sC_1' \\[2ex] g_m - sC_{gd} - \dfrac{1}{sL} & sC_2' + sC_{gd} + \dfrac{1}{sL} + \dfrac{1}{r_d} & -sC_2' - g_m - \dfrac{1}{r_d} \\[2ex] -sC_1' - g_m & -sC_2' - \dfrac{1}{r_d} & sC_1' + sC_2' + g_m + \dfrac{1}{r_d} \end{bmatrix} \tag{6.11}$$

where $C_1' = C_1 + C_{gs}$ and $C_2' = C_2 + C_{ds}$. The network determinant is found to be

$$Y_{33} = \frac{L(C_1'C_2' + C_1'C_{gd} + C_2'C_{gd})s^3 + L(C_1'/r_d + C_{gd}/r_d + C_{gd}g_m)s^2}{sL}$$

$$+ \frac{(C_1' + C_2')s + g_m + 1/r_d}{sL} \tag{6.12a}$$

which is simplified to

$$Y_{33} \approx \frac{C_1C_2Ls^3 + (LC_1/r_d)s^2 + (C_1 + C_2)s + g_m + 1/r_d}{sL} \tag{6.12b}$$

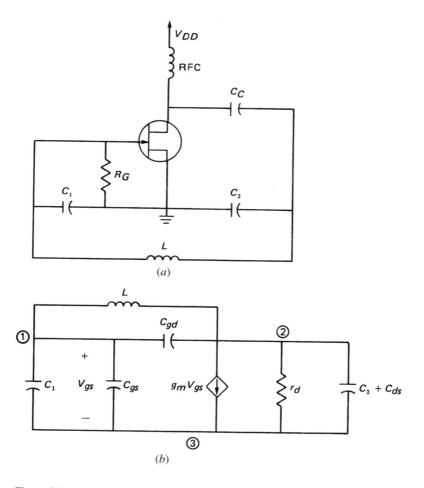

Figure 6.1 (a) A practical version of a FET Colpitts oscillator and (b) its equivalent network.

if the capacitances of the FET are ignored. Let

$$C_1 = 750\,\text{pF} \quad C_2 = 2500\,\text{pF}$$
$$L = 40\,\mu\text{H} \quad r_d = 50\,\text{k}\Omega \tag{6.13}$$

Substituting these in (6.12b) yields the characteristic equation

$$75p^3 + 0.006p^2 + 0.325p + 2 \cdot 10^{-5} + g_m = 0 \tag{6.14}$$

where $p = s/10^8$.

To investigate the stability performance of the feedback network, we apply the Routh criterion to (6.14). This leads to the following array of elements:

$$
\begin{array}{c|cc}
p^3 & 75 & 0.325 \\
p^2 & 0.006 & 2 \cdot 10^{-5} + g_m \\
p^1 & 0.075 - 12.5 \cdot 10^3 g_m & 0 \\
p^0 & 2 \cdot 10^{-5} + g_m & 0
\end{array}
\tag{6.15}
$$

According to the Routh criterion, all the zeros of (6.14) will be restricted to the open LHS if all the elements of the first column in the array (6.15) are positive. This requires that

$$
0.075 - 12.5 \cdot 10^3 g_m > 0 \tag{6.16}
$$

giving

$$
g_m < 6 \, \mu\text{mho} \tag{6.17}
$$

If we choose $g_m = 6 \, \mu$mho, the third row of the array (6.15) will vanish, and the network will have a pair of natural frequencies on the real-frequency axis, which are the roots of the equation

$$
0.006 p^2 + 2 \cdot 10^{-5} + g_m = 0 \tag{6.18}
$$

which results when the auxiliary polynomial of the second row is set to zero, giving

$$
s = 10^8 p = \pm j 6.583 \cdot 10^6 \tag{6.19}
$$

or $f = 1.048 \, \text{MHz}$. Thus, for $g_m = 6 \, \mu$mho, the network will oscillate with a frequency of 1.048 MHz.

Suppose that the capacitances of the FET are not ignored and let

$$
C_{gd} = C_{ds} = 1 \, \text{pF} \quad C_{gs} = 5 \, \text{pF} \tag{6.20}
$$

The corresponding characteristic equation is determined from (6.12a), and is given by

$$
75.66 p^3 + (0.006 + 0.4 g_m) p^2 + 0.326 p + 2 \cdot 10^{-5} + g_m = 0 \tag{6.21}
$$

To investigate stability, in this case it is convenient to apply the Liénard-Chipart criterion. The Hurwitz determinant is found to be

$$
\Delta_3 = \det \begin{bmatrix} 0.006 + 0.4 g_m & 2 \cdot 10^{-5} + g_m & 0 \\ 75.66 & 0.326 & 0 \\ 0 & 0.006 + 0.4 g_m & 2 \cdot 10^{-5} + g_m \end{bmatrix} \tag{6.22}
$$

To avoid the evaluation of Δ_3, we choose condition (2) of Theorem 6.4. For the zeros of Eq. (6.21) to be confined to the open LHS, we require that

$$a_3 = 2 \cdot 10^{-5} + g_m > 0 \tag{6.23a}$$

$$a_1 = 0.006 + 0.4 g_m > 0 \tag{6.23b}$$

$$\Delta_2 = \det \begin{bmatrix} 0.006 + 0.4 g_m & 2 \cdot 10^{-5} + g_m \\ 75.66 & 0.326 \end{bmatrix} = 0.443 \cdot 10^{-3} - 75.53 g_m > 0$$

$$\tag{6.23c}$$

yielding

$$g_m < 5.86 \, \mu\text{mho} \tag{6.24}$$

Let $g_m = 5 \, \mu$mho; substituting it in (6.21) yields

$$75.66 p^3 + 0.006 p^2 + 0.326 p + 25 \cdot 10^{-6} = 0 \tag{6.25}$$

To apply Theorem 6.3, we compute the Stieltjes continued fraction (6.9), which is found to be

$$\frac{P_1}{P_2} = \frac{75.66 p^3 + 0.326 p}{0.006 p^2 + 25 \cdot 10^{-6}} = 12,610 p + \cfrac{1}{0.546 p + 1/440 p} \tag{6.26}$$

Since all the coefficients are positive and the continued-fraction expansion does not terminate prematurely, all the roots of (6.25) are confined to the open LHS, as expected.

On the other hand, suppose that we let $g_m = 7 \, \mu$mho. We expect that the natural frequencies will move to the closed RHS and the network becomes unstable. For $g_m = 7 \, \mu$mho, (6.21) becomes

$$75.66 p^3 + 0.006 p^2 + 0.326 p + 27 \cdot 10^{-6} = 0 \tag{6.27}$$

giving

$$\frac{P_1}{P_2} = 12,610 p + \cfrac{1}{-0.429 p - 1/518.52 p} \tag{6.28}$$

Since two coefficients are negative, (6.27) has at least one zero in the closed RHS and the network is unstable.

6.3 THE NYQUIST CRITERION

The stability criteria described in the preceding section are useful if the network determinant is known analytically to be a function of the complex-frequency variable s. Even if we have the network determinant, the roots alone do not tell us the degree of stability when the feedback amplifier is stable, nor do they provide us with any information as to how to stabilize an unstable amplifier. Furthermore, for a physical network there remains the difficulty of getting an accurate formulation of the network determinant itself, since every equivalent network is, to a greater or lesser extent, an idealization of the physical reality. Also, the parasitic effects, which are usually ignored in the formulation of network determinant, play an important role for the stability study. These limitations are overcome by applying the Nyquist criterion to the return difference to be described below.

Let $f(s)$ be an analytic function that is regular within and on a given closed contour C in the s-plane, with the exception of a finite number of poles inside the contour. Assume that $f(s)$ has a zero or pole of order n at a point s_1 and write

$$f(s) = (s - s_1)^{\pm n} f_1(s) \tag{6.29}$$

where the plus sign denotes a zero and the minus sign a pole. Write $f'(s) = df(s)/ds$ and obtain

$$f'(s) = \pm n(s - s_1)^{\pm n - 1} f_1(s) + (s - s_1)^{\pm n} f_1'(s) \tag{6.30}$$

which when combined with (6.29) gives

$$\frac{f'(s)}{f(s)} = \frac{\pm n}{s - s_1} + \frac{f_1'(s)}{f_1(s)} \tag{6.31}$$

This result indicates that the function $f'(s)/f(s)$ has a simple pole at the point s_1, which is a zero or pole of $f(s)$, with the residue $+n$ or $-n$. The same process can now be repeated for the other zeros and poles of $f(s)$ by considering the function $f_1(s)$ which has the same zeros and poles as in $f(s)$ except for the one at s_1, multiplicity included. Thus, if n_{zi} and n_{pi} are the orders of the zeros s_{zi} and poles s_{pi} of $f(s)$, respectively, (6.31) can be expanded as

$$\frac{f'(s)}{f(s)} = \sum \frac{n_{zi}}{s - s_{zi}} - \sum \frac{n_{pi}}{s - s_{pi}} \tag{6.32}$$

where the summations are taken over all zeros and poles of $f(s)$. Let C be a closed curve in the s-plane. Consider the closed line integral

$$\oint_c \frac{f'(s)}{f(s)} ds = \oint_c \left(\sum \frac{n_{zi}}{s - s_{zi}} - \sum \frac{n_{pi}}{s - s_{pi}} \right) ds \tag{6.33}$$

Since the only singularities of the function $f'(s)/f(s)$ inside the closed contour C are the zeros and poles of $f(s)$ that lie inside C, by the residue theorem, the value of the closed contour integral is given by

$$\oint_c \frac{f'(s)}{f(s)} ds = 2\pi j \left(\sum n_{zi} - \sum n_{pi} \right) = 2\pi j (n_z - n_p) \qquad (6.34)$$

where the summations are now taken only over the zeros and poles of $f(s)$ that lie within the contour C, and n_z and n_p denote, respectively, the numbers of zeros and poles of $f(s)$ within the contour C, counting each zero and pole according to its multiplicity. Since

$$\frac{f'(s)}{f(s)} = \frac{d \ln f(s)}{ds} \qquad (6.35)$$

we find that with s_a and s_b symbolizing the initial and final points on C, (6.34) becomes

$$\ln |f(s)| \Big|_{s_a}^{s_b} + j \arg f(s) \Big|_{s_a}^{s_b} = 2\pi j (n_z - n_p) \qquad (6.36)$$

As s starts at a point on the contour C, traverses along the contour in the positive direction (counterclockwise), and returns to the starting point, the locus of $f(s)$ in the f-plane will also return to its starting point. Thus, the value of $\ln |f(s)|$ will be the same at the starting point as at the end of the closed contour C; only the value of $\arg f(s)$ may be different because of the multiple-valuedness of the function $\ln f(s)$. As a result, the first contour integral on the left-hand side of (6.36) vanishes, and the value of the second integral equals j times the increase in phase angle of $f(s)$ as s traverses the contour C. If we now divide by 2π on both sides of (6.36), the result on the left-hand side should be the number of times the mapping of the contour C in the f-plane goes around its origin counterclockwise. If n_{cw} denotes the number of *clockwise* encirclements of the origin by the mapping of the closed contour C, then we have

$$-n_{cw} = n_z - n_p \qquad (6.37)$$

If, instead, s traverses the contour C in the negative direction (clockwise), we introduce a minus sign on the right-hand side of (6.36), and (6.37) becomes

$$n_{cw} = n_z - n_p \qquad (6.38)$$

In words, (6.38) states that the number of clockwise encirclements of the origin by the mapping of the closed contour C in the f-plane is equal to the number of zeros minus the number of poles of $f(s)$ within the contour C as s traverses the contour C in a clockwise direction under the general assumption that $f(s)$ has

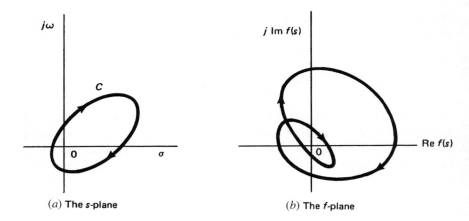

(a) **The s-plane** (b) **The f-plane**

Figure 6.2 (a) A closed contour C in the s-plane and (b) the mapping of the closed contour C in the f-plane.

neither zeros nor poles on the contour C. The situation is as depicted in Fig. 6.2. As the variable s traverses a closed contour C in the s-plane in a clockwise direction, we find a closed curve representing the mapping of C in the f-plane, as illustrated in Fig. 6.2b. The number of clockwise encirclements of the origin by the closed curve in the f-plane is equal to the number of zeros minus the number of poles of $f(s)$ within C. By encirclement we mean a complete revolution of a radius vector drawn from the origin to a moving point on the closed curve in the f-plane.

The Nyquist criterion is based on (6.38). Our objective is to determine whether or not the network determinant has any poles in the closed RHS. We therefore choose the closed contour C containing the $j\omega$-axis and a semicircle of infinite radius, as indicated in Fig. 6.3, thereby enclosing the entire closed RHS. As required by (6.38), the contour C is traversed in a clockwise direction, which corresponds to increasing frequency. The contour of Fig. 6.3 is known as the *Nyquist contour.*

Suppose that the function to be tested is the return difference, written as $F(s)$ to exhibit its dependence on the complex frequency s, of a single-loop feedback amplifier with respect to the controlling parameter x of an active device:

$$F(s) = \frac{Y_{uv}(x)}{Y_{uv}(0)} \tag{6.39}$$

Then, according to (6.1), $Y_{uv}(0)$ is independent of the controlling parameters of the active devices. This implies that $Y_{uv}(0)$ is the network determinant of the amplifier when all of the active devices are made completely dead, and the resulting network is strictly passive. Therefore, the zeros of $Y_{uv}(0)$ are confined to the open

LHS, and we conclude from (6.39) that the zeros of the return difference in the closed RHS are precisely those natural frequencies of the feedback amplifier that lie inside the closed RHS. To guarantee stability, the mapping of the Nyquist contour C in the F-plane should not encircle the origin. The mapping of the Nyquist contour C in the F-plane is called the *Nyquist plot*.

In practice, it is more convenient to work with the return ratio $T(s)$, which is related to $F(s)$ by $F(s) = 1 + T(s)$. Thus, the difference between a Nyquist plot of $F(s)$ and that of $T(s)$ is simply a shift in the position of the imaginary axis. To obtain a plot in the T-plane from that of the F-plane, we simply shift the imaginary axis of the F-plane to the right by a unit distance. Since the origin in the F-plane corresponds to the point $-1 + j0$ in the T-plane, we see that the amplifier will be stable if the Nyquist plot does not encircle the *critical point* $-1 + j0$ in the T-plane.

We summarize the above results by stating the following theorem, known as the *Nyquist stability criterion* [Nyquist (1932)].

Theorem 6.5 Assume that a feedback amplifier is stable when the controlling parameter of a chosen active device is set to zero. Then a necessary and sufficient condition for the feedback amplifier to be stable under normal operating conditions is that the Nyquist plot of the return ratio with respect to this controlling parameter should not encircle the critical point $-1 + j0$ in the T-plane.

Relation (6.38) was derived under the assumption that the function under test has no zeros and poles on the boundary of the contour C. For the Nyquist plot of the return ratio $T(s)$, Theorem 6.5 is valid only if $T(s)$ has no zeros and poles on the Nyquist contour of Fig. 6.3. These are not really serious restrictions, because we can

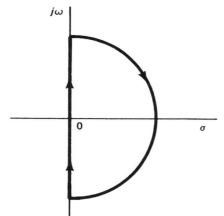

Figure 6.3 The Nyquist contour.

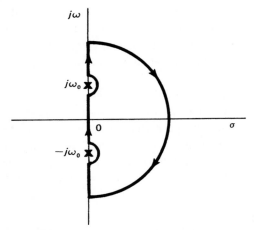

Figure 6.4 The indented Nyquist contour.

tell from the Nyquist plot whether or not the assumptions hold in a given situation. If the Nyquist plot of $T(s)$ intersects the critical point $-1 + j0$ in the T-plane, or, equivalently, if the Nyquist plot of $F(s)$ intersects the origin in the F-plane, then $F(s)$ must have a zero somewhere on the Nyquist contour, either on the $j\omega$-axis or at infinity, depending on the value of s at the intersection point. This tells us that not all of the zeros of $F(s)$ are confined to the open LHS, at least one being on the real-frequency axis, and the feedback amplifier under investigation is therefore unstable. If the Nyquist plot of $T(s)$ in the T-plane becomes unbounded at some real frequency ω_0, then $F(s)$ must have a pole at $j\omega_0$. To avoid this difficulty, we can modify the Nyquist contour by indentations into the RHS with vanishingly small semicircular arcs centered at these poles, as illustrated in Fig. 6.4, because these small indentations do not affect the number of zeros of $F(s)$ computed by (6.38). This situation does not occur if all the zeros of $Y_{uv}(0)$ are confined to the open LHS, as in the case for the single-loop feedback amplifiers discussed above.

We now investigate the Nyquist plot of $T(s)$. From physical considerations, as frequency increases, the available gain of the active devices decreases. As frequency approaches infinity, the forward transmission through the active devices is reduced to zero, and the return difference becomes unity. Therefore the mapping of the points over the semicircular arc at infinity of the Nyquist contour is simply the origin in the T-plane or $F = 1$ in the F-plane. To complete the Nyquist plot, we need only check the behavior of $T(s)$ on the real-frequency axis, that is, $T(j\omega)$. Since we are concerned only with networks with real elements and since $T(s)$ is analytic on the $j\omega$-axis, we have

$$\overline{T}(j\omega) = T(\overline{j\omega}) = T(-j\omega) \tag{6.40}$$

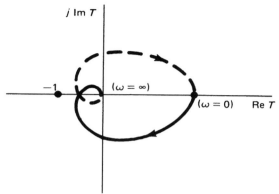

Figure 6.5 A typical Nyquist plot.

meaning that the behavior of T on the negative $j\omega$-axis is simply the mirror image of T on the positive $j\omega$-axis with respect to the real axis of the T-plane. In other words, the Nyquist plot of the return ratio T is completely specified by its behavior for frequencies on the positive $j\omega$-axis. In fact, the Nyquist plot can be achieved experimentally, since the return ratio of an amplifier can be obtained from a frequency response measurement, which includes all parasitic effects. A typical Nyquist plot is shown in Fig. 6.5. The solid line indicates the behavior of the return ratio $T(j\omega)$ for nonnegative frequencies from $\omega = 0$ to $\omega = \infty$. The remaining plot for frequencies from $\omega = 0$ to $\omega = -\infty$ is obtained as the mirror image of the solid line with respect to the real axis of the T-plane. In the plot, since the locus does not encircle the critical point $-1 + j0$, the amplifier is stable. On the other hand, the Nyquist plot of Fig. 6.6 is unstable, because the locus encircles the critical point $-1 + j0$ twice.

6.4 APPLICATIONS OF THE NYQUIST CRITERION TO SINGLE-LOOP FEEDBACK AMPLIFIERS

In this section, we study the relations between stability and feedback for single-loop feedback amplifiers in terms of the Nyquist plot of the return ratio.

The Nyquist plot discussed in the foregoing gives a practical way for the stability study. It also provides a quantitative measure of the degree of stability of the feedback amplifier. Consider the stable plot of Fig. 6.5, which is enlarged around the origin as shown in Fig. 6.7. The feedback amplifier is stable and additional feedback is allowed until the value T_m, as indicated in Fig. 6.7, passes through the critical point $-1 + j0$, corresponding to a pair of conjugate natural frequencies on the real-frequency axis. This condition provides a measure of the maximum

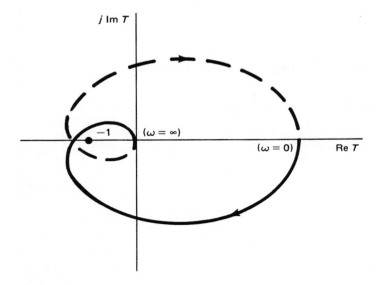

Figure 6.6 An unstable Nyquist plot.

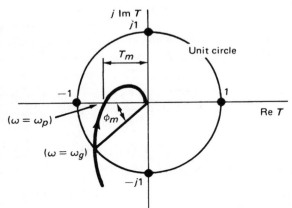

Figure 6.7 A stable Nyquist plot enlarged around the origin.

allowable feedback permitted to ensure amplifier stability. For this we define the following.

Definition 6.4: Gain margin At the frequency when the phase angle of the return ratio equals $-180°$, the magnitude in decibels of the return ratio below the 0-db level is called the *gain margin*. The corresponding frequency is termed the *phase-crossover frequency*.

In Fig. 6.7, the phase-crossover frequency is indicated by ω_p, and the gain margin is defined by the equation

$$\text{Gain margin} = -20 \log |T_m| = 20 \log \left| \frac{1}{T_m} \right| \qquad (6.41)$$

Thus, the gain margin is the additional amount of feedback in decibels that is permitted before the amplifier becomes unstable. In a similar way, we define the phase margin.

Definition 6.5: Phase margin At the frequency when the magnitude of the return ratio is unity or 0 dB, the phase difference in degrees of the return ratio above $-180°$ is called the *phase margin*. The corresponding frequency is termed the *gain-crossover frequency*.

In Fig. 6.7, at the frequency $\omega = \omega_g$, the gain-crossover frequency, the magnitude of the return ratio equals unity and its phase angle is $(-180° + \phi_m)$. Thus, the angle ϕ_m is the phase margin and represents the additional phase lag at unit magnitude permitted for the return ratio before the amplifier becomes unstable. Thus, the gain and phase margins are margins of safety against instability and are the generally accepted measures of the degree of stability of a feedback amplifier. In practice, a gain margin of about 10 dB and a phase margin of about 30° are considered adequate to guard against the effect of variations of the parameters of the active devices, the power supply, the temperature, and the others.

For an absolutely stable single-loop feedback amplifier, the gain-crossover frequency ω_g is necessarily smaller than the phase-crossover frequency ω_p. In Fig. 6.5, if the curve represents the Nyquist plot of a single-loop feedback amplifier under normal operating conditions, the amplifier must be absolutely stable because any reduction of the controlling parameter will only "shrink" the curve radially about the origin. On the other hand, the Nyquist plot of Fig. 6.8a shows that the amplifier is stable under normal operating conditions. However, by varying the controlling parameters of some active devices between 0 and their nominal values, the Nyquist plot may shrink radially about the origin, as shown in Fig. 6.8b. By applying Nyquist criterion to Fig. 6.8b, we see that the plot encircles the critical point $-1 + j0$ twice, and the amplifier is therefore unstable. We remark that in Fig. 6.8 only the plots corresponding to positive frequencies are shown. The plots corresponding to the negative real-frequency axis, being the mirror images of those shown in Fig. 6.8, are not given, for simplicity. If the plot of Fig. 6.8b is shrunk further, the amplifier may become stable again. We conclude that the single-loop feedback amplifier is conditionally stable. In terms of the natural frequencies, the above analysis indicates that, for a conditionally stable amplifier, as the midband

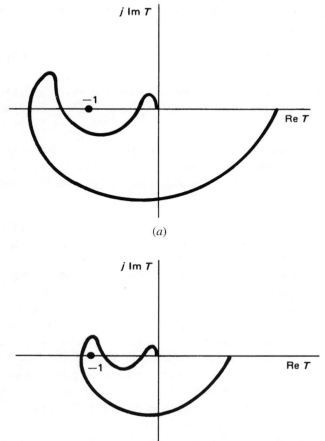

(a)

(b)

Figure 6.8 (a) A stable Nyquist plot under normal operating conditions and (b) an unstable Nyquist plot during transition.

value of the return ratio $T(0)$ is reduced, some of the natural frequencies cross over into the RHS. If, however, $T(0)$ is reduced further, the natural frequencies that were in the RHS return to the open LHS, thereby making the amplifier stable again.

Example 6.2 The network of Fig. 6.9 shows a three-stage common-emitter amplifier with a resistor R_f connected from the output to the input to provide a single external feedback. Assume that the three transistors are identical and are

Figure 6.9 A three-stage common-emitter feedback amplifier.

represented by their hybrid-pi equivalent network of Fig. 4.10, with

$$g_m = 0.2 \text{ mho} \quad r_x = 50\,\Omega \quad r_\pi = 150\,\Omega$$

$$r_o = \infty \quad C_\pi = 195 \text{ pF} \quad C_\mu = 5 \text{ pF} \tag{6.42}$$

$$R_1 = 100\,\Omega \quad R_2 = 75\,\Omega \quad R_3 = R_4 = 1\,\text{k}\Omega$$

As demonstrated in Sec. 4.2.2, the return ratio of the amplifier is found from Eqs. (4.54b) and (4.67) to be

$$T(s) = -\mu\beta = \frac{T(0)}{(1 + s/s_1)(1 + s/s_2)(1 + s/s_3)} \tag{6.43}$$

where

$$T(0) = \frac{R_1 R_2 R_3 R_4 R_f r_\pi^3 g_m^3}{(R_2 + R_f)[(r_x + r_\pi)(R_1 + R_f) + R_1 R_f]} \tag{6.44a}$$
$$\times (r_x + r_\pi + R_3)(r_x + r_\pi + R_4)$$

$$s_k = \frac{r_x + r_\pi + \hat{R}_k}{C_k r_\pi (r_x + \hat{R}_k)} \quad k = 1, 2, 3 \tag{6.44b}$$

in which $\hat{R}_1 = R_1 R_f/(R_1 + R_f)$, $\hat{R}_2 = R_3$ and $\hat{R}_3 = R_4$, and

$$C_1 = C_\pi + C_\mu \left[1 + \frac{g_m R_3 (r_\pi + r_x)}{r_\pi + r_x + R_3} \right] \tag{6.44c}$$

$$C_2 = C_\pi + C_\mu \left[1 + \frac{g_m R_4 (r_\pi + r_x)}{r_\pi + r_x + R_4} \right] \tag{6.44d}$$

$$C_3 = C_\pi + C_\mu \left(1 + \frac{g_m R_2 R_f}{R_2 + R_f} \right) \tag{6.44e}$$

Using (6.42) in conjunction with the fact that $R_f \gg R_1, R_2$, we obtain

$$C_1 = C_2 = 367 \, \text{pF} \quad C_3 \approx 275 \, \text{pF}$$

$$s_1 \approx 3.633 \cdot 10^7 \quad s_2 = 2.076 \cdot 10^7 \tag{6.45}$$

$$s_3 = 2.770 \cdot 10^7 \quad T(0) \approx \frac{469 \cdot 10^3}{R_f}$$

giving

$$T(s) = \frac{T(0)}{(1 + p/3.633)(1 + p/2.076)(1 + p/2.770)} \tag{6.46}$$

where $p = s/10^7$. To make the Nyquist plot of the return ratio, we rewrite (6.46) as

$$T(j\lambda) = \frac{T(0)}{(1 + j\lambda/3.633)(1 + j\lambda/2.076)(1 + j\lambda/2.770)} \tag{6.47}$$

where $\lambda = \omega/10^7$. The Nyquist plots of $T(j\lambda)$ are presented in Fig. 6.10 for two different midband values of $T(0)$: $T(0) = 10$ and $T(0) = 5$. For $T(0) = 10$, the feedback resistance is found to be $R_f = 46.9 \, \text{k}\Omega$; and for $T(0) = 5$, $R_f = 93.8 \, \text{k}\Omega$. In Fig. 6.10, as before, only the loci corresponding to the nonnegative λ are shown, for simplicity, with λ chosen at the points

$$\lambda_k = \frac{k}{2} \quad k = 0, 1, \ldots, 9 \tag{6.48}$$

As can be seen from the plots, for $T(0) = 10$ the amplifier is unstable, because the Nyquist locus encircles the critical point $-1 + j0$. For $T(0) = 5$ the amplifier is stable with a gain margin $= 4.6 \, \text{dB}$, because

$$T(j4.84) = 0.589 \underline{/180°} \tag{6.49}$$

The phase-crossover frequency is $\omega_p = 48.4 \cdot 10^6$ rad/s or $f_p = 7.7 \, \text{MHz}$. The phase margin of the amplifier is about $20°$, because

$$T(j3.75) = 1 \underline{/160.5°} \tag{6.50}$$

The gain-crossover frequency is $\omega_g = 37.5 \cdot 10^6$ rad/s or $f_g = 5.97 \, \text{MHz}$.

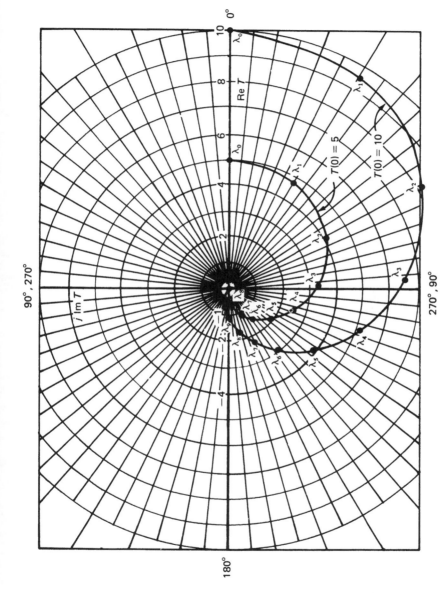

Figure 6.10 The Nyquist plots of the return ratio $T(s)$ of the amplifier of Fig. 6.9 for $T(0) = 5$ and $T(0) = 10$.

The Nyquist plot of the return ratio $T(s)$ is nothing more than the plot of $T(j\omega)$ as a function of ω in polar form. It is sometimes more useful to plot the magnitude of $T(j\omega)$ in decibels and its phase in degrees each against the logarithm of frequency instead of putting them in a single curve. Thus, the critical point in the Nyquist plot corresponds to the lines of unit magnitude or zero-decibel level and $-180°$ phase in the separate frequency response plots, which are usually referred to as the *Bode plot.* The frequencies at which these lines are crossed by their respective response curves are the gain-crossover and phase-crossover frequencies.

Example 6.3 The Bode plot of Eq. (6.47) is presented in Fig. 6.11 for $T(0) = 10$ and $T(0) = 5$. For $T(0) = 10$ we see that the gain-crossover frequency ω_g, which is about $52.1 \cdot 10^6$ rad/s, is larger than the phase-crossover frequency $\omega_p = 48.4 \cdot 10^6$ rad/s, and the amplifier is unstable. For $T(0) = 5$, we have $\omega_g = 37.5 \cdot 10^6$ rad/s, which is smaller than $\omega_p = 48.4 \cdot 10^6$ rad/s, and the amplifier is therefore stable. The definitions of the gain and phase margins are indicated in the Bode plot of Fig. 6.11. From the plot, we see that for $T(0) = 5$ the gain margin is about 4.6 dB and the phase margin is about $20°$, confirming Eqs. (6.49) and (6.50).

6.5 THE ROOT-LOCUS METHOD

In the foregoing, we showed that by changing the midband value of the return ratio $T(0)$, we change the locations of the natural frequencies of the feedback amplifier. As $T(0)$ is increased, some of the natural frequencies may cross into the closed RHS, thereby rendering the amplifier unstable. It would be extremely useful if the locus of the natural frequencies can be plotted as a function of $T(0)$ in the s-plane. The resulting locus is called the *root locus* and was first given by Evans (1948). The root-locus method is important in design because the root loci may be sketched rather quickly with the aid of an angle measuring device called a *spirule*, permitting study of many possible designs in a short period of time. By knowing the trajectories of the natural frequencies, the designer has a considerable amount of insight and understanding of the stability, time-, and frequency-domain behaviors of the network. In this section, we study this technique.

As indicated in (6.39), the natural frequencies of a feedback amplifier are the zeros of the return difference with respect to the controlling parameter of an active device, provided that the amplifier is stable when the controlling parameter vanishes. Thus, the root locus can be determined by the roots of the equation

$$F(s) = 1 + T(s) = 0 \qquad\qquad (6.51)$$

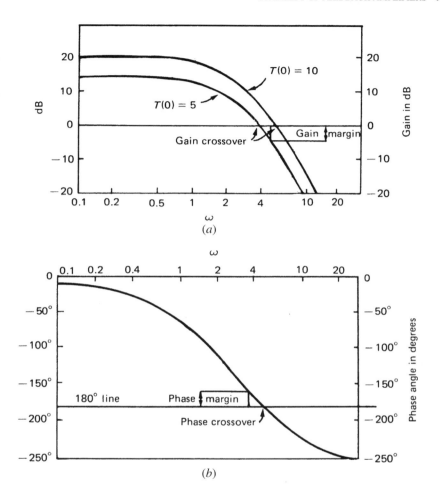

Figure 6.11 (*a*) The Bode plot of the gain of the return ratio $T(s)$ of the amplifier of Fig. 6.9 for $T(0) = 5$ and $T(0) = 10$ and (*b*) the Bode plot of the phase of the return ratio $T(s)$ of the amplifier of Fig. 6.9.

where $T(s)$ is the return ratio. The trajectories of the roots of (6.51) are therefore the points s in the s-plane satisfying the conditions

$$|T(s)| = 1 \qquad (6.52a)$$

$$\arg T(s) = (2k + 1)\pi \qquad (6.52b)$$

for any integer k. We write

$$T(s) = T(0) \frac{\prod_{i=1}^{m}(1 - s/z_i)}{\prod_{i=1}^{n}(1 - s/p_j)} = T(0) \frac{N(s)}{D(s)} \qquad (6.53a)$$

$$F(s) = F(0) \frac{\prod_{j=1}^{n}(1 - s/s_j)}{\prod_{j=1}^{n}(1 - s/p_j)} = \frac{P(s)}{D(s)} \qquad (6.53b)$$

where z_i and s_j are the zeros of $T(s)$ and $F(s)$, respectively, and p_j the poles of $T(s)$ and $F(s)$. From physical consideration, the order of the denominator polynomial of $T(s)$ is usually greater than that of the numerator polynomial, and consequently $T(s)$ has $n - m$ zeros at infinity. In the following, we study the loci of the natural frequencies s_j as a function of the midband value of the return ratio $T(0)$.

We first illustrate the root-locus method by the series-parallel and parallel-series feedback amplifiers of Secs. 4.2.3 and 4.2.4. As shown in these sections, the return ratio can generally be expressed in the form

$$T(s) = -\mu\beta = \frac{T(0)}{(1 + s/\sigma_1)(1 + s/\sigma_2)} \qquad (6.54)$$

where σ_1 and σ_2 are real and positive with $\sigma_1 \leqslant \sigma_2$. The locations of the natural frequencies of the amplifier as a function of $T(0)$ are determined by setting $T(s) = -1$, yielding

$$s^2 + (\sigma_1 + \sigma_2)s + \sigma_1\sigma_2 [1 + T(0)] = 0 \qquad (6.55)$$

whose roots are given by

$$s_1, s_2 = -\frac{1}{2}(\sigma_1 + \sigma_2) \pm \frac{1}{2}\sqrt{(\sigma_1 - \sigma_2)^2 - 4\sigma_1\sigma_2 T(0)} \qquad (6.56)$$

As $T(0)$ varies from zero to infinity, the two branches of the root locus traced out by the natural frequencies s_1 and s_2 are shown in Fig. 6.12. For $T(0) \leqslant (\sigma_1 - \sigma_2)^2/4\sigma_1\sigma_2$, the two branches, starting at points $-\sigma_1$ and σ_2, meet at $-\frac{1}{2}(\sigma_1 - \sigma_2)$. For $T(0) > (\sigma_1 - \sigma_2)^2/4\sigma_1\sigma_2$, s_1 and s_2 become complex and their loci move up and down along the vertical line Re $T = -\frac{1}{2}(\sigma_1 + \sigma_2)$, as depicted in Fig. 6.12. Since the locus remains in the open LHS for all nonnegative values of $T(0)$, the amplifier is therefore always stable. We remark that this conclusion is reached based on the assumption that, in the frequency band of interest, the series-parallel and parallel-series feedback amplifiers can be approximated by the unilateralized models of Figs. 4.26 and 4.32. This results in two dominant poles

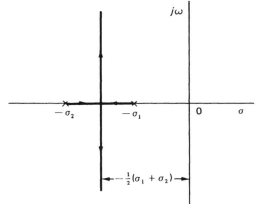

Figure 6.12 The root locus of a simplified parallel-series or series-parallel feedback amplifier.

in $T(s)$ as shown in Eq. (6.54). However, if we use the complete hybrid-pi model for the transistors, we introduce additional nondominant poles in $T(s)$ and the root locus is modified so that some of the branches will bend toward the $j\omega$-axis, showing the possibility of instability for large amounts of feedback. The root-locus technique can be used to predict the stability of an amplifier only if we have an accurate, realistic representation of the return ratio. This is one of the drawbacks of testing stability using models. We shall illustrate this after presenting rules for the construction of the root locus.

6.5.1 Rules for the Construction of the Root Locus

As demonstrated in the foregoing, the construction of the root locus is relatively simple if the return ratio $T(s)$ has only two poles. The procedure becomes more complicated if $T(s)$ has three or more poles, because the solution of the characteristic equation is much more difficult. It is therefore important to develop simple rules that will enable us readily to sketch the root locus. Very often, a rough sketch will provide us sufficient information to undertake a preliminary design. The rules together with their justifications will now be presented.

By using the expression (6.53a), the trajectories of the roots of (6.51) are the points s in the s-plane satisfying the conditions

$$T(0) \prod_{i=1}^{m} \left| 1 - \frac{s}{z_i} \right| = \prod_{j=1}^{n} \left| 1 - \frac{s}{p_j} \right| \qquad (6.57a)$$

$$\sum_{i=1}^{m} \arg\left(1 - \frac{s}{z_i}\right) - \sum_{j=1}^{n} \arg\left(1 - \frac{s}{p_j}\right) = (2k+1)\pi \qquad (6.57b)$$

From (6.57a) we see that as $T(0)$ approaches zero, s approaches p_j, whereas as $T(0)$ approaches infinity, s approaches z_i. Thus, the branches of the locus start at the poles of $T(s)$ and terminate at the zeros of $T(s)$.

Rule 1 The root loci start from the poles and terminate at the zeros of $T(s)$.

We remark that for a rational function the number of poles is always equal to the number of its zeros if poles or zeros at the infinity are included. For $n > m$, $n - m$ branches of the locus will terminate at infinity.

Since we deal only with networks with real elements, the return ratio $T(s)$ has only real coefficients. As a result, the roots of the characteristic equation must occur in complex conjugate if they are complex.

Rule 2 The root loci are symmetric with respect to the σ-axis.

From (6.57b) we see that on the σ-axis, the complex conjugate poles and zeros of $T(s)$ contribute nothing to the phase of $T(s)$, and the phase of $T(s)$ will be $(2k+1)\pi$ if s is to the left of an odd number of poles and/or zeros. The poles and zeros of $T(s)$ are referred to as the *critical frequencies*.

Rule 3 The parts of the σ-axis, which are to the left of an odd number of real critical frequencies of $T(s)$, are parts of the loci.

From (6.53a) we can write, for large values of $|s|$,

$$T(s) = (-1)^{m-n} \frac{T(0) p_1 p_2 \cdots p_n [s^m - (z_1 + z_2 + \cdots + z_m)s^{m-1} + \cdots]}{z_1 z_2 \cdots z_m [s^n - (p_1 + p_2 + \cdots + p_n)s^{n-1} + \cdots]}$$

$$= (-1)^{m-n} T(0) \frac{p_1 p_2 \cdots p_n}{z_1 z_2 \cdots z_m} \left(s^m - \sum_{i=1}^{m} z_i s^{m-1} + \cdots\right)$$

$$\times \left(s^n - \sum_{j=1}^{n} p_j s^{n-1} + \cdots\right)^{-1}$$

$$= (-1)^{m-n} T(0) \frac{p_1 p_2 \cdots p_n}{z_1 z_2 \cdots z_m} \left(s^m - \sum_{i=1}^{m} z_i s^{m-1} + \cdots\right)$$

$$\times \left(s^{-n} + \sum_{j=1}^{n} p_j s^{-n-1} + \cdots \right)$$

$$= T(0) \frac{p_1 p_2 \cdots p_n}{z_1 z_2 \cdots z_m} (-s)^{m-n} \left(1 - \frac{\sum_{i=1}^{m} z_i - \sum_{j=1}^{n} p_j}{s} + \cdots \right)$$

(6.58)

The third equation is obtained after appealing to the binomial expansion. To determine the locus for large values of $|s|$, we set (6.58) to -1 or

$$T(s) = e^{j(2k+1)\pi}$$

(6.59)

for any integer k, yielding, for $n > m$,

$$-s \left(1 - \frac{\sum_{i=1}^{m} z_i - \sum_{j=1}^{n} p_j}{s} + \cdots \right)^{1/(m-n)}$$

$$= \left[\frac{z_1 z_2 \cdots z_m}{T(0) p_1 p_2 \cdots p_n} \right]^{1/(m-n)} e^{-j\theta_k}$$

(6.60)

where

$$-\theta_k = \frac{(2k+1)\pi}{m-n} \quad k = 0, 1, \ldots, n-m-1$$

(6.61)

Applying the binomial expansion again to the left-hand side of (6.60) gives

$$-s - \frac{\sum_{i=1}^{m} z_i - \sum_{j=1}^{n} p_j}{n-m} \approx \left[\frac{z_1 z_2 \cdots z_m}{T(0) p_1 p_2 \cdots p_n} \right]^{1/(m-n)} e^{-j\theta_k}$$

(6.62)

or, more compactly,

$$s = s_0 - M_0 e^{-j\theta_k} \quad k = 0, 1, \ldots, n-m-1$$

(6.63)

where

$$s_0 = \frac{\sum_{j=1}^{n} p_j - \sum_{i=1}^{m} z_i}{n-m} \quad n > m$$

(6.64)

$$M_0 = \left[\frac{z_1 z_2 \cdots z_m}{T(0) p_1 p_2 \cdots p_n} \right]^{1/(m-n)}$$

(6.65)

Thus, we can state the following rule.

Rule 4 For large values of $|s|$, the root loci become asymptotic to the straight lines that form angles of $-\theta_k$, as given in (6.61), with the σ-axis of the s-plane. Furthermore, the asymptotes intersect on the σ-axis at s_0 of (6.64).

We now proceed to determine the angles of departure from a pole and arrival at a zero. The *angle of departure (arrival)* from a point on the root locus is defined as the angle made at the point by the tangent vector for increasing (decreasing) the value of $T(0)$. Since not all of the poles and zeros of $T(s)$ will be distinct, for our purposes we express (6.53a) as

$$T(s) = T(0)\frac{N(s)}{D(s)} = \tilde{T}(0)\frac{\prod_{i=1}^{a}(s - z_i)^{m_i}}{\prod_{j=1}^{b}(s - p_j)^{n_j}} \tag{6.66}$$

where m_i and n_j denote the orders of the zero z_i and the pole p_j, respectively, and a and b are the numbers of distinct zeros and poles of the return ratio $T(s)$.

Let s be a point on the locus that is arbitrarily close to a pole p_j of $T(s)$ of order n_j. As $T(0)$ approaches zero, according to Rule 1 the point s approaches p_j. Substituting (6.66) in (6.52b) yields

$$\sum_{i=1}^{a} m_i \arg(s - z_i) - \sum_{j=1}^{b} n_j \arg(s - p_j) = (2k + 1)\pi \tag{6.67}$$

for any integer k. Then we have

$$n_j \arg(s - p_j) = -(2k + 1)\pi + \sum_{i=1}^{a} m_i \arg(s - z_i)$$
$$- \sum_{\substack{i=1 \\ i\neq j}}^{b} n_i \arg(s - p_i) \tag{6.68}$$

As $T(0)$ approaches 0,

$$\phi_{pj} = \lim_{T(0)\to 0} \arg(s - p_j) \tag{6.69}$$

becomes the angles of departure at the pole p_j, and we obtain

$$\phi_{pj} = \frac{1}{n_j}\left[-(2k + 1)\pi + \sum_{i=1}^{a} m_i \arg(p_j - z_i) - \sum_{\substack{i=1 \\ i\neq j}}^{b} n_i \arg(p_j - p_i)\right] \tag{6.70}$$

for $k = 0, 1, \ldots, n_j - 1$.

In a similar manner, by letting $T(0)$ approach infinity, the angles of arrival at a zero z_j of $T(s)$ of order m_j are found to be

$$\Phi_{zj} = \frac{1}{m_j} \left[(2k+1)\pi + \sum_{i=1}^{b} n_i \arg(z_j - p_i) - \sum_{\substack{i=1 \\ i \neq j}}^{a} m_i \arg(z_j - z_i) \right] \quad (6.71)$$

for $k = 0, 1, \ldots, m_j - 1$. We summarize these results as follows.

Rule 5 The angles of departure from a pole p_j of order n_j of $T(s)$ are given by ϕ_{pj} of (6.70), and the angles of arrival at a zero z_j of order m_j are given by ϕ_{zj} of (6.71).

Finally, we determine the breakaway points on the root locus. A breakaway point s_b on the root locus is a zero of the return difference $F(s)$ of order at least 2 for some $T(0)$. Hence, at the breakaway point s_b the derivative of $F(s)$ must also be zero, showing that

$$\left. \frac{d}{ds} F(s) \right|_{s=s_b} = \left. \frac{d}{ds} T(s) \right|_{s=s_b} = 0 \quad (6.72)$$

Conversely, the roots of the equation $dT(s)/ds = 0$ are the zeros of $F(s)$ to the order of at least 2, and we get the following rule.

Rule 6 The breakaway points on the root locus are determined by solving the equation $dT(s)/ds = 0$ for its roots.

We remark that, barring cancellations, the roots of $dT(s)/ds = 0$ are the same as those of $dT^{-1}(s)/ds = 0$. The latter is more convenient to apply.

We illustrate the above results by the following examples.

Example 6.4 Consider the three-stage common-emitter feedback amplifier of Fig. 6.9. As computed in Example 6.2, the return ratio of the amplifier is found to be

$$T(s) = \frac{T(0)}{(1 + p/3.633)(1 + p/2.076)(1 + p/2.770)} \quad (6.73)$$

where $p = s/10^7$. We wish to construct its root locus.

The root locus will contain three branches, starting from the poles $p_1 = -3.633$, $p_2 = -2.076$, and $p_3 = -2.770$ and terminating at the three zeros z_i at infinity, all being scaled down by a factor of 10^7. According to Rule 3, the parts of the σ-axis from $\sigma = -2.076$ to -2.770 and from $\sigma = -3.633$ to $-\infty$ are parts of the loci.

For large values of $|s|$, the root loci become asymptotic to straight lines that form angles of

$$-\theta_k = -\frac{(2k+1)\pi}{3-0} \quad k = 0, 1, 2 \tag{6.74}$$

giving $-\theta_0 = -\pi/3$, $-\theta_1 = -\pi$, and $-\theta_2 = -5\pi/3$, with the σ-axis of the s-plane. Furthermore, the asymptotes intersect on the σ-axis at the point

$$s_0 = \frac{-(3.633 + 2.076 + 2.770) + 0}{3-0} = -2.826 \tag{6.75}$$

The angles of departure at the poles p_j are obtained from (6.70) as

$$\phi_{pj} = -\pi + 0 - \sum_{\substack{i=1 \\ i \neq j}}^{3} \arg(p_j - p_i) \quad j = 1, 2, 3 \tag{6.76}$$

giving $\phi_{p1} = -3\pi$, $\phi_{p2} = -\pi$, and $\phi_{p3} = -2\pi$.

Finally, the breakaway point s_b on the locus is determined by solving for the roots of $dT^{-1}(s)/ds = 0$, which is equivalent to solving

$$\frac{d}{ds}[(p+3.633)(p+2.076)(p+2.770)] = 0 \tag{6.77a}$$

or

$$3p^2 + 16.958p + 23.356 = 0 \tag{6.77b}$$

giving $p = -2.376$ and -3.277. Since the breakaway point must lie between -2.076 and -2.770, we have $s_b = -2.376$. By summarizing all these, the root locus plot of $T(s)$ is presented in Fig. 6.13.

To determine the value of $T(0)$ for which the locus will intersect the $j\omega$-axis, we consider the numerator of $F(s) = 1 + T(s)$, yielding

$$p^3 + 8.479p^2 + 23.356p + 20.892[1 + T(0)] = 0 \tag{6.78}$$

To apply the Routh criterion, we form the array of the elements of (6.78):

$$
\begin{array}{c|cc}
p^3 & 1 & 23.356 \\
p^2 & 8.479 & 20.892[1 + T(0)] \\
p^1 & 20.892 - 2.464T(0) & 0 \\
p^0 & 20.892[1 + T(0)] & 0
\end{array} \tag{6.79}
$$

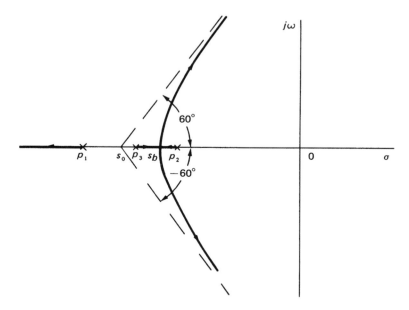

Figure 6.13 The root locus of a three-stage common-emitter feedback amplifier.

Setting the third row to zero yields $T(0) = 8.479$. The corresponding frequencies at which the loci cross the $j\omega$-axis are obtained by setting the auxiliary polynomial of the second row to zero:

$$8.479\,p^2 + 198.035 = 0 \tag{6.80}$$

giving $p = \pm j4.833$.

Example 6.5 Consider the voltage-shunt feedback amplifier of Fig. 6.14. Assume that the transistor is represented by its hybrid-pi model, with

$$g_m = 0.4 \text{ mho} \quad r_x = 50\ \Omega \quad r_\pi = 250\ \Omega$$
$$C_\pi = 195 \text{ pF} \quad C_\mu = 5 \text{ pF} \quad r_o = 50\ \text{k}\Omega = \frac{1}{g_o} \tag{6.81a}$$

For the other network elements, let

$$R_1 = 10\ \text{k}\Omega \quad R_f = 40\ \text{k}\Omega \quad R_2 = 4\ \text{k}\Omega \tag{6.81b}$$

As demonstrated in Sec. 4.2.2, the open-loop current gain can be computed by using the network of Fig. 4.17, which after applying the Miller effect can be approximated

Figure 6.14 A voltage-shunt feedback amplifier.

by the unilateralized model of Fig. 4.18. In Example 4.4, the open-loop current gain was found to be

$$\mu(s) = \frac{I_2}{I_s} = \frac{81.69}{1 + p/0.059} \tag{6.82}$$

where $p = s/10^7$. The transfer current ratio of the feedback network from (4.54b) is given by $\beta = -R_2/R_f = -0.1$. Thus, we obtain the return ratio of the amplifier as

$$T(s) = -\mu\beta = \frac{8.169}{1 + p/0.059} \tag{6.83}$$

Since $T(s)$ has a simple pole at -0.059, the root locus is a straight line moving from the point -0.059 to $-\infty$ along the $-\sigma$-axis as $T(0)$ is increased. This implies that the amplifier is always stable.

Instead of using the approximate unilateralized network of Fig. 4.18, we use the hybrid-pi model for the transistor in the network of Fig. 4.17. The resulting equivalent network is shown in Fig. 6.15, whose indefinite-admittance matrix is obtained by inspection as

$$\mathbf{Y}(p) = 10^{-3} \begin{bmatrix} 20.125 & 0 & -20 & -0.125 \\ 0 & 0.295 + 0.05p & 400 - 0.05p & -400.295 \\ -20 & -0.05p & 24 + 2p & -4 - 1.95p \\ -0.125 & -0.295 & -404 - 1.95p & 404.42 + 1.95p \end{bmatrix}$$
$$\tag{6.84}$$

Figure 6.15 A high-frequency, small-signal equivalent network of the voltage-shunt feedback amplifier of Fig. 6.14.

By appealing to Eq. (2.94), the open-loop current gain is found to be

$$\mu(s) = \frac{I_2}{I_s} = -\frac{G_2 V_{24}}{I_s} = -\frac{G_2 Y_{12,44}}{Y_{33}}$$

$$= \frac{5(400 - 0.05p)}{1.962p^2 + 418.524p + 24.485} = \frac{2.548(400 - 0.05p)}{(p + 0.0585)(p + 213.256)} \quad (6.85)$$

giving the return ratio

$$T(s) = \frac{8.17(1 - p/8000)}{(1 + p/0.0585)(1 + p/213.256)} \quad (6.86)$$

The new return ratio now has two poles at $p_1 = -0.0585$ and $p_2 = -213.256$ and a zero at $z_1 = 8000$. If we ignore the effect of the nondominant pole p_2 and zero z_1, (6.86) reduces to (6.83). The root locus of (6.86) has two branches. The sections of the real axis from -0.0585 to 8000 and from -213.256 to $-\infty$ are parts of the loci. One of the branches will eventually terminate at $\sigma = 8000$. This indicates that for some value of $T(0)$, one of the natural frequencies will cross over the $j\omega$-axis, thereby rendering the feedback amplifier unstable. This example emphasizes the fact that the root-locus technique can be used to predict the stability of an amplifier only if we have an accurate, realistic representation of the return ratio.

Example 6.6 The return ratio of a feedback amplifier is described by

$$T(s) = \frac{3T(0)(s + 2)}{(s + 1)(s + 3)(s^2 + 2s + 2)} \quad (6.87)$$

The function has two real poles at $p_1 = -1$ and $p_2 = -3$, one pair of complex conjugate poles at $p_3 = -1 + j1$ and $p_4 = -1 - j1$, and one finite real zero

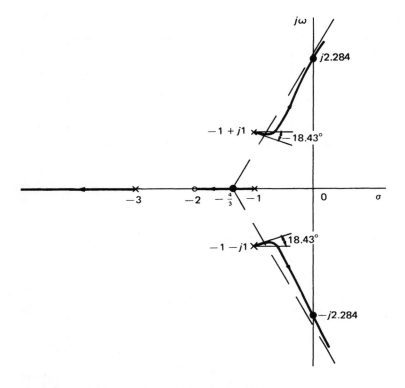

Figure 6.16 The root locus for the return ratio function (6.87).

at $z_1 = -2$, as indicated in Fig. 6.16. The root locus therefore consists of four branches.

As s approaches infinity, the root loci become asymptotic to straight lines that form angles of

$$-\theta_k = -\frac{(2k+1)\pi}{4-1} \quad k = 0, 1, 2 \tag{6.88}$$

giving $-\theta_0 = -\pi/3$, $-\theta_1 = -\pi$ and $-\theta_2 = -5\pi/3$, with the σ-axis of the s-plane. Furthermore, the asymptotes intersect on the σ-axis at the point

$$s_0 = \frac{-6+2}{4-1} = -\frac{4}{3} \tag{6.89}$$

According to Rule 3, the sections of the real axis from -1 to -2 and from -3 to $-\infty$ constitute part of the loci. The other two branches of loci start from $-1 \pm j1$ and tend to infinity along the $\pm 60°$ asymptotes, as shown in Eq. (6.88).

The angles of departure at the poles $-1 \pm j1$ are obtained from (6.70) as

$$\phi_{p3}, \phi_{p4} = -\pi + \arg(1 \pm j1) - \arg(\pm j) - \arg(2 \pm j1) - \arg(\pm j2)$$

$$= -\pi \pm \frac{\pi}{4} \mp \frac{\pi}{2} \mp 26.57° \mp \frac{\pi}{2} = -341.57°, -18.43° \qquad (6.90)$$

as indicated in Fig. 6.16. The angle of arrival at the zero $z_1 = -2$ is found to be

$$\phi_{z1} = \pi + \arg(-1) + \arg 1 + \arg(-1 - j1) + \arg(-1 + j1) = 2\pi \quad (6.91)$$

as expected. Summarizing all these yields the root locus plot of $T(s)$, as shown in Fig. 6.16.

Finally, to determine the value of $T(0)$ for which the locus will intersect the $j\omega$-axis, we consider the numerator of $F(s) = 1 + T(s)$, yielding

$$s^4 + 6s^3 + 13s^2 + [14 + 3T(0)]s + 6 + 6T(0) = 0 \qquad (6.92)$$

The first three rows of the Routh array are given by

$$
\begin{array}{c|ccc}
s^4 & 1 & 13 & 6 + 6T(0) \\
s^3 & 6 & 14 + 3T(0) & 0 \\
s^2 & 10.667 - 0.5T(0) & 6 + 6T(0) & 0
\end{array} \qquad (6.93)
$$

Equation (6.92) has a pair of imaginary zeros when

$$6[6 + 6T(0)] = [10.667 - 0.5T(0)][14 + 3T(0)] \qquad (6.94)$$

giving $T(0) = 5.768$. The corresponding frequencies at which the loci cross the $j\omega$-axis are obtained by setting the auxiliary polynomial of the third row to zero, yielding $s = \pm j2.284$, which are also indicated in Fig. 6.16.

6.5.2 Stabilization Techniques

To desensitize a feedback amplifier to parameter variations, we should increase the amount of negative feedback. However, as indicated above, there is a limit to the amount of feedback that is permitted before the amplifier becomes unstable. In this section, we demonstrate by using the parallel-parallel feedback configuration that the stability performance of a feedback amplifier can be improved by introducing a phase advance into the feedback network to counteract part of the phase-retarding effect of the basic amplifier.

Let the open-loop transfer function of a feedback amplifier be expressed explicitly as the ratio of two polynomials:

$$\mu(s) = \frac{N_\mu(s)}{D_\mu(s)} \qquad (6.95)$$

Likewise, the transfer function of the feedback network is written as

$$\beta(s) = \frac{N_\beta(s)}{D_\beta(s)} \qquad (6.96)$$

The return ratio and the closed-loop transfer function of the amplifier are obtained as

$$T(s) = -\mu(s)\beta(s) = -\frac{N_\mu(s)N_\beta(s)}{D_\mu(s)D_\beta(s)} \qquad (6.97)$$

$$w(s) = \frac{\mu(s)}{1 + T(s)} = \frac{N_\mu(s)D_\beta(s)}{D_\mu(s)D_\beta(s) - N_\mu(s)N_\beta(s)} \qquad (6.98)$$

Observe that the zeros of $w(s)$ are either zeros of $\mu(s)$ or poles of $\beta(s)$ and that the zeros of $\beta(s)$ do not show explicitly in $w(s)$. If some of the zeros of $N_\beta(s)$ are chosen to coincide with those of $D_\mu(s)$, these cancelled zeros will not appear in $T(s)$ but they will appear as the poles of $w(s)$, as can be seen from (6.98). For this reason, the zeros of the transfer function of the feedback network are called *phantom zeros*.

Consider the three-stage common-emitter feedback amplifier of Fig. 6.9. As demonstrated in Sec. 4.2.2, the open-loop current gain can be expressed in the form

$$\mu(s) = \frac{I_2}{I_s} = \frac{T(0)}{(1 - s/p_1)(1 - s/p_2)(1 - s/p_3)} \qquad (6.99)$$

with $p_i = -s_i (i = 1, 2, 3)$. The transfer current ratio of the feedback network, as defined in (4.52a), is given by

$$\beta(s) = y_{12}R_2 \approx y_{f12}R_2 = -\frac{R_2}{R_f} \qquad (6.100)$$

where y_{12} and y_{f12} are the reverse-transfer admittance parameters of the composite two-port network and the feedback two-port network, respectively.

One way to introduce a phantom zero in $\beta(s)$ is to add a capacitor C_f across the feedback resistor R_f of Fig. 6.9. The resulting network is presented in Fig. 6.17. The amplifier will have the same open-loop current gain as in Eq. (6.99), but the

Figure 6.17 A three-stage common-emitter feedback amplifier possessing a phantom zero in its feedback circuit.

current ratio of the feedback network is changed to

$$\beta(s) = -R_2 C_f \left(s + \frac{1}{R_f C_f} \right) = -\frac{R_2(1 - s/z_1)}{R_f} \qquad (6.101)$$

where

$$z_1 = -\frac{1}{R_f C_f} \qquad (6.102)$$

The corresponding return ratio becomes

$$T(s) = -\mu(s)\beta(s) = \frac{T(0)(1 - s/z_1)}{(1 - s/p_1)(1 - s/p_2)(1 - s/p_3)} \qquad (6.103)$$

The location of the phantom zero z_1 is of critical importance in the nature of the root locus. In general, one branch of the root locus will terminate at the real zero z_1, giving rise to one natural frequency in the neighborhood of the phantom zero for any appreciable amount of feedback. Suppose that we choose z_1 such that

$$z_1 = p_3 \qquad (6.104)$$

The resulting return ratio will have only two real poles, whose root locus as computed in (6.54)–(6.56) is shown in Fig. 6.18. As a result, the amplifier is always stable. This is to be compared with the root-locus plot of (6.73), as presented in Fig. 6.13, in which one branch of the locus bends toward the $j\omega$-axis. This means

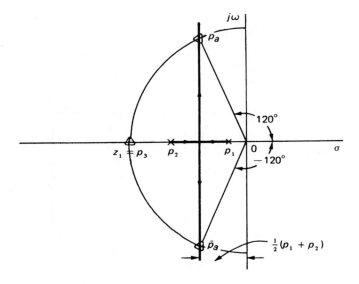

Figure 6.18 The root locus for the return ratio function (6.103) with $z_1 = p_3$.

that for sufficiently large $T(0)$, the amplifier will be unstable. The closed-loop current gain is found to be

$$w(s) = \frac{\mu(s)}{F(s)} = \frac{w(0)}{(1 - s/z_1)\,[(1 - s/p_1)(1 - s/p_2) + T(0)]} \qquad (6.105)$$

Suppose that we wish to design a three-pole feedback amplifier having a maximally flat magnitude for its closed-loop current gain. This is equivalent to selecting an appropriate value of $T(0)$ so that the natural frequencies will lie on the circle of radius $|z_1|$, centered at the origin, as depicted in Fig. 6.18. The desired natural frequencies are indicated by the small triangles and are given by $z_1 = -1/R_f C_f$ and

$$p_a, \bar{p}_a = \frac{(1 \pm j\sqrt{3})z_1}{2} \qquad (6.106)$$

in which we implicitly assume that $|p_3| > |p_2| \geqslant |p_1|$, giving

$$w(s) = \frac{w(0)}{(1 - s/z_1)(1 - s/p_a)(1 - s/\bar{p}_a)} \qquad (6.107)$$

At $s = j1/R_f C_f$, $|w(s)| = |w(0)|/\sqrt{2}$, showing that the -3-dB bandwidth of $|w(j\omega)|$ is $|z_1|$, or

$$\omega_{3-\mathrm{dB}} = |z_1| = |p_a| \tag{6.108}$$

The necessary value of $T(0)$ to achieve the desired pole locations is found by using the fact that $|T(s)| = 1$ on the locus and in particular at the point $s = p_a$. From (6.103) we obtain

$$T(0) = \left|\left(1 - \frac{p_a}{p_1}\right)\right|\left|\left(1 - \frac{p_a}{p_2}\right)\right| \tag{6.109}$$

We illustrate the above by considering the following example.

Example 6.7 Consider the return ratio

$$T(s) = \frac{T(0)}{(1 + p/2.076)(1 + p/2.770)(1 + p/3.633)} \tag{6.110}$$

as discussed in Example 6.2. Following Eq. (6.103), we have $p_1 = -2.076$, $p_2 = -2.770$, and $p_3 = -3.633$. Choose

$$z_1 = p_3 = -3.633 \tag{6.111}$$

giving the desired locations of the natural frequencies of the amplifier at

$$z_1 = p_3 = -\frac{1}{R_f C_f} = -3.633 \tag{6.112a}$$

$$p_a, \bar{p}_a = -1.8165 \mp j3.1463 \tag{6.112b}$$

The -3-dB bandwidth of the magnitude of the closed-loop current gain $w(j\omega)$ from (6.108) is therefore

$$\omega_{3\text{-dB}} = 36.33 \cdot 10^6 \,\mathrm{rad/s} \tag{6.113}$$

after denormalization, or $f_{3\text{-dB}} = 5.78\,\mathrm{MHz}$.

The necessary value of $T(0)$ to attain the desired pole locations is determined from (6.109):

$$T(0) = |(0.125 - j1.516)(0.344 - j1.136)| = 1.806 \tag{6.114}$$

Using a capacitor in shunt with the feedback resistor R_f, we produce a real phantom zero in the transfer function of the feedback network. It is possible to produce complex phantom zeros in $\beta(s)$ by using the bridged-T network, as shown in Fig. 6.19. In this case, the transfer current ratio of the feedback network is determined as

$$\beta(s) = y_{12}R_2 \approx y_{f12}R_2 = -R_2 C_{f1} \frac{s^2 + as + b}{s + c} \qquad (6.115)$$

where

$$a = \frac{2R_{f1}C_{f1} + R_{f2}C_{f2}}{R_{f1}^2 C_{f1}C_{f2} + 2R_{f1}R_{f2}C_{f1}C_{f2}} \qquad (6.116a)$$

$$b = \frac{1}{R_{f1}^2 C_{f1}C_{f2} + 2R_{f1}R_{f2}C_{f1}C_{f2}} \qquad (6.116b)$$

$$c = \frac{2}{2R_{f2}C_{f2} + R_{f1}C_{f1}} \qquad (6.116c)$$

With the same three-stage common-emitter basic amplifier described by Eq. (6.99), the return ratio of the feedback amplifier of Fig. 6.19 can be written as

$$T(s) = \frac{T(0)(1 - s/z_1)(1 - s/z_2)}{(1 - s/p_1)(1 - s/p_2)(1 - s/p_3)(1 - s/p_4)} \qquad (6.117)$$

where $z_1, z_2 = \bar{z}_1$, and p_4 are the zeros and pole of $\beta(s)$, as given in (6.115). A typical pole-zero pattern of (6.117) together with the resulting root locus are presented in Fig. 6.20. Two branches of the locus terminate at the phantom zeros z_1 and z_2. If the other two branches of the locus are far from z_1 and z_2 for a reasonably large value of $T(0)$, the two natural frequencies near z_1 and z_2, which are poles of the closed-loop current gain $w(s)$, together with the pole p_4 of $\beta(s)$, which is the finite zero of $w(s)$, are dominant for the low-pass response. For design purposes, we need only concentrate on the loci terminating at z_1 and z_2. For example, we can choose the locations of z_1 and p_4 so that a maximally flat-magnitude or maximally flat-delay type of low-pass response is achieved. For a large value of $T(0)$, the dominant poles of $w(s)$ will be very close to z_1 and z_2. For any decrease of $T(0)$ caused by aging or environmental effect, the dominant poles move away from z_1 and z_2 only slightly. Hence, only a slight change in the dominant low-pass response is obtained. Furthermore, as will be seen in the next section, the natural frequencies in the neighborhood of phantom zeros z_1 and z_2 are relatively insensitive to the variations of the parameters. At any rate, as in the case of the real phantom zero, the technique provides a substantially large $T(0)$ without causing instability.

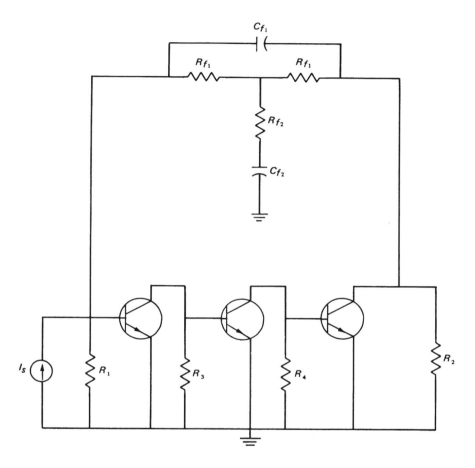

Figure 6.19 A three-stage common-emitter feedback amplifier possessing a pair of complex phantom zeros in its feedback circuit.

6.6 ROOT SENSITIVITY

In Chap. 3 we defined the sensitivity function as the ratio of the fractional change in a transfer function to the fractional change in an element (Definition 3.1). In the present situation, we are interested in knowing the change in the position of a natural frequency resulting from the variation of an element. A measure of this variation is called the root sensitivity and was first introduced by Truxal and Horowitz (1956).

Definition 6.6: Root sensitivity The *root sensitivity* is the sensitivity of a natural frequency s_k of a network with respect to an element x of interest and is

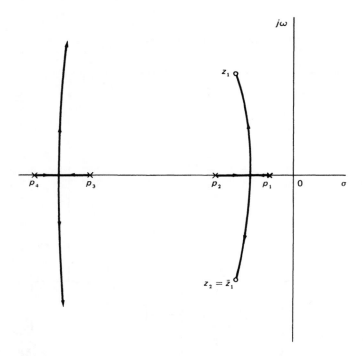

Figure 6.20 A typical root locus for the return ratio function (6.117).

defined by the relation

$$\mathcal{S}(s_k; x) = x \frac{\partial s_k}{\partial x} \tag{6.118}$$

A comparison of (6.118) with (3.21) shows that the root sensitivity is similar to the sensitivity function when In w is under consideration. First we derive a useful relation between root sensitivity and the sensitivity function.

Refer to Eq. (6.53). The numerator polynomial of the return difference can be written as

$$P(s) = D(s) + T(0)N(s) = K \prod_{k=1}^{n} (s - s_k) \tag{6.119}$$

where $s_k (k = 1, 2, \ldots, n)$ are the natural frequencies of the feedback amplifier. On taking the logarithm, we obtain

$$\ln P(s) = \ln K + \sum_{k=1}^{n} \ln(s - s_k) \tag{6.120}$$

We compute the root sensitivity with respect to the element $x = T(0)$, the midband value of return ratio. To this end, we differentiate (6.120) with respect to x, which is $T(0)$. Assuming that s_k is simple, we get

$$\frac{\partial \ln P(s)}{\partial x} = \frac{N(s)}{P(s)} = \frac{\partial K/\partial x}{K} - \sum_{k=1}^{n} \frac{\partial s_k/\partial x}{s - s_k} \qquad (6.121)$$

This shows that $-\partial s_k/\partial x$ is the residue of the function $N(s)/P(s)$ evaluated at the pole s_k. Thus, we can write

$$\frac{\partial s_k}{\partial x} = -(s - s_k)\left.\frac{N(s)}{P(s)}\right|_{s=s_k} = -\left.\frac{N(s)}{P'(s)}\right|_{s=s_k} = -\frac{N(s_k)}{P'(s_k)} \qquad (6.122)$$

where the prime denotes the derivative with respect to s. Thus, from (6.119) the root sensitivity with respect to $T(0)$ can be written as

$$\mathcal{S}(s_k; T(0)) = T(0)\frac{\partial s_k}{\partial T(0)} = -\frac{T(0)N(s_k)}{P'(s_k)} = \frac{D(s_k)}{P'(s_k)} \qquad (6.123)$$

From (6.53b), the right-hand side of (6.123) is recognized as the residue of the reciprocal of the return difference evaluated at the pole s_k. Thus, we conclude that the root sensitivity of the natural frequency s_k of a network with respect to the midband value of the return ratio is equal to the residue of the reciprocal of the return difference evaluated at its pole s_k.

As discussed in Sec. 5.1, the sensitivity function $S(x)$ with respect to an element x is related to the return difference and the transfer function $w(x)$ by

$$S(x) = \frac{1}{F(x)}\left[1 - \frac{w(0)}{w(x)}\right] \qquad (6.124)$$

In the case where $w(0) = 0$, the sensitivity becomes equal to the reciprocal of the return difference. For practical amplifiers, $|w(0)|$ is usually much smaller than $|w(x)|$ in the passband, and $S(x) \approx 1/F$ may be used as a good estimate of the sensitivity in the same frequency band. For the situation where a feedback amplifier is represented by its ideal feedback model, the sensitivity function of the closed-loop transfer function $w(\mu)$ with respect to the forward amplifier gain is given by

$$S(\mu) = \frac{1}{1 - \mu(s)\beta(s)} = \frac{1}{F(\mu)} \qquad (6.125)$$

Since $\beta(0)$ is considered fixed as far as the sensitivity function is concerned, we can use the variable parameter

$$T(0) = -\mu(0)\beta(0) \qquad (6.126)$$

rather than $\mu(0)$, and (6.125) becomes

$$S(T(0)) = \frac{1}{F(T(0))} = \frac{D(s)}{P(s)} \tag{6.127}$$

Comparing (6.123) with (6.127) yields

$$S(s_k; T(0)) = \zeta_{s_k} \tag{6.128}$$

where ζ_{s_k} denotes the residue of the sensitivity function with respect to $T(0)$ evaluated at the pole s_k. We remark that in the above analysis the return difference is written in two different ways: $F(x)$ and $F(s)$. The variable s, when shown explicitly, is used to exhibit its dependence on s, although F is dependent on both x and s. This should not create any difficulty. In the following, we express the root sensitivity in terms of the poles and zeros of the return ratio.

Consider the reciprocal of the root sensitivity, as given in (6.123), which can be manipulated into the form

$$
\begin{aligned}
\frac{1}{S(s_k; T(0))} &= \frac{P'(s_k)}{D(s_k)} = \frac{D'(s_k) + T(0)N'(s_k)}{D(s_k)} \\
&= \frac{D'(s_k)}{D(s_k)} + T(0)\frac{N(s_k)N'(s_k)}{D(s_k)N(s_k)} \\
&= \frac{D'(s_k)}{D(s_k)} - \frac{N'(s_k)}{N(s_k)} \\
&= \frac{d}{ds}\left[\ln D(s) - \ln N(s)\right]\big|_{s=s_k} \\
&= \sum_{j=1}^{n}(s_k - p_j)^{-1} - \sum_{i=1}^{m}(s_k - z_i)^{-1}
\end{aligned}
\tag{6.129}
$$

The third line follows directly from (6.119) and the fifth line from (6.53a). The prime, as before, denotes the derivative with respect to s.

Since s_k is a natural frequency, it must lie on the root-locus plot of $T(s)$. In fact, from (6.129) we can determine the angle ϕ_k of the tangent to the locus at any point s_k on the locus, as follows:

$$
\begin{aligned}
\phi_k &= \arg\left[\frac{\partial s_k}{\partial T(0)}\right] = \arg S(s_k; T(0)) = -\arg\left[\frac{1}{S(s_k; T(0))}\right] \\
&= -\arg\left[\sum_{j=1}^{n}(s_k - p_j)^{-1} - \sum_{i=1}^{m}(s_k - z_i)^{-1}\right]
\end{aligned}
\tag{6.130}
$$

In addition to the root sensitivity with respect to the changes of $T(0)$, in many problems we wish to evaluate the sensitivity of the natural frequency s_k to small changes in the locations of the poles and zeros of the return ratio $T(s)$. Assume first that z_i is a zero of $T(s)$. With z_i at its nominal value, by the definition of s_k we have

$$P(s_k) = D(s_k) + T(0)N(s_k) = 0 \tag{6.131}$$

Now let z_i be changed by a small amount from z_i to $z_i + \delta z_i$. The polynomial $P(s)$ will be changed from $P(s)$ to $P(s) + \delta P(s)$. Since the change is small, to a first-order approximation, the change in $P(s)$ can be written as

$$\delta P(s) \approx \frac{\partial P(s)}{\partial z_i} \delta z_i \tag{6.132}$$

With $P(s)$ changed to $P(s) + \delta P(s)$, the natural frequency s_k will correspondingly move to $s_k + \delta s_k$, so that

$$P(s_k + \delta s_k) + \delta P(s_k + \delta s_k) = 0 \tag{6.133}$$

By expanding $P(s_k + \delta s_k)$ and $\delta P(s_k + \delta s_k)$ by Taylor series, the first-order approximations are found to be

$$P(s_k + \delta s_k) \approx P(s_k) + \left.\frac{\partial P(s)}{\partial s}\right|_{s=s_k} \delta s_k = \left.\frac{\partial P(s)}{\partial s}\right|_{s=s_k} \delta s_k \tag{6.134a}$$

$$\delta P(s_k + \delta s_k) \approx \delta P(s_k) + \left.\frac{\partial \delta P(s)}{\partial s}\right|_{s=s_k} \delta s_k \tag{6.134b}$$

Substituting (6.134) in (6.133) in conjunction with (6.132) yields

$$\left.\frac{\partial P(s)}{\partial s}\right|_{s=s_k} \delta s_k + \left.\frac{\partial P(s)}{\partial z_i}\right|_{s=s_k} \delta z_i \approx 0 \tag{6.135}$$

giving

$$\mathcal{S}(s_k; z_i) = z_i \frac{\partial s_k}{\partial z_i} = -z_i \left.\frac{\partial P(s)/\partial z_i}{\partial P(s)/\partial s}\right|_{s=s_k}$$

$$= -z_i \left.\frac{T(0)\partial N(s)/\partial z_i}{\partial P(s)/\partial s}\right|_{s=s_k} \tag{6.136}$$

To evaluate the numerator on the right-hand side of (6.136), we write

$$N(s) = \left(1 - \frac{s}{z_i}\right) N_1(s) \tag{6.137}$$

assuming that the zero z_i is simple. Then we have

$$\frac{\partial N(s)}{\partial z_i} = \frac{s N(s)}{z_i(z_i - s)} \tag{6.138}$$

Substituting (6.138) in (6.136) and using (6.123), we obtain

$$\mathcal{S}(s_k; z_i) = -\frac{s_k T(0) N(s_k)}{(z_i - s_k) P'(s_k)} = \mathcal{S}(s_k; T(0)) \frac{s_k}{z_i - s_k} \tag{6.139}$$

In a similar manner, we can show that the root sensitivity with respect to a pole p_j of $T(s)$ is given by

$$\mathcal{S}(s_k; p_j) = \mathcal{S}(s_k; T(0)) \frac{s_k}{p_j - s_k} \tag{6.140}$$

The details are left as an exercise (see Prob. 6.11). Thus, knowing the root sensitivity with respect to the midband value of the return ratio, we can compute the root sensitivities with respect to the zeros and poles of the return ratio directly from (6.139) and (6.140).

Example 6.8 Consider the return ratio function

$$T(s) = \frac{T(0)}{s^2 + 2s + 2} = \frac{T(0)}{(s + 1 + j1)(s + 1 - j1)} \tag{6.141}$$

Then we have

$$F(s) = 1 + T(s) = \frac{s^2 + 2s + 2 + T(0)}{s^2 + 2s + 2} \tag{6.142}$$

$$S(s) = S(T(0)) = \frac{s^2 + 2s + 2}{s^2 + 2s + 2 + T(0)} \tag{6.143}$$

Assume that $T(0) = 3$. The natural frequencies are zeros of $F(s)$ or the roots of the equation

$$P(s) = s^2 + 2s + 5 = 0 \tag{6.144}$$

giving $S_1 = -1 + j2$ and $s_2 = -1 - j2$. To compute the root sensitivities of s_k with respect to $T(0)$, we apply (6.123) and obtain

$$\mathcal{S}(s_k; T(0)) = \frac{s^2 + 2s + 2}{2s + 2}\bigg|_{s=s_k} = \pm\frac{j3}{4} \quad k = 1, 2 \tag{6.145}$$

The partial fraction expansion of $S(s)$ is given by

$$S(s) = \frac{s^2 + 2s + 2}{s^2 + 2s + 5} = 1 + \frac{j3/4}{s+1-j2} - \frac{j3/4}{s+1+j2} \qquad (6.146)$$

giving $\zeta_{s_k} = \pm j3/4$. Comparing this with (6.145) confirms (6.128). Also, $\mathcal{S}(s_k; T(0))$ can be computed directly by formula (6.129), as follows:

$$\frac{1}{\mathcal{S}(s_k; T(0))} = \frac{1}{-1 \pm j2 - (-1+j1)} + \frac{1}{-1 \pm j2 - (-1-j1)} \mp \frac{j4}{3}$$

$$(6.147)$$

Thus, the angle of the tangent to the root locus at the point s_1 is $90°$.

For illustrative purposes, we compute the root sensitivity directly from its definition. For a general value of $T(0)$, the natural frequencies are found to be

$$s_1, s_2 = -1 \pm j\sqrt{1 + T(0)} \qquad (6.148)$$

Differentiating (6.148) with respect to $T(0)$ and applying the definition (6.118), we get

$$\mathcal{S}(s_k; T(0)) = T(0)\frac{ds_k}{dT(0)} = \pm j\frac{T(0)}{2\sqrt{1+T(0)}} \qquad (6.149)$$

For $T(0) = 3$, $\mathcal{S}(s_k; T(0)) = \pm j3/4$, confirming (6.145).

To compute the root sensitivities of s_k with respect to the poles p_j of the return ratio (6.141), we use (6.140); the results are given by

$$\mathcal{S}(s_k; p_k) = \mathcal{S}(s_k; T(0))\frac{s_k}{p_k - s_k} = \frac{3(1 \mp j2)}{4} \quad k = 1, 2 \qquad (6.150a)$$

$$\mathcal{S}(s_1; p_2) = \overline{\mathcal{S}}(s_2; p_1) = \mathcal{S}(s_1; T(0))\frac{s_1}{p_2 - s_1} = \frac{1 - j2}{4} \qquad (6.150b)$$

where $p_1 = -1 + j1$ and $p_2 = -1 - j1$. Observe that

$$\mathcal{S}(s_1; p_1) + \mathcal{S}(s_1; p_2) = \overline{\mathcal{S}}(s_2; p_1) + \overline{\mathcal{S}}(s_2; p_2) = 1 - j2 \qquad (6.151)$$

This result is expected, because the natural frequencies s_k and poles p_j must move in complex conjugate pairs.

6.7 BODE FORMULAS

In Sec. 6.4, we demonstrated that the gain and phase of the return ratio are important in stability considerations. In this section, we show that they are not entirely independent; one dictates the other. Specifically, we present a number of

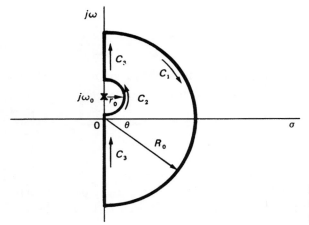

Figure 6.21 A closed contour indented around the point $j\omega_0$ in the s-plane.

relationships between the real and imaginary parts of a network function and its natural logarithm. These relations are well known in mathematics and are referred to as the *Hilbert transforms*. However, because Bode (1945) first applied them to network theory, we shall call them the *Bode formulas*.

Let $w(s)$ be a network function that is analytic in the entire closed RHS, and write

$$w(j\omega) = R(\omega) + jX(\omega) \qquad (6.152)$$

To develop the desired relations, we shall integrate the function $w(s)/(s-j\omega_0)$ around the closed contour C as shown in Fig. 6.21, where ω_0 is any value of ω. The small indentation to the right has been introduced to avoid the pole of the function $w(s)/(s - j\omega_0)$ at the point $s = j\omega_0$, so that the integrand is analytic on the boundary and within the closed contour C. Applying Cauchy's integral theorem, we have

$$\oint_C \frac{w(s)}{s - j\omega_0}\,ds = 0 \qquad (6.153)$$

The complete contour consists of three parts: the large semicircle C_1 of radius R_0, the small semicircular indentation C_2 of radius r_0 about the point $j\omega_0$, and the imaginary axis C_3. The contour integral on the left-hand side of (6.153) can be expressed as the sum of three line integrals along the paths C_1, C_2, and C_3. For contour C_1, we have $s = R_0 e^{j\theta}$, as indicated in Fig. 6.21, and we take the limit as

R_0 approaches infinity:

$$\lim_{R_0 \to \infty} \int_{C_1} \frac{w(s)}{s - j\omega_0} ds = \lim_{R_0 \to \infty} \int_{\pi/2}^{-\pi/2} \frac{w(R_0 e^{j\theta})}{R_0 e^{j\theta} - j\omega_0} j R_0 e^{j\theta} d\theta$$

$$= \int_{\pi/2}^{-\pi/2} j w(\infty) d\theta = -j\pi w(\infty) = -j\pi R(\infty)$$

$$(6.154)$$

in which we have used the relation $w(\infty) = R(\infty)$ because the imaginary part of $w(s)$, being an odd function, must be zero at infinity.

Likewise, for the small semicircular contour C_2 we have

$$\lim_{r_0 \to 0} \int_{C_2} \frac{w(s)}{s - j\omega_0} ds = \lim_{r_0 \to 0} \int_{-\pi/2}^{\pi/2} \frac{w(j\omega_0 + r_0 e^{j\phi})}{r_0 e^{j\phi}} j r_0 e^{j\phi} d\phi$$

$$= \int_{-\pi/2}^{\pi/2} j w(j\omega_0) d\phi = j\pi w(j\omega_0) \qquad (6.155)$$

Finally, for the imaginary-axis contour C_3 we have

$$\int_{C_3} \frac{w(s)}{s - j\omega_0} ds = \lim_{\substack{R_0 \to \infty \\ r_0 \to \infty}} \left[\int_{-R_0}^{\omega_0 - r_0} \frac{w(j\omega)}{\omega - \omega_0} d\omega + \int_{\omega_0 + r_0}^{R_0} \frac{w(j\omega)}{\omega - \omega_0} d\omega \right]$$

$$\equiv \int_{-\infty}^{\infty} \frac{w(j\omega)}{\omega - \omega_0} d\omega \qquad (6.156)$$

The integration in (6.156) must avoid the pole at $s = j\omega_0$ in a symmetric manner and will give the *principal value* of the integral on the right. In all the subsequent analysis, we must keep this point in mind. Now, substituting (6.154), (6.155), and (6.156) in (6.153) yields

$$\int_{-\infty}^{\infty} \frac{w(j\omega)}{\omega - \omega_0} d\omega = j\pi \left[R(\infty) - w(j\omega_0) \right] \qquad (6.157)$$

If we equate the reals and imaginaries, we get

$$R(\omega_0) = R(\infty) - \frac{1}{\pi} \int_{-\infty}^{\infty} \frac{X(\omega)}{\omega - \omega_0} d\omega \qquad (6.158a)$$

$$X(\omega_0) = \frac{1}{\pi} \int_{-\infty}^{\infty} \frac{R(\omega)}{\omega - \omega_0} d\omega \qquad (6.158b)$$

Since ω in (6.158) is only a dummy variable of integration, and since ω_0 is an arbitrary point on the $j\omega$-axis, it is convenient to rewrite (6.158) as

$$R(\omega) = R(\infty) - \frac{1}{\pi} \int_{-\infty}^{\infty} \frac{X(u)}{u - \omega} du \qquad (6.159a)$$

$$X(\omega) = \frac{1}{\pi} \int_{-\infty}^{\infty} \frac{R(u)}{u - \omega} du \qquad (6.159b)$$

The above result has two important implications. It states that if a network function is devoid of poles on the $j\omega$-axis, then its imaginary part is completely determined by the behavior of its real part on the $j\omega$-axis. Conversely, if the imaginary part is specified for all ω, its real part is completely determined within an additive constant. It is significant to note that the real or imaginary part need not be a realizable rational function — its corresponding imaginary or real part can be computed from the integral. In fact, the real or imaginary part can even be specified in graphic form.

6.7.1 Reactance and Resistance Integral Theorems

We consider two special cases of the integral relations of (6.159). First, in (6.159a) we set $\omega = 0$. Since $X(u)/u$ is an even function, we get

$$\int_0^{\infty} \frac{X(u)}{u} du = \frac{\pi}{2} [R(\infty) - R(0)] \qquad (6.160)$$

which is known as the *reactance-integral theorem*. It states that the behavior of the real part at the two extreme frequencies determines the integral of the imaginary part, weighted by the reciprocal of frequency, over all frequencies. Alternatively, if the left-hand side is plotted against the logarithmic frequency, then the area under the imaginary part is proportional to the net change in the real part between the two extreme frequencies: zero and infinite frequencies.

Next, let us multiply both sides of (6.159b) by ω and then take the limit as ω approaches infinity, giving

$$\lim_{\omega \to \infty} \omega X(\omega) = \frac{1}{\pi} \lim_{\omega \to \infty} \lim_{R_0 \to \infty} \int_{-R_0}^{R_0} \frac{\omega R(u)}{u - \omega} du \qquad (6.161)$$

There are two limiting operations involved on the right-hand side. The interchange of limits is permissible only if the integral is uniformly convergent for all ω. This condition is satisfied if $R(\infty) = 0$. If $R(\infty) \neq 0$, we can consider $R(\omega) - R(\infty)$ as the given real part of a network function. With this manipulation, together with the fact that $R(\infty)$ is even, we get

$$\int_0^\infty [R(u) - R(\infty)]\,du = -\frac{\pi}{2}\lim_{\omega\to\infty}\omega X(\omega) \qquad (6.162)$$

which is referred to as the *resistance-integral theorem*.

Consider a special situation where $w(s)$ has a simple zero at infinity. Then $R(\infty) = 0$ and after appealing to the initial value theorem we get

$$\int_0^\infty R(u)\,du = \frac{\pi}{2}\lim_{s\to\infty} s w(s) = w_\delta(0^+) \qquad (6.163)$$

where the limit is to be taken in the sector $|\arg s| \leq \pi/2 - \epsilon$, and $w_\delta(t)$ is the inverse Laplace transformation of $w(s)$ or the impulse response.

Example 6.9 Consider a one-port network that is constrained a priori by a leading shunt capacitor of value C, which may represent certain inevitable parasitic effects, as shown in Fig. 6.22. Let $Z_1(s)$ be the impedance facing the capacitor C. The total input impedance can be written as

$$Z(s) = \frac{Z_1(s)}{1 + Cs Z_1(s)} \qquad (6.164)$$

Assume that the one-port N is passive and that $Z_1(\infty) \neq 0$. Then $Z(s)$ must have a simple zero at infinity and (6.163) applies. Inserting (6.164) in (6.163) gives

$$\int_0^\infty R(u)\,du = \frac{\pi}{2}\lim_{s\to\infty}\frac{s Z_1(s)}{1 + Cs Z_1(s)} = \frac{\pi}{2C} \qquad (6.165)$$

As indicated at the beginning of this section, the resistance-integral theorem is applicable only to functions that are analytic in the closed RHS. If a function has simple poles on the $j\omega$-axis, the contour of integration must be indented to the right around these poles like the one shown in Fig. 6.21 for the pole at $s = j\omega_0$. If we go through the same procedure as in the preceding development, we find that additional terms, being proportional to the residues at these poles, must be

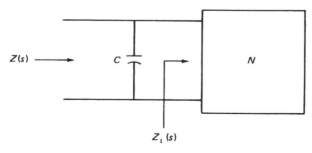

$Z(s) \longrightarrow$ C

$Z_1(s)$

N

Figure 6.22 A constrained one-port network.

subtracted from the right-hand side of (6.165). Since the residues are positive, for all cases, whether $Z(s)$ has simple poles on the $j\omega$-axis or not and whether $Z_1(\infty) \neq 0$ or not, the result can be written as

$$\int_0^\infty R(\omega)d\omega \leq \frac{\pi}{2C} \tag{6.166}$$

Note that the dummy variable has been changed to ω to suggest the physical meaning. Since power gain is usually associated with $R(\omega)$, the integral constraint (6.166) provides a basic limitation on the gain-bandwidth product introduced by the presence of the shunt capacitance C.

6.7.2 Phase-Integral and Gain-Integral Theorems

Consider the logarithm of a network function $w(s)$ and write

$$\ln w(j\omega) = A(\omega) + jB(\omega) \tag{6.167}$$

where $A(\omega)$ is the *gain function* and $B(\omega)$ is the *angle* or *phase function*. Assume that $w(s)$ is devoid of zeros and poles in the open RHS. We therefore permit $w(s)$ to have poles on the $j\omega$-axis, but it cannot have zero in the open RHS. Since poles and zeros of $w(s)$ are the logarithmic singularities of $\ln w(s)$, to apply Cauchy's integral theorem to $\ln w(s)$, as in (6.153), we must again make small semicircular indentations to the right around these singularities in selecting the contour of integration. If we can show that the small semicircular indentations around the logarithmic singularities will not contribute anything to the contour integral, identical results apply if we replace $w(s)$ by $\ln w(s)$, $R(\omega)$ by $A(\omega)$, and $X(\omega)$ by $B(\omega)$.

To confirm our assertion, let $j\omega_1$ be a zero or pole of order k of $w(s)$ and write

$$w(s) = (s - j\omega_1)^k w_1(s) \tag{6.168}$$

where k is either a positive or negative integer, giving

$$\ln w(s) = k \ln(s - j\omega_1) + \ln w_1(s) \tag{6.169}$$

It is sufficient to show that contour integral

$$\int_P \frac{\ln(s - j\omega_1)}{s - j\omega_0} ds = 0, \quad \omega_1 \neq \omega_0 \tag{6.170}$$

where P is an infinitesimal semicircular indentation to the right around the point $s = j\omega_1$. Any point s on P is expressible as $s = j\omega_1 + re^{j\theta}$, and we have the following estimate:

$$\left| \int_P \frac{\ln(s - j\omega_1)}{s - j\omega_0} ds \right| = \left| \int_P \frac{(\ln r + j\theta) j r e^{j\theta}}{r e^{j\theta} + j(\omega_1 - \omega_0)} d\theta \right|$$

$$\leq \int_P \frac{r |\ln r|}{|\omega_1 - \omega_0|} d\theta + \int_P \frac{r\theta}{|\omega_1 - \omega_0|} d\theta = \frac{2\theta |r \ln r| + r\theta^2}{2|\omega_1 - \omega_0|}$$

$$(6.171)$$

Since

$$\lim_{r \to 0} r \ln r = 0 \tag{6.172}$$

we obtain

$$\lim_{r \to 0} \int_P \frac{\ln(s - j\omega_1)}{s - j\omega_0} ds = 0 \tag{6.173}$$

This establishes (6.170) and justifies our assertion that the logarithmic singularities lying on the path of the contour integral (6.153), with $\ln w(s)$ replacing $w(s)$, do not contribute anything to the integral.

From (6.159), (6.160), and (6.162), we can write

$$A(\omega) = A(\infty) - \frac{1}{\pi} \int_{-\infty}^{\infty} \frac{B(u)}{u - \omega} du \tag{6.174a}$$

$$B(\omega) = \frac{1}{\pi} \int_{-\infty}^{\infty} \frac{A(u)}{u - \omega} du \tag{6.174b}$$

giving

$$\int_0^{\infty} \frac{B(u)}{u} du = \frac{\pi}{2} [A(\infty) - A(0)] \tag{6.175}$$

which is called the *phase-integral theorem*, and

$$\int_0^{\infty} [A(u) - A(\infty)] du = -\frac{\pi}{2} \lim_{\omega \to \infty} \omega B(\omega) \tag{6.176}$$

which is called the *gain-integral theorem*.

6.7.3 Gain-Phase Theorem

In this section, we express the phase at a particular frequency in terms of the gain specified at all frequencies. This result is useful in the design of feedback amplifiers.

Our starting point is (6.174b), which, for our purposes, is rewritten as

$$B(\omega_a) = \frac{1}{\pi}\left[\int_{-\infty}^{0}\frac{A(\omega)}{\omega-\omega_a}d\omega + \int_{0}^{\infty}\frac{A(\omega)}{\omega-\omega_a}d\omega\right] \qquad (6.177)$$

Notice that the dummy variable u has been changed to ω to indicate its physical significance. In the first integral on the right-hand side of (6.177), if we replace ω by $-\omega$ and change the limits accordingly, we obtain

$$\int_{-\infty}^{0}\frac{A(\omega)}{\omega-\omega_a}d\omega = -\int_{\infty}^{0}\frac{A(-\omega)}{-\omega-\omega_a}d\omega = -\int_{0}^{\infty}\frac{A(\omega)}{\omega+\omega_a}d\omega \qquad (6.178)$$

Substituting this in (6.177) in conjunction with the fact that

$$\int_{0}^{\infty}\frac{A(\omega_a)}{\omega^2-\omega_a^2}d\omega = A(\omega_a)\int_{0}^{\infty}\frac{1}{\omega^2-\omega_a^2}d\omega = 0 \qquad (6.179)$$

we get

$$B(\omega_a) = \frac{2\omega_a}{\pi}\int_{0}^{\infty}\frac{A(\omega)-A(\omega_a)}{\omega^2-\omega_a^2}d\omega \qquad (6.180)$$

A more convenient expression is obtained if a change to logarithmic frequency is made. Define

$$v = \ln\left(\frac{\omega}{\omega_a}\right) \quad \text{or} \quad \frac{\omega}{\omega_a} = e^v \qquad (6.181)$$

By using (6.181), (6.180) can be expressed as

$$B(\omega_a) = \frac{1}{\pi}\int_{-\infty}^{\infty}\frac{A(\omega)-A(\omega_a)}{\sinh v}dv \qquad (6.182)$$

The argument of $A(\omega)$ has been retained as ω for simplicity, although it should be written as $A(\omega_a e^v)$. Integrating the right-hand side of (6.182) by parts, we obtain

$$\begin{aligned}B(\omega_a) &= -\frac{1}{\pi}\left\{[A(\omega)-A(\omega_a)]\ln\coth\frac{v}{2}\right\}\Big|_{-\infty}^{\infty} + \frac{1}{\pi}\int_{-\infty}^{\infty}\frac{dA(\omega)}{dv}\ln\coth\frac{v}{2}dv\\&= \frac{1}{\pi}\int_{-\infty}^{\infty}\frac{dA(\omega)}{dv}\ln\coth\frac{|v|}{2}dv\end{aligned} \qquad (6.183)$$

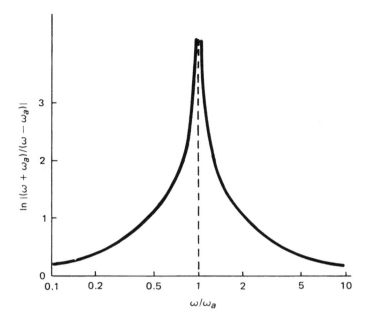

Figure 6.23 A plot of the function (6.184).

which is known as the *gain-slope theorem*. It states that the phase at any frequency depends on the slope of the gain at all frequencies when plotted against logarithmic frequency, the relative importance of any gain slope being determined by the weighting factor

$$\ln \coth \frac{|v|}{2} = \ln \left| \frac{\omega + \omega_a}{\omega - \omega_a} \right| \tag{6.184}$$

A plot of the function (6.184) is presented in Fig. 6.23. Observe that it rises sharply in the vicinity of $v = 0$ or $\omega = \omega_a$. For frequencies much higher than ω_a, the weighting factor is approximately $2\omega_a/\omega$, whereas at frequencies much lower than ω_a it is approximately $2\omega/\omega_a$. Therefore, at ω_a most of the contribution to the phase comes from the slope of the gain characteristic in the vicinity of ω_a. In terms of (6.184), (6.183) becomes

$$B(\omega_a) = \frac{1}{\pi} \int_{-\infty}^{\infty} \frac{dA(\omega)}{d\omega} \ln \left| \frac{\omega + \omega_a}{\omega - \omega_a} \right| d\omega \tag{6.185}$$

The gain slope $dA(\omega_a e^v)/dv$ in formula (6.183) is given in terms of *nepers per log radian*. This is to say that, for a unit gain slope, A will change by 1 neper between frequencies that are in the ratio $e = 2.7183$. A unit gain slope is evidently the same as a change of 6 dB per octave or 20 dB per decade.

Example 6.10 Assume that the gain characteristic has a constant slope on a logarithmic frequency scale at all frequencies. Then we have

$$A(\omega) = A(\omega_a e^v) = kv \tag{6.186}$$

k being the constant slope, and $dA(\omega)/dv = k$. From (6.183), the phase at any frequency ω_a is given by

$$B(\omega_a) = \frac{k}{\pi} \int_{-\infty}^{\infty} \ln \coth \frac{|v|}{2} dv = \frac{1}{2} k \pi \tag{6.187}$$

Thus, the phase characteristic is constant and equal to $\frac{1}{2}\pi$ times the slope of the gain characteristic.

Example 6.11 Assume that the gain characteristic is everywhere constant except for a discontinuity at frequency ω_0, as shown in Fig. 6.24. The slope is

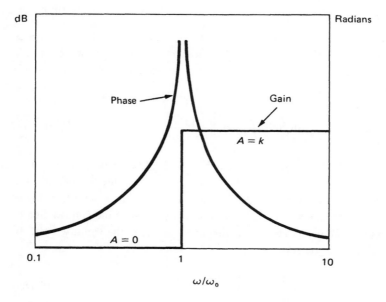

Figure 6.24 A prescribed gain characteristic and its associated phase plot.

obviously an impulse function and is given by

$$\frac{dA(\omega_a e^v)}{dv} = k\delta(\omega - \omega_0) \tag{6.188}$$

Substituting it in (6.185) yields the phase characteristic as

$$B(\omega_a) = \frac{k}{\pi} \ln \left| \frac{\omega_0 + \omega_a}{\omega_0 - \omega_a} \right| \tag{6.189}$$

A plot of (6.189) is also presented in Fig. 6.24.

6.7.4 Phase for the Semi-Infinite Constant-Slope Characteristic

One of the gain characteristics of considerable interest in feedback amplifier design is the *semi-infinite constant slope* as shown in Fig. 6.25, where the gain is zero up to $\omega = \omega_0$ and thereafter has a constant slope k on a logarithmic frequency scale. In this section, we present approximate formulas for computing its phase.

The constant slope k, as discussed in Example 6.10, can be viewed as the sum of two semi-infinite characteristics with the same slope, but running in opposite directions from ω_0, as depicted in Fig. 6.26, with the semi-infinite characteristics being identified by the solid and broken lines. The sum of the accompanying phase characteristics must therefore be equal to $k\pi/2$, as indicated in (6.187). This shows that the phase shift at ω_0 for the semi-infinite gain slope of Fig. 6.25 is equal to

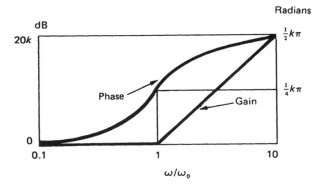

Figure 6.25 The semi-infinite constant slope and its associated phase plot.

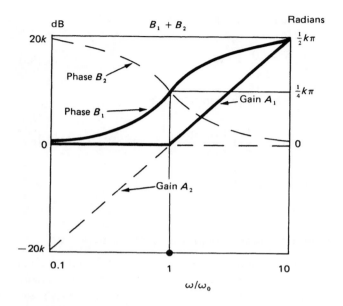

Figure 6.26 The decomposition of the constant slope into the sum of two semi-infinite characteristics with the same slope but running in the opposite directions and their associated phase plots.

half of the asymptotic value $k\pi/2$, and we can write

$$B(\infty) = \frac{k\pi}{2} \quad \text{and} \quad B(\omega_0) = \frac{k\pi}{4} \tag{6.190}$$

Furthermore, we see from the plot of Fig. 6.26 that the phase characteristic of the semi-infinite gain slope exhibits odd symmetry on a logarithmic frequency scale about the point determined by $\omega = \omega_0$ and $B(\omega) = k\pi/4$. This results in the general properties of the phase characteristic as shown in Fig. 6.25, which allows us to restrict the computation of the phase to values of ω_α below ω_a.

We now proceed to develop approximate formulas for the computation of the phase at any frequency ω_a below ω_0 for the semi-infinite constant slope k of Fig. 6.25. For our purposes, we define

$$x = \frac{\omega_a}{\omega} \quad \text{and} \quad x_a = \frac{\omega_a}{\omega_0} \tag{6.191}$$

In the range of frequencies below ω_0, the gain slope is zero; for frequencies above ω_0, the gain slope is k on a logarithmic frequency scale. These are equivalent to $dA(\omega)/d\omega = 0$, $\omega < \omega_0$, and $dA(\omega)/d\omega = k/\omega$, $\omega \geq \omega_0$. Using these in (6.185), we obtain

$$B(\omega_a) = \frac{k}{\pi} \int_{\omega_0}^{\infty} \ln \left| \frac{\omega + \omega_a}{\omega - \omega_a} \right| \frac{d\omega}{\omega} = \frac{k}{\pi} \int_{0}^{x_a} \ln \left| \frac{1 + x}{1 - x} \right| \frac{dx}{x} \qquad (6.192)$$

To develop the desired formulas, we set

$$y = \frac{1 - x}{1 + x} \quad \text{and} \quad y_a = \frac{1 - x_a}{1 + x_a} \qquad (6.193)$$

In terms of the new variable y, (6.192) can be written as

$$B(x_a) = -\frac{k}{\pi} \int_{x=0}^{x=x_a} \ln y \, d(\ln x) \qquad (6.194)$$

The above equation can be integrated by parts, and the result is given by

$$B(x_a) = -\frac{k}{\pi} (\ln x \ln y) \Big|_{x=0}^{x=x_a} + \frac{k}{\pi} \int_{x=0}^{x=x_a} \ln x \, d(\ln y) \qquad (6.195)$$

which is transformed into (see Prob. 6.17)

$$
\begin{aligned}
B(x_a) &= -\frac{k}{\pi} \ln x_a \ln y_a - \frac{k}{\pi} \int_{y=y_a}^{y=1} \ln x \, d(\ln y) \\
&= -\frac{k}{\pi} \ln x_a \ln y_a - \frac{k}{\pi} \int_{y=0}^{y=1} \ln x \, d(\ln y) + \frac{k}{\pi} \int_{y=0}^{y=y_a} \ln x \, d(\ln y)
\end{aligned}
$$
$$(6.196)$$

By using (6.192), it is straightforward to demonstrate that the second term on the right-hand side of (6.169) represents the phase of a semi-infinite constant slope k at the point ω_a, which according to (6.190) is equal to $k\pi/4$. Likewise, from (6.192), the third term represents the phase at the point y_a. Substituting these in (6.196) yields

$$B(x_a) + B(y_a) = \frac{k\pi}{4} - \frac{k}{\pi} \ln x_a \ln y_a \qquad (6.197)$$

where x_a and y_a are related by (6.193).

If in (6.193) x_a varies from 0 to 0.414, y_a changes from 1 to 0.414. Likewise, if y_a varies from 0 to 0.414, x_a changes from 1 to 0.414. Thus, by using (6.197), the phase characteristic can be computed at all frequencies if we know it only between zero and 0.414. Within this range, we can expect a power series expansion for B

to converge rapidly. Substituting the power series expansion

$$\ln \frac{1+x}{1-x} = 2\left(x + \frac{x^3}{3} + \frac{x^5}{5} + \cdots\right) \quad x^2 < 1 \tag{6.198}$$

in (6.192) and integrating term by term, we obtain

$$B(\omega_a) = \frac{2k}{\pi}\left(x_a + \frac{x_a^3}{9} + \frac{x_a^5}{25} + \cdots\right) \tag{6.199}$$

If we use only the first term in (6.199) in conjunction with (6.197), we obtain the desired approximate formulas for the phase characteristic as

$$B(\omega_a) \approx \frac{2k\omega_a}{\pi\omega_0} \quad 0 \le \omega_a \le 0.414\omega_0 \tag{6.200a}$$

$$\approx \frac{k\pi}{4} - \frac{k}{\pi}\ln\left(\frac{\omega_a}{\omega_0}\right)\ln\left(\frac{\omega_0 - \omega_a}{\omega_0 + \omega_a}\right)$$
$$- \frac{2k(\omega_0 - \omega_a)}{\pi(\omega_0 + \omega_a)} \quad 0.414\omega_0 \le \omega_a \le \omega_0 \tag{6.200b}$$

The maximum error in B as computed from these expressions is about 2 percent. If the first two terms in (6.199) are used, the result is almost exact. These formulas were first given by Bode (1945) and are very useful in the design of feedback amplifiers, to be discussed in the following section.

One of the major applications of the approximate formulas is the graphic evaluation of the phase shift $B(\omega)$ when the gain characteristic is known either analytically or graphically. The gain characteristic is first approximated by a series of straight-line segments. The straight-line approximation is next resolved into the sum of semi-infinite constant slopes of various steepness. With (6.200), the phase characteristics of the component semi-infinite constant slopes are then computed, and these are summed up to give the total phase characteristic of the gain response.

6.8 BODE'S DESIGN THEORY

A basic problem in the design of a feedback amplifier is the determination of the maximum amount of feedback in decibels for a given frequency band. To guard against instability, certain gain and phase margins are also specified. As discussed in this and the preceding two chapters, the return ratio T is essential for stability study and controls the amount of feedback, sensitivity, and the input and output impedance levels. In the following, we present Bode's design theory in terms of the return ratio $T(j\omega)$. We shall confine our discussion to low-pass characteristics of

$T(j\omega)$, because the other types of characteristics can be obtained from the low-pass by the standard techniques of frequency transformation.

An ideal low-pass characteristic for the return ratio $T(j\omega)$ is one whose gain characteristic is completely flat from zero to cutoff frequency ω_0 and whose phase is constant beyond the cutoff frequency. In this way, the flat portion of the gain characteristic provides a constant loop transmission in the passband and the flat portion of the phase characteristic gives the desired phase margin. To this end, we introduce *Bode's ideal cutoff characteristic*, expressed by the return ratio

$$\ln T(j\omega) = A + jB = \ln\left[\frac{T(0)}{(\sqrt{1 - \omega^2/\omega_0^2} + j\omega/\omega_0)^{2(1-\gamma)}}\right]$$

$$= \ln T(0) - 2(1-\gamma)\ln\left(\sqrt{1 - \frac{\omega^2}{\omega_0^2}} + \frac{j\omega}{\omega_0}\right) \qquad (6.201)$$

The exponent is written in the form $2(1-\gamma)$ to suit the feedback amplifier problem. A plot of A and B for the choice, $\gamma = \frac{1}{6}$, is presented in Fig. 6.27. Note that A is plotted in relation to the midband value $T(0)$ for simplicity. Thus, the 0-dB level in Fig. 6.27 corresponds to 20 log $T(0)$. Observe that the gain characteristic is completely flat from 0 to the cutoff frequency ω_0, and it drops off at the rate of $12(1-\gamma)$ dB/octave at high frequencies. The phase shift is changed from 0 to

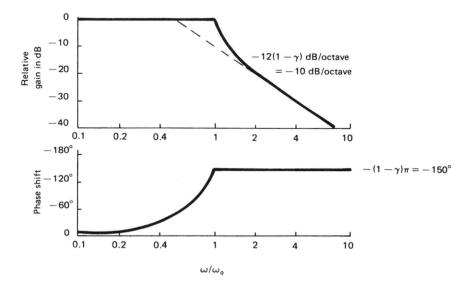

Figure 6.27   The gain and phase plots of Bode's ideal cutoff characteristic for $\gamma = \frac{1}{6}$.

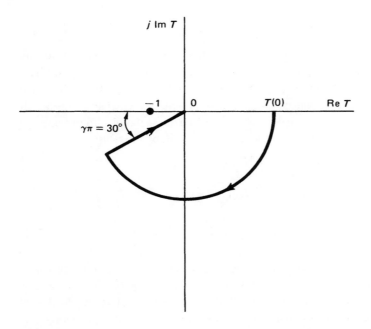

Figure 6.28 The Nyquist plot of Bode's ideal cutoff characteristic for $\gamma = \frac{1}{6}$.

$-(1-\gamma)\pi$ when the frequency is increased from 0 to ω_0, and it remains a constant phase $-(1-\gamma)\pi$ beyond ω_0. The corresponding Nyquist plot of Fig. 6.27 is shown in Fig. 6.28. For the choice of $\gamma = \frac{1}{6}$, the corresponding phase margin is 30°. In general, the phase margin is $\gamma\pi$ in radians. In Fig. 6.28, if we set $\gamma = 0$, then the gain falls off at the rate of 12 dB/octave and the asymptotic phase reaches $-180°$ while the gain of the return ratio is still larger than unity. The amplifier is therefore unstable.

We remark that in applying the formulas developed in the preceding sections, we implicitly assume that the transfer functions are devoid of zeros in the open RHS. Such transfer functions are termed the *minimum-phase transfer functions*. In feedback amplifiers, nonminimum-phase transfer functions are seldom used because excess phase shift is never desired for stability reasons. We almost always use the minimum-phase feedback and interstage networks.

6.8.1 Single-Loop Absolutely Stable Transistor Feedback Amplifiers

As demonstrated in Sec. 4.2, if a one-pole approximation is used for a common-emitter stage, its logarithmic gain falls off ultimately at the rate of at least

6 dB/octave. If n such stages are connected in cascade, the gain cuts off at a rate of $6n$ dB/octave. For broadband feedback amplifier design, we should include the nondominant pole neglected in the simplified equivalent network obtained by unilateralizing the hybrid-pi model used for the transistors. These nondominant effects are important in stability study, because they introduce an additional phase shift at low frequencies. In general, the return ratio $T(s)$ is a rational function, being the ratio of two polynomials, having m finite zeros and n finite poles. From physical considerations, $m < n$. At high frequencies, the return ratio behaves as s^{m-n} and cuts off at a rate of $-6(n-m)$ dB/octave. The asymptote with this slope in the Bode plot is called the *final asymptote*. The corresponding phase reaches $-(n-m)\pi/2$ rad asymptotically. Therefore, in practice the return ratio $T(j\omega)$ can be shaped to follow Bode's ideal characteristic only up to a certain frequency.

Consider a typical set of Bode plots as shown in Fig. 6.29, where ω_g is the gain-crossover frequency and ω_p is the phase-crossover frequency. Since $\omega_p < \omega_g$, the amplifier is unstable, and we must reduce the gain and control the phase shift by introducing the corrective networks.

Suppose that the desired shape of the return ratio $T(j\omega)$ is Bode's ideal cutoff characteristic of Fig. 6.27. This ideal cutoff characteristic is superimposed on the gain characteristic of Fig. 6.29, and the resulting curves are shown in Fig. 6.30. The shaded area is therefore the loss characteristic that should be introduced by the shaping networks. At high frequencies, the overall gain characteristic is still governed by the high-frequency asymptotic behavior of the gain curve of Fig. 6.29. Therefore, the final asymptote has a slope of $-6k$ dB/octave, as indicated in Fig. 6.30, at the frequency ω_c. To obtain the corresponding phase for this modified gain characteristic, denoted by the bottom curves in each part of Fig. 6.30, we simply add the phase contributions resulting from the ideal cutoff characteristic and a semi-infinite constant slope equal to

$$-6k + 12(1-\gamma)\text{dB/octave} \qquad (6.202)$$

which begins at the frequency ω_c. This added semi-infinite slope modifies the total phase to the shape as shown in Fig. 6.30. Observe that the much-celebrated constant-phase characteristic of the ideal cutoff characteristic has been completely destroyed. Furthermore, the plots indicate that the amplifier is unstable because of the additional phase contribution at low frequencies by the added semi-infinite slope. In the following, we present Bode's theory for obtaining a stable characteristic.

Bode's approach is by introducing a horizontal step in the gain characteristic as shown in Fig. 6.31a. The horizontal step is located below the 0-dB level by an

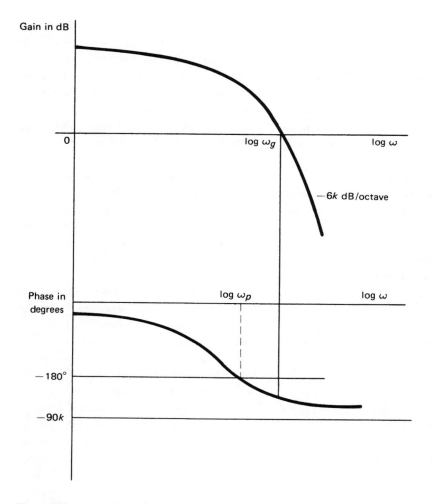

Figure 6.29 A typical set of Bode plots.

amount equal to the desired gain margin G_m. The characteristic of Fig. 6.31a can be represented as the sum of three gain characteristics: the ideal cutoff characteristic with an asymptotic slope of $-12(1 - \gamma)$ dB/octave and two semi-infinite constant slopes, one of which starts at the frequency ω_c with a slope $-6k$ dB/octave, and the other of which starts at ω_b with a slope $12(1 - \gamma)$ dB/octave. The results are presented in Fig. 6.31b. From Eq. (6.200a), we see that, at low frequencies, the phase shift $B_b(\omega)$ of the semi-infinite constant slope equal to $12(1 - \gamma)$ dB/octave

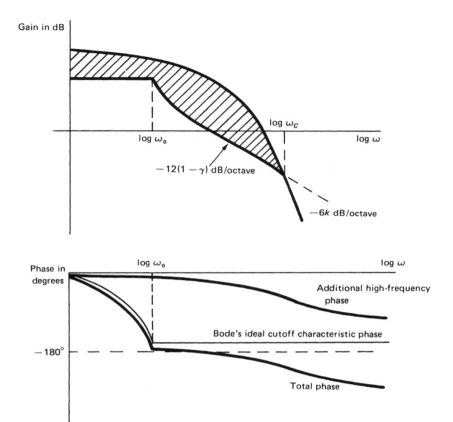

Figure 6.30 The superposition of the gain and phase plots of Bode's ideal cutoff characteristic and the Bode plots of Fig. 6.29.

is given by

$$B_b(\omega) \approx \frac{4(1-\gamma)\omega}{\pi\omega_b} \quad 0 \leqslant \omega \leqslant 0.414\omega_b \tag{6.203}$$

Note that the slope used in formula (6.200a) is stated in terms of k nepers per log radian, which is equivalent to $6k$ dB per octave or $20k$ dB per decade. Likewise, the phase shift $B_c(\omega)$ of the semi-infinite constant slope equal to $-6k$ dB/octave is obtained as

$$B_c(\omega) \approx -\frac{2k\omega}{\pi\omega_c} \quad 0 \leqslant \omega \leqslant 0.414\omega_c \tag{6.204}$$

Figure 6.31 (a) The introduction of a horizontal step in the gain characteristic of Fig. 6.30 and (b) its decomposition into the sum of three gain characteristics of various asymptotic slopes.

To obtain the phase characteristic that follows the shape of Bode's ideal cutoff characteristic at low frequencies, we choose ω_b so that

$$B_b(\omega) + B_c(\omega) = 0 \tag{6.205}$$

giving

$$\omega_b = \frac{2(1-\gamma)\omega_c}{k} \tag{6.206}$$

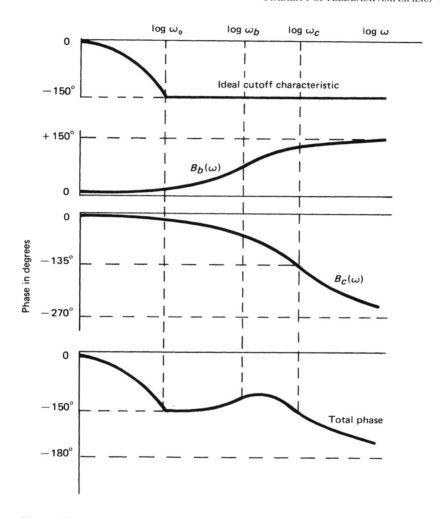

Figure 6.32 The phase plots of the three gain characteristics of Fig. 6.31b, together with their sum.

The phase plots of the three gain characteristics of Fig. 6.31b, together with their sum, are presented in Fig. 6.32 for $k = 3$ and $\gamma = \frac{1}{6}$. In this case, $\omega_b = 5\omega_c/9$. Observe that at low frequencies the total phase shift is practically constant and equal to that of the ideal cutoff characteristic phase, and it becomes less negative around ω_b. At high frequencies, the phase associated with the final asymptote takes over and dominates the others. The gain characteristic of Fig. 6.31a is referred to as *Bode's ideal loop gain characteristic*.

To determine the maximum amount of feedback within the useful frequency band and with preassigned gain margin G_m and phase margin $\gamma\pi$, we set

$$20\log|T(j\omega_b)| = 20\log\left|\frac{T(0)}{(\sqrt{1 - \omega_b^2/\omega_0^2} + j\omega_b/\omega_0)^{2(1-\gamma)}}\right| = -G_m$$

(6.207)

giving

$$10^{G_m/20}|T(0)| = \left|\sqrt{\frac{\omega_b^2}{\omega_0^2} - 1} + \frac{\omega_b}{\omega_0}\right|^{2(1-\gamma)}$$

(6.208)

Since $\omega_b \gg \omega_0$, Eq. (6.208) can be simplified. In terms of decibels, we obtain the maximum amount of feedback as

$$20\log|T(0)| \approx 40(1-\gamma)\log\left(\frac{2\omega_b}{\omega_0}\right) - G_m$$

(6.209)

It is useful to express this feedback in terms of the gain-crossover frequency ω_g, the ideal cutoff band-edge frequency ω_0, and the specified gain and phase margins G_m and $\gamma\pi$. From Fig. 6.31a we find that ω_c and ω_g are related by

$$-20k(\log\omega_c - \log\omega_g) = -G_m$$

(6.210)

yielding

$$\omega_c = \omega_g \cdot 10^{G_m/20k}$$

(6.211)

By using this in conjunction with (6.206), the maximum amount of feedback is found to be (see Prob. 6.18)

$$20\log|T(0)| = 40(1-\gamma)\log\frac{4(1-\gamma)\omega_g}{k\omega_0} + \frac{2(1-\gamma)G_m}{k} - G_m$$

(6.212)

As can be seen from Figs. 6.31a and 6.32, the modified gain characteristic is clearly stable. The maximum amount of allowed feedback is given by (6.212) for a prescribed gain margin G_m. For illustrative purposes, we also present the corresponding Nyquist plot in Fig. 6.33.

Example 6.12 We again consider the three-stage common-emitter feedback amplifier of Fig. 6.9. As demonstrated in Example 6.2, the return ratio can be

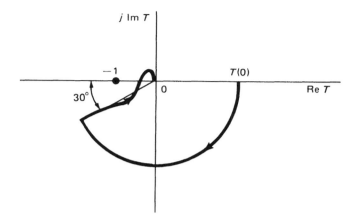

Figure 6.33 The Nyquist plot of the modified gain characteristic of Fig. 6.31a.

approximated by the expression

$$T(j\lambda) = \frac{T(0)}{(1 + j\lambda/3.633)(1 + j\lambda/2.076)(1 + j\lambda/2.770)} \tag{6.213}$$

where $\lambda = \omega/10^7$. The function has only three finite poles, giving $m = 0, n = 3$, and $k = 3$. The gain-crossover frequency was computed in Example 6.2 and was found to be $\omega_g = 37.5 \cdot 10^6$ rad/s. Suppose that the desired gain margin is 10 dB and a phase margin is about $30°$, giving

$$G_m = 10 \text{ dB} \qquad \gamma = \frac{30}{180} = \frac{1}{6} \tag{6.214}$$

Assume, for simplicity, that $\omega_0 = 10^6$ rad/s. Substituting these in (6.212), we obtain the maximum amount of allowed feedback:

$$20 \log |T(0)| = 54 + 5.56 - 10 = 49.56 \text{ dB} \tag{6.215}$$

6.8.2 Corrective Networks

Once the loss characteristic of the interstage shaping network is known, the desired corrective network can be synthesized. A widely used technique in practice is to employ a one-port impedance Z in one of the inter-stages, as shown in Fig. 6.34, because shaping in the overall feedback path will affect the overall transfer function. A typical one-port corrective network can be realized by a series tuned RLC network as illustrated in Fig. 6.35. For more complicated feedback amplifiers, more

Figure 6.34 The location of a one-port corrective network.

Figure 6.35 A typical one-port corrective network.

elaborate one-port networks are required. The design of these networks would, of course, differ from amplifier to amplifier. We shall terminate this discussion rather abruptly at this point, because a more detailed discussion would take us far afield into network synthesis, the insertion-loss synthesis in particular.

6.9 SUMMARY

We began this chapter by introducing the concept of a single-loop feedback amplifier and the notions of absolute stability and conditional stability. In a single-loop feedback amplifier, the return differences with respect to the controlling parameters of the active devices under normal operating conditions are the same.

For stability analysis, we reviewed briefly the Routh-Hurwitz criterion and Liénard-Chipart criterion. These criteria are of limited practical value, because the network determinant or its equivalent is not usually given analytically. In the case where the polynomial under consideration has numerical coefficients, the

Routh-Hurwitz criterion can be put in Guillemin form in terms of the Stieltjes continued fraction. The Guillemin form is relatively simple to test. In applying the Hurwitz criterion, we are required to evaluate the Hurwitz determinants of all orders. To reduce this number, we presented the Liénard-Chipart criterion, which requires that one evaluate only about half of the Hurwitz determinants, every other one.

To study system stability when the network determinant is not known analytically, we described in some detail the Nyquist criterion. The Nyquist criterion is important in that it provides a quantitative measure of the degree of stability of a feedback amplifier. Furthermore, the parasitic effects, which play an important role in stability study, can also be included. To provide a measure of the maximum amount of feedback permitted to ensure amplifier stability, we introduced the terms of gain margin and phase margin. The gain margin is the additional amount of feedback in decibels that is permitted before the amplifier becomes unstable. Likewise, the phase margin is the additional phase lagging that is permitted before the amplifier becomes unstable. Thus, they are margins of safety against instability and are the generally accepted measures of the degree of stability of a feedback amplifier.

By changing the amount of feedback, we change the locations of the natural frequencies of a feedback amplifier. It would be extremely useful if the locus of the natural frequencies were plotted as a function of the midband value of the return ratio. For this we discussed in detail the root-locus technique. By knowing the trajectories of the natural frequencies, the designer has a considerable amount of insight and understanding of the stability, time-, and frequency-domain behaviors of the network. The construction of the root locus is relatively simple if the return ratio has only two poles. The procedure becomes more complicated if it has three or more poles. To this end, we developed simple rules that enable us readily to sketch the root locus. Very often, a rough sketch will provide sufficient information for a preliminary design to be undertaken.

In addition to the sensitivity function defined previously, we find it useful to know the change in the position of a natural frequency as a result of the variation of an element. For this we introduced the concept of root sensitivity. It was shown that the root sensitivity with respect to the midband value of the return ratio is equal to the residue of the reciprocal of the return difference evaluated at the pole that is the natural frequency under consideration. In addition, we derived formulas relating the root sensitivities with respect to poles and zeros of the return ratio and its midband value.

The gain and phase of the return ratio are important in stability considerations. However, they are not entirely independent; one dictates the other. In this chapter, we studied a number of relations between the real and imaginary parts of a network function and its natural logarithm. We showed that if a network function is devoid of poles in the closed RHS, then its imaginary part is completely determined by the behavior of its real part on the $j\omega$-axis. Conversely, if the imaginary part is specified for all ω, its real part is completely determined within an additive constant. In terms of the natural logarithm of a network function, the restrictions are that the network function be devoid of poles and zeros in the open RHS. Under these constraints, the phase is completely determined by the behavior of the gain on the $j\omega$-axis, or the gain is completely determined within an additive constant by the behavior of the phase on the $j\omega$-axis. We also derived a useful formula relating the phase at a particular frequency in terms of the gain specified at all frequencies known as the gain-phase theorem.

Finally, we discussed Bode's design theory by introducing Bode's ideal cutoff characteristic. Specifically, we demonstrated how to modify a gain characteristic to follow Bode's ideal cutoff characteristic up to certain frequencies for the preassigned gain and phase margins. A formula giving the maximum amount of feedback within the useful frequency band was derived in terms of the gain-crossover frequency, the ideal cutoff band-edge frequency, and the prescribed gain and phase margins.

PROBLEMS

6.1 Identify the polynomials whose roots are restricted to the open LHS:

(a) $s^4 + 3s^3 + 6s^2 + 3s + 2$
(b) $2s^4 + 3s^3 + 3s^2 + 3s + 3$
(c) $s^5 + 7s^4 + 14s^3 + 15s^2 + 15s + 9$
(d) $s^5 + 10s^4 + 16s^3 + 16s^2 + 14s + 9$
(e) $s^7 + 2s^6 + 6s^5 + 14s^4 + 11s^3 + 18s^2 + 6s + 6$

6.2 A practical version of a FET Hartley oscillator is shown in Fig. 6.36. Assume that the following parameters are used for the oscillator:

$$L_1 = 2 \text{ mH} \quad L_2 = 2 \text{ mH} \quad M = 0.5 \text{ mH}$$
$$C = 200 \text{ pF} \quad r_d = 50 \text{ k}\Omega \tag{6.216}$$

ignoring the capacitances of the FET. Determine the frequency of oscillation and the required value of transconductance g_m.

6.3 A transistor Colpitts oscillator is shown in Fig. 6.37. Determine the frequency of oscillation and the minimum value of the transistor forward current gain

Figure 6.36 A practical version of a FET Hartley oscillator.

h_{fe} for the following network parameters:

$$C_1 = 0.01 \ \mu\text{F} \quad C_2 = 0.02 \ \mu\text{F}$$
$$L = 80 \ \mu\text{H} \quad h_{ie} = 1.1 \ \text{k}\Omega \qquad (6.217)$$
$$h_{oe} = h_{re} = 0$$

6.4 Show that the angles of arrival at a zero z_j of the return ratio $T(s)$ on the root locus are given by (6.71).

6.5 Use the Nyquist criterion to study the stability of feedback amplifiers having the following return ratios:

(a)

$$T(s) = \frac{s+2}{(s+1)(s+3)} \qquad (6.218a)$$

(b)

$$T(s) = \frac{s^2 + 2s + 2}{s^2(s^2 + s + 2)} \qquad (6.218b)$$

(c)

$$T(s) = \frac{s^2 + 3s + 2}{s^3 + 2s^2 + s + 1} \qquad (6.218c)$$

Figure 6.37 A transistor Colpitts oscillator.

6.6 Repeat Example 6.2 for the following set of network parameters:

$$g_m = 0.3 \text{ mho} \quad r_x = 50 \, \Omega \quad r_\pi = 125 \, \Omega$$
$$r_0 = \infty \quad c_\pi = 200 \text{ pF} \quad C_\mu = 6 \text{ pF} \quad (6.219)$$
$$R_1 = R_2 = 50 \, \Omega \quad R_3 = R_4 = 1 \text{ k}\Omega$$

6.7 Consider the three-stage common-emitter feedback amplifier of Fig. 6.17. As indicated in Sec. 6.5.2, the return ratio has the general form shown in Eq. (6.103). Let $p_1 = 1$, $p_2 = 2$, $p_3 = 3$, and $z_1 = 2.5$. Sketch the root locus for this return ratio function.

6.8 Sketch the root locus for the following return ratio functions and determine the range of the values of the midband return ratio $T(0)$ for which the feedback amplifier is stable:

(a)

$$T(s) = \frac{4T(0)(s+1)}{(s+2)(s^2+2s+2)} \qquad (6.220a)$$

(b)

$$T(s) = \frac{2T(0)}{(s+1)^2(s+2)} \qquad (6.220b)$$

(c)

$$T(s) = \frac{1.5T(0)(s^2+2s+2)}{(s+1)(s^2+s+3)} \qquad (6.220c)$$

(d)

$$T(s) = \frac{2T(0)(s+2)}{(s^2+s+1)(s^2+4s+4)} \qquad (6.220d)$$

6.9 Calculate the root sensitivities of the return ratio functions, as given in (6.220), with respect to the zeros and poles of $T(s)$ and also with respect to the midband value $T(0)$ of the return ratios.

6.10 Sketch the root locus for a negative feedback amplifier having the return ratio

$$T(s) = \frac{3T(0)(s+2)}{(s+1)(s+3)(s^2+2s+2)} \qquad (6.221)$$

6.11 Derive formula (6.140).

6.12 By integrating the function $[\ln w(s)]/(s^2+\omega_a^2)$ around the basic contour with small semicircular indentations to the right at $s = \pm j\omega_a$, demonstrate that (6.180) can be obtained in this way.

6.13 By integrating the function $s[w(s) - R(\infty)]/(s^2+\omega_a^2)$ around the basic contour with small semicircular indentations to the right at $s = \pm j\omega_a$, show that

$$R(\omega) = R(\infty) - \frac{2}{\pi}\int_0^\infty \frac{\omega X(\omega)}{\omega^2 - \omega_a^2}d\omega \qquad (6.222)$$

where, as before, $w(j\omega) = R(\omega) + jX(\omega)$.

6.14 Demonstrate that the gain-integral theorem (6.176) can be derived by integrating the function $\ln[w(s)/w(\infty)]$ around the basic contour composed of the $j\omega$-axis and the infinite semicircle to the right.

Figure 6.38 A given feedback amplifier.

6.15 By using the hybrid-pi model for the transistor in Fig. 6.38 with parameters as specified in Eq. (6.81), investigate the effect of varying the capacitance C_f on the natural frequencies of the amplifier.

6.16 A three-stage common-emitter feedback amplifier has the return ratio

$$T(s) = \frac{T(0)}{(1 + s/\sigma_1)(1 + s/\sigma_2)(1 + s/\sigma_3)} \tag{6.223}$$

where ($\sigma_i = 1, 2, 3$) are real and positive. Determine the midband value $T(0)$ of the return ratio under which the amplifier is on the verge of instability.

6.17 In (6.195), demonstrate that

$$\lim_{x \to 0} \ln x \ln y = 0 \tag{6.224}$$

where y is defined in (6.193).

6.18 Assuming that γ is small when compared with unity, we can expand $\log(1-\gamma)$ in a power series and ignore powers of γ higher than the first. Show that (6.212) can be approximated by the expression

$$A_m - A = (A_m + 17.4)\gamma + \frac{k-2}{k}G_m + \frac{2}{k}\gamma G_m \tag{6.225}$$

where $A = 20 \log |T(0)|$ and

$$A_m = A\big|_{\gamma=0}^{G_m=0} = 0 \tag{6.226}$$

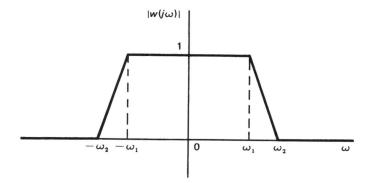

Figure 6.39 The prescribed magnitude of a minimum-phase transfer function.

6.19 Demonstrate that the phase-integral theorem (6.175) can be derived by integrating the function $[\ln w(s)]/s$ around the basic contour with a small semicircular indentation to the right at the origin.

6.20 Assume that the magnitude of a transfer function is given by the nth-order maximally flat response

$$|w(j\omega)| = (1 + \omega^{2n})^{-1/2} \qquad (6.227)$$

Show that the phase of the function is given by

$$B(\omega) = -\frac{n}{\pi} \int_0^\infty u^{-1}(1 + u^{-2n})^{-1} \ln \left| \frac{u + \omega}{u - \omega} \right| du \qquad (6.228)$$

6.21 The magnitude of a minimum-phase transfer function is sketched in Fig. 6.39. By making appropriate approximations, sketch the corresponding phase as a function of ω.

6.22 As discussed in Sec. 4.2.1, Fig. 4.7 is a current-series feedback amplifier. By using the hybrid-pi model of Fig. 4.10 for the transistor, with

$$g_m = 0.4 \text{ mho} \quad r_x = 50\,\Omega \quad r_\pi = 100\,\Omega$$
$$C_\pi = 200 \text{ pF} \quad C_\mu = 5 \text{ pF} \quad r_o = 50 \text{ k}\Omega \qquad (6.229)$$
$$R_1 = 500\,\Omega \quad R_2 = 5 \text{ k}\Omega \quad R_e = 50\,\Omega$$

calculate the return ratio with respect to the controlling parameter g_m over a sufficiently large range of frequencies, so that the return ratio locus can be plotted. By using the Nyquist criterion, determine the stability of this feedback amplifier.

6.23 Sketch the root locus for the return ratio considered in Prob. 6.22, assuming that its midband value can be varied independently.

6.24 The closed-loop transfer function of a feedback amplifier is written as

$$w(s) = K \frac{\Pi_{i=1}^{m}(s + z_i)}{\Pi_{j=1}^{n}(s + s_j)} \tag{6.230}$$

Assume that K, z_j, and s_j are functions of some parameter x. Show that

$$S_x^w = S_x^K + \sum_{i=1}^{m} \frac{\mathcal{S}(z_i; x)}{s + z_i} - \sum_{j=1}^{n} \frac{\mathcal{S}(s_j; x)}{s + s_j} \tag{6.231}$$

where S_x^w and S_x^K denote the sensitivity functions of $w(s)$ and K with respect to x, respectively.

BIBLIOGRAPHY

Barnett, S. and D. D. Šiljak: Routh's Algorithm: A Centennial Survey, *SIAM Review*, vol. 19, no. 3, pp. 472–489, 1977.

Blecher, F. H.: Design Principles for Single Loop Transistor Feedback Amplifiers, *IRE Trans. Circuit Theory*, vol. CT-4, no. 3, pp. 145–156, 1957.

Bode, H. W.: "Network Analysis and Feedback Amplifier Design," Princeton, N.J.: Van Nostrand, 1945.

Boylestad, R. and L. Nashelsky: "Electronic Devices and Circuit Theory," 2d ed., Englewood Cliffs, N.J.: Prentice-Hall, 1978.

Chen, C. F. and C. Hsu: The Determination of Root Loci Using Routh's Algorithm, *J. Franklin Inst.*, vol. 281, no. 2, pp. 114–121, 1966.

Chen, C. F. and L. S. Shieh: Continued Fraction Inversion by Routh's Algorithm, *IEEE Trans. Circuit Theory*, vol. CT-16, no. 2, pp. 197–202, 1969.

Chen, W. K.: Indefinite-Admittance Matrix Formulation of Feedback Amplifier Theory, *IEEE Trans. Circuits and Systems*, vol. CAS-23, no. 8, pp. 498–505, 1976.

Cherry, E. M.: A New Result in Negative-Feedback Theory, and Its Application to Audio Power Amplifiers, *Int. J. Circuit Theory and Applications*, vol. 6, no. 3, pp. 265–288, 1978.

Evans, W. R.: Graphical Analysis of Control Systems, *AIEE Trans.*, vol. 67, part II, pp. 547–551, 1948.

Evans, W. R.: "Control System Dynamics," New York: McGraw-Hill, 1954.

Fuller, A. T., ed.: "Stability of Motion," London: Taylor & Francis, 1975.

Gantmacher, F. R.: "The Theory of Matrices," vol. 2, New York: Chelsea, 1959.

Ghausi, M. S.: "Electronic Circuits," New York: Van Nostrand Reinhold, 1971.

Ghausi, M. S. and D. O. Pederson: A New Design Approach for Feedback Amplifiers, *IRE Trans. Circuit Theory*, vol. CT-8, no. 3, pp. 274–284, 1961.

Guillemin, E. A.: "The Mathematics of Circuit Analysis," Cambridge, Mass.: The MIT Press, 1949.

Haykin, S. S.: "Active Network Theory," Reading, Mass.: Addison-Wesley, 1970.

Horowitz, I. M.: "Synthesis of Feedback Systems," New York: Academic, 1963.

Hurwitz, A.: Über die Bedingungen, unter welchen eine Gleichung nur Wurzeln mit negativen reellen Teilen besitzt, *Math. Ann.*, vol. 46, pp. 273–284, 1895.

Krall, A. M.: An Extension and Proof of the Root-Locus Method, *J. Soc. Indust. Appl. Math.*, vol. 9, no. 4, pp. 644–653, 1961.

Kuh, E. S. and R. A. Rohrer: "Theory of Linear Active Networks," San Francisco, Calif.: Holden-Day, 1967.

Liénard, A. and M. H. Chipart: Sur la Signe de la Partie Réelle des Racines d'une Equation Algébrique, *J. Math. Pures Appl.*, vol. 10, pp. 291–346, 1914.

Mulligan, J. H.: Transient Response and the Stabilization of Feedback Amplifiers, *AIEE Trans.*, vol. 78, part II, pp. 495–503, 1960.

Nyquist, H.: Regeneration Theory, *Bell Syst. Tech. J.*, vol. 11, no. 1, pp. 126–147, 1932.

Porter, B.: "Stability Criteria for Linear Dynamical Systems," New York: Academic, 1968.

Routh, E. J.: "A Treatise on the Stability of a Given State of Motion," London: Macmillan, 1877.

Routh, E. J.: "Advanced Part of a Treatise on Advanced Rigid Dynamics," Cambridge: Cambridge University Press, 1930.

Truxal, J. G.: "Automatic Feedback Control System Synthesis," New York: McGraw-Hill, 1955.

Truxal, J. G. and I. M. Horowitz: Sensitivity Considerations in Active Network Synthesis, *Proc. Second Midwest Symp. Circuit Theory, Michigan State University, East Lansing, Mich., Dec.*, pp. 6-1-6-11, 1956.

Ur, H.: Root Locus Properties and Sensitivity Relations in Control Systems, *IRE Trans. Automatic Control*, vol. AC-5, no. 1, pp. 57–65, 1960.

SEVEN

MULTIPLE–LOOP FEEDBACK AMPLIFIERS

In the preceding three chapters, we studied the theory of single-loop feedback amplifiers. The concept of feedback was introduced in terms of return difference. We found that return difference plays an important role in the study of amplifier stability, its sensitivity to the variations of the parameters, and the determination of its transfer and driving-point impedances. The fact that return difference can be measured experimentally for many practical amplifiers indicates that we can include all the parasitic effects in the stability study, and that stability problem can be reduced to a Nyquist plot.

In this chapter, we study multiple-loop feedback amplifiers, which contain a multiplicity of inputs, outputs, and feedback loops. We first review briefly the rules of the matrix signal-flow graph, and then generalize the concept of return difference for a controlled source to the notion of return difference matrix for a multiplicity of controlled sources. For measurement situations, we introduce the null return difference matrix and discuss its physical significance. In particular, we show that the determinant of the overall transfer matrix can be expressed explicitly in terms of the determinants of the return difference and the null return difference matrices, thus generalizing Blackman's formula for the input impedance. This is followed by the derivations of the generalized feedback formulas and the formulation of the multiple-loop feedback theory in terms of the hybrid matrix. The problem of multiparameter sensitivity together with its relation to the return difference matrix is discussed. Finally, we develop formulas for computing multiparameter sensitivity functions.

7.1 MATRIX SIGNAL-FLOW GRAPHS

As demonstrated in Sec. 5.7, the feedback amplifier theory can be formulated equivalently in terms of a signal-flow graph representing the signal transmissions among the various ports. To formulate the multiple-loop feedback amplifier theory, we need the matrix version of the scalar signal-flow graph. In this section, we review briefly the matrix signal-flow graph and present some of its properties.

A scalar signal-flow graph is a graphic representation of a system of linear algebraic equations, which is put into the form

$$\sum_{j=0}^{n} a_{ij}x_j = x_i \quad (i = 1, 2, \ldots, n) \tag{7.1}$$

where x_0 is the only excitation source and x_i are the dependent variables. The scalar signal-flow graph associated with the system (7.1) is a weighted directed graph in which the nodes correspond to the dependent variables x_i and the independent variable x_0 and the weights to the *transmittances* a_{ij}. The construction of the associated scalar signal-flow graph for the ith equation is shown in Fig. 7.1. The *graph transmission* w_{0k} from x_0 to x_k is defined by the equation $x_k = w_{0k}x_0$. Application of Mason's gain formula yields the graph transmission w_{0k} by inspection of the loops and paths of the graph.

In a similar fashion, a *matrix signal-flow graph* is a directed-graph representation of a system of linear algebraic matrix equations, which is put into the form

$$\sum_{j=0}^{n} \mathbf{A}_{ij}\mathbf{X}_j = \mathbf{X}_i \quad (i = 1, 2, \ldots, n) \tag{7.2}$$

where $\mathbf{X}_0$ is the only source vector and $\mathbf{X}_i$ are the dependent vectors. The nodes of the directed graph correspond to the independent vector $\mathbf{X}_0$ and the dependent

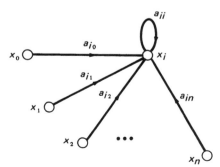

Figure 7.1 The scalar signal-flow graph for the ith equation of (7.1).

vectors $\mathbf{X}_i$ and the weights of the directed edges to the matrices $\mathbf{A}_{ij}$. To each node $\mathbf{X}_i$, there corresponds the matrix equation (7.2). The construction of the associated matrix signal-flow graph is as illustrated in Fig. 7.2 for the ith equation in Eq. (7.2). The *graph transmission* $\mathbf{W}_{0k}$ from the source node $\mathbf{X}_0$ to the output node $\mathbf{X}_k$ is defined by the matrix equation

$$\mathbf{X}_k = \mathbf{W}_{0k}\mathbf{X}_0 \tag{7.3}$$

In general, Mason's gain formula cannot be utilized to obtain the graph transmission $\mathbf{W}_{0k}$.

As an example of a matrix signal-flow graph, consider the following system of linear matrix equations:

$$\mathbf{X}_1 = \mathbf{AX}_2 + \mathbf{BX}_0 \tag{7.4a}$$

$$\mathbf{X}_3 = \mathbf{CX}_2 + \mathbf{DX}_0 \tag{7.4b}$$

$$\mathbf{X}_2 = \mathbf{EX}_1 \tag{7.4c}$$

The associated matrix signal-flow graph is presented in Fig. 7.3.

One way to compute the graph transmission is to apply the basic reduction rules repeatedly until it can be written down directly by inspection. A direct approach is to develop a topological procedure to obtain the graph transmission without

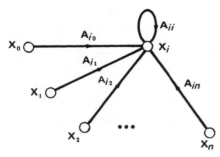

Figure 7.2 The matrix signal-flow graph for the ith equation of (7.2).

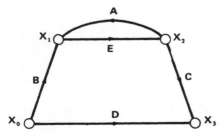

Figure 7.3 The matrix signal-flow graph associated with the system (7.4).

graph reduction. The best approach is probably a combination of both, so that the problem at hand can be solved in the most satisfying manner. In the following, we shall describe these two approaches.

7.1.1 Basic Reduction Rules

The basic reduction rules for a matrix signal-flow graph are given below.

1. *Series reduction.* As indicated in Fig. 7.4a, the directed edges with transmittances $\mathbf{A}_{ij}$ and $\mathbf{A}_{jk}$ are replaced by a single directed edge with transmittance $\mathbf{A}_{ij}\mathbf{A}_{jk}$.
2. *Parallel reduction.* As indicated in Fig. 7.4b, the directed edges with transmittances $\mathbf{A}_{ij}$ and $\mathbf{B}_{ij}$ are replaced by a single directed edge with transmittance $\mathbf{A}_{ij} + \mathbf{B}_{ij}$.
3. *Removal of a self-loop.* The removal of a self-loop with transmittance $\mathbf{A}_{ii}$ at node $\mathbf{X}_i$ is equivalent to replacing the transmittances $\mathbf{A}_{ij}$ of the edges directed from nodes $\mathbf{X}_j$ (where $j \neq i$; $j = 0, 1, \ldots, n$) to node $\mathbf{X}_i$ by $(\mathbf{1} - \mathbf{A}_{ii})^{-1}\mathbf{A}_{ij}$, where $\mathbf{1}$, as before, denotes the identity matrix of appropriate order. To justify this operation, we observe that the ith equation of (7.2) can be rewritten as

$$\sum_{\substack{j=0 \\ j \neq i}}^{n}(\mathbf{1} - \mathbf{A}_{ii})^{-1}\mathbf{A}_{ij}\mathbf{X}_j = \mathbf{X}_i \qquad (7.5)$$

and the result follows. The procedure is illustrated in Fig. 7.5.

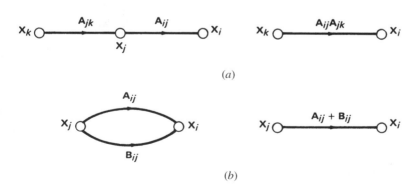

(a)

(b)

Figure 7.4 The basic matrix signal-flow graph reduction rules, (a) Series reduction and (b) parallel reduction.

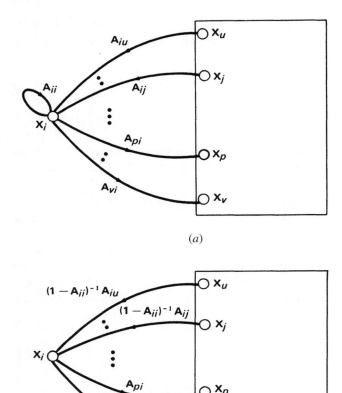

(a)

(b)

Figure 7.5 The removal of a self-loop in a matrix signal-flow graph. (a) Before the removal and (b) after the removal.

4. *Absorption of a node.* To absorb a node $\mathbf{X}_k$, we replace the transmittances $\mathbf{A}_{ij}$ of the edges directed from $\mathbf{X}_j$ to $\mathbf{X}_i$ by

$$\tilde{\mathbf{A}}_{ij} = \mathbf{A}_{ij} + \mathbf{A}_{ik}(\mathbf{1} - \mathbf{A}_{kk})^{-1}\mathbf{A}_{kj} \qquad (7.6)$$

for all i and j. The operation is depicted in Fig. 7.6, and is equivalent to eliminating the vector $\mathbf{X}_k$ in (7.2). To see this, we replace i by k in (7.5) and substitute the resulting equation for $\mathbf{X}_k$ in (7.2),

giving

$$\sum_{\substack{j=0 \\ j \neq k}}^{n} \tilde{\mathbf{A}}_{ij} \mathbf{X}_j = \mathbf{X}_i \quad (i = 1, 2, \ldots, k-1, k+1, \ldots, n) \tag{7.7}$$

The operations described in Fig. 7.6 follow directly from Eq. (7.7).

As an illustration, we compute the graph transmission from node $\mathbf{X}_0$ to node $\mathbf{X}_3$ in the matrix flow graph of Fig. 7.3. To this end, we first remove node $\mathbf{X}_2$ by the procedure outlined in Fig. 7.6; the resulting matrix flow graph is shown in Fig. 7.7a. Repeating the process once more by removing node $\mathbf{X}_1$ in Fig. 7.7a yields the reduced matrix flow graph of Fig. 7.7b. The graph transmission from node $\mathbf{X}_0$ to node $\mathbf{X}_3$ is found to be

$$\mathbf{W}_{03} = \mathbf{D} + \mathbf{CE}(\mathbf{1} - \mathbf{AE})^{-1}\mathbf{B} \tag{7.8}$$

(a)

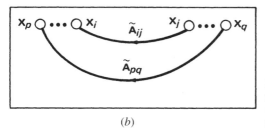

(b)

Figure 7.6 The absorption of a node in a matrix signal-flow graph. (a) Before the absorption and (b) after the absorption.

(a)

(b)

Figure 7.7 (a) The resulting matrix signal-flow graph after absorbing node $\mathbf{X}_2$ in that of Fig. 7.3 and (b) the resulting matrix signal-flow graph after absorbing node $\mathbf{X}_1$, in (a).

Alternatively, in Fig. 7.7a we can first remove the self-loop at node $\mathbf{X}_1$ by the procedure outlined in Fig. 7.5 and then apply the series and parallel reduction rules. The results are presented in Fig. 7.8, and we obtain, of course, the same graph transmission.

7.1.2 Topological Procedure

In this section, we describe a topological procedure that avoids the necessity of repeated graph reduction to obtain the desired graph transmission. To facilitate our discussion, we introduce several terms.

A node is said to be *split* when it is replaced by two nodes, a source node and a sink node, such that all edges terminating at the original node are made to terminate on the new sink node and all edges outgoing from the original node are made to originate at the new source. The *node transmission* $\mathbf{W}_i$ of a node $\mathbf{X}_i$ is the graph transmission from the source node to the sink node created by splitting node $\mathbf{X}_i$ after the original source node $\mathbf{X}_0$ is removed. The *directed path product* of a directed path from node $\mathbf{X}_i$ to node $\mathbf{X}_j$ is the product of the transmittances associated with the edges of the directed path written in the order from node $\mathbf{X}_j$ to node $\mathbf{X}_i$. Finally, the *node factor* of a node $\mathbf{X}_i$ on a directed path P is $(1 - \mathbf{W}_i)^{-1}$, where $\mathbf{W}_i$ is the node transmission calculated under the condition that all nodes on the directed path P between node $\mathbf{X}_i$ and the output node are split. Clearly, the node factor is dependent on the choice of the directed path.

(a)

(b)

(c)

Figure 7.8 The reduction of the matrix signal-flow graph of Fig. 7.7a.

With these preliminaries, the following steps yield the graph transmission from the source node X_0 to any dependent node X_j in a matrix signal-flow graph:

1. Determine all the directed paths from node X_0 to node X_j.
2. For each directed path P, compute the node factors of the nodes X_i on P. A *modified directed path product* is formed from the directed path product of P by inserting the node factor of every node X_i on P between the transmittances of the edges incident at node X_i in P.
3. The graph transmission from node X_0 to node X_j is equal to the sum of the modified directed path products of all the directed paths from node X_0 to node X_j.

We illustrate the above procedure by the following example.

Example 7.1 We compute the graph transmission from node X_0 to node X_3 in the matrix signal-flow graph of Fig. 7.3 by the procedure outlined above.

1. There are two directed paths from X_0 to X_3. The directed path products are given by **D** and **CEB**.
2. As shown in Fig. 7.9, for the directed path X_0, X_1, X_2, and X_3 the node factors of nodes X_1 and X_2 are given by $(\mathbf{1} - \mathbf{0})^{-1}$ and $(\mathbf{1} - \mathbf{EA})^{-1}$, respectively. The

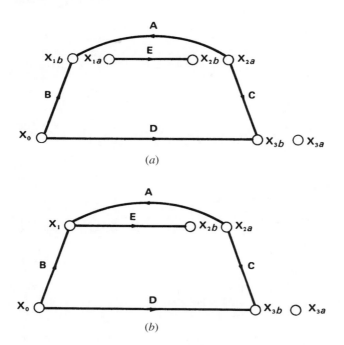

Figure 7.9 The computation of node factors for the directed path $\mathbf{X}_0, \mathbf{X}_1, \mathbf{X}_2$, and $\mathbf{X}_3$ for (a) node $\mathbf{X}_1$ and (b) node $\mathbf{X}_2$.

modified directed path product becomes $\mathbf{C}(\mathbf{1} - \mathbf{EA})^{-1}\mathbf{EB}$. The other modified directed path product is simply $\mathbf{D}$.

3. The graph transmission is found to be

$$\mathbf{W}_{03} = \mathbf{D} + \mathbf{C}(\mathbf{1} - \mathbf{EA})^{-1}\mathbf{EB} \qquad (7.9)$$

where $\mathbf{X}_3 = \mathbf{W}_{03}\mathbf{X}_0$. It is straightforward to demonstrate that (7.8) and (7.9) are equivalent (see Prob. 7.1). We remark that the identity matrix in (7.8) is of order equal to the number of rows of $\mathbf{A}$, whereas the one in (7.9) is of order equal to the number of rows of $\mathbf{E}$. Thus, their orders are generally different but the final products are the same.

7.2 THE MULTIPLE-LOOP FEEDBACK AMPLIFIER THEORY

The general configuration of a multiple-input, multiple-output, and multiple-feedback loop amplifier is shown in Fig. 7.10, in which the input, output,

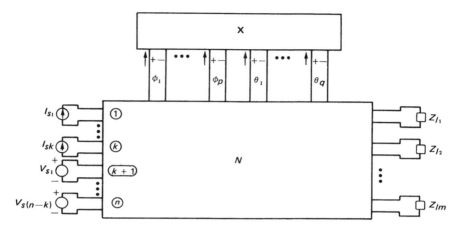

Figure 7.10 The general configuration of a multiple-input, multiple-output, and multiple-feedback loop amplifier.

and feedback variables may be either currents or voltages. For the specific arrangement as indicated in Fig. 7.10, the input and output variables can be represented by an n-dimensional vector $\mathbf{u}$ and an m-dimensional vector $\mathbf{y}$ as

$$\mathbf{u}(s) = \begin{bmatrix} u_1 \\ u_2 \\ \vdots \\ u_k \\ u_{k+1} \\ u_{k+2} \\ \vdots \\ u_n \end{bmatrix} = \begin{bmatrix} I_{s1} \\ I_{s2} \\ \vdots \\ I_{sk} \\ V_{s1} \\ V_{s2} \\ \vdots \\ V_{s(n-k)} \end{bmatrix} \qquad \mathbf{y}(s) = \begin{bmatrix} y_1 \\ y_2 \\ \vdots \\ y_r \\ y_{r+1} \\ y_{r+2} \\ \vdots \\ \mathbf{y}_m \end{bmatrix} = \begin{bmatrix} I_1 \\ I_2 \\ \vdots \\ I_r \\ V_{r+1} \\ V_{r+2} \\ \vdots \\ V_m \end{bmatrix} \qquad (7.10)$$

respectively. The elements of interest can be represented by a rectangular matrix $\mathbf{X}$ of order $q \times p$ relating the controlled and controlling variables by the matrix

Figure 7.11 A feedback network containing two current-controlled voltage sources.

equation

$$\boldsymbol{\theta} = \begin{bmatrix} \theta_1 \\ \theta_2 \\ \vdots \\ \theta_q \end{bmatrix} = \begin{bmatrix} x_{11} & x_{12} & \cdots & x_{1p} \\ x_{21} & x_{22} & \cdots & x_{2p} \\ \vdots & \cdots & \cdots & \cdots \\ x_{q1} & x_{q2} & \cdots & x_{qp} \end{bmatrix} \begin{bmatrix} \phi_1 \\ \phi_2 \\ \vdots \\ \phi_p \end{bmatrix} = \mathbf{X}\boldsymbol{\phi} \qquad (7.11)$$

where the p-dimensional vector $\boldsymbol{\phi}$ is called the *controlling vector*, and the q-dimensional vector $\boldsymbol{\theta}$ the *controlled vector*. As before, the controlled variables θ_k and the controlling variables ϕ_k can be either currents or voltages. The matrix $\mathbf{X}$ can represent either a transfer-function matrix or a driving-point function matrix. If $\mathbf{X}$ represents a driving-point function matrix, the vectors $\boldsymbol{\theta}$ and $\boldsymbol{\phi}$ are of the same dimension ($q = p$) and their components are the currents and voltages of a p-port network.

As an illustration, consider the feedback network of Fig. 7.11, which was considered in Example 5.16. Suppose that we are interested in the effects of element Y_1 and the controlling parameters $\alpha_1, \alpha_2, \beta_1$, and β_2 on the whole network. We choose $Y_1, \alpha_1, \alpha_2, \beta_1$, and β_2 as the elements of $\mathbf{X}$, and Eq. (7.11) becomes (see Probs. 7.10 and 7.20)

$$\boldsymbol{\theta} = \begin{bmatrix} I_1 \\ V_a \\ V_b \end{bmatrix} = \begin{bmatrix} Y_1 & 0 & 0 \\ 0 & \beta_1 & \beta_2 \\ 0 & \alpha_1 & \alpha_2 \end{bmatrix} \begin{bmatrix} V_1 \\ I_2 \\ I_3 \end{bmatrix} = \mathbf{X}\boldsymbol{\phi} \qquad (7.12)$$

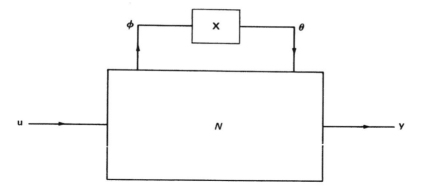

Figure 7.12 The block diagram of the general feedback configuration of Fig. 7.10.

The general configuration of Fig. 7.10 can be represented equivalently by the block diagram of Fig. 7.12, in which N is a $(p + q + m + n)$-port network and the elements of interest are exhibited explicitly by the block **X**. Consider the $(p + q + m + n)$-port network N itself. The vectors **u** and $\boldsymbol{\theta}$ are the inputs to N, whereas the vectors $\boldsymbol{\phi}$ and **y** are the outputs of N. Since N is linear, the input and output vectors are related by the matrix equations

$$\boldsymbol{\phi} = \mathbf{A}\boldsymbol{\theta} + \mathbf{B}\mathbf{u} \tag{7.13a}$$

$$\mathbf{y} = \mathbf{C}\boldsymbol{\theta} + \mathbf{D}\mathbf{u} \tag{7.13b}$$

where **A**, **B**, **C**, and **D** are transfer-function matrices of orders $p \times q$, $p \times n$, $m \times q$, and $m \times n$, respectively. The vectors $\boldsymbol{\theta}$ and $\boldsymbol{\phi}$ are not independent and, from (7.11), are related by

$$\boldsymbol{\theta} = \mathbf{X}\boldsymbol{\phi} \tag{7.14}$$

The above system of three linear matrix equations can be represented by a matrix signal-flow graph as shown in Fig. 7.13, which is known as the *fundamental matrix feedback-flow graph*. The graph is identical to that shown in Fig. 7.3 used to illustrate the matrix signal-flow graph reduction rules. The overall closed-loop *transfer-function matrix* of the multiple-loop feedback amplifier is defined by the equation

$$\mathbf{y} = \mathbf{W}(\mathbf{X})\mathbf{u} \tag{7.15}$$

where $\mathbf{W}(\mathbf{X})$ is of order $m \times n$. As before, to emphasize the importance of **X**, the matrix **W** is written as $\mathbf{W}(\mathbf{X})$ for the present discussion, even though it is also

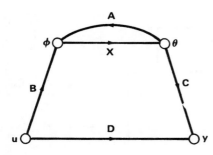

Figure 7.13 The fundamental matrix feedback-flow graph.

a function of the complex-frequency variable s. By using the matrix flow-graph reduction rules or the topological procedure discussed in the preceding section, the transfer-function matrix is found to be

$$\mathbf{W}(\mathbf{X}) = \mathbf{D} + \mathbf{CX}(\mathbf{1}_p - \mathbf{AX})^{-1}\mathbf{B} \qquad (7.16a)$$

or

$$\mathbf{W}(\mathbf{X}) = \mathbf{D} + \mathbf{C}(\mathbf{1}_q - \mathbf{XA})^{-1}\mathbf{XB} \qquad (7.16b)$$

where $\mathbf{1}_p$ denotes the identity matrix of order p. Clearly, we have

$$\mathbf{W}(\mathbf{0}) = \mathbf{D} \qquad (7.17)$$

In the particular situation where $\mathbf{X}$ is square and nonsingular, (7.16) can be written as

$$\mathbf{W}(\mathbf{X}) = \mathbf{D} + \mathbf{C}(\mathbf{X}^{-1} - \mathbf{A})^{-1}\mathbf{B} \qquad (7.18)$$

We illustrate the above procedure by the following example.

Example 7.2 Consider the voltage-series feedback amplifier of Fig. 7.14, which was discussed in Example 4.11. Assume that the two transistors are identical, with $h_{ie} = 1.1$ kΩ, $h_{fe} = 50$, $h_{re} = h_{oe} = 0$. The equivalent network of the amplifier is presented in Fig. 7.15. Let the controlling parameters of the two controlled current sources be the elements of interest. Then we have

$$\theta = \begin{bmatrix} I_a \\ I_b \end{bmatrix} = 10^{-4} \begin{bmatrix} 455 & 0 \\ 0 & 455 \end{bmatrix} \begin{bmatrix} V_{13} \\ V_{45} \end{bmatrix} = \mathbf{X}\phi \qquad (7.19)$$

Figure 7.14 A voltage-series feedback amplifier.

Assume that the output voltage V_{25} and input current I_{51} are the output variables. Then the seven-port network N defined by the variables V_{13}, V_{45}, V_{25}, I_{51}, I_a, I_b, and V_s can be characterized by the matrix equations

$$\boldsymbol{\phi} = \begin{bmatrix} V_{13} \\ V_{45} \end{bmatrix} = \mathbf{A}\boldsymbol{\theta} + \mathbf{B}\mathbf{u} = \begin{bmatrix} a_{11} & a_{12} \\ a_{21} & a_{22} \end{bmatrix} \begin{bmatrix} I_a \\ I_b \end{bmatrix} + \begin{bmatrix} b_{11} \\ b_{21} \end{bmatrix} [V_s] \qquad (7.20a)$$

$$\mathbf{y} = \begin{bmatrix} V_{25} \\ V_{51} \end{bmatrix} = \mathbf{C}\boldsymbol{\theta} + \mathbf{D}\mathbf{u} = \begin{bmatrix} c_{11} & c_{12} \\ c_{21} & c_{22} \end{bmatrix} \begin{bmatrix} I_a \\ I_b \end{bmatrix} + \begin{bmatrix} d_{11} \\ d_{21} \end{bmatrix} [V_s] \qquad (7.20b)$$

To compute a_{11}, a_{21}, c_{11}, and c_{21}, we set $I_b = 0$, $V_s = 0$, and $I_a = 1$ A. The voltages V_{13}', V_{45}', and V_{25}' and current I_{51}' in the resulting network as shown in Fig. 7.16 are numerically equal to a_{11}, a_{21}, c_{11}, and c_{21}, respectively. The results are given by

$$a_{11} = -90.782\ \Omega \quad a_{21} = -942.507\ \Omega$$
$$c_{11} = 45.391\ \Omega \quad c_{21} = -0.08252 \qquad (7.21)$$

For a_{12}, a_{22}, c_{12}, and c_{22}, we set $I_a = 0$, $V_s = 0$, and $I_b = 1$ A. The voltages V_{13}'', V_{45}'', and V_{25}'' and current I_{51}'' in the resulting network of Fig. 7.17 are numerically equal to a_{12}, a_{22}, c_{12}, and c_{22}, respectively. The element values are found to be

$$a_{12} = 45.391\ \Omega \quad a_{22} = 0$$
$$c_{12} = -2372.32\ \Omega \quad c_{22} = 0.04126 \qquad (7.22)$$

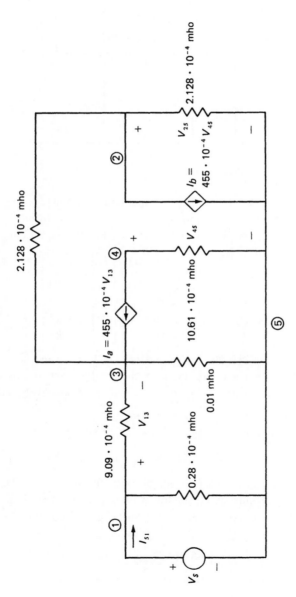

Figure 7.15   An equivalent network of the voltage-series feedback amplifier of Fig. 7.14.

Figure 7.16 The network used for the calculation of a_{11}, a_{21}, c_{11} , and c_{21}.

Figure 7.17 The network used for the calculation of a_{12}, a_{22}, c_{12}, and c_{22}.

Finally, to compute b_{11}, b_{21}, d_{11}, and d_{21}, we set $I_a = I_b = 0$ and $V_s = 1$ V, as shown in Fig. 7.18, and calculate $\tilde{V}_{13}, \tilde{V}_{45}, \tilde{V}_{25}$, and $\tilde{I}_{51}$, respectively. The results are given by

$$b_{11} = 0.91748 \quad b_{21} = 0$$
$$d_{11} = 0.04126 \quad d_{21} = 0.000862 \text{ mho}$$

(7.23)

Substituting these in (7.20), we obtain

$$\begin{bmatrix} V_{13} \\ V_{45} \end{bmatrix} = \begin{bmatrix} -90.782 & 45.391 \\ -942.507 & 0 \end{bmatrix} \begin{bmatrix} I_a \\ I_b \end{bmatrix} + \begin{bmatrix} 0.91748 \\ 0 \end{bmatrix} [V_s] \qquad (7.24a)$$

$$\begin{bmatrix} V_{25} \\ I_{51} \end{bmatrix} = \begin{bmatrix} 45.391 & -2372.32 \\ -0.08252 & 0.04126 \end{bmatrix} \begin{bmatrix} I_a \\ I_b \end{bmatrix} + \begin{bmatrix} 0.041260 \\ 0.000862 \end{bmatrix} [V_s] \qquad (7.24b)$$

Figure 7.18 The network used for the calculation of b_{l1}, b_{21}, d_{11}, and d_{21}.

According to (7.15), the transfer-function matrix of the amplifier is defined by the matrix equation

$$\mathbf{y} = \begin{bmatrix} V_{25} \\ I_{51} \end{bmatrix} = \begin{bmatrix} w_{11} \\ w_{21} \end{bmatrix} [V_s] = \mathbf{W}(\mathbf{X})\mathbf{u} \tag{7.25}$$

Since $\mathbf{X}$ is square and nonsingular, we can use (7.18) to calculate $\mathbf{W}(\mathbf{X})$, giving

$$\mathbf{W}(\mathbf{X}) = \mathbf{D} + \mathbf{C}(\mathbf{X}^{-1} - \mathbf{A})^{-1}\mathbf{B}$$

$$= \begin{bmatrix} 0.041260 \\ 0.000862 \end{bmatrix} + \begin{bmatrix} 45.391 & -2372.32 \\ -0.08252 & 0.04126 \end{bmatrix} \begin{bmatrix} 4.856 & 10.029 \\ -208.245 & 24.914 \end{bmatrix}$$

$$\times \begin{bmatrix} 0.91748 \\ 0 \end{bmatrix} \cdot 10^{-4} = \begin{bmatrix} 45.387 \\ 0.369 \cdot 10^{-4} \end{bmatrix} = \begin{bmatrix} w_{11} \\ w_{21} \end{bmatrix} \tag{7.26a}$$

where

$$(\mathbf{X}^{-1} - \mathbf{A})^{-1} = \begin{bmatrix} 4.856 & 10.029 \\ -208.245 & 24.914 \end{bmatrix} \cdot 10^{-4} \tag{7.26b}$$

Thus, the closed-loop voltage gain of the amplifier is given by $w_{11} = V_{25}/V_s = 45.387$, confirming Eq. (4.139). The input impedance facing the voltage source V_s, being the reciprocal of w_{21}, is given by

$$\frac{V_s}{I_{51}} = \frac{1}{w_{21}} = 27.1 \text{ k}\Omega \tag{7.27}$$

To verify this result, we refer to the indefinite-admittance matrix $\mathbf{Y}$ given in Eq. (4.138). By applying formula (2.95), the input impedance facing V_s is found to be

$$z_{11,55} = \frac{V_s}{I_{51}} = \frac{Y_{11,55}}{Y_{uv}} = \frac{466,069 \cdot 10^{-12}}{171,983 \cdot 10^{-16}} = 27.1 \text{ k}\Omega \tag{7.28}$$

confirming (7.27), where

$$Y_{uv} = Y_{55} = 171,983 \cdot 10^{-16} \tag{7.29}$$

7.2.1 The Return Difference Matrix

In the study of a single-loop feedback amplifier, we usually single out an element for particular attention. The element is generally one that is either crucial in terms of its effect on the entire system or of main concern to the designer. For a multiple-loop feedback amplifier, instead of singling out an element, we pay particular attention to a group of elements represented by the matrix $\mathbf{X}$ and study its effects on the whole system. In this section, we generalize the concepts of return difference by introducing the return difference matrix.

As indicated in Sec. 5.7, if we break the branch with transmittance x in the fundamental feedback-flow graph of Fig. 5.17, the negative of the signal returned to the left of the breaking mark as shown in Fig. 5.18 when a unit signal is applied to the right of the breaking mark and when the input excitation is set to zero is the return ratio for the element x. Instead of applying a unit signal, we can apply any signal g to the right of the breaking mark. Then the ratio of the returned signal h to the applied signal g must be the negative of the return ratio $T(x)$ for the element x, or $h = -T(x)g$ because the network is assumed to be linear. The difference between the applied signal g and the returned signal h defines the return difference $F(x)$ with respect to the element x:

$$g - h = [1 + T(x)]g = F(x)g \tag{7.30}$$

giving $F(x) = 1 + T(x)$.

With the above interpretations, the extension of these concepts to multiple-loop feedback amplifiers is now clear. Refer to the fundamental matrix feedback-flow graph of Fig. 7.13. Suppose that we break the input of the branch with transmittance $\mathbf{X}$, set the input excitation vector $\mathbf{u}$ to zero, and apply a signal p-vector $\mathbf{g}$ to the right of the breaking mark, as depicted in Fig. 7.19. Then the returned signal p-vector $\mathbf{h}$ to the left of the breaking mark is found to be

$$\mathbf{h} = \mathbf{AXg} \tag{7.31}$$

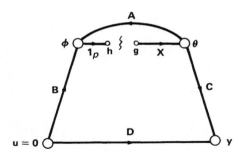

Figure 7.19 The physical interpretation of the loop-transmission matrix.

The square matrix $\mathbf{AX}$ is called the *loop-transmission matrix* and its negative is referred to as the *return ratio matrix*, denoted by the symbol $\mathbf{T(X)}$:

$$\mathbf{T(X)} = -\mathbf{AX} \tag{7.32}$$

The difference between the applied signal vector $\mathbf{g}$ and the returned signal vector $\mathbf{h}$ is given by

$$\mathbf{g} - \mathbf{h} = (\mathbf{1}_p - \mathbf{AX})\mathbf{g} \tag{7.33}$$

The square matrix $\mathbf{1}_p - \mathbf{AX}$ is defined as the *return difference matrix* with respect to $\mathbf{X}$ and is denoted by the symbol $\mathbf{F(X)}$:

$$\mathbf{F(X)} = \mathbf{1}_p - \mathbf{AX} \tag{7.34}$$

Combining this with (7.32) yields

$$\mathbf{F(X)} = \mathbf{1}_p + \mathbf{T(X)} \tag{7.35}$$

As an illustration, we compute the return difference matrix with respect to the matrix $\mathbf{X}$, as given in (7.19). From (7.24a), the return ratio matrix is found to be

$$\mathbf{T(X)} = -\mathbf{AX} = \begin{bmatrix} 4.131 & -2.065 \\ 42.884 & 0 \end{bmatrix} \tag{7.36}$$

giving the return difference matrix as

$$\mathbf{F(X)} = \mathbf{1}_2 + \mathbf{T(X)} = \begin{bmatrix} 5.131 & 2.065 \\ 42.884 & 1 \end{bmatrix} \tag{7.37}$$

As demonstrated in Sec. 4.3.1, the scalar return ratio $T(x)$ for a one-port admittance x is equal to the ratio of the admittance x to the admittance y that x faces, and the scalar return difference $F(x)$ with respect to a one-port admittance

x is equal to the ratio of the total admittance looking into the node pair where x is connected to the admittance that x faces:

$$F(x) = 1 + T(x) = 1 + \frac{x}{y} = y^{-1}(y + x) \qquad (7.38)$$

A similar interpretation can be made for the return ratio matrix and the return difference matrix if $\mathbf{X}$ denotes the admittance matrix of a p-port network. In this case, the controlling vector $\boldsymbol{\phi}$ denotes the p-port voltage vector and the controlled vector $\boldsymbol{\theta}$ the p-port current vector. By writing $\boldsymbol{\phi} = \mathbf{V}$ and $\boldsymbol{\theta} = \mathbf{I}$ with $\phi_j = V_j$ and $\theta_j = I_j = (j = 1, 2, \ldots, p)$ denoting the port voltages and currents, (7.11) can be written as

$$I_j = x_{j1}V_l + x_{j2}V_2 + \cdots + x_{jp}V_p \ (j = 1, 2, \ldots, p) \qquad (7.39)$$

As discussed in Sec. 4.3.1, a one-port admittance x can be represented equivalently by a voltage-controlled current source $I = xV$, the controlling voltage V being the terminal voltage of the current source. In a similar way, a p-port admittance matrix $\mathbf{X}$ can be represented equivalently by a vector voltage-controlled current source $\mathbf{I} = \mathbf{XV}$ whose jth component is given by (7.39). After removing the p-port network, we insert at its jth port a voltage-controlled current source $I_j = x_{j1}V_1 + x_{j2}V_2 + \cdots + x_{jp}V_p$ for $j = 1, 2, \ldots, p$. Now we break the input of each of the controlling branches, as depicted in Fig. 7.20, and apply a voltage source of g_j volts just above the breaking mark, the voltage $\tilde{V}_j$ appearing at the jth port beneath the breaking mark being the returned voltage. Denoting the applied voltages g_j by the p-vector $\mathbf{g}$ and the returned voltages $\tilde{V}_j$ by $\mathbf{h}$, we have

$$\mathbf{I} = \mathbf{Xg} \qquad (7.40a)$$

$$-\mathbf{I} = \mathbf{Yh} \qquad (7.40b)$$

where $\mathbf{Y}$ is the admittance matrix facing the current sources I_j of Fig. 7.20. This leads to

$$\mathbf{h} = -\mathbf{Y}^{-1}\mathbf{Xg} \qquad (7.41a)$$

$$\mathbf{g} - \mathbf{h} = (\mathbf{1}_p + \mathbf{Y}^{-1}\mathbf{X})\mathbf{g} \qquad (7.41b)$$

provided that $\mathbf{Y}$ is nonsingular, giving

$$\mathbf{T}(\mathbf{X}) = \mathbf{Y}^{-1}\mathbf{X} \qquad (7.42a)$$

$$\mathbf{F}(\mathbf{X}) = \mathbf{Y}^{-1}(\mathbf{Y} + \mathbf{X}) \qquad (7.42b)$$

Figure 7.20 The physical interpretation of the return difference matrix with respect to a p-port admittance matrix.

This is therefore a direct generalization of the scalar return difference of (7.38). The matrix $\mathbf{Y} + \mathbf{X}$ represents the total admittance matrix looking into the junctions of the p-port network and the multiport network N.

In the situation where $\mathbf{X}$ denotes the general hybrid matrix of a p-port network, a similar result can be obtained. Without loss of generality, we assume that $\mathbf{X}$ is characterized by the partitioned matrix equation

$$\boldsymbol{\theta} = \begin{bmatrix} \mathbf{V}_\alpha \\ \mathbf{I}_\beta \end{bmatrix} = \begin{bmatrix} \mathbf{X}_{11} & \mathbf{X}_{12} \\ \mathbf{X}_{21} & \mathbf{X}_{22} \end{bmatrix} \begin{bmatrix} \mathbf{I}_\alpha \\ \mathbf{V}_\beta \end{bmatrix} = \mathbf{X}\boldsymbol{\phi} \qquad (7.43)$$

where $[\mathbf{I}'_\alpha, \mathbf{I}'_\beta]'$ and $[\mathbf{V}'_\alpha, \mathbf{V}'_\beta]'$ denote the port current vector and the port voltage vector of the p-port network, respectively, and the primes, as before, denote the matrix transpose. As in the admittance case, the p-port network with hybrid matrix $\mathbf{X}$ can be replaced by a controlled vector source. For each controlled variable $\theta_j = V_j$ in $\mathbf{V}_\alpha$, we insert at the jth port a controlled voltage source $V_j = x_{j1}\phi_1 + x_{j2}\phi_2 + \cdots + x_{jp}\phi_p$, and for each $\theta_j = I_j$ in $\mathbf{I}_\beta$ we insert a controlled current source $I_j = x_{j1}\phi_1 + x_{j2}\phi_2 + \cdots + x_{jp}\phi_p$. We then break the input of each of the controlling branches and apply a signal of g_j volts or amperes, as the case may be, just above the breaking mark. The current or voltage appearing at the jth port beneath the breaking mark is the returned signal $h_j = \tilde{I}_j$ or $\tilde{V}_j$, as shown in

Figure 7.21 The physical interpretation of the return difference matrix with respect to a p-port general hybrid matrix.

Fig. 7.21. Denote by the partitioned p-vectors

$$\mathbf{g} = \begin{bmatrix} \tilde{\mathbf{I}}_\alpha \\ \tilde{\mathbf{V}}_\beta \end{bmatrix} \quad \mathbf{h} = \begin{bmatrix} \hat{\mathbf{I}}_\alpha \\ \hat{\mathbf{V}}_\beta \end{bmatrix} \tag{7.44}$$

the applied signals g_j and the returned signals h_j, respectively. Also, denote by the p-vector $\tilde{\boldsymbol{\theta}}$ the signals at the p ports resulting from the controlled sources, and write

$$\tilde{\boldsymbol{\theta}} = \begin{bmatrix} \tilde{\mathbf{V}}_\alpha \\ \tilde{\mathbf{I}}_\beta \end{bmatrix} \tag{7.45}$$

With these symbols, we have

$$\tilde{\boldsymbol{\theta}} = \begin{bmatrix} \tilde{\mathbf{V}}_\alpha \\ \tilde{\mathbf{I}}_\beta \end{bmatrix} = \begin{bmatrix} \mathbf{X}_{11} & \mathbf{X}_{12} \\ \mathbf{X}_{21} & \mathbf{X}_{22} \end{bmatrix} \begin{bmatrix} \tilde{\mathbf{I}}_\alpha \\ \tilde{\mathbf{V}}_\beta \end{bmatrix} = \mathbf{Xg} \tag{7.46}$$

For the multiport network N, we open-circuit all the independent current sources and short-circuit all the independent voltage sources. The p ports facing the p-port network $\mathbf{X}$ can be characterized by its general hybrid matrix in partitioned form,

$$\begin{bmatrix} \tilde{\mathbf{V}}_\alpha \\ -\tilde{\mathbf{I}}_\beta \end{bmatrix} = \begin{bmatrix} \mathbf{H}_{\alpha\alpha} & \mathbf{H}_{\alpha\beta} \\ \mathbf{H}_{\beta\alpha} & \mathbf{H}_{\beta\beta} \end{bmatrix} \begin{bmatrix} -\hat{\mathbf{I}}_\alpha \\ \hat{\mathbf{V}}_\beta \end{bmatrix} \tag{7.47}$$

which can be rewritten as

$$\tilde{\boldsymbol{\theta}} = \begin{bmatrix} \tilde{\mathbf{V}}_\alpha \\ \tilde{\mathbf{I}}_\beta \end{bmatrix} = - \begin{bmatrix} \mathbf{H}_{\alpha\alpha} & -\mathbf{H}_{\alpha\beta} \\ -\mathbf{H}_{\beta\alpha} & \mathbf{H}_{\beta\beta} \end{bmatrix} \begin{bmatrix} \hat{\mathbf{I}}_\alpha \\ \hat{\mathbf{V}}_\beta \end{bmatrix} = -\tilde{\mathbf{H}}\mathbf{h} \tag{7.48a}$$

where

$$\tilde{\mathbf{H}} = \begin{bmatrix} \mathbf{H}_{\alpha\alpha} & -\mathbf{H}_{\alpha\beta} \\ -\mathbf{H}_{\beta\alpha} & \mathbf{H}_{\beta\beta} \end{bmatrix} \tag{7.48b}$$

Combining (7.46) and (7.48a) yields

$$\mathbf{h} = -\tilde{\mathbf{H}}^{-1}\mathbf{X}\mathbf{g} \tag{7.49a}$$

$$\mathbf{g} - \mathbf{h} = (\mathbf{1}_p + \tilde{\mathbf{H}}^{-1}\mathbf{X})\mathbf{g} \tag{7.49b}$$

provided that $\hat{\mathbf{H}}$ is nonsingular, giving

$$\mathbf{T}(\mathbf{X}) = \tilde{\mathbf{H}}^{-1}\mathbf{X} \tag{7.50a}$$

$$\mathbf{F}(\mathbf{X}) = \mathbf{1}_p + \mathbf{T}(\mathbf{X}) = \tilde{\mathbf{H}}^{-1}(\tilde{\mathbf{H}} + \mathbf{X}) \tag{7.50b}$$

Alternatively, Eq. (7.50b) can be interpreted as follows. Consider an $(n + p)$-port network N_1 terminated in a p-port network N_2 as shown in Fig. 7.22. The $(n + p)$-port network N_1 can be characterized by its general hybrid matrix equation, again in partitioned form:

$$\begin{bmatrix} \mathbf{I}_1 \\ \mathbf{V}_2 \\ \mathbf{I}_3 \\ \mathbf{V}_4 \end{bmatrix} = \begin{bmatrix} \mathbf{H}_{11} & \mathbf{H}_{12} & \mathbf{H}_{13} & \mathbf{H}_{14} \\ \mathbf{H}_{21} & \mathbf{H}_{22} & \mathbf{H}_{23} & \mathbf{H}_{24} \\ \mathbf{H}_{31} & \mathbf{H}_{32} & \mathbf{H}_{33} & \mathbf{H}_{34} \\ \mathbf{H}_{41} & \mathbf{H}_{42} & \mathbf{H}_{43} & \mathbf{H}_{44} \end{bmatrix} \begin{bmatrix} \mathbf{V}_1 \\ \mathbf{I}_2 \\ \mathbf{V}_3 \\ \mathbf{I}_4 \end{bmatrix} \tag{7.51}$$

where the port current vectors $\mathbf{I}_k$ and the port voltage vectors $\mathbf{V}_k$ $(k = 1, 2, 3,$ and $4)$ are defined in Fig. 7.22. The p-port network N_2 is described by the general hybrid matrix equation

$$\begin{bmatrix} \mathbf{V}_3 \\ -\mathbf{I}_4 \end{bmatrix} = \begin{bmatrix} \mathbf{X}_{11} & \mathbf{X}_{12} \\ \mathbf{X}_{21} & \mathbf{X}_{22} \end{bmatrix} \begin{bmatrix} -\mathbf{I}_3 \\ \mathbf{V}_4 \end{bmatrix} \tag{7.52}$$

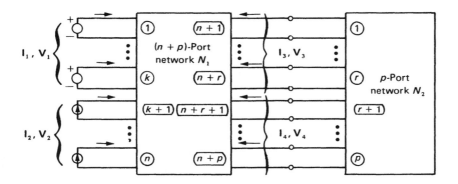

Figure 7.22 The $(n + p)$-port imbedding of a p-port network N_2.

From (7.51) and (7.52), we can solve $\mathbf{I}_1$ and $\mathbf{V}_2$ in terms of $\mathbf{V}_1$ and $\mathbf{I}_2$, and the result is given by

$$\begin{bmatrix} \mathbf{I}_1 \\ \mathbf{V}_2 \end{bmatrix} = \left(\begin{bmatrix} \mathbf{H}_{11} & \mathbf{H}_{12} \\ \mathbf{H}_{21} & \mathbf{H}_{22} \end{bmatrix} + \begin{bmatrix} \mathbf{H}_{13} & -\mathbf{H}_{14} \\ \mathbf{H}_{23} & -\mathbf{H}_{24} \end{bmatrix} \begin{bmatrix} \mathbf{M}_{11} & \mathbf{M}_{12} \\ \mathbf{M}_{21} & \mathbf{M}_{22} \end{bmatrix}^{-1} \right.$$

$$\left. \times \begin{bmatrix} -\mathbf{H}_{31} & -\mathbf{H}_{32} \\ \mathbf{H}_{41} & \mathbf{H}_{42} \end{bmatrix} \right) \begin{bmatrix} \mathbf{V}_1 \\ \mathbf{I}_2 \end{bmatrix} \tag{7.53a}$$

where

$$\begin{bmatrix} \mathbf{M}_{11} & \mathbf{M}_{12} \\ \mathbf{M}_{21} & \mathbf{M}_{22} \end{bmatrix} = \begin{bmatrix} \mathbf{X}_{11} & \mathbf{X}_{12} \\ \mathbf{X}_{21} & \mathbf{X}_{22} \end{bmatrix}^{-1} + \begin{bmatrix} \mathbf{H}_{33} & -\mathbf{H}_{34} \\ -\mathbf{H}_{43} & \mathbf{H}_{44} \end{bmatrix} \tag{7.53b}$$

provided that the various inverses exist. The coefficient matrix of (7.53a) defines the general hybrid matrix of the n-port network as shown in Fig. 7.22. Comparing this coefficient matrix with (7.18), we can make the following identifications:

$$\mathbf{D} = \begin{bmatrix} \mathbf{H}_{11} & \mathbf{H}_{12} \\ \mathbf{H}_{21} & \mathbf{H}_{22} \end{bmatrix} \quad \mathbf{C} = \begin{bmatrix} \mathbf{H}_{13} & -\mathbf{H}_{14} \\ \mathbf{H}_{23} & -\mathbf{H}_{24} \end{bmatrix} \tag{7.54a}$$

$$\mathbf{A} = - \begin{bmatrix} \mathbf{H}_{33} & -\mathbf{H}_{34} \\ -\mathbf{H}_{43} & \mathbf{H}_{44} \end{bmatrix} \quad \mathbf{B} = \begin{bmatrix} -\mathbf{H}_{31} & -\mathbf{H}_{32} \\ \mathbf{H}_{41} & \mathbf{H}_{42} \end{bmatrix} \tag{7.54b}$$

From (7.34) the return difference matrix is found to be

$$\mathbf{F(X)} = \mathbf{1}_p - \mathbf{AX} = \tilde{\mathbf{H}}^{-1} (\tilde{\mathbf{H}} + \mathbf{X}) \tag{7.55}$$

where

$$\tilde{\mathbf{H}} = \begin{bmatrix} \mathbf{H}_{\alpha\alpha} & -\mathbf{H}_{\alpha\beta} \\ -\mathbf{H}_{\beta\alpha} & \mathbf{H}_{\beta\beta} \end{bmatrix} = \begin{bmatrix} \mathbf{H}_{33} & -\mathbf{H}_{34} \\ -\mathbf{H}_{43} & \mathbf{H}_{44} \end{bmatrix}^{-1} \tag{7.56}$$

since, as can be seen from (7.51), by setting $\mathbf{V}_1 = \mathbf{0}$ and $\mathbf{I}_2 = \mathbf{0}$ we have

$$\begin{bmatrix} \mathbf{V}_3 \\ \mathbf{I}_4 \end{bmatrix} = \begin{bmatrix} \mathbf{H}_{33} & \mathbf{H}_{34} \\ \mathbf{H}_{43} & \mathbf{H}_{44} \end{bmatrix}^{-1} \begin{bmatrix} \mathbf{I}_3 \\ \mathbf{V}_4 \end{bmatrix}\Bigg|_{\substack{\mathbf{V}_1 = \mathbf{0} \\ \mathbf{I}_2 = \mathbf{0}}} \tag{7.57}$$

The coefficient matrix of (7.57) is precisely the hybrid matrix looking into the p ports of N_1 when all the independent current sources are open-circuited and all the independent voltage sources are short-circuited.

We illustrate the above results by the following example.

Example 7.3 A three-terminal active device N_2 is used in a feedback amplifier as shown in Fig. 7.23a. The feedback is provided by the impedances Z_a and Z_b. The input excitation is the voltage source V_s, whereas the output is the voltage across the impedance Z_2. The complete amplifier can be viewed as the four-port N imbedding of the active device N_2, as shown in Fig. 7.23b. Suppose that the active device N_2 is the element of interest and is characterized by its hybrid parameters h_{ij}, with

$$\boldsymbol{\theta} = \begin{bmatrix} V_3 \\ I_4 \end{bmatrix} = \begin{bmatrix} h_{11} & h_{12} \\ h_{21} & h_{22} \end{bmatrix} \begin{bmatrix} I_3 \\ V_4 \end{bmatrix} = \mathbf{X}\boldsymbol{\phi} \tag{7.58}$$

The hybrid matrix looking into the two ports facing N_2 with V_s short-circuited is described by the hybrid matrix equation

$$\begin{bmatrix} V_3 \\ -I_4 \end{bmatrix} = \lambda_1 \begin{bmatrix} Z_2 & 1 \\ -1 & \lambda_2 \end{bmatrix} \begin{bmatrix} -I_3 \\ V_4 \end{bmatrix} \tag{7.59}$$

where

$$\lambda_1 = \frac{Z_b}{Z_b + Z_2} \tag{7.60a}$$

$$\lambda_2 = \frac{Z_a + Z_b + Z_2}{Z_a Z_b} \tag{7.60b}$$

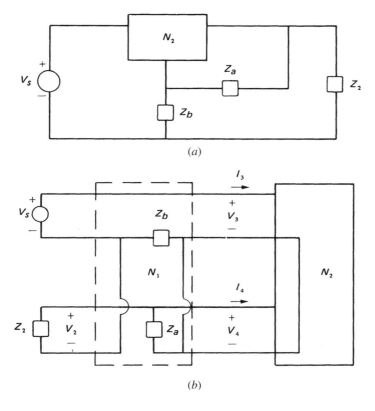

Figure 7.23 (a) A feedback amplifier employing a three-terminal active device and (b) the representation of the complete amplifier as the four-port imbedding of the active device.

giving

$$\tilde{\mathbf{H}} = \lambda_1 \begin{bmatrix} Z_2 & -1 \\ 1 & \lambda_2 \end{bmatrix} \tag{7.61}$$

whose inverse is found to be

$$\tilde{\mathbf{H}}^{-1} = \lambda_3 \begin{bmatrix} \lambda_2 & 1 \\ -1 & z_2 \end{bmatrix} \tag{7.62a}$$

where

$$\lambda_3 = \frac{Z_a}{Z_a + Z_2} \tag{7.62b}$$

Substituting $\tilde{\mathbf{H}}^{-1}$ in (7.50b) yields the return difference matrix with respect to the driving-point hybrid matrix $\mathbf{X}$:

$$
\mathbf{F}(\mathbf{X}) = \tilde{\mathbf{H}}^{-1}(\tilde{\mathbf{H}} + \mathbf{X}) = \begin{bmatrix} 1 + \lambda_3(h_{11}\lambda_2 + h_{21}) & \lambda_3(h_{12}\lambda_2 + h_{22}) \\ \lambda_3(h_{21}Z_2 - h_{11}) & 1 + \lambda_3(h_{22}Z_2 - h_{12}) \end{bmatrix}
$$
(7.63)

The four-port network N_1 in Fig. 7.23b can be described by the matrix equations

$$
\boldsymbol{\phi} = \begin{bmatrix} I_3 \\ V_4 \end{bmatrix} = \lambda_3 \begin{bmatrix} -\lambda_2 & -1 \\ 1 & -Z_2 \end{bmatrix} \begin{bmatrix} V_3 \\ I_4 \end{bmatrix} + \lambda_3 \begin{bmatrix} \lambda_2 \\ -1 \end{bmatrix} [V_s] = \mathbf{A}\boldsymbol{\theta} + \mathbf{B}\mathbf{u} \quad (7.64a)
$$

$$
\mathbf{y} = [V_2] = \lambda_3 \begin{bmatrix} -\dfrac{Z_2}{Z_a} - Z_2 \end{bmatrix} \begin{bmatrix} V_3 \\ I_4 \end{bmatrix} \lambda_3 \begin{bmatrix} \dfrac{Z_2}{Z_a} \end{bmatrix} [V_s] = \mathbf{C}\boldsymbol{\theta} + \mathbf{D}\mathbf{u} \quad (7.64b)
$$

giving

$$
\mathbf{A} = \lambda_3 \begin{bmatrix} -\lambda_2 & -1 \\ 1 & -Z_2 \end{bmatrix}
$$
(7.65)

Substituting (7.65) in (7.34), we obtain

$$
\mathbf{F}(\mathbf{X}) = \mathbf{1}_2 - \mathbf{A}\mathbf{X} = \mathbf{1}_2 + \tilde{\mathbf{H}}^{-1}\mathbf{X} = \tilde{\mathbf{H}}^{-1}(\mathbf{H} + \mathbf{X})
$$
(7.66)

confirming (7.63).

To compute the overall voltage gain function, we appeal to (6.15) with $\mathbf{y} = [V_2]$, $\mathbf{u} = [V_s]$, and $\mathbf{W}(\mathbf{X}) = [w(\mathbf{X})]$, giving

$$
w(\mathbf{X}) = \frac{V_2}{V_s} = \mathbf{D} + \mathbf{C}(\mathbf{X}^{-1} - \mathbf{A})^{-1}\mathbf{B}
$$

$$
= \frac{\lambda_3 Z_2}{Z_a} + \lambda_3^2 \begin{bmatrix} -\dfrac{Z_2}{Z_a} - Z_2 \end{bmatrix} \begin{bmatrix} p_{11} & p_{12} \\ p_{21} & p_{22} \end{bmatrix} \begin{bmatrix} \lambda_2 \\ -1 \end{bmatrix}
$$

$$
= \frac{\lambda_3 Z_2 \{1 - \lambda_3 [\lambda_2(p_{21}Z_a + p_{11}) - p_{22}Z_a - p_{12}]\}}{Z_a}
$$
(7.67a)

where

$$
\begin{bmatrix} p_{11} & p_{12} \\ p_{21} & p_{22} \end{bmatrix} = \frac{1}{\Delta_\lambda} \begin{bmatrix} \lambda_3 Z_2 + \dfrac{h_{11}}{\Delta_h} & -\lambda_3 + \dfrac{h_{12}}{\Delta_h} \\ \lambda_3 + \dfrac{h_{21}}{\Delta_h} & \lambda_2 \lambda_3 + \dfrac{h_{22}}{\Delta_h} \end{bmatrix}
$$
(7.67b)

in which $\Delta_h = h_{11}h_{22} - h_{12}h_{21}$ and

$$\Delta_\lambda = \left(\lambda_2\lambda_3 + \frac{h_{22}}{\Delta_h}\right)\left(\lambda_3 Z_2 + \frac{h_{11}}{\Delta_h}\right) + \left(\lambda_3 - \frac{h_{12}}{\Delta_h}\right)\left(\lambda_3 + \frac{h_{21}}{\Delta_h}\right)$$

(7.67c)

Alternatively, by appealing to formula (3.25a) in conjunction with the impedance matrix (3.75), the voltage gain function (7.67a) can be obtained, the details being omitted.

7.2.2 The Null Return Difference Matrix

As a direct extension of the null return difference for the single-loop feedback amplifiers, in the present section we introduce the notion of the null return difference matrix for multiple-loop feedback networks.

Referring to the fundamental matrix feedback-flow graph of Fig. 7.13, we break, as before, the branch with transmittance $\mathbf{X}$, as illustrated in Fig. 7.24, and apply a signal p-vector $\mathbf{g}$ to the right of the breaking mark. We then adjust the input excitation n-vector $\mathbf{u}$ so that the total output m-vector $\mathbf{y}$ resulting from the inputs $\mathbf{g}$ and $\mathbf{u}$ is zero. From Fig. 7.24, the input vectors $\mathbf{g}$ and $\mathbf{u}$ are related by

$$\mathbf{Du} + \mathbf{CXg} = \mathbf{0} \tag{7.68}$$

giving

$$\mathbf{u} = -\mathbf{D}^{-1}\mathbf{CXg} \tag{7.69}$$

provided that the matrix $\mathbf{D}$ is square and nonsingular. This requires that the output $\mathbf{y}$ be of the same dimension as the input $\mathbf{u}$, that is, $m = n$. Physically, this is rather obvious, because the effects at the output caused by $\mathbf{g}$ can be neutralized by

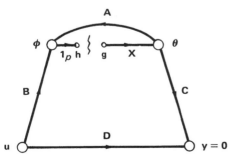

Figure 7.24 The physical interpretation of the null return difference matrix.

a unique input excitation $\mathbf{u}$ only when $\mathbf{u}$ and $\mathbf{y}$ are of the same dimension. With inputs $\mathbf{g}$ and $\mathbf{u}$, as given in Eq. (7.69), the returned signal to the left of the breaking mark in Fig. 7.24 is found to be

$$\mathbf{h} = \mathbf{Bu} + \mathbf{AXg} = (-\mathbf{BD}^{-1}\mathbf{CX} + \mathbf{AX})\mathbf{g} \tag{7.70a}$$

giving

$$\mathbf{g} - \mathbf{h} = (\mathbf{1}_p - \mathbf{AX} + \mathbf{BD}^{-1}\mathbf{CX})\mathbf{g} \tag{7.70b}$$

The square matrix

$$\hat{\mathbf{F}}(\mathbf{X}) = \mathbf{1}_p + \hat{\mathbf{T}}(\mathbf{X}) = \mathbf{1}_p - \mathbf{AX} + \mathbf{BD}^{-1}\mathbf{CX} = \mathbf{1}_p - \hat{\mathbf{A}}\mathbf{X} \tag{7.71}$$

is called the *null return difference matrix* with respect to $\mathbf{X}$, where

$$\hat{\mathbf{T}}(\mathbf{X}) = -\mathbf{AX} + \mathbf{BD}^{-1}\mathbf{CX} = -\hat{\mathbf{A}}\mathbf{X} \tag{7.72a}$$

$$\hat{\mathbf{A}} = \mathbf{A} - \mathbf{BD}^{-1}\mathbf{C} \tag{7.72b}$$

The square matrix $\hat{\mathbf{T}}(\mathbf{X})$ is referred to as the *null return ratio matrix.*

In the case where $\mathbf{X}$ denotes the admittance matrix of a p-port network, we have a simple interpretation of the null return difference matrix $\hat{\mathbf{F}}(\mathbf{X})$. Refer to the network of Fig. 7.25. An $(n + p)$-port network N_1 is terminated in a p-port network N_2. The port voltage vectors $\mathbf{V}_1$ and $\mathbf{V}_2$ and port current vectors $\mathbf{I}_s$ and $\mathbf{I}_2$ are defined in Fig. 7.25. As in the case for the return difference matrix, the p-port network N_2 can be replaced by an equivalent voltage-controlled current source vector

$$\hat{\mathbf{I}}_2 = \mathbf{Xg} \tag{7.73}$$

We then adjust the input source vector $\mathbf{I}_s$ to the n-port until the output vector $\mathbf{V}_1$ is identically zero. Under this situation, the voltage vector $\hat{\mathbf{h}}$ at the p-port network, being the returned voltage, is related to $\hat{\mathbf{I}}_2$ by

$$-\hat{\mathbf{I}}_2 = \hat{\mathbf{Y}}\hat{\mathbf{h}} \tag{7.74}$$

where $\hat{\mathbf{Y}}$ is the admittance matrix facing the p-port network N_2 when $\mathbf{V}_1$ is set to zero. In other words, $\hat{\mathbf{Y}}$ is the admittance matrix facing $\mathbf{X}$ when the input ports are short-circuited. Combining (7.73) and (7.74) yields

$$\hat{\mathbf{h}} = -\hat{\mathbf{Y}}^{-1}\mathbf{Xg} \tag{7.75}$$

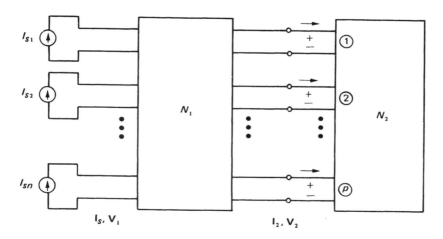

Figure 7.25 The physical interpretation of the null return difference matrix with respect to a p-port admittance matrix.

provided that $\hat{\mathbf{Y}}$ is nonsingular, giving

$$\mathbf{g} - \hat{\mathbf{h}} = (\mathbf{l}_p + \hat{\mathbf{Y}}^{-1}\mathbf{X})\mathbf{g} \qquad (7.76)$$

The null return difference matrix is given by

$$\hat{\mathbf{F}}(\mathbf{X}) = \mathbf{l}_p + \hat{\mathbf{T}}(\mathbf{X}) = \mathbf{l}_p + \hat{\mathbf{Y}}^{-1}\mathbf{X} = \hat{\mathbf{Y}}^{-1}(\hat{\mathbf{Y}} + \mathbf{X}) \qquad (7.77)$$

where $\hat{\mathbf{T}}(\mathbf{X}) = \hat{\mathbf{Y}}^{-1}\mathbf{X}$ is the null return ratio matrix.

We illustráte the above results by the following example.

Example 7.4 Consider the voltage-series feedback amplifier of Fig. 7.14, as discussed in Example 7.2. Assume that the voltage V_{25} is the output variable. Then, from (7.24), we have

$$\boldsymbol{\phi} = \begin{bmatrix} V_{13} \\ V_{45} \end{bmatrix} \begin{bmatrix} -90.782 & 45.391 \\ -942.507 & 0 \end{bmatrix} \begin{bmatrix} I_a \\ I_b \end{bmatrix}$$

$$+ \begin{bmatrix} 0.91748 \\ 0 \end{bmatrix} [V_s] = \mathbf{A}\boldsymbol{\theta} + \mathbf{B}\mathbf{u} \qquad (7.78a)$$

$$\mathbf{y} = [V_{25}] = [45.391 \quad -2372.32] \begin{bmatrix} I_a \\ I_b \end{bmatrix}$$

$$+ [0.04126][V_s] = \mathbf{C}\boldsymbol{\theta} + \mathbf{D}\mathbf{u} \qquad (7.78b)$$

Substituting the coefficient matrices in (7.72*b*), we obtain

$$\hat{\mathbf{A}} = \mathbf{A} - \mathbf{B}\mathbf{D}^{-1}\mathbf{C} = \begin{bmatrix} -1100.12 & 52,797.6 \\ -942.507 & 0 \end{bmatrix} \tag{7.79}$$

giving the null return difference matrix with respect to **X** as

$$\hat{\mathbf{F}}(\mathbf{X}) = \mathbf{1}_2 - \hat{\mathbf{A}}\mathbf{X} = \begin{bmatrix} 51.055 & -2402.29 \\ 42.884 & 1 \end{bmatrix} \tag{7.80}$$

Suppose that the input current I_{51} is chosen as the output variable. From (7.24*b*) we have

$$\mathbf{y} = [I_{51}] = [-0.08252 \quad 0.04126] \begin{bmatrix} I_a \\ I_b \end{bmatrix} + [0.000862][V_s] \tag{7.81}$$

and the corresponding null return difference matrix becomes

$$\hat{\mathbf{F}}(\mathbf{X}) = \mathbf{1}_2 - \hat{\mathbf{A}}\mathbf{X} = \begin{bmatrix} 1.13426 & -0.06713 \\ 42.8841 & 1 \end{bmatrix} \tag{7.82a}$$

where

$$\hat{\mathbf{A}} = \begin{bmatrix} -2.95085 & 1.47543 \\ -942.507 & 0 \end{bmatrix} \tag{7.82b}$$

7.2.3 The Transfer-Function Matrix and Feedback

In this section, we apply the results developed in the foregoing to show the effect of feedback on the transfer-function matrix $\mathbf{W}(\mathbf{X})$. Specifically, we express det $\mathbf{W}(\mathbf{X})$ in terms of det $\mathbf{X}(0)$ and the determinants of the return difference and null return difference matrices, thereby generalizing Blackman's impedance formula for a single input to a multiplicity of inputs.

Before we proceed to develop the desired relation, we first establish the following determinantal identity:

$$\det (\mathbf{1}_m + \mathbf{M}\mathbf{N}) = \det (\mathbf{1}_n + \mathbf{N}\mathbf{M}) \tag{7.83}$$

where **M** and **N** are two arbitrary matrices of orders $m \times n$ and $n \times m$, respectively. To verify this result, we first show that (7.83) is valid when **M** and **N** are square matrices of order n. Then if $\mathbf{M}$ is nonsingular, we have

$$\det (\mathbf{1}_n + \mathbf{MN}) = \det [\mathbf{M}(\mathbf{1}_n + \mathbf{NM})\mathbf{M}^{-1}] = \det \mathbf{M} \det (\mathbf{1}_n + \mathbf{NM}) \det \mathbf{M}^{-1}$$

$$= \det (\mathbf{1}_n + \mathbf{NM}) \tag{7.84}$$

If $\mathbf{M}$ is singular, it has a zero eigenvalue, and hence for some λ the matrix $\lambda\mathbf{1}_n + \mathbf{M}$ is nonsingular. Thus, from (7.84), we have

$$\det [\mathbf{1}_n + (\lambda\mathbf{1}_n + \mathbf{M})\mathbf{N}] = \det [\mathbf{1}_n + \mathbf{N}(\lambda\mathbf{1}_n + \mathbf{M})] \tag{7.85}$$

Observe that both sides of (7.85) are polynomials in λ of at most degree n. They must be identical for all λ for which $\lambda\mathbf{1}_n + \mathbf{M}$ is nonsingular. Since $\lambda = 0$ is a zero of $\det (\lambda\mathbf{1}_n + \mathbf{M})$, there exists a positive number λ_0 such that $\lambda\mathbf{1}_n + \mathbf{M}$ is nonsingular for all real λ satisfying $0 < \lambda < \lambda_0$. Therefore, (7.85) is valid when $\lambda = 0$.

Consider now the situation where $\mathbf{M}$ and $\mathbf{N}$ are not square. Let

$$\tilde{\mathbf{M}} = \begin{bmatrix} \mathbf{M} & \mathbf{0} \\ \mathbf{0} & \mathbf{0} \end{bmatrix} \quad \tilde{\mathbf{N}} = \begin{bmatrix} \mathbf{N} & \mathbf{0} \\ \mathbf{0} & \mathbf{0} \end{bmatrix} \tag{7.86}$$

be the augmented square matrices of order $n + m$. Then we have

$$\det (\mathbf{1}_{n+m} + \tilde{\mathbf{M}}\tilde{\mathbf{N}}) = \det \begin{bmatrix} \mathbf{1}_m + \mathbf{MN} & \mathbf{0} \\ \mathbf{0} & \mathbf{1}_n \end{bmatrix}$$

$$= \det (\mathbf{1}_n + \mathbf{MN}) \tag{7.87a}$$

$$\det (\mathbf{1}_{n+m} + \tilde{\mathbf{N}}\tilde{\mathbf{M}}) = \det \begin{bmatrix} \mathbf{1}_n + \mathbf{NM} & \mathbf{0} \\ \mathbf{0} & \mathbf{1}_m \end{bmatrix}$$

$$= \det (\mathbf{1}_n + \mathbf{NM}) \tag{7.87b}$$

But $\tilde{\mathbf{M}}$ and $\tilde{\mathbf{N}}$ are square, and from the above discussion for square matrices we conclude that $\det(\mathbf{1}_{n+m} + \tilde{\mathbf{M}}\tilde{\mathbf{N}}) = \det (\mathbf{1}_{n+m} + \tilde{\mathbf{N}}\tilde{\mathbf{M}})$. The identity (7.83) follows. This completes the derivation of the identity.

Using (7.83), we next establish the following generalization of Blackman's formula for input impedance.

Theorem 7.1 In a multiple-loop feedback amplifier, if $\mathbf{W}(0) = \mathbf{D}$ is nonsingular, then the determinant of the transfer-function matrix $\mathbf{W}(\mathbf{X})$ is related to the determinants of the return difference matrix $\mathbf{F}(\mathbf{X})$ and the null return difference matrix $\hat{\mathbf{F}}(\mathbf{X})$ by

$$\det \mathbf{W}(\mathbf{X}) = \det \mathbf{W}(\mathbf{0}) \frac{\det \hat{\mathbf{F}}(\mathbf{X})}{\det \mathbf{F}(\mathbf{X})} \tag{7.88}$$

PROOF. From $(7.16a)$ we have

$$\mathbf{W}(\mathbf{X}) = \mathbf{D}[\mathbf{1}_n + \mathbf{D}^{-1}\mathbf{C}\mathbf{X}(\mathbf{1}_p - \mathbf{A}\mathbf{X})^{-1}\mathbf{B}] \tag{7.89}$$

giving

$$\det \mathbf{W}(\mathbf{X}) = \det \mathbf{W}(\mathbf{0}) \det [\mathbf{1}_n + \mathbf{D}^{-1}\mathbf{C}\mathbf{X}(\mathbf{1}_p - \mathbf{A}\mathbf{X})^{-1}\mathbf{B}]$$

$$= \det \mathbf{W}(\mathbf{0}) \det [\mathbf{1}_p + \mathbf{B}\mathbf{D}^{-1}\mathbf{C}\mathbf{X}(\mathbf{1}_p - \mathbf{A}\mathbf{X})^{-1}]$$

$$= \det \mathbf{W}(\mathbf{0}) \det (\mathbf{1}_p - \mathbf{A}\mathbf{X} + \mathbf{B}\mathbf{D}^{-1}\mathbf{C}\mathbf{X}) \det (\mathbf{1}_p - \mathbf{A}\mathbf{X})^{-1}$$

$$= \frac{\det \mathbf{W}(\mathbf{0}) \det \hat{\mathbf{F}}(\mathbf{X})}{\det \mathbf{F}(\mathbf{X})} \tag{7.90}$$

The second line follows directly from the identity (7.83).

The relation (7.88) is a direct extension of (4.176) for the scalar transfer function $w(x)$. As indicated in (4.177), the input impedance $Z(x)$ looking into a terminal pair can be conveniently expressed as

$$Z(x) = Z(0) \frac{F(\text{input short-circuited})}{F(\text{input open-circuited})} \tag{7.91}$$

A similar expression can be derived from (7.88) if $\mathbf{W}(\mathbf{X})$ denotes the impedance matrix of an n-port network as shown in Fig. 7.10. In this case, $\mathbf{F}(\mathbf{X})$ is the return difference matrix with respect to $\mathbf{X}$ for the situation when the n ports where the impedance matrix is defined are left open without any sources, and we write $\mathbf{F}(\mathbf{X}) = \mathbf{F}(\text{input open-circuited})$. Likewise, $\hat{\mathbf{F}}(\mathbf{X})$ is the return difference matrix with respect to $\mathbf{X}$ for the input port-current vector $\mathbf{I}_s$ and the output port-voltage vector $\mathbf{V}$ under the condition that $\mathbf{I}_s$ is adjusted so that the port-voltage vector $\mathbf{V}$ is identically zero. In other words, $\hat{\mathbf{F}}(\mathbf{X})$ is the return difference matrix for the situation when the n ports where the impedance matrix is defined are short-circuited, and we write $\hat{\mathbf{F}}(\mathbf{X}) = \mathbf{F}(\text{input short-circuited})$. Consequently, the determinant of the impedance matrix $\mathbf{Z(X)}$ of an n-port network can be expressed, from (7.88), as

$$\det \mathbf{Z(X)} = \det \mathbf{Z(0)} \frac{\det \mathbf{F}(\text{input short-circuited})}{\det \mathbf{F}(\text{input open-circuited})} \tag{7.92}$$

As an application of Eq. (7.92), let $\mathbf{X}$ be the admittance matrix of a p-port network N_2 that is imbedded in an $(n + p)$-port network N_1, as shown in Fig. 7.25. From (7.42b) and (7.77), we have

$$\mathbf{F(X)} = \mathbf{Y}^{-1}(\mathbf{Y} + \mathbf{X}) = \mathbf{1}_p + \mathbf{Y}^{-1}\mathbf{X} \tag{7.93a}$$

$$\mathbf{\hat{F}(X)} = \mathbf{\hat{Y}}^{-1}(\mathbf{\hat{Y}} + \mathbf{X}) = \mathbf{1}_p + \mathbf{\hat{Y}}^{-1}\mathbf{X} \tag{7.93b}$$

where $\mathbf{Y}$ and $\mathbf{\hat{Y}}$ denote the admittance matrices facing the p-port network N_2 when the input n ports of N_1 are open-circuited and short-circuited, respectively. The determinant of the impedance matrix $\mathbf{Z(X)}$ of the n-port network of Fig. 7.25 is given by

$$\det \mathbf{Z(x)} = \det \mathbf{Z(0)} \frac{\det(\mathbf{1}_p + \mathbf{\hat{Y}}^{-1}\mathbf{X})}{\det(\mathbf{1}_p + \mathbf{Y}^{-1}\mathbf{X})} \tag{7.94}$$

In particular, for $n = 1$ and $p = 1$, Eq. (7.94) reduces to

$$Z_{\text{in}} = z_{11} \frac{1 + x/y_{22}}{1 + xz_{22}} = z_{11} - \frac{xz_{12}z_{21}}{1 + xz_{22}} \tag{7.95}$$

where Z_{in} is the input impedance of the two-port network N_1 when its output is terminated in the admittance x, and z_{ij} and y_{ij} are the open-circuit impedance and the short-circuit admittance parameters of N_1.

Example 7.5 Refer again to the voltage-series feedback amplifier of Fig. 7.14. As computed in (7.37), the return difference matrix with respect to the two controlling parameters is given by

$$\mathbf{F(X)} = \begin{bmatrix} 5.131 & -2.065 \\ 42.884 & 1 \end{bmatrix} \tag{7.96}$$

whose determinant is found to be

$$\det \mathbf{F(X)} = 93.68646 \tag{7.97}$$

If V_{25} of Fig. 7.15 is chosen as the output and V_s as the input, the null return difference matrix is, from (7.80),

$$\mathbf{\hat{F}(X)} = \begin{bmatrix} 51.055 & -2402.29 \\ 42.884 & 1 \end{bmatrix} \tag{7.98}$$

giving

$$\det \hat{\mathbf{F}}(\mathbf{X}) = 103, 071 \qquad (7.99)$$

By appealing to (7.88), the voltage gain V_{25}/V_s of the amplifier can be written as

$$w(\mathbf{X}) = \frac{V_{25}}{V_s} = w(\mathbf{0})\frac{\det \hat{\mathbf{F}}(\mathbf{X})}{\det \mathbf{F}(\mathbf{X})} = 0.04126\frac{103.071}{93.68646} = 45.39 \qquad (7.100)$$

confirming (4.139), where $w(\mathbf{0}) = 0.04126$, as given in (7.78b).

Suppose, instead, that the input current I_{51} is chosen as the output and V_s as the input. From (7.82a) the null return difference matrix becomes

$$\hat{\mathbf{F}}(\mathbf{X}) = \begin{bmatrix} 1.13426 & -0.06713 \\ 42.8841 & 1 \end{bmatrix} \qquad (7.101)$$

with

$$\det \hat{\mathbf{F}}(\mathbf{X}) = 4.01307 \qquad (7.102)$$

By invoking (7.88), the amplifier input admittance is obtained as

$$w(\mathbf{X}) = \frac{I_{51}}{V_s} = w(\mathbf{0})\frac{\det \hat{\mathbf{F}}(\mathbf{X})}{\det \mathbf{F}(\mathbf{X})} = 8.62 \cdot 10^{-4}\frac{4.01307}{93.6865} = 36.92\,\mu\text{mho}$$

$$(7.103)$$

or $27.1\,\text{k}\Omega$, confirming (7.27) or (7.28), where $w(\mathbf{0}) = 862\,\mu\text{mho}$ is found from (7.81).

Another useful application of the generalized Blackman's formula (7.88) is that it provides the basis of a procedure for the indirect measurement of return difference. Refer to Fig. 5.23. Suppose that we wish to measure the return difference $F(y_{21})$ with respect to the forward short-circuit transfer admittance y_{21} of a two-port device, which is represented by its admittance parameters y_{ij}. Choose the controlling parameters y_{21} and y_{12} as the elements of interest; then, from Fig. 5.23, we have

$$\boldsymbol{\theta} = \begin{bmatrix} I_a \\ I_b \end{bmatrix} = \begin{bmatrix} y_{21} & 0 \\ 0 & y_{12} \end{bmatrix}\begin{bmatrix} V_1 \\ V_2 \end{bmatrix} = \mathbf{X}\boldsymbol{\phi} \qquad (7.104)$$

where I_a and I_b are the currents of the voltage-controlled current sources. By appealing to formula (7.92), the impedance looking into terminals a and b of

Fig. 5.23 can be written as

$$z_{aa,bb}(y_{12}, y_{21}) = z_{aa,bb}(0, 0)\frac{\det \mathbf{F}(\text{input short-circuited})}{\det \mathbf{F}(\text{input open-circuited})} \qquad (7.105)$$

When the input terminals a and b are open-circuited, the resulting return difference matrix is exactly the same as that found under normal operating conditions, and we have

$$\mathbf{F}(\text{input open-circuited}) = \mathbf{F}(\mathbf{X}) = \begin{bmatrix} F_{11} & F_{12} \\ F_{21} & F_{22} \end{bmatrix} \qquad (7.106)$$

Since $\mathbf{F}(\mathbf{X}) = \mathbf{1}_2 - \mathbf{AX}$, the elements F_{11} and F_{21} are calculated with $y_{12} = 0$, whereas F_{12} and F_{22} are evaluated with $y_{21} = 0$. When the input terminals a and b are short-circuited, the feedback loop is interrupted and only the (2, 1)-element of the matrix $\mathbf{A}$ is nonzero, giving

$$\det \mathbf{F}(\text{input short-circuited}) = 1 \qquad (7.107)$$

As will be shown in Eq. (7.206a), for diagonal $\mathbf{X}$ the return difference function $F(y_{21})$ can be expressed in terms of $\det \mathbf{F}(\mathbf{X})$ and its (1, 1)-cofactor:

$$F(y_{21}) = \frac{\det \mathbf{F}(\mathbf{X})}{F_{22}} \qquad (7.108)$$

Substituting these in (7.105) yields

$$F(y_{12})|_{y_{21}=0}F(y_{21}) = \frac{z_{aa,bb}(0, 0)}{z_{aa,bb}(y_{12}, y_{21})} \qquad (7.109)$$

where

$$F_{22} = 1 - a_{22}y_{12}|_{y_{21}=0} = F(y_{12})|_{y_{21}=0} \qquad (7.110)$$

and a_{22} is the (2, 2)-element of $\mathbf{A}$. Relation (7.109) was derived earlier in Chap. 5 and is given by Eq. (5.194). As demonstrated in Chap. 5, the elements $F(y_{12})|_{y_{21}=0}$ and $z_{aa,bb}(0, 0)$ can be measured by the arrangements of Figs. 5.31 and 5.32, respectively.

7.2.4 The Complementary Return Difference Matrix

In deriving the null return difference matrix $\hat{\mathbf{F}}(\mathbf{X})$ of (7.71), we assume that the matrix $\mathbf{D}$ is square and nonsingular. In the case that $\mathbf{D}$ is square and singular, the null return difference matrix is not defined. To facilitate our discussion, we introduce the concept of the complementary return difference matrix.

Refer to the matrix feedback-flow graph of Fig. 7.24. As before, we apply a signal p-vector $\mathbf{g}$ to the right of the breaking mark. We then adjust the input excitation n-vector $\mathbf{u}$ so that

$$\mathbf{u} + \check{\mathbf{y}} = \mathbf{0} \tag{7.111}$$

assuming that $\mathbf{u}$ and $\check{\mathbf{y}}$ are of the same dimension ($m = n$), where $\check{\mathbf{y}}$ denotes the output m-vector resulting from the input $\mathbf{g}$ alone with $\mathbf{u} = \mathbf{0}$. The required input vector $\mathbf{u}$ becomes

$$\mathbf{u} = -\check{\mathbf{y}} = -\mathbf{CXg} \tag{7.112}$$

and the returned signal to the left of the breaking mark is found to be

$$\mathbf{h} = \mathbf{Bu} + \mathbf{AXg} = (-\mathbf{BCX} + \mathbf{AX})\mathbf{g} \tag{7.113}$$

giving

$$\mathbf{g} - \mathbf{h} = (\mathbf{1}_p - \mathbf{AX} + \mathbf{BCX})\mathbf{g} = (\mathbf{1}_p - \check{\mathbf{A}}\mathbf{X})\mathbf{g} \tag{7.114}$$

where

$$\check{\mathbf{A}} = \mathbf{A} - \mathbf{BC} \tag{7.115}$$

The square matrix (see Prob. 7.29)

$$\check{\mathbf{F}}(\mathbf{X}) = \mathbf{1}_p - \check{\mathbf{A}}\mathbf{X} \tag{7.116}$$

relating the difference between the applied and returned signals to the applied signal under the condition (7.111) is called the *complementary return difference matrix*. Comparing this with (7.71), we see that $\check{\mathbf{F}}(\mathbf{X})$ can be obtained from $\hat{\mathbf{F}}(\mathbf{X})$ by setting $\mathbf{D} = \mathbf{1}_n$ in (7.71).

Theorem 7.1 does not apply when $\mathbf{W}(0) = \mathbf{D}$ is singular, because in all such cases the null return difference matrix does not exist. As an application of (7.116), consider the situation where $m = n = 1$ with $w(0) = D = 0$. Suppose that in Fig. 7.13 we replace the branch $\mathbf{D}$ with one having unit transmittance, as shown in Fig. 7.26. The graph transmission from u to y in Fig. 7.26 becomes $1 + w(\mathbf{X})$ and the null return difference matrix with respect to $\mathbf{X}$ is obtained from (7.71) by setting $\mathbf{D} = 1$, the latter being recognized as the complementary return difference

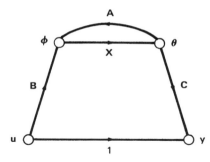

Figure 7.26 The fundamental matrix feedback-flow graph with $\mathbf{D} = 1$.

matrix with respect to $\mathbf{X}$ in Fig. 7.13. Substituting these in (7.88) gives

$$1 + w(\mathbf{X}) = [1 + w(\mathbf{0})] \frac{\det \check{\mathbf{F}}(\mathbf{X})}{\det \mathbf{F}(\mathbf{X})} \tag{7.117}$$

or (see Prob. 7.31)

$$w(\mathbf{X}) = \frac{\det \check{\mathbf{F}}(\mathbf{X}) - \det \mathbf{F}(\mathbf{X})}{\det \mathbf{F}(\mathbf{X})} \tag{7.118}$$

Example 7.6 Refer again to the voltage-series feedback amplifier of Fig. 7.14. Assume that the voltage V_{45} is the output variable and V_s is the input excitation. The required equations can be obtained from (7.24a) and are given by

$$\boldsymbol{\phi} = \begin{bmatrix} V_{13} \\ V_{45} \end{bmatrix} \begin{bmatrix} -90.782 & 45.391 \\ -942.507 & 0 \end{bmatrix} \begin{bmatrix} I_a \\ I_b \end{bmatrix} \begin{bmatrix} 0.91748 \\ 0 \end{bmatrix} [V_s] = \mathbf{A}\boldsymbol{\theta} + \mathbf{B}\mathbf{u} \tag{7.119a}$$

$$\mathbf{y} = [V_{45}] = [-942.507 \quad 0] \begin{bmatrix} I_a \\ I_b \end{bmatrix} + [0][V_s] = \mathbf{C}\boldsymbol{\theta} + \mathbf{D}\mathbf{u} \tag{7.119b}$$

Since $\mathbf{D} = 0$, the null return difference matrix $\hat{\mathbf{F}}(\mathbf{X})$ is not defined. However, the complementary return difference matrix from (7.116) is found to be

$$\check{\mathbf{F}}(\mathbf{X}) = \mathbf{1}_2 - \check{\mathbf{A}}\mathbf{X} = \begin{bmatrix} -34.21468 & -2.06529 \\ 42.88407 & 1 \end{bmatrix} \tag{7.120}$$

where

$$\check{\mathbf{A}} = \mathbf{A} - \mathbf{B}\mathbf{C} = \begin{bmatrix} 773.949 & 45.391 \\ -942.507 & 0 \end{bmatrix} \tag{7.121}$$

giving

$$\det \check{\mathbf{F}}(\mathbf{X}) = 54.35336 \qquad (7.122)$$

By appealing to (7.118), the voltage gain V_{45}/V_s is calculated as

$$w(\mathbf{X}) = \frac{V_{45}}{V_s} = \frac{\det \check{\mathbf{F}}(\mathbf{X}) - \det \mathbf{F}(\mathbf{X})}{\det \mathbf{F}(\mathbf{X})} = -0.42 \qquad (7.123)$$

where $\det \mathbf{F}(\mathbf{X}) = 93.68646$ is given in (7.97). Comparing this with (4.142a) verifies our computation.

7.2.5 Physical Significance of the Matrices A, Â, and Ǎ

As indicated in (7.34), (7.71), and (7.116), the return difference matrix, the null return difference matrix, and the complementary return difference matrix are defined by the square matrices (see Prob. 7.30)

$$\mathbf{F}(\mathbf{X}) = \mathbf{1}_p - \mathbf{A}\mathbf{X} \qquad (7.124a)$$

$$\hat{\mathbf{F}}(\mathbf{X}) = \mathbf{1}_p - \hat{\mathbf{A}}\mathbf{X} \qquad (7.124b)$$

$$\check{\mathbf{F}}(\mathbf{X}) = \mathbf{1}_p - \check{\mathbf{A}}\mathbf{X} \qquad (7.124c)$$

respectively, where

$$\hat{\mathbf{A}} = \mathbf{A} - \mathbf{B}\mathbf{D}^{-1}\mathbf{C} \qquad (7.125a)$$

$$\check{\mathbf{A}} = \mathbf{A} - \mathbf{B}\mathbf{C} \qquad (7.125b)$$

In this section, we present physical interpretations of the matrices $\mathbf{A}$, $\hat{\mathbf{A}}$, and $\check{\mathbf{A}}$ and demonstrate how they can be evaluated directly from the network.

From (7.13a) we see that if the controlled sources of θ are treated as the independent sources, the p-vector ϕ of the controlling variables is equal to $\mathbf{A}\theta$ when the input n-vector $\mathbf{u}$ is set to zero. In fact, we used this interpretation to calculate the elements of $\mathbf{A}$ in Example 7.2. As discussed in Sec. 7.2.2, the null return difference matrix $\hat{\mathbf{F}}(\mathbf{X})$ is the return difference matrix with respect to $\mathbf{X}$ when the input $\mathbf{u}$ is adjusted so that the output $\mathbf{y}$ is identically zero. Thus, $\hat{\mathbf{A}}$ is the matrix relating ϕ to θ when $\mathbf{u}$ is adjusted so that $\mathbf{y} = \mathbf{0}$. Let this $\mathbf{u}$ be designated as $\hat{\mathbf{u}}$, and we have

$$\phi = \hat{\mathbf{A}}\theta|_{\mathbf{u}=\hat{\mathbf{u}}} \qquad (7.126)$$

Likewise, the complementary return difference matrix $\check{\mathbf{F}}(\mathbf{X})$ is the return difference matrix with respect to $\mathbf{X}$ when the input $\mathbf{u}$ is adjusted so that $\mathbf{u} + \check{\mathbf{y}} = \mathbf{0}$, where $\check{\mathbf{y}}$ is defined in (7.111). Therefore, $\check{\mathbf{A}}$ is the matrix relating $\boldsymbol{\phi}$ to $\boldsymbol{\theta}$ when $\mathbf{u}$ is adjusted so that $\mathbf{u} + \check{\mathbf{y}} = \mathbf{0}$. Let this $\mathbf{u}$ be designated as $\check{\mathbf{u}}$, and we get

$$\boldsymbol{\phi} = \check{\mathbf{A}}\boldsymbol{\theta}|_{\mathbf{u}=\check{\mathbf{u}}} \tag{7.127}$$

The evaluation of these matrices is simplified considerably by the fact that the controlled sources are treated as independent sources. We illustrate the above procedure by the following example.

Example 7.7 We use the equivalent network of Fig. 7.15 of the feedback amplifier of Fig. 7.14 to illustrate the evaluation of the matrices $\hat{\mathbf{A}}$ and $\check{\mathbf{A}}$. The evaluation of $\mathbf{A}$ was outlined in Example 7.2.

We first replace the two voltage-controlled current sources by two independent current sources as shown in Fig. 7.27. Assume that V_{25} is the output variable. To compute $\hat{\mathbf{A}}$, we must adjust the input voltage V_s so that the output $V_{25} = 0$. The indefinite-admittance matrix of Fig. 7.27 can be written down by inspection and is given by

$$\mathbf{Y} = 10^{-4} \begin{bmatrix} 9.37 & 0 & -9.09 & 0 & -0.28 \\ 0 & 4.256 & -2.128 & 0 & -2.128 \\ -9.09 & -2.128 & 111.218 & 0 & -100 \\ 0 & 0 & 0 & 10.61 & -10.61 \\ -0.28 & -2.128 & -100 & -10.61 & 113.02 \end{bmatrix} \tag{7.128}$$

Figure 7.27 The network used to calculate the matrices $\hat{\mathbf{A}}$ and $\check{\mathbf{A}}$.

By appealing to formulas (2.94) and (2.97), the output voltage V_{25} resulting from the three sources V_s, I_a, and I_b is found to be

$$V_{25} = \frac{Y_{12,55}}{Y_{11,55}} V_s + \frac{Y_{32,45,11}}{Y_{11,55}} I_a + \frac{Y_{52,25,11}}{Y_{11,55}} I_b$$

$$= 0.04126 V_s + 45.391 I_a - 2372.33 I_b \qquad (7.129)$$

confirming (7.78b). We remark that in applying (2.94) for the contributions of I_a and I_b, we must first short-circuit terminals 1 and 5. This is equivalent to the operations indicated in (7.129). Setting V_{25} to zero yields the desired input excitation

$$\hat{u} = \hat{V}_s = -1100.121 I_a + 57,497.1 I_b \qquad (7.130)$$

By using this value for V_s, the voltages V_{13} and V_{45} and currents I_a and I_b in Fig. 7.27 are related by the equation (see Prob. 7.5)

$$\boldsymbol{\phi} = \begin{bmatrix} V_{13} \\ V_{45} \end{bmatrix} = \begin{bmatrix} -1100.126 & 52,797.8 \\ -942.5071 & 0 \end{bmatrix} \begin{bmatrix} I_a \\ I_b \end{bmatrix} = \hat{\mathbf{A}}\boldsymbol{\theta} \qquad (7.131)$$

The coefficient matrix is the same as that given in (7.79).

To compute $\check{\mathbf{A}}$, we must adjust the input V_s so that $V_s + \check{V}_{25} = 0$, where $\check{V}_{25}$ denotes the contributions to V_{25} by the current sources I_a and I_b. From (7.129) we have

$$\check{V}_{25} = 45.391 I_a - 2372.33 I_b \qquad (7.132)$$

requiring that

$$\check{u} = \check{V}_s = -\check{V}_{25} = -45.391 I_a + 2372.33 I_b \qquad (7.133)$$

Setting $V_s = \check{V}_s$ in Fig. 7.27, we obtain (see Prob. 7.6)

$$\boldsymbol{\phi} = \begin{bmatrix} V_{13} \\ V_{45} \end{bmatrix} = \begin{bmatrix} -132.427 & 2221.96 \\ -942.507 & 0 \end{bmatrix} \begin{bmatrix} I_a \\ I_b \end{bmatrix} = \hat{\mathbf{A}}\boldsymbol{\theta} \qquad (7.134)$$

For illustrative purposes, we also compute $\check{\mathbf{A}}$ by (7.125b). From (7.78) we get

$$\check{\mathbf{A}} = \mathbf{A} - \mathbf{BC} = \begin{bmatrix} -132.427 & 2221.95 \\ -942.507 & 0 \end{bmatrix} \qquad (7.135)$$

giving

$$\check{\mathbf{F}}(\mathbf{X}) = \mathbf{1}_2 - \check{\mathbf{A}}\mathbf{X} = \begin{bmatrix} 7.0254 & -101.099 \\ 42.884 & 1 \end{bmatrix} \tag{7.136}$$

with $\det \check{\mathbf{F}}(\mathbf{X}) = 4342.555$. By invoking (7.118), the voltage gain is found to be

$$w(\mathbf{X}) = \frac{V_{25}}{V_s} = \frac{\det \check{\mathbf{F}}(\mathbf{X}) - \det \mathbf{F}(\mathbf{X})}{\det \mathbf{F}(\mathbf{X})} = 45.35 \tag{7.137}$$

confirming (7.100) within computational accuracy.

7.2.6 Invariance of the Determinants of Feedback Matrices

So far we have introduced three feedback matrices $\mathbf{F}(\mathbf{X})$, $\hat{\mathbf{F}}(\mathbf{X})$, and $\check{\mathbf{F}}(\mathbf{X})$ and studied some of their properties. These matrices generally depend on the choice of the elements of interest $\mathbf{X}$. In this section, we demonstrate that their determinants are invariant with respect to the ordering of the components of $\boldsymbol{\theta}$ and $\boldsymbol{\phi}$ in $\boldsymbol{\theta} = \mathbf{X}\boldsymbol{\phi}$.

Let $\boldsymbol{\theta}_\gamma$ and $\boldsymbol{\phi}_\gamma$ respectively denote the vectors obtained from $\boldsymbol{\theta}$ and $\boldsymbol{\phi}$ by any reordering of the components. Then there exist nonsingular matrices $\mathbf{Q}$ and $\mathbf{P}$ of orders q and p such that $\boldsymbol{\theta}_\gamma = \mathbf{Q}\boldsymbol{\theta}$ and $\boldsymbol{\phi}_\gamma = \mathbf{P}\boldsymbol{\phi}$. In terms of $\boldsymbol{\theta}_\gamma$ and $\boldsymbol{\phi}_\gamma$, the controlled-source constraints become $\boldsymbol{\theta}_\gamma = \mathbf{X}_\gamma \boldsymbol{\phi}_\gamma$, where $\mathbf{X}_\gamma = \mathbf{Q}\mathbf{X}\mathbf{P}^{-1}$, and the equations corresponding to (7.13) can be written as

$$\boldsymbol{\phi}_\gamma = \mathbf{A}_\gamma \boldsymbol{\theta}_\gamma + \mathbf{B}_\gamma \mathbf{u} = \mathbf{P}\mathbf{A}\mathbf{Q}^{-1}\boldsymbol{\theta}_\gamma + \mathbf{P}\mathbf{B}\mathbf{u} \tag{7.138a}$$

$$\mathbf{y} = \mathbf{C}_\gamma \boldsymbol{\theta}_\gamma + \mathbf{D}_\gamma \mathbf{u} = \mathbf{C}\mathbf{Q}^{-1}\boldsymbol{\theta}_\gamma + \mathbf{D}\mathbf{u} \tag{7.138b}$$

where

$$\mathbf{A}_\gamma = \mathbf{P}\mathbf{A}\mathbf{Q}^{-1} \quad \mathbf{B}_\gamma = \mathbf{P}\mathbf{B} \quad \mathbf{C}_\gamma = \mathbf{C}\mathbf{Q}^{-1} \quad \mathbf{D}_\gamma = \mathbf{D} \tag{7.139}$$

By applying (7.83), the determinant of the return difference matrix $\mathbf{F}(\mathbf{X}_\gamma)$ can be manipulated into the form

$$\det \mathbf{F}(\mathbf{X}_\gamma) = \det(\mathbf{1}_p - \mathbf{A}_\gamma \mathbf{X}_\gamma) = \det(\mathbf{1}_p - \mathbf{P}\mathbf{A}\mathbf{Q}^{-1}\mathbf{Q}\mathbf{X}\mathbf{P}^{-1})$$

$$= \det(\mathbf{1}_p - \mathbf{P}\mathbf{A}\mathbf{X}\mathbf{P}^{-1}) = \det(\mathbf{1}_p - \mathbf{P}^{-1}\mathbf{P}\mathbf{A}\mathbf{X})$$

$$= \det(\mathbf{1}_p - \mathbf{A}\mathbf{X}) = \det \mathbf{F}(\mathbf{X}) \tag{7.140}$$

Likewise, the determinant of the null return difference matrix $\hat{\mathbf{F}}(\mathbf{X}_\gamma)$ can be written as

$$
\begin{aligned}
\det \hat{\mathbf{F}}(\mathbf{X}_\gamma) &= \det\left[\mathbf{1}_p - (\mathbf{A}_\gamma - \mathbf{B}_\gamma \mathbf{D}_\gamma^{-1}\mathbf{C}_\gamma)\mathbf{X}_\gamma\right] \\
&= \det\left[\mathbf{1}_q - \mathbf{X}_\gamma(\mathbf{A}_\gamma - \mathbf{B}_\gamma \mathbf{D}_\gamma^{-1}\mathbf{C}_\gamma)\right] \\
&= \det\left[\mathbf{1}_q - \mathbf{QXP}^{-1}(\mathbf{PAQ}^{-1} - \mathbf{PBD}^{-1}\mathbf{CQ}^{-1})\right] \\
&= \det\left[\mathbf{1}_q - \mathbf{QX}(\mathbf{AQ}^{-1} - \mathbf{BD}^{-1}\mathbf{CQ}^{-1})\right] \\
&= \det\left[\mathbf{1}_p - (\mathbf{AQ}^{-1} - \mathbf{BD}^{-1}\mathbf{CQ}^{-1})\mathbf{QX}\right] \\
&= \det\left[\mathbf{1}_p - (\mathbf{A} - \mathbf{BD}^{-1}\mathbf{C})\mathbf{X}\right] = \det \hat{\mathbf{F}}(\mathbf{X}) \qquad (7.141)
\end{aligned}
$$

The above property is evidently shared by $\det \check{\mathbf{F}}(\mathbf{X}_\gamma)$, because $\check{\mathbf{F}}(\mathbf{X}_\gamma)$ can be obtained from $\hat{\mathbf{F}}(\mathbf{X}_\gamma)$ by setting $\mathbf{D} = \mathbf{1}_n$. Thus, $\det \mathbf{F}(\mathbf{X})$, $\det \hat{\mathbf{F}}(\mathbf{X})$, and $\det \check{\mathbf{F}}(\mathbf{X})$ are invariant with respect to the ordering of the components of the controlling and controlled variables.

Example 7.8 Suppose that we reorder the elements of $\boldsymbol{\phi}$ in (7.19) as given below:

$$
\boldsymbol{\theta}_\gamma = \begin{bmatrix} I_a \\ I_b \end{bmatrix} = 10^{-4}\begin{bmatrix} 0 & 455 \\ 455 & 0 \end{bmatrix}\begin{bmatrix} V_{45} \\ V_{13} \end{bmatrix} = \mathbf{X}_\gamma \boldsymbol{\phi}_\gamma \qquad (7.142)
$$

The equations corresponding to (7.24) are given by

$$
\begin{bmatrix} V_{45} \\ V_{13} \end{bmatrix} = \begin{bmatrix} -942.507 & 0 \\ -90.782 & 45.391 \end{bmatrix}\begin{bmatrix} I_a \\ I_b \end{bmatrix} + \begin{bmatrix} 0 \\ 0.91748 \end{bmatrix}[V_s] \qquad (7.143a)
$$

$$
[V_{25}] = [45.391 \quad -2372.32]\begin{bmatrix} I_a \\ I_b \end{bmatrix} + [0.04126][V_s] \qquad (7.143b)
$$

From (7.124) the feedback matrices are calculated as follows:

$$\mathbf{F}(\mathbf{X}_\gamma) = \mathbf{1}_2 - \mathbf{A}_\gamma \mathbf{X}_\gamma = \begin{bmatrix} 1 & 42.8841 \\ -2.06529 & 5.13058 \end{bmatrix} \qquad (7.144a)$$

$$\hat{\mathbf{F}}(\mathbf{X}_\gamma) = \mathbf{1}_2 - (\mathbf{A}_\gamma - \mathbf{B}_\gamma \mathbf{D}_\gamma^{-1} \mathbf{C}_\gamma)\mathbf{X}_\gamma = \begin{bmatrix} 1 & 42.8841 \\ -2402.29 & 51.0555 \end{bmatrix} \qquad (7.144b)$$

$$\check{\mathbf{F}}(\mathbf{X}_\gamma) = \mathbf{1}_2 - (\mathbf{A}_\gamma - \mathbf{B}_\gamma \mathbf{C}_\gamma)\mathbf{X}_\gamma = \begin{bmatrix} 1 & 42.8841 \\ -101.0986 & 7.02544 \end{bmatrix} \qquad (7.144c)$$

giving

$$\det \mathbf{F}(\mathbf{X}_\gamma) = 93.69868 \qquad (7.145a)$$

$$\det \hat{\mathbf{F}}(\mathbf{X}_\gamma) = 103,071.1 \qquad (7.145b)$$

$$\det \check{\mathbf{F}}(\mathbf{X}_\gamma) = 4342.548 \qquad (7.145c)$$

The matrices (7.144) are different from those given in (7.96), (7.98), and (7.136), but their determinants are the same.

7.3 EXTENSIONS TO FEEDBACK MATRICES

In this section, the concepts of return difference and null return difference matrices are generalized. Their properties, physical significance, and interrelations are presented.

As discussed in Sec. 5.4, the general return difference $F_k(x)$ with respect to the controlling parameter x of a voltage-controlled current source $I = xV$ can be interpreted as follows: Replace the controlled current source $I = xV$ by the parallel combination of two controlled sources $I_1 = kV_1$ and $I_2 = x'V_2$, with $x' = x - k$, as shown in Fig. 5.10. Break the controlling branch of the controlled current source $I_2 = x'V_2$ and apply a voltage source of 1 V to the right of the breaking mark. The difference of the 1-V excitation and the returned voltage appearing at the left of the breaking mark under the condition that the input excitation to the amplifier is set to zero is the general return difference $F_k(x)$. The significance of this interpretation is that it corresponds to the situation where the feedback amplifier under study is made partially active rather than completely dead, as in the original interpretation of $F(x)$. A similar procedure can now be employed to define the general return difference and null return difference matrices.

Refer to the fundamental matrix feedback-flow graph of Fig. 7.13. The controlled source vector $\theta = \mathbf{X}\boldsymbol{\phi}$ corresponding to the branch $\mathbf{X}$ is replaced by two controlled source p-vectors:

$$\theta = \mathbf{X}\boldsymbol{\phi} = (\mathbf{X} - \mathbf{K})\boldsymbol{\phi} + \mathbf{K}\boldsymbol{\phi} = \theta_1 + \mathbf{K}\boldsymbol{\phi} \qquad (7.146)$$

where $\mathbf{K}$ is an arbitrary reference matrix of order $q \times p$, and

$$\theta_1 = \mathbf{X}_1\boldsymbol{\phi}_1 \quad \text{and} \quad \mathbf{X}_1 = \mathbf{X} - \mathbf{K} \qquad (7.147)$$

with $\boldsymbol{\phi}_1 = \boldsymbol{\phi}$. By using these in conjunction with (7.13), the associated matrix signal-flow graph is presented in Fig. 7.28. To emphasize the effect of $\mathbf{X}_1$, we apply the procedure outlined in Sec. 7.1 to absorb the nodes $\boldsymbol{\phi}$ and θ in Fig. 7.28. After removing node $\boldsymbol{\phi}$, the resulting graph is shown in Fig. 7.29a. Repeat the process once more by removing node θ in Fig. 7.29a. The final graph is presented in Fig. 7.29b with

$$\mathbf{A}_1 = \mathbf{A}(\mathbf{1}_q, -\mathbf{KA})^{-1} = (\mathbf{1}_p - \mathbf{AK})^{-1}\mathbf{A} \qquad (7.148a)$$

$$\mathbf{B}_1 = \mathbf{B} + \mathbf{A}(\mathbf{1}_q - \mathbf{KA})^{-1}\mathbf{KB} = (\mathbf{1}_p - \mathbf{AK})^{-1}\mathbf{B} \qquad (7.148b)$$

$$\mathbf{C}_1 = \mathbf{C}(\mathbf{1}_q - \mathbf{KA})^{-1} \qquad (7.148c)$$

$$\mathbf{D}_1 = \mathbf{D} + \mathbf{C}(\mathbf{1}_q - \mathbf{KA})^{-1}\mathbf{KB} = \mathbf{D} + \mathbf{CK}(\mathbf{1}_p - \mathbf{AK})^{-1}\mathbf{B} \qquad (7.148d)$$

In Fig. 7.29b, we break the input of the branch with transmittance $\mathbf{X}_1$, set the input excitation vector $\mathbf{u} = \mathbf{0}$, and apply a signal p-vector $\mathbf{g}$ to the right of the breaking mark, as shown in Fig. 7.30. The returned signal p-vector $\mathbf{h}$ to the left of the breaking mark is found to be $\mathbf{h} = \mathbf{A}_1\mathbf{X}_1\mathbf{g}$. The square matrix $-\mathbf{A}_1\mathbf{X}_1$ is called the *general return ratio matrix* with respect to $\mathbf{X}$ for a general reference $\mathbf{K}$ and is

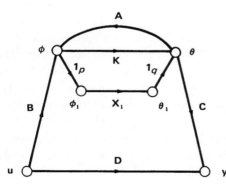

Figure 7.28 The matrix signal-flow graph associated with the systems (7.13), (7.146), and (7.147).

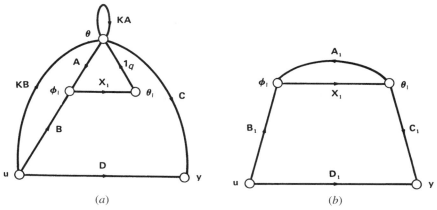

Figure 7.29 (*a*) The resulting matrix signal-flow graph after absorbing the node ϕ in that of Fig. 7.28 and (*b*) the resulting matrix signal-flow graph after absorbing the node θ in (*a*).

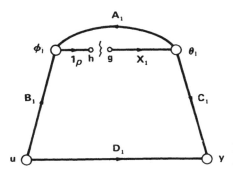

Figure 7.30 The physical interpretation of the general return difference matrix $\mathbf{F_K(X)}$.

denoted by

$$\mathbf{T_K(X)} = -\mathbf{A}_1\mathbf{X}_1 \tag{7.149}$$

The difference between the applied signal vector **g** and the returned signal **h** is given by

$$\mathbf{g} - \mathbf{h} = (\mathbf{1}_p - \mathbf{A}_1\mathbf{X}_1)\mathbf{g} \tag{7.150}$$

The square matrix

$$\mathbf{F_K(X)} = \mathbf{1}_p - \mathbf{A}_1\mathbf{X}_1 \tag{7.151}$$

is defined as the *general return difference matrix* with respect to $\mathbf{X}$ for a reference $\mathbf{K}$. For convenience, we write $\mathbf{T_0}(\mathbf{X}) = \mathbf{T}(\mathbf{X})$ and $\mathbf{F_0}(\mathbf{X}) = \mathbf{F}(\mathbf{X})$.

Likewise, if, in Fig. 7.30, we adjust the input vector $\mathbf{u}$ so that the total output vector $\mathbf{y}$ resulting from $\mathbf{u}$ and $\mathbf{g}$ is zero, the returned signal becomes

$$\mathbf{h} = (\mathbf{A_1 X_1} - \mathbf{B_1 D_1^{-1} C_1 X_1})\mathbf{g} = \hat{\mathbf{A}}_1 \mathbf{X}_1 \mathbf{g} \tag{7.152}$$

where

$$\hat{\mathbf{A}}_1 = \mathbf{A}_1 - \mathbf{B_1 D_1^{-1} C_1} \tag{7.153}$$

provided that $\mathbf{D}_1$ is square and nonsingular. The square matrix

$$\hat{\mathbf{T}}_K(\mathbf{X}) = -\hat{\mathbf{A}}_1 \mathbf{X}_1 \tag{7.154}$$

is referred to as the *general null return ratio matrix* with respect to $\mathbf{X}$ for a general reference $\mathbf{K}$. The difference between the applied signal vector $\mathbf{g}$ and the returned signal vector $\mathbf{h}$ is given by

$$\mathbf{g} - \mathbf{h} = (\mathbf{1}_p - \hat{\mathbf{A}}_1 \mathbf{X}_1)\mathbf{g} = [\mathbf{1}_p + \hat{\mathbf{T}}_K(\mathbf{X})]\mathbf{g} \tag{7.155}$$

The square matrix

$$\hat{\mathbf{F}}_K(\mathbf{X}) = \mathbf{1}_p - \hat{\mathbf{A}}_1 \mathbf{X}_1 = \mathbf{1}_p + \hat{\mathbf{T}}_K(\mathbf{X}) \tag{7.156}$$

is the *general null return difference matrix* with respect to $\mathbf{X}$ for a general reference $\mathbf{K}$. For simplicity, we write $\hat{\mathbf{T}}_0(\mathbf{X}) = \hat{\mathbf{T}}(\mathbf{X})$ and $\hat{\mathbf{F}}_0(\mathbf{X}) = \hat{\mathbf{F}}(\mathbf{X})$.

Alternatively, we can break the branch $\mathbf{X}_1$ of Fig. 7.28 and apply a signal p-vector $\mathbf{g}$ to the right of the breaking mark, as shown in Fig. 7.31. We then adjust the input vector $\mathbf{u}$ so that the total output $\mathbf{y}$ resulting from $\mathbf{u}$ and $\mathbf{g}$ is zero, giving

$$\boldsymbol{\theta} = \mathbf{X}_{1g} + \mathbf{K}\boldsymbol{\phi} \tag{7.157a}$$

$$\boldsymbol{\phi} = \mathbf{A}\boldsymbol{\theta} + \mathbf{Bu} \tag{7.157b}$$

$$\mathbf{y} = \mathbf{C}\boldsymbol{\theta} + \mathbf{Du} = \mathbf{0} \tag{7.157c}$$

Substituting $(7.157a)$ in $(7.157b)$ and $(7.157c)$ and then eliminating $\mathbf{u}$ yield

$$(\mathbf{1}_p - \mathbf{AK} + \mathbf{BD^{-1}CK})\boldsymbol{\phi} = (\mathbf{A} - \mathbf{BD^{-1}C})\mathbf{X}_1 \mathbf{g} \tag{7.158}$$

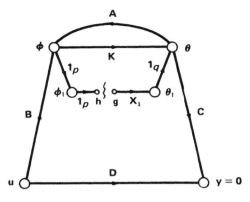

Figure 7.31 An alternative physical interpretation of the general return difference matrix $\mathbf{F_K(X)}$.

Since $\mathbf{h} = \boldsymbol{\phi}$, the above equation can be rewritten as

$$\mathbf{h} = (\mathbf{1}_p - \hat{\mathbf{A}}\mathbf{K})^{-1}\hat{\mathbf{A}}\mathbf{X}_1\mathbf{g} \tag{7.159}$$

where $\hat{\mathbf{A}} = \mathbf{A} - \mathbf{B}\mathbf{D}^{-1}\mathbf{C}$, as given in (7.72$b$). Comparing (7.159) with (7.152), we obtain

$$\hat{\mathbf{A}}_1 = (\mathbf{1}_p - \hat{\mathbf{A}}\mathbf{K})^{-1}\hat{\mathbf{A}} \tag{7.160}$$

We now proceed to establish relations similar to those given in (5.85) and (5.97). To this end, we substitute (7.147) and (7.148a) in (7.151), yielding

$$\begin{aligned}
\mathbf{F_K(X)} = \mathbf{1}_p - \mathbf{A}_1\mathbf{X}_1 &= \mathbf{1}_p - (\mathbf{1}_p - \mathbf{A}\mathbf{K})^{-1}\mathbf{A}(\mathbf{X} - \mathbf{K}) \\
&= (\mathbf{1}_p - \mathbf{A}\mathbf{K})^{-1}(\mathbf{1}_p - \mathbf{A}\mathbf{K} - \mathbf{A}\mathbf{X} + \mathbf{A}\mathbf{K}) \\
&= (\mathbf{1}_p - \mathbf{A}\mathbf{K})^{-1}(\mathbf{1}_p - \mathbf{A}\mathbf{X}) \tag{7.161}
\end{aligned}$$

or

$$\mathbf{F_K(X)} = \mathbf{F}^{-1}(\mathbf{K})\mathbf{F}(\mathbf{X}) \tag{7.162}$$

In words, this states that the general return difference matrix with respect to $\mathbf{X}$ for a general reference $\mathbf{K}$ is equal to the product of the inverse of the return difference matrix with respect to $\mathbf{K}$ and the return difference matrix with respect to $\mathbf{X}$, both being defined for the zero reference.

For the null return difference matrix, we substitute (7.160) in (7.156) and obtain

$$\hat{\mathbf{F}}_{\mathbf{K}}(\mathbf{X}) = \mathbf{1}_P - \hat{\mathbf{A}}_1 \mathbf{X}_1 = \mathbf{1}_p - (\mathbf{1}_p - \hat{\mathbf{A}}\mathbf{K})^{-1}\hat{\mathbf{A}}\mathbf{X},$$

$$= (\mathbf{1}_p - \hat{\mathbf{A}}\mathbf{K})^{-1}(\mathbf{1}_p - \hat{\mathbf{A}}\mathbf{K} - \hat{\mathbf{A}}\mathbf{X}_1)$$

$$= (\mathbf{1}_p - \hat{\mathbf{A}}\mathbf{K})^{-1}(\mathbf{1}_p - \hat{\mathbf{A}}\mathbf{X}) \tag{7.163}$$

or

$$\hat{\mathbf{F}}_{\mathbf{K}}(\mathbf{X}) = \hat{\mathbf{F}}^{-1}(\mathbf{K})\hat{\mathbf{F}}(\mathbf{X}) \tag{7.164}$$

In fact, by repeated applications of (7.162) and (7.164), the return difference and null return difference matrices can be factored in finite product forms. To see this, let $\mathbf{K}_1$ denote the reference condition for $\mathbf{X}$, where $\mathbf{K}_1$ is obtained from $\mathbf{X}$ by setting some of its elements to particular values. Then from (7.162) we have

$$\mathbf{F}(\mathbf{X}) = \mathbf{F}(\mathbf{K}_1)\mathbf{F}_{\mathbf{K}_1}(\mathbf{X}) \tag{7.165a}$$

We next consider $\mathbf{K}_1$ and a reference matrix $\mathbf{K}_2$, where $\mathbf{K}_2$ is obtained from $\mathbf{K}_1$ by setting some of the elements of $\mathbf{K}_1$ to particular values. This gives

$$\mathbf{F}(\mathbf{K}_1) = \mathbf{F}(\mathbf{K}_2)\mathbf{F}_{\mathbf{K}_2}(\mathbf{K}_1) \tag{7.165b}$$

Continuing this process, $\mathbf{F}(\mathbf{X})$ can be factored into the form

$$\mathbf{F}(\mathbf{X}) = \mathbf{F}(\mathbf{K}_n) \prod_{i=n}^{1} \mathbf{F}_{\mathbf{K}_i}(\mathbf{K}_{i-1}) \tag{7.166}$$

where $\mathbf{K}_i$ is obtained from $\mathbf{K}_{i-1}$ by setting some of the elements of $\mathbf{K}_{i-1}$ to special values for $i = 1, 2, \ldots, n$ with $\mathbf{K}_0 = \mathbf{X}$. Likewise, we can show that

$$\hat{\mathbf{F}}(\mathbf{X}) = \hat{\mathbf{F}}(\mathbf{K}_n) \prod_{i=n}^{1} \hat{\mathbf{F}}_{\mathbf{K}_i}(\mathbf{K}_{i-1}) \tag{7.167}$$

Example 7.9 From Example 7.2, the equations for the voltage-series feedback amplifier of Fig. 7.15 are given by

$$
\begin{bmatrix} V_{13} \\ V_{45} \end{bmatrix} = \begin{bmatrix} -90.782 & 45.391 \\ -942.507 & 0 \end{bmatrix} \begin{bmatrix} I_a \\ I_b \end{bmatrix} + \begin{bmatrix} 0.91748 \\ 0 \end{bmatrix} [V_s] = \mathbf{A}\boldsymbol{\theta} + \mathbf{B}\mathbf{u}
$$

(7.168a)

$$
[I_{51}] = [-0.08252 \quad 0.04126] \begin{bmatrix} I_a \\ I_b \end{bmatrix} + [0.000862][V_s] = \mathbf{C}\boldsymbol{\theta} + \mathbf{D}\mathbf{u}
$$

(7.168b)

The elements of interest are the transconductances of the two controlled sources, each having the value 0.0455 mho. Assume that their reference values are 0.01 mho and 0.02 mho. Then we have

$$
\mathbf{X} = \begin{bmatrix} 0.0455 & 0 \\ 0 & 0.0455 \end{bmatrix} \quad \mathbf{K} = \begin{bmatrix} 0.01 & 0 \\ 0 & 0.02 \end{bmatrix}
$$

(7.169)

giving

$$
\mathbf{X}_1 = \mathbf{X} - \mathbf{K} = \begin{bmatrix} 0.0355 & 0 \\ 0 & 0.0255 \end{bmatrix}
$$

(7.170)

The corresponding matrices of (7.148) are computed as follows:

$$
\mathbf{A}_1 = (\mathbf{1}_2 - \mathbf{A}\mathbf{K})^{-1}\mathbf{A} = \begin{bmatrix} -90.44353 & 4.33779 \\ 90.06962 & -40.88412 \end{bmatrix}
$$

(7.171a)

$$
\mathbf{B}_1 = (\mathbf{1}_2 - \mathbf{A}\mathbf{K})^{-1}\mathbf{B} = \begin{bmatrix} 0.08768 \\ -0.82638 \end{bmatrix}
$$

(7.171b)

$$
\mathbf{C}_1 = \mathbf{C}(\mathbf{1}_2 - \mathbf{K}\mathbf{A})^{-1} = [-0.08221 \quad 0.003943]
$$

(7.171c)

$$
\mathbf{D}_1 = \mathbf{D} + \mathbf{C}\mathbf{K}(\mathbf{1}_2 - \mathbf{A}\mathbf{K})^{-1}\mathbf{B} = [0.0001077]
$$

(7.171d)

The general return difference matrix with respect to $\mathbf{X}$ for the reference $\mathbf{K}$ is found to be

$$
\mathbf{F}_\mathbf{K}(\mathbf{X}) = \mathbf{1}_2 - \mathbf{A}_1\mathbf{X}_1 = \begin{bmatrix} 4.21075 & -0.11061 \\ 3.19747 & 2.04254 \end{bmatrix}
$$

(7.172)

Alternatively, using (7.162) in conjunction with (7.37), we get

$$\mathbf{F_K(X)} = \mathbf{F}^{-1}(\mathbf{K})\mathbf{F(X)} = \begin{bmatrix} 4.21078 & -0.11059 \\ 3.19707 & 2.04229 \end{bmatrix} \qquad (7.173)$$

confirming (7.172) within computational error, where

$$\mathbf{F(K)} = \begin{bmatrix} 1.90782 & -0.90782 \\ 9.42507 & 1 \end{bmatrix} \qquad (7.174)$$

For the null return difference matrix with respect to $\mathbf{X}$ for the reference value $\mathbf{K}$, we first compute

$$\hat{\mathbf{A}}_1 = \mathbf{A}_1 - \mathbf{B}_1 \mathbf{D}_1^{-1} \mathbf{C}_1 = \begin{bmatrix} -23.51528 & 1.12774 \\ -720.8654 & -10.62956 \end{bmatrix} \qquad (7.175)$$

This matrix can also be determined from (7.160) and is found to be

$$\hat{\mathbf{A}}_1 = (\mathbf{1}_2 - \hat{\mathbf{A}}\mathbf{K})^{-1}\hat{\mathbf{A}} = \begin{bmatrix} -23.52616 & 1.12831 \\ -720.7764 & -10.63441 \end{bmatrix} \qquad (7.176)$$

where $\hat{\mathbf{A}}$ is given in (7.82b). From (7.156) we obtain

$$\hat{\mathbf{F}}_\mathbf{K}(\mathbf{X}) = \mathbf{1}_2 - \hat{\mathbf{A}}_1 \mathbf{X}_1 = \begin{bmatrix} 1.83518 & -0.02877 \\ 25.58756 & 1.27118 \end{bmatrix} \qquad (7.177)$$

or from (7.164) in conjunction with (7.82a) we have

$$\hat{\mathbf{F}}_\mathbf{K}(\mathbf{X}) = \hat{\mathbf{F}}^{-1}(\mathbf{K})\hat{\mathbf{F}}(\mathbf{X}) = \begin{bmatrix} 1.83517 & -0.02877 \\ 25.58771 & 1.27116 \end{bmatrix} \qquad (7.178)$$

where

$$\hat{\mathbf{F}}(\mathbf{K}) = \mathbf{1}_2 - \hat{\mathbf{A}}\mathbf{K} = \begin{bmatrix} 1.02951 & -0.02951 \\ 9.42507 & 1 \end{bmatrix} \qquad (7.179)$$

7.3.1 The Transfer-Function Matrix and the General Feedback Matrices

As a direct extension of (7.88), we demonstrate that the determinant of the transfer-function matrix can be expressed in terms of the determinants of the general return difference matrix and the general null return difference matrix. We also give a simple physical interpretation of the general return difference and null return difference matrices for the admittance matrix of a p-port network.

Let $\mathbf{W}(\mathbf{X})$ be the transfer-function matrix of a multiple-loop feedback amplifier. Assume that $\mathbf{W}(\mathbf{K})$ is nonsingular, where $\mathbf{K}$ is the reference matrix. Then from (7.88) we have

$$\det \mathbf{W}(\mathbf{X}) = \det \mathbf{W}(\mathbf{0}) \frac{\det \hat{\mathbf{F}}(\mathbf{X})}{\det \mathbf{F}(\mathbf{X})} \qquad (7.180a)$$

$$\det \mathbf{W}(\mathbf{K}) = \det \mathbf{W}(\mathbf{0}) \frac{\det \hat{\mathbf{F}}(\mathbf{K})}{\det \mathbf{F}(\mathbf{K})} \qquad (7.180b)$$

Dividing $(7.180a)$ by $(7.180b)$ yields

$$\det \mathbf{W}(\mathbf{X}) = \det \mathbf{W}(\mathbf{K}) \frac{\det \hat{\mathbf{F}}(\mathbf{X}) \det \mathbf{F}(\mathbf{K})}{\det \hat{\mathbf{F}}(\mathbf{K}) \det \mathbf{F}(\mathbf{X})} \qquad (7.181)$$

By appealing to (7.162) and (7.164), the above equation can be simplified and is given by

$$\det \mathbf{W}(\mathbf{X}) = \det \mathbf{W}(\mathbf{K}) \frac{\det \hat{\mathbf{F}}_{\mathbf{K}}(\mathbf{K})}{\det \mathbf{F}_{\mathbf{K}}(\mathbf{K})} \qquad (7.182)$$

Consider the situation where $\mathbf{W}(\mathbf{X})$ denotes the impedance matrix of an n-port network of Fig. 7.10. In this case, $\mathbf{F}_{\mathbf{K}}(\mathbf{X})$ is the general return difference matrix for the situation when the n ports where the impedance matrix is defined are left open without any sources, and we write $\mathbf{F}_{\mathbf{K}}(\mathbf{X}) = \mathbf{F}_{\mathbf{K}}(\text{input open-circuited})$. Likewise, $\hat{\mathbf{F}}_{\mathbf{K}}(\mathbf{X})$ is the general return difference matrix for the situation when the n ports where the impedance matrix is defined are short-circuited, and we write $\hat{\mathbf{F}}_{\mathbf{K}}(\mathbf{X}) = \mathbf{F}_{\mathbf{K}}(\text{input short-circuited})$. As a result, the determinant of the impedance matrix $\mathbf{Z}(\mathbf{X})$ of an n-port network can be expressed, from (7.182), as

$$\det \mathbf{Z}(\mathbf{X}) = \det \mathbf{Z}(\mathbf{K}) \frac{\det \mathbf{F}_{\mathbf{K}}(\text{input short-circuited})}{\det \mathbf{F}_{\mathbf{K}}(\text{input open-circuited})} \qquad (7.183)$$

As an application of (7.182), let $\mathbf{X}$ be the admittance matrix of a p-port network N_2 imbedded in an $(n + p)$-port network N_1, as shown in Fig. 7.25. From $(7.93a)$ we have

$$\mathbf{F}(\mathbf{X}) = \mathbf{Y}^{-1}(\mathbf{Y} + \mathbf{X})$$

$$\mathbf{F}(\mathbf{K}) = \mathbf{Y}^{-1}(\mathbf{Y} + \mathbf{K}) \qquad (7.184a)$$

where $\mathbf{Y}$ is the admittance matrix facing the p-port network N_2 when the input n ports of N_1 are open-circuited, giving

$$\mathbf{F}_{\mathbf{K}}(\mathbf{X}) = \mathbf{F}^{-1}(\mathbf{K})\mathbf{F}(\mathbf{X}) = (\mathbf{Y} + \mathbf{K})^{-1}(\mathbf{Y} + \mathbf{X}) \qquad (7.185)$$

This is a direct extension of Eq. (5.104). In a similar manner, from (7.93b) we can show that

$$\hat{\mathbf{F}}_{\mathbf{K}}(\mathbf{X}) = \hat{\mathbf{F}}^{-1}(\mathbf{K})\hat{\mathbf{F}}(\mathbf{X}) = (\hat{\mathbf{Y}} + \mathbf{K})^{-1}(\hat{\mathbf{Y}} + \mathbf{X}) \tag{7.186}$$

where $\hat{\mathbf{Y}}$ is the admittance matrix facing the p-port N_2 when the input n ports of N_1 are short-circuited. Substituting these in (7.183) gives

$$\det \mathbf{Z}(\mathbf{X}) = \det \mathbf{Z}(\mathbf{K}) \frac{\det (\mathbf{Y} + \mathbf{K})\det (\hat{\mathbf{Y}} + \mathbf{X})}{\det (\hat{\mathbf{Y}} + \mathbf{K})\det (\mathbf{Y} + \mathbf{X})} \tag{7.187}$$

Example 7.10 Consider the same problem as discussed in Example 7.9. From (7.172) and (7.178) we have

$$\det \mathbf{F}_{\mathbf{K}}(\mathbf{X}) = 8.95430 \tag{7.188a}$$

$$\det \hat{\mathbf{F}}_{\mathbf{K}}(\mathbf{X}) = 3.06895 \tag{7.188b}$$

The network function of interest is the driving-point admittance facing the voltage source V_s in Fig. 7.15, that is, $w(\mathbf{X}) = I_{51}/V_s$. To apply (7.182), we first compute $w(\mathbf{K})$, which is the input admittance when the controlling parameters of the two controlled sources in Fig. 7.15 are replaced by their reference values: $I_a = 0.01\, V_{13}$ and $I_b = 0.02\, V_{45}$. The corresponding indefinite-admittance matrix can be written down by inspection and is given by

$$\mathbf{Y} = 10^{-4} \begin{bmatrix} 9.37 & 0 & -9.09 & 0 & -0.28 \\ 0 & 4.256 & -2.128 & 200 & -202.128 \\ -109.09 & -2.128 & 211.218 & 0 & -100 \\ 100 & 0 & -100 & 10.61 & -10.61 \\ -0.28 & -2.128 & -100 & -210.61 & 313.02 \end{bmatrix} \tag{7.189}$$

By appealing to formula (2.94), the input admittance $w(\mathbf{K})$ is found to be

$$w(\mathbf{K}) = \frac{Y_{uv}(\mathbf{K})}{Y_{11,55}(\mathbf{K})} = \frac{56,057.62 \cdot 10^{-16}}{52,049.75 \cdot 10^{-12}} = 107.70\,\mu\text{mho} \tag{7.190}$$

Substituting these in (7.182), we obtain the input admittance

$$w(\mathbf{X}) = w(\mathbf{K})\frac{\det \hat{\mathbf{F}}_{\mathbf{K}}(\mathbf{X})}{\det \mathbf{F}_{\mathbf{K}}(\mathbf{X})} = 36.91\,\mu\text{mho} \tag{7.191}$$

or the input impedance is $27.09\,\text{k}\Omega$, confirming (7.27) or (7.28).

7.3.2   Relations to Scalar Return Difference and Null Return Difference

In Chaps. 4 and 5, we defined the return difference and the null return difference with respect to the controlling parameter x of a voltage-controlled current source $I = xV$. These concepts have been extended to the return difference and null return difference matrices with respect to a matrix $\mathbf{X}$. In the present section, we develop relations between the scalar return difference and null return difference with respect to an element of $\mathbf{X}$ and the return difference matrix and the null return difference matrix with respect to $\mathbf{X}$. The relations are useful in that we can express the sensitivity function in terms of the determinants of the return difference and null return difference matrices.

Let $\theta_i'' = x_{ij}\phi_j$ be the controlled source of interest, where x_{ij} is the (i, j)-element of $\mathbf{X}$, as defined in (7.11). The return difference $F(x_{ij})$ with respect to the controlling parameter x_{ij} can be defined as follows: Replace the controlled source $\theta_i'' = x_{ij}\phi_j$ by an independent source of strength $x_{ij}g_j$, and set all other independent sources to zero. The signal h_j appearing at the controlling branch ϕ_j is the returned signal. The return difference $F(x_{ij})$ is defined by the equation

$$g_j - h_j = F(x_{ij})g_j \tag{7.192}$$

Let $\mathbf{K}$ be the matrix obtained from $\mathbf{X}$ by setting $x_{ij} = 0$, that is,

$$\mathbf{K} = \mathbf{X}|_{x_{ij}=0} \tag{7.193}$$

Then $\mathbf{X}_1 = \mathbf{X} - \mathbf{K}$ has the form that all of its elements are zero except the one in the ith row and jth column, which is x_{ij}. Let $\mathbf{g}$ be a p-vector, all of its elements being zero except the jth row, which is g_j. The operation of replacing the controlled source $\theta_i'' = x_{ij}\phi_j$ by an independent source of strength $x_{ij}g_j$ is equivalent to replacing the controlled source vector $\boldsymbol{\theta} = \mathbf{X}\boldsymbol{\phi}$ by the sum of an independent source vector $\mathbf{X}_1\mathbf{g}$ and a controlled source vector $\boldsymbol{\theta}'' = \mathbf{K}\boldsymbol{\phi}$. In terms of the fundamental matrix feedback-flow graph of Fig. 7.28, the operation is the same as that depicted in Fig. 7.30 or Fig. 7.31. The returned signal vector $\mathbf{h}$ is therefore found to be $\mathbf{h} = \mathbf{A}_1\mathbf{X}_1\mathbf{g}$. The signal h_j, being the jth component of $\mathbf{h}$, is given by

$$h_j = a_{ji}^1 x_{ij} g_j \tag{7.194}$$

where a_{ij}^1 are the elements of $\mathbf{A}_1$. Combining this with (7.192) yields

$$F(x_{ij}) = 1 - a_{ji}^1 x_{ij} \tag{7.195}$$

The return difference matrix with respect to $\mathbf{X}$ for the reference $\mathbf{K}$ is, from (7.151),

$$\mathbf{F_K(X)} = \mathbf{1}_p - \mathbf{A}_1\mathbf{X}_1 \tag{7.196}$$

whose determinant is found to be

$$\det \mathbf{F_K(X)} = 1 - a_{ji}^1 x_{ij} \tag{7.197}$$

showing that

$$F(x_{ij}) = \det \mathbf{F_K(X)} \tag{7.198}$$

or, after appealing to (7.162), we have

$$F(x_{ij}) = \frac{\det \mathbf{F(X)}}{\det \mathbf{F(X)}|_{x_{ij}=0}} \tag{7.199}$$

In words, this states that the scalar return difference of a multiple-loop feedback amplifier with respect to an element x_{ij} of a matrix $\mathbf{X}$ is equal to the ratio of the determinants of the return difference matrix with respect to $\mathbf{X}$ and that with respect to $\mathbf{K}$, where $\mathbf{K}$ is obtained from $\mathbf{X}$ by setting x_{ij} to zero. In a similar manner, we can show that the scalar null return difference with respect to x_{ij} is equal to the ratio of the determinants of the null return difference matrix with respect to $\mathbf{X}$ and that with respect to $\mathbf{K}$:

$$\hat{F}(x_{ij}) = \frac{\det \hat{\mathbf{F}}(\mathbf{X})}{\det \hat{\mathbf{F}}(\mathbf{X})|_{x_{ij}=0}} \tag{7.200}$$

The justification of (7.200) is left as an exercise (see Prob. 7.8).

The above relations can be generalized to a general reference value. Let k_{ij} be a general reference value of x_{ij}. Then, from (7.199), we have

$$F(k_{ij}) = \frac{\det \mathbf{F(X)}|_{x_{ij}=k_{ij}}}{\det \mathbf{F(X)}|_{x_{ij}=0}} \tag{7.201}$$

Invoking (5.85) in conjunction with (7.199) and (7.201), we obtain the scalar return difference with respect to x_{ij} for a general reference value k_{ij} as

$$F_{k_{ij}}(x_{ij}) = \frac{F(x_{ij})}{F(k_{ij})} = \frac{\det \mathbf{F(X)}}{\det \mathbf{F(X)}|_{x_{ij}=k_{ij}}} \tag{7.202}$$

Likewise, we can show that the scalar null return difference with respect to x_{ij} for a general reference value k_{ij} can be written as (see Prob. 7.9)

$$\hat{F}_{k_{ij}}(x_{ij}) = \frac{\det \hat{\mathbf{F}}(\mathbf{X})}{\det \hat{\mathbf{F}}(\mathbf{X})|_{x_{ij} = k_{ij}}} \qquad (7.203)$$

In the special situation where $\mathbf{X}$ is diagonal, the denominators of (7.199) and (7.200) can be simplified. In this case, let x_{ii} of $\mathbf{X}$ be the element of interest, and write

$$\Delta = \det \mathbf{F}(\mathbf{X}) \qquad (7.204a)$$

$$\hat{\Delta} = \det \hat{\mathbf{F}}(\mathbf{X}) \qquad (7.204b)$$

Then we have

$$\det \mathbf{F}(\mathbf{X})|_{x_{ii} = 0} = \det(\mathbf{1}_p - \mathbf{AX})|_{x_{ii} = 0} = \Delta_{ii} \qquad (7.205a)$$

$$\det \hat{\mathbf{F}}(\mathbf{X})|_{x_{ii} = 0} = \det(\mathbf{1}_p - \hat{\mathbf{A}}\mathbf{X})|_{x_{ii} = 0} = \hat{\Delta}_{ii} \qquad (7.205b)$$

where Δ_{ii}, and $\hat{\Delta}_{ii}$ are the cofactors of the (i, i)-element of $\mathbf{F}(\mathbf{X})$ and $\hat{\mathbf{F}}(\mathbf{X})$, respectively. This leads to

$$F(x_{ii}) = \frac{\Delta}{\Delta_{ii}} \qquad (7.206a)$$

$$\hat{F}(x_{ii}) = \frac{\hat{\Delta}}{\hat{\Delta}_{ii}} \qquad (7.206b)$$

In Sec. 5.6, we derived the expression for the relative sensitivity function in terms of the general return difference and null return difference. For a scalar transfer function w with respect to the element x_{ij}, the expression is given by

$$S'(x'_{ij}) = \frac{1}{F_{k_{ij}}(x_{ij})} - \frac{1}{\hat{F}_{k_{ij}}(x_{ij})} \qquad (7.207)$$

where $x'_{ij} = x_{ij} - k_{ij}$. From (7.202) and (7.203), we obtain

$$S'(x'_{ij}) = \frac{\det \mathbf{F}(\mathbf{X})|_{x_{ij} = k_{ij}}}{\det \mathbf{F}(\mathbf{X})} - \frac{\det \hat{\mathbf{F}}(\mathbf{X})|_{x_{ij} = k_{ij}}}{\det \hat{\mathbf{F}}(\mathbf{X})} \qquad (7.208)$$

Thus, the relative sensitivity function can be computed directly from the general return difference and null return difference matrices. For $k_{ij} = 0$, the relative sensitivity function becomes the ordinary sensitivity function:

$$S(x_{ij}) = \frac{\det \mathbf{F}(\mathbf{X})|_{x_{ij}=0}}{\det \mathbf{F}(\mathbf{X})} - \frac{\det \hat{\mathbf{F}}(\mathbf{X})|_{x_{ij}=0}}{\det \hat{\mathbf{F}}(\mathbf{X})} \tag{7.209}$$

We illustrate the above results by the following example.

Example 7.11 We wish to compute the return differences and the null return differences with respect to the controlling parameters $\tilde{\alpha}_j = 0.0455$ ($j = 1, 2$) of the two controlled sources of Fig. 7.15. For $F(\tilde{\alpha}_1)$ and $\hat{F}(\tilde{\alpha}_1)$, we choose

$$\mathbf{K}_1 = \begin{bmatrix} 0 & 0 \\ 0 & 0.0455 \end{bmatrix} \tag{7.210}$$

The return difference with respect to $\mathbf{K}_1$ is found to be

$$\mathbf{F}(\mathbf{K}_1) = \mathbf{1}_2 - \mathbf{A}\mathbf{K}_1 = \begin{bmatrix} 1 & -2.06529 \\ 0 & 1 \end{bmatrix} \tag{7.211}$$

where $\mathbf{A}$ is given in (7.168a), giving $\det \mathbf{F}(\mathbf{K}_1) = 1$. Using this in conjunction with (7.97), we obtain from (7.199)

$$F(\tilde{\alpha}_1) = \frac{\det \mathbf{F}(\mathbf{X})}{\det \mathbf{F}(\mathbf{K}_1)} = 93.68646 \tag{7.212}$$

confirming (4.141a). Alternatively, since $\mathbf{X}$ is diagonal, (7.206a) applies and from (7.96) we have $\Delta = 93.68646$ and $\Delta_{11} = 1$, which when substituted in (7.206a) yields (7.212). By using (7.79), the null return difference with respect to $\mathbf{K}_1$ is given by

$$\hat{\mathbf{F}}(\mathbf{K}_1) = \mathbf{1}_2 - \hat{\mathbf{A}}\mathbf{K}_1 = \begin{bmatrix} 1 & 2402.29 \\ 0 & 1 \end{bmatrix} \tag{7.213}$$

with $\det \hat{\mathbf{F}}(\mathbf{K}_1) = 1$. From (7.99) we get

$$\hat{F}(\tilde{\alpha}_1) = \frac{\det \hat{\mathbf{F}}(\mathbf{X})}{\det \hat{\mathbf{F}}(\mathbf{K}_1)} = 103.07 \cdot 10^3 \tag{7.214}$$

confirming (4.171). We can also apply (7.206b) to compute $\hat{F}(\tilde{\alpha}_1)$ with $\hat{\Delta} = 103.07 \cdot 10^3$ and $\hat{\Delta}_{11} = 1$, as given by (7.80).

In a similar manner, to compute $F(\tilde{\alpha}_2)$ and $\hat{F}(\tilde{\alpha}_2)$ we choose

$$\mathbf{K}_2 = \begin{bmatrix} 0.0455 & 0 \\ 0 & 0 \end{bmatrix} \tag{7.215}$$

From (7.168a) and (7.79), $\mathbf{F}(\mathbf{K}_2)$ and $\hat{\mathbf{F}}(\mathbf{K}_2)$ are determined as

$$\mathbf{F}(\mathbf{K}_2) = \begin{bmatrix} 5.13058 & 0 \\ 42.8841 & 1 \end{bmatrix} \tag{7.216a}$$

$$\hat{\mathbf{F}}(\mathbf{K}_2) = \begin{bmatrix} 51.0555 & 0 \\ 42.8841 & 1 \end{bmatrix} \tag{7.216b}$$

which in conjunction with (7.97) and (7.99) yield

$$F(\tilde{\alpha}_2) = \frac{\det \mathbf{F}(\mathbf{X})}{\det \mathbf{F}(\mathbf{K}_2)} = 18.26 \tag{7.217a}$$

$$\hat{F}(\tilde{\alpha}_2) = \frac{\det \hat{\mathbf{F}}(\mathbf{X})}{\det \hat{\mathbf{F}}(\mathbf{K}_2)} = 2018.80 \tag{7.217b}$$

confirming (4.141b) and (4.172), respectively. As before, we can also apply (7.206) directly to compute $F(\tilde{\alpha}_2)$ and $\hat{F}(\tilde{\alpha}_2)$ with $\Delta_{22} = 5.131$ and $\hat{\Delta}_{22} = 51.055$, as computed from (7.96) and (7.98), yielding, of course, the same results. For the sensitivity functions, we use (7.209) and obtain

$$S(\tilde{\alpha}_1) = \frac{\det \mathbf{F}(\mathbf{K}_1)}{\det \mathbf{F}(\mathbf{X})} - \frac{\det \hat{\mathbf{F}}(\mathbf{K}_1)}{\det \hat{\mathbf{F}}(\mathbf{X})} = 0.01066 \tag{7.218a}$$

$$S(\tilde{\alpha}_2) = \frac{\det \mathbf{F}(\mathbf{K}_2)}{\det \mathbf{F}(\mathbf{X})} - \frac{\det \hat{\mathbf{F}}(\mathbf{K}_2)}{\det \hat{\mathbf{F}}(\mathbf{X})} = 0.05427 \tag{7.218b}$$

which are approximately equal to the inverses of $F(\tilde{\alpha}_1)$ and $F(\tilde{\alpha}_2)$, respectively.

7.4 THE HYBRID-MATRIX FORMULATION OF MULTIPLE-LOOP FEEDBACK THEORY

In this section, we first explore the relation between the determinant of the hybrid matrix and that of the conventional cutset-admittance, node-admittance, or loop-impedance matrix, and then demonstrate how the determinant of the return difference matrix can be expressed elegantly and compactly in terms of the ratio of the two functional values assumed by the determinant of the hybrid matrix under the condition that the elements of interest assume their nominal and reference values.

7.4.1 Relations between the Hybrid Determinant and Other Network Determinants

In Sec. 5.9.1, we presented relations among the nodal, cutset, and loop determinants. In the following, we introduce the hybrid matrix and indicate how its determinant is related to the other network determinants. We assume that the reader is familiar with some of the basic terms of graph theory. For a detailed account of this subject, the reader is referred to Chen (1976a).

In a feedback network N, a branch may be taken to be either a simple resistor, inductor, capacitor, or independent or dependent generator, or a series and/or parallel connection of these elements, as we choose. For our purposes, we choose the most general branch representation of Fig. 7.32, containing both an independent and/or dependent voltage generator in series with a network branch and an independent and/or dependent current generator in parallel with this combination. The independent sources V_{sk} and I_{sk} in Fig. 7.32 may be replaced by the controlled voltage and current sources, respectively.

The starting point is the three primary systems of network equations. Let t be a tree in the directed graph G_d representing the feedback network N. Let $\mathbf{Q}_f$ and $\mathbf{B}_f$ be the fundamental cutset and fundamental circuit matrices of G_d with respect to the tree t. Partition $\mathbf{Q}_f$ and $\mathbf{B}_f$ according to the branches and chords of the tree t, and partition the branch-current vector $\mathbf{I}$ and branch-voltage vector $\mathbf{V}$ accordingly. Then we have

$$\begin{bmatrix} \mathbf{1}_r & \mathbf{Q}_{\bar{t}} \end{bmatrix} \begin{bmatrix} \mathbf{I}_t \\ \mathbf{I}_{\bar{t}} \end{bmatrix} = \mathbf{Q}_f \mathbf{I}_s \qquad (7.219a)$$

$$\begin{bmatrix} -\mathbf{Q}'_{\bar{t}} & \mathbf{1}_m \end{bmatrix} \begin{bmatrix} \mathbf{V}_t \\ \mathbf{V}_{\bar{t}} \end{bmatrix} = \mathbf{B}_f \mathbf{V}_s \qquad (7.219b)$$

where $\mathbf{I}_s$ and $\mathbf{V}_s$ are vectors of generator currents and voltages, and r and m denote the rank and nullity of G_d, respectively. The branch relations are characterized by the hybrid equations in the following form:

$$\begin{bmatrix} \mathbf{V}_{\bar{t}} \\ \mathbf{I}_t \end{bmatrix} = \begin{bmatrix} \mathbf{H}_{\bar{t}\bar{t}} & \mathbf{H}_{\bar{t}t} \\ \mathbf{H}_{t\bar{t}} & \mathbf{H}_{tt} \end{bmatrix} \begin{bmatrix} \mathbf{I}_{\bar{t}} \\ \mathbf{V}_t \end{bmatrix} \qquad (7.220)$$

The subscript t signifies that the quantity is defined in the tree t, and $\bar{t}$ is the complement of t. Thus, the elements of $\mathbf{H}_{tt}$ have the dimension of an admittance

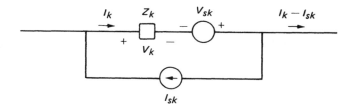

Figure 7.32 The most general branch representation.

and are defined by the tree branches, and those of $\mathbf{H}_{\bar{t}\bar{t}}$ have the dimension of an impedance and are defined by the chords of t. The elements of $\mathbf{H}_{\bar{t}t}$ and $\mathbf{H}_{t\bar{t}}$ are dimensionless and represent the coupling between the tree branches and the branches in their complement. The three systems of equations (7.219) and (7.220) can be written as a single matrix equation by considering the branch-current and branch-voltage vectors $\mathbf{I}$ and $\mathbf{V}$ as the unknowns:

$$\begin{bmatrix} -\mathbf{Q}'_{\bar{t}} & \mathbf{1}_m & 0 & 0 \\ 0 & 0 & \mathbf{1}_r & \mathbf{Q}_{\bar{t}} \\ \mathbf{H}_{tt} & 0 & -\mathbf{1}_r & \mathbf{H}_{t\bar{t}} \\ \mathbf{H}_{\bar{t}t} & -\mathbf{1}_m & 0 & \mathbf{H}_{\bar{t}\bar{t}} \end{bmatrix} \begin{bmatrix} \mathbf{V}_t \\ \mathbf{V}_{\bar{t}} \\ \mathbf{I}_t \\ \mathbf{I}_{\bar{t}} \end{bmatrix} = \begin{bmatrix} \mathbf{B}_f \mathbf{V}_s \\ \mathbf{Q}_f \mathbf{I}_s \\ 0 \\ 0 \end{bmatrix} \tag{7.221}$$

Using the elementary row operations to eliminate the variables $\mathbf{V}_{\bar{t}}$ and $\mathbf{I}_t$, we obtain a hybrid system of network equations

$$\begin{bmatrix} \mathbf{H}_{tt} & \mathbf{H}_{t\bar{t}} + \mathbf{Q}_{\bar{t}} \\ \mathbf{H}_{\bar{t}t} - \mathbf{Q}'_{\bar{t}} & \mathbf{H}_{\bar{t}\bar{t}} \end{bmatrix} \begin{bmatrix} \mathbf{V}_t \\ \mathbf{I}_{\bar{t}} \end{bmatrix} = \begin{bmatrix} \mathbf{Q}_f \mathbf{I}_s \\ \mathbf{B}_f \mathbf{V}_s \end{bmatrix} \tag{7.222}$$

or, more compactly,

$$\mathbf{H} \begin{bmatrix} \mathbf{V}_t \\ \mathbf{I}_{\bar{t}} \end{bmatrix} = \begin{bmatrix} \mathbf{Q}_f \mathbf{I}_s \\ \mathbf{B}_f \mathbf{V}_s \end{bmatrix} \tag{7.223}$$

The coefficient matrix $\mathbf{H}$ is called the *hybrid matrix* of the feedback network N, and its determinant is known as the *hybrid determinant*.

As in Sec. 5.9.1, let Δ_c, Δ_n, and Δ_m denote the cutset, nodal, and loop determinants of N, respectively. It can be shown that the hybrid determinant is related to Δ_c and Δ_m by the equations (see App. III)

$$\det \mathbf{H} = \lambda_{hc}(\det \mathbf{H}_{\bar{t}\bar{t}})\Delta_c \qquad (7.224a)$$

$$\det \mathbf{H} = \lambda_{hm}(\det \mathbf{H}_{tt})\Delta_m \qquad (7.224b)$$

where λ_{hc} and λ_{hm} are real constants, depending only on the choices of the cutsets and loops. In deriving (7.224), we implicitly assumed that the feedback amplifier N possesses the admittance and impedance representations. A formal proof of these relations can be found in Chen (1977a) and is outlined in App. III. In (7.220), if we express the branch-voltage vector $\mathbf{V}$ in terms of the branch-current vector $\mathbf{I}$, the coefficient matrix is the branch-impedance matrix and is given by

$$\mathbf{Z}_b(s) = \begin{bmatrix} \mathbf{H}_{tt}^{-1} & -\mathbf{H}_{tt}^{-1}\mathbf{H}_{t\bar{t}} \\ \mathbf{H}_{\bar{t}t}\mathbf{H}_{tt}^{-1} & \mathbf{H}_{\bar{t}\bar{t}} - \mathbf{H}_{\bar{t}t}\mathbf{H}_{tt}^{-1}\mathbf{H}_{t\bar{t}} \end{bmatrix} \qquad (7.225)$$

On the other hand, if we express $\mathbf{I}$ in terms of $\mathbf{V}$, we obtain the branch-admittance matrix as

$$\mathbf{Y}_b(s) = \begin{bmatrix} \mathbf{H}_{tt} - \mathbf{H}_{t\bar{t}}\mathbf{H}_{\bar{t}\bar{t}}^{-1}\mathbf{H}_{\bar{t}t} & \mathbf{H}_{t\bar{t}}\mathbf{H}_{\bar{t}\bar{t}}^{-1} \\ -\mathbf{H}_{\bar{t}\bar{t}}^{-1}\mathbf{H}_{\bar{t}t} & \mathbf{H}_{\bar{t}\bar{t}}^{-1} \end{bmatrix} \qquad (7.226)$$

We illustrate the above results by the following example.

Example 7.12 Consider the feedback network of Fig. 7.33, whose associated directed graph G_d is presented in Fig. 7.34. Choose the tree $t = e_1 e_3$. The

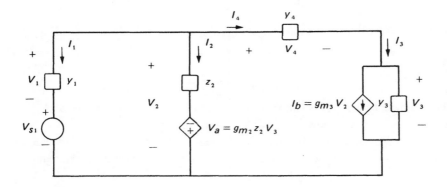

Figure 7.33 A feedback network containing two controlled sources.

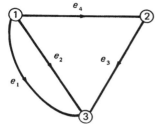

Figure 7.34 The associated directed graph of the network of Fig. 7.33.

corresponding fundamental cutset matrix is found to be

$$\mathbf{Q}_f = \begin{bmatrix} \mathbf{1}_2 & \mathbf{Q}_{\bar{t}} \end{bmatrix} = \begin{matrix} e_1 & e_3 & e_2 & e_4 \\ \begin{bmatrix} 1 & 0 & 1 & 1 \\ 0 & 1 & 0 & -1 \end{bmatrix} \end{matrix} \tag{7.227}$$

The branch vi-relations are described by the hybrid equation

$$\begin{bmatrix} V_2 \\ V_4 \\ \hline I_1 \\ I_3 \end{bmatrix} = \begin{bmatrix} z_2 & 0 & 0 & z_2 g_{m2} \\ 0 & z_4 & 0 & 0 \\ \hline 0 & 0 & y_1 & 0 \\ g_{m3} z_2 & 0 & 0 & y_3 - z_2 g_{m2} g_{m3} \end{bmatrix} \begin{bmatrix} I_2 \\ I_4 \\ \hline V_1 \\ V_3 \end{bmatrix} \tag{7.228}$$

where $z_i = 1/y_i$ $(i = 1, 2, 3, 4)$. Substituting the appropriate quantities in (7.222) yields the hybrid system of equations

$$\begin{bmatrix} y_1 & 0 & 1 & 1 \\ 0 & y_3 - g_{m2} g_{m3} z_2 & g_{m3} z_2 & -1 \\ -1 & -g_{m2} z_2 & z_2 & 0 \\ -1 & 1 & 0 & z_4 \end{bmatrix} \begin{bmatrix} V_1 \\ V_3 \\ I_2 \\ I_4 \end{bmatrix} = \begin{bmatrix} 0 \\ 0 \\ V_{s1} \\ V_{s1} \end{bmatrix} \tag{7.229}$$

The hybrid determinant is found to be

$$\det \mathbf{H} = z_2 z_4 \left[(y_1 + y_2)(y_3 + y_4) + y_4(y_3 + g_{m2} + g_{m3}) - g_{m2} g_{m3} \right] \tag{7.230}$$

The cutset-admittance matrix $\mathbf{Y}_c(s)$ is defined by the matrix triple product $\mathbf{Q}_f \mathbf{Y}_b \mathbf{Q}'_f$ and is given by

$$\mathbf{Y}_c(s) = \mathbf{Q}_f \mathbf{Y}_b \mathbf{Q}'_f = \begin{bmatrix} y_1 + y_2 + y_4 & g_{m2} - y_4 \\ g_{m3} - y_4 & y_3 + y_4 \end{bmatrix} \qquad (7.231)$$

where the branch-admittance matrix is computed by means of (7.226), giving

$$\mathbf{Y}_b(s) = \begin{bmatrix} y_1 & 0 & 0 & 0 \\ 0 & y_3 & g_{m3} & 0 \\ 0 & g_{m2} & y_2 & 0 \\ 0 & 0 & 0 & y_4 \end{bmatrix} \qquad (7.232)$$

It is straightforward to verify that

$$\det \mathbf{H} = \det \mathbf{H}_{\bar{i}\bar{i}} \det \mathbf{Y}_c = (\det \mathbf{H}_{\bar{i}\bar{i}}) \Delta_c \qquad (7.233)$$

where $\det \mathbf{H}_{\bar{i}\bar{i}} = z_2 z_4$. This confirms (7.224a) with $\lambda_{hc} = 1$. To demonstrate (7.224b), we compute the fundamental circuit matrix

$$\mathbf{B}_f = [-\mathbf{Q}'_i \quad \mathbf{1}_2] = \begin{matrix} e_1 \quad e_3 \quad e_2 \quad e_4 \\ \begin{bmatrix} -1 & 0 & 1 & 0 \\ -1 & 1 & 0 & 1 \end{bmatrix} \end{matrix} \qquad (7.234)$$

yielding the loop determinant

$$\Delta_m = \det \mathbf{B}_f \mathbf{Z}_b \mathbf{B}'_f = \det \begin{bmatrix} z_1 + \dfrac{y_3}{P} & z_1 - \dfrac{g_{m2}}{P} \\ z_1 - \dfrac{g_{m3}}{P} & z_1 + z_4 + \dfrac{y_2}{P} \end{bmatrix}$$

$$= \frac{z_1 z_4}{P} [(y_1 + y_2)(y_3 + y_4) + y_4(y_3 + g_{m2} + g_{m3}) - g_{m2} g_{m3}] \qquad (7.235)$$

where $P = y_2 y_3 - g_{m2} g_{m3}$, and the branch-impedance matrix is obtained from (7.225) as

$$
\mathbf{Z}_b(s) = \begin{bmatrix}
z_1 & 0 & 0 & 0 \\[2mm]
0 & \dfrac{y_2}{P} & -\dfrac{g_{m3}}{P} & 0 \\[4mm]
0 & -\dfrac{g_{m2}}{P} & \dfrac{y_3}{P} & 0 \\[4mm]
0 & 0 & 0 & z_4
\end{bmatrix}
\tag{7.236}
$$

Comparing (7.235) with (7.230) gives

$$
\det \mathbf{H} = (\det \mathbf{H}_{tt}) \Delta_m
\tag{7.237}
$$

where $\det \mathbf{H}_{tt} = y_1 (y_3 - g_{m2} g_{m3} z_2) = y_1 z_2 P$.

7.4.2 Hybrid Formulation of the Determinant of the Return Difference Matrix

In this section, we show that, like the scalar return difference, the determinant of the return difference matrix can be expressed elegantly and compactly as the ratio of the two functional values assumed by the hybrid determinant under the condition that the elements of interest assume their nominal values and the condition that they assume their reference values.

Let the element x_{ij} of $\mathbf{X}$ be the controlling parameter of a voltage- or current-controlled current source. Then the return difference with respect to x_{ij} for a reference value k_{ij} can be expressed as the ratio of the two functional values assumed by the cutset determinant, as given in (5.218). Using this in conjunction with (7.202), we obtain

$$
F_{k_{ij}}(x_{ij}) = \frac{\Delta_c(\mathbf{X})}{\Delta_c(\mathbf{X})|_{x_{ij}=k_{ij}}} = \frac{\det \mathbf{F}(\mathbf{X})}{\det \mathbf{F}(\mathbf{X})|_{x_{ij}=k_{ij}}}
\tag{7.238}
$$

Appealing to (7.224a) and assuming that $\det \mathbf{H}_{\bar{t}\bar{t}}$ is independent of x_{ij} yield

$$
F_{k_{ij}}(x_{ij}) = \frac{\det \mathbf{F}(\mathbf{X})}{\det \mathbf{F}(\mathbf{X})|_{x_{ij}=k_{ij}}} = \frac{\det \mathbf{H}(\mathbf{X})}{\det \mathbf{H}(\mathbf{X})|_{x_{ij}=k_{ij}}}
\tag{7.239}
$$

Likewise, if x_{ij} of $\mathbf{X}$ is the controlling parameter of a voltage- or current- controlled voltage source, then the return difference with respect to x_{ij} for a reference value k_{ij} can be expressed from (5.219) and (7.202) as

$$F_{k_{ij}}^{m}(x_{ij}) = \frac{\Delta_m(\mathbf{X})}{\Delta_m(\mathbf{X})|_{x_{ij}=k_{ij}}} = \frac{\det \mathbf{F}(\mathbf{X})}{\det \mathbf{F}(\mathbf{X})|_{x_{ij}=k_{ij}}} \qquad (7.240)$$

Applying (7.224b) and assuming that $\det \mathbf{H}_{tt}$ is independent of x_{ij}, we have

$$F_{k_{ij}}^{m}(x_{ij}) = \frac{\det \mathbf{F}(\mathbf{X})}{\det \mathbf{F}(\mathbf{X})|_{x_{ij}=k_{ij}}} = \frac{\det \mathbf{H}(\mathbf{X})}{\det \mathbf{H}(\mathbf{X})|_{x_{ij}=k_{ij}}} \qquad (7.241)$$

Observe that (7.239) and (7.241) are identical in form, meaning that either one is valid for any one of the four types of controlled sources if the hybrid matrix $\mathbf{H}$ of (7.220) can be partitioned so that $\det \mathbf{H}_{\bar{i}\bar{i}}$ is independent of the controlling parameters of the controlled current sources appearing in $\mathbf{X}$ and $\det \mathbf{H}_{tt}$ is independent of the controlling parameters of the controlled voltage sources appearing in $\mathbf{X}$. These two conditions can easily be fulfilled by simple manipulation in the formulation of the hybrid matrix $\mathbf{H}$.

Suppose that $\mathbf{X}_i$ is obtained from $\mathbf{X}_{i-1}$ by setting a certain element to its reference value for $i = 1, 2, \ldots, pq$ with $\mathbf{X}_0 = \mathbf{X}$, where $\mathbf{X}$ is of order $q \times p$. Then from (7.239) and (7.241) we have the equality

$$\frac{\det \mathbf{F}(\mathbf{X}_{i-1})}{\det \mathbf{F}(\mathbf{X}_i)} = \frac{\det \mathbf{H}(\mathbf{X}_{i-1})}{\det \mathbf{H}(\mathbf{X}_i)} \qquad (7.242)$$

for all i. Forming the product of all the terms of the above type yields

$$\frac{\det \mathbf{F}(\mathbf{X})}{\det \mathbf{F}(\mathbf{K})} = \frac{\det \mathbf{H}(\mathbf{X})}{\det \mathbf{H}(\mathbf{K})} \qquad (7.243)$$

where $\mathbf{K}$ is the reference matrix of $\mathbf{X}$. Invoking (7.162), we get the desired relation:

$$\det \mathbf{F_K}(\mathbf{X}) = \frac{\det \mathbf{H}(\mathbf{X})}{\det \mathbf{H}(\mathbf{K})} \qquad (7.244)$$

Example 7.13 In the feedback network of Fig. 7.33, let the controlling parameters $g_{m2}z_2$ and g_{m3} be the elements of interest. Then we have

$$\theta = \begin{bmatrix} I_b \\ V_a \end{bmatrix} = \begin{bmatrix} g_{m3} & 0 \\ 0 & g_{m2}z_2 \end{bmatrix} \begin{bmatrix} V_2 \\ V_3 \end{bmatrix} = \mathbf{X}\phi \qquad (7.245)$$

$$\phi = \begin{bmatrix} V_2 \\ V_3 \end{bmatrix} = -\frac{1}{\lambda} \begin{bmatrix} y_4 & y_2(y_3 + y_4) \\ y_1 + y_2 + y_4 & y_2 y_4 \end{bmatrix} \begin{bmatrix} I_a \\ V_b \end{bmatrix}$$

$$+ \frac{1}{\lambda} \begin{bmatrix} y_1(y_3 + y_4) \\ y_1 y_4 \end{bmatrix} [V_{s1}] = \mathbf{A}\theta + \mathbf{B}u \qquad (7.246)$$

where $\lambda = (y_1 + y_2)(y_3 + y_4) + y_3 y_4$. Substituting $\mathbf{X}$ and $\mathbf{A}$ from (7.245) and (7.246) in (7.34) gives the return difference matrix

$$\mathbf{F}(\mathbf{X}) = \frac{1}{\lambda} \begin{bmatrix} \lambda + g_{m3} y_4 & g_{m2}(y_3 + y_4) \\ g_{m3}(y_1 + y_2 + y_4) & \lambda + g_{m2} y_4 \end{bmatrix} \qquad (7.247)$$

whose determinant is found to be

$$\det \mathbf{F}(\mathbf{X}) = \frac{(y_1 + y_2)(y_3 + y_4) + y_4(y_3 + g_{m2} + g_{m3}) - g_{m2}g_{m3}}{(y_1 + y_2)(y_3 + y_4) + y_3 y_4} \qquad (7.248)$$

To compute $\det \mathbf{F}(\mathbf{X})$ by means of (7.244), the coefficient matrix $\mathbf{H}$ of (7.228) is inadequate, since $\det \mathbf{H}_{tt}$ is not independent of the controlling parameter $g_{m2}z_2$ of the controlled voltage source V_a. With a simple modification, the vi-relations of the network of Fig. 7.33 can be described by the hybrid equation

$$\begin{bmatrix} V_2 \\ V_4 \\ I_1 \\ I_3 \end{bmatrix} = \begin{bmatrix} z_2 & 0 & 0 & -g_{m2}z_2 \\ 0 & z_4 & 0 & 0 \\ 0 & 0 & y_1 & 0 \\ 0 & g_{m3}z_4 & 0 & y_3 + g_{m3} \end{bmatrix} \begin{bmatrix} I_2 \\ I_4 \\ V_1 \\ V_3 \end{bmatrix} \qquad (7.249)$$

The determinant of the coefficient matrix $\mathbf{H}$ is obtained as

$$\det \mathbf{H} = z_2 z_4 \left[(y_1 + y_2)(y_3 + y_4) + y_4(y_3 + g_{m2} + g_{m3}) - g_{m2}g_{m3} \right] \qquad (7.250)$$

which when substituted in (7.244) yields

$$\det \mathbf{F}(\mathbf{X}) = \frac{(y_1 + y_2)(y_3 + y_4) + y_4(y_3 + g_{m2} + g_{m3}) - g_{m2}g_{m3}}{(y_1 + y_2)(y_3 + y_4) + y_3 y_4} \qquad (7.251)$$

confirming (7.248). We remark that det $\mathbf{H}$ of (7.250) is the same as that given in (7.230). Suppose that k_2 and k_3 are the reference values for $g_{m2}z_2$ and g_{m3}. Then, from (7.162) and (7.244) we have

$$
\begin{aligned}
\det \mathbf{F_K}(\mathbf{X}) &= \frac{\det \mathbf{F}(\mathbf{X})}{\det \mathbf{F}(\mathbf{K})} = \frac{\det \mathbf{H}(\mathbf{X})}{\det \mathbf{H}(\mathbf{K})} \\
&= \frac{(y_1 + y_2)(y_3 + y_4) + y_4(y_3 + g_{m2} + g_{m3}) - g_{m2}g_{m3}}{(y_1 + y_2)(y_3 + y_4) + y_4(y_3 + k_2 y_2 + k_3) - k_2 k_3 y_2}
\end{aligned}
$$

$$(7.252)$$

in which

$$
\mathbf{K} = \begin{bmatrix} k_3 & 0 \\ 0 & k_2 \end{bmatrix}
\tag{7.253}
$$

With the coefficient matrix $\mathbf{H}$ as given in (7.249), the corresponding branch-admittance matrix $\mathbf{Y}_b$ and branch-impedance matrix $\mathbf{Z}_b$ are determined from (7.226) and (7.225), and are given by

$$
\mathbf{Y}_b(s) = \begin{bmatrix}
y_1 & 0 & 0 & 0 \\
0 & y_3 + g_{m3} & 0 & g_{m3} \\
0 & g_{m2} & y_2 & 0 \\
0 & 0 & 0 & y_4
\end{bmatrix}
\tag{7.254}
$$

$$
\mathbf{Z}_b(s) = \begin{bmatrix}
z_1 & 0 & 0 & 0 \\
0 & \dfrac{1}{\lambda_1} & 0 & \dfrac{-g_{m3}z_4}{\lambda_1} \\
0 & \dfrac{-g_{m2}z_2}{\lambda_1} & z_2 & \dfrac{g_{m2}g_{m3}z_2z_4}{\lambda_1} \\
0 & 0 & 0 & z_4
\end{bmatrix}
\tag{7.255}
$$

where $\lambda_1 = y_3 + g_{m3}$. Using $\mathbf{Q}_f$ and $\mathbf{B}_f$ as given in (7.227) and (7.234), we can compute the cutset and loop determinants. It is straightforward to confirm that they satisfy the identities (7.224) with $\lambda_{hc} = \lambda_{hm} = 1$, $\det \mathbf{H}_{\bar{t}\bar{t}} = z_2 z_4$, and $\det \mathbf{H}_{tt} = y_1(y_3 + g_{m3})$. The details are left as an exercise (see Prob. 7.11).

7.5 THE SENSITIVITY MATRIX AND MULTIPARAMETER SENSITIVITY

So far we have studied the sensitivity of a transfer function with respect to the change of a particular element in the network. In a multiple-loop feedback network, we are usually interested in the sensitivity of a transfer function with respect to the variation of a set of elements in the network. This set may include either elements that are inherently sensitive to variation or elements whose effect on the overall amplifier performance is of paramount importance to the designers. In this section, we introduce a sensitivity matrix for a multiple-loop feedback amplifier and develop formulas for computing multiparameter sensitivity functions.

7.5.1 The Sensitivity Matrix

Figure 7.35a is the block diagram of a multivariable open-loop control system with n inputs and m outputs, whereas Fig. 7.35b shows the general feedback structure. If all feedback signals are obtainable from the output and if the controllers are linear, there is no loss of generality by assuming the controller to be of the form shown in Fig. 7.36. Denote the set of Laplace transformed input signals by the n-vector $\mathbf{u}$, the set of inputs to the network $\mathbf{X}$ in the open-loop configuration of Fig. 7.35a by the p-vector $\boldsymbol{\phi}_o$, and the set of outputs of the network $\mathbf{X}$ in Fig. 7.35a by the m-vector y_o. Let the corresponding signals for the closed-loop configuration of Fig. 7.36 be denoted by the n-vector $\mathbf{u}$, the p-vector $\boldsymbol{\phi}_c$, and the m-vector $\mathbf{y}_c$,

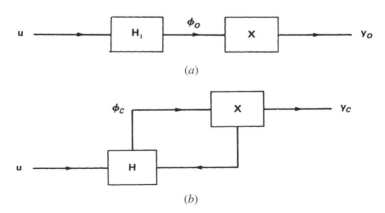

(a)

(b)

Figure 7.35 (a) The block diagram of a multivariable open-loop control system and (b) the general feedback structure.

Figure 7.36 A general feedback configuration.

respectively. Then from Figs. 7.35*a* and 7.36, we obtain the following relations:

$$\mathbf{y}_o = \mathbf{X}\boldsymbol{\phi}_o \tag{7.256a}$$

$$\boldsymbol{\phi}_o = \mathbf{H}_1\mathbf{u} \tag{7.256b}$$

$$\mathbf{y}_c = \mathbf{X}\boldsymbol{\phi}_c \tag{7.256c}$$

$$\boldsymbol{\phi}_c = \mathbf{H}_2(\mathbf{u} + \mathbf{H}_3\mathbf{y}_c) \tag{7.256d}$$

where the transfer function matrices $\mathbf{X}, \mathbf{H}_1, \mathbf{H}_2$, and $\mathbf{H}_3$ are of orders $m \times p$, $p \times n$, $p \times n$, and $n \times m$, respectively. Combining (7.256*c*) and (7.256*d*) yields

$$(\mathbf{1}_m - \mathbf{X}\mathbf{H}_2\mathbf{H}_3)\mathbf{y}_c = \mathbf{X}\mathbf{H}_2\mathbf{u} \tag{7.257}$$

or

$$\mathbf{y}_c = (\mathbf{1}_m - \mathbf{X}\mathbf{H}_2\mathbf{H}_3)^{-1}\mathbf{X}\mathbf{H}_2\mathbf{u} \tag{7.258}$$

The *closed-loop transfer function matrix* that relates the input and output is defined by the equation

$$\mathbf{y}_c = \mathbf{W}(\mathbf{X})\mathbf{u} \tag{7.259}$$

where $\mathbf{W}(\mathbf{X})$ is of order $m \times n$, giving

$$\mathbf{W}(\mathbf{X}) = (\mathbf{1}_m - \mathbf{X}\mathbf{H}_2\mathbf{H}_3)^{-1}\mathbf{X}\mathbf{H}_2 \tag{7.260}$$

Now suppose that $\mathbf{X}$ is perturbed to $\mathbf{X}^+ = \mathbf{X} + \delta\mathbf{X}$. The outputs of the open-loop and closed-loop systems of Figs. 7.35*a* and 7.36 will no longer be the same as before. Distinguishing the new from the old variables by the superscript $+$, we have

$$\mathbf{y}_o^+ = \mathbf{X}^+ \boldsymbol{\phi}_o \qquad (7.261a)$$

$$\mathbf{y}_c^+ = \mathbf{X}^+ \boldsymbol{\phi}_c^+ \qquad (7.261b)$$

$$\boldsymbol{\phi}_c^+ = \mathbf{H}_2(\mathbf{u} + \mathbf{H}_3 \mathbf{y}_c^+) \qquad (7.261c)$$

where $\boldsymbol{\phi}_o$ remains the same.

We next proceed to compare the relative effects of the variations of $\mathbf{X}$ on the performance of the open-loop and the closed-loop systems. For a meaningful comparison, we assume that $\mathbf{H}_1$, $\mathbf{H}_2$, and $\mathbf{H}_3$ are such that when there is no variation of $\mathbf{X}$, $\mathbf{y}_o = \mathbf{y}_c$. Define the error vectors resulting from perturbation of $\mathbf{X}$ as

$$\mathbf{E}_o = \mathbf{y}_o - \mathbf{y}_o^+ \qquad (7.262a)$$

$$\mathbf{E}_c = \mathbf{y}_c - \mathbf{y}_c^+ \qquad (7.262b)$$

A square matrix relating $\mathbf{E}_c$ to $\mathbf{E}_o$ is defined as the *sensitivity matrix* for the transfer function matrix $\mathbf{W}(\mathbf{X})$ with respect to the variations of $\mathbf{X}$:

$$\mathbf{E}_c = \mathbf{S}(\mathbf{X})\mathbf{E}_o \qquad (7.263)$$

In the following, we express the sensitivity matrix $\mathbf{S}(\mathbf{X})$ in terms of the system matrices $\mathbf{X}$, $\mathbf{H}_2$, and $\mathbf{H}_3$.

The output and input relation similar to that given in (7.258) for the perturbed system can be written as

$$\mathbf{y}_c^+ = (\mathbf{1}_m - \mathbf{X}^+ \mathbf{H}_2 \mathbf{H}_3)^{-1} \mathbf{X}^+ \mathbf{H}_2 \mathbf{u} \qquad (7.264)$$

Substituting (7.258) and (7.264) in (7.262b) gives

$$\begin{aligned}
\mathbf{E}_c = \mathbf{y}_c - \mathbf{y}_c^+ &= \left[(\mathbf{1}_m - \mathbf{X} \mathbf{H}_2 \mathbf{H}_3)^{-1} \mathbf{X} \mathbf{H}_2 - (\mathbf{1}_m - \mathbf{X}^+ \mathbf{H}_2 \mathbf{H}_3)^{-1} \mathbf{X}^+ \mathbf{H}_2 \right] \mathbf{u} \\
&= (\mathbf{1}_m - \mathbf{X}^+ \mathbf{H}_2 \mathbf{H}_3)^{-1} \{ [\mathbf{1}_m - (\mathbf{X} + \delta\mathbf{X}) \mathbf{H}_2 \mathbf{H}_3] \\
&\quad \times (\mathbf{1}_m - \mathbf{X} \mathbf{H}_2 \mathbf{H}_3)^{-1} \mathbf{X} \mathbf{H}_2 - (\mathbf{X} + \delta\mathbf{X}) \mathbf{H}_2 \} \mathbf{u} \\
&= (\mathbf{1}_m - \mathbf{X}^+ \mathbf{H}_2 \mathbf{H}_3)^{-1} \left[\mathbf{X} \mathbf{H}_2 - \delta\mathbf{X} \mathbf{H}_2 \mathbf{H}_3 (\mathbf{1}_m - \mathbf{X} \mathbf{H}_2 \mathbf{H}_3)^{-1} \mathbf{X} \mathbf{H}_2 \right. \\
&\quad \left. - \mathbf{X} \mathbf{H}_2 - \delta\mathbf{X} \mathbf{H}_2 \right] \mathbf{u} \\
&= -(\mathbf{1}_m - \mathbf{X}^+ \mathbf{H}_2 \mathbf{H}_3)^{-1} \delta\mathbf{X} \mathbf{H}_2 [\mathbf{1}_n + \mathbf{H}_3 \mathbf{W}(\mathbf{X})] \mathbf{u} \qquad (7.265)
\end{aligned}$$

From (7.256d) in conjunction with (7.259), we have

$$\boldsymbol{\phi}_c = \mathbf{H}_2\,[\mathbf{1}_n + \mathbf{H}_3\mathbf{W}(\mathbf{X})]\,\mathbf{u} \qquad (7.266)$$

Since by assumption $\mathbf{y}_0 = \mathbf{y}_c$,

$$\boldsymbol{\phi}_o = \boldsymbol{\phi}_c = \mathbf{H}_2\,[\mathbf{1}_n + \mathbf{H}_3\mathbf{W}(\mathbf{X})]\,\mathbf{u} \qquad (7.267)$$

yielding

$$\mathbf{E}_o = \mathbf{y}_o - \mathbf{y}_o^+ = (\mathbf{X} - \mathbf{X}^+)\boldsymbol{\phi}_o = -\delta\mathbf{X}\boldsymbol{\phi}_o = -\delta\mathbf{X}\mathbf{H}_2\,[\mathbf{1}_n + \mathbf{H}_3\mathbf{W}(\mathbf{X})]\,\mathbf{u}$$
$$(7.268)$$

Combining (7.265) and (7.268) yields the desired expression relating the error vectors $\mathbf{E}_c$ and $\mathbf{E}_o$ of the closed-loop and open-loop systems by

$$\mathbf{E}_c = (\mathbf{1}_m - \mathbf{X}^+\mathbf{H}_2\mathbf{H}_3)^{-1}\mathbf{E}_o \qquad (7.269)$$

giving the sensitivity matrix first introduced by Cruz and Perkins (1964) as

$$\mathbf{S}(\mathbf{X}) = (\mathbf{1}_m - \mathbf{X}^+\mathbf{H}_2\mathbf{H}_3)^{-1} \qquad (7.270)$$

For small variation of $\mathbf{X}$, $\mathbf{X}^+$ is approximately equal to $\mathbf{X}$. Refer to Fig. 7.36. The matrix triple product $\mathbf{X}\mathbf{H}_2\mathbf{H}_3$ may be regarded as the loop-transmission matrix, and $-\mathbf{X}\mathbf{H}_2\mathbf{H}_3$ as the return ratio matrix. The difference between the unit matrix and the loop-transmission matrix, $\mathbf{1}_m - \mathbf{X}\mathbf{H}_2\mathbf{H}_3$, can be defined as the return difference matrix (see Prob. 7.27). Therefore, Eq. (7.270) is a direct extension of the sensitivity function defined for a single-input, single-output system and for a single parameter, as shown in Eq. (5.165), where the sensitivity function of the closed-loop transfer function with respect to the forward amplifier gain is equal to the reciprocal of the return difference with respect to the forward amplifier gain. It is straightforward to demonstrate that (7.270) reduces to (5.165) for a single-input, single-output system and for a small perturbation of $\mathbf{X}$.

In the particular situation where $\mathbf{W}(\mathbf{X})$, $\delta\mathbf{X}$, and $\mathbf{X}$ are square and nonsingular, (7.270) can be put into the form similar to the scalar sensitivity function of (3.20). To this end, we use (7.262) in conjunction with (7.256) and (7.259) and obtain

$$\mathbf{E}_c = \mathbf{y}_c - \mathbf{y}_c^+ = [\mathbf{W}(\mathbf{X}) - \mathbf{W}^+(\mathbf{X})]\mathbf{u} = -\delta\mathbf{W}(\mathbf{X})\mathbf{u} \qquad (7.271a)$$

$$\mathbf{E}_o = \mathbf{y}_o - \mathbf{y}_o^+ = (\mathbf{X}\mathbf{H}_1 - \mathbf{X}^+\mathbf{H}_1)\mathbf{u} = -\delta\mathbf{X}\mathbf{H}_1\mathbf{u} \qquad (7.271b)$$

If $\mathbf{H}_1$ is nonsingular, $\mathbf{u}$ in $(7.271b)$ can be solved for and substituted in the expression of $(7.271a)$ to give

$$\mathbf{E}_c = \delta \mathbf{W}(\mathbf{X})\mathbf{H}_1^{-1}(\delta \mathbf{X})^{-1}\mathbf{E}_o \qquad (7.272)$$

As before, for meaningful comparisons, we require that $\mathbf{y}_0 = \mathbf{y}_c$ or

$$\mathbf{X}\mathbf{H}_1 = \mathbf{W}(\mathbf{X}) \qquad (7.273)$$

Using this in (7.272), we get

$$\mathbf{E}_c = \delta \mathbf{W}(\mathbf{X})\mathbf{W}^{-1}(\mathbf{X})\mathbf{X}(\delta \mathbf{X})^{-1}\mathbf{E}_o \qquad (7.274)$$

showing that (see Prob. 7.28)

$$\mathbf{S}(\mathbf{X}) = \delta \mathbf{W}(\mathbf{X})\mathbf{W}^{-1}(\mathbf{X})\mathbf{X}(\delta \mathbf{X})^{-1} \qquad (7.275)$$

This result is to be compared with the scalar sensitivity function of Eq. (3.20), which can be rewritten in the form

$$S(x) = (\delta w)w^{-1}x(\delta x)^{-1} \qquad (7.276)$$

We emphasize that the expression (7.275) is valid only when $\mathbf{W}(\mathbf{X})$, $\mathbf{X}$, and $\delta \mathbf{X}$ are square and nonsingular. However, (7.270) is well defined as long as the square matrix $\mathbf{1}_m - \mathbf{X}^+\mathbf{H}_2\mathbf{H}_3$ is nonsingular.

To demonstrate the use of (7.275), we consider the fundamental matrix feedback-flow graph of Fig. 7.13. As shown in Eq. $(7.16a)$, the overall transfer function matrix $\mathbf{W}(\mathbf{X})$ of the multiple-loop feedback amplifier of Fig. 7.12 is given by

$$\mathbf{W}(\mathbf{X}) = \mathbf{D} + \mathbf{C}\mathbf{X}(\mathbf{1}_p - \mathbf{A}\mathbf{X})^{-1}\mathbf{B} \qquad (7.277)$$

We wish to compute the sensitivity matrix $\mathbf{S}(\mathbf{X})$ for the situation when all changes concerned are differentially small:

$$\mathbf{S}(\mathbf{X}) = \delta \mathbf{W}(\mathbf{X})\mathbf{W}^{-1}(\mathbf{X})\mathbf{X}(\delta \mathbf{X})^{-1}|_{\delta \mathbf{X} \to \mathbf{0}} \qquad (7.278)$$

provided that $\mathbf{W}(\mathbf{X})$ and $\mathbf{X}$ are square and nonsingular. Then from (7.277) we have an expression for $\mathbf{W}(\mathbf{X})$ when $\mathbf{X}$ is perturbed to $\mathbf{X}^+ = \mathbf{X} + \delta \mathbf{X}$:

$$\mathbf{W}(\mathbf{X}) + \delta \mathbf{W}(\mathbf{X}) = \mathbf{D} + \mathbf{C}(\mathbf{X} + \delta \mathbf{X})(\mathbf{1}_p - \mathbf{A}\mathbf{X} - \mathbf{A}\delta \mathbf{X})^{-1}\mathbf{B} \qquad (7.279)$$

or

$$\delta \mathbf{W}(\mathbf{X}) = \mathbf{C}[(\mathbf{X} + \delta \mathbf{X})(\mathbf{1}_p - \mathbf{A}\mathbf{X} - \mathbf{A}\delta \mathbf{X})^{-1} - \mathbf{X}(\mathbf{1}_p - \mathbf{A}\mathbf{X})^{-1}]\mathbf{B} \qquad (7.280)$$

As δX approaches zero,

$$\delta W(X) = C[(X + \delta X) - X(1_p - AX)^{-1}(1_p - AX - A\delta X)]$$
$$\times (1_p - AX - A\delta X)^{-1}B$$
$$= C[\delta X + X(1_p - AX)^{-1}A\delta X](1_p - AX - A\delta X)^{-1}B$$
$$= C(1_q - XA)^{-1}(\delta X)(1_p - AX - A\delta X)^{-1}B$$
$$\approx C(1_q - XA)^{-1}(\delta X)(1_p - AX)^{-1}B \qquad (7.281)$$

Assuming that $W(X)$, X, δX, and C are nonsingular with $n = m = p = q$, Eq. (7.275) becomes

$$S(X) = C(1_n - XA)^{-1}(\delta X)(1_n - AX)^{-1}BW^{-1}(X)X(\delta X)^{-1}$$
$$= C(1_n - XA)^{-1}(\delta X)X^{-1}C^{-1}[W(X) - D]W^{-1}(X)X(\delta X)^{-1}$$
$$= C(1_n - XA)^{-1}(\delta X)X^{-1}C^{-1}[1_n - W(0)W^{-1}(X)]X(\delta X)^{-1}$$
$$= C\tilde{F}^{-1}(X)(\delta X)X^{-1}C^{-1}[1_n - W(0)W^{-1}(X)]X(\delta X)^{-1} \qquad (7.282)$$

where $W(0) = D$ and $\tilde{F}(X) = 1_n - XA$. Since from (7.83), $\det \tilde{F}(X) = \det (1_n - XA) = \det (1_n - AX) = \det F(X)$, the determinant of $S(X)$ is found to be

$$\det S(X) = \frac{\det [1_n - W(0)W^{-1}(X)]}{\det \tilde{F}(X)} = \frac{\det [1_n - W(0)W^{-1}(X)]}{\det F(X)}$$
$$(7.283)$$

This expression is a direct generalization of Eq. (5.6) for the scalar sensitivity function, and it relates the sensitivity matrix, the return difference matrix, and the transfer-function matrix of a multiple-loop feedback amplifier. Like (5.6), (7.283) indicates that if $W(0) = 0$, then $\det S(X) = 1/\det F(X)$, meaning that the determinant of the sensitivity matrix is equal to the reciprocal of the determinant of the return difference matrix $F(X)$. As to the physical significance of $\tilde{F}(X)$, we recall that in Sec. 7.2.1 the return difference matrix was defined by $F(X) = 1_n - AX$, where AX is the loop transmission matrix and is obtained by opening the loop at the input of X, as shown in Fig. 7.37a. The matrix XA has the physical significance of loop transmission if the loop is opened at the input of A, as indicated in Fig. 7.37b. Thus, $\tilde{F}(X)$ is the return difference matrix with respect to the branch A.

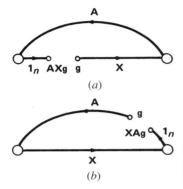

Figure 7.37 (*a*) The physical interpretation of the loop transmission **AX** and (*b*) the physical interpretation of the loop transmission **XA**.

7.5.2 The Multiparameter Sensitivity

In this section, we derive formulas for the effect of change of $\mathbf{X}$ on a scalar transfer function $w(\mathbf{X})$.

Let $x_k, k = 1, 2, \ldots, pq$, be the elements of $\mathbf{X}$. The multivariable Taylor series expansion for $w(\mathbf{X})$ with respect to x_k is given by

$$\delta w = \sum_{k=1}^{pq} \frac{\partial w}{\partial x_k} \delta x_k + \sum_{j=1}^{pq} \sum_{k=1}^{pq} \frac{\partial^2 w}{\partial x_j \partial x_k} \frac{\delta x_j \delta x_k}{2} + \cdots \qquad (7.284)$$

The first-order perturbation can then be written as

$$\delta w \approx \sum_{k=1}^{pq} \frac{\partial w}{\partial x_k} \delta x_k \qquad (7.285)$$

The single-parameter sensitivity function, as defined by Eq. (3.20), is given by

$$S(x_k) = \frac{x_k}{w} \frac{\partial w}{\partial x_k} \qquad (7.286)$$

Combining (7.285) and (7.286) yields

$$\frac{\delta w}{w} \approx \sum_{k=1}^{pq} S(x_k) \frac{\delta x_k}{x_k} \qquad (7.287)$$

This expression gives the fractional change of the transfer function w in terms of the scalar sensitivity functions $S(x_k)$.

Refer again to the fundamental matrix feedback-flow graph of Fig. 7.13. Assume that the feedback amplifier has a single input and a single output. Then from

(7.281) we have

$$\delta w(\mathbf{X}) \approx \mathbf{C}(\mathbf{1}_q - \mathbf{XA})^{-1}(\delta \mathbf{X})(\mathbf{1}_p - \mathbf{AX})^{-1}\mathbf{B} \qquad (7.288)$$

where $\mathbf{C}$ is a row q-vector and $\mathbf{B}$ is a column p-vector. Write

$$\mathbf{C} = [c_1, c_2, \ldots, c_q] \qquad (7.289a)$$

$$\mathbf{B}' = [b_1, b_2, \ldots, b_p] \qquad (7.289b)$$

$$\widetilde{\mathbf{W}} = \mathbf{X}(\mathbf{1}_p - \mathbf{AX})^{-1} = (\mathbf{1}_q - \mathbf{XA})^{-1}\mathbf{X} = \left[\tilde{w}_{ij}\right] \qquad (7.289c)$$

The increment $\delta w(\mathbf{X})$ can be expressed in terms of the elements of (7.289) and those of $\mathbf{X}$. In the case where $\mathbf{X}$ is diagonal with

$$\mathbf{X} = \operatorname{diag}(x_1, x_2, \ldots, x_p) \qquad (7.290)$$

where $p = q$, the expression for $\delta w(\mathbf{X})$ can be greatly simplified and is found to be

$$\delta w(\mathbf{X}) = \sum_{i=1}^{p}\sum_{k=1}^{p}\sum_{j=1}^{p} c_i \left(\frac{\tilde{w}_{ik}}{x_k}\right) (\delta x_k) \left(\frac{\tilde{w}_{kj}}{x_k}\right) b_j$$

$$= \sum_{i=1}^{p}\sum_{j=1}^{p}\sum_{k=1}^{p} \frac{c_i \tilde{w}_{ik} \tilde{w}_{kj} b_j}{x_k} \frac{\delta x_k}{x_k} \qquad (7.291)$$

Comparing this with (7.287), we obtain an explicit form for the single-parameter sensitivity function as

$$S(x_k) = \sum_{i=1}^{p}\sum_{j=1}^{p} \frac{c_i \tilde{w}_{ik} \tilde{w}_{kj} b_j}{x_k w(\mathbf{X})} \qquad (7.292)$$

Thus, knowing (7.289) and (7.290), we can calculate the multiparameter sensitivity function for the scalar transfer function $w(\mathbf{X})$ immediately.

Example 7.14 Consider the same voltage-series feedback amplifier of Fig. 7.14, as discussed in Example 7.2. Assume that V_s is the input and V_{25} the output. The transfer function of interest is the amplifier voltage gain V_{25}/V_s. The elements of main concern are the two controlling parameters of the controlled sources. Thus, from Example 7.2 we have

$$\mathbf{X} = \begin{bmatrix} \tilde{\alpha}_1 & 0 \\ 0 & \tilde{\alpha}_2 \end{bmatrix} = \begin{bmatrix} 0.0455 & 0 \\ 0 & 0.0455 \end{bmatrix} \tag{7.293a}$$

$$\mathbf{A} = \begin{bmatrix} -90.782 & 45.391 \\ -942.507 & 0 \end{bmatrix} \tag{7.293b}$$

$$\mathbf{B}' = [0.91748 \quad 0] \tag{7.293c}$$

$$\mathbf{C} = [45.391 \quad -2372.32] \tag{7.293d}$$

yielding

$$\widetilde{\mathbf{W}} = \mathbf{X}(\mathbf{1}_2 - \mathbf{A}\mathbf{X})^{-1} = 10^{-4} \begin{bmatrix} 4.85600 & 10.02904 \\ -208.245 & 24.91407 \end{bmatrix} \tag{7.294}$$

Also, from (7.100) we have

$$w(\mathbf{X}) = \frac{V_{25}}{V_s} = 45.39 \tag{7.295}$$

To compute the sensitivity functions with respect to $\tilde{\alpha}_1$ and $\tilde{\alpha}_2$, we apply (7.292), as follows:

$$S(\tilde{\alpha}_1) = \sum_{i=1}^{2} \sum_{j=1}^{2} \frac{c_i \tilde{w}_{i1} \tilde{w}_{1j} b_j}{\tilde{\alpha}_1 w(\mathbf{X})}$$

$$= \frac{c_1 \tilde{w}_{11} \tilde{w}_{11} b_1 + c_1 \tilde{w}_{11} \tilde{w}_{12} b_2 + c_2 \tilde{w}_{21} \tilde{w}_{11} b_1 + c_2 \tilde{w}_{21} \tilde{w}_{12} b_2}{\tilde{\alpha}_1 w}$$

$$= 0.01066 \tag{7.296a}$$

$$S(\tilde{\alpha}_2) = \frac{c_1 \tilde{w}_{12} \tilde{w}_{21} b_1 + c_1 \tilde{w}_{12} \tilde{w}_{22} b_2 + c_2 \tilde{w}_{22} \tilde{w}_{21} b_1 + c_2 \tilde{w}_{22} \tilde{w}_{22} b_2}{\tilde{\alpha}_2 w}$$

$$= 0.05426 \tag{7.296b}$$

The return differences and the null return differences with respect to $\tilde{\alpha}_1$ and $\tilde{\alpha}_2$ were computed earlier in Chap. 4 and are given by Eqs. (4.141), (4.171), and

(4.172). They are repeated below:

$$\hat{F}(\tilde{\alpha}_1) = 93.70 \qquad (7.297a)$$

$$\hat{F}(\tilde{\alpha}_2) = 18.26 \qquad (7.297b)$$

$$\hat{F}(\tilde{\alpha}_1) = 103.07 \cdot 10^3 \qquad (7.297c)$$

$$\hat{F}(\tilde{\alpha}_2) = 2018.70 \qquad (7.297d)$$

By appealing to Eq. (5.5), the sensitivity functions are found to be

$$S(\tilde{\alpha}_1) = \frac{1}{F(\tilde{\alpha}_1)} - \frac{1}{\hat{F}(\tilde{\alpha}_1)} = 0.01066 \qquad (7.298a)$$

$$S(\tilde{\alpha}_2) = \frac{1}{F(\tilde{\alpha}_2)} - \frac{1}{\hat{F}(\tilde{\alpha}_2)} = 0.05427 \qquad (7.298b)$$

confirming (7.296).

Assume that $\tilde{\alpha}_1$ is changed by 4% and $\tilde{\alpha}_2$ by 6%. The fractional change of the voltage gain $w(\mathbf{X})$ is found from (7.287) as

$$\frac{\delta w}{w} = S(\tilde{\alpha}_1)\frac{\delta \tilde{\alpha}_1}{\tilde{\alpha}_1} + S(\tilde{\alpha}_2)\frac{\delta \tilde{\alpha}_2}{\tilde{\alpha}_2} = 0.01066 \cdot 0.04 + 0.05427 \cdot 0.06$$

$$= 0.003683 \qquad (7.299)$$

or 0.37%.

7.6 COMPUTATION OF FEEDBACK MATRICES

In this section, we present compact formulas expressing various return ratio and return difference matrices and the null return ratio matrix of a linear multivariable and multiloop feedback network in terms of the first- and the second-order cofactors of the elements of the indefinite-admittance matrix. This requires that all the controlled sources of the network be first converted to voltage-controlled current sources. However, the excitations can either be current sources, voltage sources, or any combination of these sources, but the current sources are preferred. The formulas are useful in computing the feedback matrices in that they do not require any matrix inversion in computing the null return ratio matrix. In addition, they are suitable for symbolical analysis and are especially convenient when the direct transmission matrix is diagonal. The material is based on the work of Chen and Tong (1988).

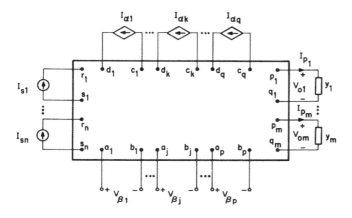

Figure 7.38 The Feedback network N_0.

Suppose that we have a feedback network N_0 containing a number of voltage-controlled current sources described by the equation

$$\mathbf{I}_\alpha = \begin{bmatrix} I_{\alpha 1} \\ I_{\alpha 2} \\ \vdots \\ I_{\alpha q} \end{bmatrix} = \begin{bmatrix} x_{11} & x_{12} & \cdots & x_{1p} \\ x_{21} & x_{22} & \cdots & x_{2p} \\ \vdots & \vdots & \ddots & \vdots \\ x_{q1} & x_{q2} & \cdots & x_{qp} \end{bmatrix} \begin{bmatrix} V_{\beta 1} \\ V_{\beta 2} \\ \vdots \\ V_{\beta p} \end{bmatrix} = \mathbf{XV}_\beta \qquad (7.300)$$

which can be represented schematically by the network N_0 of Fig. 7.38, where the current sources are used as the excitations. The results are equally valid if voltage sources or a combination of voltage and current sources are employed.
Let

$$\mathbf{I}_s = [I_{s1} \quad I_{s2} \quad \cdots \quad I_{sn}]' \qquad (7.301)$$

be the input signal vector, and

$$\mathbf{V}_o = [V_{o1} \quad V_{o2} \quad \cdots \quad V_{om}]' \qquad (7.302)$$

the output signal vector, where, as before, the prime denotes the matrix transposition. Before we proceed, we first introduce several useful relations. Assume that

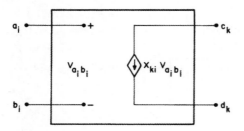

Figure 7.39 A controlled source located on nodes a, b, c and d.

the network N_0 has a voltage-controlled current source

$$I_{cd} = x V_{ab} \tag{7.303}$$

as shown in Fig. 7.39. If $\mathbf{Y}(x)$ is the indefinite-admittance matrix of the network N_0, the controlling parameter x enters $\mathbf{Y}(x)$ in a rectangular manner shown below:

$$
\mathbf{Y}(x) = \begin{array}{c} \\ \\ c \\ \\ \vdots \\ d \\ \\ \end{array}
\begin{array}{c}
\overset{\displaystyle a \qquad\qquad b}{\left[
\begin{array}{ccccc}
\cdots & \vdots & \cdots & \vdots & \cdots \\
\cdots & x & \cdots & -x & \cdots \\
\cdots & \vdots & \cdots & \vdots & \cdots \\
\cdots & -x & \cdots & x & \cdots \\
\cdots & \vdots & \cdots & \vdots & \cdots
\end{array}
\right]}
\end{array}
\tag{7.304}
$$

If $\mathbf{Y}(0)$ is the matrix obtained from $\mathbf{Y}(x)$ by setting $x = 0$, being the indefinite-admittance matrix of the subnetwork N' obtained from N_0 by removing the controlled source $I_{cd} = x v_{ab}$, any first-order cofactor $Y_{uv}(x)$ can be expanded in terms of $Y_{uv}(0)$ and the second-order cofactors $Y_{rp,sq}$ as follows:

$$Y_{uv}(x) = Y_{uv}(0) + x(Y_{uv,ca} + Y_{uv,db} - Y_{uv,cb} - Y_{uv,da}) \tag{7.305}$$

where all the second-order cofactors are independent of x. In particular, if $u = c$ and $v = a$, (7.305) reduces to

$$Y_{ca}(x) = Y_{ca}(0) + x Y_{ca,db} \tag{7.306}$$

Then we have

$$\dot{Y}_{uv}(x) = Y_{uv,ca} + Y_{uv,db} - Y_{uv,cb} - Y_{uv,da} \qquad (7.307a)$$

$$\dot{Y}_{ca}(x) = Y_{ca,db} \qquad (7.307b)$$

where, as before, $\dot{Y}_{uv}(x)$ denotes the partial derivative of the function $Y_{uv}(x)$ with respect to x.

As in (4.146), we define the third-order cofactor of the elements of $\mathbf{Y}(x)$, as follows:

$$Y_{rp,sq,ef} = \mathrm{sgn}\,(r - s)\,\mathrm{sgn}\,(p - q)\,\mathrm{sgn}\,(r - e)\,\mathrm{sgn}\,(p - f)\,\mathrm{sgn}\,(s - e)$$

$$\times \mathrm{sgn}\,(q - f)(-1)^{r+p+s+q+e+f}\det \mathbf{Y}_{rp,sq,ef} \qquad (7.308)$$

where $\mathbf{Y}_{rp,sq,ef}$ denotes the submatrix obtained from $\mathbf{Y}(x)$, $n > 3$, by deleting the rows r, s, e and columns p, q, f; and $\mathrm{sgn}\,u = 1$ if $u > 0$, $\mathrm{sgn}\,u = -1$ if $u < 0$, and $\mathrm{sgn}\,u = 0$ if $u = 0$. These lead to the following relations:

$$Y_{rp,sq}(x) = Y_{rp,sq}(0) + x(Y_{rp,sq,ca} + Y_{rp,sq,db} - Y_{rp,sq,cb} - Y_{rp,sq,da})$$

$$(7.309a)$$

$$\dot{Y}_{rp,sq}(x) = Y_{rp,sq,ca} + Y_{rp,sq,db} - Y_{rp,sq,cb} - Y_{rp,sq,da} \qquad (7.309b)$$

where $\dot{Y}_{rp,sq}(x)$ denotes the partial derivative of the function $Y_{rp,sq}(x)$ with respect to x.

7.6.1 The Return Ratio Matrix

According to (7.32), to compute the return ratio matrix $\mathbf{T}(\mathbf{X})$ we must find the matrix $\mathbf{A}$ first. In view of the physical significance of the matrix $\mathbf{A}$ as described in Sec. 7.2.5, it can be interpreted as the transfer impedance matrix of the feedback network N_0 relating the current vector $\mathbf{I}_\alpha$ to the voltage vector $\mathbf{V}_\beta$ under the condition $\mathbf{I}_s$ be set to zero and all of the controlled sources be considered as independent current sources. Let this transfer impedance matrix be denoted by $\mathbf{Z}_A$. Then $\mathbf{Y}(0) = \mathbf{Y}(\mathbf{X} = \mathbf{0})$ is the indefinite-admittance matrix of the subnetwork N_1 obtained from N_0 by removing $\mathbf{I}_\alpha$. Referring to Fig. 7.38 and (7.300) and applying formula (2.94), the (i, j)th element $Z_{ij,A}$ of $\mathbf{Z}_A$, is found to be

$$Z_{ij,A} = \frac{Y_{d_j a_i, c_j b_i}(\mathbf{0})}{Y_{uv}(\mathbf{0})}, \quad i = 1, 2, \ldots, p \quad \text{and} \quad j = 1, 2, \ldots, q \qquad (7.310)$$

By (7.32) the (i, j)th element $T_{ij,A}(\mathbf{X})$ of the return ratio matrix $\mathbf{T}(\mathbf{X})$ becomes

$$T_{ij}(\mathbf{X}) = -\sum_{k=1}^{q} Z_{ik,A} x_{kj} = -\sum_{k=1}^{q} x_{kj} \frac{Y_{d_k a_i, c_k b_i}(\mathbf{0})}{Y_{uv}(\mathbf{0})} = \sum_{k=1}^{q} x_{kj} \frac{Y_{c_k a_i, d_k b_i}(\mathbf{0})}{Y_{uv}(\mathbf{0})}$$

$$(7.311)$$

or

$$T_{ij}(\mathbf{X}) = \frac{\sum_{k=1}^{q} x_{kj} Y_{c_k a_i, d_k b_i}(\mathbf{0})}{Y_{uv}(\mathbf{0})}, \quad i, j = 1, 2, \ldots, p \qquad (7.312)$$

Invoking (7.307b) yields

$$T_{ij}(\mathbf{X}) = \frac{\sum_{k=1}^{q} x_{kj} \dot{Y}_{c_k a_i}(\mathbf{X})|_{\mathbf{X}=\mathbf{0}}}{Y_{uv}(\mathbf{0})}$$

$$= \frac{\sum_{k=1}^{q} x_{kj} \partial Y_{uv}(\mathbf{X})/\partial x_{ki}|_{\mathbf{X}=\mathbf{0}}}{Y_{uv}(\mathbf{0})}, \quad i, j = 1, 2, \ldots, p \quad (7.313)$$

As to the general return ratio matrix $\mathbf{T_K}(\mathbf{X})$ shown in (7.149), all we need to consider is the matrix $\mathbf{X}_1 = \mathbf{X} - \mathbf{K}$, the (i, j)th element x'_{ij} of which is equal to $x_{ij} - K_{ij}$, where $\mathbf{K} = [K_{ij}]$. In this case, we replace $\mathbf{Y}(\mathbf{0})$ by $\mathbf{Y}(\mathbf{K})$, and obtain

$$T_{ij,K}(\mathbf{X}) = \frac{\sum_{k=1}^{q} x'_{kj} Y_{c_k a_i, d_k b_i}(\mathbf{K})}{Y_{uv}(\mathbf{K})}, \quad i, j = 1, 2, \ldots, p \qquad (7.314a)$$

$$T_{ij,K}(\mathbf{X}) = \frac{\sum_{k=1}^{q} x'_{kj} \partial Y_{uv}(\mathbf{X})/\partial x_{ki}|_{\mathbf{X}=\mathbf{K}}}{Y_{uv}(\mathbf{K})}, \quad i, j = 1, 2, \ldots, p \,(7.314b)$$

The two sets of formulas for $\mathbf{T}(\mathbf{X})$ and $\mathbf{T_K}(\mathbf{X})$ are equally important. Provided that we can find the first-order cofactors $Y_{uv}(\mathbf{X})$ in symbolic or literal forms, the formulation with partial derivatives undoubtedly will simplify the computations, because we need only to choose those terms associated with x_{ki} from the symbolic expression of $Y_{uv}(\mathbf{X})$ when we take the partial derivatives $\partial Y_{uv}(\mathbf{X})/\partial x_{ki}$. Thus, (7.314) is more convenient than (7.149), because no matrix inversion is involved.

7.6.2   The Null Return Ratio Matrix

From (7.72) we see that in order to find the null return ratio matrix $\hat{T}(\mathbf{X})$, we have to compute the corresponding transfer impedance matrices $\mathbf{Z}_B$, $\mathbf{Z}_C$, and $\mathbf{Z}_D$ of $\mathbf{B}$, $\mathbf{C}$, and $\mathbf{D}$, respectively. For $\mathbf{Z}_D$ to be nonsingular and of order $n \times n$, the input vector and the output vector must be of the same order. The (i, j)th elements $Z_{ij,B}$,

$Z_{ij,C}$, and $Z_{ij,D}$ of $\mathbf{Z}_B$, $\mathbf{Z}_C$, and $\mathbf{Z}_D$, respectively, can be computed directly from $\mathbf{Y}(0)$ by formula (2.94), as follows:

$$Z_{ij,B} = \frac{Y_{r_j a_i, s_j b_i}(0)}{Y_{uv}(0)}, \quad i = 1, 2, \ldots, p \quad \text{and} \quad j = 1, 2, \ldots, n \qquad (7.315a)$$

$$Z_{ij,C} = \frac{Y_{d_j p_i, c_j q_i}(0)}{Y_{uv}(0)}, \quad i = 1, 2, \ldots, n \quad \text{and} \quad j = 1, 2, \ldots, q \qquad (7.315b)$$

$$Z_{ij,D} = \frac{Y_{r_j p_i, s_j q_i}(0)}{Y_{uv}(0)}, i, j = 1, 2, \ldots, n \qquad (7.315c)$$

After finding these matrices, the null return ratio matrix is computed by

$$\hat{\mathbf{T}}(\mathbf{X}) = (\mathbf{Z}_B \mathbf{Z}_D^{-1} \mathbf{Z}_C - \mathbf{Z}_A)\mathbf{X} \qquad (7.316)$$

It appears that we do not gain much in using (7.316) instead of (7.72). However, if we replace the zero matrix in (7.310) and (7.315) by the reference matrix $\mathbf{K}$, we quickly obtain the associated matrices $\mathbf{Z}_{A1}$, $\mathbf{Z}_{B1}$, $\mathbf{Z}_{C1}$, and $\mathbf{Z}_{D1}$ of $\mathbf{A}_1$, $\mathbf{B}_1$, $\mathbf{C}_1$, and $\mathbf{D}_1$, in (7.148), respectively. From (7.153) and (7.154), we obtain the general null return ratio matrix as

$$\hat{\mathbf{T}}_{\mathbf{K}}(\mathbf{X}) = (\mathbf{Z}_{B1} \mathbf{Z}_{D1}^{-1} \mathbf{Z}_{C1} - \mathbf{Z}_{A1})\mathbf{X}_1 \qquad (7.317)$$

Clearly, (7.317) is superior to (7.153) and (7.154) in that the latter requires to compute at least two matrix inversions, while the former requires only one.

7.6.3 Special Expression for the Null Return Ratio Matrix

We now consider two special cases: Single input and single output system and the n-input and n-output system.

Case 1. Single input and single output system

In this case $n = m = 1$ and $\mathbf{Z}_D$ reduces to a scalar transfer impedance. $\mathbf{Z}_B$ and $\mathbf{Z}_C$ are a column vector of order $p \times 1$ and a row vector of order $1 \times q$, respectively, and $\mathbf{Z}_A$ remains unchanged. Using (7.316) the elements of $\hat{\mathbf{T}}(\mathbf{X})$ are found to be

$$\hat{T}_{ij}(\mathbf{X}) = \frac{\sum_{k=1}^{q} x_{kj}[Y_{d_k p, c_k q}(0) Y_{r a_i, s b_i}(0) - Y_{d_k a_i, c_k b_i}(0) Y_{r p, s q}(0)]}{Y_{uv}(0) Y_{r p, s q}(0)},$$

$$i, j = 1, 2, \ldots, p \qquad (7.318)$$

If the feedback network contains only a single controlled source x_{ki}, we have

$$Y_{uv}(x_{ki}) \frac{\partial Y_{rp,sq}(x_{ki})}{\partial x_{ki}} = Y_{ra_i,sb_i} Y_{d_k p,c_k q} - Y_{d_k a_i,c_k b_i} Y_{rp,sq} \qquad (7.319)$$

Among the four second-order cofactors of (7.319), only $Y_{rp,sq}$ may contain x_{ki}. Thus (7.319) can be extended to

$$Y_{uv}(\mathbf{X}) \frac{\partial Y_{rp,sq}(\mathbf{X})}{\partial x_{ki}} = Y_{ra_i,sb_i}(\mathbf{X}) Y_{d_k p,c_k q}(\mathbf{X}) - Y_{d_k a_i,c_k b_i}(\mathbf{X}) Y_{rp,sq}(\mathbf{X}) \qquad (7.320)$$

Setting $\mathbf{X} = \mathbf{0}$, (7.320) becomes

$$Y_{uv}(\mathbf{0}) \frac{\partial Y_{rp,sq}(\mathbf{X})}{\partial x_{ki}} \bigg|_{\mathbf{X}=\mathbf{0}} = Y_{ra_i,sb_i}(\mathbf{0}) Y_{d_k p,c_k q}(\mathbf{0}) - Y_{d_k a_i,c_k b_i}(\mathbf{0}) Y_{rp,sq}(\mathbf{0}) \qquad (7.321)$$

Substituting (7.321) in (7.318) yields

$$\hat{T}_{ij}(\mathbf{X}) = \frac{\sum_{k=1}^{q} x_{kj} \partial Y_{rp,sq}(\mathbf{X})/\partial x_{ki}|_{\mathbf{X}=\mathbf{0}}}{Y_{rp,sq}(\mathbf{0})}, \quad i,j = 1, 2, \ldots, p \qquad (7.322)$$

Extending (7.309b) from x_{ki} to $\mathbf{X}$ as in (7.320) and substituting the result in (7.322) give

$$\hat{T}_{ij}(\mathbf{X})$$

$$= \frac{\sum_{k=1}^{q} x_{kj} [Y_{rp,sq,c_k a_i}(\mathbf{0}) + Y_{rp,sq,d_k b_i}(\mathbf{0}) - Y_{rp,sq,c_k b_i}(\mathbf{0}) - Y_{rp,sq,d_k a_i}(\mathbf{0})]}{Y_{rp,sq}(\mathbf{0})}$$

$$(7.323)$$

Case 2. The n-input and n-output system

In this case, $\mathbf{Z}_D$ is diagonal and can be expressed as

$$\mathbf{Z}_D = \text{diag} \left[\frac{Y_{r_1 p_1,s_1 q_1}(\mathbf{0})}{Y_{uv}(\mathbf{0})} \quad \frac{Y_{r_2 p_2,s_2 q_2}(\mathbf{0})}{Y_{uv}(\mathbf{0})} \quad \cdots \quad \frac{Y_{r_n p_n,s_n q_n}(\mathbf{0})}{Y_{uv}(\mathbf{0})} \right] \qquad (7.324)$$

and $\mathbf{Z}_B$ is the matrix of order $p \times n$, and $\mathbf{Z}_C$ is a matrix of order $n \times q$. Applying (7.310), (7.315a), (7.315b) and (7.324) to (7.316), we obtain

$$\hat{T}_{ij}(\mathbf{X}) = \sum_{k=1}^{q} x_{kj} \left[\sum_{m=1}^{n} \frac{Y_{r_m a_i,s_m b_i}(\mathbf{0}) Y_{d_k p_m,c_k q_m}(\mathbf{0})}{Y_{uv}(\mathbf{0}) Y_{r_m p_m,s_m q_m}(\mathbf{0})} - \frac{Y_{d_k a_i,c_k b_i}(\mathbf{0})}{Y_{uv}(\mathbf{0})} \right],$$

$$i,j = 1, 2, \ldots, p \qquad (7.325)$$

From (7.321) we obtain

$$
\hat{T}_{ij}(\mathbf{X}) = \sum_{k=1}^{q} \sum_{m=1}^{n} x_{kj} \frac{\partial Y_{r_m p_m, s_m q_m}(\mathbf{X})/\partial x_{ki}|_{\mathbf{X}=0}}{Y_{r_m p_m, s_m q_m}(\mathbf{0})}
$$

$$
-(n-1) \sum_{k=1}^{q} x_{kj} \frac{\partial Y_{uv}(\mathbf{X})/\partial x_{ki}|_{\mathbf{X}=0}}{Y_{uv}(\mathbf{0})} \tag{7.326a}
$$

$$
= \sum_{k=1}^{q} \sum_{m=1}^{n} \frac{x_{kj}}{Y_{r_m p_m, s_m q_m}(\mathbf{0})} [Y_{r_m p_m, s_m q_m, c_k a_i}(\mathbf{0}) + Y_{r_m p_m, s_m q_m, d_k b_i}(\mathbf{0})
$$

$$
- Y_{r_m p_m, s_m q_m, c_k b_i}(\mathbf{0}) - Y_{r_m p_m, s_m q_m, d_k a_i}(\mathbf{0})]
$$

$$
+(n-1) \sum_{k=1}^{q} x_{kj} \frac{Y_{d_k a_i, c_k b_i}(\mathbf{0})}{Y_{uv}(\mathbf{0})}, \quad i, j = 1, 2, \ldots, p \tag{7.326b}
$$

Provided that x_{kj}, $\mathbf{X} = \mathbf{0}$ and $\mathbf{0}$ are all replaced by x'_{kj}, $\mathbf{X} = \mathbf{K}$ and $\mathbf{K}$, respectively, we can obtain similar expressions for the general null return ratio matrix $\hat{T}_{\mathbf{K}}(\mathbf{X})$. To save space, they are omitted here. Thus, (7.322) and (7.326a) are suitable for symbolical analysis, and (7.323) and (7.326b) require the use of determinants of lower orders.

7.6.4 Illustrative Example

The network of Fig. 7.40 is an equivalent circuit as in Fig. 7.15 of a voltage-series feedback amplifier of Fig. 7.14. Suppose that the two controlled current sources are characterized by the equation

$$
\mathbf{I}_{\alpha} = \begin{bmatrix} I_{\alpha} \\ I_{\beta} \end{bmatrix} = \begin{bmatrix} 0.0455 & 0.01 \\ 0.02 & 0.0455 \end{bmatrix} \begin{bmatrix} V_{13} \\ V_{45} \end{bmatrix} = \mathbf{X} \mathbf{V}_{\beta} \tag{7.327}
$$

Now we compute the null return ratio matrix by (7.322) and (7.323). From Fig. 7.40 we can make the following identifications:

$$
r = 1, \quad s = 5, \quad d_1 = 3, \quad c_1 = 4, \quad d_2 = 5, \quad c_2 = 2 \tag{7.328a}
$$

$$
p = 2, \quad q = 5, \quad a_1 = 1, \quad b_1 = 3, \quad a_2 = 4, \quad b_2 = 5 \tag{7.328b}
$$

Figure 7.40 An equivalent circuit of a voltage-series feedback amplifier.

Its indefinite-admittance matrix $\mathbf{Y}$ is found to be

$$
\mathbf{Y} =
\begin{array}{c}
 \\
1 \\
2 \\
3 \\
4 \\
5
\end{array}
\begin{bmatrix}
1 & 2 & 3 & 4 & 5 \\
G_1 + G_4 & 0 & -G_4 & 0 & -G_1 \\
x_{21} & G_2 + G_f & -(G_f + x_{21}) & x_{22} & -(G_2 + x_{22}) \\
-(G_4 + x_{11}) & -G_f & G_4 + G_e + G_f + x_{11} & -x_{12} & -(G_e + x_{12}) \\
x_{11} & 0 & -x_{11} & G_5 + x_{12} & -(G_5 + x_{12}) \\
-G_1 + x_{21} & -G_2 & -G_e + x_{21} & -G_5 - x_{22} & \begin{array}{c} G_1 + G_2 + G_5 \\ +G_e + x_{22} \end{array}
\end{bmatrix}
$$

$$(7.329)$$

$$
Y_{rp,sq}(\mathbf{X}) = Y_{12,55}(\mathbf{X})
$$

$$
=
\begin{array}{c}
2 \\
3 \\
4
\end{array}
\begin{bmatrix}
1 & 3 & 4 \\
x_{21} & -(G_f + x_{21}) & x_{22} \\
-(G_4 + x_{11}) & G_4 + G_e + G_f + x_{11} & -x_{12} \\
x_{11} & -x_{11} & G_5 + x_{11}
\end{bmatrix}
\qquad (7.330)
$$

Using (7.322) and (7.323), we obtain the elements of the null return ratio matrix, as follows:

$$\hat{T}_{11}(\mathbf{X}) = \frac{x_{11}}{G_4} - \frac{x_{21}(G_e + G_f)}{G_4 G_f} = -1005.8851 \qquad (7.331a)$$

$$\hat{T}_{12}(\mathbf{X}) = \frac{x_{12}}{G_4} - \frac{x_{22}(G_e + G_f)}{G_4 G_f} = -2391.2628 \qquad (7.331b)$$

$$\hat{T}_{21}(\mathbf{X}) = \frac{x_{11}}{G_5} = 42.884071 \qquad (7.331c)$$

$$\hat{T}_{22}(\mathbf{X}) = \frac{x_{12}}{G_5} = 9.4250706 \qquad (7.331d)$$

giving the null return difference matrix with respect to $\mathbf{X}$ as

$$\hat{F}(\mathbf{X}) = \mathbf{1}_2 - \hat{\mathbf{A}}\mathbf{X} = \mathbf{1}_2 + \hat{\mathbf{T}}(\mathbf{X}) = \begin{bmatrix} -1004.8851 & -2391.2628 \\ 42.884071 & 10.4250706 \end{bmatrix} \qquad (7.332)$$

These are to be compared with the special situation worked out in Examples 7.2 and 7.4, where the coefficient matrix $\mathbf{X}$ in (7.19) is diagonal and the corresponding return difference matrix is given by (7.80).

7.7 SUMMARY

We began this chapter by reviewing the rules of the matrix signal-flow graph and presenting some of its properties. One way to obtain the graph transmission is to apply the graph reduction rules repeatedly until the final graph is composed of only one edge. The transmittance of the edge is the desired graph transmission. To avoid the necessity of repeated graph reduction, we described a topological procedure for computing the graph transmission directly from the original matrix signal-flow graph.

In the study of a single-loop feedback amplifier, we usually single out an element for particular attention. The scalar return difference and null return difference are then defined in terms of this particular element. For a multiple-loop amplifier, we pay particular attention to a group of elements and study its effects on the whole system. This leads to the concepts of return difference matrix and null return difference matrix. Like the return difference, the return difference matrix is defined as the square matrix relating the difference between the applied vector signal and the returned vector signal to the applied vector signal. Likewise, the null return difference matrix is the return difference matrix under the situation when the input excitation vector of the feedback amplifier is adjusted so that the total output resulting from this input excitation and the applied signal vector at the controlling branches of the elements of interest is identically zero. The overall transfer characteristics of a multiple-loop feedback amplifier are described by its transfer-function matrix. We showed that the determinant of the transfer-function matrix can be expressed in terms of the determinants of the return difference and null return difference matrices and its value when the elements of interest assume the zero value, thereby generalizing Blackman's impedance formula for a single input to a multiplicity of inputs.

In deriving the null return difference matrix, we assume that the effects at the output resulting from the applied signals at the controlling branches of the elements of interest can be neutralized by a unique input excitation vector of the amplifier. In the situation where the above condition is not satisfied, the null return difference matrix is not defined. To avoid this difficulty, we introduced the complementary return difference matrix. This is the return difference matrix under the situation when the input excitation vector of the amplifier is adjusted so that the sum of this input excitation and the output is identically zero. This matrix together with the return difference and null return difference matrices generally depend on the choice of the elements of interest. However, we demonstrated that their determinants are invariant with respect to the ordering of these elements.

Like the return difference and null return difference for a general reference value, the return difference and null return difference matrices were generalized for a general reference matrix. We showed that the general return difference matrix with respect to $\mathbf{X}$ for a general reference matrix $\mathbf{K}$ is equal to the product of the inverse of the return difference matrix with respect to $\mathbf{K}$ and the return difference matrix with respect to $\mathbf{X}$, both being defined for the zero reference. A similar statement can be made for the null return difference matrix for a general reference matrix. The determinant of the transfer-function matrix can therefore be expressed in terms of its value when the elements of interest assume their reference values and the determinants of the general return difference and null return difference matrices. We also developed relations between the scalar return difference and null return difference and the return difference and null return difference matrices. Specifically, we indicated that the scalar return difference of a multiple-loop feedback amplifier with respect to an element x of $\mathbf{X}$ is equal to the ratio of the determinants of the return difference matrix with respect to $\mathbf{X}$ and that with respect to $\mathbf{K}$, where $\mathbf{K}$ is obtained from $\mathbf{X}$ by setting x to zero. A similar relation was obtained between the scalar null return difference and null return difference matrix.

In Chaps. 4 and 5, the concepts of return difference and null return difference were introduced by means of the first- and second-order cofactors of the indefinite-admittance matrix. In this chapter, we demonstrated that the determinant of the return difference matrix can be expressed elegantly and compactly in terms of the ratio of the two functional values assumed by the determinant of the hybrid matrix under the condition that the elements of interest assume their nominal and reference values.

Finally, we introduced the sensitivity matrix and the multiparameter sensitivity function. We indicated that the sensitivity matrix is a direct extension of the sensitivity function defined for a single-input, single-output system and for a single parameter. In the particular situation, the sensitivity matrix reduces to the form similar to that defined for the scalar sensitivity function, and we obtained an expression relating the determinant of the sensitivity matrix and the determinants of the return difference matrix and a matrix involving only the transfer-function matrix. Formulas for the effects of changes of a group of elements on a scalar transfer function were also derived.

PROBLEMS

7.1 Show that Eqs. (7.8) and (7.9) are equivalent.

7.2 From (7.16), show that if $p = n$ and $\mathbf{B} = \mathbf{1}_n$, then

$$\mathbf{W}(\mathbf{X}) = \mathbf{W}(0)\hat{\mathbf{F}}(\mathbf{X})\mathbf{F}^{-1}(\mathbf{X}) \qquad (7.333)$$

7.3 From (7.16), show that if $q = m$ and $\mathbf{C} = \mathbf{1}_n$, then

$$\mathbf{W}(\mathbf{X}) = \mathbf{W}(0)\mathbf{F}_1^{-1}(\mathbf{X})\hat{\mathbf{F}}_1(\mathbf{X}) \qquad (7.334)$$

where

$$\mathbf{F}_1(X) = \mathbf{1}_q - \mathbf{XA} \qquad (7.335a)$$

$$\hat{\mathbf{F}}_1(\mathbf{X}) = \mathbf{1}_q - \mathbf{X}\hat{\mathbf{A}} \qquad (7.335b)$$

7.4 Consider the matrix signal-flow graph of Fig. 7.41. By using the topological procedure, confirm that the graph transmission from node $\mathbf{X}_7$ to node $\mathbf{X}_6$ is given by

$$\mathbf{W} = \mathbf{D}[\mathbf{1} - \mathbf{C}(\mathbf{1} - \mathbf{BJ})^{-1}\mathbf{K} - \mathbf{G}(\mathbf{1} - \mathbf{FH})^{-1}\mathbf{FE}(\mathbf{1} - \mathbf{JB})^{-1}\mathbf{JK}]^{-1}$$
$$\times \mathbf{G}(\mathbf{1} - \mathbf{FH})^{-1}\mathbf{FE}(\mathbf{1} - \mathbf{JB})^{-1}\mathbf{A} + \mathbf{D}[\mathbf{1} - \mathbf{C}(\mathbf{1} - \mathbf{BJ})^{-1}\mathbf{K}$$
$$- \mathbf{G}(\mathbf{1} - \mathbf{FH})^{-1}\mathbf{FE}(\mathbf{1} - \mathbf{JB})^{-1}\mathbf{JK}]^{-1}\mathbf{C}(\mathbf{1} - \mathbf{BJ})^{-1}\mathbf{BA} \qquad (7.336)$$

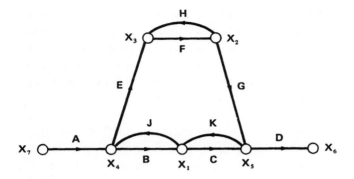

Figure 7.41 A matrix signal-flow graph.

7.5 From Fig. 7.27, show that the currents I_a and I_b and voltages V_{13} and V_{45} are related by Eq. (7.131) when the input excitation V_s is adjusted so that the output voltage $V_{25} = 0$.

7.6 In Fig. 7.27, suppose that the input V_s is adjusted so that $V_S + \check{V}_{25} = 0$, where $\check{V}_{25}$ denotes the contributions to V_{25} by the currents I_a and I_b. Show that the currents I_a and I_b and voltages V_{l3} and V_{45} are related by Eq. (7.134).

7.7 Demonstrate that (7.336) can be expressed equivalently as

$$\mathbf{W} = \mathbf{DC}(1 - \mathbf{BJ} - \mathbf{KC})^{-1}\mathbf{BA} + [\mathbf{DC}(1 - \mathbf{BJ} - \mathbf{KC})^{-1}\mathbf{KG} + \mathbf{DG}]$$

$$\cdot [1 - \mathbf{FH} - \mathbf{FEJ}(1 - \mathbf{BJ} - \mathbf{KC})^{-1}\mathbf{KG}]^{-1}$$

$$\times [\mathbf{FEA} + \mathbf{FEJ}(1 - \mathbf{BJ} - \mathbf{KC})^{-1}\mathbf{BA}] \tag{7.337}$$

7.8 Prove the identity (7.200).

7.9 Prove the identity (7.203).

7.10 In the feedback network of Fig. 7.11, let the elements of interest be represented by the matrix equation

$$\boldsymbol{\theta} = \begin{bmatrix} V_1 \\ V_a \\ V_b \end{bmatrix} = \begin{bmatrix} Z_1 & 0 & 0 \\ 0 & \beta_1 & \beta_2 \\ 0 & \alpha_1 & \alpha_2 \end{bmatrix} \begin{bmatrix} I_1 \\ I_2 \\ I_3 \end{bmatrix} = \mathbf{X}\boldsymbol{\phi} \tag{7.338}$$

Show that the return difference matrix with respect to **X** is given by

$$\mathbf{F(X)} = \frac{1}{\xi}$$

$$\begin{bmatrix} (Z_1 + Z_3)(Z_2 + Z_4) + Z_1 Z_3 & -\alpha_1 Z_3 - \beta_1(Z_2 + Z_4) & -\alpha_2 Z_3 - \beta_2(Z_2 + Z_4) \\ Z_1 Z_3 & Z_3(Z_2 + Z_4 - \alpha_1) & -\alpha_2 Z_3 \\ Z_1(Z_2 + Z_4) & -\beta_1(Z_2 + Z_4) & (Z_3 - \beta_2)(Z_2 + Z_4) \end{bmatrix}$$

$$(7.339)$$

where $\xi = Z_3(Z_2 + Z_4)$.

7.11 By applying (7.225) and (7.226) in conjunction with (7.249), verify (7.254) and (7.255). Also confirm the identities (7.224).

7.12 Show that (7.281) can also be expressed in the form

$$\delta \mathbf{W(X)} = \mathbf{C}(\mathbf{1}_q - \mathbf{X}^+\mathbf{A})^{-1}(\delta\mathbf{X})(\mathbf{1}_q - \mathbf{AX})^{-1}\mathbf{B} \qquad (7.340)$$

7.13 Repeat Example 7.14 if current I_{51} is chosen as the output variable and the input admittance I_{51}/V_S is the network function of interest.

7.14 Consider the parallel series or current-shunt feedback amplifier of Fig. 5.14, whose equivalent network is shown in Fig. 5.15b. Let the controlling parameters of the two controlled current sources be the elements of interest. Assume that the voltages V_{15} and V_{25} are the output variables. Compute the matrix equations (7.13), and determine the transfer-function matrix of the amplifier.

7.15 In Prob. 7.14, let the voltage V_{15} be the output variable. Compute the following with respect to the two controlling parameters:

(a) The return difference matrix $\mathbf{F(X)}$
(b) The null return difference matrix $\check{\mathbf{F}}(\mathbf{X})$
(c) The complementary return difference matrix $\hat{F}(\mathbf{X})$
(d) The voltage gain of the amplifier
(e) The amplifier input impedance

7.16 In Prob. 7.14, let the voltage V_{25} be the output variable. By using physical interpretations outlined in Sec. 7.2.5, compute the matrices $\mathbf{A}$, $\hat{\mathbf{A}}$, and $\check{\mathbf{A}}$ directly from the equivalent network of Fig. 5.15b. Invoke (7.118) and calculate the voltage gain V_{25}/V_S.

7.17 In Prob. 7.14, assume that the reference values for the two controlling parameters are 0.01 and 0.02 mho. Compute the following:

(a) The matrices $\mathbf{A}_1$, $\mathbf{B}_1$, $\mathbf{C}_1$, and $\mathbf{D}_1$
(b) The return difference matrix $\mathbf{F_K(X)}$

(c) The null return difference matrix $\hat{\mathbf{F}}_K(\mathbf{X})$

(d) The amplifier input impedance

7.18 Refer to Prob. 7.14. By using (7.199), (7.200), and (7.209), calculate the return differences, the null return differences, and the sensitivity functions with respect to the two controlling parameters.

7.19 In Prob. 7.14, let the voltage V_{25} be the output variable. The transfer function of interest is the amplifier voltage gain V_{25}/V_S. By using (7.292), compute the sensitivity functions with respect to the controlling parameters of the two controlled sources. Suppose that the controlling parameters are each changed by 5%. Determine the fractional change of the voltage gain.

7.20 In the feedback network of Fig. 7.11, the elements of interest are represented by the matrix equation

$$\boldsymbol{\theta} = \begin{bmatrix} I_1 \\ V_a \\ V_b \end{bmatrix} = \begin{bmatrix} Y_1 & 0 & 0 \\ 0 & \beta_1 & \beta_2 \\ 0 & \alpha_1 & \alpha_2 \end{bmatrix} \begin{bmatrix} V_1 \\ I_2 \\ I_3 \end{bmatrix} = \mathbf{X}\boldsymbol{\phi} \tag{7.341}$$

Demonstrate that the return difference matrix with respect to $\mathbf{X}$ is given by

$$\mathbf{F}(\mathbf{X}) = \frac{1}{U}$$
$$\times \begin{bmatrix} (Z_2 + Z_4)(1 + Y_1 Z_3) + Z_3 & -\beta_1(Z_2 + Z_4) - \alpha_1 Z_3 & -\beta_2(Z_2 + Z_4) - \alpha_2 Z_3 \\ -Y_1 Z_3 & Z_2 + Z_3 + Z_4 - \alpha_1 + \beta_1 & \beta_2 - \alpha_2 \\ -Y_1(Z_2 + Z_4) & \alpha_1 - \beta_1 & Z_2 + Z_3 + Z_4 + \alpha_2 - \beta_2 \end{bmatrix}$$

$$\tag{7.342}$$

where $U = Z_2 + Z_3 + Z_4$. Compute the matrix $\mathbf{A}$ by its physical interpretation.

7.21 In the feedback network of Fig. 7.11, let the voltage V_2 be the output variable. Assume that $\mathbf{X}$ of (7.338) is the matrix of interest. Show that the null return difference matrix with respect to $\mathbf{X}$ is given by the expression

$$\hat{\mathbf{F}}(\mathbf{X}) = \frac{1}{Z_3} \begin{bmatrix} Z_3 & \alpha_1 - \beta_1 & \alpha_2 - \beta_2 \\ 0 & Z_3 & 0 \\ 0 & \alpha_1 - \beta_1 & Z_3 + \alpha_2 - \beta_2 \end{bmatrix} \tag{7.343}$$

Compute the matrix $\hat{\mathbf{A}}$ by its physical interpretation.

7.22 In the feedback network of Fig. 7.11, let the voltage V_4 be the output variable. Assume that the matrix $\mathbf{X}$ of (7.338) is the matrix of interest. Show that the null return difference matrix with respect to $\mathbf{X}$ is given by (7.343).

7.23 A three-stage common-emitter feedback amplifier is shown in Fig. 6.9. Assume that the three transistors are identical, with $h_{ie} = 1.1\,k\Omega$, $h_{fe} = 50$, and $h_{re} = h_{oe} = 0$. The values of other network elements are given below:

$$R_1 = 100\,\Omega \quad R_2 = 75\,\Omega \quad R_3 = 1\,k\Omega$$
$$R_4 = 1\,k\Omega \quad R_f = 100\,k\Omega$$

(7.344)

Let the controlling parameters h_{fe} of the three transistors be the elements of interest. The transfer function of interest is the amplifier voltage gain. Compute the following:

(a) The return difference matrix $\mathbf{F}(\mathbf{X})$
(b) The null return difference matrix $\hat{\mathbf{F}}(\mathbf{X})$
(c) The complementary return difference matrix $\check{\mathbf{F}}(\mathbf{X})$
(d) The voltage gain of the amplifier
(e) The fractional change of the voltage gain for a 10% change of each of the three controlling parameters h_{fe}

7.24 For the multiple-loop feedback amplifier of Fig. 7.42, assume that transistors Q_1 and Q_3 are identical, with $h_{ie} = 1\,k\Omega$, $h_{fe} = 50$, and $h_{re} = h_{oe} = 0$, and that transistors Q_2 and Q_4 are identical, with $h_{ie} = 1.2\,k\Omega$, $h_{fe} = 75$, and $h_{re} = h_{oe} = 0$. Compute the return difference matrix, the null return difference matrix, and the complementary return difference matrix with respect to the four controlling parameters h_{fe} of the transistors by ignoring the biasing circuitry (Ramey and White, 1971).

7.25 In the fundamental matrix feedback-flow graph of Fig. 7.13, assume that the system has a single input u and a single output y. Suppose that the output node y is connected to the input node u by a branch with transmittance k as shown in Fig. 7.43. Show that the return difference matrix, written as $\mathbf{F}(\mathbf{X}, k)$, with respect to $\mathbf{X}$ in Fig. 7.43 can be written as

$$\mathbf{F}(\mathbf{X}, k) = \mathbf{1}_p - \left(\mathbf{A} + \frac{k}{1 - kd}\mathbf{BC} \right) \mathbf{X}$$

(7.345)

By using this result, show that

$$\mathbf{F}(\mathbf{X}) = \mathbf{F}(\mathbf{X}, 0)$$ (7.346a)

$$\hat{\mathbf{F}}(\mathbf{X}) = \mathbf{F}(\mathbf{X}, \infty)$$ (7.346b)

$$\check{\mathbf{F}}(\mathbf{X}) = \mathbf{F}\left(\mathbf{X}, \frac{1}{d - 1} \right)$$ (7.346c)

Figure 7.42 A multiple-loop feedback amplifier together with its biasing circuitry.

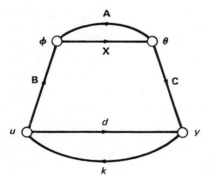

Figure 7.43 A matrix signal-flow graph associated with a new return difference matrix.

This indicates that the return difference matrix $\mathbf{F}(\mathbf{X})$, the null return difference matrix $\hat{\mathbf{F}}(\mathbf{X})$, and the complementary return difference matrix $\check{\mathbf{F}}(\mathbf{X})$ are special values of the return difference matrix $\mathbf{F}(\mathbf{X}, k)$.

7.26 Instead of invoking (7.83), show that we can apply the following formulas for the determinants of the partitioned matrices to derive the identity (7.88) directly from (7.16a):

$$\det \begin{bmatrix} \mathbf{W} & \mathbf{X} \\ \mathbf{Y} & \mathbf{Z} \end{bmatrix} = \det \mathbf{W} \det (\mathbf{Z} - \mathbf{YW}^{-1}\mathbf{X}) \quad \text{if } \det \mathbf{W} \neq 0 \quad (7.347a)$$

$$\det \begin{bmatrix} \mathbf{W} & \mathbf{X} \\ \mathbf{Y} & \mathbf{Z} \end{bmatrix} = \det \mathbf{Z} \det (\mathbf{W} - \mathbf{XZ}^{-1}\mathbf{Y}) \quad \text{if } \det \mathbf{Z} \neq 0 \quad (7.347b)$$

7.27 Refer to the block diagram of Fig. 7.36. As indicated in the paragraph following Eq. (7.270), the matrix $\mathbf{1}_m - \mathbf{XH_2H_3}$ can be defined as the return difference matrix, since, by opening the output of $\mathbf{X}$, it is the matrix relating the difference between the applied vector signal and the returned vector signal to the applied vector signal. Likewise, by opening the output of $\mathbf{H_2}$, the matrix $\mathbf{1}_p - \mathbf{H_2H_3X}$ may be regarded as the return difference matrix, and by opening the input of $\mathbf{H_2}$, the matrix $\mathbf{1}_n - \mathbf{H_3XH_2}$ is the return difference matrix appropriate to this break point. Show that all three return difference matrices have the same determinant:

$$\det (\mathbf{1}_m - \mathbf{XH_2H_3}) = \det (\mathbf{1}_p - \mathbf{H_2H_3X}) = \det(\mathbf{1}_n - \mathbf{H_3XH_2})$$
$$(7.348)$$

As a result, the determinant of the sensitivity matrix (7.270) is independent of which return difference matrix is used, and we may work any convenient return difference matrix.

7.28 In (7.260), assume that $\mathbf{XH_2}$ and $\mathbf{W(X)}$ are square and nonsingular. Suppose that $\mathbf{XH_2}$ is perturbed to $\mathbf{XH_2} + \delta(\mathbf{XH_2})$ and that $\mathbf{W(X)}$ is correspondingly perturbed to $\mathbf{W(X)} + \delta\mathbf{W(X)}$. Demonstrate that

$$\mathbf{W}^{-1}(\mathbf{X})\delta\mathbf{W(X)} \approx \mathbf{S(X)}(\mathbf{XH_2})^{-1}\delta(\mathbf{XH_2}) \quad (7.349)$$

Compare this result with (7.275) and (7.276).

7.29 Refer to the matrix feedback-flow graph of Fig. 7.24. As in Sec. 7.2.4, we apply a signal p-vector g to the right of the breaking mark. Suppose that we adjust the input excitation n-vector $\mathbf{u}$ so that $\mathbf{u} + \mathbf{y} = \mathbf{0}$, assuming that $\mathbf{u}$ and $\mathbf{y}$ are of the same dimension. Show that

$$\mathbf{g} - \mathbf{h} - \overset{\circ}{\mathbf{F}}(\mathbf{X})\mathbf{g} \quad (7.350)$$

where

$$\overset{\circ}{\mathbf{F}}(\mathbf{X}) = \mathbf{1}_p - \overset{\circ}{\mathbf{A}}\mathbf{X} \qquad (7.351a)$$

$$\overset{\circ}{\mathbf{A}} = \mathbf{A} - \mathbf{B}(\mathbf{1}_n + \mathbf{D})^{-1}\mathbf{C} \qquad (7.351b)$$

The square matrix $\mathbf{F}(\mathbf{X})$ relating the difference between the applied and the returned signals to the applied signal under the condition that $\mathbf{u} + \mathbf{y} = \mathbf{0}$ may also be defined as the *complementary return difference matrix*. For a single input and single output, show that $\overset{\circ}{\mathbf{F}}(\mathbf{X})$ can be expressed as

$$\overset{\circ}{\mathbf{F}}(\mathbf{X}) = \mathbf{F}(\mathbf{X}, -1) \qquad (7.352)$$

where $\mathbf{F}(\mathbf{X}, k)$ is defined in (7.312).

7.30 Refer to Prob. 7.29 and Sec. 7.2.5. Give a physical interpretation of the matrix $\overset{\circ}{\mathbf{A}}$ of Eq. (7.350b). By applying this interpretation to the problem considered in Example 7.7, compute $\overset{\circ}{\mathbf{A}}$ and $\overset{\circ}{\mathbf{F}}(\mathbf{X})$.

7.31 Refer to Prob. 7.29 and consider the situation where $m = n = 1$ with $w(0) = D = 0$. Express the transfer function $w(\mathbf{X})$ in terms of the determinants of the matrices $\mathbf{F}(\mathbf{X})$ and $\overset{\circ}{\mathbf{F}}(\mathbf{X})$, and compare this with Eq. (7.118).

BIBLIOGRAPHY

Acar, C.: New Return Difference Matrix, *Int. J. Circuit Theory and Applications*, vol. 3, no. 1, pp. 87–94, 1975.

Arbel, A. F.: Identification of the Return Difference Matrix of a Multivariable Linear Control System, in Terms of Its Inverse Closed-Loop Transfer Matrix, *Int. J. Circuit Theory and Applications*, vol. 1, no. 2, pp. 187–190, 1973a.

Arbel, A. F.: Return Ratio and Sensitivity Computation for Multivariable Control Systems and Feedback-Stabilized Multiple Amplifier Active Circuits, *Int. J. Circuit Theory and Applications*, vol. 1, no. 2, pp. 191–199, 1973b.

Biswas, R. N. and E. S. Kuh: Optimum Synthesis of a Class of Multiple-Loop Feedback Systems, *IEEE Trans. Circuit Theory*, vol. CT-18, no. 6, pp. 582–587, 1971a.

Biswas, R. N. and E. S. Kuh: A Multiparameter Sensitivity Measure for Linear Systems, *IEEE Trans. Circuit Theory*, vol. CT-18, no. 6, pp. 718–719, 1971b.

Blecher, F. H.: Transistor Multiple Loop Feedback Amplifiers, *Proc. Natl. Elect. Conf.*, vol. 13, pp. 19–34, 1957.

Chen, W. K.: "Applied Graph Theory: Graphs and Electrical Networks," 2d rev. ed., New York: American Elsevier, and Amsterdam: North-Holland, 1976a, chaps. 1 and 2.

Chen, W. K.: Topological Analysis of Multiple-Loop Feedback Networks, *Proc. Tenth Asilomar Conf. on Circuits, Systems, and Computers*, Pacific Grove, Calif., pp. 192–197, 1976b.

Chen, W. K.: The Hybrid Matrix in Linear Multiple-Loop Feedback Networks, *IEEE Trans. Circuits and Systems*, vol. CAS-24, no. 9, pp. 469–474, 1977a.

Chen, W. K.: Topological Evaluation of the Null Return Difference in Feedback Amplifier Theory, *Proc. Eleventh Asilomar Conf. on Circuits, Systems, and Computers*, Pacific Grove, Calif., pp. 388–395, 1977b.

Chen, W. K.: Topological Analysis of Feedback Matrices, *Proc. Twelfth International Symposium on Circuits and Systems*, Tokyo, Japan, pp. 837–840, July 17–19, 1979a. IEEE catalog no. 79CH1421-7 CAS.

Chen W. K.: Topological Evaluation of Feedback Matrices in Multiple-Loop Feedback Amplifiers, *J. Franklin Inst.*, vol. 308, no. 2, pp. 125–139, 1979b.

Chen, W. K.: Physical Interpretation of the Multiport Return Difference Matrix, *IEEE Trans. Automatic Control*, vol. AC-25, no. 3, pp. 598–600, 1980.

Chen, W. K. and H. M. Elsherif: Determinant of the Null Return-Difference Matrix, *Electron. Lett.*, vol. 13, no. 10, pp. 306–307, 1977.

Chen, W. K. and M. D. Tong: On Indefinite Impedance Matrix, *IEEE Trans. Circuits and Systems*, vol. CAS-32, no. 8, pp. 840–847, 1985.

Chen, W. K. and M. D. Tong: Analysis and Computation of Feedback Matrices in Multivariable and Multiloop Feedback Networks, *IEEE Trans. Circuits and Systems*, vol. CAS-35, no. 9, pp. 1123–1128, 1988.

Cruz, J. B., Jr. and W. R. Perkins: A New Approach to the Sensitivity Problem in Multivariable Feedback System Design, *IEEE Trans. Automatic Control*, vol. AC-9, no. 3, pp. 216–223, 1964.

Elsherif, H. M.: The Return Difference Matrix in the Multiple-Loop Feedback Systems, M.S. thesis, Ohio University, Athens, Ohio, 1974.

Elsherif, H. M. and W. K. Chen: The Return Difference Matrix in Multiple-Loop Feedback Systems, *Proc. 17th Midwest Symp. Circuits and Systems*, University of Kansas, Lawrence, Kansas, pp. 95–103, 1974.

Goldstein, A. J. and F. F. Kuo: Multiparameter Sensitivity, *IRE Trans. Circuit Theory*, vol. CT-8, no. 2, pp. 177–178, 1961.

Hakim, S. S.: Multiple-Loop Feedback Circuits, *Proc. IEE (London)*, vol. 110, no. 11, pp. 1955–1959, 1963.

Hakim, S. S.: Evaluation of Sensitivity in Transistor Multiple-Loop Feedback Amplifiers, *Proc. IEE (London)*, vol. 113, no. 2, pp. 219–224, 1966.

Kuh, E. S.: Some Results in Linear Multiple Loop Feedback Systems, *Proc. Allerton Conf on Circuit and System Theory*, University of Illinois, Urbana, Illinois, vol. 1, pp. 471–487, 1963.

Kuh, E. S. and R. A. Rolirer: "Theory of Linear Active Networks," San Francisco, Calif.: Holden-Day, 1967.

MacFarlane, A. G. J.: Return-Difference and Return-Ratio Matrices and Their Use in Analysis and Design of Multivariable Feedback Control Systems, *Proc. IEE (London)*, vol. 117, no. 10, pp. 2037–2049, 1970.

MacFarlane, A. G. J.: Return-Difference and Return-Ratio Matrices and Their Use in Analysis and Design of Multivariable Feedback Control Systems, *Proc. IEE (London)*, vol. 118, no. 7, pp. 946–947, 1971.

Ramey, R. L. and E. J. White: "Matrices and Computers in Electronic Circuit Analysis," New York: McGraw-Hill, 1971.

Riegle, D. E. and P. M. Lin: Matrix Signal Flow Graphs and an Optimum Topological Method for Evaluating Their Gains, *IEEE Trans. Circuit Theory*, vol. CT-19, no. 5, pp. 427–435, 1972.

Sandberg, I. W.: On the Theory of Linear Multi-Loop Feedback Systems, *Bell Sys. Tech. J.*, vol. 42, no. 2, pp. 355–382, 1963.

Tassny-Tschiassny, L.: The Return Difference Matrix in Linear Networks, *Proc. IEE (London)*, vol. 100, part IV, pp. 39–46, 1953.

Truxal, J. G.: "Automatic Feedback Control System Synthesis," New York: McGraw-Hill, 1955.

EIGHT

STATE-SPACE ANALYSIS
AND FEEDBACK THEORY

In this chapter, we formulate the network equations in the time-domain as a system of first-order differential equations that govern the dynamic behavior of a network. The advantages of representing the network equations in this form are numerous. First of all, such a system has been widely studied in mathematics and its solution, both analytical and numerical, is known and readily available. Secondly, the representation can easily and naturally be extended to time-varying and nonlinear networks. In fact, nearly all time-varying and nonlinear networks are characterized by this approach. Finally, the first-order differential equations are easily programmed for a digital computer or simulated on an analog computer.

We begin the chapter by first presenting procedures for the systematic formulation of network equations in the form of a system of first-order differential equations known as the *state equations*, and then discuss the number of dynamically independent *state variables* required in the formulation of these state equations. We shall demonstrate how this number or its upper bounds can be determined from network topology alone. Finally, we express the familiar feedback matrices in terms of the coefficient parameter matrices of the state equations, and discuss their physical significance. In this chapter, we consider only linear lumped networks that may be passive or active, reciprocal or nonreciprocal, and time-invariant or time-varying.

8.1 STATE EQUATIONS IN NORMAL FORM

Our objective is to choose an appropriate set of real functions $x_1(t), x_2(t), \ldots,$ $x_k(t)$ of the time t called the *state variables* so that the network equations can be put in the form of a system of k first-order differential equations as

$$\dot{x}_i(t) = \sum_{j=1}^{k} a_{ij}(t)x_j(t) + \sum_{j=1}^{h} b_{ij}(t)u_j(t) \quad (i = 1, 2, \ldots, k) \qquad (8.1)$$

where $\dot{x}_i(t)$ denotes the time derivative of $x_i(t)$, $u_j(t)$ are the known forcing or excitation functions, and $a_{ij}(t)$ and $b_{ij}(t)$ are network parameters. In matrix form, (8.1) can be compactly written as

$$\dot{\mathbf{x}}(t) = \mathbf{A}_s \mathbf{x}(t) + \mathbf{B}_s \mathbf{u}(t) \qquad (8.2)$$

where the k-vector $\mathbf{x}(t)$ formed by the state variables $x_j(t)$ is called the *state vector*, the h-vector $\mathbf{u}(t)$ formed by the h known forcing functions $u_j(t)$ is referred to as the *input vector*, and $\dot{\mathbf{x}}(t)$ denotes the k-vector whose elements are the time derivatives of those of $\mathbf{x}(t)$. The coefficient matrices $\mathbf{A}_s$ and $\mathbf{B}_s$ depend only on the network parameters and are of orders $k \times k$ and $k \times h$, respectively. For a linear time-invariant network, they are constant matrices, being independent of time t.

The state variables $x_j(t)$ may or may not be the desired output variables. For this reason, another matrix equation is required to express the desired output variable $y_j(t)(j = 1, 2, \ldots, w)$ in terms of the state variables $x_i(t)(i = 1, 2, \ldots, k)$ and the input excitations $u_j(j = 1, 2, \ldots, h)$ as

$$\mathbf{y}(t) = \mathbf{C}_s \mathbf{x}(t) + \mathbf{D}_s \mathbf{u}(t) \qquad (8.3)$$

where the w-vector $\mathbf{y}(t)$ whose elements are the w output variables $y_j(t)$ is called the *output vector*, and the known coefficient matrices $\mathbf{C}_s$ and $\mathbf{D}_s$ of network parameters are of orders $w \times k$ and $w \times h$, respectively.

Equation (8.2) is referred to as the *state equation in normal form*, Equation (8.3) is called the *output equation*. Together they are known as the *state equations*. We remark that in order to distinguish the coefficient matrices of (8.2) and (8.3) from those of (7.13), a subscript s, standing for the state equations, is used here. The symbols for the input vector $\mathbf{u}$ and output vector $\mathbf{y}$ are retained. However, they are used in two different contexts. In (7.13) they denote the Laplace transforms of the input and output variables, whereas in (8.2) and (8.3) they represent these variables in the time domain. This should not create any confusion as the context will reveal.

Our immediate problem is to choose the network variables as the state variables in order to formulate the state equations. If we call the set of instantaneous values of all the branch currents and voltages as the "state" of the network, then the knowledge of the instantaneous values of all these variables determines this instantaneous state. However, not all of these instantaneous values are required in order to determine the instantaneous state, since some can be calculated from the others by linear combinations or by means of linear algebraic equations. For example, the instantaneous voltage of a resistor can be obtained from its instantaneous current through Ohm's law. The question arises as to how many branch voltages and currents whose instantaneous values are sufficient to determine completely the instantaneous state of the network. This leads to the following definition.

Definition 8.1: Complete set of state variables In a given network, a minimal set of its branch variables is said to be a *complete set of state variables* if their instantaneous values are sufficient to determine completely the instantaneous values of all the branch variables.

In general, the complete set of state variables is not unique. In most situations, the voltages across the capacitors and the currents through the inductors will constitute a complete set of state variables for the network. However, in degenerate cases, not all these variables are independent. For example, in a network containing two capacitors connected in parallel, the voltages of these two capacitors cannot be chosen independently, one being determined by the other. Hence, at most one can be a member of a complete set, because any complete set must be minimal. For our purposes, we shall choose the capacitor voltages and inductor currents or their linear combinations as the state variables in the case of linear time-invariant networks. For linear time-varying networks, it is convenient to select the capacitor charges and the inductor flux linkages or their linear combinations as the state variables.

After selecting the state variables, we now proceed to develop the state equations. Our starting point is the three primary systems of network equations composed of the Kirchhoff's current law (KCL) equations, Kirchhoff's voltage law (KVL) equations, and element defining equations governed by the generalized Ohm's law, as follows: Let N be a network represented by a directed graph G. For each edge of G, there associate two real functions $v(t)$ and $i(t)$ of time t called the *branch voltage* and *branch current* of the edge. They satisfy the following constraints:

1. Kirchhoff's current law equations.
2. Kirchhoff's voltage law equations.
3. Element defining equations governed by the generalized Ohm's law.

A *capacitive edge* obeys a differential equation like

$$i(t) = \frac{d}{dt}[C(t)v(t)] = \frac{dq(t)}{dt} \tag{8.4a}$$

where $C(t)$ and $q(t)$ are the capacitance and charges of the corresponding network element. An *inductive edge* obeys a differential equation like

$$\dot{v}(t) = \frac{d}{dt}[L(t)i(t)] = \frac{d\lambda(t)}{dt} \tag{8.4b}$$

where $L(t)$ and $\lambda(t)$ are the inductance and flux linkages of the corresponding network element. A *resistive edge* obeys an algebraic equation like

$$v(t) = R_k(t)i_k(t) \tag{8.4c}$$

or

$$i(t) = G_k(t)v_k(t) \tag{8.4d}$$

where $i_k(t)$ and $v_k(t)$ denote the current and voltage of the edge e_k, which may or may not be the same edge whose current and voltage are $i(t)$ and $v(t)$. Thus, (8.4c) or (8.4d) represents either a controlled source whose controlling parameter is $R_k(t)$ or $G_k(t)$, or a one-port resistor whose resistance is $R_k(t)$ or whose conductance is $G_k(t)$.

To simplify our notation, from here on we shall drop the time variable t in writing the variables, vectors or matrices with the understanding that they are functions of time t. We shall include t for situations where confusion may arise or, occasionally, for emphasis.

Example 8.1 Consider the network N of Fig. 8.1a the associated directed graph G of which is shown in Fig. 8.1b. Let $i_k(t)$ and $v_k(t)$ be the current and voltage of edge $e_k (k = 1, 2, \ldots, 6)$. The KCL and KVL equations for the network are found to be

$$i_2 + i_3 - i_5 = 0 \tag{8.5a}$$

$$-i_1 - i_2 + i_4 = 0 \tag{8.5b}$$

$$i_1 + i_5 + i_6 = 0 \tag{8.5c}$$

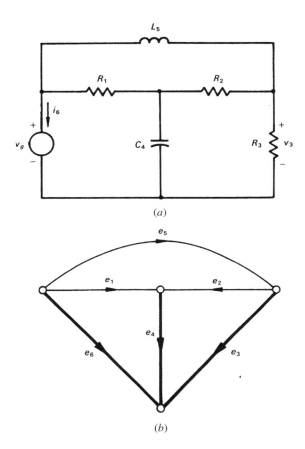

(a)

(b)

Figure 8.1 (a) A network and (b) its associated directed graph.

and

$$v_1 + v_4 - v_6 = 0 \qquad (8.6a)$$

$$v_2 - v_3 + v_4 = 0 \qquad (8.6b)$$

$$v_3 + v_5 - v_6 = 0 \qquad (8.6c)$$

respectively. The element v-i equations are obtained as

$$v_1 = R_1 i_1 \qquad (8.7a)$$

$$v_2 = R_2 i_2 \qquad (8.7b)$$

$$i_3 = \frac{v_3}{R_3} \qquad (8.7c)$$

$$i_4 = C_4 \frac{dv_4}{dt} = C_4 \dot{v}_4 \qquad (8.7d)$$

$$v_5 = L_5 \frac{di_5}{dt} = L_5 \dot{i}_5 \qquad (8.7e)$$

$$v_6 = v_g \qquad (8.7f)$$

Our objective is to express $\dot{v}_4$ and $\dot{i}_5$ in terms of the state variables v_4 and i_5, and the known source v_g. To this end, we substitute (8.7) in (8.5) and (8.6) to obtain

$$i_5 - i_2 = \frac{v_3}{R_3} \qquad (8.8a)$$

$$C_4 \frac{dv_4}{dt} = C_4 \dot{v}_4 = i_1 + i_2 \qquad (8.8b)$$

$$i_6 = -i_1 - i_5 \qquad (8.8c)$$

$$R_1 i_1 = v_g - v_4 \qquad (8.8d)$$

$$R_2 i_2 = -v_4 + v_3 \qquad (8.8e)$$

$$L_5 \frac{di_5}{dt} = L_5 \dot{i}_5 = v_g - v_3 \qquad (8.8f)$$

From (8.8a) and (8.8e), we can solve i_2 and v_3 in terms of i_5 and v_4 and obtain

$$i_2 = \frac{R_3 i_5 - v_4}{R_2 + R_3} \qquad (8.9a)$$

$$v_3 = \frac{R_2 R_3 i_5 + R_3 v_4}{R_2 + R_3} \qquad (8.9b)$$

Finally, substituting i_1, i_2 and v_3 from (8.8d) and (8.9) in (8.8b) and (8.8f) yields

$$\dot{v}_4 = \frac{dv_4}{dt} = -\frac{v_4}{R_1 C_4} - \frac{v_4}{C_4(R_2 + R_3)} + \frac{R_3 i_5}{C_4(R_2 + R_3)} + \frac{v_g}{R_1 C_4} \qquad (8.10a)$$

$$\dot{i}_5 = \frac{di_5}{dt} = -\frac{R_3 v_4}{L_5(R_2 + R_3)} - \frac{R_2 R_3 i_5}{L_5(R_2 + R_3)} + \frac{v_g}{L_5} \qquad (8.10b)$$

In matrix notation, (8.10) can be written in the form of (8.2), as follows:

$$
\begin{bmatrix} \dot{v}_4 \\ \dot{i}_5 \end{bmatrix} = \begin{bmatrix} -\dfrac{1}{R_1 C_4} - \dfrac{1}{C_4(R_2 + R_3)} & \dfrac{R_3}{C_4(R_2 + R_3)} \\ -\dfrac{R_3}{L_5(R_2 + R_3)} & -\dfrac{R_2 R_3}{L_5(R_2 + R_3)} \end{bmatrix} \begin{bmatrix} v_4 \\ i_5 \end{bmatrix} + \begin{bmatrix} \dfrac{1}{R_1 C_4} \\ \dfrac{1}{L_5} \end{bmatrix} [v_g]
$$

$$(8.11)$$

Suppose that we choose the resistor voltage v_3 and generator current i_6 as the desired output variables. Then from (8.8c) and (8.9b) in conjunction with (8.8d) we obtain

$$
v_3 = \frac{R_2 R_3 i_5 + R_3 v_4}{R_2 + R_3} \tag{8.12a}
$$

$$
i_6 = \frac{v_4 - v_g}{R_1} - i_5 \tag{8.12b}
$$

Writing these in matrix form gives the desired output equation

$$
\begin{bmatrix} v_3 \\ i_6 \end{bmatrix} = \begin{bmatrix} \dfrac{R_3}{R_2 + R_3} & \dfrac{R_2 R_3}{R_2 + R_3} \\ \dfrac{1}{R_1} & -1 \end{bmatrix} \begin{bmatrix} v_4 \\ i_5 \end{bmatrix} + \begin{bmatrix} 0 \\ \dfrac{1}{R_1} \end{bmatrix} [v_g] \tag{8.13}
$$

Equations (8.11) and (8.13) constitute the state equations for the network of Fig. 8.1(a). The complete set of state variables are the capacitor voltage v_4 and inductor current i_5.

The above example demonstrates that with appropriate elimination of 'unwanted' variables the primary system of network equations can be reduced to the state equations. Before we proceed to develop procedures for writing the state equations systematically, we define three matrices associated with the directed graph of a network, and show how they can be applied to the formulation of Kirchhoff's current and voltage equations.

8.2 GRAPH MATRICES AND KIRCHHOFF'S EQUATIONS

A network must satisfy Kirchhoff's two laws regardless of whether it is linear or nonlinear and time varying or time invariant. Kirchhoff's law equations are therefore independent of the constituents of network elements and can be determined solely from the network topology.

Consider the transistor amplifier of Fig. 8.2a. After the biasing circuitry is removed, its small–signal equivalent network is presented in Fig. 8.2b.

Figure 8.2 (*a*) A transistor amplifier together with its biasing circuitry, (*b*) its small-signal equivalent network and (*c*) the associated directed graph.

The associated directed graph G of the network N of Fig. 8.2b is shown in Fig. 8.2c. Let i_k and v_k be the current and voltage of the edge e_k ($k = 1, 2, \ldots, 8$). Applying Kirchhoff's current law to the nodes yields

$$\text{node 1: } i_1 + i_2 \qquad\qquad\qquad\qquad\qquad = 0 \qquad (8.14a)$$

$$\text{node 2: } \quad -i_2 + i_3 + i_4 + i_5 \qquad\qquad\quad = 0 \qquad (8.14b)$$

$$\text{node 3: } \qquad\qquad\qquad -i_5 + i_6 + i_7 \quad = 0 \qquad (8.14c)$$

$$\text{node 4: } \qquad\qquad\qquad\qquad\quad -i_7 + i_8 = 0 \qquad (8.14d)$$

$$\text{node 5: } -i_1 \quad -i_3 - i_4 \quad -i_6 \quad -i_8 = 0 \qquad (8.14e)$$

which can be written in matrix notation as

$$
\begin{array}{c}
\\
\text{node 1} \\
\text{node 2} \\
\text{node 3} \\
\text{node 4} \\
\text{node 5}
\end{array}
\begin{array}{cccccccc}
e_1 & e_2 & e_3 & e_4 & e_5 & e_6 & e_7 & e_8 \\
\left[\begin{array}{cccccccc}
1 & 1 & 0 & 0 & 0 & 0 & 0 & 0 \\
0 & -1 & 1 & 1 & 1 & 0 & 0 & 0 \\
0 & 0 & 0 & 0 & -1 & 1 & 1 & 0 \\
0 & 0 & 0 & 0 & 0 & 0 & -1 & 1 \\
-1 & 0 & -1 & -1 & 0 & -1 & 0 & -1
\end{array}\right]
\end{array}
\begin{bmatrix} i_1 \\ i_2 \\ i_3 \\ i_4 \\ i_5 \\ i_6 \\ i_7 \\ i_8 \end{bmatrix}
\begin{bmatrix} 0 \\ 0 \\ 0 \\ 0 \\ 0 \end{bmatrix} \qquad (8.15)
$$

This equation can be expressed more compactly as

$$\mathbf{A}_a \mathbf{i}(t) = \mathbf{0} \qquad (8.16)$$

The coefficient matrix $\mathbf{A}_a$ is called the *complete incidence matrix* of G and $\mathbf{i}(t)$ is the *branch-current vector*. In general if G is a directed graph of n nodes and b edges, the *complete incidence matrix* of G is a matrix of order $n \times b$ such that if

$$\mathbf{A}_a = [a_{ij}] \qquad (8.17)$$

then

$a_{ij} = 1$ if edge e_j is incident at node i and is directed away from node i
$a_{ij} = -1$ if edge e_j is incident at node i and is directed toward node i
$a_{ij} = 0$ if edge e_j is not incident at node i

Thus, the complete incidence matrix $\mathbf{A}_a$ of the directed graph G of Fig. 8.2c is a matrix of order 5×8. Each row corresponds to a node and each column corresponds to an edge. The edge e_1 is directed from node 1 to node 5. Hence there is a 1 in the first row position of column 1 corresponding to e_1 and a -1 in the fifth row position. Likewise, edges e_3 and e_4 are directed from node 2 to node 5 and there are 1's in second row and -1's in fifth row positions of the columns corresponding to e_3 and e_4. The complete incidence matrix of G is found to be

$$
\mathbf{A}_a = \begin{array}{c} \\ \text{node 1} \\ \text{node 2} \\ \text{node 3} \\ \text{node 4} \\ \text{node 5} \end{array}
\begin{array}{c} e_1 \quad e_2 \quad e_3 \quad e_4 \quad e_5 \quad e_6 \quad e_7 \quad e_8 \\
\left[\begin{array}{cccccccc}
1 & 1 & 0 & 0 & 0 & 0 & 0 & 0 \\
0 & -1 & 1 & 1 & 1 & 0 & 0 & 0 \\
0 & 0 & 0 & 0 & -1 & 1 & 1 & 0 \\
0 & 0 & 0 & 0 & 0 & 0 & -1 & 1 \\
-1 & 0 & -1 & -1 & 0 & -1 & 0 & -1
\end{array}\right]
\end{array} \tag{8.18}
$$

Observe that each column of $\mathbf{A}_a$ contains exactly two nonzero elements, a 1 and a -1. Hence the sum of all the rows of $\mathbf{A}_a$ is a row of zeros, showing that the rows of $\mathbf{A}_a$ are linearly dependent. This implies that the rank of $\mathbf{A}_a$ is at most $n - 1$ and equations of (8.16) are not linearly independent, one being the linear combinations of the others. It can be shown that the rank of $\mathbf{A}_a$ is also at least $n - 1$ if G is connected. A proof of this can be found in Chen (1976).

Theorem 8.1 The rank of the complete incidence matrix $\mathbf{A}_a$ of an n-node connected directed graph is equal to $n - 1$.

As a result, there is no need to consider all the rows of $\mathbf{A}_a$; $n - 1$ linearly independent ones are sufficient. A submatrix of $\mathbf{A}_a$ is called a *basis incidence matrix* of G if it is of order $(n - 1) \times b$ and of rank $n - 1$. Thus, if G is connected, a basis incidence matrix can be obtained from $\mathbf{A}_a$ simply by deleting any one of its rows. The deleted row in $\mathbf{A}_a$ corresponds to the reference potential point in the network N. If $\mathbf{A}_\varsigma$ is such a basis incidence matrix, an appropriate set of linearly independent KCL equations may be compactly written as

$$
\mathbf{A}_\varsigma \mathbf{i}(t) = \mathbf{0} \tag{8.19}
$$

From the directed graph G of Fig. 8.2c, suppose that node 5 is chosen as the reference-potential point for the network. Then a basis incidence matrix can be obtained from $\mathbf{A}_a$ of (8.18) by deleting the row corresponding to node 5, which is row 5, giving

$$A_\varsigma = \begin{bmatrix} 1 & 1 & 0 & 0 & 0 & 0 & 0 & 0 \\ 0 & -1 & 1 & 1 & 1 & 0 & 0 & 0 \\ 0 & 0 & 0 & 0 & -1 & 1 & 1 & 0 \\ 0 & 0 & 0 & 0 & 0 & 0 & -1 & 1 \end{bmatrix} \qquad (8.20)$$

A generalized version of Kirchhoff's current law equations written for individual nodes can be stated for groups of nodes represented by the so-called cutsets.

Definition 8.2: Cutset A *cutset* of a connected directed graph is a subgraph consisting of a minimal collection of edges whose removal cuts the directed graph into two connected pieces. One or two of these connected pieces may each consist of an isolated node.

In Fig. 8.2c, the subgraph $e_5 e_6 e_8$ is a cutset because it is a minimal collection of edges whose removal cuts the directed graph G into two connected pieces $e_1 e_2 e_3 e_4$ and e_7. The subgraph $e_4 e_5 e_6 e_8$ is not a cutset because if we remove the subgraph $e_5 e_6 e_8$ from G, G will also be cut into two connected pieces, showing that $e_4 e_5 e_6 e_8$ is not minimal. There are obviously many cutsets besides $e_5 e_6 e_8$. Some of them may simply be the *incidence cutsets* formed by the edges incident to a node. In most cases, a cutset can be represented by a line drawn across the edges of the cutset. Also, we assign *orientation* to a cutset by placing an arrow near the line defining the cutset. In this way, Kirchhoff's current law equations can be written for individual cutsets, as follows: the algebraic sum of currents crossing the cutset is zero at all times, or equivalently the sum of currents crossing the cutset in one direction is equal to the sum of currents crossing the cutset in the other direction at every instance. In particular, this is true for the incidence cutsets. Thus, the above procedure is a generalization of Kirchhoff's current law. Two oriented cutsets $e_1 e_3 e_4 e_5$ and $e_5 e_6 e_8$ of the directed graph G of Fig. 8.2c are depicted in Fig. 8.3, the KCL equations of which are given by

$$i_1 + i_3 + i_4 + i_5 = 0 \qquad (8.21a)$$

$$i_5 - i_6 - i_8 = 0 \qquad (8.21b)$$

As another example, consider the network of Fig. 8.4a. Its associated directed graph G is presented in Fig. 8.4b. All the cutsets of G together with their orientations are shown by the broken lines in Fig. 8.4b. The KCL equations obtained

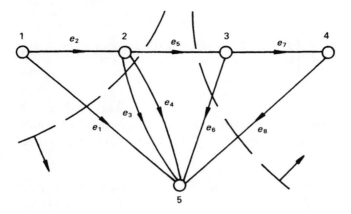

Figure 8.3 Physical interpretation of total current crossing the broken lines in the direction of the arrow associated with a cutset.

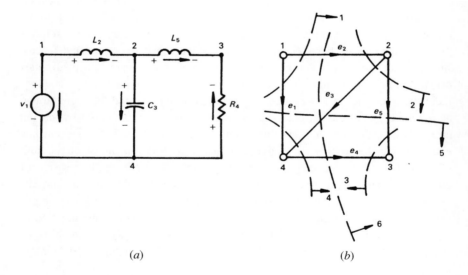

$$(a) \qquad\qquad (b)$$

Figure 8.4 (a) A network and (b) its associated directed graph.

for these cutsets are given by

$$\text{cutset 1:} \quad i_1 + i_2 \qquad\qquad\qquad = 0 \qquad\qquad (8.22a)$$

$$\text{cutset 2:} \quad - i_2 + i_3 \quad + i_5 = 0 \qquad\qquad (8.22b)$$

$$\text{cutset 3:} \qquad\qquad -i_4 - i_5 = 0 \qquad\qquad (8.22c)$$

$$\text{cutset 4:} \; -i_1 \qquad -i_3 + i_4 \qquad = 0 \qquad\qquad (8.22d)$$

$$\text{cutset 5:} \; i_1 \qquad +i_3 \qquad +i_5 = 0 \qquad\qquad (8.22e)$$

$$\text{cutset 6:} \qquad\quad i_2 - i_3 + i_4 \qquad = 0 \qquad\qquad (8.22f)$$

which can be written in matrix notation as

$$
\begin{array}{c}
\quad \begin{array}{ccccc} e_1 & e_2 & e_3 & e_4 & e_5 \end{array} \\
\begin{array}{c}
\text{cutset 1} \\ \text{cutset 2} \\ \text{cutset 3} \\ \text{cutset 4} \\ \text{cutset 5} \\ \text{cutset 6}
\end{array}
\left[\begin{array}{ccccc}
1 & 1 & 0 & 0 & 0 \\
0 & -1 & 1 & 0 & 1 \\
0 & 0 & 0 & -1 & -1 \\
-1 & 0 & -1 & 1 & 0 \\
1 & 0 & 1 & 0 & 1 \\
0 & 1 & -1 & 1 & 0
\end{array}\right]
\left[\begin{array}{c}
i_1 \\ i_2 \\ i_3 \\ i_4 \\ i_5
\end{array}\right]
=
\left[\begin{array}{c}
0 \\ 0 \\ 0 \\ 0 \\ 0 \\ 0
\end{array}\right]
\qquad (8.23)
\end{array}
$$

or, more compactly, as

$$\mathbf{Q}_a \mathbf{i}(t) = \mathbf{0} \qquad\qquad (8.24)$$

The coefficient matrix $\mathbf{Q}_a$ is called the *complete cutset matrix* of G and as before $\mathbf{i}(t)$ is the branch-current vector. In general, if G has q cutsets and b edges, the complete cutset matrix $\mathbf{Q}_a$ is a matrix of order $q \times b$ such that if

$$\mathbf{Q}_a = [q_{ij}] \qquad\qquad (8.25)$$

then

$q_{ij} = 1$ if edge e_j is in cutset i and the orientations of the cutset and the edge coincide

$q_{ij} = -1$ if edge e_j is in cutset i and the orientations of the cutset and the edge are opposite

$q_{ij} = 0$ if edge e_j is not in cutset i

Thus, each row of $\mathbf{Q}_a$ is associated with a cutset of G and each column is associated with an edge. In Fig. 8.4b the edge e_3 is contained in cutsets 2, 4, 5 and 6, so there are nonzero elements in rows 2, 4, 5 and 6 of column 3 corresponding to the cutsets 2, 4, 5 and 6. The element in row 5 is 1 because the orientation of cutset 5 coincides with that of edge e_3, and the element in row 6 is –1 because the orientation of cutset 6 is opposite to that of edge e_3.

Observe that (8.23) has six equations in five unknown currents. Evidently they are not linearly independent since some of these equations are obtainable from the others by linear combinations. Then the question is "What is the maximum number of linearly independent equations?" or, equivalently, "What is the rank of the matrix $\mathbf{Q}_a$?" Since $\mathbf{Q}_a$ contains $\mathbf{A}_a$ as a submatrix, we know right away that the rank of $\mathbf{Q}_a$ is at least $n - 1$ for an n-node connected directed graph G. It turns out that $n - 1$ is also the rank of $\mathbf{Q}_a$, a proof of which may be found in Chen (1976).

Theorem 8.2 The rank of the complete cutset matrix $\mathbf{Q}_a$ of an n-node connected directed graph is equal to $n - 1$.

The number $n - 1$ is also called the *rank* of the connected directed graph G. This shows that we need only consider $n - 1$ linearly independent equations for a connected network. For this we say that a submatrix of $\mathbf{Q}_a$, denoted by $\mathbf{Q}_\varsigma$ is a *basis cutset matrix* if $\mathbf{Q}_\varsigma$ is of order $(n - 1) \times b$ and of rank $n - 1$. All the information contained in (8.24) is now contained in

$$\mathbf{Q}_\varsigma \mathbf{i}(t) = \mathbf{0} \tag{8.26}$$

The 5-edge directed graph G of Fig. 8.4b has four nodes. Thus, the rank of the coefficient matrix $\mathbf{Q}_a$ of (8.23) is 3. The submatrix formed by the first three rows of $\mathbf{Q}_a$ as given below

$$\mathbf{Q}_\varsigma = \begin{bmatrix} 1 & 1 & 0 & 0 & 0 \\ 0 & -1 & 1 & 0 & 1 \\ 0 & 0 & 0 & -1 & -1 \end{bmatrix} \tag{8.27}$$

is a basis cutset matrix.

We remark that confining to connected directed graphs should not be deemed a restriction, because in dealing with electrical networks each connected subnetwork can be treated separately.

Finally, we turn our attention to the systematic formulation of Kirchhoff's voltage law. The approach is parallel to those discussed above for KCL.

Consider again the low-pass filter of Fig. 8.4a, the associated directed graph G of which is redrawn in Fig. 8.5. As before, let i_k and v_k be the current and voltage of the edges $e_k (k = 1, 2, 3, 4, 5)$. Using the three loops as indicated in Fig. 8.5, KVL equations can be written as

$$\text{loop 1:} \quad -v_1 + v_2 + v_3 \qquad\qquad = 0 \tag{8.28a}$$

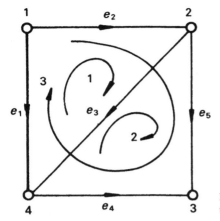

Figure 8.5 Three loops used to write KVL equations.

$$\text{loop 2:} \qquad -v_3 - v_4 + v_5 = 0 \qquad (8.28b)$$

$$\text{loop 3:} \ -v_1 + v_2 \qquad -v_4 + v_5 = 0 \qquad (8.28c)$$

which can be written in matrix notation as

$$
\begin{array}{c}
\\
\text{loop 1} \\
\text{loop 2} \\
\text{loop 3}
\end{array}
\begin{array}{ccccc}
e_1 & e_2 & e_3 & e_4 & e_5
\end{array}
\left[
\begin{array}{ccccc}
-1 & 1 & 1 & 0 & 0 \\
0 & 0 & -1 & -1 & 1 \\
-1 & 1 & 0 & -1 & 1
\end{array}
\right]
\begin{bmatrix}
v_1 \\ v_2 \\ v_3 \\ v_4 \\ v_5
\end{bmatrix}
=
\begin{bmatrix}
0 \\ 0 \\ 0
\end{bmatrix}
\qquad (8.29)
$$

or, more compactly, as

$$\mathbf{B}_a \mathbf{v}(t) = \mathbf{0} \qquad (8.30)$$

The coefficient matrix $\mathbf{B}_a$ is called the *complete circuit matrix* of G and $\mathbf{v}(t)$ is the *branch-voltage vector*. In general, if G is a directed graph of n nodes and b edges and contains κ circuits, the *complete circuit matrix* of G is a matrix of order $\kappa \times b$ such that if

$$\mathbf{B}_a = [b_{ij}] \qquad (8.31)$$

then

$b_{ij} = 1$ if edge e_j is in circuit i and the orientations of the circuit and the edge coincide

$b_{ij} = -1$ if edge e_j is in circuit i and the orientations of the circuit and the edge are opposite

$b_{ij} = 0$ if edge e_j is not in circuit i

Therefore, each row of $\mathbf{B}_a$ is associated with a loop of G and each column is associated with an edge. In Fig. 8.5 edge e_3 is contained in loops 1 and 2, so that the nonzero elements appear in rows 1 and 2 of column 3 corresponding to loops 1 and 2. The element in row 1 is 1 because the orientation of loop 1 coincides with that of edge e_3, whereas the element in row 2 is -1 because the orientation of loop 2 is opposite to that of edge e_3.

Observe that in (8.28) the third equation (8.28c) is the sum of the first two. Thus, the equations in (8.30) in general are not linearly independent. Then what is the maximum number of linearly independent equations? Or, equivalently, what is the rank of the matrix $\mathbf{B}_a$? The answer to this question is given below, a proof of which may be found in Chen (1976).

Theorem 8.3 The rank of the complete circuit matrix $\mathbf{B}_a$ of an n-node connected directed graph having b edges is equal to $b - n + 1$.

The number $b - n + 1$ is also referred to as the *nullity* of the connected directed graph G. A submatrix of $\mathbf{B}_a$, denoted by $\mathbf{B}_\varsigma$, is called a *basis circuit matrix* of G if $\mathbf{B}_\varsigma$ is of order $(b - n + 1) \times b$ and of rank $b - n + 1$. All the information contained in (8.30) is now contained in

$$\mathbf{B}_\varsigma \mathbf{v}(t) = \mathbf{0} \tag{8.32}$$

For the directed graph G of Fig. 8.5, its nullity is $b - n + 1 = 5 - 4 + 1 = 2$. The submatrix formed by the last two rows of $\mathbf{B}_a$ in (8.29), as given below,

$$\mathbf{B}_\varsigma = \begin{bmatrix} 0 & 0 & -1 & -1 & 1 \\ -1 & 1 & 0 & -1 & 1 \end{bmatrix} \tag{8.33}$$

is a basis circuit matrix of G.

8.3 TREES AND FUNDAMENTAL CUTSETS AND CIRCUITS

The tree is perhaps the single most important subgraph in electrical network theory because a number of fundamental results in network theory can be related to it.

In particular, the methods of choosing the independent KCL and KVL equations and the formulation of the state equations may all be stated in terms of the single concept of a tree.

Definition 8.3: Circuit A *circuit* of a directed graph is a subgraph composed of a sequence of edges, each having one node in common with the preceding edge in the sequence and the other node in common with the succeeding edge. Furthermore, each node is incident with exactly two edges in the sequence. An *oriented circuit* is a circuit with an orientation assigned to it by a cyclic ordering of nodes along the circuit.

An oriented circuit is frequently referred to as a *loop*. In Fig. 8.5, for example, the subgraph $e_1e_2e_5e_4$ is a circuit, which can be oriented as (1, 2, 3, 4). The orientation can be represented pictorially by an arrowhead as shown in Fig. 8.5.

Definition 8.4: Tree A *tree* of a directed graph is a connected subgraph that contains all the nodes of the graph but does not contain any circuits. An edge of a tree is called a *branch*.

The word *tree* signifies a treelike structure, a structure in one piece with branches connecting to other branches. Thus, a tree possesses three important attributes: (1) connectedness, (2) all nodes, and (3) no circuits. If any one of these characteristics is missing, the subgraph is not a tree.

For instance, the directed graph G of Fig. 8.6 has many trees; four of them are shown in Fig. 8.7. The subgraphs $e_1e_2e_4e_9$, $e_1e_3e_7e_9$, and $e_1e_2e_3e_4e_7e_9$ shown in Fig. 8.8 are not trees because $e_1e_2e_4e_9$ does not contain all the nodes of G, $e_1e_3e_7e_9$ is not connected, and $e_1e_2e_3e_4e_7e_9$ contains a circuit.

Definition 8.5: Cotree The complement of a tree in a directed graph is called a *cotree*. The edges of a cotree are called *links*.

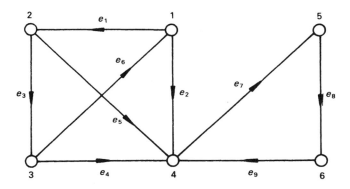

Figure 8.6 A directed graph used to illustrate the concepts of tree and cotree.

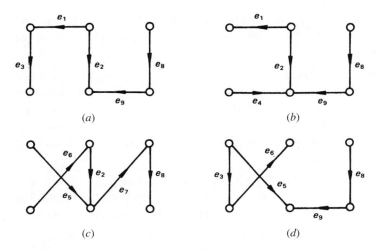

Figure 8.7 Four of the trees of the directed graph of Fig. 8.6.

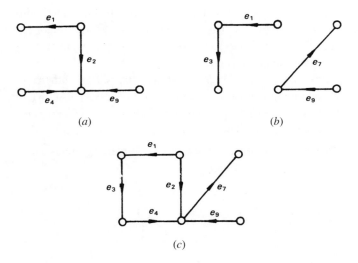

Figure 8.8 Subgraphs that are not trees of the directed graph of Fig. 8.6.

The word *cotree* is an abbreviation of the phrase "the complement of a *tree*." Thus, a cotree is defined only with respect to a chosen tree. Figure 8.9 shows the four cotrees with respect to the trees of Fig. 8.7 in the directed graph G of Fig. 8.6. Each edge in Fig. 8.9 is a link of some cotree. Observe that cotrees need not contain all the nodes of G. In fact, a cotree may not be connected and may contain circuits.

(a)

(b)

(c)

(d)

Figure 8.9 Four cotrees with respect to the trees of Fig. 8.7 in the directed graph of Fig. 8.6.

When we speak of links and branches, it is with respect to a chosen tree. Recall that the currents and voltages of network elements are also referred to as the branch currents and branch voltages. The word *branch* has nothing to do with the tree branch defined here and should not create any difficulty. Since a tree is connected and contains no circuits, there is a unique path between any two nodes in a tree. If we add one link to the tree, the resulting graph is of course no longer a tree. The link and the unique path in the tree between the two endpoints of the link constitute a circuit. If the orientation of the circuit is chosen to agree with that of the defining link, the resulting oriented circuit is called a *fundamental circuit* or an *f-circuit* for short. The number of f-circuits constructed in this way with respect to a chosen tree is clearly equal to the number of links in the corresponding cotree.

For the directed graph G of Fig. 8.10, let $T = e_2e_5e_6e_7e_8$ of Fig. 8.7c be the chosen tree. The corresponding cotree $\bar{T} = e_1e_3e_4e_9$ is shown in Fig. 8.9c. Suppose that we add the link e_3 to T. The resulting subgraph $e_2e_3e_5e_6e_7e_8$ contains the f-circuit $e_2e_3e_5e_6$, as depicted in Fig. 8.11. This f-circuit is formed by the defining link e_3 and the unique tree path composed of the branches e_2, e_5 and e_6 connecting the two endpoints of the link e_3. The three other f-circuits are $e_2e_4e_6$, $e_1e_2e_5$, and $e_7e_8e_9$, and are shown in Fig. 8.12. Observe that the orientations of these f-circuits are chosen to agree with those of the defining links.

Denote by $\mathbf{B}_f$ the submatrix of the complete circuit matrix of a connected directed graph G corresponding to a set of f-circuits. If G has n nodes and b

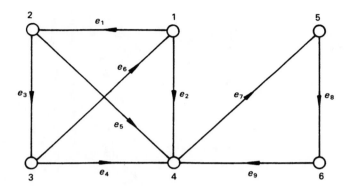

Figure 8.10 A directed graph used for illustrations.

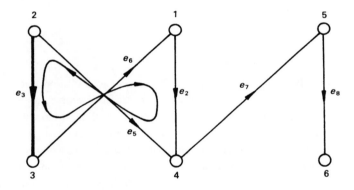

Figure 8.11 The f-circuit $e_2e_3e_5e_6$ defined by the link e_3.

edges, a tree of G contains $n - 1$ branches and a cotree of G contains $b - n + 1$ links. Since each f-circuit contains an edge — namely, the defining link — that is not contained in any other f-circuits, $\mathbf{B}_f$ must contain the identity matrix of order $b - n + 1$. Hence $\mathbf{B}_f$, being of order $(b - n + 1) \times b$ and of rank $b - n + 1$, is a basis circuit matrix. In particular, if the f-circuits are numbered in some arbitrary manner as $1, 2, \ldots, b - n + 1$, and if the link that appears in circuit i is numbered as edge e_i for $i = 1, 2, \ldots, b - n + 1$, then $\mathbf{B}_f$ can be partitioned as

$$\mathbf{B}_f = [\mathbf{1}_{b-n+1} \quad \mathbf{B}_{f12}] \tag{8.34}$$

where the columns of $\mathbf{B}_f$ have been arranged in the order of links and branches with respect to the defining tree.

Figure 8.12 The f-circuits (a) $e_2 e_4 e_6$, (b) $e_1 e_2 e_5$ and (c) $e_7 e_8 e_9$ defined by the links e_4, e_1 and e_9, respectively.

In Fig. 8.10, the f-circuits with respect to the tree $T = e_2e_5e_6e_7e_8$ are presented in Figs. 8.11 and 8.12. The f-circuit matrix is given by

$$
\mathbf{B}_f = \begin{array}{c}
\begin{array}{ccccccccc} e_1 & e_3 & e_4 & e_9 & e_2 & e_5 & e_6 & e_7 & e_8 \end{array} \\
\left[\begin{array}{cccc:ccccc}
1 & 0 & 0 & 0 & -1 & 1 & 0 & 0 & 0 \\
0 & 1 & 0 & 0 & 1 & -1 & 1 & 0 & 0 \\
0 & 0 & 1 & 0 & -1 & 0 & -1 & 0 & 0 \\
0 & 0 & 0 & 1 & 0 & 0 & 0 & 1 & 1
\end{array} \right]
\end{array} \tag{8.35}
$$

This matrix contains the identity matrix of order 4 formed by the first four columns corresponding to the links of the cotree $e_1e_3e_4e_9$, confirming that $\mathbf{B}_f$ is a basis circuit matrix.

So far we have been concentrating on circuits. A dual operation can be applied to cutsets. Let T be a tree of G and let e be a branch of T. Since T is connected and contains no circuits, the removal of e from T results in a subgraph consisting of two connected subgraphs. If S_1 and S_2 denote the node sets of these subgraphs, then S_1 and S_2 are mutually exclusive and together include all the nodes of G. Therefore, the branch e of T defines a partition of the nodes of G in a unique way. The set of edges of G connecting a node in S_1 and a node in S_2 is clearly a cutset of G, as depicted symbolically in Fig. 8.13 in which the branch e is assumed to be directed from a node in S_1 to a node in S_2. Such a cutset is termed a *fundamental cutset* or an *f-cutset* for short. An important property of an f-cutset is that it contains only one branch — namely, the defining tree branch — and some links of the cotree with respect to the same tree.

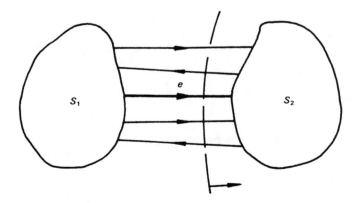

Figure 8.13 Symbolic representation of a cutset of a directed graph.

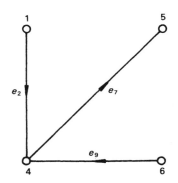

Figure 8.14 Subgraph resulting from the tree $e_2e_3e_6e_7e_9$ after the removal of the branch e_6 from the tree.

To illustrate we consider the directed graph G of Fig. 8.10. Let $T = e_2e_3e_6e_7e_9$ be the chosen tree. Suppose that e_6 of T is the branch of interest. Removing e_6 from T results in two connected subgraphs e_3 and $e_2e_7e_9$, as shown in Fig. 8.14. The node sets of these two subgraphs are respectively $S_1 = \{2, 3\}$ and $S_2 = \{1, 4, 5, 6\}$. Thus, the branch e_6 of T partitions the nodes of G into two mutually exclusive sets S_1 and S_2. The set of edges of G connecting a node in S_1 and a node in S_2 defines an f-cutset $e_1e_4e_5e_6$ of G. The operation is depicted in Fig. 8.15. The other four f-cutsets defined by the branches e_2, e_3, e_7 and e_9 are given by $e_2e_4e_5$, $e_1e_3e_5$, e_7e_8, and e_8e_9, respectively. The complete set of f-cutsets is presented pictorially in Fig. 8.16 by drawing a broken line across each f-cutset.

Recall that a cutset "cuts" a connected directed graph G into two connected subgraphs. Let S_1 and S_2 be the node sets of these two subgraphs. Then the cutset is said to be *oriented* if the node sets S_1 and S_2 are ordered either as (S_1, S_2) or (S_2, S_1). In most cases the orientation of a cutset can be represented pictorially as an arrow by placing an arrowhead near the broken line defining the cutset. The orientation of an f-cutset is chosen to coincide with the direction of the defining tree branch. In Fig. 8.15 the cutset $e_1e_4e_5e_6$ can be oriented either as $(\{2, 3\}, \{1, 4, 5, 6\})$ from left to right or as $(\{1, 4, 5, 6\}, \{2, 3\})$ from right to left. Since $e_1e_4e_5e_6$ is an f-cutset, we choose its orientation to coincide with the direction of the defining tree branch e_6. Thus it is oriented from left to right. The orientations of all the f-cutsets are indicated in Fig. 8.16.

Denote by $\mathbf{Q}_f$ the submatrix of the complete cutset matrix $\mathbf{Q}_a$ of an n-node connected directed graph G corresponding to a set of f-cutsets. Since each f-cutset contains a branch — namely, the defining tree branch — that is not contained in any

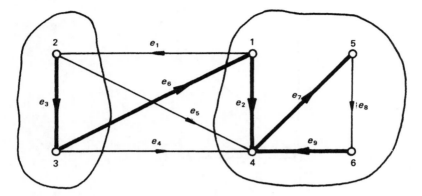

Figure 8.15   The f-cutset $e_1 e_4 e_5 e_6$ defined by the branch e_6 for the tree $e_2 e_3 e_6 e_7 e_9$.

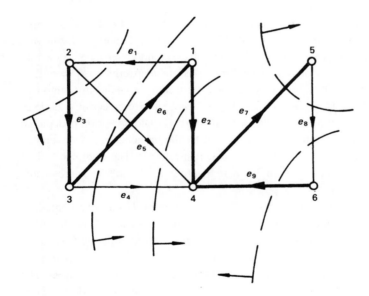

Figure 8.16   Pictorial representation of the f-cutsets defined for the tree $e_2 e_3 e_6 e_7 e_9$ by drawing a broken line across each f-cutset.

other f-cutsets and since there are $n - 1$ f-cutsets, $\mathbf{Q}_f$ must contain the identity matrix of order $n - 1$. Hence $\mathbf{Q}_f$, being of order $(n - 1) \times b$ and of rank $n - 1$, is a basis cutset matrix, where b is the number of edges of G.

Like the f-circuit matrix, if the edges of G are numbered in such a way that the last $n - 1$ columns of $\mathbf{Q}_f$ correspond to the branches of the defining tree, and if the f-cutsets are numbered correspondingly, the f-cutset matrix $\mathbf{Q}_f$ can be partitioned as

$$\mathbf{Q}_f = [\mathbf{Q}_{f11} \quad \mathbf{1}_{n-1}] \tag{8.36}$$

For the directed graph G of Fig. 8.16, the f-cutsets with respect to the tree $e_2e_3e_6e_7e_9$ are indicated in the figure. The f-cutset matrix is found to be

$$\mathbf{Q}_f = \begin{array}{c} \begin{array}{ccccccccc} e_1 & e_4 & e_5 & e_8 & e_2 & e_3 & e_6 & e_7 & e_9 \end{array} \\ \begin{bmatrix} 0 & 1 & 1 & 0 & 1 & 0 & 0 & 0 & 0 \\ -1 & 0 & 1 & 0 & 0 & 1 & 0 & 0 & 0 \\ -1 & 1 & 1 & 0 & 0 & 0 & 1 & 0 & 0 \\ 0 & 0 & 0 & -1 & 0 & 0 & 0 & 1 & 0 \\ 0 & 0 & 0 & -1 & 0 & 0 & 0 & 0 & 1 \end{bmatrix} \end{array} \tag{8.37}$$

This matrix contains the identity matrix of order 5 formed by the last five columns corresponding to the branches of the tree $e_2e_3e_6e_7e_9$, confirming that $\mathbf{Q}_f$ is a basis cutset matrix.

Thus, a systematic way in writing down a basis circuit matrix and a basis cutset matrix is first to pick a tree and then to generate the f-circuits and the f-cutsets. The f-circuit matrix $\mathbf{B}_f$ and the f-cutset matrix $\mathbf{Q}_f$ obtained for these subgraphs can be partitioned as in (8.34) and (8.36). If the branch current vector $\mathbf{i}(t)$ of (8.26) and the branch voltage vector $\mathbf{v}(t)$ of (8.32) are partitioned accordingly, the KCL and KVL equations can be partitioned from (8.26) and (8.32) as

$$\mathbf{Q}_f\mathbf{i}(t) = [\mathbf{Q}_{f11} \quad \mathbf{1}_{n-1}]\begin{bmatrix} \mathbf{i}_{\bar{T}} \\ \mathbf{i}_T \end{bmatrix} = \mathbf{0} \tag{8.38}$$

$$\mathbf{B}_f\mathbf{v}(t) = [\mathbf{1}_{b-n+1} \quad \mathbf{B}_{f12}]\begin{bmatrix} \mathbf{v}_{\bar{T}} \\ \mathbf{v}_T \end{bmatrix} = \mathbf{0} \tag{8.39}$$

or, in expanded form,

$$\mathbf{i}_T = -\mathbf{Q}_{f11}\mathbf{i}_{\bar{T}} \tag{8.40}$$

$$\mathbf{v}_{\bar{T}} = -\mathbf{B}_{f12}\mathbf{v}_T \tag{8.41}$$

where $\mathbf{i}_T$ and $\mathbf{i}_{\bar{T}}$ are the subvectors of the branch-current vector $\mathbf{i}(t)$ corresponding to the branches and links of the defining tree T, and $\mathbf{v}_T$ and $\mathbf{v}_{\bar{T}}$ are the subvectors of the branch-voltage vector $\mathbf{v}(t)$ corresponding to the branches and links of T, respectively.

The implication of (8.40) is that if the currents of the links are arbitrarily chosen and if the currents of the branches of T are calculated by (8.40), the set of currents will always satisfy KCL. Similarly, if the voltages of the branches of T are arbitrarily chosen and if the voltages of the links are obtained by (8.41), the set of voltages will always satisfy KVL. Thus, we have $b - n + 1$ degrees of freedom in choosing the currents, and $n - 1$ degrees of freedom in choosing the voltages if they are only required to satisfy KCL and KVL. However, we have b more element v-i constraints specified by (8.4). In other words, there are $2b$ unknowns in $\mathbf{i}(t)$ and $\mathbf{v}(t)$ and $2b$ equations:

$$(b - n + 1) + (n - 1) + b = 2b \tag{8.42}$$

If these equations are consistent, we are assured of a solution. If, furthermore, they are linearly independent, the solution obtained will be unique. However, for most practical networks, this number is still too large to handle. The state-variable technique is a systematic way to organize these variables, so that the number of the resulting equations called the state equations will be greatly reduced.

8.4 SYSTEMATIC PROCEDURE IN WRITING STATE EQUATIONS

Equations (8.40) and (8.41) indicate that with respect to a chosen tree, the tree-branch voltages determine all other voltages by means of the f-circuits, and the cotree-link currents determine all other currents by way of the f-cutsets. Thus, to develop state equations using capacitor voltages and inductor currents as the state variables, we should place all the voltage sources and as many capacitors as possible in a tree and all the current sources and as many inductors as possible in a cotree. This leads to the following definition.

Definition 8.6: **Normal tree** In the connected directed graph associated with a network, a *normal tree* is a tree that contains all the independent-voltage-source edges, the maximum number of capacitive edges, the minimum number of inductive edges, and none of the independent-current-source edges.

For unconnected networks, we consider the normal trees of the connected subnetworks individually. For the network N of Fig. 8.17a, its associated directed graph G is presented in Fig. 8.17b. The tree $e_3 e_4 e_6$ shown by the heavy lines in the figure is a normal tree. If the tree-branch voltages v_3, v_4 and v_6 are known, all other voltages v_1, v_2 and v_5 are determined by means of the f-circuits, as given by (8.41). Likewise, if the cotree-link currents i_1, i_2 and i_5 are specified, all other currents i_3, i_4 and i_6 can be obtained by way of the f-cutsets, as shown in (8.40).

The name *normal tree* is used because it is the tree that will enable us to write the state equation in normal form. The normal tree may or may not be unique.

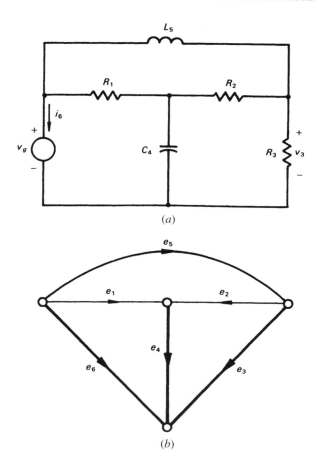

Figure 8.17 (*a*) A network and (*b*) its associated directed graph.

For instance, the subgraph $e_2e_4e_6$ is another normal tree of the directed graph G of Fig. 8.17*b*. As a more complicated example, consider the network and its associated directed graph G of Fig. 8.18. Observe that the independent current-source edge e_7 and the inductive edges e_1 and e_2 constitute a cutset. Therefore at least one of these edges must be contained in every tree of G, since a tree contains all the nodes of G. To construct a normal tree, we must exclude e_7 and include e_3 and e_4 as tree branches, yielding two normal trees $e_2e_3e_4e_5$ and $e_1e_3e_4e_5$ as shown in Fig. 8.19. The two trees $e_1e_3e_5e_6$ and $e_1e_2e_3e_4$ of Fig. 8.20 are not normal trees because the former does not contain the maximum number of capacitive edges that can be put in a tree, which is two, and the latter contains one more inductive edge

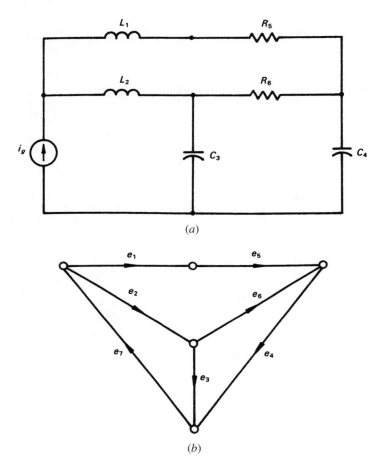

(a)

(b)

Figure 8.18 (a) A network and (b) its associated directed graph used to illustrate the selection of a normal tree.

than the required absolute minimum, which is one. For some networks, not all the capacitive edges can be made part of a tree. In Fig. 8.21b, the three capacitive edges e_1, e_2 and e_3 form a circuit, making it impossible to include them all in any tree. Thus the maximum number of capacitive edges that can be included in any tree is two. Figure 8.22 shows the three normal trees associated with the network of Fig. 8.21a.

In general, a subgraph can be made part of a tree if and only if it does not contain any circuit. Likewise a subgraph can be made part of a cotree if and only if it does not contain any cutset of the given directed graph. In Fig. 8.21b, the

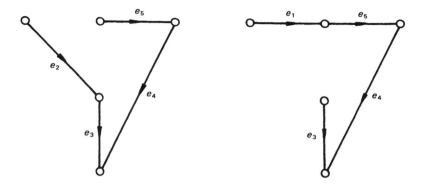

Figure 8.19   Normal trees of the network of Fig. 8.18*a*.

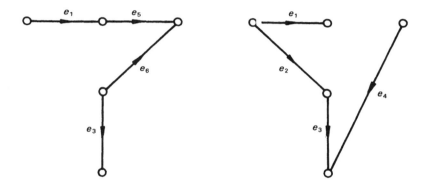

Figure 8.20   Two nonnormal trees of the network of Fig. 8.18*a*.

subgraphs e_3e_6, $e_1e_2e_6$ and e_1e_3 do not contain any circuit, and therefore each can be made part of a tree as shown in Fig. 8.22. The subgraph e_4e_5 does not contain any cutset of G of Fig. 8.21*b*, so it can be made part of a cotree, being excluded from a tree. Some of the trees that exclude e_4e_5 are given in Fig. 8.22. As another example, consider the directed graph G of Fig. 8.18*b*. The subgraph formed by e_7 and either e_1 or e_2 contains no cutset of G, and thus can be included in a cotree, as shown in Fig. 8.23. The subgraph $e_3e_4e_5$ does not contain any circuit, so there is at least one tree containing it as a subgraph. Two such trees are found in Fig. 8.19. In fact, the inclusion of a subgraph g_1 in some tree T_1 and the exclusion of a subgraph g_2 with respect to some other tree T_2 can be achieved with respect to the same tree T provided that g_1 and g_2 do not have edges in common. We state this formally below, a proof of which may be found in Chen (1976).

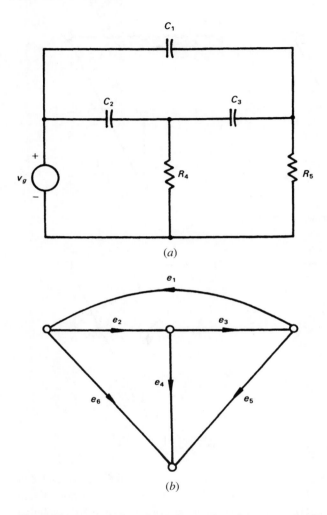

Figure 8.21 (*a*) A network and (*b*) its associated directed graph used to illustrate the selection of a normal tree.

Theorem 8.4 Let g_1 and g_2 be two edge-disjoint subgraphs of a connected directed graph G. If there is a tree T_1 containing g_1 as a subgraph and, furthermore, if there is a cotree $\bar{T}_2$ containing g_2, then there is a tree T of G for which g_1 is contained in T and g_2 is contained in its complement $\bar{T}$.

The result is extremely useful in determining the maximum number of capacitive edges and the minimum number of inductive edges that can be included

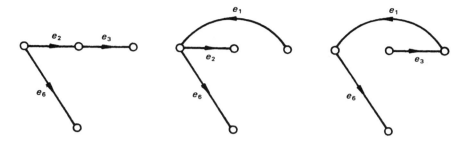

Figure 8.22 Normal trees of the network of Fig. 8.21*a*.

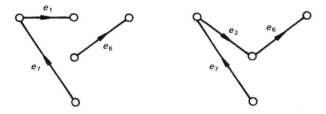

Figure 8.23 Cotrees of the directed graph of Fig. 8.18*b*.

in a tree. For instance, in Fig. 8.24*b* edges e_1, e_2 and e_7 constitute a cutset and edges e_3, e_4 and e_8 a circuit. Hence at least one of the edges e_1, e_2 and e_7 will be in every tree, and at most two of the edges e_3, e_4 and e_8 can be made part of any tree. Figure 8.25*a* is a tree T_1 containing the subgraph $e_3 e_8$, and Fig. 8.25*b* shows a tree $T_2 = e_1 e_4 e_5 e_6$ for which the subgraph $e_2 e_7$ is contained in its complement $\bar{T}_2 = e_2 e_3 e_7 e_8$. According to Theorem 8.4, there exists a tree T for which $e_3 e_8$ is contained in T and $e_2 e_7$ is contained in its complement $\bar{T}$. Such a tree T is presented in Fig. 8.25*c*, showing that a normal tree must contain one inductive edge and one capacitive edge.

Based on the discussions in this and preceding sections, we now outline steps in writing the state equation for a network. These steps are a systematic way to eliminate the unwanted variables in the primary system of equations, so that the resulting equations are the state equations and can be put in normal form.

Step 1. From a given network, construct a directed graph G representing its interconnection and indicating the voltage and current references of its elements — namely, the directions of the edges.

Step 2. In G select a normal tree T and choose as the state variables the capacitor voltages of T and the inductor currents of its complement $\bar{T}$.

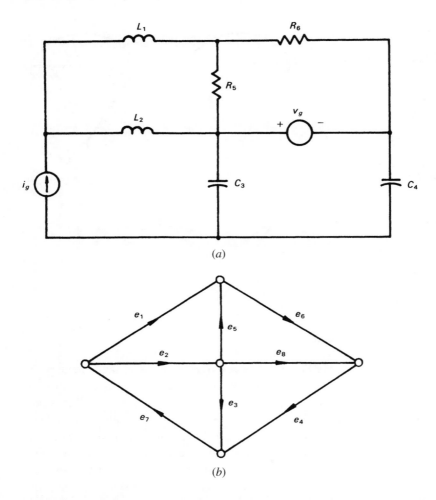

(a)

(b)

Figure 8.24 (a) A network and (b) its associated directed graph used to illustrate the determination of the maximum number of capacitive edges and the minimum number of inductive edges that can be included in a tree.

Step 3. Assign each branch of $\bar{T}$ a voltage symbol, and assign each link of $\bar{T}$ a current symbol.

Step 4. Use KCL to express each tree-branch current as a sum of cotree-link currents by way of the f-cutsets defined by $\bar{T}$, and indicate it in G if necessary.

Step 5. Use KVL to express each cotree-link voltage as a sum of tree-branch voltages by means of the f-circuits defined by $\bar{T}$, and indicate it in G if necessary.

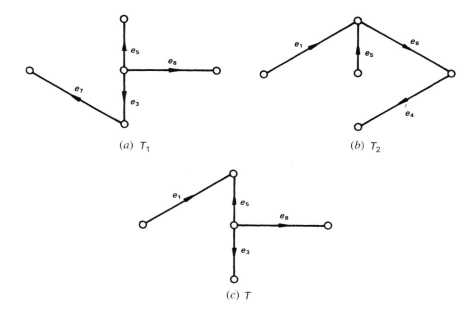

Figure 8.25 (*a*) A tree containing edges e_3 and e_8, (*b*) a tree for which $e_2 e_7$ is contained in its complement, and (*c*) the existence of a tree T in the directed graph of Fig. 8.24*b* for which $e_3 e_8$ is contained in T and $e_2 e_7$ is contained in its complement.

Step 6. Write the element *v-i* equations for the passive elements and separate these equations into two groups:

(a) those element *v-i* equations for the tree-branch capacitors and the cotree-link inductors.
(b) those element *v-i* equations for all other passive elements.

Step 7. Eliminate the nonstate variables among the equations obtained in the preceding step. *Nonstate variables* are defined as those variables that are neither state variables nor known independent sources.

Step 8. Rearrange the terms and write the resulting equations in normal form.

We illustrate the above steps by the following examples.

Example 8.2 We follow the above eight steps to write the state equations for the active network N of Fig. 8.26*a*.

Step 1. The associated directed graph G of N is presented in Fig. 8.26*b*.

Step 2. Select a normal tree T consisting of the edges e_1, e_3 and e_7. Edge e_5 is excluded from the tree because it is usually simpler to put controlled current

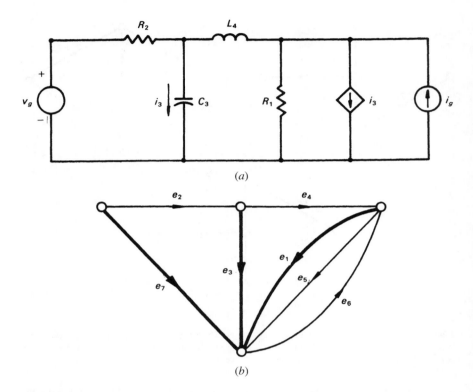

(a)

(b)

Figure 8.26 (a) An active network and (b) its associated directed graph used to illustrate the procedure of writing the state equation in normal form.

sources in the cotree. The subgraph $e_3e_5e_7$, nevertheless, is also a normal tree. The normal tree $T = e_1e_3e_7$ is exhibited explicitly in Fig. 8.26b by the heavy lines.

Step 3. As illustrated in Fig. 8.27a, the tree-branches e_1, e_3 and e_7 are assigned the voltage symbols v_1, v_3 and v_g; and the cotree-links e_2, e_4, e_5 and e_6 are assigned the current symbols i_2, i_4, i_3 and i_g, respectively. Edge e_5 is given the symbol i_3 because its current is controlled by the current of e_3, which is i_3. Edges e_6 and e_7 are assigned i_g and v_g because they represent the known independent sources.

Step 4. By writing KCL equations for the f-cutsets defined by T, the branch currents i_1, i_3 and i_7 are found to be

$$i_1 = i_4 + i_g - i_3 \qquad (8.43a)$$

$$i_3 = i_2 - i_4 \qquad (8.43b)$$

$$i_7 = -i_2 \qquad (8.43c)$$

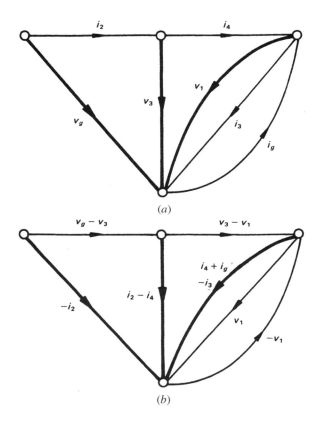

Figure 8.27 (*a*) The assignment of the voltage symbols to the tree branches and the current symbols to the cotree links, and (*b*) the expressions for the tree-branch currents in terms of cotree-link currents and the cotree-link voltages in terms of the tree-branch voltages.

These are also indicated in Fig. 8.27*b*.

Step 5. By writing KVL equations for the f-circuits defined by T, the voltages v_2, v_4, v_5 and v_6 are obtained as

$$v_2 = v_g - v_3 \qquad (8.44a)$$

$$v_4 = v_3 - v_1 \qquad (8.44b)$$

$$v_5 = v_1 \qquad (8.44c)$$

$$v_6 = -v_1 \qquad (8.44d)$$

These are also indicated in Fig. 8.27*b*.

Step 6. The element v-i equations for the tree-branch capacitor cotree-link inductor are given by

$$C_3\dot{v}_3 = i_3 = i_2 - i_4 \tag{8.45a}$$

$$L_4\dot{i}_4 = v_4 = v_3 - v_1 \tag{8.45b}$$

For the other passive elements, we have

$$v_1 = R_1 i_1 = R_1(i_4 + i_g - i_3) \tag{8.46a}$$

$$i_2 = \frac{v_2}{R_2} = \frac{v_g - v_3}{R_2} \tag{8.46b}$$

In fact, these equations can be obtained directly from Fig. 8.27 by writing the v-i equations for the voltage and current variables of the corresponding edges of the two directed graphs, depending of course on the character of the individual elements.

Step 7. The state variables are v_3 and i_4 and the nonstate variables to be eliminated in (8.45) are i_2 and v_1. For this we express i_2 and v_1 in terms of the state variables. From (8.43b) and (8.46), we obtain

$$v_1 = R_1\left(2i_4 + i_g + \frac{v_3}{R_2} - \frac{v_g}{R_2}\right) \tag{8.47a}$$

$$i_2 = \frac{v_g - v_3}{R_2} \tag{8.47b}$$

Substituting these in (8.45) yields

$$C_3\dot{v}_3 = \frac{v_g - v_3}{R_2} - i_4 \tag{8.48a}$$

$$L_4\dot{i}_4 = \left(1 - \frac{R_1}{R_2}\right)v_3 - 2R_1 i_4 - R_1 i_g + \frac{R_1 v_g}{R_2} \tag{8.48b}$$

Step 8. Writing these equations in matrix form gives the state equation in normal form for the active network N of Fig. 8.26a:

$$\begin{bmatrix} \dot{v}_3 \\ \dot{i}_4 \end{bmatrix} = \begin{bmatrix} -\dfrac{1}{R_2 C_3} & -\dfrac{1}{C_3} \\ \dfrac{1}{L_4} - \dfrac{R_1}{R_2 L_4} & -\dfrac{2R_1}{L_4} \end{bmatrix} \begin{bmatrix} v_3 \\ i_4 \end{bmatrix} + \begin{bmatrix} \dfrac{1}{R_2 C_3} & 0 \\ \dfrac{R_1}{R_2 L_4} & -\dfrac{R_1}{L_4} \end{bmatrix} \begin{bmatrix} v_g \\ i_g \end{bmatrix} \tag{8.49}$$

Example 8.3 We shall write the state equation for the active network N of Fig. 8.28a, which contains two mutually coupled inductors with mutual inductance M. Again we follow the eight steps with very brief comments.

Step 1. The associated directed graph G of N is shown in Fig. 8.28b.

Step 2. Select a normal tree T in G as shown by the heavy lines in Fig. 8.28b.

Step 3. The tree-branches are assigned the voltages v_3, v_4 and v_g; and the cotree-links are assigned the currents i_1, i_2, αi_3 and i_5. These are indicated by the first elements of the pairs adjacent to the edges of G in Fig. 8.28b.

Steps 4 and 5. By writing KCL and KVL equations for the f-cutsets and f-circuits defined by T, the branch currents are expressed in terms of the link

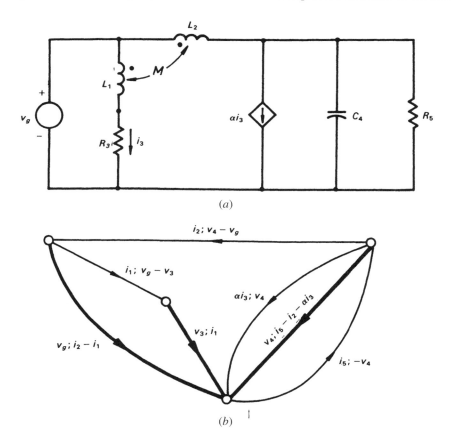

Figure 8.28 (*a*) A network containing two mutually coupled inductors and (*b*) its associated directed graph used to illustrate the procedure for writing the state equation in normal form.

currents and the link voltages are expressed in terms of the branch voltages. These are shown as the second elements of the pairs adjacent to the edges of G in Fig. 8.28b.

Step 6. The element v-i equations can now be written down directly from Fig. 8.28 using the relation $i_3 = i_1$. We obtain

$$C_4 \dot{v}_4 = i_5 - i_2 - \alpha i_1 \qquad (8.50a)$$

$$L_1 \dot{i}_1 - M \dot{i}_2 = v_g - v_3 \qquad (8.50b)$$

$$L_2 \dot{i}_2 - M \dot{i}_1 = v_4 - v_g \qquad (8.50c)$$

$$v_3 = R_3 i_1 \qquad (8.51a)$$

$$i_5 = -\frac{v_4}{R_5} \qquad (8.51b)$$

Step 7. The state variables are v_4, i_1 and i_2 and the nonstate variables to be eliminated in (8.50) are v_3 and i_5. Substituting (8.51) in (8.50), we obtain three equations involving only the state variables and the known source v_g. Using matrix notation these three equations can be written as

$$\begin{bmatrix} C_4 & 0 & 0 \\ 0 & L_1 & -M \\ 0 & -M & L_2 \end{bmatrix} \begin{bmatrix} \dot{v}_4 \\ \dot{i}_1 \\ \dot{i}_2 \end{bmatrix} = \begin{bmatrix} -1/R_5 & -\alpha & -1 \\ 0 & -R_3 & 0 \\ 1 & 0 & 0 \end{bmatrix} \begin{bmatrix} v_4 \\ i_1 \\ i_2 \end{bmatrix} + \begin{bmatrix} 0 \\ v_g \\ -v_g \end{bmatrix} \qquad (8.52)$$

Step 8. Premultiplying both sides of (8.52) by the matrix

$$\begin{bmatrix} C_4 & 0 & 0 \\ 0 & L_1 & -M \\ 0 & -M & L_2 \end{bmatrix}^{-1} = \frac{1}{\Delta} \begin{bmatrix} \Delta/C_4 & 0 & 0 \\ 0 & L_2 & M \\ 0 & M & L_1 \end{bmatrix} \qquad (8.53)$$

where $\Delta = L_1 L_2 - M^2 \neq 0$ by assuming that the inductors are not perfectly coupled, we obtain the state equation for the active network N of Fig. 8.28a:

$$\begin{bmatrix} \dot{v}_4 \\ \dot{i}_1 \\ \dot{i}_2 \end{bmatrix} = \begin{bmatrix} -\dfrac{1}{C_4 R_5} & -\dfrac{\alpha}{C_4} & -\dfrac{1}{C_4} \\[2mm] \dfrac{M}{\Delta} & -\dfrac{L_2 R_3}{\Delta} & 0 \\[2mm] \dfrac{L_1}{\Delta} & -\dfrac{M R_3}{\Delta} & 0 \end{bmatrix} \begin{bmatrix} v_4 \\ i_1 \\ i_2 \end{bmatrix} + \begin{bmatrix} 0 \\[2mm] \dfrac{L_2 - M}{\Delta} \\[2mm] \dfrac{M - L_1}{\Delta} \end{bmatrix} [v_g] \qquad (8.54)$$

Finally, we remark that if the output variables are specified, they can be expressed in terms of the state variables and the known source through the weighted directed graph of Fig. 8.28b and (8.50).

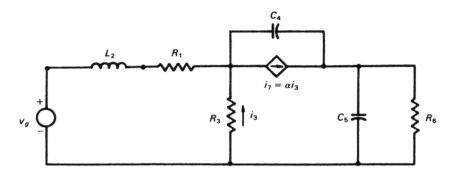

Figure 8.29 An equivalent network of a transistor amplifier.

Example 8.4 Consider the equivalent network N of a transistor amplifier shown in Fig. 8.29. We shall follow the eight steps to develop the state equation. For illustrative purposes, we assume that the capacitances C_4 and C_5 and resistance R_1 are time-varying, and are governed by the following equations:

$$C_4(t) = \frac{1}{2 + \sin 2t} \tag{8.55a}$$

$$C_5(t) = \frac{1}{1 + \frac{1}{2}\cos t} \tag{8.55b}$$

$$R_1(t) = 2 + \sin 3t \tag{8.55c}$$

Step 1. The associated directed graph G of N is presented in Fig. 8.30.

Step 2. Select a normal tree $T = e_1 e_4 e_5 e_8$ in G as shown by the heavy lines in Fig. 8.30.

Step 3. As indicated in Fig. 8.31, the tree branches are assigned the voltages v_1, v_4, v_5 and v_8 and the cotree links are assigned the currents i_2, i_3, i_6 and i_7.

Step 4. Using KCL express each tree-branch current as a sum of cotree-link currents by means of the f-cutsets defined by T:

$$i_1 = i_2 \tag{8.56a}$$

$$i_4 = i_2 + i_3 - i_7 \tag{8.56b}$$

$$i_5 = i_2 + i_3 - i_6 \tag{8.56c}$$

$$i_8 = -i_2 \tag{8.56d}$$

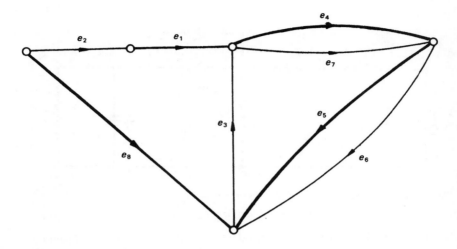

Figure 8.30 The associated directed graph of the network of Fig. 8.29.

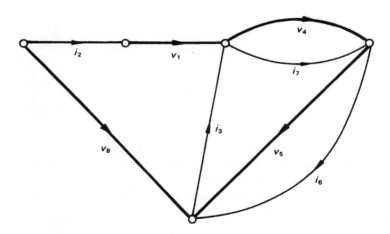

Figure 8.31 The assignment of voltage symbols for the tree branches and the current symbols for the cotree links for the directed graph of Fig. 8.30.

Step 5. Using KVL express each cotree-link voltage as a sum of tree-branch voltages by way of the f-circuits defined by T:

$$v_2 = v_8 - v_1 - v_4 - v_5 \qquad (8.57a)$$

$$v_3 = -v_4 - v_5 \qquad (8.57b)$$

$$v_6 = v_5 \tag{8.57c}$$

$$v_7 = v_4 \tag{8.57d}$$

Step 6. For convenience, we choose the capacitor charges

$$q_4 = C_4 v_4 \tag{8.58a}$$

$$q_5 = C_5 v_5 \tag{8.58b}$$

and inductor flux linkage

$$\lambda_2 = L_2 i_2 \tag{8.58c}$$

to be the state variables. The element defining equations become

$$\dot{q}_4 = i_2 + i_3 - i_7 \tag{8.59a}$$

$$\dot{q}_5 = i_2 + i_3 - i_6 \tag{8.59b}$$

$$\dot{\lambda}_2 = v_8 - v_1 - v_4 - v_5 \tag{8.59c}$$

$$v_1 = R_1 i_2 \tag{8.60a}$$

$$i_3 = -\frac{v_4 + v_5}{R_3} \tag{8.60b}$$

$$i_6 = \frac{v_5}{R_6} \tag{8.60c}$$

$$i_7 = -\frac{\alpha(v_4 + v_5)}{R_3} \tag{8.60d}$$

where $v_8 = v_g$.

Step 7. To eliminate the unwanted variables v_1, i_3, i_6 and i_7 in (8.59), we express them in terms of the state variables q_4, q_5 and λ_2 by making use of (8.58) and (8.60). This gives

$$v_1 = \frac{R_1 \lambda_2}{L_2} \tag{8.61a}$$

$$i_3 = -\frac{q_4}{R_3 C_4} - \frac{q_5}{R_3 C_5} \tag{8.61b}$$

$$i_6 = \frac{q_5}{R_6 C_5} \tag{8.61c}$$

$$i_7 = -\frac{\alpha q_4}{R_3 C_4} - \frac{\alpha q_5}{R_3 C_5} \tag{8.61d}$$

Substituting these in (8.59) in conjunction with (8.55) and (8.58) yields

$$\dot{\lambda}_2 = v_g - \frac{R_1 \lambda_2}{L_2} - (2 + \sin 2t)q_4 - \left(1 + \frac{1}{2}\cos t\right)q_5 \tag{8.62a}$$

$$\dot{q}_4 = \frac{\lambda_2}{L_2} + \frac{(2 + \sin 2t)(\alpha - 1)q_4}{R_3} + \frac{(1 + \frac{1}{2}\cos t)(\alpha - 1)q_5}{R_3} \tag{8.62b}$$

$$\dot{q}_5 = \frac{\lambda_2}{L_2} - \frac{(2 + \sin 2t)q_4}{R_3} - \frac{(1 + \frac{1}{2}\cos t)(R_3 + R_6)q_5}{R_3 R_6} \tag{8.62c}$$

Step 8. Writing these equations in matrix gives

$$\dot{\mathbf{x}}(t) = \mathbf{A}_s(t)\mathbf{X}(t) + \mathbf{B}_s(t)\mathbf{u}(t) \tag{8.63}$$

where

$$\mathbf{A}_s(t) = \begin{bmatrix} -\dfrac{R_1}{L_2} & -(2 + \sin 2t) & -\left(1 + \dfrac{1}{2}\cos t\right) \\[2ex] \dfrac{1}{L_2} & \dfrac{(2 + \sin 2t)(\alpha - 1)}{R_3} & \dfrac{(1 + \frac{1}{2}\cos t)(\alpha - 1)}{R_3} \\[2ex] \dfrac{1}{L_2} & -\dfrac{2 + \sin 2t}{R_3} & -\dfrac{(1 + \frac{1}{2}\cos t)(R_3 + R_6))}{R_3 R_6} \end{bmatrix} \tag{8.64a}$$

$$\mathbf{B}_s = \begin{bmatrix} 1 \\ 0 \\ 0 \end{bmatrix}, \quad \mathbf{x}(t) = \begin{bmatrix} \lambda_2 \\ q_4 \\ q_5 \end{bmatrix}, \quad \mathbf{u}(t) = [v_g(t)] \tag{8.64b}$$

Observe that since the network is time-varying, the coefficient matrix $\mathbf{A}_s(t)$ is a function of time t.

Suppose that the resistor voltage v_6 and inductor current i_2 are selected as the outputs. Then from (8.57c), (8.58b) and (8.58c) we obtain

$$L_2 i_2 = \lambda_2 \tag{8.65a}$$

$$C_5 v_6 = q_5 \tag{8.65b}$$

or, in matrix notation, the output equation becomes

$$\begin{bmatrix} i_2 \\ v_6 \end{bmatrix} = \begin{bmatrix} \frac{1}{L_2} & 0 & 0 \\ 0 & 0 & \frac{1}{C_5} \end{bmatrix} \begin{bmatrix} \lambda_2 \\ q_4 \\ q_5 \end{bmatrix} \tag{8.66}$$

Equations (8.63) and (8.66) constitute the state equations for the active network N of Fig. 8.29.

8.5 STATE EQUATIONS FOR DEGENERATE NETWORKS

A network is said to be *degenerate* if it satisfies one or both of the following conditions:

1. It contains a circuit composed only of capacitors and/or independent or dependent voltage sources.
2. It contains a cutset composed only of inductors and/or independent or dependent current sources.

In the present section we demonstrate by simple examples that a degenerate network may possess any one of the following attributes:

1. It has no solution.
2. It has a solution but has no unique solution.
3. If capacitor voltages and inductor currents are routinely chosen as the state variables, its state equation cannot be put in the normal form of (8.2).

Any linear physical network always has a solution — a unique solution. However, the networks that we are dealing with are models representing the physical networks. The models are made up of the idealized elements such as ideal resistors, capacitors, inductors, and independent or dependent sources. As a result we cannot always assume the existence and uniqueness of their solutions. Consider, for example, a circuit composed of the series connection of independent and/or dependent voltage sources. For the network to have a solution, KVL requires that the algebraic sum of these voltages be zero. Otherwise the network will have no solution at all. In reality, of course there are no such things as independent or dependent voltage sources. Physical voltage sources always contain other elements such as resistance. Even the existence of a solution in such a network is not unique. For instance, in Fig. 8.32 a resistor R is connected in parallel with two identical batteries. There is a circuit composed only of independent voltage sources v_1 and v_2. It is straightforward to verify that the complete solution of the network is given by

$$\mathbf{v} = \begin{bmatrix} v_1 \\ v_2 \\ v_R \end{bmatrix} = \begin{bmatrix} V \\ V \\ V \end{bmatrix} \tag{8.67}$$

Figure 8.32 A simple network that does not possess a unique solution.

$$\mathbf{i} = \begin{bmatrix} i_1 \\ i_2 \\ i_R \end{bmatrix} = \begin{bmatrix} k \\ V/R - k \\ V/R \end{bmatrix} \tag{8.68}$$

where k is an arbitrary real constant, showing that the network possesses infinitely many solutions.

Likewise, if a degenerate network contains a cutset comprised only of independent and/or dependent current sources, such a network either has no solution or no unique solution if one exists. For our purposes, we shall exclude these networks from our consideration because they cannot be adequately represented by the state equations.

The network of Fig. 8.24a is degenerate because there is a circuit composed only of the capacitors C_3 and C_4 and the independent voltage source v_g. In addition, it also contains a cutset composed only of the inductors L_1 and L_2 and the independent current source i_g. Such a network cannot be described by the normal-form state equation (8.2) if capacitor voltages and inductor currents are used as the state variables, and its state equation assumes the general form

$$\dot{\mathbf{x}}(t) = \mathbf{A}_s \mathbf{x}(t) + \mathbf{B}_{s1} \mathbf{u}(t) + \mathbf{B}_{s2} \dot{\mathbf{u}}(t) \tag{8.69}$$

and the output equation becomes

$$\mathbf{y}(t) = \mathbf{C}_s \mathbf{x}(t) + \mathbf{D}_{s1} \mathbf{u}(t) + \mathbf{D}_{s2} \dot{\mathbf{u}}(t) \tag{8.70}$$

where $\mathbf{B}_{s1}, \mathbf{B}_{s2}, \mathbf{D}_{s1}$ and $\mathbf{D}_{s2}$ are the known parameter matrices of appropriate orders. We remark that in some situations the output equation (8.70) may also involve the second- and higher-order derivatives of $\mathbf{u}(t)$. Equation (8.69) may be regarded as the *modified normal-form state equation* because it can be expressed in the form

$$\dot{\mathbf{x}}(t) = \mathbf{A}_s \mathbf{x}(t) + \mathbf{F}\mathbf{u}(t) \tag{8.71}$$

by letting

$$\mathbf{F} = \mathbf{B}_{s1} + \mathbf{B}_{s2}p \tag{8.72}$$

where p is the *linear differential operator* for any vector function of time t,

$$p\mathbf{u}(t) = \frac{d\mathbf{u}(t)}{dt} = \dot{\mathbf{u}}(t) \tag{8.73}$$

If, however, one would insist on considering (8.2) as the only acceptable normal form of the state equation, the derivative of the input vector can be removed by the transformation

$$\mathbf{x}(t) \rightarrow \mathbf{x}(t) + \mathbf{B}_{s2}\mathbf{u}(t) \tag{8.74}$$

which, when substituted in (8.69), yields

$$\dot{\mathbf{x}}(t) = \mathbf{A}_s \mathbf{x}(t) + \mathbf{B}_s \mathbf{u}(t) \tag{8.75}$$

where

$$\mathbf{B}_s = \mathbf{B}_{s1} + \mathbf{A}_s \mathbf{B}_{s2} \tag{8.76}$$

The price we paid in achieving this is that, instead of using $\mathbf{x}(t)$ as the state vector, we use $\mathbf{x}(t) + \mathbf{B}_{s2}\mathbf{u}(t)$ as the new state vector.

We illustrate the above results by the following examples.

Example 8.5 Consider the network N of Fig. 8.33, which is obtained from the network of Fig. 8.29 by inserting an inductor L_{10}, a capacitor C_9 and an independent current source $i_g(t)$. We follow the eight steps outlined in the preceding section to write its state equation in normal form.

Step 1. The directed graph G representing the network N of Fig. 8.33 is presented in Fig. 8.34.

Step 2. Select $T = e_1 e_4 e_5 e_8 e_{10}$ to be the normal tree, as depicted in Fig. 8.34 by the heavy lines.

Step 3. As indicated in Fig. 8.34, the tree-branch voltages are assigned the symbols v_1, v_4, v_5, v_8 and v_{10} and the cotree-link currents are assigned the symbols i_2, i_3, i_6, i_7, i_9 and i_{11}. The state variables are chosen to be v_4, v_5 and i_2.

Figure 8.33 An active network used to illustrate the formulation of its state equation.

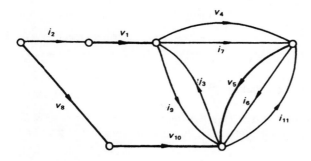

Figure 8.34 The associated directed graph of the active network of Fig. 8.33.

Step 4. Using KCL express each tree-branch current as a sum of the cotree-link currents:

$$i_1 = i_2 \tag{8.77a}$$

$$i_4 = i_2 + i_3 - i_7 - i_9 \tag{8.77b}$$

$$i_5 = i_2 + i_3 - i_6 - i_9 + i_{11} \tag{8.77c}$$

$$i_8 = -i_2 \tag{8.77d}$$

$$i_{10} = -i_2 \tag{8.77e}$$

Step 5. Using KVL express each cotree-link voltage as a sum of the tree-branch voltages:

$$v_2 = v_8 + v_{10} - v_1 - v_4 - v_5 \tag{8.78a}$$

$$v_3 = -v_4 - v_5 \tag{8.78b}$$

$$v_6 = v_5 \tag{8.78c}$$

$$v_7 = v_4 \tag{8.78d}$$

$$v_9 = v_4 + v_5 \tag{8.78e}$$

$$v_{11} = -v_5 \tag{8.78f}$$

Step 6. Applying the v-i equations to the edges of G in conjunction with (8.77) and (8.78) yields

$$C_4 \dot{v}_4 = i_2 + i_3 - i_7 - i_9 \tag{8.79a}$$

$$C_5 \dot{v}_5 = i_2 + i_3 - i_6 - i_9 + i_{11} \tag{8.79b}$$

$$L_2 \dot{i}_2 = v_8 + v_{10} - v_1 - v_4 - v_5 \tag{8.79c}$$

$$v_1 = R_1 i_2 \tag{8.80a}$$

$$v_8 = v_g \tag{8.80b}$$

$$v_{10} = -L_{10} \dot{i}_2 \tag{8.80c}$$

$$i_3 = -\frac{v_4}{R_3} - \frac{v_5}{R_3} \tag{8.80d}$$

$$i_6 = \frac{v_5}{R_6} \tag{8.80e}$$

$$i_7 = \frac{\alpha v_3}{R_3} = \frac{\alpha(-v_4 - v_5)}{R_3} \tag{8.80f}$$

$$i_9 = C_9 \dot{v}_4 + C_9 \dot{v}_5 \tag{8.80g}$$

$$i_{11} = i_g \tag{8.80h}$$

Step 7. Substituting (8.80) in (8.79) to eliminate the nonstate variables $v_1, v_8, v_{10}, i_3, i_6, i_7, i_9$ and i_{11} gives

$$
\begin{bmatrix} L_2 + L_{10} & 0 & 0 \\ 0 & C_4 + C_9 & C_9 \\ 0 & C_9 & C_5 + C_9 \end{bmatrix}
\begin{bmatrix} \dot{i}_2 \\ \dot{v}_4 \\ \dot{v}_5 \end{bmatrix}
=
\begin{bmatrix} -R_1 & -1 & -1 \\ 1 & \dfrac{\alpha - 1}{R_3} & \dfrac{\alpha - 1}{R_3} \\ 1 & -\dfrac{1}{R_3} & -\dfrac{1}{R_3} - \dfrac{1}{R_6} \end{bmatrix}
\begin{bmatrix} i_2 \\ v_4 \\ v_5 \end{bmatrix}
$$

$$
+ \begin{bmatrix} 0 & 1 \\ 0 & 0 \\ 1 & 0 \end{bmatrix}
\begin{bmatrix} i_g \\ v_g \end{bmatrix} \tag{8.81}
$$

Step 8. Writing the equation (8.81) in normal form yields

$$\dot{\mathbf{x}}(t) = \mathbf{A}_s \mathbf{x}(t) + \mathbf{B}_s \mathbf{u}(t) \tag{8.82}$$

where

$$\mathbf{A}_s = \frac{1}{\Delta}\begin{bmatrix} -\dfrac{R_1\Delta}{L_2+L_{10}} & -\dfrac{\Delta}{L_2+L_{10}} & -\dfrac{\Delta}{L_2+L_{10}} \\[2mm] C_5 & \dfrac{(\alpha-1)C_5+\alpha C_9}{R_3} & \dfrac{(\alpha-1)C_5+\alpha C_9}{R_3}+\dfrac{C_9}{R_6} \\[2mm] C_4 & -\dfrac{\alpha C_9+C_4}{R_3} & -\dfrac{C_4+\alpha C_9}{R_3}-\dfrac{C_4+C_9}{R_6} \end{bmatrix} \tag{8.83a}$$

$$\mathbf{B}_s = \frac{1}{\Delta}\begin{bmatrix} 0 & \dfrac{\Delta}{L_2+L_{10}} \\[2mm] -C_9 & 0 \\[2mm] C_4+C_9 & 0 \end{bmatrix} \tag{8.83b}$$

$$\mathbf{x}(t) = \begin{bmatrix} i_2 \\ v_4 \\ v_5 \end{bmatrix}, \quad \mathbf{u}(t) = \begin{bmatrix} i_g \\ v_g \end{bmatrix} \tag{8.83c}$$

$$\Delta = C_4 C_5 + C_9(C_4+C_5) \tag{8.83d}$$

Example 8.6 We follow the eight steps outlined in the preceding section to write the state equation in normal form for the degenerate network N of Fig. 8.24a.

Step 1. The associated directed graph G of the network is shown in Fig. 8.24b.

Step 2. Select a normal tree $T = e_1 e_3 e_5 e_8$, as exhibited explicitly in Fig. 8.35 by the heavy lines.

Step 3. The tree branches e_1, e_3, e_5 and e_8 are assigned the voltages v_1, v_3, v_5 and v_g; and the cotree-links e_2, e_4, e_6 and e_7 are assigned the currents i_2, i_4, i_6 and i_g, as indicated by the first members of the pairs of variables located adjacent to the edges.

Step 4. Each tree-branch current is expressed as a sum of cotree-link currents:

$$i_1 = i_g - i_2 \tag{8.84a}$$
$$i_3 = i_g - i_4 \tag{8.84b}$$
$$i_5 = i_2 + i_6 - i_g \tag{8.84c}$$
$$i_8 = i_4 - i_6 \tag{8.84d}$$

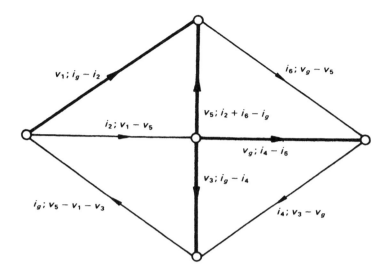

Figure 8.35 The associated directed graph of the degenerate network of Fig. 8.24a with a selected normal tree exhibited explicitly by the heavy lines.

Step 5. Each cotree-link voltage is expressed as a sum of tree-branch voltages:

$$v_2 = v_1 - v_5 \tag{8.85a}$$

$$v_4 = v_3 - v_g \tag{8.85b}$$

$$v_6 = v_g - v_5 \tag{8.85c}$$

$$v_7 = v_5 - v_1 - v_3 \tag{8.85d}$$

Step 6. The element v-i equations are found to be

$$L_2\dot{i}_2 = v_1 - v_5 \tag{8.86a}$$

$$C_3\dot{v}_3 = i_g - i_4 \tag{8.86b}$$

$$v_5 = R_5(i_2 + i_6 - i_g) \tag{8.87a}$$

$$i_6 = \frac{v_g - v_5}{R_6} \tag{8.87b}$$

$$L_1(\dot{i}_g - \dot{i}_2) = v_1 \tag{8.87c}$$

$$C_4(\dot{v}_3 - \dot{v}_g) = i_4 \tag{8.87d}$$

Step 7. Use (8.87) to eliminate the nonstate variables v_1, i_4 and v_5 in (8.86) and obtain

$$(L_1 + L_2)\dot{i}_2 = L_1\dot{i}_g - \frac{R_5 R_6}{R_5 + R_6}\left(i_2 - i_g + \frac{v_g}{R_6}\right) \tag{8.88a}$$

$$(C_3 + C_4)\dot{v}_3 = i_g + C_4\dot{v}_g \tag{8.88b}$$

Observe that the two equations involve not only i_2, v_3, i_g and v_g but also the derivatives of i_g and v_g, which may not exist.

Step 8. Writing (8.88) in matrix notation yields

$$\begin{bmatrix} \dot{i}_2 \\ \dot{v}_3 \end{bmatrix} = \begin{bmatrix} -\dfrac{R_0}{L_1 + L_2} & 0 \\ 0 & 0 \end{bmatrix} \begin{bmatrix} i_2 \\ v_3 \end{bmatrix} + \begin{bmatrix} \dfrac{R_0}{L_1 + L_2} & -\dfrac{R_0}{R_6(L_1 + L_2)} \\ \dfrac{1}{C_3 + C_4} & 0 \end{bmatrix} \begin{bmatrix} i_g \\ v_g \end{bmatrix}$$

$$+ \begin{bmatrix} \dfrac{L_1}{L_1 + L_2} & 0 \\ 0 & \dfrac{C_4}{C_3 + C_4} \end{bmatrix} \begin{bmatrix} \dot{i}_g \\ \dot{v}_g \end{bmatrix} \tag{8.89}$$

where $\dot{i}_g$ and $\dot{v}_g$ are assumed to exist and

$$R_0 = \frac{R_5 R_6}{R_5 + R_6} \tag{8.90}$$

To transform (8.89) to the form (8.75), we compute first the matrix $\mathbf{B}_s$ from (8.76)

$$\mathbf{B}_s = \mathbf{B}_{s1} + \mathbf{A}_s \mathbf{B}_{s2} = \begin{bmatrix} \dfrac{R_0 L_2}{(L_1 + L_2)^2} & -\dfrac{R_0}{R_6(L_1 + L_2)} \\ \dfrac{1}{C_3 + C_4} & 0 \end{bmatrix} \tag{8.91}$$

The new state vector is determined from (8.74) as

$$\mathbf{x}(t) = \begin{bmatrix} x_1 \\ x_2 \end{bmatrix} = \begin{bmatrix} i_2 \\ v_3 \end{bmatrix} + \begin{bmatrix} \dfrac{L_1 i_g}{L_1 + L_2} \\ \dfrac{C_4 v_g}{C_3 + C_4} \end{bmatrix} = \begin{bmatrix} i_2 + \dfrac{L_1 i_g}{L_1 + L_2} \\ v_3 + \dfrac{C_4 v_g}{C_3 + C_4} \end{bmatrix} \tag{8.92}$$

Substituting these in (8.75) yields the state equation in normal form for the network N of Fig. 8.24a:

$$\begin{bmatrix} \dot{x}_1 \\ \dot{x}_2 \end{bmatrix} = \begin{bmatrix} -\dfrac{R_0}{L_1 + L_2} & 0 \\ 0 & 0 \end{bmatrix} \begin{bmatrix} x_1 \\ x_2 \end{bmatrix} + \begin{bmatrix} \dfrac{R_0 L_2}{(L_1 + L_2)^2} & -\dfrac{R_0}{R_6(L_1 + L_2)} \\ \dfrac{1}{C_3 + C_4} & 0 \end{bmatrix} \begin{bmatrix} i_g \\ v_g \end{bmatrix}$$

(8.93)

Observe that the state variables in (8.93) are no longer the inductor current i_2 and the capacitor voltage v_3. Instead they are the linear combinations of i_2 and i_g and v_3 and v_g, respectively, as defined in (8.92).

8.6   STATE-SPACE FORMULATION OF FEEDBACK THEORY

In this section, we formulate the general feedback theory in terms of the coefficient matrices of the state equations.

Let N be a network N characterized by its state equations

$$\dot{\mathbf{x}}(t) = \mathbf{A}_s \mathbf{x}(t) + \mathbf{B}_s \mathbf{u}(t) \tag{8.94a}$$

$$\mathbf{y}(t) = \mathbf{C}_s \mathbf{x}(t) + \mathbf{D}_s \mathbf{u}(t) \tag{8.94b}$$

Taking the Laplace transforms on both sides gives

$$s\tilde{\mathbf{x}}(s) = \mathbf{A}_s \tilde{\mathbf{x}}(s) + \mathbf{B}_s \tilde{\mathbf{u}}(s) + \mathbf{x}(0) \tag{8.95a}$$

$$\tilde{\mathbf{y}}(s) = \mathbf{C}_s \tilde{\mathbf{x}}(s) + \mathbf{D}_s \tilde{\mathbf{u}}(s) \tag{8.95b}$$

where the symbol tilde $\sim$ is used to denote the Laplace transform $\tilde{\mathbf{x}}(s)$ of the corresponding time-domain function $\mathbf{x}(t)$. Note that s is the complex-frequency variable, whereas the subscript s in $\mathbf{A}_s$, $\mathbf{B}_s$, $\mathbf{C}_s$ and $\mathbf{D}_s$, stands for state equations and is used to distinguish the coefficient matrices $\mathbf{A}$, $\mathbf{B}$, $\mathbf{C}$ and $\mathbf{D}$ in (7.13). From (8.95a) we obtain

$$\tilde{\mathbf{x}}(s) = (s\mathbf{1} - \mathbf{A}_s)^{-1} [\mathbf{x}(0) + \mathbf{B}_s \tilde{\mathbf{u}}(s)] \tag{8.96}$$

Substituting this in (8.95b) gives

$$\tilde{\mathbf{y}}(s) = \mathbf{C}_s (s\mathbf{1} - \mathbf{A}_s)^{-1} \mathbf{x}(0) + [\mathbf{C}_s (s\mathbf{1} - \mathbf{A}_s)^{-1} \mathbf{B}_s + \mathbf{D}_s] \tilde{\mathbf{u}}(s) \tag{8.97}$$

The first term on the right-hand side is the zero-input response, and the second term is the zero-state response.

For a single-input and single-output network N, the quantities $\tilde{\mathbf{y}}(s)$, $\tilde{\mathbf{u}}(s)$ and $\mathbf{D}_s$ all become scalars and the transfer function $w(s)$ of the network N is obtained after setting the initial conditions $\mathbf{x}(0)$ to zero:

$$w(s) = \frac{\tilde{y}(s)}{\tilde{u}(s)} = D_s + \mathbf{C}_s(s\mathbf{1} - \mathbf{A}_s)^{-1}\mathbf{B}_s \qquad (8.98)$$

In feedback theory, we always focus our attention on a group of elements of particular interest. For our purposes, we choose the elements represented by the coefficient matrix $\mathbf{A}_s$. Following Fig. 7.13, the *fundamental matrix feedback-flow graph* for the state equations (8.94) of a single-input and single-output network N is shown in Fig. 8.36. To compute the return difference matrix $\mathbf{F}(\mathbf{A}_s)$ with respect to $\mathbf{A}_s$, we set $\tilde{u}(s) = 0$ and consider only the feedback loop in the resulting matrix feedback-flow graph of Fig. 8.37. We first break the loop at the input of the edge $\mathbf{A}_s$, then apply a signal vector $\mathbf{g}$, as shown in Fig. 8.38, and calculate the returned signal $\mathbf{h}$. This gives

$$\mathbf{h} = \frac{1}{s}\mathbf{A}_s\mathbf{g} \qquad (8.99)$$

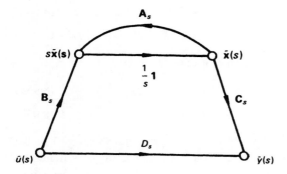

Figure 8.36   The fundamental matrix feedback-flow graph.

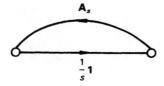

Figure 8.37   The feedback loop of the fundamental matrix feedback-flow graph of Fig. 8.36.

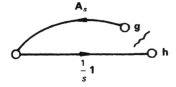

Figure 8.38 The physical interpretation of loop transmission.

The difference between the applied signal vector **g** and the returned signal vector **h** is given by

$$\mathbf{g} - \mathbf{h} = \left[1 - \frac{1}{s}\mathbf{A}_s \right] \mathbf{g} \tag{8.100}$$

The coefficient matrix relating **g** to **g** − **h** is the *return difference matrix* with respect to $\mathbf{A}_s$:

$$\mathbf{F}(\mathbf{A}_s) = 1 - \frac{1}{s}\mathbf{A}_s = \frac{1}{s}(s\mathbf{1} - \mathbf{A}_s) \tag{8.101}$$

the determinant of which is found to be

$$\det \mathbf{F}(\mathbf{A}_s) = \frac{\det (s\mathbf{1} - \mathbf{A}_s)}{s^n} \tag{8.102}$$

where n is the order of $\mathbf{A}_s$. Observe that the numerator of $\det \mathbf{F}(\mathbf{A}_s)$ is the characteristic polynomial of $\mathbf{A}_s$, whose roots are the eigenvalues of $\mathbf{A}_s$.

To compute the null return difference matrix $\hat{\mathbf{F}}(\mathbf{A}_s)$ with respect to $\mathbf{A}_s$, we again refer to the fundamental matrix feedback-flow graph of Fig. 8.36. As before, we break the edge with transmittance $\mathbf{A}_s$ at its input, as illustrated in Fig. 8.39, and

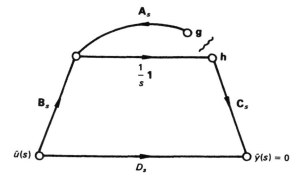

Figure 8.39 The physical interpretation of the null return difference matrix.

apply a signal n-vector $\mathbf{g}$. We then adjust the input excitation $\tilde{u}(s)$, so that the total output $\tilde{y}(s)$ due to the excitations $\mathbf{g}$ and $\tilde{u}(s)$ is identically zero. We obtain from Fig. 8.39

$$\tilde{y}(s) = D_s\tilde{u}(s) + \frac{1}{s}C_s[\mathbf{A}_s\mathbf{g} + \mathbf{B}_s\tilde{u}(s)] = 0 \tag{8.103}$$

from which we can solve $\tilde{u}(s)$ in terms of $\mathbf{g}$ and obtain

$$\tilde{u}(s) = -\frac{\mathbf{C}_s\mathbf{A}_s\mathbf{g}}{sD_s + \mathbf{C}_s\mathbf{B}_s} \tag{8.104}$$

The returned signal $\mathbf{h}$, with the presence of both sources $\tilde{u}(s)$ and $\mathbf{g}$, is found to be

$$\mathbf{h} = \frac{1}{s}[\mathbf{A}_s\mathbf{g} + \mathbf{B}_s\tilde{u}(s)] = \frac{1}{s}\left(\mathbf{1}_n - \frac{\mathbf{B}_s\mathbf{C}_s}{sD_s + \mathbf{C}_s\mathbf{B}_s}\right)\mathbf{A}_s\mathbf{g} \tag{8.105}$$

The difference between the applied signal $\mathbf{g}$ and the returned signal $\mathbf{h}$ under the condition that the input excitation $\tilde{u}(s)$ assumes the special value of (8.104) is given by

$$\mathbf{g} - \mathbf{h} = \left[\mathbf{1}_n - \frac{1}{s}\left(\mathbf{1}_n - \frac{\mathbf{B}_s\mathbf{C}_s}{sD_s + \mathbf{C}_s\mathbf{B}_s}\right)\mathbf{A}_s\right]\mathbf{g} \tag{8.106}$$

The coefficient matrix

$$\begin{aligned}
\hat{\mathbf{F}}(\mathbf{A}_s) &= \mathbf{1}_n - \frac{1}{s}\left(\mathbf{1}_n - \frac{\mathbf{B}_s\mathbf{C}_s}{sD_s + \mathbf{C}_s\mathbf{B}_s}\right)\mathbf{A}_s \\
&= \mathbf{1}_n - \frac{1}{s}\hat{\mathbf{A}}_s = \frac{1}{s}(s\mathbf{1}_n - \hat{\mathbf{A}}_s)
\end{aligned} \tag{8.107}$$

where $sD_s + \mathbf{C}_s\mathbf{B}_s$ is assumed to be nonzero and

$$\hat{\mathbf{A}}_s = \left(\mathbf{1}_n - \frac{\mathbf{B}_s\mathbf{C}_s}{sD_s + \mathbf{C}_s\mathbf{B}_s}\right)\mathbf{A}_s \tag{8.108}$$

is the *null return difference matrix* with respect to $\mathbf{A}_s$.

We are now in a position to establish the following theorem, which represents a generalization of Blackman's impedance formula (4.177) by expressing the transfer function $w(s)$ in terms of the determinants of the return difference and the null return difference matrices. As before we write $w(s)$ as $w(\mathbf{A}_s)$ to emphasize the important dependence of the transfer function w on the state matrix $\mathbf{A}_s$, even though it is also a function of the complex-frequency variable s.

Theorem 8.5 In a single-input and single-output feedback network, the transfer function $w(\mathbf{A}_s)$ is related to the determinants of the return difference matrix $\mathbf{F}(\mathbf{A}_s)$ and the null return difference matrix $\hat{\mathbf{F}}(\mathbf{A}_s)$ by

$$w(\mathbf{A}_s) = w(\mathbf{0})\frac{\det \hat{\mathbf{F}}(\mathbf{A}_s)}{\det \mathbf{F}(\mathbf{A}_s)} \tag{8.109}$$

where $w(\mathbf{0})$ is the transfer function under the condition that the coefficient $\mathbf{A}_s$ matrix in the state equation of the feedback network is set to zero.

PROOF. Let the matrix $\mathbf{A}_s$ be of order n. From (8.98) we obtain

$$
\begin{aligned}
w(\mathbf{A}_s) &= D_s + \mathbf{C}_s(s\mathbf{1}_n - \mathbf{A}_s)^{-1}\mathbf{B}_s \\
&= \frac{sD_s + \mathbf{C}_s\mathbf{B}_s}{s} - \frac{\mathbf{C}_s[\mathbf{1}_n - s(s\mathbf{1}_n - \mathbf{A}_s)^{-1}]\mathbf{B}_s}{s} \\
&= \frac{sD_s + \mathbf{C}_s\mathbf{B}_s}{s}\det\left\{1 - \frac{\mathbf{C}_s[\mathbf{1}_n - s(s\mathbf{1}_n - \mathbf{A}_s)^{-1}\mathbf{B}_s]}{sD_s + \mathbf{C}_s\mathbf{B}_s}\right\} \\
&= w(\mathbf{0})\det\left\{\mathbf{1}_n - \frac{\mathbf{B}_s\mathbf{C}_s[\mathbf{1}_n - s(s\mathbf{1}_n - \mathbf{A}_s)^{-1}]}{sD_s + \mathbf{C}_s\mathbf{B}_s}\right\} \\
&= w(\mathbf{0})\det\left[\left(\mathbf{1}_n - \frac{\mathbf{B}_s\mathbf{C}_s}{sD_s + \mathbf{C}_s\mathbf{B}_s}\right)(s\mathbf{1}_n - \mathbf{A}_s)\right. \\
&\qquad \left. + \frac{s\mathbf{B}_s\mathbf{C}_s}{sD_s + \mathbf{C}_s\mathbf{B}_s}\right][\det(s\mathbf{1}_n - \mathbf{A}_s)^{-1}] \\
&= w(\mathbf{0})[\det(s\mathbf{1}_n - \hat{\mathbf{A}}_s)][\det (s\mathbf{1}_n - \mathbf{A}_s)^{-1}] \\
&= w(\mathbf{0})\frac{\det (s\mathbf{1}_n - \hat{\mathbf{A}}_s)}{\det (s\mathbf{1}_n - \mathbf{A}_s)} = w(\mathbf{0})\frac{\det \hat{\mathbf{F}}(\mathbf{A}_s)}{\det \mathbf{F}(\mathbf{A}_s)} \tag{8.110}
\end{aligned}
$$

The fourth line follows from (7.83), and the fifth line follows from the fact that the determinant of the product of two square matrices is equal to the product of their determinants. So the theorem is proved.

The significance of formula (8.109) is that when $D_s = 0$, the matrix

$$\hat{\mathbf{A}}_s = \left(\mathbf{1}_n - \frac{\mathbf{B}_s\mathbf{C}_s}{\mathbf{C}_s\mathbf{B}_s}\right)\mathbf{A}_s \tag{8.111}$$

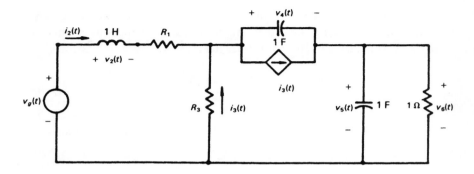

Figure 8.40 An active network used to illustrate the computation of the feedback matrices.

is a constant matrix. Then the determinant of the null return difference matrix can be written from (8.107) as

$$\det \hat{\mathbf{F}}(\mathbf{A}_s) = \frac{\det (s\mathbf{1}_n - \hat{\mathbf{A}}_s)}{s^n} \tag{8.112}$$

Using this in conjunction with (8.102) in (8.109) shows that the eigenvalues of $\hat{\mathbf{A}}_s$ are zeros of the transfer function $w(\mathbf{A}_s)$ and the eigenvalues of $\mathbf{A}_s$ are poles of $w(\mathbf{A}_s)$. These results are consistent with those obtained by Brocken (1965), using the concept of the inverse system.

We illustrate the above result by the following example.

Example 8.7 For the active network N of Fig. 8.40, we wish to compute its feedback matrices and the input admittance facing the voltage source v_g. The resistances R_1 and R_3 are chosen so that the coefficient $\mathbf{A}_s$ matrix of the state equations possesses an eigenvalue of multiplicity 3 at -1. All the network element values have been scaled so that they appear in ohms, farads, and henrys.

Applying the procedure of the preceding section yields the state equations

$$\begin{bmatrix} \dot{i}_2 \\ \dot{v}_4 \\ \dot{v}_5 \end{bmatrix} = \begin{bmatrix} -R_1 & -1 & -1 \\ 1 & 0 & 0 \\ 1 & -1/R_3 & -1 - 1/R_3 \end{bmatrix} \begin{bmatrix} i_2 \\ v_4 \\ v_5 \end{bmatrix} + \begin{bmatrix} 1 \\ 0 \\ 0 \end{bmatrix} [v_g] \tag{8.113a}$$

$$[i_2] = [1 \quad 0 \quad 0] \begin{bmatrix} i_2 \\ v_4 \\ v_5 \end{bmatrix} [0] [v_g] \tag{8.113b}$$

where the output variable is the inductor current i_2. The characteristic equation of the $\mathbf{A}_s$ matrix is found to be

$$\det(\lambda\mathbf{1}_3 - \mathbf{A}_s) = \lambda^3 + \left(1 + R_1 + \frac{1}{R_3}\right)\lambda^2 + \left(\frac{R_1}{R_3} + R_1 + 2\right)\lambda + 1 = 0$$

(8.114)

To produce a root of multiplicity 3 at $\lambda = -1$, we set

$$1 + R_1 + \frac{1}{R_3} = 3 \tag{8.115a}$$

$$\frac{R_1}{R_3} + R_1 + 2 = 3 \tag{8.115b}$$

yielding

$$R_1 = 0.382\ \Omega, \quad R_3 = 0.618\ \Omega \tag{8.116}$$

Using these values in (8.113a), the $\mathbf{A}_s$ matrix becomes

$$\mathbf{A}_s = \begin{bmatrix} -0.382 & -1 & -1 \\ 1 & 0 & 0 \\ 1 & -1.618 & -2.618 \end{bmatrix} \tag{8.117a}$$

The other coefficient matrices are identified as

$$\mathbf{B}_s = \begin{bmatrix} 1 \\ 0 \\ 0 \end{bmatrix}, \quad \mathbf{C}_s = [1\ \ 0\ \ 0], \quad \mathbf{D}_s = [0] \tag{8.117b}$$

The return difference matrix with respect to $\mathbf{A}_s$ is computed from (8.101) as

$$\mathbf{F}(\mathbf{A}_s) = \frac{1}{s}(s\mathbf{1}_3 - \mathbf{A}_s) = \frac{1}{s}\begin{bmatrix} s + 0.382 & 1 & 1 \\ -1 & s & 0 \\ -1 & 1.618 & s + 2.618 \end{bmatrix} \tag{8.118}$$

the determinant of which is given by

$$\det \mathbf{F}(\mathbf{A}_s) = \frac{(s+1)^3}{s^3} \tag{8.119}$$

as expected from (8.114).

To compute the null return difference matrix with respect to $\mathbf{A}_s$, we first compute from (8.108) and (8.117)

$$\hat{\mathbf{A}}_s = \left(\mathbf{1}_3 - \frac{\mathbf{B}_s \mathbf{C}_s}{s D_s + \mathbf{C}_s \mathbf{B}_s}\right) \mathbf{A}_s = (\mathbf{1}_3 - \mathbf{B}_s \mathbf{C}_s)\mathbf{A}_s$$

$$= \begin{bmatrix} 0 & 0 & 0 \\ 0 & 1 & 0 \\ 0 & 0 & 1 \end{bmatrix} \begin{bmatrix} -0.382 & -1 & -1 \\ 1 & 0 & 0 \\ 1 & -1.618 & -2.618 \end{bmatrix}$$

$$= \begin{bmatrix} 0 & 0 & 0 \\ 1 & 0 & 0 \\ 1 & -1.618 & -2.618 \end{bmatrix} \tag{8.120}$$

Substituting this in (8.107) gives the null return difference matrix

$$\hat{\mathbf{F}}(\mathbf{A}_s) = \frac{1}{s}(s\mathbf{1}_3 - \hat{\mathbf{A}}_s) = \frac{1}{s}\begin{bmatrix} s & 0 & 0 \\ -1 & s & 0 \\ -1 & 1.618 & s+2.618 \end{bmatrix} \tag{8.121}$$

the determinant of which is given by

$$\det \hat{\mathbf{F}}(\mathbf{A}_s) = \frac{s^2(s+2.618)}{s^3} = \frac{s+2.618}{s} \tag{8.122}$$

The desired input admittance is found by substituting (8.119) and (8.122) in (8.109):

$$w(\mathbf{A}_s) = \frac{I_2(s)}{V_g(s)} = w(0)\frac{\det \hat{\mathbf{F}}(\mathbf{A}_s)}{\det \mathbf{F}(\mathbf{A}_s)} = \frac{s(s+2.618)}{(s+1)^3} \tag{8.123}$$

where $I_2(s)$ and $V_g(s)$ are the Laplace transforms of $i_2(t)$ and $v_g(t)$, and

$$w(0) = D_s + \frac{1}{s}\mathbf{C}_s\mathbf{B}_s = 0 + \frac{1}{s} = \frac{1}{s} \tag{8.124}$$

Q.E.D.

Recall that in formulating the return difference and null return difference matrices by considering the edge $\mathbf{A}_s$ in the fundamental matrix feedback-flow graph of Fig. 8.36, one difficulty is that when $D_s \neq 0$ the matrix $\hat{\mathbf{A}}_s$ of (8.108) is no longer a constant matrix, being a function of s. An alternate formulation is possible by considering the edge $(1/s)\mathbf{1}$ in the matrix feedback-flow graph G of Fig. 8.36. To compute the return difference matrix $\mathbf{F}(s)$, we set $\tilde{u}(s) = 0$, break the

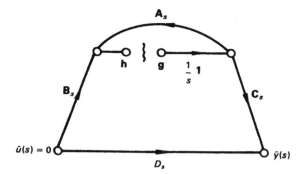

Figure 8.41 The physical interpretation of the return difference matrix.

edge $(1/s)\mathbf{1}$ at its input, apply a signal vector $\mathbf{g}$, and calculate the returned signal $\mathbf{h}$, as depicted in Fig. 8.41, yielding

$$\mathbf{g} - \mathbf{h} = \left[\mathbf{1} - \frac{1}{s}\mathbf{A}_s\right]\mathbf{g} \tag{8.125}$$

The *return difference matrix* becomes

$$\mathbf{F}(s) = \frac{1}{s}(s\mathbf{1} - \mathbf{A}_s) \tag{8.126}$$

which is identical to (8.101).

To compute the null return difference matrix, we use the matrix feedback-flow graph G of Fig. 8.42. As before, we break the edge $(1/s)\mathbf{1}$ at its input, apply a signal vector $\mathbf{g}$, and then adjust the input $\tilde{u}(s)$ so that the total output $\bar{y}(s)$ resulting from the inputs $\mathbf{g}$ and $\tilde{u}(s)$ is identically zero. We obtain from Fig. 8.42

$$\bar{y}(s) = \frac{1}{s}\mathbf{C}_s\mathbf{g} + D_s\tilde{u}(s) = 0 \tag{8.127}$$

giving the desired special value of the input $\tilde{u}(s)$ for $D_s \neq 0$ as

$$\tilde{u}(s) = -\frac{\mathbf{C}_s\mathbf{g}}{sD_s} \tag{8.128}$$

The returned signal $\mathbf{h}$ becomes

$$\mathbf{h} = \frac{1}{s}\mathbf{A}_s\mathbf{g} + \mathbf{B}_s\tilde{u}(s) = \frac{1}{s}\left(\mathbf{A}_s - \frac{\mathbf{B}_s\mathbf{C}_s}{D_s}\right)\mathbf{g} \tag{8.129}$$

or

$$\mathbf{g} - \mathbf{h} = \left[\mathbf{1} - \frac{1}{s}\left(\mathbf{A}_s - \frac{\mathbf{B}_s\mathbf{C}_s}{D_s}\right)\right]\mathbf{g} \tag{8.130}$$

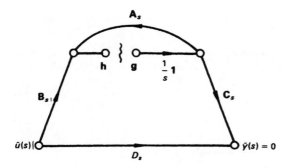

Figure 8.42 The physical interpretation of the null return difference matrix.

The coefficient matrix on the right-hand side of (8.130) is the *null return difference matrix*

$$\hat{\mathbf{F}}(s) = 1 - \frac{1}{s}\left(\mathbf{A}_s - \frac{\mathbf{B}_s\mathbf{C}_s}{D_s}\right)$$

$$= 1 - \frac{1}{s}\hat{\mathbf{A}}_0 = \frac{1}{s}(s1 - \hat{\mathbf{A}}_0) \qquad (8.131)$$

where D_s is assumed to be nonzero and

$$\hat{\mathbf{A}}_0 = \mathbf{A}_s - \frac{\mathbf{B}_s\mathbf{C}_s}{D_s} \qquad (8.132)$$

A relation similar to that of (8.109) can be derived directly from (8.98), as follows:

$$
\begin{aligned}
w(s) &= D_s + \mathbf{C}_s(s1 - \mathbf{A}_s)^{-1}\mathbf{B}_s \\
&= D_s\left[1 + \frac{\mathbf{C}_s(s1 - \mathbf{A}_s)^{-1}\mathbf{B}_s}{D_s}\right] \\
&= D_s\,\det\left[1 + \frac{\mathbf{C}_s(s1 - \mathbf{A}_s)^{-1}\mathbf{B}_s}{D_s}\right] \\
&= D_s\,\det\left[1 + \frac{\mathbf{B}_s\mathbf{C}_s(s1 - \mathbf{A}_s)^{-1}}{D_s}\right] \\
&= D_s\left[\det\left(s1 - \mathbf{A}_s + \frac{\mathbf{B}_s\mathbf{C}_s}{D_s}\right)\right][\det(s1 - \mathbf{A}_s)^{-1}] \\
&= D_s\frac{\det(s1 - \hat{\mathbf{A}}_0)}{\det(s1 - \mathbf{A}_s)} = D_s\frac{\det\hat{\mathbf{F}}(s)}{\det\mathbf{F}(s)} \qquad (8.133)
\end{aligned}
$$

provided that $D_s \neq 0$, where the fourth line follows from (7.83).

Equation (8.133) states that the eigenvalues of the matrix $\hat{\mathbf{A}}_0$ are zeros of the transfer function $w(s)$, and the eigenvalues of $\mathbf{A}_s$ are the poles of $w(s)$. In other words, the zeros of the determinant of the null return difference matrix $\hat{\mathbf{F}}(s)$ are the zeros of the transfer function $w(s)$, and the zeros of the determinant of the return difference matrix $\mathbf{F}(s)$ are the poles of $w(s)$.

Example 8.8 Consider the state equations

$$
\dot{\mathbf{x}}(t) = \begin{bmatrix} \dot{i}_2 \\ \dot{v}_4 \\ \dot{v}_5 \end{bmatrix} = \begin{bmatrix} -0.382 & -1 & -1 \\ 1 & 0 & 0 \\ 1 & -1.618 & -2.618 \end{bmatrix} \begin{bmatrix} i_2 \\ v_4 \\ v_5 \end{bmatrix} + \begin{bmatrix} 1 \\ 0 \\ 0 \end{bmatrix} [v_g]
$$

$$
= \mathbf{A}_s \mathbf{x}(t) + \mathbf{B}_s \mathbf{u}(t) \tag{8.134a}
$$

$$
y(t) = [v_2] = [-0.382 \quad -1 \quad -1] \begin{bmatrix} i_2 \\ v_4 \\ v_5 \end{bmatrix} + [1][v_g]
$$

$$
= \mathbf{C}_s \mathbf{x}(t) + \mathbf{D}_s \mathbf{u}(t) \tag{8.134b}
$$

as previously considered in Example 8.7 except that the output variable is the voltage v_2 across the inductor in the network of Fig. 8.40.

To compute the return difference and the null return difference matrices, we substitute the appropriate coefficient matrices of (8.134) in (8.126) and (8.131), and obtain

$$
\mathbf{F}(s) = \frac{1}{s}(s\mathbf{1}_3 - \mathbf{A}_s) = \frac{1}{s} \begin{bmatrix} s + 0.382 & 1 & 1 \\ -1 & s & 0 \\ -1 & 1.618 & s + 2.618 \end{bmatrix} \tag{8.135}
$$

$$
\hat{\mathbf{F}}(s) = \frac{1}{s}(s\mathbf{1}_3 - \hat{\mathbf{A}}_0) = \frac{1}{s} \begin{bmatrix} s & 0 & 0 \\ -1 & s & 0 \\ -1 & 1.618 & s + 2.618 \end{bmatrix} \tag{8.136}
$$

where

$$
\hat{\mathbf{A}}_0 = \mathbf{A}_s - \frac{\mathbf{B}_s \mathbf{C}_s}{D_s} = \begin{bmatrix} 0 & 0 & 0 \\ 1 & 0 & 0 \\ 1 & -1.618 & -2.618 \end{bmatrix} \tag{8.137}
$$

The determinants of $\mathbf{F}(s)$ and $\hat{\mathbf{F}}(s)$ are found to be

$$\det \mathbf{F}(s) = \frac{(s+1)^3}{s^3} \tag{8.138}$$

$$\det \hat{\mathbf{F}}(s) = \frac{s+2.618}{s} \tag{8.139}$$

obtaining from (8.133)

$$w(s) = \frac{V_2(s)}{V_g(s)} = D_s \frac{\det \hat{\mathbf{F}}(s)}{\det \mathbf{F}(s)} = \frac{s^2(s+2.618)}{(s+1)^3} \tag{8.140}$$

where $V_2(s)$ is the Laplace transform of $v_2(t)$.   Q.E.D.

We now extend the preceding discussion to the situation where only a portion of the $\mathbf{A}_s$ matrix is of interest. Thus, we decompose the matrix as a sum of two matrices:

$$\mathbf{A}_s = \mathbf{A}_{s1} + \mathbf{K} \tag{8.141}$$

where as in (7.147) $\mathbf{A}_{s1}$ represents the elements of interest and $\mathbf{K}$ is called the *reference matrix*.

In Fig. 8.36 we split the edge with transmittance $\mathbf{A}_s$ into two edges $\mathbf{A}_{s1}$ and $\mathbf{K}$, as in Fig. 7.28. For ease of reduction, we insert two edges with unity transmittance at the input and the output of the edge $\mathbf{A}_{s1}$. The new equations governing the matrix feedback-flow graph of Fig. 8.43 can now be written as

$$(s\mathbf{1} - \mathbf{K})\tilde{\mathbf{x}}(s) = \mathbf{A}_{s1}\tilde{\mathbf{x}}(s) + \mathbf{B}_s\tilde{\mathbf{u}}(s) \tag{8.142a}$$

$$\tilde{\mathbf{y}}(s) = \mathbf{C}_s\tilde{\mathbf{x}}(s) + D_s\tilde{\mathbf{u}}(s) \tag{8.142b}$$

$$\tilde{\mathbf{x}}(s) = (s\mathbf{1} - \mathbf{K})^{-1}(s\mathbf{1} - \mathbf{K})\tilde{\mathbf{x}}(s) \tag{8.142c}$$

These equations can now be represented by the new matrix feedback-flow graph of Fig. 8.44, where we have focused our attention on the edge with transmittance $\mathbf{A}_{s1}$ after combining the edges with transmittances $\mathbf{K}$ and $(1/s)\mathbf{1}$ in Fig. 8.43. The combined edge has transmittance $(s\mathbf{1} - \mathbf{K})^{-1}$.

To compute the return difference matrix, we set $\tilde{\mathbf{u}}(s) = 0$ and consider only the feedback loop. We first break the loop at the input of the edge $\mathbf{A}_{s1}$, and then

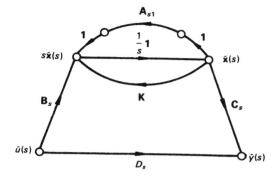

Figure 8.43   The fundamental matrix feedback-flow graph with reference matrix exhibited explicitly.

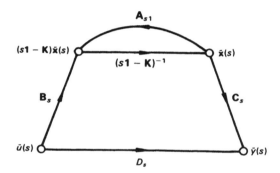

Figure 8.44   The resulting matrix feedback-flow graph after absorbing two nodes in the graph of Fig. 8.43.

Figure 8.45   Computation of the general return difference matrix for a general reference.

apply a signal vector **g**, as depicted in Fig. 8.45. The returned signal is given by

$$\mathbf{h} = (s\mathbf{1} - \mathbf{K})^{-1}\mathbf{A}_{s1}\mathbf{g} \tag{8.143}$$

giving

$$\mathbf{g} - \mathbf{h} = [\mathbf{1} - (s\mathbf{1} - \mathbf{K})^{-1}\mathbf{A}_{s1}]\mathbf{g} \tag{8.144}$$

The resulting return difference matrix is called the *general return difference matrix* with respect to $\mathbf{A}_s$ for the general reference $\mathbf{K}$, and is given by

$$\mathbf{F_K}(\mathbf{A}_s) = \mathbf{1} - (s\mathbf{1} - \mathbf{K})^{-1}\mathbf{A}_{s1} \qquad (8.145)$$

Clearly, for $\mathbf{K} = \mathbf{0}$ then

$$\mathbf{F_0}(\mathbf{A}_s) = \mathbf{F}(\mathbf{A}_s) \qquad (8.146)$$

the original return difference matrix of (8.101). As in (7.162), $\mathbf{F_K}(\mathbf{A}_s)$ can also be written as

$$\begin{aligned}
\mathbf{F_K}(\mathbf{A}_s) &= (s\mathbf{1} - \mathbf{K})^{-1}(s\mathbf{1} - \mathbf{K} - \mathbf{A}_{s1}) \\
&= (s\mathbf{1} - \mathbf{K})^{-1}(s\mathbf{1} - \mathbf{A}_s) \\
&= \mathbf{F}^{-1}(\mathbf{K})\mathbf{F}(\mathbf{A}_s)
\end{aligned} \qquad (8.147)$$

where from (8.101)

$$\mathbf{F}(\mathbf{K}) = \mathbf{1} - \frac{1}{s}\mathbf{K} \qquad (8.148)$$

Likewise, to compute the null return difference matrix, we refer to the matrix feedback-flow graph of Fig. 8.46. As before, we break the edge with transmittance $\mathbf{A}_{s1}$ at its input, apply a signal vector $\mathbf{g}$, and then adjust the input $\tilde{u}(s)$, so that the total output $\tilde{y}(s)$ resulting from both inputs $\mathbf{g}$ and $\tilde{u}(s)$ is identically zero. This requires that

$$\tilde{y}(s) = D_s\tilde{u}(s) + \mathbf{C}_s(s\mathbf{1} - \mathbf{K})^{-1}[\mathbf{B}_s\tilde{u}(s) + \mathbf{A}_{s1}\mathbf{g}] = 0 \qquad (8.149)$$

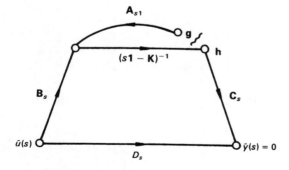

Figure 8.46 Computation of the general null return difference matrix for a general reference.

obtaining the special input

$$\tilde{u}(s) = -\frac{\mathbf{C}_s(s\mathbf{1} - \mathbf{K})^{-1}\mathbf{A}_{s1}\mathbf{g}}{D_s + \mathbf{C}_s(s\mathbf{1} - \mathbf{K})^{-1}\mathbf{B}_s} \qquad (8.150)$$

The returned signal **h**, with the presence of both sources $\tilde{u}(s)$ and **g**, is found to be

$$\mathbf{h} = (s\mathbf{1} - \mathbf{K})^{-1}\left[\mathbf{B}_s\tilde{u}(s) + \mathbf{A}_{s1}\mathbf{g}\right] \qquad (8.151a)$$

giving

$$\mathbf{g} - \mathbf{h} = \left\{\mathbf{1} - \left[\mathbf{1} - \frac{(s\mathbf{1} - \mathbf{K})^{-1}\mathbf{B}_s\mathbf{C}_s}{D_s + \mathbf{C}_s(s\mathbf{1} - \mathbf{K})^{-1}\mathbf{B}_s}\right](s\mathbf{1} - \mathbf{K})^{-1}\mathbf{A}_{s1}\right\}\mathbf{g} \qquad (8.151b)$$

The coefficient matrix

$$\hat{\mathbf{F}}_{\mathbf{K}}(\mathbf{A}_s) = \mathbf{1} - \left[\mathbf{1} - \frac{(s\mathbf{1} - \mathbf{K})^{-1}\mathbf{B}_s\mathbf{C}_s}{D_s + \mathbf{C}_s(s\mathbf{1} - \mathbf{K})^{-1}\mathbf{B}_s}\right](s\mathbf{1} - \mathbf{K})^{-1}\mathbf{A}_{s1} \qquad (8.152)$$

is called the *general null return difference matrix* with respect to $\mathbf{A}_s$ for the general reference **K**.

To show that a relation similar to (7.164) exists for $\hat{\mathbf{F}}_{\mathbf{K}}(\mathbf{A}_s)$, we first compute from (8.107) the null return difference matrix

$$\hat{\mathbf{F}}(\mathbf{K}) = \mathbf{1}_n - \frac{1}{s}\left(\mathbf{1}_n - \frac{\mathbf{B}_s\mathbf{C}_s}{sD_s + \mathbf{C}_s\mathbf{B}_s}\right)\mathbf{K} \qquad (8.153a)$$

$$= \hat{\mathbf{F}}(\mathbf{A}_s) + \frac{1}{s}\left(\mathbf{1}_n - \frac{\mathbf{B}_s\mathbf{C}_s}{sD_s + \mathbf{C}_s\mathbf{B}_s}\right)\mathbf{A}_{s1} \qquad (8.153b)$$

$$= \frac{[(sD_s + \mathbf{C}_s\mathbf{B}_s)\mathbf{1}_n - \mathbf{B}_s\mathbf{C}_s]}{s(sD_s + \mathbf{C}_s\mathbf{B}_s)}(s\mathbf{1}_n - \mathbf{K}) + \frac{s\mathbf{B}_s\mathbf{C}_s}{s(sD_s + \mathbf{C}_s\mathbf{B}_s)} \qquad (8.153c)$$

with respect to **K**. Then we use (8.153b) to multiply the first term on the right-hand side of (8.152) and use (8.153c) to multiply the second term of (8.152) to yield the matrix product

$$\hat{\mathbf{F}}(\mathbf{K})\hat{\mathbf{F}}_{\mathbf{K}}(\mathbf{A}_s) = \hat{\mathbf{F}}(\mathbf{A}_s) + \frac{1}{s}\left(\mathbf{1}_n - \frac{\mathbf{B}_s\mathbf{C}_s}{sD_s + \mathbf{C}_s\mathbf{B}_s}\right)\mathbf{A}_{s1}$$

$$-\frac{1}{s}\left(\mathbf{1}_n - \frac{\mathbf{B}_s\mathbf{C}_s}{sD_s + \mathbf{C}_s\mathbf{B}_s}\right)\mathbf{A}_{s1} - \frac{s\mathbf{B}_s\mathbf{C}_s(s\mathbf{1}_n - \mathbf{K})^{-1}}{s(sD_s + \mathbf{C}_s\mathbf{B}_s)}\mathbf{A}_{s1}$$

$$+ \frac{[(sD_s + C_sB_s)\mathbf{1}_n - \mathbf{B}_s\mathbf{C}_s] + s\mathbf{B}_s\mathbf{C}_s(s\mathbf{1}_n - \mathbf{K})^{-1}}{s(sD_s + \mathbf{C}_s\mathbf{B}_s)[D_s + \mathbf{C}_s(s\mathbf{1}_n - \mathbf{K})^{-1}\mathbf{B}_s]}$$

$$\times \mathbf{B}_s\mathbf{C}_s(s\mathbf{1}_n - \mathbf{K})^{-1}\mathbf{A}_{s1}$$

$$= \hat{\mathbf{F}}(\mathbf{A}_s) + \frac{\mathbf{B}_s e\mathbf{C}_s(s\mathbf{1}_n - \mathbf{K})^{-1}\mathbf{A}_{s1} - e\mathbf{B}_s\mathbf{C}_s(s\mathbf{1}_n - \mathbf{K})^{-1}\mathbf{A}_{s1}}{s(sD_s + \mathbf{C}_s\mathbf{B}_s)[D_s + \mathbf{C}_s(s\mathbf{1}_n - \mathbf{K})^{-1}\mathbf{B}_s]}$$

$$= \hat{\mathbf{F}}(\mathbf{A}_s) + \mathbf{0} = \hat{\mathbf{F}}(\mathbf{A}_s) \qquad (8.154a)$$

where

$$e = \mathbf{C}_s[s(s\mathbf{1}_n - \mathbf{K})^{-1} - \mathbf{1}_n]\mathbf{B}_s \qquad (8.154b)$$

is a scalar, showing that

$$\hat{\mathbf{F}}_{\mathbf{K}}(\mathbf{A}_s) = \hat{\mathbf{F}}^{-1}(\mathbf{K})\hat{\mathbf{F}}(\mathbf{A}_s) \qquad (8.155)$$

Finally to compute the determinants of the general return difference matrix and the general null return difference matrix, we take determinants on both sides of (8.147) and (8.155) and obtain

$$\det \mathbf{F}_{\mathbf{K}}(\mathbf{A}_s) = \frac{\det \mathbf{F}(\mathbf{A}_s)}{\det \mathbf{F}(\mathbf{K})} \qquad (8.156a)$$

$$\det \hat{\mathbf{F}}_{\mathbf{K}}(\mathbf{A}_s) = \frac{\det \hat{\mathbf{F}}(\mathbf{A}_s)}{\det \hat{\mathbf{F}}(\mathbf{K})} \qquad (8.156b)$$

Alternatively, we can express $\hat{\mathbf{F}}(\mathbf{K})$ in the form of (8.148) by letting

$$\hat{\mathbf{K}} = \left(\mathbf{1}_n - \frac{\mathbf{B}_s\mathbf{C}_s}{sD_s + \mathbf{C}_s\mathbf{B}_s}\right)\mathbf{K} \qquad (8.157a)$$

Substituting this in (8.153a) yields

$$\hat{\mathbf{F}}(\mathbf{K}) = \mathbf{1}_n - \frac{1}{s}\hat{\mathbf{K}} \qquad (8.157b)$$

Clearly, for $\mathbf{K} = \mathbf{0}$, then $\hat{\mathbf{K}} = \mathbf{0}$, $\hat{\mathbf{F}}(\mathbf{K}) = \mathbf{1}_n$ and from (8.155)

$$\hat{\mathbf{F}}_{\mathbf{0}}(\mathbf{A}_s) = \hat{\mathbf{F}}(\mathbf{A}_s) \qquad (8.158)$$

the original null return difference matrix of (8.107).

Equation (8.147) states that in a single-input and single-output feedback network, the general return difference matrix $\mathbf{F}_{\mathbf{K}}(\mathbf{A}_s)$ with respect to the $\mathbf{A}_s$ matrix of the linear system (8.94) for the general reference matrix $\mathbf{K}$ is equal to the product of the inverse of the return difference matrix $\mathbf{F}(\mathbf{K})$ with respect to the general reference matrix $\mathbf{K}$ and the return difference matrix $\mathbf{F}(\mathbf{A}_s)$ with respect to the matrix $\mathbf{A}_s$. Similarly, from (8.155) we can state that in a single-input and single-output feedback network, the general null return difference matrix $\hat{\mathbf{F}}_{\mathbf{K}}(\mathbf{A}_s)$ with respect to the $\mathbf{A}_s$ matrix of the linear system (8.94) for the general reference matrix $\mathbf{K}$ is equal to the product of the inverse of the null return difference matrix $\hat{\mathbf{F}}(\mathbf{K})$ with respect to $\mathbf{K}$ and the null return difference matrix $\hat{\mathbf{F}}(\mathbf{A}_s)$ with respect to $\mathbf{A}_s$.

Suppose that we are interested in the scalar return difference $F(a_{ij})$ and null return difference $\hat{F}(a_{ij})$ with respect to an element a_{ij} of $\mathbf{A}_s$. Then as shown in Sec. 7.3.2 we set

$$\mathbf{K} = \mathbf{A}_s|_{a_{ij}=0} \tag{8.159}$$

and obtain from (7.199)

$$F(a_{ij}) = \frac{\det \mathbf{F}(\mathbf{A}_s)}{\det \mathbf{F}(\mathbf{A}_s)|_{a_{ij}=0}} \tag{8.160}$$

or from (8.156a)

$$F(a_{ij}) = \det \mathbf{F}_{\mathbf{K}}(\mathbf{A}_s) \tag{8.161}$$

Likewise from (7.200) and (8.156b) we have

$$\hat{F}(a_{ij}) = \det \hat{\mathbf{F}}_{\mathbf{K}}(\mathbf{A}_s) = \frac{\det \hat{\mathbf{F}}(\mathbf{A}_s)}{\det \hat{\mathbf{F}}(\mathbf{A}_s)|_{a_{ij}=0}} \tag{8.162}$$

To obtain the sensitivity of the transfer function $w(\mathbf{A}_s)$ with respect to a_{ij}, we appeal to formula (5.5) and obtain

$$\begin{aligned} S(a_{ij}) &= \frac{1}{F(a_{ij})} - \frac{1}{\hat{F}(a_{ij})} \\ &= \frac{\det \mathbf{F}(\mathbf{A}_s)|_{a_{ij}=0}}{\det \mathbf{F}(\mathbf{A}_s)} - \frac{\det \hat{\mathbf{F}}(\mathbf{A}_s)|_{a_{ij}=0}}{\det \hat{\mathbf{F}}(\mathbf{A}_s)} \end{aligned} \tag{8.163}$$

Thus, it is only necessary to calculate the determinants of the return difference matrix and the null return difference matrix of the network under the nominal operating condition and under the condition $a_{ij} = 0$ in order to determine the sensitivity of the transfer function with respect to the element a_{ij} of the $\mathbf{A}_s$ matrix.

We illustrate the above results by the following example.

Example 8.9 Consider the same problem as in Example 8.7. For our purposes, we write the matrix $\mathbf{A}_s$ in terms of the element a_{31} as

$$
\mathbf{A}_s = \begin{bmatrix} -0.382 & -1 & -1 \\ 1 & 0 & 0 \\ a_{31} & -1.618 & -2.618 \end{bmatrix}
\tag{8.164}
$$

where $a_{31} = 1$. The other matrices are obtained from (8.117b) as

$$
\mathbf{B}_s = \begin{bmatrix} 1 \\ 0 \\ 0 \end{bmatrix}, \quad \mathbf{C}_s = [1 \ \ 0 \ \ 0], \quad D_s = [0]
\tag{8.165}
$$

Substituting these in (8.101) and (8.107) yields

$$
\mathbf{F}(\mathbf{A}_s) = \frac{1}{s} \begin{bmatrix} s + 0.382 & 1 & 1 \\ -1 & s & 0 \\ -a_{31} & 1.618 & s + 2.618 \end{bmatrix}
\tag{8.166}
$$

$$
\hat{\mathbf{F}}(\mathbf{A}_s) = \frac{1}{s} \begin{bmatrix} s & 0 & 0 \\ -1 & s & 0 \\ -a_{31} & 1.618 & s + 2.618 \end{bmatrix}
\tag{8.167}
$$

the determinants of which are found to be

$$
\det \mathbf{F}(\mathbf{A}_s) = \frac{s^3 + 3s^2 + (a_{31} + 2)s + 1}{s^3}
\tag{8.168}
$$

$$
\det \hat{\mathbf{F}}(\mathbf{A}_s) = \frac{s + 2.618}{s}
\tag{8.169}
$$

Thus, the sensitivity of the input admittance facing the voltage source in the active network of Fig. 8.40 can be calculated directly from (8.163), as follows:

$$
S(a_{31}) = \frac{\det \mathbf{F}(\mathbf{A}_s)|_{a_{31}=0}}{\det \mathbf{F}(\mathbf{A}_s)} - \frac{\det \hat{\mathbf{F}}(\mathbf{A}_s)|_{a_{31}=0}}{\det \hat{\mathbf{F}}(\mathbf{A}_s)}
$$

$$
= \frac{s^3 + 3s^2 + 2s + 1}{(s + 1)^3} - 1 = -\frac{s}{(s + 1)^3}
\tag{8.170}
$$

To verify (8.147), (8.155) and (8.156), let

$$
\mathbf{K} = \begin{bmatrix} -0.382 & -1 & -1 \\ 1 & 0 & 0 \\ 0 & -1.618 & -2.618 \end{bmatrix}
\tag{8.171}
$$

Then from (8.148) we obtain the return difference matrix with respect to $\mathbf{K}$ as

$$\mathbf{F}(\mathbf{K}) = \frac{1}{s}(s\mathbf{1}_3 - \mathbf{K}) = \frac{1}{s}\begin{bmatrix} s+0.382 & 1 & 1 \\ -1 & s & 0 \\ 0 & 1.618 & s+2.618 \end{bmatrix} \tag{8.172}$$

the determinant and inverse of which are found to be

$$\det \mathbf{F}(\mathbf{K}) = \frac{s^3 + 3s^2 + 2s + 1}{s^3} \tag{8.173}$$

$$\mathbf{F}^{-1}(\mathbf{K}) = \frac{1}{s^2 \det \mathbf{F}(\mathbf{K})}$$

$$\times \begin{bmatrix} s(s+2.618) & -s-1 & -s \\ s+2.618 & (s+0.382)(s+2.618) & -1 \\ -1.618 & -1.618(s+0.382) & s^2+0.382s+1 \end{bmatrix} \tag{8.174}$$

Substituting these in (8.145) or (8.147) gives the general return difference matrix with respect to $\mathbf{A}_s$ for the reference value $\mathbf{K}$ as

$$\mathbf{F_K}(\mathbf{A}_s) = \mathbf{1}_3 - (s\mathbf{1}_3 - \mathbf{K})^{-1}\mathbf{A}_{s1} = \mathbf{F}^{-1}(\mathbf{K})\mathbf{F}(\mathbf{A}_s)$$

$$= \frac{1}{P(s)}\begin{bmatrix} (s+1)^3 & 0 & 0 \\ 1 & P(s) & 0 \\ -s^2-0.382s-1 & 0 & P(s) \end{bmatrix} \tag{8.175}$$

where

$$P(s) = s^3 + 3s^2 + 2s + 1 \tag{8.176}$$

$$\mathbf{A}_{s1} = \mathbf{A}_s - \mathbf{K} = \begin{bmatrix} 0 & 0 & 0 \\ 0 & 0 & 0 \\ 1 & 0 & 0 \end{bmatrix} \tag{8.177}$$

The determinant of $\mathbf{F_K}(\mathbf{A}_s)$ is given by

$$\det \mathbf{F_K}(\mathbf{A}_s) = \frac{(s+1)^3}{s^3+3s^2+2s+1} = \frac{\det \mathbf{F}(\mathbf{A}_s)}{\det \mathbf{F}(\mathbf{K})} \tag{8.178}$$

confirming (8.156a).

To calculate the general null return difference matrix with respect to $\mathbf{A}_s$ for the reference value $\mathbf{K}$, we apply (8.152) and obtain

$$\hat{\mathbf{F}}_{\mathbf{K}}(\mathbf{A}_s) = \mathbf{1}_3 - \left[\mathbf{1}_3 - \frac{(s\mathbf{1}_3 - \mathbf{K})^{-1}\mathbf{B}_s\mathbf{C}_s}{D_s + \mathbf{C}_s(s\mathbf{1}_3 - \mathbf{K})^{-1}\mathbf{B}_s} \right] (s\mathbf{1}_3 - \mathbf{K})^{-1}\mathbf{A}_{s1}$$

$$= \frac{1}{s(s + 2.618)} \begin{bmatrix} s(s + 2.618) & 0 & 0 \\ 0 & s(s + 2.618) & 0 \\ -s & 0 & s(s + 2.618) \end{bmatrix}$$

$$(8.179a)$$

the determinant of which is found to be

$$\det \hat{\mathbf{F}}_{\mathbf{K}}(\mathbf{A}_s) = 1 \tag{8.179b}$$

To verify (8.156b), we first calculate the null return difference matrix with respect to $\mathbf{K}$ from (8.157b) and obtain

$$\hat{\mathbf{F}}(\mathbf{K}) = \mathbf{1}_3 - \frac{1}{s}\hat{\mathbf{K}} = \frac{1}{s} \begin{bmatrix} s & 0 & 0 \\ -1 & s & 0 \\ 0 & 1.618 & s + 2.618 \end{bmatrix} \tag{8.180a}$$

where from (8.157a)

$$\hat{\mathbf{K}} = \left(\mathbf{1}_3 - \frac{\mathbf{B}_s\mathbf{C}_s}{sD_s + \mathbf{C}_s\mathbf{B}_s} \right) \mathbf{K} = \begin{bmatrix} 0 & 0 & 0 \\ 1 & 0 & 0 \\ 0 & -1.618 & -2.618 \end{bmatrix} \tag{8.180b}$$

The inverse and determinant of $\hat{\mathbf{F}}(\mathbf{K})$ are found to be

$$\hat{\mathbf{F}}^{-1}(\mathbf{K}) = \frac{1}{s + 2.618} \begin{bmatrix} s + 2.618 & 0 & 0 \\ \dfrac{s + 2.618}{s} & s + 2.618 & 0 \\ -\dfrac{1.618}{s} & -1.618 & s \end{bmatrix} \tag{8.181a}$$

$$\det \hat{\mathbf{F}}(\mathbf{K}) = \frac{s + 2.618}{s} \tag{8.181b}$$

Substituting (8.181a) and (8.167) in (8.155) yields

$$\hat{\mathbf{F}}^{-1}(\mathbf{K})\hat{\mathbf{F}}(\mathbf{A}_s) = \frac{1}{s(s+2.618)}\begin{bmatrix} s+2.618 & 0 & 0 \\ \dfrac{s+2.618}{s} & s+2.618 & 0 \\ -\dfrac{1.618}{s} & -1.618 & s \end{bmatrix}$$

$$\times \begin{bmatrix} s & 0 & 0 \\ -1 & s & 0 \\ -1 & 1.618 & s+2.618 \end{bmatrix}$$

$$= \frac{1}{s(s+2.618)}\begin{bmatrix} s(s+2.618) & 0 & 0 \\ 0 & s(s+2.618) & 0 \\ -s & 0 & s(s+2.618) \end{bmatrix}$$

$$= \hat{\mathbf{F}}_\mathbf{K}(\mathbf{A}_s) \qquad (8.182a)$$

confirming (8.155), the determinant of which is found to be

$$\det \hat{\mathbf{F}}_\mathbf{K}(\mathbf{A}_s) = 1 \qquad (8.182b)$$

This determinant can also be calculated from (8.156b) using (8.169) and (8.181b), as follows:

$$\det \hat{\mathbf{F}}_\mathbf{K}(\mathbf{A}_s) = \frac{\det \hat{\mathbf{F}}(\mathbf{A}_s)}{\det \hat{\mathbf{F}}(\mathbf{K})} = \frac{(s+2.618)/s}{(s+2.618)/s} = 1 \qquad (8.182c)$$

Instead of using the output equation (8.113b), suppose that we select the inductor voltage v_2 as the output variable by using the output equation (8.134b), everything else being the same. Then from (8.101), (8.107), (8.145), (8.148), (8.152), (8.153), and (8.157) we obtain

$$\mathbf{F}(\mathbf{A}_s) = \mathbf{1}_3 - \frac{1}{s}\mathbf{A}_s = \frac{1}{s}\begin{bmatrix} s+0.382 & 1 & 1 \\ -1 & s & 0 \\ 1 & 1.618 & s+2.618 \end{bmatrix} \qquad (8.183a)$$

$$\mathbf{F(K)} = \mathbf{1}_3 - \frac{1}{s}\mathbf{K} = \frac{1}{s}\begin{bmatrix} s + 0.382 & 1 & 1 \\ -1 & s & 0 \\ 0 & 1.618 & s + 2.618 \end{bmatrix} \quad (8.183b)$$

$$\mathbf{F_K(A_s)} = \mathbf{1}_3 - (s\mathbf{1}_3 - \mathbf{K})^{-1}\mathbf{A}_{s1}$$

$$= \frac{1}{p(s)}\begin{bmatrix} s + P(s) & 0 & 0 \\ 1 & P(s) & 0 \\ -s^2 - 0.382s - 1 & 0 & P(s) \end{bmatrix} \quad (8.183c)$$

$$\hat{\mathbf{F}}(\mathbf{A}_s) = \mathbf{1}_3 - \frac{1}{s}\left(\mathbf{1}_3 - \frac{\mathbf{B}_s\mathbf{C}_s}{s\mathbf{D}_s + \mathbf{C}_s\mathbf{B}_s}\right)\mathbf{A}_s$$

$$= \frac{1}{s}\begin{bmatrix} \dfrac{s^2 - 2}{s - 0.382} & \dfrac{s + 1.618}{s - 0.382} & \dfrac{s + 2.618}{s - 0.382} \\ -1 & s & 0 \\ -1 & 1.618 & s + 2.618 \end{bmatrix} \quad (8.183d)$$

$$\hat{\mathbf{F}}(\mathbf{K}) = \mathbf{1}_3 - \frac{1}{s}\left(\mathbf{1}_3 - \frac{\mathbf{B}_s\mathbf{C}_s}{s\mathbf{D}_s + \mathbf{C}_s\mathbf{B}_s}\right)\mathbf{K} = \mathbf{1}_3 - \frac{1}{s}\hat{\mathbf{K}}$$

$$= \frac{1}{s}\begin{bmatrix} \dfrac{s^2 - 1}{s - 0.382} & \dfrac{s + 1.618}{s - 0.382} & \dfrac{s + 2.618}{s - 0.382} \\ -1 & s & 0 \\ 0 & 1.618 & s + 2.618 \end{bmatrix} \quad (8.183e)$$

the inverse of which is obtained as

$$\hat{\mathbf{F}}^{-1}(\mathbf{K}) = \begin{bmatrix} \dfrac{s - 0.382}{s} & -\dfrac{1}{s} & -\dfrac{1}{s} \\ \dfrac{s - 0.382}{s^2} & \dfrac{s^2 - 1}{s^2} & -\dfrac{1}{s^2} \\ -\dfrac{1.618(s - 0.382)}{s^2(s + 2.618)} & -\dfrac{1.618(s^2 - 1)}{s^2(s + 2.618)} & \dfrac{s^3 + 1.618}{s^2(s + 2.618)} \end{bmatrix}$$

$$(8.183f)$$

$$\hat{\mathbf{F}}_{\mathbf{K}}(\mathbf{A}_s) = \mathbf{1}_3 - \left[\mathbf{1}_3 - \frac{(s\mathbf{1}_3 - \mathbf{K})^{-1}\mathbf{B}_s\mathbf{C}_s}{D_s + \mathbf{C}_s(s\mathbf{1}_3 - \mathbf{K})^{-1}\mathbf{B}_s} \right] (s\mathbf{1}_3 - \mathbf{K})^{-1}\mathbf{A}_{s1}$$

$$= \begin{bmatrix} 1 & 0 & 0 \\ 0 & 1 & 0 \\ -\dfrac{1}{s+2.618} & 0 & 1 \end{bmatrix} \tag{8.183g}$$

where

$$\mathbf{C}_s = [-0.382 \quad -1 \quad -1] \tag{8.183h}$$

$$D_s = [1] \tag{8.183i}$$

$$(s\mathbf{1}_3 - \mathbf{K})^{-1} = \frac{1}{P(s)}$$

$$\times \begin{bmatrix} s(s+2.618) & -s-1 & -s \\ s+2.618 & (s+0.382)(s+2.618) & -1 \\ -1.618 & -1.618(s+0.382) & s^2+0.382s+1 \end{bmatrix} \tag{8.183j}$$

and $\mathbf{K}$, $P(s)$ and $\mathbf{A}_{s1}$ are given by (8.171), (8.176), and (8.177), respectively.
Substituting these in (8.147) and (8.155) we obtain

$$\mathbf{F}_{\mathbf{K}}(\mathbf{A}_s) = \mathbf{F}^{-1}(\mathbf{K})\mathbf{F}(\mathbf{A}_s) = \frac{1}{p(s)} \begin{bmatrix} s+P(s) & 0 & 0 \\ 1 & P(s) & 0 \\ -s^2-0.382s-1 & 0 & P(s) \end{bmatrix} \tag{8.184a}$$

confirming (8.183c), and

$$\hat{\mathbf{F}}_{\mathbf{K}}(\mathbf{A}_s) = \hat{\mathbf{F}}^{-1}(\mathbf{K})\hat{\mathbf{F}}(\mathbf{A}_s) = \begin{bmatrix} 1 & 0 & 0 \\ 0 & 1 & 0 \\ -\dfrac{1}{s+2.618} & 0 & 1 \end{bmatrix} \tag{8.184b}$$

confirming (8.183g), where

$$
\mathbf{F}^{-1}(\mathbf{K}) = \frac{s}{P(s)}
\begin{bmatrix}
s(s+2.618) & -s-1 & -s \\
s+2.618 & (s+0.382)(s+2.618) & -1 \\
-1.618 & -1.618(s+0.382) & s^2+0.382s+1
\end{bmatrix}
$$

(8.184c)

8.7 STATE-SPACE FORMULATION OF MULTIPLE-LOOP FEEDBACK NETWORKS

In the preceding section, we derived the feedback matrices with respect to the coefficient $\mathbf{A}_s$ matrix of the state equation for a single-input and single-output feedback network, and showed how their determinants are related to the poles and zeros of the transfer function and their use in sensitivity analysis of the transfer function with respect to any element of the $\mathbf{A}_s$ matrix. In the present section, we shall extend these concepts to multiple-input, multiple-output and multiple-loop feedback networks.

The block diagram of a multiple-input, multiple-output and multiple-loop feedback network N_f is shown in Fig. 7.12, and is redrawn in Fig. 8.47, where, as before, the elements of interest are exhibited explicitly by the block $\mathbf{X}$. The rectangular matrix $\mathbf{X}$ can be either a transfer function matrix or a driving-point immittance matrix characterized by the relation

$$
\tilde{\boldsymbol{\theta}}(s) = \mathbf{X}\tilde{\boldsymbol{\varphi}}(s)
$$

(8.185)

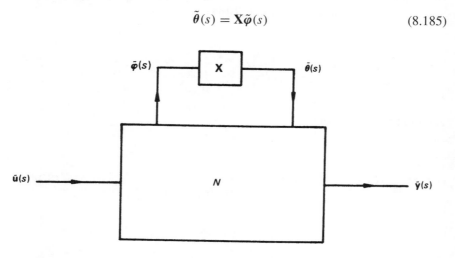

Figure 8.47 The block diagram of a general feedback network.

where $\mathbf{X}$ is of order $q \times p$, and $\tilde{\boldsymbol{\theta}}$ and $\tilde{\boldsymbol{\varphi}}$ are the q-vector and p-vector, respectively. In the feedback network N_f of Fig. 8.47, the block N can be considered as a multi-port network, in which the n-vector $\tilde{\mathbf{u}}$ and the q-vector $\tilde{\boldsymbol{\theta}}$ are inputs and the m-vector $\tilde{\mathbf{y}}$ and the p-vector $\tilde{\boldsymbol{\varphi}}$ are outputs of N. Since the network N is linear, it can be characterized by the equations

$$\tilde{\boldsymbol{\varphi}}(s) = \mathbf{A}\tilde{\boldsymbol{\theta}}(s) + \mathbf{B}\tilde{\mathbf{u}}(s) \tag{8.186a}$$

$$\tilde{\mathbf{y}}(s) = \mathbf{C}\tilde{\boldsymbol{\theta}}(s) + \mathbf{D}\tilde{\mathbf{u}}(s) \tag{8.186b}$$

as in (7.13), where $\mathbf{A}$, $\mathbf{B}$, $\mathbf{C}$ and $\mathbf{D}$ are transfer-function matrices of orders $p \times q$, $p \times n$, $m \times q$ and $m \times n$, respectively.

Also, the feedback network N_f of Fig. 8.47 can be characterized by its state equations

$$\dot{\mathbf{x}}(t) = \mathbf{A}_s\mathbf{x}(t) + \mathbf{B}_s\mathbf{u}(t) \tag{8.187a}$$

$$\mathbf{y}(t) = \mathbf{C}_s\mathbf{x}(t) + \mathbf{D}_s\mathbf{u}(t) \tag{8.187b}$$

as in (8.2) and (8.3). Equations (8.185) and (8.186) are the frequency-domain characterization of N_f whereas its time-domain description is given by (8.187). Recall that we use the symbol tilde $\sim$ to denote the Laplace transform $\tilde{\mathbf{u}}(s)$ of the corresponding time-domain function $\mathbf{u}(t)$. Assume that the matrix $\mathbf{X}$ is independent of the time t. Then the multi-port network N can be described by the state equations*

$$\dot{\mathbf{x}}(t) = \mathbf{A}_o\mathbf{x}(t) + \mathbf{B}_o\mathbf{u}(t) + \mathbf{B}_1\boldsymbol{\theta}(t) \tag{8.188a}$$

$$\mathbf{y}(t) = \mathbf{C}_o\mathbf{x}(t) + \mathbf{D}_o\mathbf{u}(t) + \mathbf{D}_1\boldsymbol{\theta}(t) \tag{8.188b}$$

$$\boldsymbol{\varphi}(t) = \mathbf{C}_1\mathbf{x}(t) + \mathbf{D}_2\mathbf{u}(t) + \mathbf{D}_3\boldsymbol{\theta}(t) \tag{8.188c}$$

Assume that the feedback network N_f is initially relaxed. Taking the Laplace transform of (8.187) and (8.188) yields

$$s\tilde{\mathbf{x}}(s) = \mathbf{A}_s\tilde{\mathbf{x}}(s) + \mathbf{B}_s\tilde{\mathbf{u}}(s) \tag{8.189a}$$

$$\tilde{\mathbf{y}}(s) = \mathbf{C}_s\tilde{\mathbf{x}}(s) + \mathbf{D}_s\tilde{\mathbf{u}}(s) \tag{8.189b}$$

$$s\tilde{\mathbf{x}}(s) = \mathbf{A}_o\tilde{\mathbf{x}}(s) + \mathbf{B}_o\tilde{\mathbf{u}}(s) + \mathbf{B}_1\tilde{\boldsymbol{\theta}}(s) \tag{8.190a}$$

---

*The matrices $\mathbf{B}_1$, $\mathbf{C}_1$, and $\mathbf{D}_1$, are different from those defined in (7.148).

$$\tilde{\mathbf{y}}(s) = \mathbf{C}_o\tilde{\mathbf{x}}(s) + \mathbf{D}_o\tilde{\mathbf{u}}(s) + \mathbf{D}_1\tilde{\boldsymbol{\theta}}(s) \tag{8.190b}$$

$$\tilde{\boldsymbol{\varphi}}(s) = \mathbf{C}_1\tilde{\mathbf{x}}(s) + \mathbf{D}_2\tilde{\mathbf{u}}(s) + \mathbf{D}_3\tilde{\boldsymbol{\theta}}(s) \tag{8.190c}$$

Substituting (8.185) in (8.190c) gives

$$\tilde{\boldsymbol{\varphi}}(s) = (\mathbf{1}_p - \mathbf{D}_3\mathbf{X})^{-1}[\mathbf{C}_1\tilde{\mathbf{x}}(s) + \mathbf{D}_2\tilde{\mathbf{u}}(s)] \tag{8.191}$$

provided that the square matrix $(\mathbf{1}_p - \mathbf{D}_3\mathbf{X})$ is nonsingular. Using this in (8.185) we obtain

$$\tilde{\boldsymbol{\theta}}(s) = \mathbf{X}(\mathbf{1}_p - \mathbf{D}_3\mathbf{X})^{-1}[\mathbf{C}_1\tilde{\mathbf{x}}(s) + \mathbf{D}_2\tilde{\mathbf{u}}(s)] \tag{8.192}$$

Substituting (8.192) in (8.190a) and (8.190b) to eliminate $\tilde{\boldsymbol{\theta}}$ results in the equations

$$s\tilde{\mathbf{x}}(s) = [\mathbf{A}_o + \mathbf{B}_1\mathbf{X}(\mathbf{1}_p - \mathbf{D}_3\mathbf{X})^{-1}\mathbf{C}_1]\tilde{\mathbf{x}}(s)$$
$$+ [\mathbf{B}_o + \mathbf{B}_1\mathbf{X}(\mathbf{1}_p - \mathbf{D}_3\mathbf{X})^{-1}\mathbf{D}_2]\tilde{\mathbf{u}}(s) \tag{8.193a}$$

$$\tilde{\mathbf{y}}(s) = [\mathbf{C}_o + \mathbf{D}_1\mathbf{X}(\mathbf{1}_p - \mathbf{D}_3\mathbf{X})^{-1}\mathbf{C}_1]\tilde{\mathbf{x}}(s)$$
$$+ [\mathbf{D}_o + \mathbf{D}_1\mathbf{X}(\mathbf{1}_p - \mathbf{D}_3\mathbf{X})^{-1}\mathbf{D}_2]\tilde{\mathbf{u}}(s) \tag{8.193b}$$

Comparing this with (8.189), we can make the following identifications:

$$\mathbf{A}_s = \mathbf{A}_o + \mathbf{B}_1\mathbf{X}(\mathbf{1}_p - \mathbf{D}_3\mathbf{X})^{-1}\mathbf{C}_1 \tag{8.194a}$$

$$\mathbf{B}_s = \mathbf{B}_o + \mathbf{B}_1\mathbf{X}(\mathbf{1}_p - \mathbf{D}_3\mathbf{X})^{-1}\mathbf{D}_2 \tag{8.194b}$$

$$\mathbf{C}_s = \mathbf{C}_o + \mathbf{D}_1\mathbf{X}(\mathbf{1}_p - \mathbf{D}_3\mathbf{X})^{-1}\mathbf{C}_1 \tag{8.194c}$$

$$\mathbf{D}_s = \mathbf{D}_o + \mathbf{D}_1\mathbf{X}(\mathbf{1}_p - \mathbf{D}_3\mathbf{X})^{-1}\mathbf{D}_2 \tag{8.194d}$$

These equations express the coefficient matrices of the state equations of the feedback network N_f in terms of the matrix $\mathbf{X}$, considered to be independent of the complex frequency s, and the coefficient matrices of the state equations of the subnetwork N.

Recall that for the network equations (8.186), the return difference matrix $\mathbf{F}(\mathbf{X})$ with respect to $\mathbf{X}$ is given from (7.34) by

$$\mathbf{F}(\mathbf{X}) = \mathbf{1}_p - \mathbf{A}\mathbf{X} \tag{8.195}$$

To express $\mathbf{F}(\mathbf{X})$ in terms of the coefficient matrices of (8.190), we solve for $\tilde{\mathbf{x}}(s)$ in (8.190a) and obtain

$$\tilde{\mathbf{x}}(s) = (s\mathbf{1}_k - \mathbf{A}_o)^{-1}[\mathbf{B}_1\tilde{\boldsymbol{\theta}}(s) + \mathbf{B}_o\tilde{\mathbf{u}}(s)] \tag{8.196}$$

Substituting this in (8.190c) yields

$$\tilde{\varphi}(s) = [\mathbf{D}_3 + \mathbf{C}_1(s\mathbf{1}_k - \mathbf{A}_o)^{-1}\mathbf{B}_1]\tilde{\theta}(s)$$
$$+ [\mathbf{D}_2 + \mathbf{C}_1(s\mathbf{1}_k - \mathbf{A}_o)^{-1}\mathbf{B}_o]\tilde{u}(s) \qquad (8.197)$$

Comparing this with (8.186a) identifies

$$\mathbf{A} = \mathbf{D}_3 + \mathbf{C}_1(s\mathbf{1}_k - \mathbf{A}_o)^{-1}\mathbf{B}_1 \qquad (8.198a)$$

$$\mathbf{B} = \mathbf{D}_2 + \mathbf{C}_1(s\mathbf{1}_k - \mathbf{A}_o)^{-1}\mathbf{B}_o \qquad (8.198b)$$

Combining (8.195) and (8.198a) gives the return difference matrix with respect to $\mathbf{X}$ as

$$\mathbf{F}(\mathbf{X}) = \mathbf{1}_p - [\mathbf{D}_3 + \mathbf{C}_1(s\mathbf{1}_k - \mathbf{A}_o)^{-1}\mathbf{B}_1]\mathbf{X} \qquad (8.199)$$

We now proceed to the calculation of the determinant of $\mathbf{F}(\mathbf{X})$, as follows:

$$\det \mathbf{F}(\mathbf{X}) = \det \{[\mathbf{1}_p - \mathbf{C}_1(s\mathbf{1}_k - \mathbf{A}_o)^{-1}\mathbf{B}_1\mathbf{X}(\mathbf{1}_p - \mathbf{D}_3\mathbf{X})^{-1}](\mathbf{1}_p - \mathbf{D}_3\mathbf{X})\}$$
$$= [\det (\mathbf{1}_p - \mathbf{D}_3\mathbf{X})]\{\det [\mathbf{1}_p - \mathbf{C}_1(s\mathbf{1}_k - \mathbf{A}_o)^{-1}\mathbf{B}_1\mathbf{X}(\mathbf{1}_p - \mathbf{D}_3\mathbf{X})^{-1}]\}$$
$$= [\det (\mathbf{1}_p - \mathbf{D}_3\mathbf{X})]\{\det [\mathbf{1}_k - (s\mathbf{1}_k - \mathbf{A}_o)^{-1}\mathbf{B}_1\mathbf{X}(\mathbf{1}_p - \mathbf{D}_3\mathbf{X})^{-1}\mathbf{C}_1]\}$$
$$= [\det (\mathbf{1}_p - \mathbf{D}_3\mathbf{X})][\det (s\mathbf{1}_k - \mathbf{A}_o)^{-1}]$$
$$\times \det [s\mathbf{1}_k - \mathbf{A}_o - \mathbf{B}_1\mathbf{X}(\mathbf{1}_p - \mathbf{D}_3\mathbf{X})^{-1}\mathbf{C}_1]$$
$$= [\det (\mathbf{1}_p - \mathbf{D}_3\mathbf{X})][\det (s\mathbf{1}_k - \mathbf{A}_o)^{-1}][\det (s\mathbf{1}_k - \mathbf{A}_s)]$$
$$= [\det (\mathbf{1}_p - \mathbf{D}_3\mathbf{X})]\frac{\det (s\mathbf{1}_k - \mathbf{A}_s)}{\det s\mathbf{1}_k - \mathbf{A}_o} \qquad (8.200)$$

The third line follows from (7.83), and the sixth line from (8.194a). Since from (8.194a)

$$\mathbf{A}_o = \mathbf{A}_s|_{\mathbf{x}=\mathbf{0}} \qquad (8.201)$$

we have

$$\det \mathbf{F}(\mathbf{X}) = [\det (\mathbf{1}_p - \mathbf{D}_3\mathbf{X})]\frac{\det (s\mathbf{1}_k - \mathbf{A}_s)}{\det (s\mathbf{1}_k - \mathbf{A}_s)|_{\mathbf{x}=\mathbf{0}}} \qquad (8.202)$$

Thus, the eigenvalues of the $\mathbf{A}_s$ matrix under the nominal condition and under the condition $\mathbf{X} = \mathbf{0}$ are the zeros and poles of the determinant of the return difference matrix with respect to $\mathbf{X}$.

To extend this to a general reference $\mathbf{K}$, we appeal to (7.162)

$$\mathbf{F_K}(\mathbf{X}) = \mathbf{F}^{-1}(\mathbf{K})\mathbf{F}(\mathbf{X}) \tag{8.203}$$

where from (8.199)

$$\mathbf{F}(\mathbf{K}) = \mathbf{1}_p - [\mathbf{D}_3 + \mathbf{C}_1(s\mathbf{1}_k - \mathbf{A}_o)^{-1}\mathbf{B}_1]\mathbf{K} \tag{8.204}$$

Equation (8.203) expresses the general return difference matrix with respect to $\mathbf{X}$ for a general reference $\mathbf{K}$ in terms of the product of the inverse of the return difference matrix with respect to $\mathbf{K}$ and the return difference matrix with respect to $\mathbf{X}$, both being defined for the zero reference.

To find an expression for the determinant of the general return difference matrix, we replace $\mathbf{X}$ by $\mathbf{K}$ in (8.202) and obtain

$$\det \mathbf{F}(\mathbf{K}) = [\det (\mathbf{1}_p - \mathbf{D}_3\mathbf{K})]\frac{\det (s\mathbf{1}_k - \mathbf{A}_s)|_{\mathbf{X}=\mathbf{K}}}{\det (s\mathbf{1}_k - \mathbf{A}_s)|_{\mathbf{X}=\mathbf{0}}} \tag{8.205}$$

Using this and (8.202), the determinant of the general return difference matrix is found from (8.203) to be

$$\det \mathbf{F_K}(\mathbf{X}) = \frac{\det \mathbf{F}(\mathbf{X})}{\det \mathbf{F}(\mathbf{K})} = \frac{[\det (\mathbf{1}_p - \mathbf{D}_3\mathbf{X})][\det (s\mathbf{1}_k - \mathbf{A}_s)]}{[\det (\mathbf{1}_p - \mathbf{D}_3\mathbf{K})][\det (s\mathbf{1}_k - \mathbf{A}_s)]|_{\mathbf{X}=\mathbf{K}}} \tag{8.206}$$

Thus, the eigenvalues of the $\mathbf{A}_s$ matrix under the nominal condition and under the condition $\mathbf{X} = \mathbf{K}$ are the zeros and poles of the determinant of the general return difference matrix with respect to $\mathbf{X}$ for the general reference $\mathbf{K}$.

We illustrate the above results by the following example.

Example 8.10 Consider a series-parallel or voltage-series feedback amplifier of Fig. 8.48. Assume that the two transistors are identical and each can be represented by the simplified equivalent network of Fig. 1.14 with $R_1 = 1\,\text{k}\Omega$, $R_2 = \infty$, $C = 100\,\text{pF}$ and $g_m = 0.05\,\text{mhos}$. A small-signal equivalent network of the amplifier is presented in Fig. 8.49, in which $I_5 = 0.05V_3$, $I_6 = 0.05V_4$ and a second current source I_{g2} is connected at the output port, so that the input and output impedances of the amplifier can be computed later from its transfer function matrix.

Let the controlling parameters of the two controlled current sources be the elements of interest. Then we have

$$\tilde{\boldsymbol{\theta}}(s) = \begin{bmatrix} I_5 \\ I_6 \end{bmatrix} = \begin{bmatrix} 0.05 & 0 \\ 0 & 0.05 \end{bmatrix}\begin{bmatrix} V_3 \\ V_4 \end{bmatrix} = \mathbf{X}\tilde{\boldsymbol{\varphi}}(s) \tag{8.207}$$

Figure 8.48 A voltage-series feedback amplifier.

Figure 8.49 A small-signal equivalent network of the voltage-series feedback amplifier of Fig. 8.48.

Assume that the input port voltage V_1 and the output port voltage V_2 are the output variables. Then the eight-port network N defined by the variables I_5, I_6, V_3, V_4, I_{g1}, I_{g2}, V_1 and V_2 can be characterized by the matrix equations

$$\tilde{\varphi}(s) = \begin{bmatrix} V_3 \\ V_4 \end{bmatrix} = \mathbf{A}\tilde{\theta}(s) + \mathbf{B}\tilde{u}(s) = -\frac{1}{3} \begin{bmatrix} \dfrac{520}{p+1.827} & -\dfrac{400}{p+1.827} \\ \dfrac{3000}{p+1.100} & 0 \end{bmatrix} \begin{bmatrix} I_5 \\ I_6 \end{bmatrix}$$

$$+ \frac{1}{3(p+1.827)} \begin{bmatrix} 2480 & -400 \\ 0 & 0 \end{bmatrix} \begin{bmatrix} I_{g1} \\ I_{g2} \end{bmatrix} \qquad (8.208a)$$

$$\tilde{\mathbf{y}}(s) = \begin{bmatrix} V_1 \\ V_2 \end{bmatrix} = \frac{1}{3(p + 1.827)} \begin{bmatrix} 520(p + 1) & -400(p + 1) \\ 400(p + 2) & -1000(p + 1.88) \end{bmatrix} \begin{bmatrix} I_5 \\ I_6 \end{bmatrix}$$

$$+ \frac{1}{3(p + 1.827)} \begin{bmatrix} 520(p + 5.769) & 400(p + 1) \\ 400(p + 1) & 1000(p + 1.88) \end{bmatrix} \begin{bmatrix} I_{g1} \\ I_{g2} \end{bmatrix}$$

$$= \mathbf{C}\tilde{\theta}(s) + \mathbf{D}\tilde{\mathbf{u}}(s) \tag{8.208b}$$

where $p = s/10^7$. Substituting $\mathbf{A}$ from (8.208a) and $\mathbf{X}$ from (8.207) in (8.195) gives the return difference matrix with respect to $\mathbf{X}$ as

$$\mathbf{F}(\mathbf{X}) = \mathbf{1}_2 - \mathbf{A}\mathbf{X} = \begin{bmatrix} \dfrac{p + 10.494}{p + 1.827} & -\dfrac{6.667}{p + 1.827} \\ \dfrac{50}{p + 1.1} & 1 \end{bmatrix} \tag{8.209}$$

the determinant of which is given by

$$\det \mathbf{F}(\mathbf{X}) = \frac{s^2 + 11.594 \times 10^7 s + 344.893 \times 10^{14}}{(s + 1.827 \times 10^7)(s + 1.100 \times 10^7)}$$

$$= \frac{\left[s + (5.797 + j17.643)10^7\right]\left[s + (5.797 - j17.643)10^7\right]}{(s + 1.827 \times 10^7)(s + 1.100 \times 10^7)} \tag{8.210}$$

Applying the procedure outlined in Sec. 8.4, the state equations of the feedback amplifier network N_f of Fig. 8.49 are found to be

$$\dot{\mathbf{x}}(t) = \begin{bmatrix} \dot{v}_3(t) \\ \dot{v}_4(t) \end{bmatrix} = 10^7 \begin{bmatrix} -10.494 & 20/3 \\ -50 & -1.1 \end{bmatrix} \begin{bmatrix} v_3(t) \\ v_4(t) \end{bmatrix}$$

$$+ \frac{10^8}{3} \begin{bmatrix} 248 & -40 \\ 0 & 0 \end{bmatrix} \begin{bmatrix} i_{g1}(t) \\ i_{g2}(t) \end{bmatrix}$$

$$= \mathbf{A}_s \mathbf{x}(t) + \mathbf{B}_s \mathbf{u}(t) \tag{8.211a}$$

$$\mathbf{y}(t) = \begin{bmatrix} v_1(t) \\ v_2(t) \end{bmatrix} = \frac{1}{3} \begin{bmatrix} 28.48 & -20 \\ 19.6 & -50 \end{bmatrix} \begin{bmatrix} v_3(t) \\ v_4(t) \end{bmatrix}$$

$$+ \frac{100}{3} \begin{bmatrix} 5.2 & 4 \\ 4 & 10 \end{bmatrix} \begin{bmatrix} i_{g1}(t) \\ i_{g2}(t) \end{bmatrix}$$

$$= \mathbf{C}_s \mathbf{x}(t) + \mathbf{D}_s \mathbf{u}(t) \tag{8.211b}$$

where the lower-case variables $v_1, v_2, v_3, v_4, i_{g1}$ and i_{g2} denote the time-domain functions of their Laplace transforms $V_1, V_2, V_3, V_4, I_{g1}$ and I_{g2}, respectively.

If $\boldsymbol{\theta}$ is also considered as an input vector, the state equations for the subnetwork N become

$$
\dot{\mathbf{x}}(t) = \begin{bmatrix} \dot{v}_3 \\ \dot{v}_4 \end{bmatrix} = 10^7 \begin{bmatrix} -1.827 & 0 \\ 0 & -1.1 \end{bmatrix} \begin{bmatrix} v_3 \\ v_4 \end{bmatrix} + \frac{10^9}{3} \begin{bmatrix} 24.8 & -4 \\ 0 & 0 \end{bmatrix} \begin{bmatrix} i_{g1} \\ i_{g2} \end{bmatrix}
$$

$$
+ \frac{10^9}{3} \begin{bmatrix} -5.2 & 4 \\ -30 & 0 \end{bmatrix} \begin{bmatrix} i_5 \\ i_6 \end{bmatrix} = \mathbf{A}_o \mathbf{x}(t) + \mathbf{B}_o \mathbf{u}(t) + \mathbf{B}_1 \boldsymbol{\theta}(t) \qquad (8.212a)
$$

$$
\mathbf{y}(t) = \begin{bmatrix} v_1 \\ v_2 \end{bmatrix} = \frac{1}{3} \begin{bmatrix} 2.48 & 0 \\ -0.4 & 0 \end{bmatrix} \begin{bmatrix} v_3 \\ v_4 \end{bmatrix} + \frac{100}{3} \begin{bmatrix} 5.2 & 4 \\ 4 & 10 \end{bmatrix} \begin{bmatrix} i_{g1} \\ i_{g2} \end{bmatrix}
$$

$$
+ \frac{100}{3} \begin{bmatrix} 5.2 & -4 \\ 4 & -10 \end{bmatrix} \begin{bmatrix} i_5 \\ i_6 \end{bmatrix} = \mathbf{C}_o \mathbf{x}(t) + \mathbf{D}_o \mathbf{u}(t) + \mathbf{D}_1 \boldsymbol{\theta}(t) \qquad (8.212b)
$$

$$
\boldsymbol{\varphi} = \begin{bmatrix} v_3 \\ v_4 \end{bmatrix} = \begin{bmatrix} 1 & 0 \\ 0 & 1 \end{bmatrix} \begin{bmatrix} v_3 \\ v_4 \end{bmatrix} + \begin{bmatrix} 0 & 0 \\ 0 & 0 \end{bmatrix} \begin{bmatrix} i_{g1} \\ i_{g2} \end{bmatrix}
$$

$$
+ \begin{bmatrix} 0 & 0 \\ 0 & 0 \end{bmatrix} \begin{bmatrix} i_5 \\ i_6 \end{bmatrix} = \mathbf{C}_1 \mathbf{x}(t) + \mathbf{D}_2 \mathbf{u}(t) + \mathbf{D}_3 \boldsymbol{\theta}(t) \qquad (8.212c)
$$

where i_5 and i_6 are the time-domain functions of the Laplace transform currents I_5 and I_6, respectively. Substituting the appropriate coefficient matrices of (8.212) in (8.199) yields the return difference matrix with respect to $\mathbf{X}$:

$$
\mathbf{F}(\mathbf{X}) = \mathbf{1}_2 - [\mathbf{D}_3 + \mathbf{C}_1(s\mathbf{1}_2 - \mathbf{A}_o)^{-1}\mathbf{B}_1]\mathbf{X}
$$

$$
= \begin{bmatrix} \dfrac{s + 10.494 \times 10^7}{s + 1.827 \times 10^7} & -\dfrac{6.667 \times 10^7}{s + 1.827 \times 10^{10}} \\[3mm] \dfrac{50 \times 10^7}{s + 1.1 \times 10^7} & 1 \end{bmatrix} \qquad (8.213)
$$

confirming (8.209), the determinant of which can be written as

$$
\det \mathbf{F}(\mathbf{X}) = \frac{\det (s\mathbf{1}_2 - \mathbf{A}_s)}{\det (s\mathbf{1}_2 - \mathbf{A}_o)}
$$

$$
= \frac{[s + (5.797 + j17.643)10^7][s + (5.797 - j17.643)10^7]}{(s + 1.827 \times 10^7)(s + 1.1 \times 10^7)} \qquad (8.214)
$$

showing that the eigenvalues of $\mathbf{A}_s$ are zeros of $\det \mathbf{F}(\mathbf{X})$, and those of $\mathbf{A}_o$ are poles of $\det \mathbf{F}(\mathbf{X})$. Q.E.D.

We now proceed to express the null return difference matrix in terms of the coefficient matrices of the state equations. Refer to (7.71). The square matrix

$$\hat{\mathbf{F}}(\mathbf{X}) = \mathbf{1}_p - \hat{\mathbf{A}}\mathbf{X} \tag{8.215}$$

where

$$\hat{\mathbf{A}} = \mathbf{A} - \mathbf{B}\mathbf{D}^{-1}\mathbf{C} \tag{8.216}$$

is called the null return difference matrix with respect to $\mathbf{X}$, provided that $\mathbf{D}$ is square and nonsingular. For our purposes, we express the feedback matrices $\mathbf{A}$, $\mathbf{B}$, $\mathbf{C}$ and $\mathbf{D}$ in (8.186) in terms of the coefficient matrices in the equations (8.188). We have already done so for the matrices $\mathbf{A}$ and $\mathbf{B}$ as given in (8.198). For $\mathbf{C}$ and $\mathbf{D}$, we substitute (8.196) in (8.190b) and obtain

$$\tilde{\mathbf{y}}(s) = [\mathbf{D}_o + \mathbf{C}_o(s\mathbf{1}_k - \mathbf{A}_o)^{-1}\mathbf{B}_o]\tilde{\mathbf{u}}(s)$$
$$+[\mathbf{D}_1 + \mathbf{C}_o(s\mathbf{1}_k - \mathbf{A}_o)^{-1}\mathbf{B}_1]\tilde{\boldsymbol{\theta}}(s) \tag{8.217}$$

Comparing this with (8.186b) identifies

$$\mathbf{C} = \mathbf{D}_1 + \mathbf{C}_o(s\mathbf{1}_k - \mathbf{A}_o)^{-1}\mathbf{B}_1 \tag{8.218a}$$

$$\mathbf{D} = \mathbf{D}_o + \mathbf{C}_o(s\mathbf{1}_k - \mathbf{A}_o)^{-1}\mathbf{B}_o \tag{8.218b}$$

Substituting $\mathbf{A}$ and $\mathbf{B}$ from (8.198) and $\mathbf{C}$ and $\mathbf{D}$ from (8.218) in (8.216) gives the desired matrix

$$\hat{\mathbf{A}} = \mathbf{D}_3 + \mathbf{C}_1(s\mathbf{1}_k - \mathbf{A}_o)^{-1}\mathbf{B}_1 - [\mathbf{D}_2 + \mathbf{C}_1(s\mathbf{1}_k - \mathbf{A}_o)^{-1}\mathbf{B}_o]$$
$$\times [\mathbf{D}_o + \mathbf{C}_o(s\mathbf{1}_k - \mathbf{A}_o)^{-1}\mathbf{B}_o]^{-1}[\mathbf{D}_1 + \mathbf{C}_o(s\mathbf{1}_k - \mathbf{A}_o)^{-1}\mathbf{B}_1] \tag{8.219}$$

The null return difference matrix with respect to $\mathbf{X}$ becomes

$$\hat{\mathbf{F}}(\mathbf{X}) = \mathbf{1}_p - \hat{\mathbf{A}}\mathbf{X}$$
$$= \mathbf{1}_p - \{\mathbf{D}_3 + \mathbf{C}_1(s\mathbf{1}_k - \mathbf{A}_o)^{-1}\mathbf{B}_1 - [\mathbf{D}_2 + \mathbf{C}_1(s\mathbf{1}_k - \mathbf{A}_o)^{-1}\mathbf{B}_o]$$
$$\times [\mathbf{D}_o + \mathbf{C}_o(s\mathbf{1}_k - \mathbf{A}_o)^{-1}\mathbf{B}_o]^{-1}[\mathbf{D}_1 + \mathbf{C}_o(s\mathbf{1}_k - \mathbf{A}_o)^{-1}\mathbf{B}_1]\}\mathbf{X}$$
$$\tag{8.220}$$

and can be simplified and expressed in a form similar to (8.199):

$$\hat{\mathbf{F}}(\mathbf{X}) = \mathbf{1}_p - [\hat{\mathbf{D}}_3 + \hat{\mathbf{C}}_1(s\mathbf{1}_k - \hat{\mathbf{A}}_o)^{-1}\hat{\mathbf{B}}_1]\mathbf{X} \tag{8.221}$$

provided that $\mathbf{D}_o$ is square and nonsingular, where

$$\hat{\mathbf{A}}_o = \mathbf{A}_o - \mathbf{B}_o \mathbf{D}_o^{-1} \mathbf{C}_o \qquad (8.222a)$$

$$\hat{\mathbf{B}}_1 = \mathbf{B}_1 - \mathbf{B}_o \mathbf{D}_o^{-1} \mathbf{D}_1 \qquad (8.222b)$$

$$\hat{\mathbf{C}}_1 = \mathbf{C}_1 - \mathbf{D}_2 \mathbf{D}_o^{-1} \mathbf{C}_o \qquad (8.222c)$$

$$\hat{\mathbf{D}}_3 = \mathbf{D}_3 - \mathbf{D}_2 \mathbf{D}_o^{-1} \mathbf{D}_1 \qquad (8.222d)$$

Following Sec. 7.3.1, let $\mathbf{W}(\mathbf{X})$ be the transfer-function matrix of the feedback network N_f relating the input vector $\tilde{\mathbf{u}}$ to the output vector $\tilde{\mathbf{y}}$ by the equation

$$\tilde{\mathbf{y}}(s) = \mathbf{W}(\mathbf{X})\tilde{\mathbf{u}}(s) \qquad (8.223)$$

By eliminating the vector $\tilde{\mathbf{x}}$ in (8.189), we obtain

$$\tilde{\mathbf{y}}(s) = [\mathbf{D}_s + \mathbf{C}_s(s\mathbf{1}_k - \mathbf{A}_s)^{-1}\mathbf{B}_s]\tilde{\mathbf{u}}(s) \qquad (8.224)$$

giving

$$\mathbf{W}(\mathbf{X}) = \mathbf{D}_s + \mathbf{C}_s(s\mathbf{1}_k - \mathbf{A}_s)^{-1}\mathbf{B}_s \qquad (8.225)$$

If $\mathbf{W}(\mathbf{X})$ is square, its determinant is found to be

$$\begin{aligned}
\det \mathbf{W}(\mathbf{X}) &= \det [\mathbf{D}_s + \mathbf{C}_s(s\mathbf{1}_k - \mathbf{A}_s)^{-1}\mathbf{B}_s] \\
&= (\det \mathbf{D}_s)\{\det [\mathbf{1}_n + \mathbf{C}_s(s\mathbf{1}_k - \mathbf{A}_s)^{-1}\mathbf{B}_s \mathbf{D}_s^{-1}]\} \\
&= (\det \mathbf{D}_s)\{\det [\mathbf{1}_k + (s\mathbf{1}_k - \mathbf{A}_s)^{-1}\mathbf{B}_s \mathbf{D}_s^{-1}\mathbf{C}_s]\} \\
&= (\det \mathbf{D}_s)[\det (s\mathbf{1}_k - \mathbf{A}_s)^{-1}]\{\det[s\mathbf{1}_k - \mathbf{A}_s + \mathbf{B}_s \mathbf{D}_s^{-1}\mathbf{C}_s]\} \\
&= (\det \mathbf{D}_s) [\det (s\mathbf{1}_k - \mathbf{A}_s)^{-1}][\det(s\mathbf{1}_k - \hat{\mathbf{A}}_s)] \\
&= (\det \mathbf{D}_s)\frac{\det(s\mathbf{1}_k - \hat{\mathbf{A}}_s)}{\det(s\mathbf{1}_k - \mathbf{A}_s)} \qquad (8.226)
\end{aligned}$$

provided that $\mathbf{D}_5$ is square and nonsinguiar, where

$$\hat{\mathbf{A}}_s = \mathbf{A}_s - \mathbf{B}_s \mathbf{D}_s^{-1}\mathbf{C}_s \qquad (8.227)$$

and the third line follows from (7.83). Thus, we obtain

$$\det \mathbf{W}(\mathbf{X}) = (\det \mathbf{D}_s)\frac{\det(s\mathbf{1}_k - \hat{\mathbf{A}}_s)}{\det(s\mathbf{1}_k - \mathbf{A}_s)} \qquad (8.228)$$

It states that if $\mathbf{D}_s$ is nonsingular, the eigenvalues of the matrix $\hat{\mathbf{A}}_s$ are the zeros of the determinant of the transfer-function matrix $\mathbf{W}(\mathbf{X})$, and the eigenvalues of $\mathbf{A}_s$ are the poles of det $\mathbf{W}(\mathbf{X})$.

To express the determinant of the null return difference matrix in terms of the eigenvalues of certain matrices, we appeal to (7.88) in conjunction with (8.202) and (8.228) and obtain

$$\det \hat{\mathbf{F}}(\mathbf{X}) = [\det(\mathbf{1}_p - \mathbf{D}_3\mathbf{X})] \frac{[\det(s\mathbf{1}_k - \hat{\mathbf{A}}_s)](\det \mathbf{D}_s)}{[\det(s\mathbf{1}_k - \hat{\mathbf{A}}_s)]|_{\mathbf{x}=\mathbf{0}}(\det \mathbf{D}_s)|_{\mathbf{x}=\mathbf{0}}} \qquad (8.229)$$

provided that $\mathbf{D}_s$ is nonsingular when $\mathbf{X}$ assumes its nominal value and when $\mathbf{X}$ assumes the value $\mathbf{X} = \mathbf{0}$. Thus, the eigenvalues of $\hat{\mathbf{A}}_s$ when $\mathbf{X}$ assumes its nominal value and when it assumes the value $\mathbf{X} = \mathbf{0}$ are the zeros and poles of the determinant of the null return difference matrix with respect to $\mathbf{X}$, respectively.

Finally, we consider the general null return difference matrix $\hat{\mathbf{F}}_{\mathbf{K}}(\mathbf{X})$ with respect to $\mathbf{X}$ for a general reference $\mathbf{K}$. From (7.164) we have

$$\hat{\mathbf{F}}_{\mathbf{K}}(\mathbf{X}) = \hat{\mathbf{F}}^{-1}(\mathbf{K})\hat{\mathbf{F}}(\mathbf{X}) \qquad (8.230)$$

where from (8.220)

$$\hat{\mathbf{F}}(\mathbf{K}) = \mathbf{1}_p - \hat{\mathbf{A}}\mathbf{K} \qquad (8.231)$$

obtaining from (8.229)

$$\det \hat{\mathbf{F}}(\mathbf{K}) = [\det(\mathbf{1}_p - \mathbf{D}_3\mathbf{K})] \frac{[\det(s\mathbf{1}_k - \hat{\mathbf{A}}_s)]|_{\mathbf{x}=\mathbf{K}} (\det \mathbf{D}_s)|_{\mathbf{x}=\mathbf{K}}}{[\det(s\mathbf{1}_k - \hat{\mathbf{A}}_s)]|_{\mathbf{x}=\mathbf{0}} (\det \mathbf{D}_s)|_{\mathbf{x}=\mathbf{0}}} \qquad (8.232)$$

Substituting this and (8.229) in

$$\det \hat{\mathbf{F}}_{\mathbf{K}}(\mathbf{X}) = \frac{\det \hat{\mathbf{F}}(\mathbf{X})}{\det \hat{\mathbf{F}}(\mathbf{K})} \qquad (8.233)$$

gives

$$\det \hat{\mathbf{F}}_{\mathbf{K}}(\mathbf{X}) = \frac{[\det(s\mathbf{1}_k - \hat{\mathbf{A}}_s)][\det(\mathbf{1}_p - \mathbf{D}_3\mathbf{X})](\det \mathbf{D}_s)}{[\det(s\mathbf{1}_k - \hat{\mathbf{A}}_s)]|_{\mathbf{x}=\mathbf{K}}[\det(\mathbf{1}_p - \mathbf{D}_3\mathbf{K})](\det \mathbf{D}_s)|_{\mathbf{x}=\mathbf{K}}} \qquad (8.234)$$

It states that the eigenvalues of $\mathbf{A}_s$ when $\mathbf{X}$ assumes its nominal value and when $\mathbf{X}$ assumes the reference value $\mathbf{X} = \mathbf{K}$ are the zeros and poles of the determinant of the general null return difference matrix with respect to $\mathbf{X}$ for the general reference $\mathbf{K}$.

As in (7.163), the general null return difference matrix can also be expressed from (8.215) and (8.231) as

$$\hat{\mathbf{F}}_\mathbf{K}(\mathbf{X}) = (\mathbf{1}_p - \hat{\mathbf{A}}\mathbf{K})^{-1}(\mathbf{1}_p - \hat{\mathbf{A}}\mathbf{X}) \qquad (8.235)$$

We illustrate the above results by the following examples.

Example 8.11 Consider the same feedback amplifier network N_f of Fig. 8.49. To compute the null return difference matrix $\hat{\mathbf{F}}(\mathbf{X})$ with respect to $\mathbf{X}$ in (8.207), we first compute the matrix $\hat{\mathbf{A}}$ of (8.216) using the coefficient matrices of (8.208), and obtain

$$\hat{\mathbf{A}} = \mathbf{A} - \mathbf{B}\mathbf{D}^{-1}\mathbf{C} = -10^3 \begin{bmatrix} \dfrac{1}{p+8.332} & 0 \\ \dfrac{1}{p+1.1} & 0 \end{bmatrix} \qquad (8.236)$$

giving from (8.215)

$$\hat{\mathbf{F}}(\mathbf{X}) = \mathbf{1}_2 - \hat{\mathbf{A}}\mathbf{X} = \begin{bmatrix} \dfrac{p+58.332}{p+8.332} & 0 \\ \dfrac{50}{p+1.1} & 1 \end{bmatrix} \qquad (8.237)$$

where, as before, $p = s/10^7$. The determinant of the null return difference matrix becomes

$$\det \hat{\mathbf{F}}(\mathbf{X}) = \frac{s + 58.332 \times 10^7}{s + 8.332 \times 10^7} \qquad (8.238)$$

The transfer-function matrix $\mathbf{W}(\mathbf{X})$ relating the input vector $\tilde{\mathbf{u}}(s)$ and the output vector $\tilde{\mathbf{y}}(s)$ is determined from (7.16a) as

$$\mathbf{W}(\mathbf{X}) = \mathbf{D} + \mathbf{C}\mathbf{X}(\mathbf{1}_2 - \mathbf{A}\mathbf{X})^{-1}\mathbf{B}$$

$$= \frac{1}{q(p)} \begin{bmatrix} w_{11} & w_{12} \\ w_{21} & w_{22} \end{bmatrix} \qquad (8.239)$$

where

$$q(p) = 3(p + 5.797 + j17.643)(p + 5.797 - j17.643) \qquad (8.240a)$$

$$w_{11} = 520(p + 28.435 + j34.291)(p + 28.435 - j34.291) \qquad (8.240b)$$

$$w_{12} = 400(p+1)(p+1.1) \tag{8.240c}$$

$$w_{21} = 400(p+26.05+j69.839)(p+26.05-j69.839) \tag{8.240d}$$

$$w_{22} = 1000(p+1.1)(p+7.895) \tag{8.240e}$$

Its determinant is found to be

$$\det \mathbf{W}(\mathbf{X}) = \frac{12 \times 10^4 (p+1.1)(p+58.332)}{q(p)} \tag{8.241}$$

which can also be computed from (7.88) in conjunction with (8.214), (8.238) and (7.17), where

$$\det \mathbf{W}(\mathbf{0}) = \det \mathbf{D} = \frac{4 \times 10^4 (p+8.332)}{p+1.827} \tag{8.242}$$

We remark that w_{11} is the input impedance of the amplifier, w_{22} is the output impedance, and w_{12} and w_{21} are the transfer impedances. For the amplifier voltage-ratio function, we have

$$
\begin{aligned}
\frac{V_2}{V_1}\bigg|_{I_{g2}=0} &= \frac{w_{21}}{w_{11}} \\
&= \frac{10(p+26.05+j69.839)(p+26.05-j69.839)}{13(p+28.435+j34.291)(p+28.435-j34.291)}
\end{aligned} \tag{8.243}
$$

To compute the null return difference matrix using (8.221), we substitute the coefficient matrices of (8.212) in (8.222), and first obtain the desired matrices

$$\hat{\mathbf{A}}_o = \mathbf{A}_o - \mathbf{B}_o \mathbf{D}_o^{-1} \mathbf{C}_o = 10^7 \begin{bmatrix} -8.332 & 0 \\ 0 & -1.1 \end{bmatrix} \tag{8.244a}$$

$$\hat{\mathbf{B}}_1 = \mathbf{B}_1 - \mathbf{B}_o \mathbf{D}_o^{-1} \mathbf{D}_1 = -10^{10} \begin{bmatrix} 1 & 0 \\ 1 & 0 \end{bmatrix} \tag{8.244b}$$

$$\hat{\mathbf{C}}_1 = \mathbf{C}_1 - \mathbf{D}_2 \mathbf{D}_o^{-1} \mathbf{C}_o = \mathbf{C}_1 = \mathbf{1}_2 = \begin{bmatrix} 1 & 0 \\ 0 & 1 \end{bmatrix} \tag{8.244c}$$

$$\hat{\mathbf{D}}_3 = \mathbf{D}_3 - \mathbf{D}_2 \mathbf{D}_o^{-1} \mathbf{D}_1 = \mathbf{0} = \begin{bmatrix} 0 & 0 \\ 0 & 0 \end{bmatrix} \tag{8.244d}$$

Substituting these in (8.221) yields

$$\hat{\mathbf{F}}(\mathbf{X}) = \mathbf{1}_2 - [\hat{\mathbf{D}}_3 + \hat{\mathbf{C}}_1(s\mathbf{1}_2 - \hat{\mathbf{A}}_o)^{-1}\hat{\mathbf{B}}_1]\mathbf{X}$$

$$= \begin{bmatrix} \dfrac{p + 58.332}{p + 8.332} & 0 \\[2ex] \dfrac{50}{p + 1.1} & 1 \end{bmatrix} \tag{8.245}$$

To verify (8.228), we first compute from (8.227) the matrix

$$\hat{\mathbf{A}}_s = \mathbf{A}_s - \mathbf{B}_s\mathbf{D}_s^{-1}\mathbf{C}_s = 10^7 \begin{bmatrix} -58.332 & 0 \\ -50 & -1.1 \end{bmatrix} \tag{8.246}$$

giving

$$\det(s\mathbf{1}_2 - \hat{\mathbf{A}})_s = (s + 1.1 \times 10^7)(s + 58.332 \times 10^7) \tag{8.247}$$

Also, from (8.211a) we obtain

$$s\mathbf{1}_2 - \mathbf{A}_s = \begin{bmatrix} s + 10.494 \times 10^7 & -\dfrac{2 \times 10^8}{3} \\[2ex] 50 \times 10^7 & s + 1.1 \times 10^7 \end{bmatrix} \tag{8.248}$$

the determinant of which is given by

$$\det(s\mathbf{1}_2 - \mathbf{A}_s) = [s + (5.797 + j17.643)10^7][s + (5.797 - j17.643)10^7] \tag{8.249}$$

Substituting these in (8.228) yields

$$\det \mathbf{W}(\mathbf{X}) = (\det \mathbf{D}_s)\frac{\det(s\mathbf{1}_2 - \hat{\mathbf{A}}_s)}{\det(s\mathbf{1}_2 - \mathbf{A}_s)}$$

$$= \frac{4 \times 10^4(s + 1.1 \times 10^7)(s + 58.332 \times 10^7)}{[s + (5.797 + j17.643)10^7][s + (5.797 - j17.643)10^7]} \tag{8.250}$$

where $\det \mathbf{D}_s = 4 \times 10^4$, confirming (8.241). Thus, the eigenvalues of $\hat{\mathbf{A}}_s$ are zeros of $\det \mathbf{W}(\mathbf{X})$ and the eigenvalues of $\mathbf{A}_s$ are poles of $\det \mathbf{W}(\mathbf{X})$.

To apply (8.229) we first compute from (8.194) and (8.227) the matrices

$$s\mathbf{1}_2 - \hat{\mathbf{A}}_s|_{\mathbf{x}=0} = s\mathbf{1}_2 - (\mathbf{A}_o - \mathbf{B}_o\mathbf{D}_o^{-1}\mathbf{C}_o) = \begin{bmatrix} s + 8.332 \times 10^7 & 0 \\ 0 & s + 1.1 \times 10^7 \end{bmatrix}$$

$$(8.251)$$

$$\mathbf{D}_s|_{\mathbf{x}=0} = \mathbf{D}_o = \frac{1}{3}\begin{bmatrix} 520 & 400 \\ 400 & 1000 \end{bmatrix} \qquad (8.252)$$

the determinants of which are given by

$$\det (s\mathbf{1}_2 - \hat{\mathbf{A}}_s)|_{\mathbf{x}=0} = (s + 8.332 \times 10^7)(s + 1.1 \times 10^7) \qquad (8.253)$$

$$\det \mathbf{D}_s|_{\mathbf{x}=0} = \det \mathbf{D}_o = 4 \times 10^4 \qquad (8.254)$$

Substituting these in (8.229) yields

$$\det \hat{\mathbf{F}}(\mathbf{X}) = [\det (\mathbf{1}_2 - \mathbf{D}_3\mathbf{X})] \frac{[\det (s\mathbf{1}_2 - \hat{\mathbf{A}}_s)](\det \mathbf{D}_s)}{[\det (s\mathbf{1}_2 - \hat{\mathbf{A}}_s)]|_{\mathbf{x}=0}(\det \mathbf{D}_o)}$$

$$= \frac{s + 58.332 \times 10^7}{s + 8.332 \times 10^7} \qquad (8.255)$$

confirming (8.238). Q.E.D.

Example 8.12 We wish to compute the general return difference and the general null return difference matrices of the feedback network of Fig. 8.49 with respect to the controlling parameters of the two voltage-controlled current sources for the general reference

$$\mathbf{K} = \begin{bmatrix} 0.01 & 0 \\ 0 & 0.02 \end{bmatrix} \qquad (8.256)$$

Then from (8.208a) the return difference matrix with respect to **K** is obtained as

$$\mathbf{F}(\mathbf{K}) = \mathbf{1}_2 - \mathbf{AK} = \begin{bmatrix} \dfrac{3p + 10.681}{3(p + 1.827)} & -\dfrac{8}{3(p + 1.827)} \\ \dfrac{30}{3(p + 1.1)} & 1 \end{bmatrix} \qquad (8.257)$$

the inverse of which is found to be

$$\mathbf{F}^{-1}(\mathbf{K}) = \frac{1}{f(p)} \begin{bmatrix} 3(p+1.1)(p+1.827) & 8(p+1.1) \\ -30(p+1.827) & (3p+10.681)(p+1.1) \end{bmatrix}$$

$$(8.258)$$

where $p = s/10^7$ and

$$f(p) = 3p^2 + 13.981p + 91.7491$$

$$= 3(p+2.3302+j5.0153)(p+2.3302-j5.0153) \quad (8.259)$$

Substituting (8.209) and (8.258) in (8.203) yields the general return difference matrix with respect to $\mathbf{X}$ in (8.207) for the reference $\mathbf{K}$:

$$\mathbf{F_K}(\mathbf{X}) = \mathbf{F}^{-1}(\mathbf{K})\mathbf{F}(\mathbf{X})$$

$$= \frac{3}{f(p)} \begin{bmatrix} p^2+11.594p+144.8767 & -4p-4.4 \\ 40p+73.077 & p^2+4.6603p+70.5864 \end{bmatrix}$$

$$(8.260)$$

the determinant of which is found to be

$$\det \mathbf{F_K}(\mathbf{X}) = \frac{p^4+16.2543p^3+429.4946p^2+1961.8556p+10547.8635}{(p^2+4.6603p+30.5863)^2}$$

$$= \frac{p^2+11.594p+344.881}{p^2+4.6603p+30.5863}$$

$$= \frac{(p+5.797+j17.643)(p+5.797-j17.643)}{(p+2.3302+j5.0153)(p+2.3302-j5.0153)} \quad (8.261)$$

To verify (8.206) we compute from (8.194a) the matrix

$$\mathbf{A}_s|_{\mathbf{X}=\mathbf{K}} = \mathbf{A}_o + \mathbf{B}_1\mathbf{K}(\mathbf{1}_2 - \mathbf{D}_3\mathbf{K})^{-1}\mathbf{C}_1$$

$$= 10^7 \begin{bmatrix} -3.5603 & 2.667 \\ -10 & -1.1 \end{bmatrix} \quad (8.262)$$

giving

$$s\mathbf{1}_2 - \mathbf{A}_s|_{\mathbf{X}=\mathbf{K}} = 10^7 \begin{bmatrix} p+3.5603 & -2.667 \\ 10 & p+1.1 \end{bmatrix} \quad (8.263)$$

the determinant of which is given by

$$\det (s\mathbf{1}_2 - \mathbf{A}_s)|_{\mathbf{X}=\mathbf{K}} = (p^2+4.6603p+30.5863)10^{14} \quad (8.264)$$

Substituting this and (8.249) in (8.206) gives

$$\det \mathbf{F_K(X)} = \frac{[\det (\mathbf{1}_2 - \mathbf{D}_3\mathbf{X})] [\det (s\mathbf{1}_2 - \mathbf{A}_s)]}{[\det (\mathbf{1}_2 - \mathbf{D}_3\mathbf{K})][\det (s\mathbf{1}_2 - \mathbf{A}_s)]|_{\mathbf{X}=\mathbf{K}}}$$

$$= \frac{p^2 + 11.594p + 344.881}{p^2 + 4.6603p + 30.5863} \tag{8.265}$$

confirming (8.261), where $\mathbf{D}_3 = \mathbf{0}$. Thus, the eigenvalues of the coefficient $\mathbf{A}_s$ matrix of the state equation when $\mathbf{X}$ assumes its nominal value and when $\mathbf{X}$ assumes the reference value $\mathbf{X} = \mathbf{K}$ are the zeros and poles of the determinant of the general return difference matrix with respect to $\mathbf{X}$ for the reference $\mathbf{K}$, respectively.

To compute the general null return difference matrix with respect to $\mathbf{X}$ for the reference $\mathbf{K}$, we first compute $\hat{\mathbf{F}}(\mathbf{K})$ using (8.231) and (8.236) and obtain

$$\hat{\mathbf{F}}(\mathbf{K}) = \mathbf{1}_2 - \hat{\mathbf{A}}\mathbf{K} = \begin{bmatrix} \dfrac{p + 18.332}{p + 8.332} & 0 \\ \dfrac{10}{p + 1.1} & 1 \end{bmatrix} \tag{8.266}$$

the inverse of which is found to be

$$\hat{\mathbf{F}}^{-1}(\mathbf{K}) = \frac{p + 8.332}{p + 18.332} \begin{bmatrix} 1 & 0 \\ -\dfrac{10}{p + 1.1} & \dfrac{p + 18.332}{p + 8.332} \end{bmatrix} \tag{8.267}$$

Substituting this and (8.245) in (8.230) yields

$$\hat{\mathbf{F}}_\mathbf{K}(\mathbf{X}) = \hat{\mathbf{F}}^{-1}(\mathbf{K})\hat{\mathbf{F}}(\mathbf{X})$$

$$= \frac{1}{p + 18.332} \begin{bmatrix} p + 58.332 & 0 \\ \dfrac{40p + 333.28}{p + 1.1} & p + 18.332 \end{bmatrix} \tag{8.268}$$

the determinant of which is found to be

$$\det \hat{\mathbf{F}}_\mathbf{K}(\mathbf{X}) = \det \hat{\mathbf{F}}^{-1}(\mathbf{K})\hat{\mathbf{F}}(\mathbf{X}) = \frac{\det \hat{\mathbf{F}}(\mathbf{X})}{\det \hat{\mathbf{F}}(\mathbf{K})} = \frac{p + 58.332}{p + 18.332} \tag{8.269}$$

To verify (8.234) we first compute the matrices of (8.194) by setting $\mathbf{X} = \mathbf{K}$ and obtain

$$\mathbf{B}_s|_{\mathbf{X}=\mathbf{K}} = \mathbf{B}_o + \mathbf{B}_1\mathbf{K}(\mathbf{1}_2 - \mathbf{D}_3\mathbf{K})^{-1}\mathbf{D}_2 = \mathbf{B}_o \tag{8.270a}$$

$$\mathbf{C}_s|_{\mathbf{X}=\mathbf{K}} = \mathbf{C}_o + \mathbf{D}_1\mathbf{K}(\mathbf{1}_2 - \mathbf{D}_3\mathbf{K})^{-1}\mathbf{C}_1 = \mathbf{C}_o + \mathbf{D}_1\mathbf{K}$$

$$= \frac{1}{3}\begin{bmatrix} 7.68 & -8 \\ 3.6 & -20 \end{bmatrix} \tag{8.270b}$$

$$\mathbf{D}_s|_{\mathbf{X}=\mathbf{K}} = \mathbf{D}_o + \mathbf{D}_1\mathbf{K}(\mathbf{1}_2 - \mathbf{D}_3\mathbf{K})^{-1}\mathbf{D}_2 = \mathbf{D}_o \tag{8.270c}$$

Using these in conjunction with (8.212) and (8.262) yields

$$\hat{\mathbf{A}}_s|_{\mathbf{X}=\mathbf{K}} = (\mathbf{A}_s - \mathbf{B}_s\mathbf{D}_s^{-1}\mathbf{C}_s)|_{\mathbf{X}=\mathbf{K}}$$

$$= 10^7 \begin{bmatrix} -18.3336 & 0 \\ -10 & -1.1 \end{bmatrix} \tag{8.271}$$

the characteristic polynominal of which is found to be

$$\det(s\mathbf{1}_2 - \hat{\mathbf{A}}_s)|_{\mathbf{X}=\mathbf{K}} = (s + 18.3336 \times 10^7)(s + 1.1 \times 10^7) \tag{8.272}$$

Substituting (8.247) and (8.272) in (8.234) gives

$$\det \hat{\mathbf{F}}_\mathbf{K}(\mathbf{X}) = \frac{[\det(s\mathbf{1}_2 - \hat{\mathbf{A}}_s)][\det(\mathbf{1}_2 - \mathbf{D}_3\mathbf{X})](\det \mathbf{D}_s)}{[\det(s\mathbf{1}_2 - \hat{\mathbf{A}}_s)]|_{\mathbf{X}=\mathbf{K}}[\det(\mathbf{1}_2 - \mathbf{D}_3\mathbf{K})](\det \mathbf{D}_s)|_{\mathbf{X}=\mathbf{K}}}$$

$$= \frac{\det(s\mathbf{1}_2 - \hat{\mathbf{A}}_s)}{\det(s\mathbf{1}_2 - \hat{\mathbf{A}}_s)|_{\mathbf{X}=\mathbf{K}}} = \frac{s + 58.332 \times 10^7}{s + 18.333 \times 10^7} \tag{8.273}$$

confirming (8.269), where $\mathbf{D}_3 = \mathbf{0}$ and

$$\det \mathbf{D}_s = \det \mathbf{D}_s|_{\mathbf{X}=\mathbf{K}} = 4 \times 10^4 \tag{8.274}$$

Equation (8.273) states that the eigenvalues of the matrix $\hat{\mathbf{A}}_s$ when $\mathbf{X}$ assumes its nominal value and when $\mathbf{X}$ assumes the reference value $\mathbf{X} = \mathbf{K}$ are the zeros and poles of the determinant of the general null return difference matrix with respect to $\mathbf{X}$ for the reference $\mathbf{K}$. Note that there was a cancellation of the common factor $(s + 1.1 \times 10^7)$ in (8.273), resulting in one zero and one pole in $\det \hat{\mathbf{F}}_\mathbf{K}(\mathbf{X})$. Thus, not all the eigenvalues of $\hat{\mathbf{A}}_s$ are observable.

8.8 SUMMARY

In this chapter we introduced the concept of state. For electric networks the set of instantaneous values of all the branch currents and voltages is called the state of the network. A minimal set of branch currents and voltages the instantaneous values of which are sufficient to determine the state of a network is termed a complete set of state variables. The inductor currents and capacitor voltages are usually chosen as the state variables, since the equations associated with these variables are dynamical as opposed to variables associated with resistive edges, which are static. The objective was to frame the network equations as a system of first-order differential equations involving only the state variables and independent sources called the state equation. Apart from the state equation, the output variables can be expressed algebraically in terms of the state variables and independent sources called the output equation. The state equation and the output equation together are referred to as the state equations.

A network must satisfy Kirchhoff's two laws regardless of whether it is linear or nonlinear and time varying or time invariant. Therefore, Kirchhoff's law equations are independent of the constituents of network elements and can be determined solely from the network topology. For this we introduced the complete incidence matrix, the complete cutset matrix and the complete circuit matrix. These matrices can be compactly and elegantly used to describe the Kirchhoff's laws. However, not all the network equations written for these matrices are linearly independent. To ascertain the maximum number of linearly independent equations associated with these matrices, we had to determine their ranks. Specifically, we showed that the rank of the complete incidence matrix and the complete cutset matrix for an n-node connected network is $n - 1$, and the rank of the complete circuit matrix is $b - n + 1$, where b is the number of branches of the network. Thus, we need only consider the basis incidence, cutset and circuit matrices, which contain the same amount of information as the original ones. To write down these matrices systematically, we introduced the notions of tree and cotree and their associated fundamental cutset and circuit matrices. Since the fundamental cutset and circuit matrices contain the identity matrix, they are basis cutset and circuit matrices.

With respect to a chosen tree, the tree-branch voltages determine all other voltages by means of the f-circuits, and the cotree-link currents determine all other currents by way of the f-cutsets. Thus, to develop state equations using capacitor voltages and inductor currents as the state variables, we should place all the voltage sources and as many capacitors as possible in a tree and all the current sources and as many inductors as possible in a cotree. Such a tree is termed a normal tree. Based on the foregoing, we outlined eight steps in writing the state equation in normal form for a network. These steps are a systematic way to eliminate the unwanted variables in the primary system of network equations, so that the resulting equations are the state equation and can be put in normal form.

A network is degenerate if it contains a circuit composed only of capacitors and/or independent or dependent voltage sources, or a cutset composed only of inductors and/or independent or dependent current sources. A degenerate network may not possess a solution, a unique solution, or its state equation cannot be put in normal form if capacitor voltages and inductor currents are routinely chosen as the state variables.

We next formulated the feedback theory in terms of the coefficient matrices of the state equations using the fundamental matrix feedback-flow graph as a tool. Specifically, we expressed the return difference matrix and the null return difference matrix in terms of the coefficient matrices of the state equations, and showed that for a single-input and single-output feedback network, the transfer function is equal to the ratio of the determinants of the null return difference matrix and the return difference matrix times the transfer function when the elements of interest vanish. These results were then extended to the general reference, where we expressed the general return difference matrix and the general null return difference matrix for a general reference in terms of the coefficient matrices of the state equations, and were applied to the calculation of the sensitivity of a network function to an element.

Finally, we considered the general situation where a feedback network may have multiple inputs, multiple outputs and multiple feedback loops. We derived the expressions for the return difference matrix and the null return difference matrix, and showed that eigenvalues of the $\mathbf{A}_s$ or $\hat{\mathbf{A}}_s$ matrix under the nominal condition and under the condition that the elements of interest vanish give the zeros and poles of the determinant of the return difference matrix or the null return difference matrix. For a general reference, we found that the eigenvalues of the $\mathbf{A}_s$ or $\hat{\mathbf{A}}_s$ matrix under the nominal condition and under the condition that the elements of interest assume their reference values give the zeros and poles of the determinant of the general return difference matrix or the general null return difference matrix. In addition, the eigenvalues of $\hat{\mathbf{A}}_s$ and $\mathbf{A}_s$ matrices are zeros and poles of the determinant of the transfer-function matrix.

PROBLEMS

8.1 Verify that the state equations for the active network of Fig. 8.40 are those given by (8.113).

8.2 Verify that the state equations for the feedback amplifier of Fig. 8.49 are those given by (8.211).

8.3 In the network of Fig. 8.49, let the controlling parameters of the two voltage-controlled current sources be the elements of interest. Show that the eight-port network defined by the variables I_5, I_6, V_3, V_4, I_{g1}, I_{g2}, V_1 and V_2 can be characterized by the matrix equations (8.208).

8.4 Verify equations (8.194) for the active network of Fig. 8.49.

Figure 8.50 An active network the state equation of which is to be ascertained.

8.5 Obtain the state equations for the network of Fig. 8.50. The output variables are the input current i and output voltage v_4. Verify that, for $R_4 = 4\,\Omega$, $R_5 = 2\,\Omega$, $C_1 = 1\,\text{F}$, $C_2 = 3\,\text{F}$, $L_3 = 5\,\text{H}$ and $\alpha = 6$ the state equation reduces to

$$\begin{bmatrix} \dot{v}_1 \\ \dot{v}_2 \\ \dot{i}_3 \end{bmatrix} = \begin{bmatrix} -6.5 & 0 & -1 \\ 2 & 0 & 1/3 \\ 0.2 & -0.2 & -0.8 \end{bmatrix} \begin{bmatrix} v_1 \\ v_2 \\ i_3 \end{bmatrix} + \begin{bmatrix} 0.5 \\ 0 \\ 0 \end{bmatrix} [v_g] \qquad (8.275)$$

8.6 Show that the state equation for the network of Fig. 8.51 is given by

$$\begin{bmatrix} \dot{i}_1 \\ \dot{i}_2 \\ \dot{v}_3 \end{bmatrix} = \frac{1}{\Delta} \begin{bmatrix} 0 & -MR_4 & M - L_2 \\ 0 & -L_1 R_4 & L_1 - M \\ \Delta/C_3 & -\Delta/C_3 & 0 \end{bmatrix} \begin{bmatrix} i_1 \\ i_2 \\ v_3 \end{bmatrix}$$

$$+ \frac{1}{\Delta} \begin{bmatrix} L_2 - M & -R_4 M \\ M - L_1 & -L_1 R_4 \\ 0 & 0 \end{bmatrix} \begin{bmatrix} v_g \\ i_g \end{bmatrix} \qquad (8.276)$$

where $\Delta = L_1 L_2 - M^2 \neq 0$.

8.7 For the state equation of (8.275), compute the return difference matrix and general return difference matrix with respect to the coefficient matrix $\mathbf{A}_s$ of the state vector, where the reference matrix $\mathbf{K}$ is obtained from $\mathbf{A}_s$ by setting the third row and first column element a_{31} of $\mathbf{A}_s$ to zero.

8.8 Consider the active network of Fig. 8.50, the state equation of which is given by (8.275). Suppose that the voltage v_4 is the output variable. Determine the output equation of the network, and compute the null return difference matrix and the general null return difference matrix with respect to the coefficient $\mathbf{A}_s$ matrix of the state equation, where the reference matrix $\mathbf{K}$ is obtained from $\mathbf{A}_s$ by setting the third row and first column element a_{31} of $\mathbf{A}_s$ to zero.

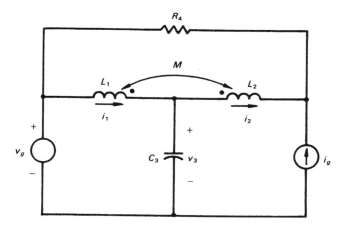

Figure 8.51 A network containing two mutually coupled inductors.

Figure 8.52 An equivalent network of a transistor amplifier.

8.9 In the equivalent network of a transistor amplifier as shown in Fig. 8.52, let the voltage V_o across the resistor R_3 be the output variable. Assume that the element of interest is the controlling parameter g_m of the voltage-controlled current source. Show that the return difference and the null return difference with respect to g_m are given by

$$F(g_m) = \frac{RC(R_1 + R_3)s + R + R_1 + R_3 + g_m R R_3}{RC(R_1 + R_3)s + R + R_1 + R_3} \tag{8.277}$$

$$\hat{F}(g_m) = \frac{RC_s + 1 + Rg_m}{RCs + 1} \tag{8.278}$$

8.10 Consider the same network as in Pro. 8.9. Compute the transfer function $w(g_m) = V_o/V_g$ by means of (8.225), and use (5.5) to calculate its sensitivity with respect to g_m.

8.11 A single-input and single-output network is characterized by the state equations

$$\begin{bmatrix} \dot{x}_1 \\ \dot{x}_2 \end{bmatrix} = \begin{bmatrix} 1 & 1 \\ -1 & 2 \end{bmatrix} \begin{bmatrix} x_1 \\ x_2 \end{bmatrix} + \begin{bmatrix} 1 \\ 1 \end{bmatrix} [u(t)] \qquad (8.279a)$$

$$y(t) = [1 \quad 1] \begin{bmatrix} x_1 \\ x_2 \end{bmatrix} \qquad (8.279b)$$

Show that the return difference matrix with respect to the coefficient $\mathbf{A}_s$ matrix of (8.279a) is given by

$$\mathbf{F}(\mathbf{A}_s) = \frac{1}{s} \begin{bmatrix} s-1 & -1 \\ -a_{21} & s-2 \end{bmatrix} \qquad (8.280)$$

where $a_{21} = -1$. If the reference matrix $\mathbf{K}$ is obtained from $\mathbf{A}_s$ by setting $a_{21} = 0$, compute the general return difference matrix with respect to $\mathbf{A}_s$ for the reference $\mathbf{K}$.

8.12 For the state equations of (8.279) and the reference matrix $\mathbf{K}$ considered in Pro. 8.11, compute the null return difference matrix and the general null return difference matrix with respect to $\mathbf{A}_s$ for the reference $\mathbf{K}$. Using (8.163) verify that the sensitivity of the transfer function with respect to the element a_{21} is given by

$$S(a_{21}) = \frac{s^2 - 3s + 2}{s^2 - 3s + 3} - \frac{s-1}{s-1.5} \qquad (8.281)$$

8.13 Use (8.98) to verify that the transfer function of the system (8.279) is given by

$$w(s) = \frac{2s-3}{s^2 - 3s + 3} \qquad (8.282)$$

8.14 Consider the parallel-series amplifier of Fig. 5.15. Use the equivalent network of Fig. 1.14 for the transistors. Let the controlling parameters of the two controlled current sources be the elements of interest. Assume that the voltages V_{15} and V_{25} are the output variables. Compute the state equations for the amplifier.

8.15 In Prob. 8.14 let the voltage V_{15} be the output variable. Compute the following:

(a) The return difference matrix $\mathbf{F}(\mathbf{A}_s)$.
(b) The null return difference matrix $\hat{\mathbf{F}}(\mathbf{A}_s)$.

(*c*) The general return difference matrix $\mathbf{F}_K(\mathbf{A}_s)$.
(*d*) The general null return difference matrix $\hat{\mathbf{F}}_K(\mathbf{A}_s)$.
(*e*) The voltage gain of the amplifier.
(*f*) The amplifier input impedance.

where $\mathbf{A}_s$ is the coefficient matrix in the state equation of the amplifier, and the reference matrix $\mathbf{K}$ is chosen so that $100\mathbf{K} = \mathbf{1}$.

8.16 In Prob. 8.14, compute the transfer-function matrix for the output variables $\mathbf{V}_{15}$ and $\mathbf{V}_{25}$ and input variables I_s.

8.17 A network is characterized by its state equations

$$\begin{bmatrix} \dot{x}_1 \\ \dot{x}_2 \end{bmatrix} = \begin{bmatrix} -1 & 2 \\ -5 & -10 \end{bmatrix} \begin{bmatrix} x_1 \\ x_2 \end{bmatrix} + \begin{bmatrix} 0 \\ 5 \end{bmatrix} [u(t)] \tag{8.283a}$$

$$y(t) = [0 \quad -2] \begin{bmatrix} x_1 \\ x_2 \end{bmatrix} + [3][u(t)] \tag{8.283b}$$

Compute the return difference matrix and the null return difference matrix with respect to the $\mathbf{A}_s$ matrix of (8.283a). Use these results to verify formula (8.109) by showing that the transfer function is given by

$$w(s) = \frac{3s^2 + 23s + 50}{s^2 + 11s + 20} \tag{8.284}$$

8.18 A network is characterized by its state equations

$$\begin{bmatrix} \dot{x}_1 \\ \dot{x}_2 \end{bmatrix} = \begin{bmatrix} 0 & 1 \\ -2 & -3 \end{bmatrix} \begin{bmatrix} x_1 \\ x_2 \end{bmatrix} + \begin{bmatrix} 1 & 0 \\ 1 & 1 \end{bmatrix} \begin{bmatrix} u_1 \\ u_2 \end{bmatrix} \tag{8.285a}$$

$$\begin{bmatrix} y_1 \\ y_2 \\ y_3 \end{bmatrix} = \begin{bmatrix} 1 & 0 \\ 1 & 1 \\ 0 & 2 \end{bmatrix} \begin{bmatrix} x_1 \\ x_2 \end{bmatrix} + \begin{bmatrix} 0 & 0 \\ 1 & 0 \\ 0 & 1 \end{bmatrix} \begin{bmatrix} u_1 \\ u_2 \end{bmatrix} \tag{8.285b}$$

Compute the return difference matrix and the null return difference matrix with respect to the coefficient $\mathbf{A}_s$ matrix of (8.285a). Compute the transfer-function matrix of the network.

BIBLIOGRAPHY

Brockett, R. W.: Poles, Zeros, and Feedback: State Space Interpretation, *IEEE Trans. Automatic Control*, vol. AC-10, no. 2, pp. 129–135, 1965.
Chen, W. K.: "Applied Graph Theory: Graphs and Electrical Networks," 2nd rev. ed., New York: American Elsevier, and Amsterdam: North-Holland, 1976.

Chen, W. K.: Physical Interpretation of the Multiport Return Difference Matrix, *IEEE Trans. Automatic Control*, vol. AC-25, no. 3, pp. 598–600, 1980.

Chen, W. K. and M. D. Tong: On Indefinite-Impedance Matrix, *IEEE Trans. Circuits and Systems*, vol. CAS-32, no. 8, pp. 840–847, 1985.

Chen, W. K. and M. D. Tong: Analysis and Computation of Feedback Matrices in Multivariable and Multiloop Feedback Networks, *IEEE Trans. Circuits and Systems*, vol. CAS-35, no. 9, pp. 1123–1128, 1988.

Elsherif, H. M.: Loop, Cutset, Hybrid and State Analyses of Linear Multiple-Loop Feedback Systems, Ph.D. dissertation, Ohio University, Athens, Ohio, 1980.

Elsherif, H. M. and W. K. Chen: State-Space Analysis of the Multiple-Loop Feedback Networks, *Proc. 23rd Midwest Symp. Circuits and Systems*, University of Toledo, Toledo, Ohio, pp. 284–288, 1980a.

Elsherif, H. M. and W. K. Chen: Analysis of Multiple-Loop Feedback Networks, *Proc. 13th International Symp. Circuits and Systems*, Port Chester, New York, pp. 808–811, 1980b. IEEE catalog no. 80CH1511-5CAS.

Kalman, R. E.: Irreducible Realizations and the Degree of a Matrix of Rational Functions, *J. Soc. Indust. Appl. Math.*, vol. 13, no. 2, pp. 520–544, 1965.

Kuh, E. S.: State Variables and Feedback Theory, *IEEE Trans. Circuit Theory*, vol. CT-16, no. 1, pp. 23–26, 1969.

Lan, J. L. and W. K. Chen: On Loop-Impedance Matrix Formulation of Feedback Amplifier Theory, *J. Franklin Inst.*, vol. 320, no. 1, pp. 1–14, 1985.

MacFarlane, A. G. J.: Return-Difference and Return-Ratio Matrices and Their Use in Analysis and Design of Multivariable Feedback Control Systems, *Proc. IEE (London)*, vol. 117, no. 10, pp. 2037–2049, 1970.

MacFarlane, A. G. J.: Return-Difference and Return-Ratio Matrices and Their Use in Analysis and Design of Multivariable Feedback Control Systems, *Proc. IEE (London)*, vol. 118, no. 7, pp. 946–947, 1971.

Riegle, D. E. and P. M. Lin: Matrix Signal Flow Graphs and an Optimum Topological Method for Evaluating Their Gains, *IEEE Trans. Circuit Theory*, vol. CT-19, no. 5, pp. 427–435, 1972.

Sandberg, I. W.: On the Theory of Linear Multi-Loop Feedback Systems, *Bell Sys. Tech. J.*, vol. 42, no. 2, pp. 355–382, 1963.

Tassny-Tschiassny, L.: The Return Difference Matrix in Linear Networks, *Proc. IEE (London)*, vol. 100, part IV, pp. 39–46, 1953.

NINE

TOPOLOGICAL ANALYSIS OF ACTIVE NETWORKS

The concept of the "natural frequencies" of a network arises from the consideration of its transient behavior. They are conventionally given as the zeros of the determinant of the loop-impedance matrix or the cutset-admittance matrix. Since in general the zeros of these two determinants are different, this definition involves some ambiguity. One of the classical problems is to count the number of natural frequencies of a network by inspection. An early solution to determine this number called the *degrees of freedom* was given by Guillemin (1931), applicable only to RLC networks that do not contain any all-capacitor or all-inductor loops. The term originates from the study of mechanical systems, in which we can attach significance to the word "position" or "configuration". Reza (1955) gave the solution for networks containing only two types of elements and adopted the term "order of complexity", which we shall follow in this book. The complete solution for RLC networks was obtained independently by Bryant (1959, 1960), Bers (1959), and Seshu and Reed (1961). The extension to active networks was recent and has been considered by many workers [see, for example, Chen (1972a)]. The main difficulty lies in the fact that, unlike the case for RLC networks, topology of the network alone is not sufficient; network parameters must also be involved, which complicates the problem considerably. However, various upper bounds on the order of complexity of a general network are available. In the first part of the chapter, we shall present a unified summary on many of the known results.

The studies and designs of electronic circuits, signals, and systems are based on a variety of models, which consist of interconnections of *idealized* physical elements such as inductors, capacitors, resistors, and generators. A physical inductor should be thought of as a coil of wire with series and/or shunt resistance and even capacitance, in addition to inductance, for its complete representation. Similar statements can be made for any physical component. Although such a procedure is physically justifiable, it is extremely inconvenient from a theoretical point of view; for, using this procedure, the equations of a simple network will become very complicated. All of the techniques of network synthesis that have been developed will be useless, and a great many procedures for designing practical networks such as filters, interstage networks, wave shaping networks etc. will be hopelessly complicated. No one is going to give them up merely to avoid the use of idealized elements. Thus, for all purposes, both in analysis and synthesis, we will continue to use models composed of idealized physical elements. A consequence of this idealization is that we cannot always assume the existence and uniqueness of network solutions. It is of great interest, therefore, to determine the conditions under which unique solutions can be obtained. These conditions are especially useful in computer-aided network analysis when a numerical solution does not converge. They help distinguish those cases where a network does not possess a unique solution from those where the fault lies with the integration technique. We note that even when a network response is unstable, a numerical solution exists. Thus when a numerical solution does not converge, it is important to distinguish network instability, divergence due to improper numerical integration, and divergence due to the lack of existence of a unique solution. For these reasons, we shall also present a unified summary of various existence conditions.

9.1 NATURAL FREQUENCIES

We first define precisely the term "order of complexity" of a network, and then show how it can be determined or its upper bounds obtained by inspection.

Definition 9.1: Natural frequencies The *natural frequencies* of a network are defined as the roots of the determinantal polynomial of the operator matrix of the network equations when these are framed as a set of first-order differential equations and/or algebraic equations for the branch currents and voltages.

Definition 9.2: Order of complexity The number of natural frequencies of a network is called the *order of complexity* of the network, counting each frequency according to its multiplicity.

In state-variable formulation of network equations, the order of complexity is the minimal number of dynamically-independent branch voltages and currents whose instantaneous values are sufficient to determine completely the instantaneous state of the network, and hence is the dimension of the state space. It is also the maximum number of linearly independent initial conditions that can be specified for a network. In mathematics, it is the number of arbitrary constants appearing in the general solution of the network equations. These are the various ways of stating the same thing.

Since natural frequencies are inherent characteristics of a network, being independent of the external excitations, for simplicity and without loss of generality, in this section we shall assume that all the networks considered are devoid of independent sources.

Example 9.1 Consider the active network N of Fig. 9.1a, the associated directed graph of which is shown in Fig. 9.1b. We now write down the network equations governed by the Kirchhoff current and voltage laws and the generalized Ohm's law applied to each branch.

Kirchhoff's current law becomes

$$[\mathbf{Q}_p \ \ \mathbf{Q}_d]\begin{bmatrix}\mathbf{I}_p \\ \mathbf{I}_d\end{bmatrix} = \mathbf{0} \tag{9.1}$$

where

$$\mathbf{Q}_f = [\mathbf{Q}_p \ \ \mathbf{Q}_d] = \begin{array}{cccc} \alpha_1 & \alpha_2 & \alpha_3 & \alpha_4 \\ \begin{bmatrix} -1 & 0 & 1 & -1 \\ 1 & 1 & 0 & 1 \end{bmatrix} \end{array} \tag{9.2}$$

is the fundamental cutset matrix with respect to the tree $\alpha_2\alpha_3$, and

$$\mathbf{I}' = \begin{bmatrix}\mathbf{I}_p' & \mathbf{I}_d'\end{bmatrix} = \begin{bmatrix} i_1 & i_2 & i_3 & i_4 \end{bmatrix} \tag{9.3}$$

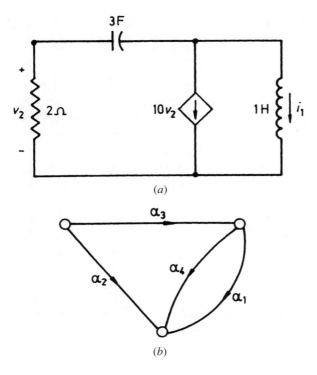

(a)

(b)

Figure 9.1 (a) An active network, the order of complexity of which is 2, and (b) its associated digraph.

is the transpose of the branch-current vector. Kirchhoff's voltage law becomes

$$[\mathbf{B}_p \quad \mathbf{B}_d] \begin{bmatrix} \mathbf{V}_p \\ \mathbf{V}_d \end{bmatrix} = \mathbf{0} \qquad (9.4)$$

where

$$\mathbf{B}_f = [\mathbf{B}_p \quad \mathbf{B}_d] = \begin{array}{c} \begin{array}{cccc} \alpha_1 & \alpha_2 & \alpha_3 & \alpha_4 \end{array} \\ \begin{bmatrix} 1 & -1 & 1 & \vdots & 0 \\ 0 & -1 & 1 & \vdots & 1 \end{bmatrix} \end{array} \qquad (9.5)$$

is the fundamental circuit matrix with respect to the same tree $\alpha_2\alpha_3$, and

$$\mathbf{V}' = \begin{bmatrix} \mathbf{V}'_p & \mathbf{V}'_d \end{bmatrix} = \begin{bmatrix} v_1 & v_2 & v_3 & \vdots & v_4 \end{bmatrix} \tag{9.6}$$

is the transpose of the branch-voltage vector. Denote the differential operator d/dt by D, the generalized Ohm's law for the branches can be written in matrix form as

$$\mathbf{Z}_{LMR}\mathbf{I}_p = \mathbf{Y}_C\mathbf{V}_p \tag{9.7}$$

$$\mathbf{I}_d = \mathbf{Y}_d\mathbf{V}_p \tag{9.8}$$

where

$$\mathbf{Z}_{LMR} = \begin{bmatrix} D & 0 & 0 \\ 0 & 2 & 0 \\ 0 & 0 & 1 \end{bmatrix} \tag{9.9}$$

$$\mathbf{Y}_C = \begin{bmatrix} 1 & 0 & 0 \\ 0 & 1 & 0 \\ 0 & 0 & 3D \end{bmatrix} \tag{9.10}$$

$$\mathbf{Y}_d = \begin{bmatrix} 0 & 10 & 0 \end{bmatrix} \tag{9.11}$$

$$\mathbf{I}_d = \begin{bmatrix} i_4 \end{bmatrix} \tag{9.12}$$

Observe that the matrices (9.1)–(9.6) have been partitioned according to the types of branches in the network. The subscript d is used to represent the branches of the dependent or controlled sources, whereas p is for all other passive branches.

Equations (9.1), (9.4), (9.7), and (9.8) can be written as a single matrix equation such as

$$\mathbf{H}(D)\begin{bmatrix} \mathbf{I} \\ \mathbf{V} \end{bmatrix} = \mathbf{0} \tag{9.13}$$

where the coefficient matrix $\mathbf{H}(D)$ called the *operator matrix* is given by

$$\mathbf{H}(D) = \begin{bmatrix} \mathbf{Q}_p & \mathbf{Q}_d & \mathbf{0} & \mathbf{0} \\ \mathbf{0} & \mathbf{0} & \mathbf{B}_p & \mathbf{B}_d \\ -\mathbf{Z}_{LMR} & \mathbf{0} & \mathbf{Y}_C & \mathbf{0} \\ \mathbf{0} & -\mathbf{1}_d & \mathbf{Y}_d & \mathbf{0} \end{bmatrix}$$

$$= \left[\begin{array}{ccc:c:ccc:c} -1 & 0 & 1 & -1 & 0 & 0 & 0 & 0 \\ 1 & 1 & 0 & 1 & 0 & 0 & 0 & 0 \\ \hdashline 0 & 0 & 0 & 0 & 1 & -1 & 1 & 0 \\ 0 & 0 & 0 & 0 & 0 & -1 & 1 & 1 \\ \hdashline -D & 0 & 0 & 0 & 1 & 0 & 0 & 0 \\ 0 & -2 & 0 & 0 & 0 & 1 & 0 & 0 \\ 0 & 0 & -1 & 0 & 0 & 0 & 3D & 0 \\ \hdashline 0 & 0 & 0 & -1 & 0 & 10 & 0 & 0 \end{array}\right] \tag{9.14}$$

and $\mathbf{1}_d$ is the identity matrix of appropriate order. Thus, according to Definition 9.1, the natural frequencies of the network are the roots of the polynomial det $\mathbf{H}(\lambda)$, λ being the complex frequency variable. Hence from (9.14) we obtain

$$\det \mathbf{H}(\lambda) = -63\lambda^2 - 6\lambda - 1 \tag{9.15}$$

giving the natural frequencies

$$\lambda_1, \lambda_2 = -0.0476 \pm j0.1166 \tag{9.16}$$

Thus, the order of complexity of the network of Fig. 9.1a is 2. We remark that the partitioned matrices (9.1)–(9.14) are valid in general, not merely for the network of Fig. 9.1a.

In the study of the order of complexity, it will be necessary to expand the determinant of the operator matrix $\mathbf{H}(\lambda)$. By applying the Laplace expansion to det $\mathbf{H}(\lambda)$, Mark (1971) has shown that for a network N composed of resistors, capacitors, self and mutual inductors, and controlled current sources, the determinant of the operator matrix $\mathbf{H}(\lambda)$ is related to the determinant of its node-admittance matrix $\mathbf{Y}_n$ by

$$\det \mathbf{H}(s) = \text{constant} \ (\det \mathbf{Z}_{LMR})(\det \mathbf{Y}_n)$$

$$= \text{constant} \ s^{\beta_L} \ (\det \mathbf{Y}_n) \tag{9.17}$$

where β_L denotes the number of inductors in N. This relationship is a slight generalization of an earlier result of Bryant (1959) and Purslow and Spence (1967). In deriving (9.17), it is implicitly assumed that the inductance matrix formed by the set of inductors of N is a positive-definite matrix. This condition is stronger than the usual requirement that the inductance matrix be nonnegative definite in that it prohibits perfectly coupled transformers.

In Fig. 9.1a, we can compute the node-admittance matrix $\mathbf{Y}_n$ whose determinant is given by

$$\det \mathbf{Y}_n = \det \begin{bmatrix} 3s + 0.5 & -3s \\ 10 - 3s & 3s + 1/s \end{bmatrix} = \frac{63s^2 + 6s + 1}{2s} \qquad (9.18)$$

confirming (9.17) and yielding the same natural frequencies.

9.2 DIGRAPH ASSOCIATED WITH AN ACTIVE NETWORK

Let

$$\mathbf{Y} = [y_{ij}] \qquad (9.19)$$

be the indefinite-admittance matrix of an n-node active network N. We associate with $\mathbf{Y}$ an n-node directed graph called the *digraph.*

Definition 9.3: Associated digraph Associate with an indefinite-admittance matrix $\mathbf{Y} = [y_{ij}]$ of order n an n-node directed graph called the *associated digraph* G, such that there is an edge directed from node i to node j with weight $-y_{ij}, i \neq j$, in G if and only if $y_{ij} \neq 0$.

We remark that the diagonal elements of $\mathbf{Y}$ have no direct bearing on the construction of G. If $\mathbf{Y}$ is the indefinite-admittance matrix of an active network N, we write $G = G(N)$. A careful study of the rules for G shows that G can be greatly simplified if we agree that an undirected edge stands for a pair of oppositely directed edges such that the weight (admittance) associated with each directed edge in the pair is the same as that of the undirected edge. With this simplification, it is clear that the resulting G is a *composite graph* containing both directed and undirected edges. For the most practical and commonly used networks consisting of resistors, capacitors, self and mutual inductors, transistors, gyrators, and voltage-controlled current sources, G can be constructed directly from the equivalent network N simply by superimposing the corresponding directed graphs of the voltage-controlled current sources of Fig. 9.2, the directed graphs

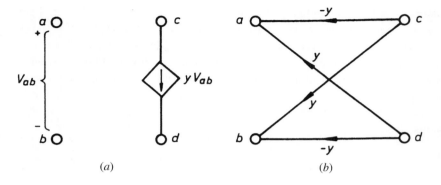

Figure 9.2 (a) A voltage-controlled current source, and (b) its associated digraph.

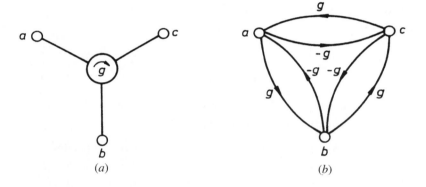

Figure 9.3 (a) A gyrator, and (b) its associated digraph.

of the gyrators of Fig. 9.3, or the associated graphs of the transformers or the mutual couplings of the coils of Fig. 9.4, upon the subnetwork of N composed of resistors, capacitors, and self-inductors. We remark that the restriction to voltage-controlled current sources in N is not really serious, since all other types of controlled sources can easily be transformed into the above type by using i- or e-shift if necessary. We shall further agree that the directed edges in the composite graph G will be called the *active edges*, and the undirected edges will be called the *passive edges*. Observe that in Fig. 9.2 if $a = c$ and $b = d$, the unilateral element corresponds to a two-terminal passive network branch.

Figure 9.4 (*a*) A transformer and (*b*) its associated graph.

Figure 9.5 A transformer-coupled transistor amplifier.

Consider, for example, the transformer-coupled transistor amplifier N of Fig. 9.5. The subnetwork of N composed of the resistors G_1, G_2, G_3, G_6 and G_7 corresponds to the edges G_1, G_2, G_3, G_6 and G_7 in the associated composite graph G of Fig. 9.6. If we superimpose on this subnetwork the associated directed graph of the equivalent voltage-controlled current source

$$I_{34} = aI_1 = aG_1V_{13} \tag{9.20}$$

where $V_{13} = V_1 - V_3$, and V_k ($k = 1, 2, 3, 4, 5$) is the potential measured from terminal k to some common but unspecified reference point, identifying from Fig. 9.2

$$a = 1, \quad b = 3, \quad c = 3, \quad d = 4, \quad y = aG_1 \tag{9.21}$$

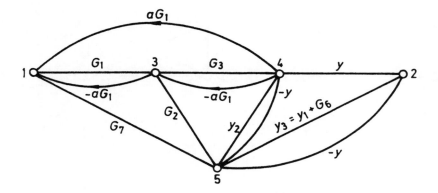

Figure 9.6 The associated digraph of the transistor amplifier of Fig. 9.5.

and the associated graph of the transformer of Fig. 9.4 with

$$a = 4, \quad b = 5, \quad c = 2, \quad d = 5 \tag{9.22}$$

$$y_1 = \frac{L_1}{s(L_1 L_2 - M^2)} \tag{9.23a}$$

$$y_2 = \frac{L_2}{s(L_1 L_2 - M^2)} \tag{9.23b}$$

$$y = \frac{M}{s(L_1 L_2 - M^2)} \tag{9.23c}$$

we obtain the associated digraph or the composite graph G of Fig. 9.6. After combining the passive parallel edges, the resulting graph is shown in Fig. 9.7.

Definition 9.4: Directed tree A *directed tree* with reference node k, denoted by T_k, in a directed graph is a tree in which each of its directed edges is directed toward the reference node k in the unique path from the initial node of the edge to k in the tree.

For the directed graph G of Fig. 9.8, the set of directed trees T_4 with reference node 4 is shown in Fig. 9.9. The corresponding set of directed trees T_3 with reference node 3 is presented in Fig. 9.10.

In the composite graph $G(N)$ of an active network N, a directed tree T_k with reference node k is a tree in which each of its active edges, if they exist, is directed

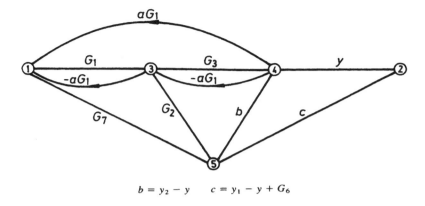

$$b = y_2 - y \qquad c = y_1 - y + G_6$$

Figure 9.7 The associated digraph $G(N)$ of the active network N of Fig. 9.5.

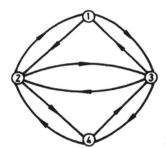

Figure 9.8 A directed graph used to illustrate the directed trees.

toward the reference node k in the unique path from the initial node of the active edge to k in the tree. Thus, trees composed only of passive edges are directed trees with respect to any node. In the composite graph $G(N)$ of Fig. 9.7, three directed trees T_4 are presented in Fig. 9.11.

For a subgraph g of $G(N)$, denote by $f(g)$ the product of the weights associated with the edges of g. Thus, if $g = T_k$ is a directed tree of $G(N)$, then $f(T_k)$ denotes the product of the weights associated with the edges of T_k. For the three directed trees T_4 of Fig. 9.11, their products $f(T_4)$ are given by

$$(-aG_1)G_7bc, \quad (-aG_1)G_7by, \quad (-aG_1)G_7cy, \tag{9.24}$$

where $b = y_2 - y$ and $c = y_1 - y + G_6$. In addition, there are 21 other directed trees T_4 in $G(N)$ composed only of passive edges. Their products $f(T_4)$

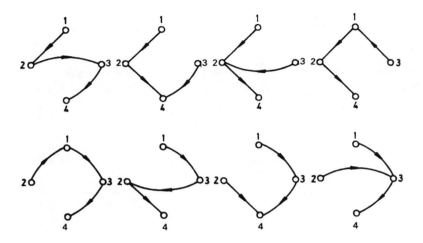

Figure 9.9 The set of directed trees T_4 with reference node 4 in the directed graph of Fig. 9.8.

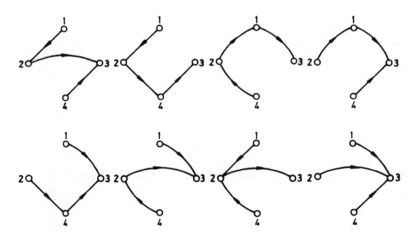

Figure 9.10 The set of directed trees T_3 with reference node 3 in the directed graph of Fig. 9.8.

are found to be

$$
\begin{array}{cccc}
G_1G_7bc, & G_1G_7by, & G_1G_7cy, & G_1G_2by, \\
G_1G_2bc, & G_2G_7by, & G_2G_7bc, & G_1G_2cy, \\
G_2G_7cy, & G_1G_2G_3y, & G_1G_2G_3c, & G_2G_3G_7y, \\
G_2G_3G_7c, & G_1G_3by, & G_1G_3bc, & G_3G_7by,
\end{array}
$$

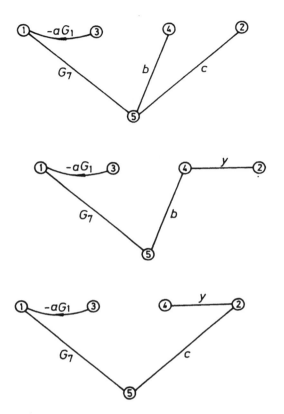

Figure 9.11   Three directed trees T_4 with reference node 4 in the associated digraph of Fig. 9.7.

$$G_3G_7bc, \quad G_1G_3G_7y, \quad G_1G_3G_7c, \quad G_1G_3yc,$$
$$G_3G_7yc,$$

(9.25)

The significance of the above formulation is contained in the following fundamental theorem, which relates the determinant of the node-admittance matrix $\mathbf{Y}_n$ of an active network N with the directed trees of $G(N)$.

Theorem 9.1   The determinant of the node-admittance matrix $\mathbf{Y}_n$ of an active network N is equal to the sum of the directed-tree admittance products of the directed trees T_k in the associated digraph $G(N)$ of N:

$$\det \mathbf{Y}_n = \sum_{T_k} f(T_k)$$

(9.26)

The proof of this theorem is omitted, and can be found in Chen (1976). We remark that since k on the right-hand side of (9.26) is arbitrary, we have the freedom to choose any node as the reference node of the directed trees. This freedom of choice is reflected in the indefinite-admittance matrix $\mathbf{Y}$ in that it is an equicofactor matrix. It emphasizes the fact that in solving an electrical network problem it is immaterial as to which node is taken to be the reference potential point, since the solution must be the same in every case. These statements are really three different ways of saying the same thing.

We illustrate the above result by the following example.

Example 9.2 Consider the transistor amplifier of Fig. 9.5, the associated digraph G of which is shown in Fig. 9.7. The node-admittance matrix $\mathbf{Y}_n$ of the amplifier with node 5 chosen as the reference node is found to be

$$
\mathbf{Y}_n = \begin{bmatrix}
G_1 + G_7 & 0 & -G_1 & 0 \\
0 & c + y & 0 & -y \\
aG_1 - G_1 & 0 & G_2 + G_3 + G_1 - aG_1 & -G_3 \\
-aG_1 & -y & aG_1 - G_3 & b + y + G_3
\end{bmatrix}
\tag{9.27}
$$

According to Theorem 9.1 we have

$$
\det \mathbf{Y}_n = \sum_{T_k} f(T_k) = \sum_{T_4} f(T_4)
\tag{9.28}
$$

The associated digraph G of Fig. 9.7 has three active directed trees T_k as shown in Fig. 9.11 and 21 passive directed trees T_k, the admittance products of which are listed in (9.25), obtaining from (9.28)

$$
\begin{aligned}
\det \mathbf{Y}_n = & (-aG_1)G_7bc + (-aG_1)G_7by + (-aG_1)G_7cy + G_1G_7bc \\
& + G_1G_7by + G_1G_7cy + G_1G_2by + G_1G_2bc + G_2G_7by + G_2G_7bc \\
& + G_1G_2cy + G_2G_7cy + G_1G_2G_3y + G_1G_2G_3c + G_2G_3G_7y \\
& + G_2G_3G_7c + G_1G_3by + G_1G_3bc + G_3G_7by + G_3G_7bc \\
& + G_1G_3G_7y + G_1G_3G_7c + G_1G_3yc + G_3G_7yc
\end{aligned}
\tag{9.29}
$$

Each term on the right-hand side of the above equation corresponds to a directed-tree admittance product $f(T_4)$.

We remark that instead of selecting node 4 as the reference node for the directed trees in (9.28), we may choose any other node as the reference node, and the final result will be the same in each and every case. This does not mean that the number of directed trees for different choices of reference node will be the same. In fact, for different choices of the reference node, the number of directed trees may be substantially different. For example, if we choose node 5 as the reference node, the number of directed trees T_5 in the digraph G is 44. This is in direct contrast to the situation where there are only 24 directed trees T_4 with reference node 4. However, the final sum of all the directed-tree admittance products remains the same.

9.3 ORDER OF COMPLEXITY

We now apply Theorem 9.1 to the study of the order of complexity of an active network. To be specific and without loss of generality, we shall assume that, in this section, an active network N is composed of resistors, capacitors, self and mutual inductors, and controlled current sources, that each controlled current source is controlled by a distinct passive branch voltage or current, and that we have not excited N by any independent sources. The second condition is not to be deemed as a restriction, because all other types of dependent sources can be transformed into the above types, using the source transformations if necessary [see, for example, Chen (1990)]. Thus, gyrators and positive and negative impedance converters are included. We also assume that each network element is considered as a separate branch and that the inductance matrix of N is positive definite, a fact that was mentioned earlier.

Observe that there are three classes of edges in the associated digraph $G(N)$ of an active network N:

(i) *Inductive edges*: edges whose associated admittances are of the form $1/Ls$, L being a real constant.
(ii) *Capacitive edges*: edges whose associated admittances are of the form Cs, C being a real constant.
(iii) *Resistive edges*: edges whose associated admittances are real constants.

Definition 9.5: *C*-**cut and** *L*-**cut** A *C-cut* (*L-cut*) of $G(N)$ is a cutset or an edge-disjoint union of cutsets composed only of capacitive (inductive) edges.

Definition 9.6: *C*-**circuit and** *L*-**circuit** A *C-circuit* (*L-circuit*) of $G(N)$ is a circuit composed only of capacitive (inductive) edges.

9.3.1 Preliminaries and Symbols

We first state a result from graph theory, the proof of which can be found in Chen (1976).

Lemma 9.1 *Let g be a subgraph of a connected graph G. Then the maximum number of edges of g that can be made part of any tree of G is given by the rank of g, and the minimum number of edges of g contained in any tree of G is given by the rank of the graph obtained from G by shorting all the edges not contained in g.*

We remark that the *rank* of a graph is equal to the number of its nodes minus the number of connected parts called the *components*. In $G = G(N)$, denote by G_L the subgraph consisting of all the inductive edges, i.e. the subgraph obtained by deleting (open-circuiting) all the noninductive edges. Similarly, G_C denotes the subgraph consisting of all the capacitive edges. By G_L^* we mean the graph formed from G by shorting all the noninductive edges. This is similarly valid for G_C^*.

For a subgraph g of G, denote by $r(g)$ and $m(g)$ the rank and nullity of g, respectively, where *nullity* of a graph is equal to the number of its edges minus its rank. Thus, by Lemma 9.1, the maximum number of edges of G_C that can be made part of any tree of G is given by $r(G_C)$, and the minimum number of edges of G_L contained in any tree of G is $r(G_L^*)$. Also, denote by $b_L(g)$ and $b_C(g)$ the numbers of the inductive edges and capacitive edges of g, respectively . For $g = T$, a tree of G, $b_L(T)$ represents the number of inductive edges of T. In particular, we write $b_L(G) = b_L$ and $b_C(G) = b_C$, for simplicity. In Fig. 9.6 we have $b_C = 0$ and $b_L = 5$, counting y_3 as the parallel connection of edges y_1 and G_6.

9.3.2 RLC Networks

An *RLC network* is a network composed of resistors, inductors, and capacitors. In this section, we show that the order of complexity of an RLC network can be determined from its topology alone by inspection. Although it is possible to obtain the general formulas for active networks first, and treat the RLC networks as a special case, formulas for the RLC networks are the simplest and are sufficiently important to be considered separately.

We note that for an RLC network N, $G(N)$ and N are isomorphic, and the inductive, capacitive, and resistive edges of $G(N)$ become the inductors, capacitors and resistors of N. Therefore, we can use $G(N)$ and N interchangeably. When we speak of a C-circuit in N, we mean a circuit consisting only of capacitors. This is similarly valid for all other terms.

Theorem 9.2 The order of complexity of an RLC network is given by the number of reactive elements, less the number of independent C-circuits and the number of independent L-cuts.

PROOF. Let $\mathbf{Y}_n$ be the node-admittance matrix of a given RLC network N. Write

$$\det \mathbf{Y}_n = a_k s^k + a_{k-1} s^{k-1} + \cdots + a_0 + a_{-1} s^{-1} + \cdots + a_{-q} s^{-q}$$

(9.30)

where $a_k \neq 0$ and $a_{-q} \neq 0$, and k and q are positive integers to be determined. From (9.17) it is evident that the order of complexity can be determined once the integers k and q are known. Hence it suffices to ascertain k and q. Since from (9.26)

$$\det \mathbf{Y}_n = \sum_T f(T)$$

(9.31)

where T is a tree of N, it is clear that

$$k = b_C(T_1) - b_L(T_1)$$

(9.32)

where the tree T_1 maximizes this difference.

Similarly, we have

$$q = b_L(T_2) - b_C(T_2)$$

(9.33)

where the tree T_2 maximizes this difference. It can be shown [see, for example Chen (1976)] that it is possible simultaneously to maximize the number of capacitors and minimize the number of inductors in the same tree, say T_1. Let $T = T_1$ be such a tree. Since

$$b_L - q \geqq 0$$

(9.34)

from (9.17) the order of complexity σ of N becomes

$$\sigma = k + q + (\beta_L - q) = k + b_L$$
$$= b_C(T) + b_L(\overline{T})$$

(9.35)

where $\overline{T}$ is the complement of T in G, a cotree of G. From Lemma 9.1, we have

$$b_C(T) = r(G_C)$$

(9.36)

$$b_L(T) = r(G_L^*)$$

(9.37)

Substituting these in (9.35) yields

$$\sigma = b_C(T) + b_L(\overline{T}) = r(G_C) + b_L - b_L(T)$$
$$= b_C - m(G_C) + b_L - r(G_L^*)$$

(9.38)

Observe that the number of linearly independent C-circuits in G_C, being equal to its nullity $m(G_C)$, is the same as that in G, and that the number of linearly independent L-cuts in G_L^*, being equal to its rank $r(G_L^*)$, is the same as that in G. The theorem follows from here, and the proof is completed.

Corollary 9.1 The number of nonzero natural frequencies of an RLC network is equal to its order of complexity, less the number of independent L-circuits and the number of independent C-cuts.

PROOF. From (9.17), (9.30), and (9.34), it is evident that the number of zero natural frequencies is given by $b_L - q$. Using the symbols defined in the proof of Theorem 9.2 yields

$$b_L - q = b_L - b_L(T_2) + b_C(T_2) \tag{9.39}$$

Appealing to Lemma 9.1, we obtain

$$b_L(T_2) = r(G_L) \tag{9.40a}$$
$$b_C(T_2) = r(G_C^*) \tag{9.40b}$$

Substituting these in (9.39) yields

$$b_L - q = b_L - r(G_L) + r(G_C^*) = m(G_L) + r(G_C^*) \tag{9.41}$$

showing that the number of zero natural frequencies is given by $m(G_L)$, which is equal to the number of linearly independent L-circuits in G, plus $r(G_C^*)$, which is equal to the number of linearly independent C-cuts in G. So the corollary is proved.

Alternative expressions for Corollary 9.1 are possible and they are stated in the following:

Corollary 9.2 The number of nonzero natural frequencies of an RLC network is given by

$$f = r(G_C) + r(G_L) - r(G_C^*) - r(G_L^*) \tag{9.42a}$$
$$= m(G_C^*) + m(G_L^*) - m(G_C) - m(G_L) \tag{9.42b}$$
$$= \left[m(G_L^*) - m(G_L) \right] + \left[r(G_C) - r(G_C^*) \right] \tag{9.42c}$$

Corollary 9.3 The number of nonzero natural frequencies of an LC network is given by

$$f = 2[m(G) - m(G_C) - m(G_L)] \tag{9.43}$$

If the RLC network contains neither the L-circuits nor the C-circuits, (9.42b) reduces to $f = m(G_C^*) + m(G_L^*)$, which gives the Guillemin's algorithm. Formula (9.42c) was conjectured by Mason and Adler [see, for example, Reza (1955)], and formula (9.43) was first given by Reza (1955) who called f the *number of dynamically independent loops*.

We shall illustrate the above results by the following example.

Example 9.3 Consider the RLC network N of Fig. 9.12. Using the symbols defined above gives

$$b_L = 6, \qquad\qquad b_C = 9 \tag{9.44}$$

$$r(G_C) = 9 - 2 = 7, \qquad r(G_L) = 6 - 1 = 5 \tag{9.45a}$$

$$r(G_C^*) = 3 - 1 = 2, \qquad r(G_L^*) = 2 - 1 = 1 \tag{9.45b}$$

$$m(G_C) = 9 - 9 + 2 = 2, \quad m(G_L) = 6 - 6 + 1 = 1 \tag{9.46a}$$

$$m(G_C^*) = 9 - 3 + 1 = 7, \quad m(G_L^*) = 6 - 2 + 1 = 5 \tag{9.46b}$$

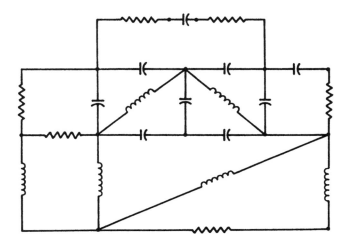

Figure 9.12 An RLC network used to illustrate the order of complexity.

Now we can state the following:

The number of linearly independent C-cuts $= r(G_C^*) = 2$
The number of linearly independent L-cuts $= r(G_L^*) = 1$
The number of linearly independent C-circuits $= m(G_C) = 2$
The number of linearly independent L-circuits $= m(G_L) = 1$

Thus, the order of complexity of N is given by

$$\sigma = b_C + b_L - r(G_L^*) - m(G_C) = 9 + 6 - 1 - 2 = 12 \qquad (9.47)$$

which is the maximum number of linearly independent initial conditions that can be specified for the network of Fig. 9.12 or, equivalently, the total number of natural frequencies of N, zero frequency and multiplicity included. The number of nonzero natural frequencies, according to Corollary 9.1, is given by

$$f = \sigma - m(G_L) - r(G_C^*) = 12 - 1 - 2 = 9 \qquad (9.48)$$

which can also be computed directly from (9.42). The reader is urged to verify this.

Physically, the natural frequencies at the origin correspond to the constant currents that circulate around the circuits consisting only of inductors or the constant voltages that apply across the cutsets consisting only of capacitors. We remark that these constant currents and voltages do not give rise to any additional algebraic constraints among the currents in the L-circuits or the voltages across the C-cuts, only on their rates of change. This is in direct contrast to an incorrect statement made by Guillemin (1931), whose argument applied to L-circuits, for example, is as follows: Such an L-circuit gives rise to a differential relation of the form

$$L_1 \frac{di_1}{dt} + L_2 \frac{di_2}{dt} + \cdots + L_k \frac{di_k}{dt} = 0 \qquad (9.49)$$

Integration of this expression from 0 to t leads to

$$L_1 i_1 + L_2 i_2 + \cdots + L_k i_k = \text{constant} \qquad (9.50)$$

It might appear that this also represents a constraint on the inductor currents. However, the constant term on the right-hand side is not specified, whose determination requires an independent relationship. A similar statement can be made about voltages across the branches of a C-cut.

9.3.3 Active Networks

As mentioned earlier, for a general active network, topology alone is not sufficient to determine its order of complexity; network parameters must also be involved. The general characterization has not been solved so far; partial solutions are available. In this section, we present some of the known solutions.

From (9.17) it is clear that the order of complexity of an active network N is equal to the number of inductors plus the highest index of s in the nodal determinant. This also shows that the finite nonzero zeros of det $\mathbf{H}(s)$ and det $\mathbf{Y}_n$ are the same. Since, according to Theorem 9.1, det $\mathbf{Y}_n$ is equal to the sum of directed-tree-admittance products $f(T_k)$ of the directed trees T_k of $G(N)$, the highest index of s in det $\mathbf{Y}_n$ corresponds to a T_k that maximizes the difference between the number of capacitive edges and the number of inductive edges contained in it. Such a directed tree T_k that possesses the highest index of s is called a *normal directed tree* of $G(N)$. This definition differs from the tree used by Bryant (1959) for an RLC network in that it is not always possible to maximize the number of capacitive edges and minimize the number of inductive edges in the same tree. Since in $G(N)$ not all of the normal directed-tree-admittance products have the same sign, in applying Theorem 9.1, cancellations may occur. Thus, let

$$w = \sum_{T_k} f(T_k) \qquad (9.51)$$

where T_k is a normal directed tree and the summation is taken over all such T_k. This leads to the following result.

Theorem 9.3 An upper bound on the order of complexity σ of an active network N containing β_L inductors is given by

$$\sigma \leqq \sigma_{\max} = \beta_L + b_C(T) - b_L(T) \qquad (9.52)$$

where T is a normal directed tree of $G(N)$, the reference node of which is arbitrary. The equality applies if and only if $w \neq 0$.

For example, in Fig. 9.6, $G_1 G_2 G_3 G_6$ is a normal directed tree T of $G(N)$. Thus, we have $\beta_L = 2$, $b_C(T) = 0$ and $b_L(T) = 0$, and according to (9.52) the order of complexity is bounded above by 2. Since $w \neq 0$, or $\sigma = 2$.

In the following, we consider the order of complexity for a restricted class of networks composed of resistors, inductors, capacitors, and voltage- and/or current-controlled current sources.

Definition 9.7: Complete tree of a network A *complete tree* of a given network N is a tree composed only of resistors, inductors, and capacitors of N

and such that the replacement of the controlling elements (passive branches whose voltages or currents control the controlled current sources) of the controlled current sources in the tree by their corresponding controlled current sources would result in a tree of N. In particular, a tree is considered complete if it does not contain any controlling elements.

This definition is essentially the same as that of a *complete tree* in the Coates-Mayeda (1958) formulation of network equations. The main difference is that it is now stated in terms of the branches of the network rather than the edges of the corresponding voltage and current graphs. We remark that a complete tree of the voltage and current graphs in the Coates-Mayeda formulation is not necessarily a complete tree of N, as given by Definition 9.7.

Before we proceed to state and prove our result, we note that in $G(N)$ the inductive edges are due to self-inductors and/or inductor-current-controlled current sources, that the capacitive edges are due to capacitors and/or capacitor-current- controlled current sources, and that the resistive edges are due to resistors and/or voltage-controlled current sources, the controlling parameters of which are assumed to be constant.

Theorem 9.4 In a network N composed of resistors, inductors, capacitors, and controlled current sources, let T be a complete tree of N chosen so that the difference between the number of capacitors and the number of inductors is maximized. Then an upper bound on the order of complexity σ of N is given by the number of capacitors contained in T plus the number of inductors contained in its cotree $\overline{T}$, i.e.

$$\sigma \leqq \sigma_{\max} = b_C(T) + b_L(\overline{T}) \tag{9.53}$$

PROOF. Let N'' be the network obtained from N by converting all the current-controlled current sources into equivalent voltage-controlled current sources. This is always possible because the current and voltage of a passive branch of N are related by a finite admittance. In $G(N'')$ let T_k'' be a normal directed tree. Our objective is to show that

$$\beta_L + b_C(T_k'') - b_L(T_k'') = b_C(T) + b_L(\overline{T}) \tag{9.54}$$

To this end, we first demonstrate that we can construct a complete tree T of N that has the same index of s as that of T_k''. Let the edge (c, b) of Fig. 9.2b be a capacitive (inductive) active edge of $G(N'')$ contained in T_k''. It can be shown [see Chen (1972b)] that $T_k'' \cup (a, b) - (c, b)$ is a directed tree of $G(N'')$ having the same index of s as that of T_k''. Likewise, we can show that there exists such a directed tree if any one of the remaining three active edges of Fig. 9.2b is in T_k''.

We remark that at most one of the four active edges of Fig. 9.2b can appear in any T_k'' that will not result in any cancellations in (9.51).

Repeating this process for each capacitive or inductive active edge of T_k'' yields a passive tree T'' of $G(N'')$ whose corresponding subnetwork in N a complete tree T possessing the property stated in the theorem. Since T_k'' and T'' have the same index of s, we have

$$b_C(T_k'') = b_C(T) \tag{9.55}$$

$$b_L(T_k'') = b_L(T) \tag{9.56}$$

Since $\beta_L - b_L(T) = b_L(\overline{T})$, formula (9.53) follows immediately. This completes the proof of the theorem.

A result very similar to (9.53) was derived by Purslow and Spence (1967), who define a complete tree of N as either a tree consisting only of passive elements or a tree containing some controlled current sources, the replacement of these sources in the tree by their corresponding controlling branches resulting in a tree consisting only of passive elements. In order to avoid confusion, let the trees defined by them be called the *common trees*. Since the set of complete trees is a subset of the set of common trees, the bound given in (9.53) is sharper than that obtained by replacing the term complete tree by common tree in Theorem 9.4.

As an illustration, consider the network N of Fig. 9.13. It is easily seen that the branches R, C and L constitute the only complete tree of N. According to Theorem 9.4, we have

$$\sigma_{\max} = 1 + 0 = 1 \tag{9.57}$$

On the other hand, the branches R, C and $a_1 i_L$ form a common tree of N, and Purslow and Spence's theorem would give $\sigma_{\max} = 1 + 1 = 2$. Therefore, if

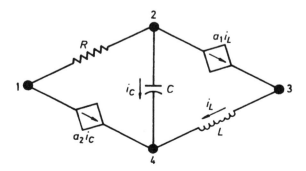

Figure 9.13 An active network containing controlled current sources.

N possesses a unique solution, $a_1 \neq 1$, $a_2 \neq -1$ and $\sigma = 1$. This shows that even if the parameters of the controlled sources do not bear special relation to the other network parameters, Purslow and Spence's theorem does not give the order of complexity of N, as claimed by the authors, only an upper bound.

Corollary 9.4 In a network N composed of resistors, inductors, capacitors, and controlled current sources, let T be a common tree of N with the associated number

$$\sigma_{\max} = b_C(T) + b_L(\overline{T}) + b_{cs}(T) - b_{is}(T) \tag{9.58}$$

where $b_{cs}(T)$ and $b_{is}(T)$ denote the numbers of capacitor-current-controlled current-sources and inductor-current-controlled current-sources in T, respectively, and $\overline{T}$ is the complement of T in N, a cotree of N. Then a bound on the order of complexity of N is given by $\sigma_{\max}$.

This corollary is a special case of a more general formula derived by Tow (1968), which also includes the situation where N may contain controlled voltage sources. However, in the latter case, the common trees cannot be defined in N alone; they are stated in terms of the associated voltage and current graphs of N. The general case is not considered here, for simplicity.

Example 9.4 Consider the network N of Fig. 9.13, in which the branches R, C and $a_1 i_L$ constitute a common tree. This common tree also maximizes the number given in (9.58). Thus, according to Corollary 9.4, an upper bound on the order of complexity σ of the network is given by

$$\sigma \leq \sigma_{\max} = 1 + 1 + 0 - 1 = 1 \tag{9.59}$$

Equivalently, we can use the common tree formed by the branches R, C and L, yielding

$$\sigma \leq \sigma_{\max} = 1 + 0 + 0 - 0 = 1 \tag{9.60}$$

Finally, we mention a very special case, as discussed by Su (1968), in which N contains a single nonreciprocal network element such as a gyrator, transistor or vacuum tube.

Corollary 9.5 The order of complexity of a network containing a gyrator (transistor) is not less than that of the network (equivalent network) with the gyrator (controlled current source) removed.

PROOF. By partitioning the set of directed trees of $G(N)$ according to those that do not contain any active edge, and those that contain at least one, the right-hand side of (9.26) can be put in two summations, each being nonnegative. Using this in conjunction with (9.17), the corollary follows immediately. The details are omitted.

9.4 UNIQUE SOLVABILITY

In this section, we present a unified summary on the existence conditions that are both necessary and sufficient for the unique solvability of networks that contain resistors, inductors, capacitors, transformers, independent sources, controlled sources of all types, and other linear devices that can be modeled by these elements such as transistors, gyrators, impedance converters, etc.

Kirchhoff (1847) was the first to prove the existence of a solution for a resistive network. His proof was elaborated by Weyl (1923) and Eckmann (1944–1945), who also gave a deeper insight into the problem. Seshu and Reed (1961) presented necessary and sufficient conditions for the existence of a unique solution for an RLC network with mutual inductances and independent sources. In an attempt to cover a more general case, Roth (1955, 1959) applied the theory of algebraic topology and derived a sufficient condition which was elaborated by Chen (1976). Malik and Hale (1967) extended Seshu and Reed's results by presenting conditions that are both necessary and sufficient for networks that also contain dependent voltage sources not controlled by short-circuit currents and dependent current sources not controlled by open-circuit voltages. Purslow (1970) and Parker and Barmes (1971) tackled the problem through the reduction process used to obtain the state equations for the networks and obtained conditions that are both necessary and sufficient for an active network to possess a unique solution. Finally, Chen and Chan (1974) extended Malik and Hale's work to networks that contain all types of controlled sources.

Unified summary of these results will now be presented below, each being in a separate section.

9.4.1 Roth's Condition

Based on the concept of ohmicness, Roth (1955, 1959) derived a condition that is sufficient for the existence of a unique solution for a general network. His condition also becomes necessary for RLCM networks.

Definition 9.8: Ohmicness A square matrix $\mathbf{F}$ over the complex field is said to be *ohmic* if the associated quadratic form

$$\mathbf{X}^*\mathbf{F}\mathbf{X} \neq 0 \tag{9.61}$$

for each nonzero vector $\mathbf{X}$ over the complex field.

For example, any positive or negative definite matrix is ohmic. However, the converse is not necessarily true.

Theorem 9.5 Let N be a network which does not contain any circuit (cutset) composed only of independent and dependent voltage (current) generators. Let N'' be the network obtained from N by removing all the independent sources. Then the network N possesses a unique solution if the branch-immittance matrix of N'' is ohmic.

The proof can be found in Chen (1976), and is omitted here.

Example 9.5 Let N be the network obtained from that of Fig. 9.1a with an independent voltage generator inserted in series with the 2-Ω resistor. Then N'' is the network shown in Fig. 9.1a. In N'' if each network element is considered as a separate branch, as indicated in Fig. 9.1b, the branch-admittance matrix $\mathbf{Y}_b$ of N'' becomes

$$\mathbf{Y}_b(s) = \begin{bmatrix} 1/s & 0 & 0 & 0 \\ 0 & 0.5 & 0 & 0 \\ 0 & 0 & 3s & 0 \\ 0 & 10 & 0 & 0 \end{bmatrix} \tag{9.62}$$

Since we can find a vector $\mathbf{X}$, the transpose of which is given by

$$\mathbf{X}' = \begin{bmatrix} 0 & 0 & 0 & 1 \end{bmatrix} \neq \mathbf{0} \tag{9.63}$$

such that $\mathbf{X}^*\mathbf{Y}_b\mathbf{X} = 0$, the matrix $\mathbf{Y}_b$ is not ohmic, and consequently we cannot apply Theorem 9.5 to determine the unique solvability of N. However, if the controlled current source $10V_2$ and the inductor are considered as a single branch, as shown in Fig. 9.14, the branch-admittance matrix becomes

$$\mathbf{Y}_b(s) = \begin{bmatrix} 1/s & 10 & 0 \\ 0 & 0.5 & 0 \\ 0 & 0 & 3s \end{bmatrix} \tag{9.64}$$

which is positive definite for positive real s. Therefore, it is ohmic and according to Theorem 9.5, the network N possesses a unique solution.

An *RLCM network* is an RLC network with mutual inductances. The branch-impedance matrix of an RLCM network N can be partitioned, according to the

Figure 9.14 A different representation of the network of Fig. 9.1*a*.

resistors, inductors and capacitors of N, into the form

$$
\mathbf{Z}_b(s) = \begin{bmatrix} \mathbf{R} & 0 & 0 \\ 0 & s\mathbf{L} & 0 \\ 0 & 0 & \dfrac{1}{s}\mathbf{D} \end{bmatrix}
\tag{9.65}
$$

by simultaneously permuting the rows and columns if necessary, where $\mathbf{R}$ is diagonal with positive diagonal entries, and so is $\mathbf{D}$, and $\mathbf{L}$ is positive definite by postulate. Thus, $\mathbf{Z}_b(s)$ is ohmic, and from Theorem 9.5 we obtain the conditions of Seshu and Reed (1961).

    Corollary 9.6 Let N be an RLCM network with independent generators. Then N possesses a unique solution if and only if there is no circuit consisting only of voltage generators and there is no cutset consisting only of current generators.

9.4.2 Chen and Chan's Conditions

To facilitate our discussion and without loss of generality, we shall consider the modified network N_m obtained from a given active network N by means of the following transformations:

  (i) Transformers and other linear devices such as gyrators and impedance converters are replaced by their equivalent controlled-source representations.

 (ii) Controlling short-circuit currents (open-circuit node-pair voltages) are considered as currents (voltages) in (across) branches with zero impedance (admittance). Each controlling short-circuit (open-circuit) branch of a controlled voltage-source (current-source) is replaced with two resistors of 1 ohm and -1 ohm in series (parallel), which are still considered as a single branch.

(iii) Each controlled source is controlled by a distinct passive branch, which is defined to be a resistor, inductor, or capacitor. This assumption is not to be deemed as a restriction, since, for example, a current-controlled current-source controlled by more than one passive branch current may be replaced by equivalent current sources connected in parallel, each being controlled by a distinct passive branch current.

(iv) Each controlled current (voltage) source controlled by the branch voltage (current) of a branch with finite nonzero impedance is replaced by an equivalent current-controlled current-source (voltage-controlled voltage-source) through Ohm's law.

Thus, the modified network N_m can be depicted symbolically as shown in Fig. 9.15. For examples, the modified networks N_m of Figs. 9.16a, 9.17a, and 9.18a are presented in Figs. 9.16b, 9.17b, and 9.18b, respectively.

In the following, we shall partition the branch currents and voltages of N_m into various classes. To simplify the notation, we shall use **I** to denote the branch-current vector and **V** the branch-voltage vector with subscripts representing various types of branches, as follows:

j: independent current sources
cc: current-controlled current-sources
vc: voltage-controlled current-sources, the controlling branches of which are made of ± 1-mho admittance pairs in parallel
ccp: controlling branches of the current-controlled current-sources
vcp: controlling branches of the voltage-controlled current-sources
p: noncontrolling passive branches
vv: voltage-controlled voltage-sources
cv: current-controlled voltage-sources, whose controlling branches are made of ± 1-ohm impedance pairs in series
vvp: controlling branches of the voltage-controlled voltage-sources
cvp: controlling branches of the current-controlled voltage-sources
e: independent voltage sources

Thus, $\mathbf{I}_{cc}$ denotes the branch-current vector corresponding to the currents in the current-controlled current-sources, and $\mathbf{V}_{vvp}$ represents the branch-voltage vector corresponding to the voltages across the controlling branches of the voltage-controlled voltage-sources. The meaning of these symbols applied to other types of branches is obvious, and no further elaboration is necessary.

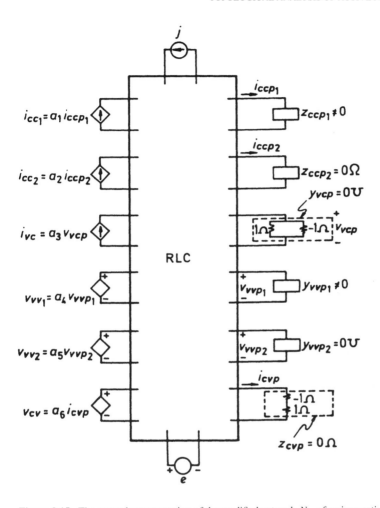

Figure 9.15   The general representation of the modified network N_m of a given active network N.

The controlled sources are described by the systems of equations as follows:

$$\mathbf{I}_{cc} = \mathbf{A}_{cc}\mathbf{I}_{ccp} \qquad (9.66a)$$

$$\mathbf{I}_{vc} = \mathbf{A}_{vc}\mathbf{V}_{vcp} \qquad (9.66b)$$

$$\mathbf{V}_{vv} = \mathbf{A}_{vv}\mathbf{V}_{vvp} \qquad (9.66c)$$

$$\mathbf{V}_{cv} = \mathbf{A}_{cv}\mathbf{I}_{cvp} \qquad (9.66d)$$

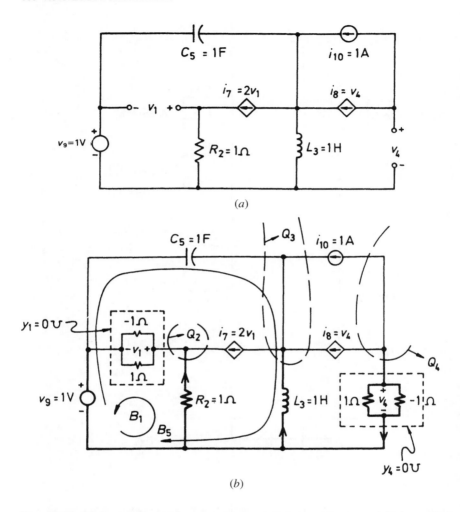

Figure 9.16 (*a*) A network containing two voltage-controlled current sources, and (*b*) its modified network N_m.

where **A**'s are nonsingular diagonal matrices of the known branch controlling parameters.

For example, in Figs. 9.16*a*, 9.17*a* and 9.18*a* the corresponding equations are given by

$$\mathbf{I}_{vc} = \begin{bmatrix} i_7 \\ i_8 \end{bmatrix} = \begin{bmatrix} 2 & 0 \\ 0 & 1 \end{bmatrix} \begin{bmatrix} v_1 \\ v_4 \end{bmatrix} = \mathbf{A}_{vc}\mathbf{I}_{vcp} \qquad (9.67a)$$

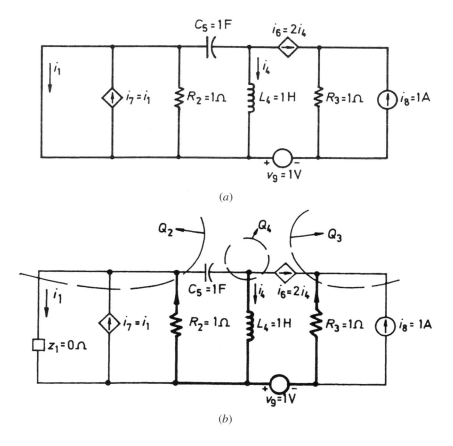

Figure 9.17 (a) A network containing two current-controlled current sources, and (b) its modified network N_m.

$$\mathbf{I}_{cc} = \begin{bmatrix} i_7 \\ i_6 \end{bmatrix} = \begin{bmatrix} 1 & 0 \\ 0 & 2 \end{bmatrix} \begin{bmatrix} i_1 \\ i_4 \end{bmatrix} = \mathbf{A}_{cc}\mathbf{I}_{ccp} \qquad (9.67b)$$

$$\mathbf{V}_{cv} = \begin{bmatrix} v_2 \\ v_3 \end{bmatrix} = \begin{bmatrix} R_2 & 0 \\ 0 & R_3 \end{bmatrix} \begin{bmatrix} i_7 \\ i_8 \end{bmatrix} = \mathbf{A}_{cv}\mathbf{I}_{cvp} \qquad (9.67c)$$

respectively.

It can be shown that in a uniquely solvable linear network N, there exists a tree T in its modified network N_m such that all the voltage sources, dependent and independent, are tree branches of T, and all the current sources, dependent

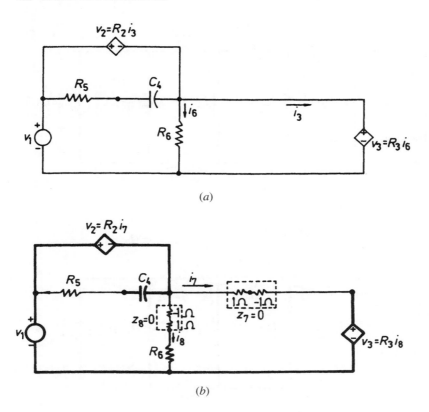

(a)

(b)

Figure 9.18 (a) A network containing two current-controlled voltage sources, and (b) its modified network N_m.

and independent, are links of its complement. A proof of this can be found in Chen and Chan (1974). Let T be such a tree in N_m containing all the voltage sources, whose complement $\overline{T}$, a cotree of N_m, contains all the current sources. Let $\mathbf{Q}_f$ and $\mathbf{B}_f$ be the fundamental cutset and circuit matrices of N_m defined with respect to the tree T, respectively. We use the superscripts T and $\overline{T}$ to indicate the rows of $\mathbf{Q}_f$ and $\mathbf{B}_f$ corresponding to the nonsource branches in T and $\overline{T}$. Thus, Q_{cv}^T denotes the submatrix of $\mathbf{Q}_f$ whose columns correspond to the current-controlled voltage-sources and whose rows correspond to the non-source branches in the tree T, and $\mathbf{B}_{cvp}^{\overline{T}}$ represents the submatrix of $\mathbf{B}_f$ whose columns correspond to the controlling branches of the current-controlled voltage-sources and whose rows correspond to the nonsource branches in the cotree $\overline{T}$. We remark that these submatrices can be obtained directly from the modified network N_m without first writing down the f-cutset matrix $\mathbf{Q}_f$ or the f-circuit matrix $\mathbf{B}_f$.

As examples for writing down these submatrices directly from N_m by inspection, it is straightforward to confirm that the matrices

$$
\left[\mathbf{Q}_{vc}^T \ \mathbf{Q}_{vcp}^T\right] = \begin{array}{c} \\ Q_4 \\ Q_2 \\ Q_3 \end{array} \overset{\begin{array}{cccc} i_7 & i_8 & y_1 & y_4 \end{array}}{\left[\begin{array}{cccc} 0 & 1 & 0 & 1 \\ 1 & 0 & -1 & 0 \\ -1 & 1 & 0 & 0 \end{array}\right]} \tag{9.68a}
$$

$$
\mathbf{Q}_{cc}^T = \begin{array}{c} \\ Q_4 \\ Q_2 \\ Q_3 \end{array} \overset{\begin{array}{cc} i_7 & i_6 \end{array}}{\left[\begin{array}{cc} 0 & 1 \\ 1 & 0 \\ 0 & 1 \end{array}\right]} \tag{9.68b}
$$

$$
\left[\mathbf{B}_{cv}^T \ \mathbf{B}_{cvp}^T\right] = \begin{array}{c} \\ R_5 \\ z_8 \\ z_7 \end{array} \overset{\begin{array}{cccc} v_2 & v_3 & z_7 & z_8 \end{array}}{\left[\begin{array}{cccc} 1 & 0 & 0 & 0 \\ 1 & 0 & 0 & 1 \\ 1 & 1 & 1 & 0 \end{array}\right]} \tag{9.68c}
$$

are the submatrices of the f-cutset matrices of Figs. 9.16b and 9.17b and the f-circuit matrix of Fig. 9.18b, respectively, with respect to the chosen trees $v_9 R_2 L_3 y_4$, $R_2 L_4 R_3 v_9$ and $C_4 R_6 v_1 v_2 v_3$ in their respective modified networks N_m. In the following, we use these symbols to state the known results on the unique solvability of N.

Denote by N_m'' the network derived from N_m by open-circuiting (removing) all the current sources and the -1-Ω resistors of the parallel connected ± 1-mho admittance pairs, and by short-circuiting all the voltage sources and the -1-Ω resistors of the series connected ± 1-Ω impedance pairs, as depicted symbolically in Fig. 9.19. Thus, N_m'' is an RLC network. The branches of N_m'' corresponding to the controlling branches of N_m will, for convenience, still be called the controlling branches, even thought it contains no controlled sources.

Theorem 9.6 An active network N is uniquely solvable if and only if its modified network N_m satisfies the following constraints:

(i) There exists a tree T such that all the dependent and independent voltage sources are tree branches of T and all the dependent and independent current sources are links of $\overline{T}$. This is equivalent to the statement that there exists

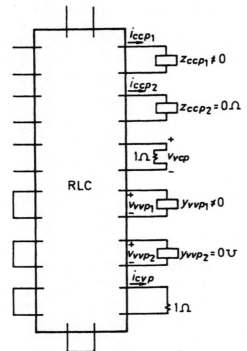

Figure 9.19 The associated sourceless network N_m'' of the modified network N_m.

neither a circuit composed only of dependent and independent voltage sources nor a cutset composed only of dependent and independent current sources.

(ii)

$$\det(\boldsymbol{\phi}^{-1} - \mathbf{WK}) \neq 0 \tag{9.69}$$

does not vanish identically in s, where

$$\boldsymbol{\phi} = \begin{bmatrix} \mathbf{A}_{vc} & \mathbf{0} & \mathbf{0} & \mathbf{0} \\ \mathbf{0} & \mathbf{A}_{cc} & \mathbf{0} & \mathbf{0} \\ \mathbf{0} & \mathbf{0} & \mathbf{A}_{vv} & \mathbf{0} \\ \mathbf{0} & \mathbf{0} & \mathbf{0} & \mathbf{A}_{cv} \end{bmatrix} \tag{9.70a}$$

$$\mathbf{K} = \begin{bmatrix} \mathbf{0} & \mathbf{0} & \mathbf{B}_{vv}^T & \mathbf{B}_{cv}^T - \mathbf{B}_{cvp}^T \mathbf{A}_{cv}^{-1} \\ \mathbf{Q}_{vc}^T - \mathbf{Q}_{vcp}^T \mathbf{A}_{vc}^{-1} & \mathbf{Q}_{cc}^T & \mathbf{0} & \mathbf{0} \end{bmatrix} \tag{9.70b}$$

and $\mathbf{W}$ is the transfer function matrix whose rows correspond to the controlling variables of the controlling branches, and whose columns correspond to the rows of $\mathbf{K}$ such that its kth row and qth column element w_{kq} is the transfer function found in N_m'' that relates the controlling variable (voltage or current) of the controlling branch corresponding to the kth column of ϕ to an excitation connected at the branch α_q corresponding to the qth row of $\mathbf{K}$, as follows:

1. If α_q is a link, insert a voltage source in series with α_q, the polarty of which is the same as that defined for α_q.
2. If α_q is a tree branch, connect a current source in parallel with α_q, the current-reference direction of which is the same as that defined by α_q.

The proof of this theorem can be found in Chen and Chan (1974) and is omitted here. We illustrate this result by the following examples.

Example 9.6 Consider the active network N of Fig. 9.16a, the modified network N_m of which is shown in Fig. 9.16b. In N_m let T be the tree consisting of the branches v_9, R_2, L_3 and y_4. Thus, condition (i) of the theorem is satisfied. Since N contains only current sources controlled by the open-circuit voltages and an independent voltage source, condition (ii) is reduced to

$$\det[\mathbf{A}_{vc}^{-1} - \mathbf{W}(\mathbf{Q}_{vc}^T - \mathbf{Q}_{vcp}^T \mathbf{A}_{vc}^{-1}] \neq 0 \qquad (9.71)$$

where $\mathbf{A}_{vc}$, $\mathbf{Q}_{vc}^T$ and $\mathbf{Q}_{vcp}^T$ are given in (9.67a) and (9.68a). The matrix $\mathbf{W}$ is the transfer-impedance matrix relating the voltages across the controlling branches of the two 1-Ω resistors to the currents of the current sources across the tree branches y_4', R_2 and L_3, as indicated in Fig. 9.20. This gives

$$\mathbf{W} = \begin{array}{c} \\ v_1 \\ v_4 \end{array} \begin{array}{ccc} j_4 & j_2 & j_3 \\ \left[\begin{array}{ccc} 0 & 0.5 & 0 \\ -1 & 0 & 0 \end{array} \right] \end{array} \qquad (9.72)$$

Substituting these in (9.71) and performing the indicated operations yield

$$\det[\mathbf{A}_{vc}^{-1} - \mathbf{W}(\mathbf{Q}_{vc}^T - \mathbf{Q}_{vcp}^T \mathbf{A}_{vc}^{-1})] = -0.25 \neq 0 \qquad (9.73)$$

showing that condition (ii) is also satisfied. Thus we conclude that the active network N is uniquely solvable.

Example 9.7 Figure 9.17a is an active network N containing two current-controlled current sources, the modified network N_m of which is given in Fig. 9.17b. Select a tree T in N_m composed of the branches R_2, L_4, R_3 and v_9. Thus, condition

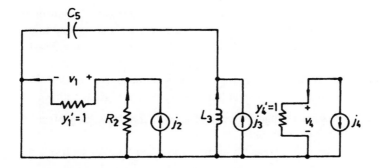

Figure 9.20 The multiport network associated with the network of Fig. 9.16b for computing the elements of the matrix **W**.

Figure 9.21 The multiport network associated with the network of Fig. 9.17b for computing the elements of the matrix **W**.

(i) is satisfied. Since in N_m only current-controlled current sources are involved, condition (ii) is reduced to

$$\det(\mathbf{A}_{cc}^{-1} - \mathbf{W}\mathbf{Q}_{cc}^{T}) \neq 0 \tag{9.74}$$

where $\mathbf{A}_{cc}$ and $\mathbf{Q}_{cc}^{T}$ are given in (9.67b) and (9.68b). **W** is the transfer-function matrix of current ratios relating the currents in the controlling branches z_1 and L_4 to the currents of the current sources connected across the tree branches L_4, R_2 and R_3, as indicated in Fig. 9.21, obtaining

$$\mathbf{W} = \begin{array}{c} \\ i_1 \\ \\ i_4 \end{array} \begin{array}{c} \overset{j_4 \qquad j_2 \quad j_3}{\left[\begin{array}{ccc} \dfrac{-s^2}{s^2+1} & 1 & 0 \\ \dfrac{-1}{s^2+1} & 0 & 0 \end{array}\right]} \end{array} \tag{9.75}$$

Substituting (9.67b), (9.68b), and (9.75) in (9.74) gives

$$\det(\mathbf{A}_{cc}^{-1} - \mathbf{W}\mathbf{Q}_{cc}^{T}) = 0 \tag{9.76}$$

showing that the network does not possess a unique solution.

Example 9.8 Consider the active network N of Fig. 9.18a, the modified network N_m of which is presented in Fig. 9.18b. Choose a tree of N_m composed of the branches C_4, R_6, v_1, v_2 and v_3. We see that condition (i) is satisfied. Condition (ii) reduces to

$$\det[\mathbf{A}_{cv}^{-1} - \mathbf{W}(\mathbf{B}_{cv}^{\overline{T}} - \mathbf{B}_{cvp}^{\overline{T}}\mathbf{A}_{cv}^{-1})] \neq 0 \tag{9.77}$$

The submatrices $\mathbf{A}_{cv}$, $\mathbf{B}_{cv}^{\overline{T}}$, and $\mathbf{B}_{cvp}^{\overline{T}}$ are given in (9.67c) and (9.68c), and from Fig. 9.22

$$\mathbf{W} = \begin{array}{c} \\ i_7 \\ i_8 \end{array} \begin{array}{ccc} e_5 & e_8 & e_7 \\ \left[\begin{array}{ccc} 0 & 0 & -1 \\ 0 & -(1+R_6)^{-1} & 0 \end{array} \right] \end{array} \tag{9.78}$$

Substituting these in (9.77) yields

$$\det[\mathbf{A}_{cv}^{-1} - \mathbf{W}(\mathbf{B}_{cv}^{\overline{T}} - \mathbf{B}_{cvp}^{\overline{T}}\mathbf{A}_{cv}^{-1})] = \frac{R_6 - R_3}{R_3(1 + R_6)} \tag{9.79}$$

Thus, N is uniquely solvable as long as R_3 and R_6 do not have the same value.

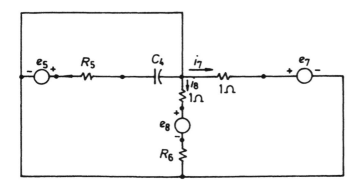

Figure 9.22 The multiport network associated with the network of Fig. 9.18b for computing the elements of the matrix $\mathbf{W}$.

9.4.3 Malik and Hale's Conditions

In this section, we discuss solvability conditions proposed by Malik and Hale (1967). Their conditions are valid only for the class of networks that do not contain any voltage sources controlled by short-circuit currents and current sources controlled by open-circuit voltages. Let N be such a network satisfying the restrictions outlined at the beginning of Sec. 9.4.2. Denote by M the network derived from N by short-circuiting all independent voltage sources and open-circuiting all independent current sources, and then by replacing every controlled current source by an independent current source and every controlled voltage source by an independent voltage source called the *replacement sources*. For convenience, the branches of M corresponding to the controlling branches of N will still be called the controlling branches, even though M does not contain any controlled sources.

Theorem 9.7 Equation (9.69) is satisfied if, and only if,

$$\det(\hat{\mathbf{W}} - \boldsymbol{\phi}^{-1}) \neq 0 \qquad (9.80)$$

identically in s where $\hat{\mathbf{W}}$ is the square transfer-function matrix relating the controlling variables of the controlling branches to the replacement sources in M such that the kth row and qth column element $\hat{w}_{kq}$ of $\hat{\mathbf{W}}$ is the transfer function found in M that relates the controlling variable of the controlled source corresponding to the kth row of $\boldsymbol{\phi}'$ to the replacement source corresponding to the qth column of $\boldsymbol{\phi}'$ when all other replacement sources have been properly removed from M, where $\boldsymbol{\phi}'$ denotes the transpose of $\boldsymbol{\phi}$.

PROOF. Under the stipulated conditions, (9.69) and (9.70) reduce to

$$\det(\boldsymbol{\phi}^{-1} - \mathbf{WK}) = \pm \det(\mathbf{WK} - \boldsymbol{\phi}^{-1}) \neq 0 \qquad (9.81)$$

where

$$\boldsymbol{\phi} = \begin{bmatrix} \mathbf{A}_{cc} & \mathbf{0} \\ \mathbf{0} & \mathbf{A}_{vv} \end{bmatrix} \qquad (9.82a)$$

$$\mathbf{K} = \begin{bmatrix} \mathbf{0} & \overline{\mathbf{B}}_{vv}^{T} \\ \mathbf{Q}_{cc}^{T} & \mathbf{0} \end{bmatrix} \qquad (9.82b)$$

Let $\mathbf{X}$ be the vector of the controlling variables. Then $\mathbf{X} = -\mathbf{WC}$, where $\mathbf{C}$ is the corresponding source vector described in Theorem 9.6. Let $\mathbf{C} = -\mathbf{KF}$, where $\mathbf{F}$ is the vector of the controlled sources. Then $\mathbf{X} = \mathbf{WKF}$.

Observe that the elements of $\mathbf{KF}$ can be identified either as the algebraic sums of the controlled voltage sources contained in the fundamental circuits or as the algebraic sums of the controlled current sources contained in the fundamental cutsets of N. These are similarly valid in M with respect to the corresponding tree of T in M. Since the actual branch distribution of the independent sources is not important in computing the response, only the circuit distribution of the voltage sources and the cutset distribution of current sources matter [see, for example, Chen (1976)]. This means that $\mathbf{KF}$ can be replaced by $-\mathbf{C}$ in M in computing $\mathbf{X}$, and vice versa. Thus $\hat{\mathbf{W}} = \mathbf{WK}$, and the theorem follows directly from (9.81). This completes the proof of the theorem.

In particular, if N contains a single controlled source, condition (9.80) becomes

$$f(s) \neq 1/\phi \text{ identically in } s \qquad (9.83)$$

where $f(s)$ is the transfer function of M relating the controlling variable to the replacement source, and ϕ is the controlling parameter of the controlled source.

Example 9.9 Consider the network of Fig. 9.13, the associated network M of which is presented in Fig. 9.23. The transfer-function matrix $\hat{\mathbf{W}}$ can be computed directly from M, and is given by

$$\begin{bmatrix} i_C \\ i_L \end{bmatrix} = \begin{bmatrix} -1 & -1 \\ 0 & 1 \end{bmatrix} \begin{bmatrix} j_C \\ j_L \end{bmatrix} = \hat{\mathbf{W}} \begin{bmatrix} j_C \\ j_L \end{bmatrix} \qquad (9.84)$$

The controlling parameter matrix $\boldsymbol{\phi}$, under the present situation, reduces to

$$\boldsymbol{\phi} = \mathbf{A}_{cc} = \begin{bmatrix} a_2 & 0 \\ 0 & a_1 \end{bmatrix} \qquad (9.85)$$

Substituting these in (9.80) yields

$$\det\left(\hat{\mathbf{W}} - \boldsymbol{\phi}^{-1}\right) = \left(1 + \frac{1}{a_2}\right)\left(\frac{1}{a_1} - 1\right) \qquad (9.86)$$

Figure 9.23 The multiport network associated with the network of Fig. 9.13 for computing the elements of the matrix $\hat{\mathbf{W}}$.

which according to Theorem 9.7 shows that the active network is uniquely solvable provided that $a_1 \neq 1$ and $a_2 \neq 1$, a fact that was mentioned in an illustration for Theorem 9.4.

9.4.4 Purslow's Conditions

Purslow (1970) stated his conditions in terms of the state equations. However, the reduction process used to obtain the required matrix is lengthy and cumbersome, and, furthermore, the matrix whose determinant we must evaluate is of order that is equal to the number of controlled sources plus the number of state variables of the corresponding passive network. In the special situation when there are no circuits composed only of the voltage sources or cutsets composed only of current sources, the matrix condition reduces to that of Malik and Hale (1967). In the following, we state a useful result of Purslow (1970), omitting the details of his reduction process. As a matter of fact, a similar approach based on the state-variable technique was reported by Parker and Barmes (1971).

Theorem 9.8 If a network N composed of resistors, inductors and capacitors and controlled and uncontrolled sources of all types is uniquely solvable,[†] then the following two conditions must be satisfied:

(i) There exists a tree such that all the uncontrolled voltage sources are tree branches and all uncontrolled current sources are links.
(ii) There exists a tree such that all the controlled and uncontrolled voltage sources, the currents of which do not control any sources, are tree branches, and all controlled and uncontrolled current sources, the voltages of which do not control any sources, are links.

PROOF. Condition (i) is obvious. To prove condition (ii), we observe that N can be obtained from its modified network N_m by open-circuiting the controlling branches made of ± 1-mho admittance pairs in parallel and by short-circuiting the controlling branches made of ± 1-Ω impedance pairs in series. If N is uniquely solvable, then according to Theorem 9.6 there is a tree T in N_m for which all the controlled and uncontrolled (dependent and independent) voltage sources are tree branches of T and all the controlled and uncontrolled current sources are links of $\overline{T}$, the complement of T in N_m. This is equivalent to stating that there exists neither a circuit composed only of controlled and uncontrolled voltage sources nor a cutset composed only of controlled and uncontrolled current sources in N_m.

[†] It is implicitly assumed that the controlling voltages of the controlled sources are voltages across the branches of the network, not simply node-pair voltages as used in the previous theorems.

Suppose that there is a circuit in N composed only of controlled and uncontrolled voltage sources, the currents of which do not control any sources. Then this circuit must also be in N_m, contradicting the assumption that since N is uniquely solvable, by Theorem 9.6, no such circuit can exist in N_m. Thus, N has neither circuits nor cutsets of the above type, and contains a desired tree. This completes the proof of the theorem.

In Fig. 9.18a, the tree composed of the branches v_1, R_5 and v_2 is such a tree. In Fig. 9.17a, one of such trees is composed of the branches R_2, L_4, R_3 and v_9. The latter example also shows that the existence of such a tree does not necessarily imply that the network is uniquely solvable since we showed in Example 9.7 that this network is not uniquely solvable.

9.5 TOPOLOGY AND THE SUMMATION OF RETURN DIFFERENCES

In this section, we show that the sum of the reciprocals of the return differences with respect to all network elements equals the nullity of the network if nodal determinant is used, and equals the rank of the network if the loop determinant is employed. Some consequences of these results in relation to the summations of driving-point and transfer immittances are discussed.

9.5.1 Nodal Formulation of Return Difference

Let $\mathbf{Y}$ be the indefinite-admittance matrix of a given network N. Then according to Definition 4.1, the return difference $F(x)$ with respect to a network element x for the zero reference value is given by

$$F(x) = \frac{Y_{uv}(x)}{Y_{uv}(0)} \qquad (9.87)$$

Let N be an n-node network of e branches b_x ($x = 1, 2, \ldots, e$) that may contain mutual couplings and controlled sources. Assume that each branch b_x has a nonzero admittance y_x connected between the nodes x_1 and x_2. Since a one-port admittance or a transformer can be represented equivalently by voltage-controlled current sources, without loss of generality, in this section we may assume that all branches of N are voltage-controlled current sources

$$I_x = y_x V_{j_1 j_2}, \quad x - 1, 2, \ldots, e \qquad (9.88)$$

where

$$V_{j_1 j_2} = V_{j_1} - V_{j_2} \qquad (9.89)$$

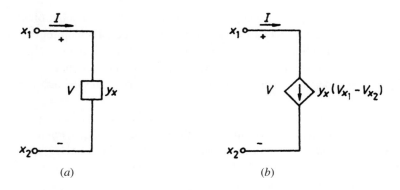

Figure 9.24 (*a*) An one-port admittance y_x, and (*b*) its equivalent controlled source representation.

and V_x is the nodal voltage from node x to some arbitrary but unspecified reference point. In the case $x_1 = j_1$ and $x_2 = j_2$, y_x becomes the one-port admittance of branch b_x.

For example, the one-port admittance y_x of Fig. 9.24*a* can be represented equivalently by the voltage-controlled current source

$$I = y_x V = y_x (V_{x_1} - V_{x_2}) \tag{9.90}$$

of Fig. 9.24*b*. The transformer of Fig. 9.25*a* is characterized by the matrix equation

$$\begin{bmatrix} V_1 \\ V_2 \end{bmatrix} = \begin{bmatrix} L_1 s & M s \\ M s & L_2 s \end{bmatrix} \begin{bmatrix} I_1 \\ I_2 \end{bmatrix} \tag{9.91}$$

the inverse of which is given by

$$\begin{bmatrix} I_1 \\ I_2 \end{bmatrix} = \begin{bmatrix} y_b & -y \\ -y & y_a \end{bmatrix} \begin{bmatrix} V_1 \\ V_2 \end{bmatrix} \tag{9.92}$$

where $L_1 L_2 - M^2 \neq 0$ and

$$y_a = \frac{L_1}{s(L_1 L_2 - M^2)} \tag{9.93a}$$

$$y_b = \frac{L_2}{s(L_1 L_2 - M^2)} \tag{9.93b}$$

$$y = \frac{M}{s(L_1 L_2 - M^2)} \tag{9.93c}$$

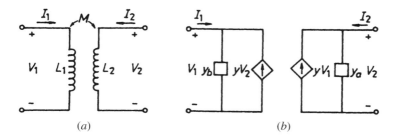

(a)   (b)

Figure 9.25 (a) An imperfectly coupled transformer, and (b) its equivalent controlled source representation.

and can be represented equivalently by the two-port network of Fig. 9.25b containing two voltage-controlled current sources.

As indicated in (2.15), the transadmittance y_x of the voltage-controlled current source

$$I_x = y_x V_{j_1 j_2} = y_x (V_{j_1} - V_{j_2}) \tag{9.94}$$

of (9.88) enters the indefinite-admittance matrix $\mathbf{Y}$ in a rectangular pattern with y_x appearing at the entries (x_1, j_1) and (x_2, j_2) and $-y_x$ at (x_1, j_2) and (x_2, j_1), as shown below:

$$
\begin{array}{c}
\quad\quad\quad V_{j_1} \quad V_{j_2} \quad V_{x_1} \quad V_{x_2} \\
\begin{array}{c} I_{j_1} \\ I_{j_2} \\ I_{x_1} \\ I_{x_2} \end{array}
\left[
\begin{array}{cccc}
0 & 0 & 0 & 0 \\
0 & 0 & 0 & 0 \\
y_x & -y_x & 0 & 0 \\
-y_x & y_x & 0 & 0
\end{array}
\right]
\end{array}
\tag{9.95}
$$

As before, we write $Y_{uv} = Y_{uv}(x)$ and $Y_{rp,sq} = Y_{rp,sq}(x)$ to focus our attention on a particular element x, even though they are also functions of the complex-frequency variable s. With these preliminaries, we state the following main result of this section.

Theorem 9.9 The sum of the reciprocals of the return differences with respect to all network elements and based on the nodal formulation equals the nullity of the network.

PROOF. Without loss of generality, we may assume that the given network N is connected, and contains n nodes, e branches and of nullity m. Let $\mathbf{Y}$ be the indefinite-admittance matrix of N. We next expand the first-order cofactor $Y_{uv}(y_x)$ along each column or row of the submatrix $\mathbf{Y}_{uv}(y_x)$ obtained from $\mathbf{Y}$ by deleting row u and column v and then add the resulting expansions. Since each column or row expansion gives $Y_{uv}(y_x)$, the sum of the above expansions equals $(n-1)Y_{uv}(y_x)$ or

$$\sum_{x=1}^{e} y_x(Y_{x_1 j_1, uv} + Y_{x_2 j_2, uv} - Y_{x_1 j_2, uv} - Y_{x_2 j_1, uv}) = (n-1)Y_{uv}(y_x)$$

$$(9.96)$$

Appealing to (4.148) yields

$$\sum_{x=1}^{e} y_x Y_{x_1 j_1, x_2 j_2} = (n-1)Y_{uv}(y_x) \tag{9.97}$$

Also, from (9.95) $Y_{uv}(y_x)$ can be expanded as

$$Y_{uv}(y_x) = Y_{uv}(0) + y_x(Y_{x_1 j_1, uv} + Y_{x_2 j_2, uv} - Y_{x_1 j_2, uv} - Y_{x_2 j_1, uv})$$

$$= Y_{uv}(0) + y_x Y_{x_1 j_1, x_2 j_2} \tag{9.98}$$

The second line follows from (4.148). Summing over all network elements yields

$$\sum_{x=1}^{e} Y_{uv}(y_x) - \sum_{x=1}^{e} Y_{uv}(0) = e Y_{uv}(y_x) - \sum_{x=1}^{e} Y_{uv}(0)$$

$$= \sum_{x=1}^{e} y_x Y_{x_1 j_1, x_2 j_2} = (n-1)Y_{uv}(y_x) \quad (9.99)$$

or

$$\sum_{x=1}^{e} \frac{Y_{uv}(0)}{Y_{uv}(y_x)} = e - n + 1 = m \tag{9.100}$$

Using (9.87) in (9.100) we obtain

$$\sum_{x=1}^{e} \frac{1}{F(y_x)} = m \tag{9.101}$$

This completes the proof of the theorem.

We illustrate the above results by the following examples.

Example 9.10 Consider the indefinite-admittance matrix

$$
\mathbf{Y} =
\begin{bmatrix}
g_1 & -g_1 & 0 & 0 & 0 \\
-g_1 & g_1 + g_2 + sC & -g_2 & 0 & -sC \\
0 & \beta g_2 - g_2 & g_2 - \beta g_2 + \dfrac{1}{sL} & -\dfrac{1}{sL} & 0 \\
0 & 0 & -\dfrac{1}{sL} & g_3 + \dfrac{1}{sL} & -g_3 \\
0 & -\beta g_2 - sC & \beta g_2 & -g_3 & g_3 + sC
\end{bmatrix}
$$

$$(9.102)$$

of the active network N of Fig. 2.5. Expanding along the first column of $\mathbf{Y}_{55}$ gives

$$Y_{55} = g_1(Y_{11,55} - Y_{21,55}) \tag{9.103}$$

Likewise, expanding along the second, third and fourth columns of $\mathbf{Y}_{55}$ we obtain

$$Y_{55} = -g_1 Y_{12,55} + (g_1 + g_2 + sC)Y_{22,55} + (\beta g_2 - g_2)Y_{32,55} \tag{9.104a}$$

$$Y_{55} = -g_2 Y_{23,55} + \left(g_2 - \beta g_2 + \frac{1}{sL}\right) Y_{33,55} - \frac{Y_{43,55}}{sL} \tag{9.104b}$$

$$Y_{55} = -\frac{Y_{34,55}}{sL} + \left(g_3 + \frac{1}{sL}\right) Y_{44,55} \tag{9.104c}$$

respectively. Adding (9.103) and (9.104) yields

$$4Y_{55} = g_1(Y_{11,55} + Y_{22,55} - Y_{12,55} - Y_{21,55}) + \beta g_2(Y_{32,55} - Y_{33,55})$$

$$\qquad + g_2(Y_{22,55} + Y_{33,55} - Y_{23,55} - Y_{32,55}) + sCY_{22,55}$$

$$\qquad + \frac{1}{sL}(Y_{33,55} + Y_{44,55} - Y_{34,55} - Y_{43,55}) + g_3 Y_{44,55} \tag{9.105}$$

verifying the identity (9.96). Appealing to (4.148), the above equation is simplified to

$$4Y_{55} = g_1 Y_{11,22} + \beta g_2 Y_{32,53} + g_2 Y_{22,33} + sCY_{22,55} + \frac{1}{sL}Y_{33,44} + g_3 Y_{44,55}$$

$$(9.106)$$

confirming (9.97).

To illustrate the theorem, we first compute the first-order cofactors Y_{uv} by setting the individual elements to zero, as follows :

$$Y_{uv}(g_1 = 0) = Y_{22}(g_1 = 0) = Y_{22}(g_1) - g_1 Y_{11,22} \qquad (9.107a)$$

$$Y_{uv}(g_2 = 0) = Y_{33}(g_2 = 0) = Y_{33}(g_2) - g_2 Y_{22,33} \qquad (9.107b)$$

$$Y_{uv}(g_3 = 0) = Y_{55}(g_3 = 0) = Y_{55}(g_3) - g_3 Y_{44,55} \qquad (9.107c)$$

$$Y_{uv}\left(\frac{1}{sL} = 0\right) = Y_{44}\left(\frac{1}{sL} = 0\right) = Y_{44}\left(\frac{1}{sL}\right) - \frac{Y_{33,44}}{sL} \qquad (9.107d)$$

$$Y_{uv}(sC = 0) = Y_{55}(sC = 0) = Y_{55}(sC) - sCY_{22,55} \qquad (9.107e)$$

$$Y_{uv}(\beta g_2 = 0) = Y_{33}(\beta g_2 = 0) = Y_{33}(\beta g_2) + \beta g_2 Y_{52,33} \qquad (9.107f)$$

in which we have used the fact that all the cofactors of the elements of the indefinite-admittance matrix $\mathbf{Y}$ are equal. We remark that in computing the return differences we assume that all the individual admittances are distinct. Thus, g_2 and βg_2 are considered as two distinct admittances. When g_2 is set to zero, βg_2 is not affected, and vice versa. Using (9.107) the sum of the reciprocals of the return differences with respect to all network elements is found to be

$$\sum_{x=1}^{6} \frac{1}{F(y_x)} = \frac{1}{F(g_1)} + \frac{1}{F(g_2)} + \frac{1}{F(g_3)} + \frac{1}{F\left(\frac{1}{sL}\right)} + \frac{1}{F(sC)} + \frac{1}{F(\beta g_2)}$$

$$= \frac{1}{Y_{uv}}\left[Y_{uv}(g_1 = 0) + Y_{uv}(g_2 = 0) + Y_{uv}(g_3 = 0)\right.$$

$$\left. + Y_{uv}\left(\frac{1}{sL} = 0\right) + Y_{uv}(sC = 0) + Y_{uv}(\beta g_2 = 0)\right]$$

$$= \frac{1}{Y_{uv}}\left[6Y_{uv} - g_1 Y_{11,22} - g_2 Y_{22,33} - g_3 Y_{44,55}\right.$$

$$\left. - \frac{Y_{33,44}}{sL} - sCY_{22,55} - \beta g_2 Y_{32,53}\right]$$

$$= \frac{6Y_{uv} - 4Y_{uv}}{Y_{uv}} = 2 \qquad (9.108)$$

where we have invoked (9.106) and the fact that

$$Y_{uv} = Y_{22}(g_1) = Y_{33}(g_2) = Y_{55}(g_3) = Y_{44}\left(\frac{1}{sL}\right)$$

$$= Y_{55}(sC) = Y_{33}(\beta g_2) \qquad (9.109)$$

We remark that after the removal of the independent current source I_g from the network of Fig. 2.5, the resulting network is of nullity 2. Thus, the sum of the reciprocals of the return differences with respect to all network elements equals the nullity of the network after the removal of all the independent sources.

Example 9.11 The indefinite-admittance matrix of the equivalent network of a transistor of Fig. 2.2 is found from (2.5) to be

$$\mathbf{Y} = \begin{bmatrix} g_1 + sC_1 + sC_2 & -sC_2 & -g_1 - sC_1 \\ g_m - sC_2 & g_2 + sC_2 & -g_2 - g_m \\ -g_1 - sC_1 - g_m & -g_2 & g_1 + g_2 + g_m + sC_1 \end{bmatrix} \tag{9.110}$$

To compute the return differences with respect to all network elements, we first compute the first-order cofactors $Y_{uv} = Y_{33}$ by setting the individual elements to zero, as follows:

$$Y_{33}(g_1 = 0) = g_2(sC_1 + sC_2) + sC_2(g_m + sC_1) \tag{9.111a}$$

$$Y_{33}(g_2 = 0) = sC_2(g_1 + sC_1 + g_m) \tag{9.111b}$$

$$Y_{33}(sC_1 = 0) = g_2(g_1 + sC_2) + sC_2(g_1 + g_m) \tag{9.111c}$$

$$Y_{33}(sC_2 = 0) = g_2(g_1 + sC_1) \tag{9.111d}$$

$$Y_{33}(g_m = 0) = g_2(g_1 + sC_1 + sC_2) + sC_2(g_1 + sC_1) \tag{9.111e}$$

The sum of the reciprocals of the return differences with respect to all network elements is found to be

$$\sum_{x=1}^{5} \frac{1}{F(x)} = \frac{1}{F(g_1)} + \frac{1}{F(g_2)} + \frac{1}{F(sC_1)} + \frac{1}{F(sC_2)} + \frac{1}{F(g_m)}$$

$$= \frac{\begin{array}{c} Y_{33}(g_1 = 0) + Y_{33}(g_2 = 0) + Y_{33}(sC_1 = 0) \\ + Y_{33}(sC_2 = 0) + Y_{33}(g_m = 0) \end{array}}{Y_{33}}$$

$$= \frac{3[(g_1 + sC_1)(g_2 + sC_2) + sC_2(g_2 + g_m)]}{Y_{33}}$$

$$= \frac{3Y_{33}}{Y_{33}} = 3 \tag{9.112}$$

the nullity of the network of Fig. 2.2.

We now discuss some consequences of Theorem 9.9 in relation to the summation of driving-point and transfer impedances.

Assume that a current source is connected between any two nodes x_1 and x_2 so that a current I_s is injected into the x_1 node and at the same time is extracted from the x_2 node. The transfer impedance, as defined in (2.89), between the node pairs x_1, x_2 and j_1, j_2 is obtained from (2.94) as

$$z_{x_1 j_1, x_2 j_2} = \frac{V_{j_1 j_2}}{I_s} = \frac{Y_{x_1 j_1, x_2 j_2}}{Y_{uv}} \qquad (9.113)$$

Substituting this in (9.97) gives

$$\frac{\sum_{x=1}^{e} y_x Y_{x_1 j_1, x_2 j_2}}{Y_{uv}} = \sum_{x=1}^{e} y_x \frac{Y_{x_1 j_1, x_2 j_2}}{Y_{uv}}$$

$$= \sum_{x=1}^{e} y_x z_{x_1 j_1, x_2 j_2} = n - 1 = r \qquad (9.114)$$

where r denotes the rank of the network, which is assumed to be connected. In the case of an unconnected network of c components, the above procedure can be applied to each component to yield the general relation

$$\sum_{x=1}^{e} y_x z_{x_1 j_1, x_2 j_2} = n - c = r \qquad (9.115)$$

Theorem 9.10 The sum of the products of all transadmittances y_x and their associated transfer impedances $z_{x_1 j_1, x_2 j_2}$ between the controlled branches from node x_1 to node x_2 and the controlling branches from node j_1 to node j_2 in a network equals the rank of the network.

In particular, when $x_1 = j_1$ and $x_2 = j_2$, $z_{x_1 x_1, x_2 x_2}$ denotes the driving-point impedance of the network across the terminals of the branch y_x, and (9.115) becomes a well-known result on the summation of driving-point impedances:

$$\sum_{x=1}^{e} y_x z_{x_1 x_1, x_2 x_2} = r \qquad (9.116)$$

Corollary 9.7 For an RLC network with no coupling, the sum of products of all branch admittances and the driving-point impedances across the terminals of the branches equals the rank of the network.

Figure 9.26 An equivalent network of a transistor.

Corollary 9.7 was first pointed out by Tsang (1954), an extension of which to RLC networks with mutual couplings was given by Hajj (1984). The general result (Theorem 9.9) was shown by Chen (1984).

We illustrate the above results by the following examples.

Example 9.12 An equivalent network N of a transistor is shown in Fig. 9.26, the indefinite-admittance matrix of which is given by

$$\mathbf{Y} = \begin{bmatrix} g_1 + sC_1 + sC_2 & -sC_2 & -g_1 - sC_1 \\ g_m - sC_2 & g_2 + sC_2 & -g_2 - g_m \\ -g_1 - sC_1 - g_m & -g_2 & g_1 + g_2 + g_m + sC_1 \end{bmatrix} \quad (9.117)$$

The driving-point impedances across the admittances g_1 or sC_1, sC_2 and g_2 are, respectively, given by

$$z_{11,33} = \frac{Y_{11,33}}{Y_{uv}} = \frac{g_2 + sC_2}{Y_{uv}} \quad (9.118a)$$

$$z_{11,22} = \frac{Y_{11,22}}{Y_{uv}} = \frac{g_1 + g_2 + g_m + sC_1}{Y_{uv}} \quad (9.118b)$$

$$z_{22,33} = \frac{Y_{22,33}}{Y_{uv}} = \frac{g_1 + sC_1 + sC_2}{Y_{uv}} \quad (9.118c)$$

The transfer impedance between the node pairs 2,3 and 1,3 of the voltage- controlled current source $g_m V$ and its controlling voltage V is obtained as

$$z_{21,33} = \frac{Y_{21,33}}{Y_{uv}} = \frac{sC_2}{Y_{uv}} \quad (9.119)$$

The sum of the products of the transadmittances and their associated transfer impedances is computed as follows:

$$\sum_{x=1}^{5} y_x z_{x_1 j_1, x_2 j_2} = g_1 z_{11,33} + s C_1 z_{11,33} + s C_2 z_{11,22} + g_2 z_{22,33} + g_m z_{21,33}$$

$$= \frac{1}{Y_{uv}} \left[g_1(g_2 + sC_2) + sC_1(g_2 + sC_2) \right.$$

$$+ sC_2(g_1 + g_2 + g_m + sC_1)$$

$$+ g_2(g_1 + sC_1 + sC_2) + g_m sC_2 \Big]$$

$$= \frac{2Y_{33}}{Y_{uv}} = \frac{2Y_{33}}{Y_{33}} = 2 \tag{9.120}$$

which is the rank of the network of Fig. 9.26, confirming Theorem 9.10.

Example 9.13 The T-equivalent network of a transistor is shown in Fig. 9.27, the indefinite-admittance matrix of which is given by

$$\mathbf{Y} = \begin{bmatrix} g_e & 0 & 0 & -g_e \\ -\alpha g_e & g_c & 0 & \alpha g_e - g_c \\ 0 & 0 & g_b & -g_b \\ \alpha g_e - g_e & -g_c & -g_b & g_e + g_b + g_c - \alpha g_e \end{bmatrix} \tag{9.121}$$

Figure 9.27 The T-equivalent network of a transistor.

as previously computed in (2.34). The driving-point impedances across the admittances g_e, g_b and g_c are, respectively, given by

$$z_{11,44} = \frac{Y_{11,44}}{Y_{uv}} = \frac{g_b g_c}{g_b g_c g_e} = \frac{1}{g_e} \qquad (9.122a)$$

$$z_{33,44} = \frac{Y_{33,44}}{Y_{uv}} = \frac{g_c g_e}{g_b g_c g_e} = \frac{1}{g_b} \qquad (9.122b)$$

$$z_{22,44} = \frac{Y_{22,44}}{Y_{uv}} = \frac{g_b g_e}{g_b g_c g_e} = \frac{1}{g_c} \qquad (9.122c)$$

The transfer impedance between the node pairs 2,4 and 1,4 of the voltage-controlled current source $\alpha g_e V_{14}$ and its controlling voltage V_{14} from terminal 1 to terminal 4 is found to be

$$z_{21,44} = \frac{Y_{21,44}}{Y_{uv}} = \frac{0}{g_b g_c g_e} = 0 \qquad (9.123)$$

The sum of the products of the transadmittances and their associated transfer impedances is obtained as

$$\sum_{x=1}^{4} y_z z_{x_1 j_1, x_2 j_2} = g_e z_{11,44} + g_b z_{33,44} + g_c z_{22,44} + \alpha g_e z_{21,44}$$

$$= 1 + 1 + 1 + 0 = 3 \qquad (9.124)$$

which is the rank of the network of Fig. 9.27, confirming Theorem 9.10.

9.5.2 Loop Formulation of Return Difference

In the foregoing, we showed that the sum of the reciprocals of the return differences with respect to all network elements equals the nullity of the network, where the return difference is defined as the ratio of the two functional values assumed by the nodal determinant under the condition that the element of interest assumes its nominal value and the condition that the element assumes the zero value. As a result, the sum of the products of all the transadmittances and their associated transfer impedances between the controlled branches and the controlling branches in a network equals the rank of the network. In the present section, we show that if the return difference is formulated in terms of the loop determinant as in (5.219), dual results on the summations of the reciprocals of the return differences and the products of transimpedances and transfer admittances are obtained for networks containing active and coupled elements.

To simplify the formulation, let N be a connected planar network of e branches $b_x(x = 1, 2, \ldots, e)$ and of nullity m. Assume that each branch b_x has a nonzero impedance z_x. Since a one-port impedance or a transformer can be equivalently represented by current-controlled voltage sources, without loss of generality we may assume that all branches of N are current-controlled voltage sources

$$V_x = z_x(I_{j_1} - I_{j_2}), \quad x = 1, 2, \ldots, e \tag{9.125}$$

where $I_u(u = 1, 2, \ldots, m + 1)$ are the mesh currents. All the mesh currents are oriented clockwise except the outer mesh I_{m+1} which is counterclockwise. For $j = x$

$$V_x = z_x(I_{x_1} - I_{x_2}) \tag{9.126}$$

and z_x becomes the one-port impedance of branch b_x. The matrix $\mathbf{Z}$ relating the mesh-current vector $\mathbf{J}$ and the mesh voltage-source vector $\mathbf{E}$

$$\mathbf{ZJ} = \mathbf{E} \tag{9.127}$$

is called the *indefinite-impedance matrix*, which is square and of order $m + 1$. In general, the impedance z_x enters the indefinite-impedance matrix $\mathbf{Z}$ in a rectangular pattern, as in (9.95), with z_x appearing at the entries (x_1, j_1) and (x_2, j_2) and z_x at (x_1, j_2) and (x_2, j_1), where branch b_x is traversed by mesh currents I_{x_1} and I_{x_2}, as shown below:

$$\begin{array}{c} \\ V_{j_1} \\ V_{j_2} \\ V_{x_1} \\ V_{x_2} \end{array} \begin{array}{cccc} I_{j_1} & I_{j_2} & I_{x_1} & I_{x_2} \\ \left[\begin{array}{cccc} & & & \\ & & & \\ z_x & -z_x & & \\ -z_x & z_x & & \end{array} \right. \end{array} \left. \begin{array}{c} \\ \\ \\ \\ \end{array} \right] \tag{9.128}$$

Denote by $\mathbf{Z}_{ij}$, the submatrix obtained from $\mathbf{Z}$ by deleting the row i and column j. Likewise, $\mathbf{Z}_{rp,sq}$ denotes the submatrix derived from $\mathbf{Z}$ by deleting the rows r and s and columns p and q. The *first-* and *second-order cofactors* of the elements of $\mathbf{Z}$ are the scalar quantities defined by the equations

$$Z_{ij} = (-1)^{i+j} \det \mathbf{Z}_{ij} \tag{9.129}$$

$$Z_{rp,sq} = \text{sgn } (r - s) \text{ sgn } (p - q)(-1)^{r+p+s+q} \det \mathbf{Z}_{rp,sq} \tag{9.130}$$

respectively. It is straightforward to show that $\mathbf{Z}$ is an equicofactor matrix

$$Z_{ij} = Z_{uv} \tag{9.131}$$

for all i, j, u and v. The *return difference* $F(x)$ with respect to a network element x is defined as the ratio of the two functional values assumed by the loop determinant under the condition that the element x assumes its nominal value and the condition that the element x assumes the zero value:

$$F(x) = \frac{Z_{uv}(x)}{Z_{uv}(0)} \tag{9.132}$$

where, as before, $\mathbf{Z} = \mathbf{Z}(x)$ to emphasize the importance of x, even though it is also a function of the complex-frequency variable s. With these preliminaries, we state the following theorem of Chen (1985).

Theorem 9.11 The sum of the reciprocals of the return differences with respect to all network elements and based on the loop formulation in a planar network equals the rank of the planar network.

PROOF. Without loss of generality, assume that the planar network is connected, and is of e branches and of rank r; for, otherwise, we consider each component separately, and invoke the fact that the rank of a network is equal to the sum of the ranks of its components.

Let N be an n-node connected planar network, the indefinite-impedance matrix of which is $\mathbf{Z}$. As in (9.96), we expand the first-order cofactor $Z_{uv}(z_x)$ along each column or row of $\mathbf{Z}_{uv}(z_x)$, and then add the resulting expansions. This gives

$$\sum_{x=1}^{e} z_x (Z_{x_1 j_1, uv} + Z_{x_2 j_2, uv} - Z_{x_1 j_2, uv} - Z_{x_2 j_1, uv}) = m Z_{uv}(z_x) \tag{9.133}$$

Since $\mathbf{Z}$ is an equicofactor matrix, from Theorem 2.2 the second-order cofactors of its elements are related by

$$Z_{rp,sq} = Z_{rp,uv} + Z_{sq,uv} - Z_{rq,uv} - Z_{sp,uv} \tag{9.134}$$

for all r, p, s, q, u, $v \leq m + 1$. Using this in (9.133) results in

$$\sum_{x=1}^{e} z_x Z_{x_1 j_1, x_2 j_2} = m Z_{uv}(z_x) \tag{9.135}$$

Following (9.98), $Z_{uv}(z_x)$ can be expanded as

$$Z_{uv}(z_x) = Z_{uv}(0) + z_x (Z_{x_1 j_1, uv} + Z_{x_2 j_2, uv} - Z_{x_1 j_2, uv} - Z_{x_2 j_1, uv})$$

$$= Z_{uv}(0) + z_x Z_{x_1 j_1, x_2 j_2} \tag{9.136}$$

Summing both sides for all x gives

$$\sum_{x=1}^{e} Z_{uv}(z_x) - \sum_{x=1}^{e} Z_{uv}(0) = \sum_{x=1}^{e} z_x Z_{x_1 j_1, x_2 j_2} \tag{9.137}$$

Combining this with (9.135) we obtain

$$e Z_{uv}(z_x) - \sum_{x=1}^{e} Z_{uv}(0) = m Z_{uv}(z_x) \tag{9.138}$$

Dividing both sides by $Z_{uv}(z_x)$ yields

$$\sum_{x=1}^{e} \frac{Z_{uv}(0)}{Z_{uv}(z_x)} = e - m = n - 1 = r \tag{9.139}$$

or

$$\sum_{x=1}^{e} \frac{1}{F(z_x)} = r \tag{9.140}$$

This completes the proof of the theorem.

Assume that the network N is excited by a voltage source V_{in}, the output of which is I_0. If V_{in} is traversed by the mesh currents I_{x_1} and I_{x_2}, and I_0 by I_{j_1} and I_{j_2}, as depicted in Fig. 9.28, the transfer admittance can be expressed compactly as

$$y_{x_1 j_1, x_2 j_2} = \frac{I_0}{V_{in}} = \frac{Z_{x_1 j_1, x_2 j_2}}{Z_{uv}(z_x)} \tag{9.141}$$

This symbol was also used to represent the short-circuit transfer admittance, as previously defined in (2.112). Formally, we should use a different symbol instead of $y_{x_1 j_1, x_2 j_2}$ in (9.141). However, in this section since we deal exclusively with the

Figure 9.28 Network used to define transfer admittance $y_{x_1 j_1, x_2 j_2}$.

indefinite-impedance matrix, the present usage is much preferred. This should not create any difficulty as the context will reveal.

Substituting (9.141) in (9.135) we obtain a dual result of Theorem 9.10 on the summation of the products of the transimpedances and their associated transfer admittances:

$$\sum_{x=1}^{e} z_x y_{x_1 j_1, x_2 j_2} = m \tag{9.142}$$

Theorem 9.12 The sum of the products of all transimpedances z_x and their associated transfer admittances $y_{x_1 j_1, x_2 j_2}$ between the controlled branches traversed by the meshes x_1 and x_2 and the controlling branches traversed by the meshes j_1 and j_2 in a planar network equals the nullity of the network.

In particular, when $x = j$, $y_{x_1 x_1, x_2 x_2}$ denotes the driving-point admittance looking into a voltage generator connected in series with the branch impedance z_x, and (9.142) becomes a well-known result on the summation of the driving-point admittances for an RLC network:

$$\sum_{x=1}^{e} z_x y_{x_1 x_1, x_2 x_2} = m \tag{9.143}$$

This result was first pointed out by Tsang (1954), and is valid for both planar and nonplanar networks, even though it was proved only for the planar case.

Corollary 9.8 For an RLC network with no coupling, the sum of the products of all branch impedances and the driving-point admittances looking into the branches including the branch impedances themselves equals the nullity of the network.

We illustrate the above results by the following examples.

Example 9.14 Consider the active planar network N of Fig. 9.29, the indefinite-impedance matrix equation of which is found to be

$$\begin{bmatrix} z_1 + z_4 & -z_1 & -z_4 \\ -z_1 - g_m z_1 z_3 & z_1 + z_2 + z_3 + g_m z_1 z_3 & -z_2 - z_3 \\ g_m z_1 z_3 - z_4 & -z_2 - z_3 - g_m z_1 z_3 & z_2 + z_3 + z_4 \end{bmatrix} \begin{bmatrix} i_1 \\ i_2 \\ i_3 \end{bmatrix} = \begin{bmatrix} V_{in} \\ 0 \\ 0 \end{bmatrix}$$

$$\tag{9.144}$$

Figure 9.29 An equivalent network of a transistor network.

Appealing to (9.142) we obtain

$$\sum_{x=1}^{5} z_x y_{x_1 j_1, x_2 j_2} = z_1 y_{11,22} + g_m z_1 z_3 y_{31,22} + z_2 y_{22,33} + z_3 y_{22,33} + z_4 y_{11,33}$$

$$= \frac{1}{Z_{33}}[z_1(z_2 + z_3 + z_4 + g_m z_3 z_4) + z_2(z_1 + z_4) + z_3(z_1 + z_4)$$

$$+ z_4(z_1 + z_2 + z_3 + g_m z_1 z_3)]$$

$$= 2 = m \qquad (9.145)$$

the nullity of N, confirming Theorem 9.12, where

$$Z_{33} = z_1(z_2 + z_3) + z_4(z_1 + z_2 + z_3 + g_m z_1 z_3) \qquad (9.146)$$

is the first-order cofactor of the third row and third column element of the coefficient matrix of (9.144).

To verify (9.140) we let $z_5 = g_m z_1 z_3$ and compute

$$\sum_{x=1}^{5} \frac{1}{F(z_x)} = \frac{1}{F(z_1)} + \frac{1}{F(z_2)} + \frac{1}{F(z_3)} + \frac{1}{F(z_4)} + \frac{1}{F(z_5)}$$

$$= \frac{Z_{33}(z_1 = 0)}{Z_{33}(z_1)} + \frac{Z_{33}(z_2 = 0)}{Z_{33}(z_2)} + \frac{Z_{33}(z_3 = 0)}{Z_{33}(z_3)}$$

$$+ \frac{Z_{33}(z_4 = 0)}{Z_{33}(z_4)} + \frac{Z_{33}(z_5 = 0)}{Z_{33}(z_5)}$$

$$= \frac{1}{Z_{33}}[z_4(z_2 + z_3 + g_m z_1 z_3) + z_1 z_3 + z_4(z_1 + z_3 + g_m z_1 z_3)$$

Figure 9.30 A passive planar network used to illustrate Theorems 9.11 and 9.12.

$$+ z_1 z_2 + z_4(z_1 + z_2 + g_m z_1 z_3) + z_1(z_2 + z_3)$$

$$+ z_1(z_2 + z_3) + z_4(z_1 + z_2 + z_3)]$$

$$= 3 = r \qquad (9.147)$$

the rank of the network of Fig. 9.29 after short-circuiting the voltage source, confirming Theorem 9.11. We remark that when we set z_1 or z_3 to zero, $z_5 = g_m z_1 z_3$ should not be set to zero.

Example 9.15 Consider the passive planar network N of Fig. 9.30, the indefinite-impedance matrix of which is found to be

$$\mathbf{Z} = \begin{bmatrix} z_1 + z_4 + z_5 + z_6 & -z_1 - z_6 & 0 & -z_4 - z_5 \\ -z_1 - z_6 & z_2 + z_2 + z_6 & -z_2 & 0 \\ 0 & -z_2 & z_2 + z_3 + z_7 & -z_3 - z_7 \\ -z_4 - z_5 & 0 & -z_3 - z_7 & z_3 + z_4 + z_5 + z_7 \end{bmatrix}$$

$$(9.148)$$

From (9.143) we obtain

$$\sum_{x=1}^{7} z_x y_{x_1 x_1, x_2 x_2} = \frac{1}{Z_{uv}} \sum_{x=1}^{7} z_x Z_{x_1 x_1, x_2 x_2}$$

$$= \frac{1}{Z_{44}} [z_1 Z_{11,22} + z_2 Z_{22,33} + z_3 Z_{33,44} + z_4 Z_{11,44}$$

$$+ z_5 Z_{11,44} + z_6 Z_{11,22} + z_7 Z_{33,44}]$$

$$= \frac{1}{Z_{44}} \{ z_1[(z_2 + z_3 + z_7)(z_3 + z_4 + z_5 + z_7) - (z_3 + z_7)^2]$$

$$+ z_2[(z_1 + z_4 + z_5 + z_6)(z_3 + z_4 + z_5 + z_7) - (z_4 + z_5)^2]$$

$$+ z_3[(z_1 + z_4 + z_5 + z_6)(z_1 + z_2 + z_6) - (z_1 + z_6)^2]$$

$$+ z_4[(z_1 + z_2 + z_6)(z_2 + z_3 + z_7) - z_2^2)]$$

$$+ z_5[(z_1 + z_2 + z_6)(z_2 + z_3 + z_7) - z_2^2)]$$

$$+ z_6[(z_2 + z_3 + z_7)(z_3 + z_4 + z_5 + z_7) - (z_3 + z_7)^2]$$

$$+ z_7[(z_1 + z_4 + z_5 + z_6)(z_1 + z_2 + z_6) - (z_1 + z_6)^2]\}$$

$$= 3 = m \qquad (9.149)$$

the nullity of N, confirming Theorem 9.12, where the first-order cofactor is obtained as

$$Z_{44} = (z_1 + z_4 + z_5 + z_6)(z_1 + z_2 + z_6)(z_2 + z_3 + z_7)$$
$$- (z_1 + z_6)^2(z_2 + z_3 + z_7) - z_2^2(z_1 + z_4 + z_5 + z_6) \qquad (9.150)$$

To verify (9.140) we calculate

$$\sum_{x=1}^{7} \frac{1}{F(z_x)} = \frac{1}{F(z_1)} + \frac{1}{F(z_2)} + \frac{1}{F(z_3)} + \frac{1}{F(z_4)} + \frac{1}{F(z_5)} + \frac{1}{F(z_6)} + \frac{1}{F(z_7)}$$

$$= \frac{1}{Z_{44}}[Z_{44}(z_1 = 0) + Z_{44}(z_2 = 0) + Z_{44}(z_3 = 0) + Z_{44}(z_4 = 0)$$

$$+ Z_{44}(z_5 = 0) + Z_{44}(z_6 = 0) + Z_{44}(z_7 = 0)]$$

$$= \frac{1}{Z_{44}}\{[(z_4 + z_5 + z_6)(z_2 + z_6)(z_2 + z_3 + z_7) - z_6^2(z_2 + z_3 + z_7)$$

$$- z_2^2(z_4 + z_5 + z_6)] + [(z_1 + z_4 + z_5 + z_6)(z_1 + z_6)(z_3 + z_7)$$

$$- (z_1 + z_6)^2(z_3 + z_7)] - [(z_1 + z_4 + z_5 + z_6)(z_1 + z_2 + z_6)$$

$$\times (z_2 + z_7) - (z_1 + z_6)^2(z_2 + z_7) - z_2^2(z_1 + z_4 + z_5 + z_6)]$$

$$+ [(z_1 + z_5 + z_6)(z_1 + z_2 + z_6)(z_2 + z_3 + z_7)$$

$$- (z_1 + z_6)^2(z_2 + z_3 + z_7) - z_2^2(z_1 + z_5 + z_6)]$$

$$+ [(z_1 + z_4 + z_6)(z_1 + z_2 + z_6)(z_2 + z_3 + z_7)$$

$$- (z_1 + z_6)^2 (z_2 + z_3 + z_7) - z_2^2(z_1 + z_4 + z_6)]$$

$$+ [(z_1 + z_4 + z_5)(z_1 + z_2)(z_2 + z_3 + z_7) - z_1^2(z_2 + z_3 + z_7)$$

$$- z_2^2(z_1 + z_4 + z_5)] + [(z_1 + z_4 + z_5 + z_6)(z_1 + z_2 + z_6)(z_2 + z_3)$$

$$- (z_1 + z_6)^2(z_2 + z_3) - z_2^2(z_1 + z_4 + z_5 + z_6)]\}$$

$$= 4 = r \tag{9.151}$$

the rank of N, confirming Theorem 9.11.

In the foregoing, we have demonstrated that if all the mesh currents are oriented clockwise except the outer mesh which is counterclockwise, then the matrix relating the mesh-current vector and the mesh voltage-source vector is the indefinite-impedance matrix of the planar network. This choice of mesh currents is not necessary in defining the indefinite-impedance matrix. In fact, if the network is nonplanar, no such choice is possible. However, it can be shown by induction [Chen and Lu (1989)] that it is always possible to assign $m + 1$ loop current sets to a network N, planar or nonplanar, of nullity m, so that the number of loop currents traversing in one direction through a branch equals the number of loop currents traversing in the opposite direction through the branch. Furthermore, any m of these $m + 1$ loop current sets are linearly independent. Then the matrix relating the loop-current set vector and the loop voltage-source set vector is the *indefinite-impedance matrix* $\mathbf{Z}$ of the network. With this extension, Theorems 9.11 and 9.12 remain valid, *mutatis mutandis*, for nonplanar networks as can be seen by making the necessary changes in the proofs. The only difference is that if z_x is traversed by more than two loop currents, the transfer admittance $y_{x_1 j_1, x_2 j_2}$ in (9.142) and (9.143) is replaced by a sum of transfer admittances involving these loop currents only [Wang and Chen (1987)].

We illustrate this generalization by the following examples.

Example 9.16 Consider the active planar network N of Fig. 9.31. Using the loop current sets shown in the figure, the indefinite-impedance matrix of the network is found to be

$$\mathbf{Z} = \begin{bmatrix} z_2 + z_4 + z_5 & -z_2 & -z_2 & z_2 - z_4 - z_5 \\ -z_2 & z_1 + z_2 + z_6 & z_2 & -z_1 - z_2 - z_6 \\ -z_2 & z_2 - z_7 & z_2 + z_3 & z_7 - z_2 - z_3 \\ z_2 - z_4 - z_5 & z_7 - z_1 - z_2 - z_6 & -z_2 - z_3 & \begin{matrix} z_1 + z_2 + z_3 + z_4 \\ + z_5 + z_6 - z_7 \end{matrix} \end{bmatrix} \tag{9.152}$$

Figure 9.31 An active planar network used to illustrate (9.140) and (9.142).

Appealing to (9.140) we obtain

$$\sum_{x=1}^{7} \frac{1}{F(z_x)} = \frac{1}{F(z_1)} + \frac{1}{F(z_2)} + \frac{1}{F(z_3)} + \frac{1}{F(z_4)} + \frac{1}{F(z_5)}$$

$$+ \frac{1}{F(z_6)} + \frac{1}{F(z_7)}$$

$$= \frac{1}{Z_{44}}[Z_{44}(z_1 = 0) + Z_{44}(z_2 = 0) + z_{44}(z_3 = 0) + Z_{44}(z_4 = 0)$$

$$+ Z_{44}(z_5 = 0) + Z_{44}(z_6 = 0) + Z_{44}(z_7 = 0)]$$

$$= \frac{1}{Z_{44}}(z_2 z_3 z_4 + z_2 z_3 z_5 + z_2 z_3 z_6 + z_2 z_4 z_6 + z_2 z_5 z_6 + z_3 z_4 z_6$$

$$+ z_3 z_5 z_6 + z_2 z_4 z_7 + z_2 z_5 z_7 + z_1 z_3 z_4 + z_1 z_3 z_5 + z_3 z_5 z_6$$

$$+ z_3 z_4 z_6 + z_1 z_2 z_4 + z_1 z_2 z_5 + z_2 z_4 z_6 + z_2 z_5 z_6 + z_2 z_4 z_7$$

$$+ z_2 z_5 z_7 + z_1 z_2 z_5 + z_1 z_2 z_3 + z_1 z_3 z_5 + z_2 z_3 z_5 + z_2 z_3 z_6$$

$$+ z_2 z_5 z_6 + z_3 z_5 z_6 + z_2 z_5 z_7 + z_1 z_2 z_4 + z_1 z_2 z_3 + z_1 z_3 z_4$$

$$+ z_2 z_3 x_4 + z_2 z_3 z_6 + z_2 z_4 z_6 + z_3 z_4 z_6 + z_2 z_4 z_7 + z_1 z_2 z_3$$

$$+ z_1 z_2 z_4 + z_1 z_2 z_5 + z_1 z_3 z_4 + z_1 z_3 z_5 + z_2 z_3 z_4 + z_2 z_3 z_5$$

$$+ z_2z_4z_7 + z_2z_5z_7 + z_1z_2z_3 + z_1z_2z_4 + z_1z_2z_5 + z_1z_3z_4$$

$$+ z_1z_3z_5 + z_2z_3z_4 + z_2z_3z_5 + z_2z_3z_6 + z_2z_4z_6 + z_2z_5z_6$$

$$+ z_3z_5z_6 + z_3z_4z_6)$$

$$= 4 = r \tag{9.153}$$

the rank of N, where the first-order cofactor Z_{44} is given by

$$Z_{44} = z_1z_2z_3 + z_1z_2z_4 + z_1z_2z_5 + z_1z_3z_4 + z_1z_3z_5 + z_2z_3z_4 + z_2z_3z_5$$

$$+ z_2z_3z_6 + z_2z_4z_6 + z_2z_5z_6 + z_3z_4z_6 + z_3z_5z_6 + z_2z_4z_7 + z_2z_5z_7 \tag{9.154}$$

To verify (9.142) we calculate

$$\sum_{x=1}^{7} z_x y_{x_1 j_1, x_2 j_2} = \frac{1}{Z_{44}} [z_1 Z_{22,44} + z_2(Z_{11,33} + Z_{44,22} + Z_{14,32} + Z_{41,23})$$

$$+ z_3 Z_{33,44} + z_4 Z_{11,44} + z_5 Z_{11,44} + z_6 Z_{22,44} + z_7 Z_{32,44}]$$

$$= \frac{1}{Z_{44}} (z_1z_2z_3 + z_1z_2z_4 + z_1z_2z_5 + z_1z_3z_4 + z_1z_3z_5 + z_1z_2z_3$$

$$+ z_1z_2z_4 + z_1z_2z_5 + z_2z_3z_4 + z_2z_3z_5 + z_2z_3z_6 + z_2z_4z_6$$

$$+ z_2z_5z_6 + z_2z_4z_7 + z_2z_5z_7 + z_1z_2z_3 + z_1z_3z_4 + z_1z_3z_5$$

$$+ z_2z_3z_4 + z_2z_3z_5 + z_2z_3z_6 + z_3z_4z_6 + z_3z_5z_6 + z_1z_2z_4$$

$$+ z_1z_3z_4 + z_2z_3z_4 + z_2z_4z_6 + z_3z_4z_6 + z_2z_4z_7 + z_1z_2z_5$$

$$+ z_2z_5z_6 + z_2z_5z_7 + z_1z_3z_5 + z_2z_3z_5 + z_3z_5z_6 + z_2z_3z_6$$

$$+ z_2z_4z_6 + z_2z_5z_6 + z_3z_4z_6 + z_3z_5z_6 + z_2z_4z_7 + z_2z_5z_7)$$

$$= 3 = m \tag{9.155}$$

the nullity of N.

Example 9.17 Figure 9.32 is a passive nonplanar network N with no mutual coupling among its elements. Using the loop currents as indicated in the figure, the

Figure 9.32 A passive nonplanar network with no mutual coupling among its elements.

indefinite-impedance matrix of the network is found to be

$$
\mathbf{z} = \begin{bmatrix}
z_1 + z_4 + z_8 & -z_4 & 0 & -z_1 \\
-z_4 & z_2 + z_4 + z_5 & -z_5 & -z_2 \\
0 & -z_5 & z_3 + z_5 + z_9 & 0 \\
-z_1 & -z_2 & 0 & z_1 + z_2 + z_6 \\
0 & -z_2 & -z_3 & z_2 \\
0 & z_2 & 0 & -z_2 - z_6 \\
-z_8 & 0 & -z_9 & 0
\end{bmatrix}
$$

$$
\begin{bmatrix}
0 & 0 & -z_8 \\
-z_2 & z_2 & 0 \\
-z_3 & 0 & -z_9 \\
z_2 & -z_2 - z_6 & 0 \\
z_2 + z_3 + z_7 & -z_2 - z_7 & 0 \\
-z_2 - z_7 & z_2 + z_6 + z_7 + z_{10} & -z_{10} \\
0 & -z_{10} & z_8 + z_9 + z_{10}
\end{bmatrix} \tag{9.156}
$$

We first calculate

$$\sum_{x=1}^{10} \frac{1}{F(z_x)} = \frac{1}{Z_{66}} \sum_{x=1}^{10} Z_{66}(z_x = 0) = \frac{P}{Z_{66}} = 4 = r \qquad (9.157)$$

the rank of N, where P denotes the sum of 500 products involving the branch impedances z_x, and Z_{66} is the first-order cofactor of the sixth row and sixth column element of $\mathbf{Z}$ and contains 125 terms. Finally, to verify (9.143), we compute

$$\sum_{x=1}^{10} z_x y_{x_1 x_1, x_2 x_2} = \frac{1}{Z_{66}} [z_1 Z_{11,44} + z_2 (Z_{22,44} + Z_{66,55} + Z_{26,45} + Z_{62,54})$$

$$+ z_3 Z_{33,55} + z_4 Z_{11,22} + z_5 Z_{22,33} + z_6 Z_{44,66}$$

$$+ z_7 Z_{55,66} + z_8 Z_{11,77} + z_9 Z_{33,77} + z_{10} Z_{66,77}]$$

$$= 6 = m \qquad (9.158)$$

the nullity of N, where there are 750 terms inside the brackets on the right-hand side of (9.158).

9.6 TOPOLOGICAL ANALYSIS OF ACTIVE NETWORKS

In Sec. 9.2 we have indicated how to construct the associated directed graphs for a large class of most commonly used networks, and shown how they can be used for the evaluation of nodal determinant. In the present section, we demonstrate that the associated digraphs themselves may be considered as physical models of the networks, and can be employed directly for the evaluation of network functions.

Like many other topics discussed in this chapter, the basic idea of the present approach is not new, and was first exploited by Kirchhoff (1847) for the mesh system of equations, and by Maxwell (1892) for the nodal system. The modem treatment follows that of Chen (1976).

Definition 9.8: Meter branch A *meter branch* is a directed edge used to represent either an *am*meter, denoted by the symbol *am*, or a *volt*meter, denoted by the symbol *vm*. The current reference is given by the direction of the edge, and the voltage-reference plus is at the tail of the current-reference arrow.

Definition 9.9: Source branch A *source branch* is a directed edge used to represent either an independent *current source*, denoted by the symbol *cs*, or an independent *voltage source*, denoted by the symbol *vs*. The current reference is given by the direction of the edge, and the voltage-reference plus is at the *head* of the current-reference arrow.

For our purposes, the associated digraph of a network is considered as its physical model. When a source branch and a meter branch are connected to the digraph G, the source causes the meter to register. If it is an ammeter, then the reading indicates the amount of current flowing through the meter branch; if it is a voltmeter, the reading indicates the voltage drop across the meter branch.

Definition 9.10: Transmission Let a meter branch directed from node p to node q and a source branch directed from s to r be inserted in the associated digraph G of a network. The *transmission*, denoted by the symbol $H_{rp,sq}$, between the node pairs rs and pq of G is defined as the ratio of meter reading to source value.

Thus, if the meter branch denotes an ammeter, and if the source branch represents a voltage source, the transmission $H_{rp,sq}$ represents the short-circuit transfer admittance $y_{rp,sq}$. On the other hand, if the meter branch denotes a voltmeter and if the source branch represents a current source, the transmission $H_{rp,sq}$ represents the open-circuit transfer impedance $z_{rp,sq}$. This is similarly valid for all other types of network functions. These results are, of course, quite general, because the letters r, s, p and q may refer to the labels of any four nodes in any linear active network. In general, we also allow the situation where a meter branch and a source branch are first connected in series and then inserted in the digraph G.

Theorem 9.13 Let G be the associated digraph of a general linear active network. Let a meter branch directed from node p to node q and a source branch directed from s to r be inserted in G. Then for $r \neq s$ and $p \neq q$ the transmission $H_{rp,sq}$ of G is given by

$$H_{rp,sq} = \frac{\sum_{P_{rs}^p} (-1)^w f(P_{rs}^p) V_k^*}{V_j''} \tag{9.159}$$

where

$$V_k^* = \sum_{T_k^*} f(T_k^*) \tag{9.160a}$$

$$V_j'' = \sum_{T_j''} f(T_j'') \tag{9.160b}$$

P_{rs}^p is a path between nodes r and s and containing the meter branch such that for each node j in P_{rs}^p, $j \neq p$, there is a directed path from p to j in P_{rs}^p. $w = 0$ if the orientations of the meter and the source branches are both opposite to or coincident with the direction of the oriented circuit formed by P_{rs}^p and the source branch, and $w = 1$ otherwise. T_k^* and T_j'' are directed trees with reference nodes k and j in G^* and G'', respectively; where G^* is the digraph obtained from G by shorting all the edges of P_{rs}^p, and G'' is the digraph obtained from G by removing the current- source and voltmeter branches and by shorting the voltage-source and ammeter branches. The weight associated with the meter branch in G is assumed to be unity.

An alternative way to determine the sign in (9.159) is as follows: $w = 0$ if the current in the positive direction of the source will cause the meter to register a positive reading in the circuit formed by the path P_{rs}^p and the source branch, and $w = 1$ otherwise. Also, in the theorem the choice of the reference nodes k and j is arbitrary since the associated matrices of the digraphs G^* and G'' are equicofactor matrices. The requirements that the voltage-source branch vs and the ammeter branch am be short-circuited, and that the current-source branch cs and the voltmeter branch vm be open-circuited are reasonable because the voltage source and ammeter have zero internal impedance and the current source and voltmeter have infinite internal impedance.

In applying formula (9.159) we need to calculate sums of directed-tree admittance products in certain modified digraphs. To this end, we shall introduce a procedure for generating these sums by an expansion on a node-pair. Thus, for a labeled digraph G if each of its subgraphs is denoted by the "product" or by juxtaposition of its edge-designation symbols or admittances, then for each choice of the nodes $i \neq k$ in G, the sum of the directed-tree admittance products $f(T_k)$ of G can be uniquely represented by the expansion

$$V_k = \sum_{T_k} f(T_k) = \sum_m f(P_{ik}^m) V_k^* \qquad (9.161)$$

where P_{ik}^m is the mth directed path from node i to node k in G, and V_k^* denotes the sum of directed-tree admittance products $f(T_k^*)$ in the digraph G^* obtained from G by shorting all the edges of P_{ik}^m with the combined node being labeled by k. Since each directed tree T_k contains a unique directed path from i to k, the set of directed trees T_k of G can be partitioned in terms of the directed path P_{ik}^m. The term $f(P_{ik}^m) V_k^*$ represents the sum of directed-tree admittance products of all those directed trees T_k containing P_{ik}^m. Thus, if $f(P_{ik}^m)$ is factored out, the remainder is the sum of directed-tree admittance products $f(T_k^*)$ in G^*. Equation (9.161) shows the expansion of the sum of directed-tree admittance products $f(T_k)$ in terms of the directed-path products $f(P_{ik}^m)$ of the directed paths P_{ik}^m from a chosen node i to the reference node k. Obviously, if V_k^* is still complicated, the same process may be repeated until all the terms can be obtained by inspection. The amount

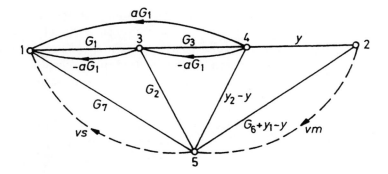

Figure 9.33 The associated digraph of the transistor amplifier of Fig. 9.5 with the meter and source branches inserted.

of work achieved by an expansion in directed paths will depend, of course, upon the structure of the particular digraph under consideration and upon the choice of node-pair for the expansion and subexpansions.

We illustrate the above results by the following examples.

Example 9.18 Consider the transformer-coupled transistor amplifier of Fig. 9.5, the associated digraph G of which is shown in Fig. 9.7. Suppose that we wish to compute the voltage-gain function $g_{12,55}$ of the amplifier. To this end, we insert a meter branch vm and a source branch vs in G and obtain the digraph of Fig. 9.33, where the dashed lines are used to denote the meter and source branches. Applying (9.159) we obtain

$$g_{12,55} = \frac{V_{25}}{V_{15}} = \frac{\sum_{P_{15}^2} (-1)^w f(P_{15}^2) V_k^*}{V_j''}$$

$$= \frac{1}{V_j''}[G_1 G_3 y + (aG_1)y(G_1 + G_2 + G_3 - aG_1) + G_1(-aG_1)y$$

$$+ (-aG_1)G_3 y + (-aG_1)(-aG_1)y]$$

$$= \frac{y(G_1 G_3 + aG_1 G_2)}{V_j''} \tag{9.162}$$

where

$$V_j'' = V_4'' = (y_2 - y)[G_3 + (G_1 + G_2) - aG_1][y + (G_6 + y_1 - y)]$$

$$+ (G_1 + G_2)G_3[y + (G_6 + y_1 - y)]$$

$$+ (G_6 + y_1 - y)y[G_3 + (G_1 + G_2) - aG_1] \tag{9.163}$$

In (9.162), $G_1 G_3 y1$ is a path P_{15}^2 between nodes 1 and 5 and containing the meter branch vm, the admittance of which is 1, such that for each node j in P_{15}^2, $j \neq 2$, there is a directed path from 2 to j in P_{15}^2. The cofactor V_k^* for this path $G_1 G_3 y1$ is to be computed in the digraph G^* obtained from that of Fig. 9.33 by shorting the edges G_1, G_3, y and the meter branch vm. The resulting digraph G^* consists of a single node. Thus, by definition, we have $V_k^* = 1$. Since the current in the positive direction of the voltage source will cause the voltmeter to register a positive reading in the circuit formed by the path $G_1 G_3 y$ and the source and meter branches vs and vm, $w = 0$.

Likewise, $(aG_1)y1$ is another such path P_{15}^2. After shorting the edges aG_1, y and the meter branch vm, the resulting digraph G^* is shown in Fig. 9.34, the sum of directed-tree admittance products $f(T_k^*)$ of which is found to be

$$V_k^* = V_3^* = V_5^* = \sum_{T_5^*} f(T_5^*) = G_1 + G_2 + G_3 - aG_1 \qquad (9.164)$$

The other paths of the type P_{15}^2 are given by $G_1(-aG_1)y1$, $(-aG_1)G_3 y1$, and $(-aG_1)(-aG_1)y1$, and their corresponding digraph G^* is the null graph composed of a single node. For all these paths P_{15}^*, $w = 0$.

The terms in V_4'' correspond to directed-tree admittance products $f(T_4'')$ in the digraph G'' obtained from that of Fig 9.33 by removing the meter branch vm and short-circuiting the source branch vs. The resulting digraph G'' is shown in Fig. 9.35. To calculate V_4'', we apply an expansion on the pair of nodes 5 and 4. A directed path from node 5 to node 4 is $y_2 - y$, the short-circuiting of which results in the digraph shown in Fig. 9.36. The sum of directed-tree admittance products

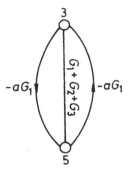

Figure 9.34 The digraph obtained from that of Fig. 9.33 after shorting the edges aG_1, y and the meter.

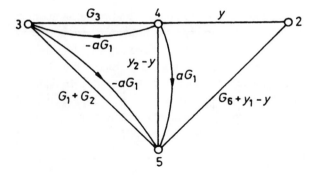

Figure 9.35 The digraph obtained from that of Fig. 9.33 by removing the meter branch vm and short-circuiting the source branch vs.

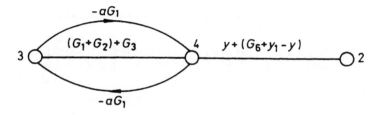

Figure 9.36 The digraph obtained from that of Fig. 9.35 after short-circuiting the directed path composed of the edge with admittance $(y_2 - y)$.

with reference node 4 in this digraph is found to be

$$[(G_1 + G_2) + G_3 - aG_1][y + (G_6 + y_1 - y)] \qquad (9.165)$$

The product of this and $(y_2 - y)$ gives the first term on the right-hand side of (9.163).

Likewise, $(G_1 + G_2)G_3$ is another directed path from node 5 to node 4, the short-circuiting of which results in a digraph composed of a single edge with admittance $y+(G_6+y_1-y)$. Thus, the product of this admittance and $(G_1+G_2)G_3$ gives the second term on the right-hand side of (9.163). Finally, by short-circuiting the directed path $(G_6 + y_1 - y)y$ from node 5 to node 4, we obtain the digraph of Fig. 9.37, the sum of directed-tree admittance products of which are given by $(G_1 + G_2) + G_3 - aG_1$. This corresponds to the third term on the right-hand side of (9.163).

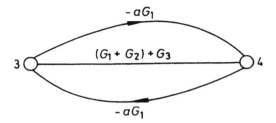

Figure 9.37 The digraph obtained from that of Fig. 9.35 after short-circuiting the directed path having admittance product $(G_6 + y_i - y)y$.

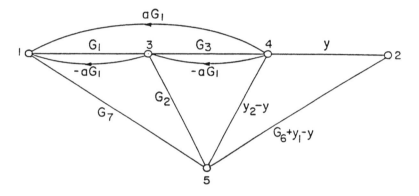

Figure 9.38 The associated digraph of the active network of Fig. 9.5.

Suppose that we wish to compute the transfer impedance $z_{12,55}$ of the amplifier. Thus, we replace the voltage-source branch vs by a current-source branch cs in the digraph of Fig. 9.33. The numerator of $z_{12,55}$ is the same as that of $g_{12,55}$ of (9.162). The denominator V''_j of $z_{12,55}$ is the sum of directed-tree admittance products in the digraph obtained from that of Fig. 9.33 by removing the meter and source branches, which, in fact, is the associated digraph G of the indefinite-admittance matrix of the amplifier itself, as shown in Fig. 9.7 and repeated in Fig. 9.38. In other words, V''_j is simply the determinant of the node-admittance matrix of the amplifier. To calculate V''_j, we apply an expansion on the pair of nodes 5 and 4, as follows:

$$V''_4 = (y_2 - y)(G_6 + y_1)[(G_2 + G_3)(G_1 + G_7) + (G_1 - aG_1)G_7]$$
$$+ G_2 G_3 (G_1 + G_7)(G_6 + y_1) + G_7 G_1 G_3 (G_6 + y_1)$$

$$+ (G_6 + y_1 - y)y[(G_2 + G_3)(G_1 + G_7) + (G_1 - aG_1)G_7]$$
$$= (y_1 + G_6)[(G_1 + G_7)(G_2 G_3 + G_2 y_2 + G_3 y_2)$$
$$+ G_1 G_7 (y_2 + G_3) - aG_1 G_7 y_2] + aG_1 G_7 y^2$$
$$- y^2 (G_2 + G_3)(G_1 + G_7) - y^2 G_1 G_7 \qquad (9.166)$$

The required transfer function is obtained directly from (9.159) as

$$z_{12,55} = \frac{V_{25}}{I_s} = \frac{y(G_1 G_3 + aG_1 G_2)}{V_4''} \qquad (9.167)$$

We emphasize that in applying formula (9.159) the choice of the reference nodes k and j is arbitrary, because the associated indefinite-admittance matrices of the digraphs are equicofactor matrices, resulting in the same sums of directed-tree admittance products. However, the number of directed trees is different for different choices of the reference nodes.

Example 9.19 Consider the ladder network of Fig. 9.39. The current-controlled voltage source $z_m I_2$ can be converted into an equivalent voltage-controlled current source as shown in Fig. 9.40. The associated digraph G of the ladder network is obtained in Fig. 9.41. Suppose that we wish to compute the voltage-gain function $g_{rq,ss}$. So we insert a meter branch vm and a source branch vs, as indicated by the dashed lines, in G. From (9.159) we obtain

$$g_{rq,ss} = \frac{V_{qs}}{V_{rs}} = \frac{\sum_{p_{rs}^q} (-1)^w f(P_{rs}^q) V_k^*}{V_j''}$$
$$= \frac{y_1 (y_3 + y)(y_5 + y) + (-y)y_1 (y_3 + y_4 + y_5 + y)}{V_j''} \qquad (9.168)$$

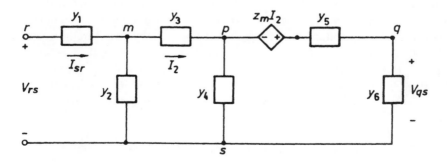

Figure 9.39 An active ladder network.

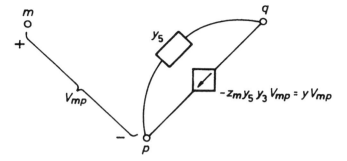

Figure 9.40 An equivalent voltage-controlled current source.

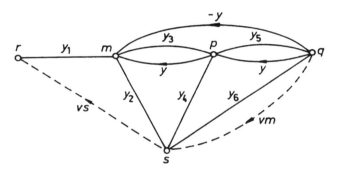

Figure 9.41 The associated digraph of the ladder network of Fig. 9.39.

where

$$V_j'' = V_q'' = y_3 y_5 (y_1 + y_2 + y_4 + y_6) + (y_1 + y_2) y_4 y_5$$
$$+ (y_1 + y_2) y_6 (y_3 + y_4 + y_5 + y) + y_3 y_4 y_6 \qquad (9.169)$$

which is obtained by an expansion on the pair of nodes m and q in the digraph
G'' obtained from that of Fig. 9.41 by removing the meter branch vm and short-
circuiting the source branch vs, as given in Fig. 9.42. The first term on the right-hand
side of (9.169) corresponds to the directed path $y_3 y_5$, the short-circuiting of which
yields the sum of directed-tree admittance products equal to $y_1 + y_2 + y_4 + y_6$. The
second term corresponds to the directed path $(y_1 + y_2) y_4 y_5$ with unity cofactor.
The third term corresponds to the directed path $(y_1 + y_2) y_6$, the short-circuiting
of which results in the digraph of Fig. 9.43. The sum of directed-tree admittance

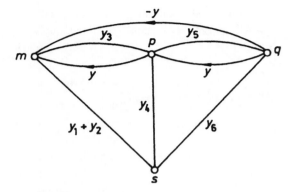

Figure 9.42 The digraph derived from that of Fig. 9.41 by short-circuiting the source branch vs and by removing the meter branch vm.

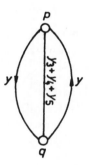

Figure 9.43 The digraph obtained from that of Fig. 9.42 after short-circuiting the directed path with admittance product $(y_1 + y_2)y_6$.

products of this digraph is given by $y_3 + y_4 + y_5 + y$. Finally, we have the directed path $y_3 y_4 y_6$ from m to q with unity cofactor.

Suppose that we wish to compute the transfer impedance $z_{rp,sq} = V_{pq}/I_{sr}$ of the ladder network of Fig. 9.39. To this end, we insert a meter branch vm and a source branch cs as shown in Fig. 9.44. Appealing to (9.159) we obtain

$$z_{rq,sq} = \frac{V_{pq}}{I_{sr}} = \frac{\sum_{P_{rs}^p} (-1)^w f(P_{rs}^p) V_k^*}{V_j''}$$

$$= \frac{y_1(y_3 + y)y_6 - (-y)y_1 y_4}{V_j''} \tag{9.170}$$

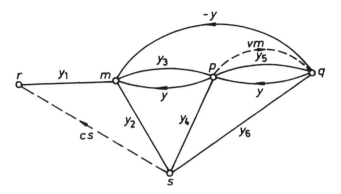

Figure 9.44 The digraph used to calculate the transfer impedance $z_{rp,sq}$.

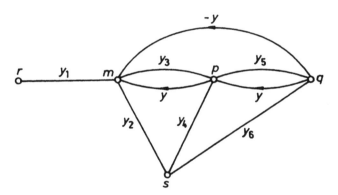

Figure 9.45 The digraph obtained from that of Fig. 9.44 after removing the source branch cs and the meter branch vm.

where I_{sr} is the input current, and

$$V_j'' = V_q'' = y_1 y_3 y_5 (y_2 + y_4 + y_6) + y_1 y_3 y_4 y_6 + y_1 y_2 y_4 y_5$$
$$+ y_1 y_2 y_6 (y_3 + y_4 + y_5 + y) \qquad (9.171)$$

The terms in (9.171) correspond to directed trees T_q'' in the digraph G'' of Fig. 9.45, which is obtained from that of Fig. 9.44 by the removal of the source branch cs and meter branch vm, and are calculated by an expansion on the pair of nodes r and q in G''. The directed paths from r to q in G'' corresponding to these terms are $y_1 y_3 y_5$, $y_1 y_3 y_4 y_6$, $y_1 y_2 y_4 y_5$, and $y_1 y_2 y_6$, respectively.

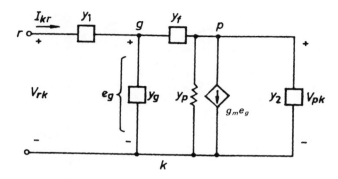

Figure 9.46 An equivalent network of a potential-feedback amplifier.

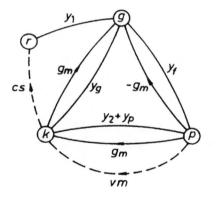

Figure 9.47 The associated digraph of the network of Fig. 9.46 with meter and source branches inserted.

Example 9.20 An equivalent network of a general potential-feedback amplifier is shown in Fig. 9.46, the associated digraph G of which is presented in Fig. 9.47. Suppose that we wish to compute the transfer impedance $z_{rp,kk} = V_{pk}/I_{kr}$ of the amplifier. So we insert a voltmeter branch vm from node p to node k, and a current source branch cs from node k to node r in G, as indicated in Fig. 9.47 by the dashed lines. Appealing to (9.159) we obtain

$$z_{rp,kk} = \frac{V_{pk}}{I_{kr}} = \frac{\sum_{P_{rk}^p}(-1)^w f(P_{rk}^p)V_k^*}{V_j''}$$

$$= \frac{y_1(y_f - g_m)(1)}{V_j''} = \frac{y_1(y_f - g_m)}{V_k''} \tag{9.172}$$

where

$$V_j'' = V_k'' = y_1 y_g [(y_f - g_m) + (y_2 + y_p + g_m)] + y_1 y_f (y_2 + y_p + g_m)$$
$$= y_1 [y_f (y_p + y_2 + g_m) + y_g (y_f + y_p + y_2)] \qquad (9.173)$$

The terms in (9.173) are obtained by an expansion on the pair of nodes r and k in the digraph G'' obtained from that of Fig. 9.47 by the removal of the source branch cs and the meter branch vm. The two directed paths from r to k are represented by their admittance products as $y_1 y_g$ and $y_1 y_f (y_2 + y_p + g_m)$.

Suppose that we wish to compute the input admittance of the amplifier of Fig. 9.46. We insert a meter branch am and a source branch vs in G, and obtain the digraph of Fig. 9.48. Applying (9.195) we obtain

$$H_{xx,kr} = \frac{1}{z_{rr,kk}} = \frac{I_{xr}}{V_{xk}} = \frac{\sum_{P_{xk}^x} (-1)^w f(P_{xk}^x) V_k^*}{V_j''}$$
$$= \frac{y_1 y_f (y_2 + y_p + g_m) 1 + y_1 y_g [(y_f - g_m) + (y_2 + y_p + g_m)]}{(y_2 + y_p)(y_1 + y_g + y_f) + (y_1 + y_g + g_m) y_f} \qquad (9.174)$$

The terms in the denominator are obtained by an expansion of the pair of nodes k and p in the digraph G'' derived from that of Fig. 9.48 by short-circuiting the meter branch am and the source branch vs, as shown in Fig. 9.49. The directed paths from k to p in G'' are represented by the admittance products $(y_2 + y_p)$ and $(y_1 + y_g + g_m) y_f$, the cofactors of which are given by $(y_1 + y_g + y_f)$ and 1, respectively.

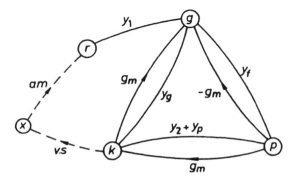

Figure 9.48 The digraph used to compute the driving-point admittance function of the amplifier of Fig. 9.46.

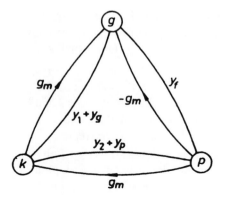

Figure 9.49 The digraph obtained from that of Fig. 9.48 after shrot-circuiting the source branch vs and meter branch am.

Each of the terms in the sum denotes a directed-tree admittance product $f(T_p'')$ in G''.

9.7 SUMMARY

We began this chapter by defining the natural frequencies of a network as the roots of the determinantal polynomial of the operator matrix of the network equations when these are framed as a set of first-order differential equations and/or algebraic equations for the branch currents and voltages. The number of these natural frequencies are called the order of complexity of the networks, counting each frequency according to its multiplicity.

As before, we characterize an active network by its indefinite-admittance matrix, from which we construct the associated digraph. We showed that the determinant of the node-admittance matrix of an active network is equal to the sum of the directed-tree admittance products of the directed trees in the associated digraph of the network. Using this result, we obtained specific bounds on the order of complexity of different classes of active networks. For an RLC network, we found that its order of complexity can be determined from its topology alone by inspection, being equal to the number of reactive elements, less the number of independent C-circuits and the number of independent L-cuts. For a general active network, topology alone is not sufficient to determine its order of complexity; network parameters must also be involved. An upper bound on the order of complexity of an active network containing β_L inductors is equal to β_L plus the number of capacitors in a normal directed tree of the associated digraph of the network, less the number of inductors in the normal directed tree. For a network composed of resistors, inductors, capacitors and controlled current sources, an upper bound on the order of complexity of the network is given by the number of capacitors contained in a complete tree of the network plus the number of inductors contained in its cotree.

We next studied the conditions on the unique solvability of networks that contain resistors, inductors, capacitors, transformers, independent sources, controlled sources of all types, and other linear devices that can be modeled by these elements such as transistors, gyrators, impedance converters, etc. Specifically, we showed that if a network contains neither a circuit composed only of independent and dependent voltage sources nor a cutset composed only of independent and dependent current sources, then it possesses a unique solution if its branch-immittance matrix is ohmic after the removal of all the independent sources. In addition, three other sets of conditions on unique solvability of an active network were discussed.

Return differences of a network and its topology are closely related. We demonstrated that the sum of the reciprocals of the return differences with respect to all network elements equals the nullity of the network, if nodal determinant is used in the formulation; and equals the rank of the network if the loop determinant is employed. As a result, the sum of the products of all transadmittances and their associated transfer impedances between the controlled branches and the controlling branches in a network equals the rank of the network, and the sum of the products of all transimpedances and their associated transfer admittances between the controlled branches and the controlling branches in a planar network equals the nullity of the network. In particular, for an RLC network with no coupling, the sum of products of all branch admittances and the driving-point impedances across the terminals of the branches equals the rank of the network, and the sum of products of all branch impedances and the driving-point admittances looking into the branches including the branch impedances themselves equals the nullity of the network.

Finally, we use the associated digraph of an active network itself to evaluate the network functions directly. We demonstrated that the digraph itself may be considered as a physical model of the network, because we can connect a meter and a source in the digraph to yield readings as if it were a physical network. In applying topological formula, we need to calculate sums of directed-tree admittance products in certain modified digraphs. To simplify the computation, we presented a procedure that expresses the sum of directed-tree admittance products in terms of the directed-path products, and is known as the node-pair expansion. This procedure may be repeated until all the terms can be obtained by inspection.

PROBLEMS

9.1 Determine the operator matrix and the order of complexity of the active network of Fig. 9.13.

9.2 Determine the operator matrix and the order of complexity of the active network of Fig. 9.23.

9.3 For the transistor amplifier of Fig. 9.26, construct the associated digraph of the amplifier and compute its nodal determinant by means of Theorem 9.1.

9.4 For the transistor amplifier of Fig. 9.27, construct the associated digraph of the amplifier and compute its nodal determinant by means of Theorem 9.1.

9.5 Determine the operator matrix and the order of complexity of the transistor amplifier of Fig. 9.26.

9.6 Applying Theorem 9.3, determine an upper bound on the order of complexity of the active network of Fig. 9.16a.

9.7 Applying Theorem 9.4, determine an upper bound on the order of complexity of the active network of Fig. 9.16a.

9.8 Determine if the active network of Fig. 9.26 is uniquely solvable.

9.9 Repeat Example 9.6 for the active network of Fig. 9.16a after inserting a resistor of resistance $R_6 = 1\ \Omega$ in parallel with the inductor L_3.

9.10 Repeat Example 9.7 for the active network of Fig. 9.17a after inserting a resistor of resistance $R_9 = 1\ \Omega$ in parallel with the inductor L_4.

9.11 Repeat Example 9.8 for the active network of Fig. 9.18a after inserting an inductor of inductance L_9 in parallel with the resistor R_6.

9.12 For the RLC network of Fig. 9.30, verify that the sum of products of all branch admittances and the driving-point impedances across the terminals of the branches equals the rank of the network.

9.13 For the RLC network of Fig. 9.32, verify that the sum of products of all branch admittances and the driving-point impedances across the terminals of the branches equals the rank of the network.

9.14 For the active ladder network of Fig. 9.39, show that the sum of the reciprocals of the return differences with respect to all network elements and based on the nodal formulation equals the nullity of the network.

9.15 For the active ladder network of Fig. 9.39, show that the sum of the reciprocals of the return differences with respect to all network elements and based on the loop formulation equals the rank of the network.

9.16 Use the active ladder network of Fig. 9.39 to verify Theorem 9.10.

9.17 Use the active ladder network of Fig. 9.39 to verify Theorem 9.12.

9.18 For the potential-feedback amplifier of Fig. 9.46, verify that the sum of the reciprocals of the return differences with respect to all network elements and based on the nodal formulation equals the nullity of the network.

9.19 For the potential-feedback amplifier of Fig. 9.46, verify that the sum of the reciprocals of the return differences with respect to all network elements and based on the loop formulation equals the rank of the network.

9.20 Use the potential-feedback amplifier of Fig. 9.46 to verify Theorem 9.10.

9.21 Use the potential-feedback amplifier of Fig. 9.46 to verify Theorem 9.12.

9.22 Applying formula (9.159), compute the transfer voltage-ratio function of the amplifier of Fig. 9.26.

9.23 Applying formula (9.159), compute the input and output impedances of the transistor amplifier of Fig. 9.26.

9.24 Applying formula (9.159), compute the transfer voltage-ratio function of the transistor amplifier of Fig. 9.27.

9.25 Applying formula (9.159), compute the input and output impedances of the transistor amplifier of Fig. 9.27, when the other port is short-circuited.

9.26 Use formula (9.159) to compute the short-circuit admittance parameters of the two-port network of Fig. 9.26.

9.27 Use formula (9.159) to compute the open-circuit impedance parameters of the two-port network of Fig. 9.26.

9.28 Use formula (9.159) to compute the short-circuit admittance parameters of the two-port network of Fig. 9.27.

9.29 Use formula (9.159) to compute the open-circuit impedance parameters of the two-port network of Fig. 9.27.

BIBLIOGRAPHY

Abdullah, K.: A Necessary Condition for Complete Solvability of RLCT Networks, *IEEE Trans. Circuit Theory*, vol. CT-19, no. 5, pp. 492–493, 1972.

Abdullah, K. and Y. Tokad: On the Existence of Mathematical Models for Multiterminal RCT Networks, *IEEE Trans. Circuit Theory*, vol. CT-19, no. 5, pp. 419–424, 1972.

Bers, A.: The Degrees of Freedom in RLC Networks, *IRE Trans. Circuit Theory*, vol. CT-6, no. 1, pp. 91–95, 1959.

Bryant, P. R.: The Order of Complexity of Electrical Networks, *Proc. IEE* (London), vol. 106C, no. 2, pp. 174–188, 1959.

Bryant, P. R.: The Degrees of Freedom in RLC Networks, *IRE Trans. Circuit Theory*, vol. CT-7, no. 2, nn. 173–174 and p. 357, 1960.

Bryant, P. R. and J. Tow: The A-matrix of Linear Passive Reciprocal Networks, *J. Franklin Inst.*, vol. 293, no. 6, pp. 401–419, 1972.

Chan, S. P.: "Introductory Topological Analysis of Electrical Networks," San Francisco, Calif.: Holt, Rinehart and Winston, 1969.

Chan, S. P., S. Y. Chan, and S. G. Chan: "Analysis of Linear Networks and Systems," Reading, Mass.: Addison-Wesley, 1972.

Chen, W. K.: Graph-Theoretic Considerations on the Invariance and Mutual Relations of the Determinants of the Generalized Network Matrices and Their Generalized Cofactors, *Quart, J. Math. Oxford* (2), vol. 21, no. 4, pp. 459–479, 1970.

Chen, W. K.: Topological Formulations and the Order of Complexity of Active Networks: A Unified Survey, *Networks,* vol. 2, pp. 237–260, 1972a.

Chen, W. K.: Characterizations of Complete Directed Trees and Two-Trees, *IEEE Trans. Circuit Theory,* vol. CT-19, no. 3, pp. 241–247, 1972b.

Chen, W. K.: Solvability and the Order of Complexity of Linear Networks, *Proc. 16th Midwest Symp. Circuit Theory,* University of Waterloo, Waterloo, Ontario, pp. XII.5.1–XII.5.10, 1973.

Chen, W. K.: "Applied Graph Theory: Graphs and Electrical Networks," 2d rev. ed., New York: American Elsevier and Amsterdam: North-Holland, 1976.

Chen, W. K.: A Theorem on the Summation of Return Differences and Some Consequences, *Proc. IEEE,* vol. 72, no. 3, pp. 396–397, 1984.

Chen, W. K.: A Dual Theorem on the Summation of Return Differences, *Proc. IEEE,* vol. 73, no. 1, pp. 157–159, 1985.

Chen, W. K.: "Linear Networks and Systems: Algorithms and Computer-Aided Implementations," Singapore: World Scientific, 1990.

Chen, W. K. and F. N. T. Chan: On the Unique Solvability of Linear Active Networks, *IEEE Trans. Circuits and Systems,* vol. CAS-21, no. 1, pp. 26–35, 1974.

Chen, W. K. and J. Lu: On the Assignment of Loops to Graphs, *Proc. 22nd International Symp. Circuits and Systems,* Portland, Oregon, pp. 331–334, 1989. IEEE catalog no. 89CH2692-2.

Coates, C. L.: General Topological Formulas for Linear Network Functions, *IRE Trans. Circuit Theory,* vol. CT-5, no. 1, pp. 30–43, 1958.

DeClaris, N. and R. Saeks: Theoretic Foundations of Finite Network and System Analysis, in "Aspects of Network and System Theory," R. E. Kalman and N. DeClaris (eds.), New York: Holt, Rinehart and Winston, pp. 3–60, 1971.

Eckmann, B.: Harmonische Funktionen und Randwert Aufgaben in einem Komplex, *Commun. Math. Helvetici,* vol. 17, pp. 240–255, 1944–1945.

Guillemin, E. A.: "Communication Networks," New York: Wiley, 1931, vol. 1.

Guillemin, E. A.: "Synthesis of Passive Networks," New York: Wiley, 1957.

Hajj, I. N.: A Note on a Theorem on the Summation of Driving-Point and Transfer Network Impedances, *Proc. IEEE,* vol. 72, no. 3, pp. 395–396, 1984.

Ho, Y. S. and P. H. Roe: Existence Theorem in the Time Domain for Linear Active Networks, *IEEE Trans. Circuits and Systems,* vol. CAS-21, no. 2, pp. 175–177, 1974a.

Ho, Y. S. and P. H. Roe: Degree of Polynomial Matrix and Explicit Formula for the Order of Complexity of Linear Active Networks, *Proc. 7th International Symp. Circuits and Systems,* San Francisco, Calif., pp. 500–504, 1974b.

Kirchhoff, G.: Über die Auflösung der Gleichungen, auf welche man bei der Untersuchungen der linearen Verteilung galvanischer Ströme geführt wird, *Poggendorf Ann. Phys.,* vol. 72, pp. 497–508, 1847.

Malik, N. R. and H. W. Haie: Equations for Active Networks: Existence of Unique Solutions, *IEEE Trans. Circuit Theory,* vol. CT-14, no. 1, pp. 37–43, 1967.

Mark, S. K.: On the Order of Complexity of Active Electrical Networks, Ph.D. dissertation, Ohio University, Athens, June 1971.

Maxwell, J. C.: "Electricity and Magnetism," Oxford, England: Clarendon Press, 1892, ch. 6 and appendix.

Mayeda, W.: Topological Formulas for Nonreciprocal Networks and Networks with Transformers, *Proc. Natl. Electronics Conf.,* vol. 14, pp. 631–643, 1958.

Milić, M. M.: General Passive Networks — Solvability, Degeneracies, and Order of Complexity, *IEEE Trans. Circuits and Systems,* vol. CAS-21, no. 2, pp. 177–183, 1974.

Parker, S. R. and V. T. Barnes: Existence of Numerical Solutions and the Order of Linear Circuits with Dependent Sources, *IEEE Trans. Circuit Theory,* vol. CT-18, no. 3, pp. 368–374, 1971.

Purslow, E. J.: Solvability and Analysis of Linear Active Networks by Use of the State Equations, *IEEE Trans. Circuit Theory,* vol. CT-17, no. 4, pp. 469–475, 1970.

Purslow, E. J. and R. Spence: Order of Complexity of Active Networks, *Proc. IEE* (London), vol. 114, Pt. 2, no. 2, pp. 195–198, 1967.

Reza, F. M.: Order of Complexity and Minimal Structures in Network Analysis, *Proc. Symp. Circuit Analysis,* University of Illinois, Urbana-Champaign, Ill., pp. 7.1–7.33, 1955.

Roth, J. P.: An Application of Algebraic Topology to Numerical Analysis: On the Existence of a Solution to the Network Problem, *Proc. Natl. Acad. Sci.,* vol. 41, pp. 518–521, 1955.

Roth, J. P.: An Application of Algebraic Topology: Kron's Method of Tearing, *Quart. Appl. Math.,* vol. 17, no. 1, pp. 1–24, 1959.

Seshu, S. and M. B. Reed: "Linear Graphs and Electrical Networks," Reading, Mass.: Addison-Wesley, 1961, chs. 6 and 8.

Su, Y. H.: Topological Formulas and the Order of Complexity for Networks with a Nonreciprocal Element, *J. Franklin Inst.,* vol. 286, no. 3, pp. 204–224, 1968.

Thulasiraman, K., R. Jayakumar, and M. N. S. Swamy: Graph-Theoretic Proof of a Network Theorem and Some Consequences, *Proc. IEEE,* vol. 71, no. 6, pp. 771–772, 1983.

Tow, J.: Order of Complexity of Linear Active Networks, *Proc. IEE* (London), vol. 115, Pt. 2, no. 9, pp. 1259–1262, 1968.

Tsang, N. F.: On Electrical Network Determinants, *J. Math. Phys.,* vol. 33, pp. 185–193, 1954.

Wang, H. Y. and W. K. Chen: A General Theorem on the Summation of Return Differences, *Proc. 20th International Symp. Circuits and Systems,* Philadelphia, Penn., pp. 160–164, 1987. IEEE catalog no. 87CH2394-5.

Weyl, H.: Repartition de corriente et uno red conductora, *Rev. Mathmatica,* vol. 5, pp. 153–164, 1923.

TEN

GENERALIZED NETWORK MATRICES
AND THEIR COFACTORS

In Chapters 4 and 5, we showed that Bode's concept of return difference plays an important role in the design of feedback amplifiers. Among the many important properties, the return difference is a generalization of the concept of the feedback factors of ideal feedback model, the sensitivity function of the amplifier is closely related to the return difference, and the return difference is basic to the study of the stability of the feedback system and to the determination of its transmission and driving-point properties.

Several important extensions and generalizations of Bode's return difference concept have been discussed. They include the concepts of the return-difference matrix and the null return difference. The former is applicable to linear feedback networks that possess a multiplicity of physical feedback loops, and the latter is useful in measurement situations and in the computation of the sensitivity. The extension of the null return difference to the notion of the null return-difference matrix is useful in the evaluation of the closed-loop gain and the driving-point impedance of a multiple-loop feedback network. Furthermore, these concepts have been modified and elaborated upon in the previous chapters.

So far we have demonstrated that although the return difference and the null return difference are not invariant with respect to the general transformations of the reference frame, they are invariant for the most common and important types of feedback networks. These results have also been extended and generalized to the multiple-loop feedback networks.

The formulation of the return difference and especially the null return difference of a feedback amplifier in terms of the first- and second-order cofactors of the elements of its indefinite-admittance matrix is significant in that it avoids the necessity of interpreting the loop gain in terms of the actual signal transmission around the closed loops in a flow graph, which in many situations is very misleading.

In the present chapter, we first explore the relation between the null return difference and the generalized cofactors, and then demonstrate how the null return difference can be expressed in terms of the ratio of the two functional values assumed by a generalized cofactor under the condition that the element assumes its nominal value and the condition that the element assumes its reference value. We shall also consider the practical problems encountered in the evaluation of these functions and present topological formulas for their determination. Topological formulas are derived which are expressed in terms of directed-tree and directed-two-tree produces in the associated directed graph of the amplifier. Extensions of topology and the summations of the products of all transimmittances and their associated transfer immittances will also be considered.

10.1 NETWORK DETERMINANTS

The invariant character of the determinant of the node-admittance matrix with respect to the choice of the reference node was first considered by Jeans (1948), and rediscovered by Percival (1954) and Shekel (1954). The same statement is also made very often about the determinant of the loop-impedance matrix [see, for example, Tsang (1954)]. It was pointed by Seshu (1955) that the determinant of the loop-impedance matrix in general is not invariant under the transformation of circuits. However, they are related by a real constant, depending only on the two choices of circuits [see, for example, Tsang (1954)]. Okada (1955) showed that the determinant of the loop-impedance matrix formed by the tree method is invariant, but the determinant of the node-admittance matrix does not remain invariant when general nodal variables are used. Cederbaum (1956a,b) considered the problem more generally and showed that the network determinants, although they are not invariant for a general loop system or for a general cutset representation, are invariant for the most common and important types of network representations such as circuits and cutsets formed by the tree method and circuits formed by the windows of a planar network. He also showed that the ratio of the determinants of the loop-impedance and the node-admittance matrices under the invariance condition is equal to the determinant of the branch-impedance matrix of the network. This is a generalization of earlier work of Tsang (1954).

In this section, we show that many of the invariant characters of the network determinants and network functions can be derived from a purely graph-theoretic viewpoint. Thus, they are valid not only for the electrical networks, but also for other systems. It will also be shown that they are invariant with respect to the choice of the incidence functions of the network, i.e. the orientation of the branches of the network.

10.1.1 The Invariance of the Determinants of the Major Submatrices

We first define the term a major submatrix of a matrix.

Definition 10.1: Major Submatrix A submatrix $\mathbf{F}_m$ of an arbitrary matrix $\mathbf{F}$ of order $p \times q$ and of rank p where $p \leq q$ is said to be a *major submatrix* of $\mathbf{F}$ if $\mathbf{F}_m$ is a nonsingular submatrix of order p in $\mathbf{F}$.

In other words, the determinant of $\mathbf{F}_m$ is a nonvanishing *major determinant* of $\mathbf{F}$. Also, let $\mathbf{B}$ and $\mathbf{Q}$ be the circuit and cutset matrices of order $n \times e$ and $r \times e$ and of ranks n and r, respectively, of a directed graph G of v nodes and e edges, where n and r denote the nullity and rank of G, respectively. It can be shown [see Chen (1976), Chapter 5] that the major submatrices $\mathbf{Q}_m$ and $\mathbf{B}_m$ of $\mathbf{Q}$ and $\mathbf{B}$, respectively, are in one-to-one correspondence with the trees t and cotrees $\bar{t}$ of G. For convenience, we shall assume that the columns of all the circuit and cutset matrices considered here have been arranged in the same edge order. As in (8.34) and (8.36), let $\mathbf{B}_f$ and $\mathbf{Q}_f$ denote the fundamental circuit and fundamental cutset matrices of G with respect to a tree t of G, respectively. Then it is evident that there exist nonsingular matrices $\mathbf{C}$ and $\mathbf{D}$ of orders n and r, respectively, such that

$$\mathbf{B}_f = \mathbf{CB} \qquad (10.1a)$$

$$\mathbf{Q}_f = \mathbf{DQ} \qquad (10.1b)$$

since the rows of $\mathbf{B}_f$ and $\mathbf{Q}_f$ are the bases of the vector spaces spanned by the rows of $\mathbf{B}$ and $\mathbf{Q}$, respectively. For convenience, let $\mathbf{Q}_{fm}$ and $\mathbf{Q}_m$ be the major submatrices of $\mathbf{Q}_f$ and $\mathbf{Q}$, respectively, corresponding to the tree t, and $\mathbf{B}_{fm}$ and $\mathbf{B}_m$ be the major submatrices of $\mathbf{B}_f$ and $\mathbf{B}$, respectively, corresponding to the cotree $\bar{t}$. It follows that $\mathbf{B}_{fm}$ and $\mathbf{Q}_{fm}$ are identity matrices of orders n and r, respectively, and

$$\mathbf{C} = \mathbf{B}_m^{-1} \tag{10.2a}$$

$$\mathbf{D} = \mathbf{Q}_m^{-1} \tag{10.2b}$$

Definition 10.2: Totally unimodular matrix A matrix with the property that the determinant of each of its square submatrices is 1, -1, or 0 is called a *totally unimodular matrix*.

Since $\mathbf{B}_f$ and $\mathbf{Q}_f$ are totally unimodular, the determinants of all the major submatrices of either $\mathbf{B}_f$ or $\mathbf{Q}_f$ have the same magnitude 1. In other words, the determinants of all the major submatrices of either the circuit matrix or the cutset matrix of a directed graph have the same magnitude. We summarize the above result as follows.

Theorem 10.1 If $\mathbf{B}$ and $\mathbf{Q}$ are the circuit and cutset matrices of a directed graph, then $\mathbf{B}_m^{-1}\mathbf{B}$ and $\mathbf{Q}_m^{-1}\mathbf{Q}$ must be totally unimodular matrices for every choice of the major submatrices $\mathbf{B}_m$ and $\mathbf{Q}_m$ of $\mathbf{B}$ and $\mathbf{Q}$, respectively.

Because the elements of $\mathbf{B}_m$ and $\mathbf{Q}_m$ are 1, -1 or 0, the determinant of these major submatrices must be an integer, not necessarily $\pm 2^i$, where i is a nonnegative integer fixed for a given $\mathbf{B}$ or $\mathbf{Q}$. As an illustration, consider the directed graph of Fig. 10.1. The circuit matrix $\mathbf{B}$ as defined by the circuits formed by the edges $e_1 e_5 e_6 e_3 e_4$, $e_1 e_2 e_6 e_7 e_4$, $e_1 e_2 e_3 e_7 e_8$ and $e_5 e_2 e_3 e_4 e_8$ is given by

$$\mathbf{B} = \begin{array}{c} e_1 \ e_5 \ e_6 \ e_3 \ e_4 \\ e_1 \ e_2 \ e_6 \ e_7 \ e_4 \\ e_1 \ e_2 \ e_3 \ e_7 \ e_8 \\ e_5 \ e_2 \ e_3 \ e_4 \ e_8 \end{array} \begin{array}{cccccccc} 1 & 2 & 3 & 4 & 5 & 6 & 7 & 8 \\ \begin{bmatrix} 1 & 0 & 1 & -1 & 1 & 1 & 0 & 0 \\ 1 & 1 & 0 & -1 & 0 & -1 & 1 & 0 \\ 1 & 1 & 1 & 0 & 0 & 0 & -1 & -1 \\ 0 & 1 & 1 & -1 & -1 & 0 & 0 & 1 \end{bmatrix} \end{array} \tag{10.3}$$

It is straightforward to verify that the determinant of all the major submatrices of $\mathbf{B}$ has a magnitude of 3, which obviously is not a power of 2.

In many applications, it is necessary to compute the magnitude square of such major determinants. For convenience let $k(\mathbf{B}) = (\det \mathbf{B}_m)^2$ and $k(\mathbf{Q}) = (\det \mathbf{Q}_m)^2$, where $\mathbf{B}_m$ and $\mathbf{Q}_m$ are the submatrices of $\mathbf{B}$ and $\mathbf{Q}$ corresponding to the cotree $\bar{t}$ and tree t of G, respectively. Also, as before let the symbols $\mathbf{A}$ and $\mathbf{B}_p$ be used to denote the incidence matrix of order $r \times e$ and of rank r of G, and the circuit matrix of order $n \times e$ and of rank n corresponding to the windows of a planar graph, respectively. Clearly, we have $k(\mathbf{A}) = 1$, $k(\mathbf{B}_p) = 1$, $k(\mathbf{B}_f) = 1$ and $k(\mathbf{Q}_f) = 1$. This is the minimum value that can be obtained for a major submatrix for any directed

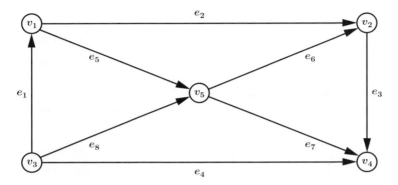

Figure 10.1 A given directed graph G.

graph G. Nevertheless, they are not the only ones for which the determinant of the major submatrices has this minimum value.

10.1.2 Generalized Network Matrices

In this section, we shall consider the matrix triple products $\mathbf{BWB}'$ and $\mathbf{QXQ}'$, where $\mathbf{W}$ and $\mathbf{X}$ are square matrices of order e, and the prime denotes the transpose of a matrix. We shall also discuss the conditions under which the determinant of the triple products is invariant with respect to the transformation from one such system to another, and their interrelationships. For our purposes, these triple products are referred to as the *generalized network matrices*. Without loss of generality, we assume that the directed graph G is connected.

Definition 10.3: Generalized Network Matrix For a given directed graph G of e edges, the matrix triple products $\mathbf{BWB}'$ and $\mathbf{QXQ}'$ are called the *generalized network matrices* of G, where $\mathbf{W}$ and $\mathbf{X}$ are square matrices of order e.

Theorem 10.2 Let $\mathbf{B}_1$ and $\mathbf{B}_2$ be two circuit matrices of rank n of a directed graph G, and let $\mathbf{Q}_1$ and $\mathbf{Q}_2$ be two cutset matrices of rank r of G. Then

$$k(\mathbf{B}_2) \det \mathbf{B}_1 \mathbf{WB}_1' = k(\mathbf{B}_1) \det \mathbf{B}_2 \mathbf{WB}_2' \qquad (10.4a)$$

$$k(\mathbf{Q}_2) \det \mathbf{Q}_1 \mathbf{XQ}_1' = k(\mathbf{Q}_1) \det \mathbf{Q}_2 \mathbf{XQ}_2' \qquad (10.4b)$$

and

$$\det \mathbf{B}_1 \mathbf{WB}_1' = \det \mathbf{B}_2 \mathbf{WB}_2' \ (\neq 0) \qquad (10.5a)$$

$$\det \mathbf{Q}_1 \mathbf{XQ}_1' = \det \mathbf{Q}_2 \mathbf{XQ}_2' \ (\neq 0) \qquad (10.5b)$$

if and only if

$$k(\mathbf{B}_1) = k(\mathbf{B}_2) \tag{10.6a}$$

$$k(\mathbf{Q}_1) = k(\mathbf{Q}_2) \tag{10.6b}$$

respectively.

PROOF. We shall prove only the circuit part since the cutset part may be proved in an entirely similar manner. Let $\mathbf{B}_{m1}$ and $\mathbf{B}_{m2}$ be the corresponding major submatrices of $\mathbf{B}_1$ and $\mathbf{B}_2$, respectively. Then from earlier discussion, we have

$$\mathbf{B}_f = \mathbf{B}_{m1}^{-1}\mathbf{B}_1 = \mathbf{B}_{m2}^{-1}\mathbf{B}_2 \tag{10.7}$$

Thus,

$$
\begin{aligned}
\det \mathbf{B}_1 \mathbf{W} \mathbf{B}_1' &= \det[(\mathbf{B}_{m1}\mathbf{B}_{m2}^{-1}\mathbf{B}_2)\mathbf{W}(\mathbf{B}_{m1}\mathbf{B}_{m2}^{-1}\mathbf{B}_2)'] \\
&= \det[(\mathbf{B}_{m1}\mathbf{B}_{m2}^{-1})(\mathbf{B}_2\mathbf{W}\mathbf{B}_2')(\mathbf{B}_{m1}\mathbf{B}_{m2}^{-1})'] \\
&= \left(\frac{\det \mathbf{B}_{m1}}{\det \mathbf{B}_{m2}}\right)^2 \det \mathbf{B}_2 \mathbf{W} \mathbf{B}_2' \\
&= \left[\frac{k(\mathbf{B}_1)}{k(\mathbf{B}_2)}\right]^2 \det \mathbf{B}_2 \mathbf{W} \mathbf{B}_2' \tag{10.8}
\end{aligned}
$$

The theorem follows immediately. This completes the proof of the theorem.

Some of the consequences of this theorem are listed below.

Corollary 10.1 The determinant of a generalized network matrix associated with a system of fundamental circuits or cutsets is invariant with respect to the transformation from one such system to another.

Corollary 10.2 For a given $\mathbf{W}$ and $\mathbf{X}$, the determinants of the generalized network matrices $\mathbf{B}\mathbf{W}\mathbf{B}'$ and $\mathbf{Q}\mathbf{X}\mathbf{Q}'$, if not identically zero, achieve their minimum absolute values if and only if $k(\mathbf{B}) = 1$ and $k(\mathbf{Q}) = 1$, respectively.

Corollary 10.3 For a given $\mathbf{W}$ and $\mathbf{X}$, the determinants of the generalized network matrices $\mathbf{B}_f\mathbf{W}\mathbf{B}_f'$, $\mathbf{B}_p\mathbf{W}\mathbf{B}_p'$, $\mathbf{A}\mathbf{X}\mathbf{A}'$ and $\mathbf{Q}_f\mathbf{X}\mathbf{Q}_f'$ all have their minimum absolute values.

For a given matrix $\mathbf{F}$, let the symbol $\mathbf{F}(I_u, J_v)$ be used to denote the submatrix of $\mathbf{F}$ consisting of the rows and columns corresponding to the integers in the sets I_u and J_v, respectively, where $I_u = \{i_1, i_2, \ldots, i_u\}$ and $J_v = \{j_1, j_2, \ldots, j_v\}$. We shall first give a formula on the number of trees of a graph before establishing an important relation between the determinants of the two types of generalized network matrices.

Theorem 10.3 The number of trees in a directed graph G is given by $\det \mathbf{BB}'/k(\mathbf{B})$ or $\det \mathbf{QQ}'/k(\mathbf{Q})$.

PROOF. Let $\mathbf{B}_m$ be a major submatrix of $\mathbf{B}$. Then $\mathbf{B}_f = \mathbf{B}_m^{-1}\mathbf{B}$. It follows that

$$\det \mathbf{BB}' = \det \mathbf{B}_m \mathbf{B}_f \mathbf{B}_f' \mathbf{B}_m' = (\det \mathbf{B}_m)(\det \mathbf{B}_f \mathbf{B}_f')(\det \mathbf{B}_m')$$

$$= (\det \mathbf{B}_m)^2 (\det \mathbf{B}_f \mathbf{B}_f') = k(\mathbf{B}) \det \mathbf{B}_f \mathbf{B}_f' \qquad (10.9)$$

Now $\det \mathbf{B}_f \mathbf{B}_f'$ is also the determinant of the loop-impedance matrix of a resistive network with resistance of each branch taken to be unity. It can be shown [see, for example, Chen (1976a,b)] that this determinant is equal to the sum of cotree-impedance products of the network. Since each of these cotree-impedance products is of value 1, it follows that $\det \mathbf{B}_f \mathbf{B}_f'$ denotes the number of cotrees or trees in G. This completes the proof of the first part of the theorem. Similarly, we can prove the second part.

Let

$$\mathbf{M}' = [\, \mathbf{Q}' \quad \mathbf{B}' \,] \qquad (10.10)$$

Then

$$\det \mathbf{MM}' = (\det \mathbf{M})^2 = \det \left(\begin{bmatrix} \mathbf{Q} \\ \mathbf{B} \end{bmatrix} [\, \mathbf{Q}' \quad \mathbf{B}' \,] \right)$$

$$= \det \begin{bmatrix} \mathbf{QQ}' & \mathbf{QB}' \\ \mathbf{BQ}' & \mathbf{BB}' \end{bmatrix}$$

$$= \det \begin{bmatrix} \mathbf{QQ}' & \mathbf{0} \\ \mathbf{0} & \mathbf{BB}' \end{bmatrix} = (\det \mathbf{QQ}') \det \mathbf{BB}'$$

$$(10.11)$$

where $\mathbf{QB}' = \mathbf{0}$ and $\mathbf{BQ}' = \mathbf{0}$. To justify this, we note that the number of edges common to a cut and a circuit is always even. If a cut has $2k$ edges in common with a circuit, then it is not difficult to see that k of these edges have the same relative orientation in the cut and in the circuit, and the other k edges have one orientation in the cut and the opposite orientation in the circuit, resulting in $\mathbf{QB}' = \mathbf{0}$ and $\mathbf{BQ}' = \mathbf{0}$.

If m denotes the number of trees of G, then from Theorem 10.3 we have

$$\det \mathbf{M}\mathbf{M}' = (\det \mathbf{M})^2 = m^2 k\,(\mathbf{Q})\,k(\mathbf{B}) \tag{10.12}$$

or

$$\det \mathbf{M} = m\,[k\,(\mathbf{Q})\,k(\mathbf{B})]^{1/2} \tag{10.13}$$

Let $\bar{I}_r$ and $\bar{J}_r$ denote the complements of the sets I_r and J_r in $\{1, 2, \ldots, e\}$, respectively, and $\mathbf{Q}(I_r)$ and $\mathbf{B}(\bar{I}_r)$ be the major submatrices of $\mathbf{Q}$ and $\mathbf{B}$ consisting of the columns corresponding to the integers in I_r and $\bar{I}_r$, respectively. Expanding $\det \mathbf{M}$ according to the minor determinant of the first r rows by Laplace's expansion, we obtain

$$\sum_{(I_r)} (-1)^{i_1+i_2+\cdots+i_r} \det \mathbf{Q}(I_r) \det \mathbf{B}(\bar{I}_r)$$

$$= \sum_{(I_r)} (-1)^{i_1+i_2+\cdots+i_r} (\pm 1)\, [k(\mathbf{Q})k(\mathbf{B})]^{1/2}$$

$$= [k(\mathbf{Q})k(\mathbf{B})]^{1/2} \sum_{(I_r)} (-1)^{i_1+i_2+\cdots+i_r} (\pm 1)$$

$$= \pm m\, [k(\mathbf{Q})k(\mathbf{B})]^{1/2} \tag{10.14}$$

Since there are exactly m terms in the summation, for the last equality to hold it is necessary that all the terms inside the summation sign agree in algebraic sign, yielding

$$(-1)^{i_1+i_2+\cdots+i_r} \det \mathbf{Q}(I_r) \det \mathbf{B}(\bar{I}_r)$$

$$= (-1)^{j_1+j_2+\cdots+j_r} \det \mathbf{Q}(J_r) \det \mathbf{B}(\bar{J}_r) \tag{10.15}$$

which leads to

$$(-1)^{i_1+i_2+\cdots+i_r+j_1+j_2+\cdots+j_r} \det \mathbf{Q}(I_r) \det \mathbf{Q}(J_r)$$

$$= \frac{k(\mathbf{Q})}{k(\mathbf{B})} \det \mathbf{B}(\bar{I}_r) \det \mathbf{B}(\bar{J}_r) \tag{10.16}$$

With these preliminaries, we are now in a position to state and prove a main theorem of this section.

Theorem 10.4 If $\mathbf{W}$ is nonsingular, then

$$k(\mathbf{Q}) \det \mathbf{B}\mathbf{W}\mathbf{B}' = k(\mathbf{B})(\det \mathbf{W}) \det \mathbf{Q}\mathbf{W}^{-1}\mathbf{Q}' \tag{10.17}$$

Proof. Let $\mathbf{X} = \mathbf{W}^{-1}$. Applying the Binet–Cauchy theorem [see, for example, Aitken (1962)] in the usual way to the expansions of det $\mathbf{BWB}'$ and det $\mathbf{QXQ}'$, we have

$$\det \mathbf{BWB}' = \sum_{(I_r)} \sum_{(J_r)} \det \mathbf{B}(\bar{I}_r) \det \mathbf{W}(\bar{I}_r, \bar{J}_r) \det \mathbf{B}(\bar{J}_r)'$$

$$= \sum_{(I_r)} \sum_{(J_r)} \det \mathbf{B}(\bar{I}_r) \det \mathbf{B}(\bar{J}_r) \det \mathbf{W}(\bar{I}_r, \bar{J}_r) \qquad (10.18)$$

and

$$\det \mathbf{QXQ}' = \sum_{(I_r)} \sum_{(J_r)} \det \mathbf{Q}(I_r) \det \mathbf{Q}(J_r) \det \mathbf{X}(I_r, J_r)$$

$$= \frac{k(\mathbf{Q})}{k(\mathbf{B})} \sum_{(I_r)} \sum_{(J_r)} (-1)^{i_1+i_2+\cdots+i_r+j_1+j_2+\cdots+j_r}$$

$$\times \det \mathbf{B}(\bar{I}_r) \det \mathbf{B}(\bar{J}_r) \det \mathbf{X}(I_r, J_r) \qquad (10.19)$$

Since $\mathbf{X}$ is the inverse of $\mathbf{W}$, by Jacobi's theorem [see Aitken (1962)], we have

$$\det \mathbf{X}(I_r, J_r) = (-1)^{i_1+i_2+\cdots+i_r+j_1+j_2+\cdots+j_r} \frac{\det \mathbf{W}(\bar{J}_r, \bar{I}_r)}{\det \mathbf{W}} \qquad (10.20)$$

Thus,

$$\det \mathbf{QXQ}' = \frac{k(\mathbf{Q})}{k(\mathbf{B})} \sum_{(I_r)} \sum_{(J_r)} \det \mathbf{B}(\bar{I}_r) \det \mathbf{B}(\bar{J}_r) \frac{\det \mathbf{W}(\bar{J}_r, \bar{I}_r)}{\det \mathbf{W}}$$

$$= \frac{k(\mathbf{Q})}{k(\mathbf{B}) \det \mathbf{W}} \sum_{(I_r)} \sum_{(J_r)} \det \mathbf{B}(\bar{I}_r)$$

$$\times \det \mathbf{B}(\bar{J}_r) \det \mathbf{W}(\bar{I}_r, \bar{J}_r)$$

$$= \frac{k(\mathbf{Q})}{k(\mathbf{B}) \det \mathbf{W}} \det \mathbf{BWB}' \qquad (10.21)$$

The theorem follows. This completes the proof of the theorem.

Corollary 10.4 If $\mathbf{W}$ is nonsingular, then

$$\det \mathbf{B}_f \mathbf{WB}'_f = (\det \mathbf{W}) \det \mathbf{Q}_f \mathbf{W}^{-1} \mathbf{Q}'_f$$

$$= (\det \mathbf{W}) \det \mathbf{AW}^{-1} \mathbf{A}' \qquad (10.22)$$

If $\mathbf{Z}_b$ and $\mathbf{Y}_b$ are the *branch-impedance* and *branch-admittance matrices* of order e of a given network N, respectively, the matrix triple products $\mathbf{BZ}_b\mathbf{B}'$ and $\mathbf{QY}_b\mathbf{Q}'$ are also referred to as the *loop-impedance matrix* $\mathbf{Z}_m$ and the *cutset-admittance matrix* $\mathbf{Y}_c$ of N, respectively. In particular, the cutset admittance matrix associated with an incidence matrix $\mathbf{A}$ of N, denoted by $\mathbf{Y}_n$, is the *node-admittance matrix*. Some consequences of the above discussion can now be stated as follows.

Corollary 10.5 The determinant of the loop-impedance matrix associated with a system of fundamental circuits or the determinant of the cutset-admittance matrix associated with a system of fundamental cutsets is invariant with respect to the transformation from one such system to another. The absolute value of such determinant is the minimum absolute value possible for a given network.

Corollary 10.6 The determinant of the cutset-admittance matrix associated with any system of r independent node-pair voltages has the minimum absolute value, this being invariant with respect to the transformations from one such system to another. If the network is planar, then the determinant of the loop-impedance matrix formed by the windows of the network has the minimum absolute value.

Corollary 10.7 The determinant of the node-admittance matrix is invariant with respect to the choice of the reference node in the network.

PROOF. Let $\mathbf{A}_{ci}$ be the incidence matrix obtained from the complete incidence matrix $\mathbf{A}_c$ by deleting the row i from $\mathbf{A}_c$. Since $\det \mathbf{A}_{ci}\mathbf{Y}_b\mathbf{A}'_{ci}$ is invariant with respect to any i, the corollary follows from the fact that $\mathbf{A}_{ci}\mathbf{Y}_b\mathbf{A}'_{ci}$ uses node i as the reference node. This completes the proof of the corollary.

A direct consequence of Theorem 10.4 is the following.

Corollary 10.8 The determinants of any two loop-impedance matrices or the determinants of any two cutset-admittance matrices are related by a real constant.

Consequently, the poles and zeros of the network determinants are independent of the choice of the loop currents or the cutset voltages, and in particular the node-pair voltages, in a network.

The relationships between the determinants of the loop-impedance matrix and the cutset-admittance matrix can be readily obtained from Theorem 10.4, and are given as follows.

Corollary 10.9 If $\mathbf{Z}_b$ is nonsingular, then

$$k(\mathbf{Q}) \det \mathbf{BZ}_b\mathbf{B}' = k(\mathbf{B})(\det \mathbf{Z}_b) \det \mathbf{QY}_b\mathbf{Q}' \tag{10.23}$$

where $\mathbf{Y}_b = \mathbf{Z}_b^{-1}$.

Corollary 10.10 If $\mathbf{Z}_b$ is nonsingular, then the ratio of the minimum absolute values of the determinants of the loop-impedance matrix and the cutset-admittance matrix is equal to the absolute value of the determinant of the branch-impedance matrix.

In the case where there are unilateral transmittances which do not form any closed loop, we may renumber the branches in such a way that $\mathbf{Z}_b$ is a triangular matrix. Thus, we obtain the following corollary.

Corollary 10.11 If the transmittances of a network do not form any closed loop, then the ratio of the minimum absolute values of the determinants of the loop-impedance matrix and the node-admittance or cutset-admittance matrix is equal to the product of all the branch driving-point impedances. The ratio is independent of the transmittances present.

A different version of Corollary 10.9 is the following.

Corollary 10.12 If $\mathbf{Z}_b$ is nonsingular, then the ratio of the determinants of the loop-impedance matrix and the cutset-admittance matrix is equal to a real constant times the determinant of the branch-impedance matrix of the network, where the real constant depends only upon the choice of the circuits and cutsets.

Corollary 10.13 The poles and zeros of the determinants of the loop-impedance matrix and the cutset-admittance matrix differ only by those contained in the determinant of the branch-impedance matrix of the network.

Corollary 10.14 For a passive RLC network without mutual inductances, the determinants of the loop-impedance matrix and the cutset-admittance matrix have the same zeros, excluding those at the origin and infinity.

10.1.3 Relations Between Loop- and Cutset-Based General Return Differences

In the early chapters, we have indicated that the general return difference is invariant with respect to the general transformation from one set of basis circuits or cutsets to another. It is commonly assumed in the literature that the general return difference is also invariant with respect to the loop and cutset formulations of the network equations. That this is not the case will be discussed in the present section.

In the loop formulation of network equations, we implicitly assume that all the dependent sources are current-controlled voltage sources, and in the cutset formulation that all the dependent sources are voltage-controlled current sources. Thus, in applying (10.23), we need to define two equivalent networks N_m and

N_c of a given feedback network N, N_m for the loop formulation and N_c for the cutset formulation, whose associated directed graphs are isomorphic, and such that if $\mathbf{Z}_b$ is the nonsingular branch-impedance matrix of N_m, then $\mathbf{Y}_b = \mathbf{Z}_b^{-1}$ is the branch-admittance matrix of N_c. Clearly, this can always be achieved by means of the Thévenin and Norton theorems and the Blakesley transformation ((e-shift or i-shift) if necessary [see, for example, Seshu and Reed (1961)]. In the following when we discuss the loop formulation, we refer to the equivalent network N_m, and the cutset formulation we refer to N_c. Write

$$\mathbf{Z}_b(k) = \mathbf{Z}_b(x)|_{x=k} \quad \text{and} \quad \mathbf{Y}_b(k) = \mathbf{Y}_b(x)|_{x=k} \qquad (10.24)$$

Then from (10.23) we find that the general return difference $F_k^m(x)$ with respect to an element x for a reference value k and based on the loop formation is related to that $F_k^c(x)$ of the same element based on the cutset formulation by the equation

$$F_k^m(x) = \frac{\det \mathbf{Z}_b(x)}{\det \mathbf{Z}_b(k)} F_k^c(x) \qquad (10.25)$$

where

$$F_k^m(x) = \frac{\det \mathbf{BZ}_b(x)\mathbf{B}'}{\det \mathbf{BZ}_b(k)\mathbf{B}'} \qquad (10.26a)$$

$$F_k^c(x) = \frac{\det \mathbf{QY}_b(x)\mathbf{Q}'}{\det \mathbf{QY}_b(k)\mathbf{Q}'} \qquad (10.26b)$$

Using (10.4) in conjunction with (10.26), we obtain a useful result that will be stated as a corollary.

Corollary 10.15 In the loop (cutset) formulation of the network determinant of a feedback network, the general return difference is invariant with respect to the transformation from one such system to another.

To facilitate our discussion, we shall consider two different situations.

Two-Port Element x

The element x of interest is assumed to be the controlling parameter of a controlled source. The controlling parameter could be transconductance, transresistance or any transfer function, which may appear more than one place in $\mathbf{Z}_b$, $\mathbf{BZ}_b(x)\mathbf{B}'$, $\mathbf{Y}_b$ or $\mathbf{QY}_b(x)\mathbf{Q}'$, as will be demonstrated in the illustrations.

Theorem 10.5 In a feedback network, assume that its branch-immittance matrix is nonsingular under the condition that the element x assumes its nominal value and the condition that the element x assumes the value k. Then the ratio of the general return difference with respect to an element x for a reference value k and based on the loop formulation to that of the same element based on the cutset formulation is equal to the ratio of the determinant of the branch-impedance matrix when the element x assumes its nominal value to that when it assumes the value k.

The theorem follows directly from (10.25) and (10.26). An important consequence of this is a result that gives the condition under which the general return difference is invariant with respect to the loop and cutset formulations.

Theorem 10.6 In a feedback network, assume that the branch-immittance matrix is nonsingular under the condition that the element x assumes the value k. Then the general return difference with respect to x for a reference value k is invariant under the general transformations between a system of basis circuits and a system of basis cutsets if and only if the determinant of the branch-immittance matrix remains unaltered when x is replaced by k.

Example 10.1 Consider the feedback network N of Fig. 10.2a. After shorting the independent voltage generator V_g, the equivalent network N_c together with its associated directed graph G is presented in Figs. 10.2b and c. Suppose that g_{m2} is the element of interest and that the reference value k is zero. Let $y_i = 1/z_i$ for $i = 1, 2, 3$ and 4. The branch-admittance matrix of N_c can easily be obtained by inspection and is given by

$$\mathbf{Y}_b(g_{m2}) = \begin{bmatrix} y_1 & 0 & 0 & 0 \\ 0 & y_2 & g_{m2} & 0 \\ 0 & g_{m3} & y_3 & 0 \\ 0 & 0 & 0 & y_4 \end{bmatrix} \tag{10.27}$$

Since $\det \mathbf{Y}_b(g_{m2})$ does not remain unaltered when g_{m2} is replaced by 0, according to Theorem 10.6, the return difference $F(g_{m2})$ depends upon the loop and cutset formulations of the network equations.

Consider the node-admittance matrix of Fig. 10.2b, which is obtained as

$$\mathbf{A}\mathbf{Y}_b(g_{m2})\mathbf{A}' = \begin{bmatrix} y_1 + y_2 + y_4 & g_{m2} - y_4 \\ g_{m3} - y_4 & y_3 + y_4 \end{bmatrix} \tag{10.28}$$

Then the return difference with respect to the element g_{m2} for the zero reference value and based on the cutset formulation can be computed from (10.26b),

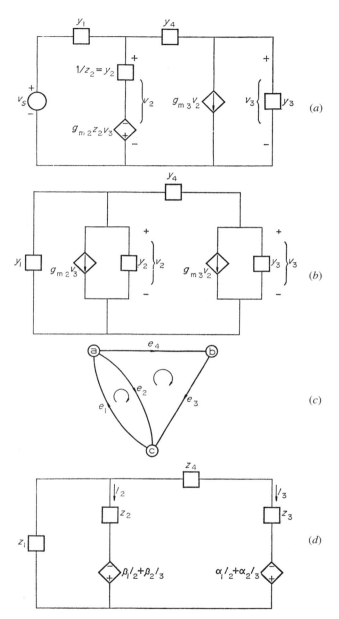

Figure 10.2 (*a*) A feedback network with two controlled sources. (*b*) The modified feedback network N_c used in the cutset formulation. (*c*) The associated directed graph G of N_c or N_m. (*d*) The modified feedback network N_m used in the loop formulation. The parameters are defined as follows: $\beta_1 = -g_{m2}g_{m3}z_2/w$, $\beta_2 = g_{m2}/w$, $\alpha_1 = g_{m3}/w$, and $\alpha_2 = -g_{m2}g_{m3}z_3/w$.

which yields

$$F^c(g_{m2}) = F_0^c(g_{m2}) = \frac{\det \mathbf{AY}_b(g_{m2})\mathbf{A}'}{\det \mathbf{AY}_b(0)\mathbf{A}'}$$

$$= \frac{(y_1 + y_2)(y_3 + y_4) + y_4(y_3 + g_{m2} + g_{m3}) - g_{m2}g_{m3}}{(y_1 + y_2)(y_3 + y_4) + y_4(y_3 + g_{m3})} \tag{10.29}$$

For illustrative purposes, we shall compute the return difference based on the loop formulation. The equivalent network N_m is given in Fig. 10.2d, which, as required, has the same associated directed graph G as that of N_c shown in Fig. 10.2c. Using the two circuits as indicated in Fig. 10.2c, the branch-impedance matrix and the loop-impedance matrix are found to be

$$\mathbf{Z}_b(g_{m2}) = \begin{bmatrix} z_1 & 0 & 0 & 0 \\ 0 & y_3/w & -g_{m2}/w & 0 \\ 0 & -g_{m3}/w & y_2/w & 0 \\ 0 & 0 & 0 & z_4 \end{bmatrix} \tag{10.30}$$

which is the inverse of $\mathbf{Y}_b(g_{m2})$, where $w = y_2y_3 - g_{m2}g_{m3}$ and

$$\mathbf{BZ}_b(g_{m2})\mathbf{B}' = \begin{bmatrix} z_1 + y_3/w & -(y_3 + g_{m2})/w \\ -(y_3 + g_{m3})/w & z_4 + (y_2 + y_3 + g_{m2} + g_{m3})/w \end{bmatrix} \tag{10.31}$$

From (10.26a), the return difference with respect to the element g_{m2} for the zero reference is computed as

$$F^m(g_{m2}) = F_0^m(g_{m2}) = \frac{\det \mathbf{BZ}_b(g_{m2})\mathbf{B}'}{\det \mathbf{BZ}_b(0)\mathbf{B}'}$$

$$= \frac{y_1y_2y_3y_4 \left[z_1(z_4w + y_2 + y_3 + g_{m2} + g_{m3}) + y_3z_4 + 1 \right]}{w \left[y_2y_3 + y_4(y_2 + y_3 + g_{m3}) + y_1y_3 + y_1y_4 \right]}$$

$$= \frac{y_2y_3 \left[y_3(y_1 + y_2 + y_4) + y_4(y_1 + y_2 + g_{m2} + g_{m3}) - g_{m2}g_{m3} \right]}{(y_2y_3 - g_{m2}g_{m3}) \left[y_3(y_1 + y_2 + y_4) + y_4(y_1 + y_2 + g_{m3}) \right]} \tag{10.32}$$

Comparing (10.29) with (10.32) gives

$$F^m(g_{m2}) = \frac{\det \mathbf{Z}_b(g_{m2})}{\det \mathbf{Z}_b(0)} F^c(g_{m2}) \tag{10.33}$$

where

$$\det \mathbf{Z}_b(g_{m2}) = \frac{z_1 z_4}{(y_2 y_3 - g_{m2} g_{m3})} \qquad (10.34a)$$

$$\det \mathbf{Z}_b(0) = z_1 z_2 z_3 z_4 \qquad (10.34b)$$

confirming our statement in Theorem 10.5. Now suppose that we set $g_{m2} = 0$. Then

$$F(g_{m2}) = F^m(g_{m2}) = F^c(g_{m2}) \qquad (10.35)$$

with $\det \mathbf{Z}_b(g_{m2}) = \det \mathbf{Z}_b(0)$, which also follows directly from Theorem 10.6.

In a feedback network N_m, a set of current-controlled voltage sources, characterized by the defining equations

$$V_{q_j} = x_{q_j} I_{u_j}, \quad j = 1, 2, \ldots, k \qquad (10.36)$$

are said to be *cyclically coupled* if $q_{\sigma+1} = u_\sigma$ for $\sigma = 1, 2, \ldots, k$, where $q_{k+1} = q_1$. For example, the following three current-controlled voltage sources

$$V_2 = x_2 I_5, \quad V_5 = x_5 I_4, \quad V_4 = x_4 I_2 \qquad (10.37)$$

are cyclically coupled. By interchanging the roles of V and I in (10.36), we can similarly define a set of *cyclically coupled voltage-controlled current sources* in N_c.

We remark that the controlled sources of nearly all practical feedback amplifiers are *not* cyclically coupled. For this large class of feedback networks, we show that the general return difference is invariant.

Corollary 10.16 In a feedback network, assume that the branch-immittance matrix is nonsingular under the condition that the element x assumes the value k, and that there are no cyclically coupled controlled sources. Then the general return difference with respect to the element x for a reference value k is invariant under the general transformations between a system of basis circuits and a system of basis cutsets.

PROOF. Since there are no cyclically coupled controlled sources, we can always renumber the controlling and controlled branches such that the index of any controlled branch is not higher than the index of its corresponding controlling branch. This results in a branch-immittance matrix with all the controlling parameters appearing above the main diagonal. Thus, $\det \mathbf{Z}_b(x)$ is independent of the controlling parameter x, and from Theorem 10.6 the corollary follows. This completes the proof of the corollary.

For example, in Fig. 10.2d if we set $g_{m2} = 0$, the network has only one controlled source, so it cannot be cyclically coupled. Thus, the general return difference with respect to the element g_{m2} is invariant, a fact that was stated earlier in (10.35) for the zero reference value.

Corollary 10.17 In a feedback network containing only a single controlled source, the general return difference with respect to the controlling parameter of the controlled source is invariant under the general transformations between a system of basis circuits and a system of basis cutsets.

One-Port Element x

Suppose that the element x of interest is a one-port network characterized by its immittance x. In the present section, we discuss how the general return difference varies with the formulation of the network equations.

We first consider the case where x is a one-port impedance. Since from Corollary 10.15 the general return difference is invariant with respect to the transformations from one system of basis circuits to another, for simplicity and without loss of generality, it is sufficient to consider a system of fundamental circuits. To this end, let t be a tree of the associated directed graph G *not* containing the edge corresponding to the branch x. This is always possible unless the edge x is a cutset, which will be considered separately. Let $\mathbf{B}_f$ be the fundamental circuit matrix of G with respect to the tree t. With this choice of circuits, we recognize that the impedance x will appear only in the diagonal element of $\mathbf{B}_f \mathbf{Z}_b(x)\mathbf{B}'_f$ corresponding to the fundamental circuit formed by the edge x. Thus, by assuming the controlling parameters do not involve x, we can write

$$\det \mathbf{B}_f \mathbf{Z}_b(x)\mathbf{B}'_f = \det \mathbf{B}_f \mathbf{Z}_b(k)\mathbf{B}'_f + (x - k) \det [\mathbf{B}_f \mathbf{Z}_b(x)\mathbf{B}'_f]_{xx} \qquad (10.38)$$

where $[\mathbf{B}_f \mathbf{Z}_b(x)\mathbf{B}'_f]_{xx}$ is the submatrix obtained from $\mathbf{B}_f \mathbf{Z}_b(x)\mathbf{B}'_f$ by deleting the row and column corresponding to the edge x. Combining (10.26a) and (10.38) yields

$$F_k^m(x) = 1 + (x - k)\frac{\det [\mathbf{B}_f \mathbf{Z}_b(x)\mathbf{B}'_f]_{xx}}{\det \mathbf{B}_f \mathbf{Z}_b(k)\mathbf{B}'_f}$$

$$= 1 + \frac{x - k}{k + z} = \frac{z + x}{z + k} \qquad (10.39)$$

where z is the driving-point impedance that the element x faces. The equation to the right of $F_k^m(x)$ follows from the fact that $\det \mathbf{B}_f \mathbf{Z}_b(k)\mathbf{B}'_f$ is the loop-impedance matrix of the feedback network N_m when the element x assumes the value k, that $[\mathbf{B}_f \mathbf{Z}_b(x)\mathbf{B}'_f]_{xx} = [\mathbf{B}_f \mathbf{Z}_b(k)\mathbf{B}'_f]_{xx}$ and that if, in N_m, with x assuming the value k, we insert a voltage generator in series with x, the ratio of $\det \mathbf{B}_f \mathbf{Z}_b(k) \mathbf{B}'_f$ to $\det [\mathbf{B}_f \mathbf{Z}_b(x) \mathbf{B}'_f]_{xx}$ is precisely the impedance looking into the generator, which is $z + k$.

In the case where the edge x of G is a cutset, then it will not belong to any circuits. Thus, $\mathbf{B} \mathbf{Z}_b(x) \mathbf{B}'$ is independent of x, and under this situation

$$F_k^m(x) = 1 \qquad (10.40)$$

which clearly is the limiting case of (10.39) as z approaches to infinity.

In the cutset formulation of the general return difference, again by Corollary 10.15 it is sufficient to consider a system of fundamental cutsets. Let t' be a tree of G containing the edge x, and let $\mathbf{Q}_f$ be the fundamental cutset matrix of G with respect to the tree t'. Then, as before, the admittance $1/x$ will appear only in the diagonal element of $\mathbf{Q}_f \mathbf{Y}_b(x) \mathbf{Q}'_f$ corresponding to the fundamental cutset defined by the edge x. Thus, for $k \neq 0$, we can write

$$\det \mathbf{Q}_f \mathbf{Y}_b(x) \mathbf{Q}'_f = \det \mathbf{Q}_f \mathbf{Y}_b(k') \mathbf{Q}'_f + (x' - k')$$
$$\times \det [\mathbf{Q}_f \mathbf{Y}_b(x) \mathbf{Q}'_f]_{xx} \qquad (10.41)$$

where

$$x' = 1/x \quad \text{and} \quad k' = 1/k \qquad (10.42)$$

Combining (10.26b) and (10.42) yields

$$F_k^c(x) = 1 + (x' - k') \frac{\det [\mathbf{Q}_f \mathbf{Y}_b(x) \mathbf{Q}'_f]_{xx}}{\det \mathbf{Q}_f \mathbf{Y}_b(k) \mathbf{Q}'_f}$$
$$= 1 + \frac{x' - k'}{y + k'} = \frac{y + x'}{y + k'} \qquad (10.43)$$

where y is the driving-point admittance that the element x faces. The equation on the right of $F_k^c(x)$ follows dually from that of (10.39). As k approaches to zero, $\det \mathbf{Q}_f \mathbf{Y}_b(k) \mathbf{Q}'_f$ approaches to

$$\det \mathbf{Q}_f \mathbf{Y}_b(k) \mathbf{Q}'_f = k' \det [\mathbf{Q}_f \mathbf{Y}_b(x) \mathbf{Q}'_f]_{xx} \qquad (10.44)$$

and consequently for $k = 0$ we have

$$F^c(x) = F_0^c(x) = 0 \qquad (10.45)$$

which may again be viewed as the limiting case of (10.43) as k' approaches to infinity or k to zero. Comparing (10.39) and (10.43) gives an equation that relates the two general return difference by

$$kF_k^m(x) = xF_k^c(x) \tag{10.46}$$

confirming the identity (10.25) in which $[\det \mathbf{Z}_b(x)] / [\det \mathbf{Z}_b(k)] = x/k$.

The treatment given above for the one-port impedance can be duplicated for the one-port admittance. In order to distinguish the two cases, we use the symbols x' and k', as defined in (10.42). In the cutset formulation, we have

$$F_{k'}^c(x') = \frac{y + x'}{y + k'} \tag{10.47}$$

and in the loop formulation and for $k' \neq 0$, we have

$$F_{k'}^m(x') = \frac{z + x}{z + k} \tag{10.48}$$

and for $k' = 0$

$$F^m(x') = F_0^m(x') = 0 \tag{10.49}$$

Finally, (10.47) and (10.48) are related by

$$x'F_{k'}^m(x') = k'F_{k'}^c(x') \tag{10.50}$$

We summarize the above results by the following theorem.

Theorem 10.7 The general return difference of a feedback network with respect to a one-port immittance x for a general reference value k and based on the loop formulation of network equations is equal to the ratio of the two impedances looking into the branch x under the condition that x assumes its nominal value and the condition that x assumes the value k, and that based on the cutset formulation is equal to the ratio of the two admittances at the node pair where x is connected and under the condition that x assumes its nominal value and the condition that x assumes the value k. It is invariant under the general transformations between a system of basis circuits and a system of basis cutsets if and only if $x = k$.

We illustrate the above results by the following example.

Example 10.2 Consider the same feedback network as in Example 10.1 except that now we wish to compute the general return difference with respect to the one-port impedance z_1 for the general reference value k. Using (10.26a) in conjunction with (10.31) with $x = z_1$, we get

$$
\begin{aligned}
F_k^m(z_1) &= \frac{z_1(z_4 w + y_2 + y_3 + g_{m2} + g_{m3}) + y_3 z_4 + 1}{k(z_4 w + y_2 + y_3 + g_{m2} + g_{m3}) + y_3 z_4 + 1} \\
&= \frac{z + z_1}{z + k}
\end{aligned}
\tag{10.51}
$$

in which the impedance that z_1 faces is given by

$$
z = \frac{y_3 z_4 + 1}{z_4 w + y_2 + y_3 + g_{m2} + g_{m3}}
\tag{10.52}
$$

Likewise, using (10.26b) in conjunction with (10.28) we have

$$
\begin{aligned}
F_k^c(z_1) &= \frac{(y_1 + y_2)(y_3 + y_4) + y_4(y_3 + g_{m2} + g_{m3}) - g_{m2} g_{m3}}{(1/k + y_2)(y_3 + y_4) + y_4(y_3 + g_{m2} + g_{m3}) - g_{m2} g_{m3}} \\
&= \frac{y + y_1}{y + 1/k}
\end{aligned}
\tag{10.53}
$$

where $y = 1/z$ is the admittance that y_1 faces.

In the case of the one-port admittance y_1, we obtain

$$
F_{k'}^c(y_1) = F_k^c(z_1)
\tag{10.54a}
$$

$$
F_{k'}^m(y_1) = F_k^m(z_1)
\tag{10.54b}
$$

where k' and k are related by (10.42).

10.2 GENERALIZED COFACTORS OF THE GENERALIZED NETWORK MATRIX DETERMINANTS

In this section, we shall look into the conditions under which some precisely defined quantities called *generalized cofactors* of the determinants of the generalized network matrices are invariant, so that they may be used to show that the network functions of a network are invariant with respect to the choice of loop currents or cutset voltages in the network.

For a given matrix $\mathbf{F}$, let the symbol $\mathbf{F}_{-i}$ be used to denote the matrix obtained from $\mathbf{F}$ by deleting the row i from $\mathbf{F}$. If $\mathbf{F}$ is of order $p \times q$, $p \leq q$, then by $\mathbf{F}(I_k)$ we mean the submatrix of order $p \times k$ consisting of the columns corresponding to the integers in I_k where $I_k = \{i_1, i_2, \ldots, i_k\}$.

Since the determinants of the major submatrices of either $\mathbf{B}_{-i}$ and $\mathbf{Q}_{-i}$, in general, do not have the same magnitude, it is important that we seek certain quantities which have this property. As an example, consider the circuit matrix **B** given by (10.3). Then after deleting the fourth row, we have

$$
\mathbf{B}_{-4} =
\begin{array}{c}
e_1\ e_5\ e_6\ e_3\ e_4 \\
e_1\ e_2\ e_6\ e_7\ e_4 \\
e_1\ e_2\ e_3\ e_7\ e_8
\end{array}
\begin{array}{cccccccc}
1 & 2 & 3 & 4 & 5 & 6 & 7 & 8 \\
\left[\begin{array}{cccccccc}
1 & 0 & 1 & -1 & 1 & 1 & 0 & 0 \\
1 & 1 & 0 & -1 & 0 & -1 & 1 & 0 \\
1 & 1 & 1 & 0 & 0 & 0 & -1 & -1
\end{array}\right]
\end{array}
\quad (10.55)
$$

It is straightforward to check that the determinant of the major submatrix consisting of the columns 1, 2 and 3 of $\mathbf{B}_{-4}$ has value 1, while the one corresponding to the column 2, 3 and 4 has value -2. Thus, they do not have the same magnitude. For convenience, let $\mathbf{B} = \left[b_{ij}\right]$ and $\mathbf{Q} = \left[q_{ij}\right]$. Define

$$
M_j(\mathbf{B}; I_{n-1}) = \sum_{i=1}^{n} (-1)^i b_{ij} \det \mathbf{B}_{-i}(I_{n-1}) \quad (10.56a)
$$

$$
M_j(\mathbf{Q}; I_{r-1}) = \sum_{i=1}^{r} (-1)^i q_{ij} \det \mathbf{Q}_{-i}(I_{r-1}) \quad (10.56b)
$$

Lemma 10.1 *Let $\mathbf{B}''$ and $\mathbf{B}^{\#}$ be two circuit matrices of a graph G of rank n, and $\mathbf{Q}''$ and $\mathbf{Q}^{\#}$ be two cutset matrices of G of rank r. Then*

$$
k(\mathbf{B}^{\#})^{\frac{1}{2}}|M_j(\mathbf{B}''; I_{n-1})| = k(\mathbf{B}'')^{\frac{1}{2}}|M_j(\mathbf{B}^{\#}; I_{n-1})| \quad (10.57a)
$$

$$
k(\mathbf{Q}^{\#})^{\frac{1}{2}}|M_j(\mathbf{Q}''; I_{r-1})| = k(\mathbf{Q}'')^{\frac{1}{2}}|M_j(\mathbf{Q}^{\#}; I_{r-1})| \quad (10.57b)
$$

if j is not in I_{n-1} and I_{r-1}; and they all vanish if j is in I_{n-1} and I_{r-1}.

PROOF. We shall only prove the circuit part since the cutset part can be proved in an entirely similar manner. If j is not in I_{n-1}, then

$$
M_j(\mathbf{B}''; I_{n-1}) = \pm \det \mathbf{B}''(I_{n-1} \cup \{j\}) \quad (10.58a)
$$

$$
M_j(\mathbf{B}^{\#}; I_{n-1}) = \pm \det \mathbf{B}^{\#}(I_{n-1} \cup \{j\}) \quad (10.58b)
$$

Without loss of generality, assume that $\mathbf{B}''(I_{n-1} \cup \{j\})$ is a major submatrix of $\mathbf{B}''$. Since

$$
\det \mathbf{B}''(I_{n-1} \cup \{j\}) = \pm k(\mathbf{B}'')^{\frac{1}{2}} \quad (10.59a)
$$

$$
\det \mathbf{B}^{\#}(I_{n-1} \cup \{j\}) = \pm k(\mathbf{B}^{\#})^{\frac{1}{2}} \quad (10.59b)
$$

the lemma follows. If j is in I_{n-1}, let us consider the matrix $\mathbf{R}$ which is obtained from $\mathbf{B}''(I_{n-1})$ by inserting the jth column of $\mathbf{B}''$ next to the left of the column corresponding to j in $\mathbf{B}''(I_{n-1})$. Now if we expand $\det \mathbf{R}$ along the inserted column, we find $\det \mathbf{R} = \pm M_j(\mathbf{B}''; I_{n-1})$. Since $\mathbf{R}$ has two identical columns, the lemma follows. This completes the proof of the lemma.

Thus, if $M_j(\mathbf{B}; I_{n-1})$ and $M_j(\mathbf{B}; J_{n-1})$ are nonzero, where $J_{n-1} = \{j_1, j_2, \ldots, j_{n-1}\}$, then they have the same magnitude. This is similarly valid for the cutsets.

Corollary 10.18 The magnitude of $M_j(\mathbf{B}_f; I_{n-1})$ and $M_j(\mathbf{Q}_f; I_{r-1})$ is invariant with respect to the transformations from one such fundamental system to another, and they are all equal to unity.

Lemma 10.2

$$M_i(\mathbf{B}''; I_{n-1}) M_j(\mathbf{B}''; J_{n-1}) = \frac{K(\mathbf{B}'')}{K(\mathbf{B}^\#)} M_i(\mathbf{B}^\#; I_{n-1}) M_j(\mathbf{B}^\#; J_{n-1}) \qquad (10.60a)$$

$$M_i(\mathbf{Q}''; I_{r-1}) M_j(\mathbf{Q}''; J_{r-1}) = \frac{K(\mathbf{Q}'')}{K(\mathbf{Q}^\#)} M_i(\mathbf{Q}^\#; I_{r-1}) M_j(\mathbf{Q}^\#; J_{r-1}) \qquad (10.60b)$$

Definition 10.4: Two-tree or 2-tree A spanning subgraph of a directed or undirected graph G is said to be a *two-tree* or *2-tree* if and only if it has two components and contains no circuits. One or, in trivial cases, both of the components may consist of an isolated node.

Very often, 2-trees in which certain designated nodes are required to be in different components, are needed. For convenience, let the subscripts be used for this purpose. For example, $t_{ab,cde}$ is the symbol for a 2-tree in which the nodes a and b are in one component, and the nodes c, d, and e are in the other component.

Corollary 10.19 $M_j(\mathbf{Q}; I_{r-1})$ is nonzero if and only if the edges in G corresponding to the integers in I_{r-1} form a 2-tree separating the two endpoints of the edge e_j corresponding to the integer j. $M_j(\mathbf{B}; I_{n-1})$ is nonzero if and only if the edges in G corresponding to the integers in I_{n-1} form a cotree in $G - e_j$.

PROOF. Since $M_j(\mathbf{Q}; I_{r-1})$ is nonzero if and only if $\mathbf{Q}(I_{r-1} \cup \{j\})$ is a major submatrix of $\mathbf{Q}$ such that j is not in I_{r-1}, and $\det \mathbf{Q}(I_{r-1} \cup \{j\}) \neq 0$ if and only if edges corresponding to the columns of $\mathbf{Q}(I_{r-1} \cup \{j\})$ form a tree containing the edge e_j in G, the first part of the corollary follows immediately. Likewise, we can prove the second part of the corollary. This completes the proof of the corollary.

Definition 10.5: Generalized Cofactor For given matrices **W** and **X** of order e, and for a given directed graph G of e edges, let

$$M_{uv}(\mathbf{B}) = \sum_{i=1}^{n} \sum_{j=1}^{n} (-1)^{i+j} b_{iu} b_{jv} \det \mathbf{B}_{-i} \mathbf{W} \mathbf{B}'_{-j} \qquad (10.61a)$$

$$M_{uv}(\mathbf{Q}) = \sum_{i=1}^{r} \sum_{j=1}^{r} (-1)^{i+j} q_{iu} q_{jv} \det \mathbf{Q}_{-i} \mathbf{X} \mathbf{Q}'_{-j} \qquad (10.61b)$$

for $u, v = 1, 2, \ldots, e$, where $\mathbf{B}'_{-j} = (\mathbf{B}_{-j})'$ and $\mathbf{Q}'_{-j} = (\mathbf{Q}_{-j})'$. The scalar quantities $M_{uv}(\mathbf{B})$ and $M_{uv}(\mathbf{Q})$ are called the *generalized cofactors* of the (u, v)-elements of $\det \mathbf{BWB}'$ and $\det \mathbf{QXQ}'$, respectively, or simply the (u, v)th *generalized cofactors*.

The reason for this name is that they reduce to the usual cofactors under additional constraints. For example, if the edges e_u and e_v of G corresponding to the columns u and v in **B** are contained only in the circuits C_k and C_m corresponding to the rows k and m in **B**, and oriented in the same directions as the circuits C_k and C_m, respectively, then we have

$$M_{uv}(\mathbf{B}) = (-1)^{k+m} \det \mathbf{B}_{-k} \mathbf{W} \mathbf{B}'_{-m} \qquad (10.62)$$

which is actually the (k, m)th cofactor in $\det \mathbf{BWB}'$. By (k, m)th cofactor in $\det \mathbf{BWB}'$, we mean the cofactor of the element z_{km} in $\det \mathbf{BWB}'$, where $\mathbf{BWB}' = [z_{ij}]$. If the edges and circuits of G are labeled in such a way that $k = u$ and $m = v$, then $M_{uv}(\mathbf{B})$ is simply the (u, v)th cofactor in $\det \mathbf{BWB}'$. This is similarly valid for the generalized cofactor $M_{uv}(\mathbf{Q})$ in $\det \mathbf{QXQ}'$.

In the following, we shall show that under certain conditions, the generalized cofactors defined above are invariant with respect to the choice of the circuits and cutsets.

Theorem 10.8 Let $\mathbf{B}''$ and $\mathbf{B}^{\#}$ be two circuit matrices of a graph G of rank n, and $\mathbf{Q}''$ and $\mathbf{Q}^{\#}$ be two cutset matrices of G of rank r. Then for any two given matrices **W** and **X** of order e, we have

$$k(\mathbf{B}^{\#}) M_{uv}(\mathbf{B}'') = k(\mathbf{B}'') M_{uv}(\mathbf{B}^{\#}) \qquad (10.63a)$$

$$k(\mathbf{Q}^{\#}) M_{uv}(\mathbf{Q}'') = k(\mathbf{Q}'') M_{uv}(\mathbf{Q}^{\#}) \qquad (10.63b)$$

for $u, v = 1, 2, \ldots, e$.

PROOF. Let $\mathbf{B}'' = [b_{ij}'']$, $\mathbf{B}^{\#} = [b_{ij}^{\#}]$, $\mathbf{Q}'' = [q_{ij}'']$ and $\mathbf{Q}^{\#} = [q_{ij}^{\#}]$. Then

$$
M_{uv}(\mathbf{B}'') = \sum_{i=1}^{n}\sum_{j=1}^{n}(-1)^{i+j}b_{iu}''b_{jv}'' \det \mathbf{B}_{-i}'' \mathbf{W}(\mathbf{B}_{-j}'')'
$$

$$
= \sum_{i=1}^{n}\sum_{j=1}^{n}(-1)^{i+j}b_{iu}''b_{jv}'' \sum_{(I_{n-1})}\sum_{(J_{n-1})} \det \mathbf{B}_{-i}''(I_{n-1})
$$

$$
\times \det \mathbf{W}(I_{n-1}; J_{n-1}) \det \mathbf{B}_{-j}''(J_{n-1})
$$

$$
= \sum_{(I_{n-1})}\sum_{(J_{n-1})} \left\{ \left[\sum_{i=1}^{n}(-1)^{i}b_{iu}'' \det \mathbf{B}_{-i}''(I_{n-1}) \right] \right.
$$

$$
\left. \times \det \mathbf{W}(I_{n-1}; J_{n-1}) \left[\sum_{j=1}^{n}(-1)^{j}b_{jv}'' \det \mathbf{B}_{-j}''(J_{n-1}) \right] \right\}
$$

$$
= \sum_{(I_{n-1})}\sum_{(J_{n-1})} M_u(\mathbf{B}''; I_{n-1}) \det \mathbf{W}(I_{n-1}; J_{n-1}) M_v(\mathbf{B}''; J_{n-1})
$$

$$
= \sum_{(I_{n-1})}\sum_{(J_{n-1})} k(\mathbf{B}'') M_u(\mathbf{B}^{\#}; I_{n-1})
$$

$$
\times \det \mathbf{W}(I_{n-1}; J_{n-1}) M_v(\mathbf{B}^{\#}; J_{n-1}) / k(\mathbf{B}^{\#})
$$

$$
= \frac{k(\mathbf{B}'')}{k(\mathbf{B}^{\#})} \sum_{(I_{n-1})}\sum_{(J_{n-1})} M_u(\mathbf{B}^{\#}; I_{n-1})
$$

$$
\times \det \mathbf{W}(I_{n-1}; J_{n-1}) M_v(\mathbf{B}^{\#}; J_{n-1})
$$

$$
= \frac{k(\mathbf{B}'')}{k(\mathbf{B}^{\#})} M_{uv}(\mathbf{B}^{\#}) \tag{10.64}
$$

Similarly, we can prove the other case. This completes the proof of the theorem.

Corollary 10.20 The generalized cofactors $M_{uv}(\mathbf{B}_f)$ and $M_{uv}(\mathbf{Q}_f)$ of $\det \mathbf{B}_f \mathbf{W} \mathbf{B}_f'$ and $\det \mathbf{Q}_f \mathbf{X} \mathbf{Q}_f'$, respectively, are invariant with respect to the transformations from one such fundamental system to another.

Corollary 10.21 For given $\mathbf{W}$ and $\mathbf{X}$, the generalized cofactors $M_{uv}(\mathbf{B})$ and $M_{uv}(\mathbf{Q})$ of $\det \mathbf{B}\mathbf{W}\mathbf{B}'$ and $\det \mathbf{Q}\mathbf{X}\mathbf{Q}'$, if not identically zero, achieve their minimum absolute values if and only if $k(\mathbf{B}) = 1$ and $k(\mathbf{Q}) = 1$, respectively.

Corollary 10.22 For given $\mathbf{W}$ and $\mathbf{X}$, the generalized cofactors $M_{uv}(\mathbf{B}_f)$, $M_{uv}(\mathbf{B}_p)$, $M_{uv}(\mathbf{Q}_f)$ and $M_{uv}(\mathbf{A})$ all have their minimum absolute values.

A direct consequence of the Theorems 10.2 and 10.8 is the following.

Corollary 10.23 The ratios $M_{uv}(\mathbf{B})/M_{ts}(\mathbf{B})$, $M_{uv}(\mathbf{B})/\det \mathbf{BWB}'$, $M_{uv}(\mathbf{Q})/M_{ts}(\mathbf{Q})$, $M_{uv}(\mathbf{Q})/\det \mathbf{QXQ}'$ are invariant with respect to the choice of circuits and cutsets.

Let $S = \{1, 2, \ldots, e\}$. For u and v in S, let $S^{uv} = S - \{u, v\}$. For convenience, the symbols I_k^{uv} and J_k^{uv} are used to denote the subsets of S^{uv}, each of which contains k elements. By $\overline{I_k^{uv}}$ and $\overline{J_k^{uv}}$ we mean the complements of I_k^{uv} and J_k^{uv} in S^{uv}, respectively. With these preliminaries, we have the following lemma.

Lemma 10.3 *If u and v are adjacent integers, then*

$$k(\mathbf{Q})M_u(\mathbf{B}; I_{n-1}^{uv})M_v(\mathbf{B}; J_{n-1}^{uv})$$
$$= k(\mathbf{B})(-1)^{a+b+1}M_v(\mathbf{Q}; \overline{I_{n-1}^{uv}})M_u(\mathbf{Q}; \overline{J_{n-1}^{uv}}) \qquad (10.65)$$

where a and b are the sums of the integers contained in $\overline{I_{n-1}^{uv}}$ and $\overline{J_{n-1}^{uv}}$, respectively.

PROOF. Without loss of generality, let us assume $v > u$. Since

$$M_u(\mathbf{B}; I_{n-1}^{uv}) = (-1)^{u'} \det \mathbf{B}(I_{n-1}^{uv} \cup \{u\}) \qquad (10.66a)$$

$$M_v(\mathbf{B}; J_{n-1}^{uv}) = (-1)^{v'} \det \mathbf{B}(J_{n-1}^{uv} \cup \{v\}) \qquad (10.66b)$$

where u' and v' denote the column indices of the columns corresponding to the integers u and v in $\mathbf{B}\left(I_{n-1}^{uv} \cup \{u\}\right)$ and $\mathbf{B}\left(J_{n-1}^{uv} \cup \{v\}\right)$, respectively, it follows from (10.16) that

$$\frac{k(\mathbf{Q})}{k(\mathbf{B})}M_u(\mathbf{B}; I_{n-1}^{uv})M_v(\mathbf{B}; J_{n-1}^{uv})$$
$$= (-1)^{u'+v'+a+b+u+v} \det \mathbf{Q}(\overline{I_{n-1}^{uv}} \cup \{v\})$$
$$\times \det \mathbf{Q}(\overline{J_{n-1}^{uv}} \cup \{u\}) \qquad (10.67)$$

Since

$$M_v(\mathbf{Q}; \overline{I_{n-1}^{uv}}) = (-1)^{v''} \det \mathbf{Q}(\overline{I_{n-1}^{uv}} \cup \{v\}) \qquad (10.68a)$$

$$M_u(\mathbf{Q}; \overline{J_{n-1}^{uv}}) = (-1)^{u''} \det \mathbf{Q}(\overline{J_{n-1}^{uv}} \cup \{u\}) \qquad (10.68b)$$

where v'' and u'' denote the column indices of the columns corresponding to the integers v and u in $\mathbf{Q}(\overline{I_{n-1}^{uv}}\cup\{v\})$ and $\mathbf{Q}(\overline{J_{n-1}^{uv}}\cup\{u\})$, respectively, and $u = u'+v''-1$

and $v = v' + u''$, it follows that

$$k(\mathbf{Q})M_u(\mathbf{B}; I_{n-1}^{uv})M_v(\mathbf{B}; J_{n-1}^{uv})$$
$$= k(\mathbf{B})(-1)^{a+b+1}M_v(\mathbf{Q}; \overline{I_{n-1}^{uv}})M_u(\mathbf{Q}; \overline{J_{n-1}^{uv}}) \qquad (10.69)$$

which completes the proof of the lemma.

Let $\mathbf{W} = [w_{ij}]$. We say that the edges e_u and e_v of G corresponding to the rows (columns) u and v of $\mathbf{W}$ are *isolated* if $w_{ui} = w_{iu} = 0$ for $i \neq u$ and $i = 1, 2, \ldots, e$ and $w_{jv} = w_{vj} = 0$ for $j \neq v$ and $j = 1, 2, \ldots, e$. In other words, if $w_{uu} \neq 0$ and $w_{vv} \neq 0$, w_{uu} and w_{vv} are the only nonzero entries in these rows and columns. By $\mathbf{W}_{uu,vv}$ we mean the matrix obtained from $\mathbf{W}$ by deleting the rows and columns u and v from $\mathbf{W}$.

Theorem 10.9 If $\mathbf{W}$ and $\mathbf{W}_{uu,vv}$ are nonsingular such that $\mathbf{X} = \mathbf{W}^{-1}$, then

$$k(\mathbf{Q})M_{uv}(\mathbf{B}) = -k(\mathbf{B})(\det \mathbf{W}_{uu,vv})M_{uv}(\mathbf{Q}) \qquad (10.70)$$

for all isolated edges e_u and e_v, $u \neq v$, of G.

PROOF. Since permuting the columns of $\mathbf{B}$ or $\mathbf{Q}$ and the corresponding rows and columns of $\mathbf{W}$ or $\mathbf{X}$ does not change the value of $M_{uv}(\mathbf{B})$ and $M_{uv}(\mathbf{Q})$, without loss of generality, let us assume that u and v are adjacent integers. Let $\mathbf{H} = \mathbf{W}_{uu,vv}$. Following a similar argument to that given in the proof of Theorem 10.8, we have

$$-M_{uv}(\mathbf{B}) = -\sum_{(I_{n-1})}\sum_{(J_{n-1})} M_u(\mathbf{B}; I_{n-1}) \det \mathbf{W}(I_{n-1}; J_{n-1}) M_v(\mathbf{B}; J_{n-1})$$

$$= -\sum_{(I_{n-1}^{uv})}\sum_{(J_{n-1}^{uv})} M_u(\mathbf{B}; I_{n-1}^{uv})M_v(\mathbf{B}; J_{n-1}^{uv}) \det \mathbf{W}(I_{n-1}^{uv}; J_{n-1}^{uv})$$

$$= \sum_{(I_{n-1}^{uv})}\sum_{(J_{n-1}^{uv})} \frac{k(\mathbf{B})}{k(\mathbf{Q})} M_v(\mathbf{Q}; \overline{I_{n-1}^{uv}})M_u(\mathbf{Q}; \overline{J_{n-1}^{uv}})(-1)^{a+b} \det \mathbf{W}(I_{n-1}^{uv}; J_{n-1}^{uv})$$

$$= \frac{k(\mathbf{B})}{k(\mathbf{Q})} \sum_{(I_{n-1}^{uv})}\sum_{(J_{n-1}^{uv})} M_v(\mathbf{Q}; \overline{I_{n-1}^{uv}})M_u(\mathbf{Q}; \overline{J_{n-1}^{uv}})(-1)^{a+b} \det \mathbf{H}(I_{n-1}^{uv}; J_{n-1}^{uv})$$

$$= \frac{k(\mathbf{B})}{k(\mathbf{Q})} \sum_{(I_{n-1}^{uv})}\sum_{(J_{n-1}^{uv})} M_v(\mathbf{Q}; \overline{I_{n-1}^{uv}})M_u(\mathbf{Q}; \overline{J_{n-1}^{uv}}) \det \mathbf{H}^{-1}(\overline{I_{n-1}^{uv}}; \overline{J_{n-1}^{uv}}) \det \mathbf{H}$$

$$= \frac{k(\mathbf{B}) \det \mathbf{H}}{k(\mathbf{Q})} \sum_{(I_{n-1}^{uv})}\sum_{(J_{n-1}^{uv})} M_v(\mathbf{Q}; \overline{I_{n-1}^{uv}})M_u(\mathbf{Q}; \overline{J_{n-1}^{uv}}) \det \mathbf{X}(\overline{J_{n-1}^{uv}}; \overline{I_{n-1}^{uv}})$$

$$= \frac{k(\mathbf{B}) \det \mathbf{H}}{k(\mathbf{Q})} \sum_{(\overline{I_{n-1}^{uv}})} \sum_{(\overline{J_{n-1}^{uv}})} M_u(\mathbf{Q}; \overline{J_{n-1}^{uv}}) \det \mathbf{X}(\overline{J_{n-1}^{uv}}; \overline{I_{n-1}^{uv}}) M_v(\mathbf{Q}; \overline{I_{n-1}^{uv}})$$

$$= \frac{k(\mathbf{B})}{k(\mathbf{Q})} (\det \mathbf{H}) M_{uv}(\mathbf{Q}) \tag{10.71}$$

The second line follows from Lemma 10.1 and the fact that the triple product is zero if v is in I_{n-1} or u is in J_{n-1} since e_u and e_v are isolated edges of G, and the fifth line is obtained by means of Jacobi's theorem as we did in the proof of Theorem 10.8. This completes the proof of the theorem.

Corollary 10.24 If e_u and e_v, $u \neq v$, are isolated edges of G, then

$$-M_{uv}(\mathbf{B}_f) = (\det \mathbf{W}_{uu,vv}) M_{uv}(\mathbf{Q}_f)$$

$$= (\det \mathbf{W}_{uu,vv}) M_{uv}(\mathbf{A}) \tag{10.72}$$

Corollary 10.25 If e_u, e_v, e_t and e_s are isolated edges of G, and if $\mathbf{W}$, $\mathbf{W}_{uu,vv}$ and $\mathbf{W}_{tt,ss}$ are nonsingular such that $\mathbf{X} = \mathbf{W}^{-1}$, then for $s \neq t$ and $u \neq v$ we have

$$\frac{M_{uv}(\mathbf{B})}{M_{ts}(\mathbf{B})} = K_1 \frac{M_{uv}(\mathbf{Q})}{M_{ts}(\mathbf{Q})} \tag{10.73a}$$

$$\frac{M_{uv}(\mathbf{B})}{\det \mathbf{BWB}'} = K_2 \frac{M_{uv}(\mathbf{Q})}{\det \mathbf{QWQ}'} \tag{10.73b}$$

where $K_1 = w_{tt} w_{ss}/w_{uu} w_{vv}$ and $K_2 = -1/w_{uu} w_{vv}$.

Thus, if $\mathbf{W}$ and $\mathbf{X}$ are the nonsingular branch-impedance and the branch-admittance matrices of a network, respectively, then it is not difficult to see that a network function can always be expressed as the ratio of a network determinant and one of its generalized cofactors or the ratio of two of its generalized cofactors. For example, if loop currents are used in the analysis, we may remove all the independent sources from the network except the one at the branch e_u, and the transfer admittance between the branches e_u and e_v corresponding to the integers u and v is given by the ratio $M_{uv}(\mathbf{B})/\det \mathbf{BWB}'$, i.e. the current in the branch e_v with the direction of e_v as reference, due to a voltage generator in e_u with a reference $+$ at the arrow head of e_u. Similarly, if cutset voltages are used, the transfer impedance between the branches e_u and e_v is given by the ratio $M_{uv}(\mathbf{Q})/\det \mathbf{QWQ}'$, i.e. the voltage at the branch e_v with a reference $+$ at the arrow head of e_v, due to a current generator at e_u with the direction of e_u as reference. Thus, from Corollary 10.23, we may conclude the following.

Theorem 10.10 The network functions of a network are invariant with respect to the ways in which the loop currents or the cutset voltages are chosen as variables, i.e. they are invariant with respect to the transformations of the reference frame.

10.3 THE GENERAL NULL RETURN DIFFERENCES

As discussed in early chapters, the null return difference $\hat{F}^c(x)$ with respect to a voltage-controlled current source $i = xV$ for the zero reference and based on the cutset formulation of network equations is defined to be one plus the negative voltage appearing at the controlling branch of the controlled source when the controlled source is replaced by an independent current source of x amperes and when the input current of the feedback network is adjusted so that its output current is identically zero. If we designate port 1 as the input of the feedback network, port 2 the output port, port 3 the port that connects to the controlling branch of the controlled source and port 4 that connects to the controlled source, as shown in Fig. 10.3, and if the port voltages are also the node-pair voltages in the cutset formulation of the network equations, then it is straightforward to show that the null return difference can be expressed

$$\hat{F}^c(x) = \frac{\Delta_{c12}}{\Delta_{c12}^0} \tag{10.74}$$

where Δ_{c12} and Δ_{c12}^0 denote the cofactors of the (1, 2)-element of the cutset admittance matrices of the feedback network under the nominal operating condition and under the condition that x vanishes, respectively. In general, the null return

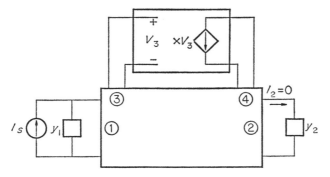

Figure 10.3 A feedback network with the port designations as indicated.

difference with respect to x for a general reference value k is given by

$$\hat{F}_k^c(x) = \frac{\Delta_{c12}}{\Delta_{c12}^k} \tag{10.75}$$

where Δ_{c12}^k is that value of Δ_{c12} when x is replaced by k. Dually, we can define the null return difference with respect to a current-controlled voltage source for a general reference value. Also, the function can be extended to the one-port case and defined in terms of the cofactors of the elements of the loop-impedance matrices.

Several points are worth mentioning at this stage. First of all, unlike the general return difference, the general null return difference is defined differently, though similar, for different types of elements. Secondly, the identity such as (10.75) cannot be used as the formal definition for the general null return difference unless some specific details of the formulation of the network equations are also stated and, furthermore, (10.75) is *variant* under the general transformations from a system of basis cutsets to another. Finally, the assumption that the port voltages be also the node-pair voltages in the cutset formulation is not always valid. For example, in the feedback network N_c of Fig. 10.4, let y_1 be the input port, y_4 the output port, y_2 the controlling branch of the controlled source $g_{m3}V_2$, and $g_{m3}V_2$ the controlled source. Clearly, these port voltages cannot be made part of any set of node-pair voltages in the cutset formulation since they contain a circuit formed by the branches y_2, y_4 and $g_{m3}V_2$. Similar statements can be made with respect to other types of elements and on the loop formulation of network equations.

In the following, we define the general null return difference in terms of the generalized cofactors that will remove all the objections raised above. Furthermore,

Figure 10.4 A feedback network N_c used in the cutset formulation.

we show that the definition is consistent with the physical interpretation outlined at the beginning of this section.

As before, let G be the associated directed graph of a feedback network. Let **B** and **Q** be a basis circuit matrix and a basis cutset matrix of G of rank r and nullity m. Set

$$\mathbf{B} = [b_{ij}] \quad \text{and} \quad \mathbf{Q} = [q_{ij}] \tag{10.76}$$

Definition 10.6: General Null Return Difference Let e_u be the edge corresponding to the independent source, which may include the source impedance, and e_v the edge corresponding to the output impedance of a feedback network. The quantities

$$\hat{F}_k^m(x) = \frac{M_{uv}(\mathbf{B})}{M_{uv}^k(\mathbf{B})} \tag{10.77a}$$

$$\hat{F}_k^c(x) = \frac{M_{uv}(\mathbf{Q})}{M_{uv}^k(\mathbf{Q})} \tag{10.77b}$$

where

$$M_{uv}^k(\mathbf{B}) = M_{uv}(\mathbf{B})|_{x=k} \tag{10.78a}$$

$$M_{uv}^k(\mathbf{Q}) = M_{uv}(\mathbf{Q})|_{x=k} \tag{10.78b}$$

are called the *general null return differences* of the feedback network with respect to a network element x, which may be the controlling parameter of a controlled source or a one-port immittance, for a general reference value k and based on loop and cutset formulations, respectively.

Observe that, like the general return difference, the general null return difference is defined as the ratio of the two functional values assumed by a network generalized cofactor under the condition that the element x assumes its nominal value and the condition that the element x assumes the value k:

$$\hat{F}_k(x) = \frac{M_{uv}(\mathbf{W})}{M_{uv}^k(\mathbf{W})} \tag{10.79}$$

where $\mathbf{W} = \mathbf{B}$ or $\mathbf{Q}$.

Let $\mathbf{B}_1$ and $\mathbf{B}_2$ be two basis circuit matrices of G, and let $\mathbf{Q}_1$ and $\mathbf{Q}_2$ be two basis cutset matrices of G. Then according to Theorem 10.8 we have

$$k(\mathbf{B}_2)M_{uv}(\mathbf{B}_1) = k(\mathbf{B}_1)M_{uv}(\mathbf{B}_2) \tag{10.80a}$$

$$k(\mathbf{Q}_2)M_{uv}(\mathbf{Q}_1) = k(\mathbf{Q}_1)M_{uv}(\mathbf{Q}_2) \tag{10.80b}$$

Using (10.80) in conjunction with (10.77), we obtain a result similar to that for the general return difference.

Theorem 10.11 The general null return difference is invariant with respect to the general transformations from one system of basis circuits or cutsets to another.

Now we show that the definition (10.79) is consistent with the physical interpretation stated at the beginning of this section. Suppose that we wish to use the cutset system of equations, and suppose that the port voltages can be made as part of the node-pair voltages, which is equivalent to stating that the ports can be made part of a tree. Since by Theorem 10.11 the null return difference is invariant with respect to the choice of the basis cutsets, without loss of generality, it is sufficient to consider the system of fundamental cutsets formed with respect to a tree containing the branches of the ports. Also, if we relabel the edges and the fundamental cutsets of G in such a way that the fundamental cutsets defined by the input and output ports have the same designations as the edges e_u and e_v of the input and output ports, respectively, then it is easy to show that

$$M_{uv}(\mathbf{Q}) = (-1)^{u+v} \det M_{uv}(\mathbf{Q}_{-u}\mathbf{Y}_b\mathbf{Q}'_{-v}) = \Delta_{c12} \qquad (10.81)$$

indicating that (10.77b) is consistent with (10.74) under the assumed constraints.

We illustrate the above results by the following examples.

Example 10.3 Consider the feedback network N_c of Fig. 10.4, whose associated directed graph G is given in Fig. 10.5. Let the source edge be e_1 and the output edge be e_4, i.e. $u = 1$ and $v = 4$. Assume that the element of interest is g_{m2}, whose controlling edge is e_3. In the cutset formulation, we consider two basis cutset matrices

$$\mathbf{Q}_1 = \begin{bmatrix} -1 & 1 & 0 & 1 \\ 0 & 0 & 1 & -1 \end{bmatrix} \qquad (10.82a)$$

$$\mathbf{Q}_2 = \begin{bmatrix} -1 & 1 & 0 & 1 \\ 1 & -1 & -1 & 0 \end{bmatrix} \qquad (10.82b)$$

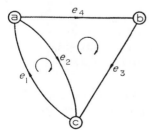

Figure 10.5 The associated directed graph G of the feedback network N_c of Fig. 10.4.

which are also the basis incidence matrices of G. Now, we compute the generalized cofactors

$$M_{14}(\mathbf{Q}_1) = -\det(\mathbf{Q}_1)_{-1}\mathbf{Y}_b(\mathbf{Q}_1)'_{-1} - \det(\mathbf{Q}_1)_{-1}\mathbf{Y}_b(\mathbf{Q}_1)'_{-2}$$

$$= -(y_3 + y_4) - (g_{m3} - y_4) = -y_3 - g_{m3} \qquad (10.83a)$$

$$M_{14}(\mathbf{Q}_2) = -\det(\mathbf{Q}_2)_{-1}\mathbf{Y}_b(\mathbf{Q}_2)'_{-1} - \det(\mathbf{Q}_2)_{-2}\mathbf{Y}_b(\mathbf{Q}_2)'_{-1}$$

$$= -(y_1 + y_2 + y_3 + g_{m2} + g_{m3}) + (y_1 + y_2 + g_{m2})$$

$$= -(y_3 + g_{m3}) \qquad (10.83b)$$

Since

$$M_{14}(\mathbf{Q}_1) = M_{14}^k(\mathbf{Q}_1) = M_{14}(\mathbf{Q}_2) = M_{14}^k(\mathbf{Q}_2) \qquad (10.84)$$

from $(10.77b)$ we obtain the general null return difference

$$\hat{F}_k^c(g_{m2}) = \frac{M_{14}(\mathbf{Q}_1)}{M_{14}^k(\mathbf{Q}_1)} = \frac{M_{14}(\mathbf{Q}_2)}{M_{14}^k(\mathbf{Q}_2)} = 1 \qquad (10.85)$$

which is expected, since, physically, if the independent current source at y_1 is adjusted so the current in y_4 is zero, the voltage across the controlling branch y_3 must be zero.

In the loop formulation, we consider the modified feedback network N_m of Fig. 10.6 and the two basis circuit matrices

$$\mathbf{B}_1 = \begin{bmatrix} 1 & 1 & 0 & 0 \\ 0 & -1 & 1 & 1 \end{bmatrix} \qquad (10.86a)$$

$$\mathbf{B}_2 = \begin{bmatrix} 1 & 1 & 0 & 0 \\ 1 & 0 & 1 & 1 \end{bmatrix} \qquad (10.86b)$$

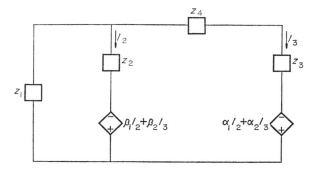

Figure 10.6 The modified feedback network N_m of Fig. 10.3.

corresponding to the basis circuits e_1e_2, $e_2e_3e_4$ and e_1e_2, $e_1e_3e_4$ of G, respectively. The required generalized cofactors are given by

$$M_{14}(\mathbf{B}_1) = -\det(\mathbf{B}_1)_{-1}\mathbf{Z}_b(\mathbf{B}_1)'_{-2} = \frac{y_3 + g_{m3}}{y_2 y_3 - g_{m2}g_{m3}} \tag{10.87a}$$

$$M_{14}(\mathbf{B}_2) = -\det(\mathbf{B}_2)_{-1}\mathbf{Z}_b(\mathbf{B}_2)'_{-2} + \det(\mathbf{B}_2)_{-2}\mathbf{Z}_b(\mathbf{B}_2)'_{-2}$$

$$= -\left(z_1 - \frac{g_{m3}}{w}\right) + \left(z_1 + z_2 + \frac{z_2 g_{m2} g_{m3}}{w}\right)$$

$$= \frac{y_3 + g_{m3}}{y_2 y_3 - g_{m2}g_{m3}} \tag{10.87b}$$

where $w = y_2 y_3 - g_{m2}g_{m3}$ and

$$\mathbf{Z}_b = \begin{bmatrix} z_1 & 0 & 0 & 0 \\ 0 & y_3/w & -g_{m2}/w & 0 \\ 0 & -g_{m3}/w & y_2/w & 0 \\ 0 & 0 & 0 & z_4 \end{bmatrix} \tag{10.88}$$

Substituting (10.87) in (10.77a) yields the general null return difference

$$\hat{F}_k^m(g_{m2}) = \frac{M_{14}(\mathbf{B}_1)}{M_{14}^k(\mathbf{B}_1)} = \frac{M_{14}(\mathbf{B}_2)}{M_{14}^k(\mathbf{B}_2)} = \frac{y_2 y_3 - kg_{m3}}{y_2 y_3 - g_{m2}g_{m3}} \tag{10.89}$$

which is different from $\hat{F}_k^c(g_{m2})$ in (10.85). However, if the two controlled sources are not cyclically coupled, which in the present case is equivalent to letting $g_{m3} = 0$, then $\hat{F}_k^m(g_{m2}) = 1$ and $\hat{F}_k^m(g_{m2}) = \hat{F}_k^c(g_{m2})$. In the following section, we show that this assertion is valid in general, and we will present conditions that are both necessary and sufficient for the general null return difference to be invariant under this type of general transformation. As expected, they are very similar to those for the general return difference. The physical interpretation for $\hat{F}_k^m(g_{m2})$ is somewhat difficult because in N_m, g_{m2} appears in all the controlling parameters of the controlled sources.

Example 10.4 Suppose that, in Example 10.3, the element of interest is g_{m3} instead of g_{m2}. Then from (10.77), (10.83) and (10.87), we have

$$\hat{F}_k^c(g_{m3}) = \frac{M_{14}(\mathbf{Q}_1)}{M_{14}^k(\mathbf{Q}_1)} = \frac{M_{14}(\mathbf{Q}_2)}{M_{14}^k(\mathbf{Q}_2)} = \frac{y_3 + g_{m3}}{y_3 + k} \tag{10.90a}$$

$$\hat{F}_k^m(g_{m3}) = \frac{M_{14}(\mathbf{B}_1)}{M_{14}^k(\mathbf{B}_1)} = \frac{M_{14}(\mathbf{B}_2)}{M_{14}^k(\mathbf{B}_2)} = \frac{y_2 y_3 - kg_{m2}}{y_2 y_3 - g_{m2}g_{m3}} \cdot \frac{y_3 + g_{m3}}{y_3 + k} \tag{10.90b}$$

Figure 10.7 A physical interpretation of the null return difference of g_{m3}.

For illustrative purposes, we interpret $\hat{F}_k^c(g_{m3})$ physically as follows. In N_c, let the controlled source $g_{m3}V_2$ be replaced by an independent current source of g_{m3} amperes, as shown in Fig. 10.7. Now we adjust the independent current source I_s so that the current in y_4 is zero, which results, after a straightforward computation, in

$$I_s = \frac{-g_{m3}(y_1 + y_2 + g_{m2})}{y_3} \tag{10.91}$$

Under this situation, the negative of the voltage across the controlling branch, which is y_2, is given by

$$T = \frac{g_{m3}}{y_3} \tag{10.92}$$

Thus, the null return difference with respect to g_{m3} for the zero reference value is obtained as

$$\hat{F}_k^c(g_{m3}) = 1 + T = \frac{y_3 + g_{m3}}{y_3} \tag{10.93}$$

confirming our result in (10.90a) for $k = 0$.

10.4 RELATIONS BETWEEN THE LOOP AND CUTSET FORMULATIONS

Like the case for the general return difference, in the present section, we present mutual relations of the general null return differences based on the loop and cutset formulations of the network equations.

Definition 10.7: Uncoupled Edges For a given square matrix $\mathbf{Z}$ of order b and for a given b-edge directed graph G, the edges e_u and e_v of G corresponding to the rows or columns u and v of $\mathbf{Z}$ are said to be *uncoupled* with respect to $\mathbf{Z}$ if $z_{ui} = z_{iu} = 0$ for all $i \neq u$ and $i = 1, 2, \ldots, b$; and $z_{vj} = z_{jv} = 0$ for all $j \neq v$ and $j = 1, 2, \ldots, b$, where $\mathbf{Z} = [z_{ij}]$.

In other words, if $z_{uu} \neq 0$ and $z_{vv} \neq 0$, for e_u and e_v to be the uncoupled edges of G, they have to be the only nonzero entries in the rows and columns u and v of $\mathbf{Z}$.

Using (10.70) in conjunction with (10.77), we obtain the desired relation

$$\hat{F}_k^m(x) = \frac{\det \mathbf{Z}_{uu,vv}(x)}{\det \mathbf{Z}_{uu,vv}(k)} \hat{F}_k^c(x) \qquad (10.94)$$

From (10.94), we can state the mutual relation of the two general null return differences as a theorem.

Theorem 10.12 In a feedback network, assume that the edges corresponding to the branches of the input and output ports are uncoupled, and that the branch-immittance matrix is nonsingular under the conditions when the element x has its nominal value and when it assumes the value k. Then the ratio of the general null return difference with respect to a two-port element x and for a general reference value k and based on the loop formulation to that of the same element based on the cutset formulation is equal to the ratio of the determinant of the branch-impedance matrix after deleting the rows and columns corresponding to the branches of the input and output ports when x assumes its nominal value to that when it assumes the value k.

The above theorem may be stated differently as follows.

Theorem 10.13 Assume that the conditions of Theorem 10.12 are satisfied. Then the general null return difference with respect to a two-port element x for a general reference value k is invariant under the general transformation between a system of basis circuits and a system of basis cutsets if and only if the determinant of the branch-immittance matrix after deleting the rows and columns corresponding to the branches of the input and output ports remains unaltered when x is replaced by k.

Consequently, like the general return difference, for nearly all of the practical feedback networks, the general null return difference is invariant.

Corollary 10.26 Assume that the two conditions of Theorem 10.12 are satisfied, and also assume that there are no cyclically coupled controlled sources. Then the general null return difference with respect to a two-port element x for a general reference value k is invariant under the general transformations between a system of basis circuits and a system of basis of cutsets.

As an illustration, consider the problem of Example 10.1 in which we have

$$\det \mathbf{Z}_{11,44} = \frac{1}{y_2 y_3 - g_{m2} g_{m3}} \tag{10.95a}$$

$$\det \mathbf{Z}_{11,44}^k = \frac{1}{y_2 y_3 - k g_{m3}} \tag{10.95b}$$

Substituting in (10.94) yields

$$\hat{F}_k^m(g_{m2}) = \frac{y_2 y_3 - k g_{m3}}{y_2 y_3 - g_{m2} g_{m3}} \hat{F}_k^c(g_{m2}) \tag{10.96}$$

confirming our earlier results given in (10.85) and (10.89). Similarly, we can confirm the relationship (10.94) by the problem discussed in Example 10.3, in which $\det \mathbf{Z}_{11,44}^k$ is given by (10.95b).

We remark that the definition (10.79) is also valid for the one-port immittance x. However, there is no straightforward physical interpretation of the general null return difference in the form of a ratio of two immittances, as in (10.39) and (10.43) for the general return difference. We note that as k approaches to zero, a property similar to (10.44) exists. Thus, for $k = 0$ the denominator of the right-hand side of (10.94) is nonzero, and the relation is well defined for all k. As a matter of fact, identities like (10.46) and (10.55) can be derived directly from (10.94) for the general null return differences. In the case where the element x is a one-port impedance, (10.94) becomes

$$k \hat{F}_k^m(x) = x \hat{F}_k^c(x) \tag{10.97}$$

and in the case where the element x' is a one-port admittance, we have

$$x' \hat{F}_{k'}^m(x') = k' \hat{F}_{k'}^c(x') \tag{10.98}$$

As an illustration, in Example 10.3, let $g_{m3} = 0$ and consider the one-port impedance $x = z_3$. From (10.77), (10.83), and (10.87) we obtain

$$\hat{F}_k^c(z_3) = k y_3 \tag{10.99a}$$

$$\hat{F}_k^m(z_3) = 1 \tag{10.99b}$$

confirming (10.97) for all values of k. Now suppose that the element x' is the one-port admittance $x' = y_3$. Then

$$\hat{F}_{k'}^c(y_3) = \frac{y_3}{k'} \tag{10.100a}$$

$$\hat{F}_{k'}^m(y_3) = 1 \tag{10.100b}$$

confirming (10.98) for all values of k'.

10.5 THE PRIMARY SYSTEMS OF EQUATIONS

So far, the network determinants used to define the general return differences are formulated in terms of the loop and cutset systems of equations, known as the secondary systems of equations. In this section, we show that the network determinant obtained from the coefficient matrix of the primary systems of network equations can also be used to define the general return difference.

In the feedback N_c or N_m, whose associated directed graph is G, let the branch-current and branch-voltage vectors be denoted by $\mathbf{I}$ and $\mathbf{V}$, respectively, and let the branch-impedance and the branch-admittance matrices by $\mathbf{Z}_b$ and $\mathbf{Y}_b$. If $\mathbf{B}$ and $\mathbf{Q}$ are a basis circuit matrix and a basis cutset matrix of N, then the primary systems of equations may be formulated as

$$\begin{bmatrix} \mathbf{Q} & \mathbf{0} \\ \mathbf{0} & \mathbf{B} \\ -\mathbf{U} & \mathbf{Y}_b \end{bmatrix} \begin{bmatrix} \mathbf{I} \\ \mathbf{V} \end{bmatrix} = \mathbf{0} \qquad (10.101a)$$

$$\begin{bmatrix} \mathbf{Q} & \mathbf{0} \\ \mathbf{0} & \mathbf{B} \\ -\mathbf{Z}_b & \mathbf{U} \end{bmatrix} \begin{bmatrix} \mathbf{I} \\ \mathbf{V} \end{bmatrix} = \mathbf{0} \qquad (10.101b)$$

$\mathbf{U}$ being the identity matrix of order e, the number of edges of G. Let the coefficient matrix of (10.101a) and (10.101b) be $\mathbf{H}_y$ and $\mathbf{H}_z$, respectively. Using $\mathbf{Q}$ or $\mathbf{B}$, we can sweep out the identity matrix without changing the values of the determinants of the coefficient matrices. This results in

$$\det \mathbf{H}_y = c \det \mathbf{Q}\mathbf{Y}_b\mathbf{Q}' \qquad (10.102a)$$

$$\det \mathbf{H}_z = \frac{1}{c} \det \mathbf{B}\mathbf{Z}_b\mathbf{B}' \qquad (10.102b)$$

c being a real constant depending only on the choice of $\mathbf{B}$ and $\mathbf{Q}$. Substituting (10.102) in (10.26) gives

$$F_k^m(x) = \frac{\det \mathbf{B}\mathbf{Z}_b(x)\mathbf{B}'}{\det \mathbf{B}\mathbf{Z}_b(k)\mathbf{B}'} = \frac{\det \mathbf{H}_z(x)}{\det \mathbf{H}_z(k)} \qquad (10.103)$$

$$F_k^c(x) = \frac{\det \mathbf{Q}\mathbf{Y}_b(x)\mathbf{Q}'}{\det \mathbf{Q}\mathbf{Y}_b(k)\mathbf{Q}'} = \frac{\det \mathbf{H}_y(x)}{\det \mathbf{H}_y(k)} \qquad (10.104)$$

Thus, all the properties derived in the foregoing for the general return difference are also valid in terms of the determinants of the coefficient matrices of the primary systems of equation instead of the secondary systems.

As an example, let us again consider the feedback network N_c and N_m of Figs. 10.4 and 10.6. Let $\mathbf{Q} = \mathbf{Q}_1$ of (10.82a) and $\mathbf{B} = \mathbf{B}_1$ of (10.86a) with $\mathbf{Y}_b$ and $\mathbf{Z}_b$ being given in (10.27) and (10.88). Substituting these in $\mathbf{H}_y$ and $\mathbf{H}_z$ gives

$$\det \mathbf{H}_y = (y_1 + y_2)(y_3 + y_4) + y_4(y_3 + g_{m2} + g_{m3}) - g_{m2}g_{m3}$$

$$= \det \mathbf{Q}_1 \mathbf{Y}_b \mathbf{Q}_1' \tag{10.105a}$$

$$\det \mathbf{H}_z = \frac{z_1(z_4 w + y_2 + y_3 + g_{m2} + g_{m3}) + (y_3 z_4 + 1)}{w}$$

$$= \det \mathbf{B}_1 \mathbf{Z}_b \mathbf{B}_1' \tag{10.105b}$$

which yield the same general return difference as in Examples 10.1 and 10.2.

10.6   INVARIANCE AND INCIDENCE FUNCTIONS

In the preceding sections, we have considered the problem of invariance of the determinants of the generalized network matrices and their generalized cofactors with respect to the choice of circuits and cutsets in a given directed graph. In this section, we shall consider the corresponding problem of invariance of these determinants with respect to the choice of the incidence functions of a graph, as discussed in Slepian (1968).

For a given directed graph G, the orientation as indicated by the arrow directions of the edges of G can be described by a function called the *incidence function* of the set of edges of G. In the following, we shall show that if the incidence function of G is altered, the determinants of the generalized network matrices $\mathbf{BWB}'$ and $\mathbf{QXQ}'$ and their generalized cofactors $M_{uv}(\mathbf{B})$ and $M_{uv}(\mathbf{Q})$, in general, *do not* remain invariant. However, if matrices $\mathbf{W}$ and $\mathbf{X}$ are the branch-impedance and branch-admittance matrices of an electrical network, respectively, then these determinants and cofactors are invariant with respect to the choice of the incidence function of the network (graph). In particular, we shall show that Slepian's result on invariance with respect to the incidence functions of a resistive network can easily be derived via the technique discussed in the preceding sections.

As an example, consider the directed graph G as shown in Fig. 10.8a. Let $\mathbf{B}$ be the circuit matrix of G corresponding to the two windows of G. Also, let $w_{ii} = w_i$ and $w_{ij} = 0$, $i \neq j$, except w_{35} which is w_6, where $\mathbf{W} = [w_{ij}]$ and

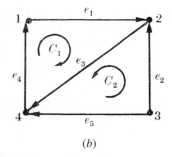

(a) (b)

Figure 10.8 (a) A directed graph G. (b) The directed graph G_1 obtained from G by reversing the orientation of the edge e_5 in G.

$i, j = 1, 2, 3, 4, 5$. Then we have

$$\mathbf{BWB}' = \begin{bmatrix} 1 & 0 & 1 & 1 & 0 \\ 0 & 1 & 1 & 0 & 1 \end{bmatrix} \begin{bmatrix} w_1 & 0 & 0 & 0 & 0 \\ 0 & w_2 & 0 & 0 & 0 \\ 0 & 0 & w_3 & 0 & w_6 \\ 0 & 0 & 0 & w_4 & 0 \\ 0 & 0 & 0 & 0 & w_5 \end{bmatrix} \begin{bmatrix} 1 & 0 \\ 0 & 1 \\ 1 & 1 \\ 1 & 0 \\ 0 & 1 \end{bmatrix}$$

$$= \begin{bmatrix} w_1 + w_3 + w_4 & w_3 + w_6 \\ w_3 & w_2 + w_3 + w_5 + w_6 \end{bmatrix} \tag{10.106}$$

yielding

$$\det \mathbf{BWB}' = (w_1 + w_4)(w_2 + w_3 + w_5 + w_6)$$
$$+ w_3(w_2 + w_5) \tag{10.107}$$

Now if we reverse the orientation of the edge e_5 in G, we obtain a new directed graph G_1 as shown in Fig. 10.8b. If $\mathbf{B}_1$ is the corresponding circuit matrix in G_1, then

$$\det \mathbf{B}_1 \mathbf{WB}'_1 = (w_1 + w_4)(w_2 + w_3 + w_5 - w_6)$$
$$+ w_3(w_2 + w_5) \tag{10.108}$$

Likewise, the generalized cofactors $M_{21}(\mathbf{B})$ and $M_{21}(\mathbf{B}_1)$ of $\mathbf{BWB}'$ and $\det \mathbf{B}_1 \mathbf{WB}'_1$, respectively, are found to be

$$M_{21}(\mathbf{B}) = -(w_3 + w_6) \quad \text{and} \quad M_{21}(\mathbf{B}_1) = -(w_3 - w_6) \tag{10.109}$$

Thus, we may conclude that the determinants of generalized network matrices and their generalized cofactors do not in general remain invariant with respect to the choice of the incidence functions of the graph. However, they remain invariant under certain conditions, and also for the most common and important type of graphs which represent the electrical networks, they are invariant.

For a given directed graph G, let G_1 be the directed graph obtained from G by reversing the orientations of certain edges of G. If $\mathbf{B}$ and $\mathbf{B}_1$ are the circuit matrices of G and G_1 corresponding to the same set of circuits, respectively, where $\mathbf{B}$ and $\mathbf{B}_1$ are of order $n \times e$ and of rank n, then

$$\mathbf{B} = \mathbf{B}_1\mathbf{D} \tag{10.110}$$

where $\mathbf{D}$ is an $e \times e$ diagonal matrix whose iith entry d_{ii} corresponding to the edge e_i of G or G_1: $d_{ii} = 1$ if the orientation of the edge e_i in G is the same as that of e_i in G_1, and $d_{ii} = -1$ otherwise. A slight generalization of this is that if $\mathbf{B}$ and $\mathbf{B}_1$ correspond to two different sets of circuits in G and G_1, it follows that

$$\mathbf{B} = \mathbf{C}\mathbf{B}_1\mathbf{D} \tag{10.111}$$

where $\mathbf{C}$ is an $n \times n$ nonsingular matrix.

We shall now consider some sufficient conditions under which the determinants of the generalized network matrices and their generalized cofactors will be invariant with respect to the choice of the incidence functions. The general characterization of this problem remains unsolved.

Corollary 10.27 If the matrices $\mathbf{W}$ and $\mathbf{X}$ are diagonal, then the determinants of the generalized network matrices $\mathbf{BWB}'$ and $\mathbf{QXQ}'$ and their generalized cofactors $M_{uv}(\mathbf{B})$ and $M_{uv}(\mathbf{Q})$ associated with a given directed graph G are invariant with respect to the change of the incidence functions of G.

PROOF. We shall only consider the circuit case, since the others can be proved in an entirely similar manner. For our purpose, it is sufficient to show that if $\mathbf{B}$ and $\mathbf{B}_1$ are the circuit matrices of G and G_1 corresponding to the same set of circuits, respectively, then

$$\det \mathbf{BWB}' = \det \mathbf{B}_1\mathbf{WB}_1' \tag{10.112}$$

where all other symbols are defined as above. This is obvious since

$$\det \mathbf{BWB}' = \det \mathbf{B}_1\mathbf{DWD}'\mathbf{B}_1' = \det \mathbf{B}_1\mathbf{DD}'\mathbf{WB}_1' \tag{10.113}$$

The corollary follows directly from the fact that $\mathbf{DD}'$ is the identity matrix of order e.

Thus, Slepian's results on the invariance of the network determinants of a resistive network with respect to the choice of the incidence functions of the network follows directly from the above corollary. As a matter of fact, we shall show that if $\mathbf{W}$ and $\mathbf{X}$ are the branch-impedance and branch-admittance matrices of a general electrical network, the network determinants will remain invariant as in the resistive case.

For a given electrical network N, let G and G_1 be two associated directed graphs, being used only to exhibit the interconnections of the branches, not the branch voltage–current relationships, of N corresponding to the two different choices of the reference directions of the branch currents or branch voltages. By convention, we assume that the reference plus for the branch voltage is at the tail of the reference arrow for the branch current. Without loss of generality, we assume that the same set of circuits is chosen for both G and G_1. Using the symbols defined above for G and G_1, we have $\mathbf{B} = \mathbf{B}_1\mathbf{D}$. Let $\mathbf{I}_b$ and $\tilde{\mathbf{I}}_b$ be the e-vectors corresponding to the branch currents and $\mathbf{V}_b$ and $\tilde{\mathbf{V}}_b$ be the e-vectors corresponding to the branch voltages of G and G_1, respectively. Then

$$\mathbf{V}_b = \mathbf{Z}_b\mathbf{I}_b \quad \text{and} \quad \tilde{\mathbf{V}}_b = \tilde{\mathbf{Z}}_b\tilde{\mathbf{I}}_b \tag{10.114}$$

where $\mathbf{Z}_b$ and $\tilde{\mathbf{Z}}_b$ are the branch-impedance matrices of N corresponding to the two different choices of reference directions of branch currents or branch voltages in G and G_1, respectively. Since

$$\mathbf{V}_b = \mathbf{D}\tilde{\mathbf{V}}_b \quad \text{and} \quad \mathbf{I}_b = \mathbf{D}\tilde{\mathbf{I}}_b \tag{10.115}$$

it follows that, after a simple substitution and the fact that $\mathbf{DD}$ is the identity matrix, we have

$$\tilde{\mathbf{Z}}_b = \mathbf{DZ}_b\mathbf{D} \tag{10.116}$$

Thus, the invariant character of the determinant of the loop-impedance matrix of a general network with respect to the choice of the incidence functions can easily be established as follows:

$$\det \mathbf{BZ}_b\mathbf{B}' = \det \mathbf{B}_1\mathbf{DZ}_b\mathbf{D}'\mathbf{B}_1' = \det \mathbf{B}_1(\mathbf{DZ}_b\mathbf{D}')\mathbf{B}_1'$$
$$= \det \mathbf{B}_1\tilde{\mathbf{Z}}_b\mathbf{B}_1' \tag{10.117}$$

This is similar valid for the determinant of the cutset-admittance matrix of a general network, and the generalized cofactors of these determinants.

Theorem 10.14 The network determinants and their generalized cofactors of a general electrical network are invariant with respect to the choice of the incidence functions of the network.

10.7 SIMPLE DERIVATIONS OF TOPOLOGICAL FORMULAS

Let Fig. 10.9 represent a passive two-port network N without mutual inductances with input nodes 1 and $1'$ and output nodes 2 and $2'$. Let T denote the sum of tree-admittance products $f(t)$ of all the trees t, and $T_{rp,sq}$ a sum of 2-tree admittance products $f(t_{rp,sq})$ of all the 2-trees $t_{rp,sq}$ with subscripts denoting nodes which are required to be in different parts, all in N_1:

$$T = \sum_t f(t) \tag{10.118a}$$

$$T_{rp,sq} = \sum_{t_{rp,sq}} f(t_{rp,sq}) \tag{10.118b}$$

Also let $\bar{T}_{rp,sq}$ denote the sum of the complements of the 2-tree impedance products $g(\bar{t}_{rp,sq})$ in N.

$$\bar{T}_{rp,sq} = \sum_{\bar{t}_{rp,sq}} g(\bar{t}_{rp,sq}) \tag{10.119}$$

For example, in Fig. 10.8 there are eight trees t: $e_1e_2e_3$, $e_3e_4e_5$, $e_1e_3e_5$, $e_2e_3e_4$, $e_1e_2e_4$, $e_1e_4e_5$, $e_1e_2e_5$ and $e_2e_4e_5$, and four 2-trees $t_{2,4}$ separating nodes 2 and 4: e_1e_2, e_1e_5, e_4e_5 and e_2e_4. If y_i is the admittance of the edge e_i and $z_i = 1/y_i$ for

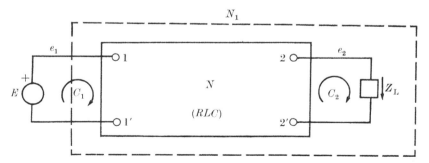

Figure 10.9 A two-port network with reference convention. Node $1'$ is used as the reference node for the nodal system.

$i = 1, 2, 3, 4, 5$, then we have

$$T = y_1 y_2 y_3 + y_3 y_4 y_5 + y_1 y_3 y_5 + y_2 y_3 y_4$$

$$+ y_1 y_2 y_4 + y_1 y_4 y_5 + y_1 y_2 y_5 + y_2 y_4 y_5 \qquad (10.120a)$$

$$T_{2,4} = \sum_{t_{2,4}} f(t_{2,4}) = y_1 y_2 + y_1 y_5 + y_4 y_5 + y_2 y_4 \qquad (10.120b)$$

$$\bar{T}_{2,4} = \sum_{\bar{t}_{2,4}} g(\bar{t}_{2,4}) = z_3 z_4 z_5 + z_2 z_3 z_4 + z_1 z_2 z_3 + z_1 z_3 z_5 \qquad (10.120c)$$

In the following, we demonstrate that, by using the identities established earlier, most of the topological formulas can easily be derived from those associated with the determinant of the node-admittance matrix $\mathbf{Y}_n$ which is relatively simple to obtain. For convenience, write

$$\Delta^n = \det \mathbf{A} \mathbf{Y}_b \mathbf{A}' \qquad (10.121a)$$

$$\Delta^n_{ij} = (-1)^{i+j} \det \mathbf{A}_{-i} \mathbf{Y}_b \mathbf{A}'_{-j} \qquad (10.121b)$$

where $\mathbf{Y}_b = \mathbf{X}$ is the branch-admittance matrix. It is well known that if node $1'$ is used as the reference node, then

$$\Delta^n = T = \sum_t f(t) \qquad (10.122a)$$

$$\Delta^n_{ij} = W_{ij,1'}(Y) = T_{ij,1'} = \sum_{t_{ij,1'}} f(t_{ij,1'}) \qquad (10.122b)$$

Theorem 10.15

$$M_{12}(\mathbf{A}) = \Delta^n_{12} - \Delta^n_{12'} = T_{12,1'2'} - T_{12',1'2} \qquad (10.123)$$

PROOF. Since

$$\Delta^n_{12} = T_{12,1'} = T_{12,1'2'} + T_{122',1'} \qquad (10.124a)$$

$$\Delta^n_{12'} = T_{12',1'} = T_{12',1'2} + T_{122',1'} \qquad (10.124b)$$

the theorem follows immediately.

Theorem 10.16 If the branches e_1 and e_2 are contained only in the cutsets 1 and 2, and oriented in the same directions as the cutsets, respectively, then the (1,2)th cofactor Δ^c_{12} in the determinant Δ^c of the cutset-admittance matrix is given by

$$\Delta^c_{12} = k(\mathbf{Q}) M_{12}(\mathbf{A}) = k(\mathbf{Q})[T_{12,1'2'} - T_{12',1'2}] \qquad (10.125)$$

PROOF. The theorem follows directly from the definition of $M_{12}(\mathbf{Q})$ and Theorem 10.15.

Theorem 10.17 If the branches e_1 and e_2 are contained only in the circuits 1 and 2, respectively, then the (1,2)th cofactor Δ_{12}^m in the determinant Δ^m of the loop-impedance matrix is given by

$$\Delta_{12}^m = k(\mathbf{B})[\bar{T}_{12,1'2'} - \bar{T}_{12',1'2}] \tag{10.126}$$

PROOF. Since $M_{12}(\mathbf{B}) = -\Delta_{12}^m$ from definition of $M_{12}(\mathbf{B})$, by Theorem 10.9, we have

$$
\begin{aligned}
-M_{12}(\mathbf{B}) &= k(\mathbf{B})(\det \mathbf{W}_{11,22})M_{12}(\mathbf{A}) \\
&= k(\mathbf{B})(\det \mathbf{W}_{11,22})[T_{12,1'2'} - T_{12',1'2}] \\
&= k(\mathbf{B})[\bar{T}_{12,1'2'} - \bar{T}_{12',1'2}]
\end{aligned} \tag{10.127}
$$

where $\mathbf{W} = \mathbf{Z}_b$. This completes the proof of the theorem.

We remark that Theorem 10.17 is the well-known Kirchhoff's rules for the mesh systems, a proof of which was given by Obermeyer as indicated in Seshu and Reed (1961). However, the proof is rather lengthy and cumbersome. It is commonly assumed that the cofactors of a network determinant are of interest only when there is at least one element, say, the ith circuit or cutset which is not in any other circuit or cutset. This is actually not necessary as can be seen from the following corollary. Of course, this assumption would greatly simplify the problem.

Corollary 10.28 If general cutsets and circuits are used, then

$$M_{12}(\mathbf{Q}) = k(\mathbf{Q})[T_{12,1'2'} - T_{12',1'2}] \tag{10.128a}$$
$$-M_{12}(\mathbf{B}) = k(\mathbf{B})[\bar{T}_{12,1'2'} - \bar{T}_{12',1'2}] \tag{10.128b}$$

10.8 TOPOLOGICAL EVALUATION OF FEEDBACK MATRICES IN MULTIPLE-LOOP FEEDBACK AMPLIFIERS

In the present section, we present topological formulas for the evaluation of the elements of the return difference and the null return difference matrices of a multiple-loop feedback amplifier. They are expressed in terms of the directed-tree and directed-two-tree admittance products in the associated digraph. In the situation where the controlling parameters of all the controlled sources are chosen as the elements of interest, the associated digraph is simply the graph representing the feedback amplifier after the independent source and controlled sources have been removed. The significance of the present approach is that it not only provides a short-cut for the evaluation of the feedback matrices, but also gives an insight into the behavior of the feedback amplifier under consideration.

10.8.1 The Return Difference Matrix

In a multiple-loop feedback network N, let the elements of interest be represented by a rectangular matrix $\mathbf{X}$, which can either be a transfer-admittance matrix or a driving-point admittance matrix characterized by

$$
\mathbf{I}_\alpha = \begin{bmatrix} I_{\alpha 1} \\ I_{\alpha 2} \\ \vdots \\ I_{\alpha q} \end{bmatrix} = \begin{bmatrix} x_{11} & x_{12} & \cdots & x_{1p} \\ x_{21} & x_{22} & \cdots & x_{2p} \\ \vdots & \vdots & \ddots & \vdots \\ x_{q1} & x_{q2} & \cdots & x_{qp} \end{bmatrix} \begin{bmatrix} V_{\beta 1} \\ V_{\beta 2} \\ \vdots \\ V_{\beta p} \end{bmatrix} = \mathbf{X}\mathbf{V}_\beta \qquad (10.129)
$$

which can be represented schematically by the network N_0 of Fig. 10.10 with the current sources being used as the excitations, where $\mathbf{X}$ is of order q by p. If $\mathbf{X}$ is a transfer-admittance matrix of the controlling parameters of the voltage-controlled current sources, $\mathbf{V}_\beta$ represents a p-dimensional vector of controlling voltages and $\mathbf{I}_\alpha$ a q-dimensional vector of controlled current sources. If $\mathbf{X}$ represents a driving-point admittance matrix, $\mathbf{I}_\alpha$ and $\mathbf{V}_\beta$ are of the same dimension, $p = q$, and they represent current and voltage vectors of a p-port network. The matrix $\mathbf{X}$ is important in terms of its effects to the whole system and is imbedded in the rest designated by

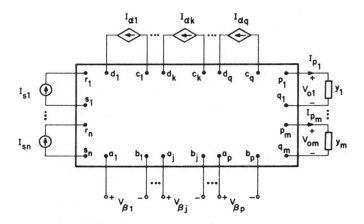

Figure 10.10 The Feedback network N_0.

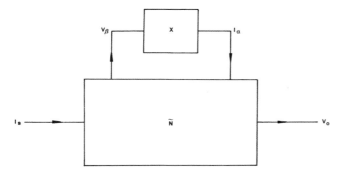

Figure 10.11 The general configuration of a multiple-loop feedback amplifier.

$\tilde{N}$ as shown in the bloc diagram of Fig. 10.11, where $\mathbf{I}_s$ denotes the input current vector and $\mathbf{V}_o$ the output voltage vector of N. Let

$$\mathbf{I}_s = [I_{s1} \quad I_{s2} \quad \cdots \quad I_{sn}]' \tag{10.130}$$

be the input signal vector, and

$$\mathbf{V}_o = [V_{o1} \quad V_{o2} \quad \cdots \quad V_{om}]' \tag{10.131}$$

the output signal vector, where, as before, the prime denotes the matrix transposition.

Since the network $\tilde{N}$ is linear, it can be characterized by

$$\mathbf{V}_\beta = \mathbf{A}\mathbf{I}_\alpha + \mathbf{B}I_s \tag{10.132a}$$
$$V_o = \mathbf{C}\mathbf{I}_\alpha + DI_s \tag{10.132b}$$

where $\mathbf{A}, \mathbf{B}, \mathbf{C}$ and D are the transfer-impedance matrices of order $p \times q$, $p \times 1$, $1 \times q$, and 1×1, respectively.

From (7.34) and (7.35), the return difference matrix with respect to $\mathbf{X}$ is given by

$$\mathbf{F}(\mathbf{X}) = \mathbf{1}_p - \mathbf{A}\mathbf{X} = \mathbf{1}_p + \mathbf{T}(\mathbf{X}) \tag{10.133}$$

where $\mathbf{T}(\mathbf{X}) = -\mathbf{A}\mathbf{X}$ is the return ratio matrix. Likewise, from (7.71) and (7.72) the null return difference matrix with respect to $\mathbf{X}$ is given by

$$\hat{\mathbf{F}}(\mathbf{X}) = \mathbf{1}_p - \hat{\mathbf{A}}\mathbf{X} = \mathbf{1}_p + \hat{\mathbf{T}}(\mathbf{X}) \tag{10.134}$$

where $\hat{\mathbf{T}}(\mathbf{X}) = -\hat{\mathbf{A}}\mathbf{X}$ is the null return ratio matrix and

$$\hat{\mathbf{A}} = \mathbf{A} - (1/D)\mathbf{B}\mathbf{C} \tag{10.135}$$

provided $D \neq 0$. Notice that here we only consider systems with a single input and a single output.

As shown in Theorem 7.1, the closed-loop transfer impedance $w(\mathbf{X})$, defined as the ratio of the output voltage V_o to the input current I_s, can be expressed in a very compact way by making use of the return difference and null return difference matrices

$$w(\mathbf{X}) = w(\mathbf{0}) \frac{\det \hat{\mathbf{F}}(\mathbf{X})}{\det \mathbf{F}(\mathbf{X})} \tag{10.136}$$

We remark that we have exhibited the importance of $\mathbf{X}$ on the closed-loop transfer impedance by writing w as $w(\mathbf{X})$.

From (8.163), the sensitivity function $S(x_{ij})$, which is defined as the ratio of the fractional change in a transfer function $w(\mathbf{X})$ to the fractional change in an element x_{ij} of $\mathbf{X}$, can be expressed as

$$S(x_{ij}) = \frac{\det \mathbf{F}(\mathbf{X})|_{x_{ij}=0}}{\det \mathbf{F}(\mathbf{X})} \left[1 - \frac{w(\mathbf{X})|_{x_{ij}=0}}{w(\mathbf{X})} \right] \tag{10.137}$$

Definition 10.8: Directed Two-Tree or Directed 2-Tree A subgraph $t_{i,j}$ of a directed graph or simply digraph G is said to be a *directed two-tree* or *directed 2-tree* of G with reference nodes i and j if and only if (1) it is a 2-tree of G, and (2) the outgoing degree of each node of $t_{i,j}$ is 1 except the nodes i and j which have outgoing degree 0. One or, in trivial case, both of the components may consist of an isolated node.

Like 2-trees, directed 2-trees in which certain designated nodes are required to be in different components are needed. For our purposes, $t_{ab,cd}$ will be used to denote a directed 2-tree in which nodes a and b are in one component, and the nodes c and d in the other, where the nodes a and c (the first subscripts) are the reference nodes of the components.

With these preliminaries, the first-order cofactor Y_{uv} and the second-order cofactor $Y_{rp,sq}$ of the indefinite-admittance matrix $\mathbf{Y} = [y_{ij}]$ of an active network can be expressed in terms of the directed tree admittance products and directed 2-tree admittance products in its associated digraph G, as defined in Definition 9.3, as follows, the proof of which can be found in Chen (1976b):

$$Y_{uv} = \sum_{t_k} f(t_k) \equiv T_k \tag{10.138}$$

$$Y_{rp,sq} = T_{rp,sq} - T_{rq,sp} \tag{10.139}$$

where the choice of the reference node k is arbitrary and

$$T_{rp,sq} = \sum_{t_{rp,sq}} f(t_{rp,sq}) \qquad (10.140)$$

t_k is a directed tree with reference node k in G, $t_{rp,sq}$ is a directed two-tree with reference nodes r and s and such that nodes r and p are in one component and nodes s and q in the other, and $f(g)$ denotes the product of the weights associated with the edges of the subgraph g with $f(\emptyset) = 0$, $\emptyset$ being the null graph.

As indicated (10.132a), $\mathbf{A}$ is the matrix relating the controlling voltage vector $\mathbf{V}_\beta$ to the controlled current vector $\mathbf{I}_\alpha$ after the input excitation I_s has been set to zero. Under this situation, the (i, j)-element of $\mathbf{AX}$ is equal to the voltage $\tilde{\mathbf{V}}_{\beta i}$ appearing at the ith controlling branch of Fig. 10.10 when the controlled current sources $I_{\alpha k}$ $(k = 1, 2, \ldots, q)$ have been replaced by the independent current sources of strengths x_{kj} as depicted in Fig. 10.12. Thus, with

$$\mathbf{F}(\mathbf{X}) = [F_{ij}] \qquad (10.141)$$

we have for the jth column elements of $\mathbf{F}(\mathbf{X})$

$$F_{ij} = -\tilde{\mathbf{V}}_{\beta i}, \quad i \neq j \qquad (10.142a)$$

$$= 1 - \tilde{\mathbf{V}}_{\beta i}, \quad i = j \qquad (10.142b)$$

Since the network is linear, the voltage $\tilde{\mathbf{V}}_{\beta i}$ due to the independent current sources x_{ij} can be obtained by considering each source separately, and the result is

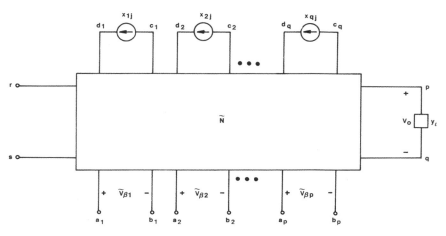

Figure 10.12 The physical interpretation of the elements of the return difference matrix with respect to the controlling parameters of the voltage-controlled current sources.

given by

$$V_{\beta i} = \frac{\sum_{k=1}^{q} x_{kj} Y_{d_k a_i, c_k b_i}(\mathbf{0})}{Y_{uv}(\mathbf{0})} \tag{10.143}$$

Appealing to (10.138), (10.139) and (10.142), we have for $i \neq j$

$$F_{ij} = \frac{\sum_{k=1}^{q} x_{kj}[T_{d_k b_i, c_k a_i}(\mathbf{0}) - T_{d_k a_i, c_k b_i}(\mathbf{0})]}{T_m(\mathbf{0})} \tag{10.144}$$

and for $i = j$

$$F_{ii} = 1 + \frac{\sum_{k=1}^{q} x_{ki}[T_{d_k b_i, c_k a_i}(\mathbf{0}) - T_{d_k a_i, c_k b_i}(\mathbf{0})]}{T_m(\mathbf{0})} \tag{10.145}$$

In the special situation where $\mathbf{X}$ is diagonal, which occurs most often in practice, (10.144) and (10.145) reduce to

$$F_{ij} = \frac{x_{jj}[T_{d_j b_i, c_j a_i}(\mathbf{0}) - T_{d_j a_i, c_j b_i}(\mathbf{0})]}{T_m(\mathbf{0})} \tag{10.146}$$

for $i \neq j$, and

$$F_{ii} = 1 + \frac{x_{ii}[T_{d_i b_i, c_i a_i}(\mathbf{0}) - T_{d_i a_i, c_i b_i}(\mathbf{0})]}{T_m(\mathbf{0})} = \frac{T_m(\mathbf{X}^{ij})}{T_m(\mathbf{0})} \tag{10.147}$$

where $\mathbf{X}^{ij}$ is the matrix derived from $\mathbf{X}$ by setting all of its elements to zero except x_{ij}. Thus, $\mathbf{X}^{ij}$ is the matrix consisting only of zeros except the ith row and jth column element, which is x_{ij}.

We remark that the directed trees and two-trees in (10.144) and (10.145) are to be evaluated in the digraph obtained from $G(\mathbf{Y})$ by removing all the directed edges corresponding to the elements of $\mathbf{X}$. If $\mathbf{X}$ is the matrix comprised of all the controlling parameters, the resulting digraph is simply the graph representing the feedback network when the input excitation and all the controlled sources have been removed, and the directed trees and two-trees become the ordinary trees and two-trees. We illustrate this by the following example.

Example 10.5 The amplifier of Fig. 10.13 makes use of voltage-series feedback by connecting the second collector to the first emitter through the voltage divider $R_f R_e$, as shown in Fig. 10.14 after the biasing and coupling circuitry have

Figure 10.13 A voltage-series feedback amplifier.

been removed. Assume that the two transistors are identical with

$$h_{ie} = 1.1\,\text{k}\Omega, \quad h_{fe} = 50, \quad h_{re} = h_{oe} = 0 \tag{10.148}$$

The equivalent network of Fig. 10.14 is presented in Fig. 10.15. Suppose that the controlling parameters of the two controlled sources are the elements of interest. Then we have

$$\mathbf{I}_\alpha = \begin{bmatrix} I_a \\ I_b \end{bmatrix} = \begin{bmatrix} 0.0455 & 0 \\ 0 & 0.0455 \end{bmatrix} \begin{bmatrix} V_{13} \\ V_{45} \end{bmatrix} = \mathbf{X}\mathbf{V}_\beta \tag{10.149}$$

Since $\mathbf{X}$ is diagonal, (10.146) and (10.147) apply. To obtain the associated digraph, we short-circuit the voltage source V_s and remove the two controlled current sources I_a and I_b. The resulting graph is shown in Fig. 10.16 with

$$a_1 = 5, \quad b_1 = 3, \quad c_1 = 4, \quad d_1 = 3 \tag{10.150a}$$

$$a_2 = 4, \quad b_2 = 5, \quad c_2 = 2, \quad d_2 = 5 \tag{10.150b}$$

Figure 10.14 The circuit of Fig. 10.13 after the biasing and coupling circuitry having been removed.

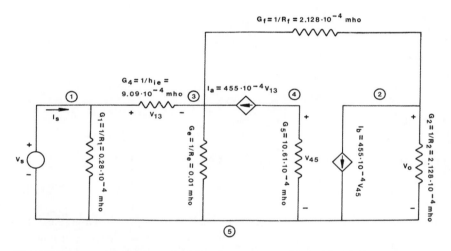

Figure 10.15 The equivalent network of the voltage-series amplifier of Fig. 10.14.

The elements of the return difference matrix can now be computed directly from (10.146) and (10.147) using the graph of Fig. 10.16, as follows[1]:

$$F_{11} = 1 + \frac{x_{11}[T_{33,45}(\mathbf{0}) - T_{35,43}(\mathbf{0})]}{T_5(\mathbf{0})}$$

$$= 1 + \frac{x_{11}(G_5 G_f + G_2 G_5 - 0)}{G_5(G_e' G_f + G_e' G_2 + G_2 G_f)} = 5.13058 \quad (10.151a)$$

---

[1] To exhibit the subscripts, it is more clear to write $T_{r,sq}$ and $T_{rp,s}$ as $T_{rr,sq}$ and $T_{rp,ss}$, respectively.

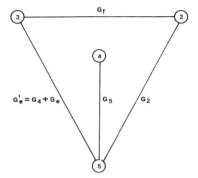

Figure 10.16 The graph used to compute the elements of the return difference matrix with respect to the controlling parameters of the network of Figure 10.15.

$$F_{21} = \frac{x_{11}[T_{35,44}(\mathbf{0}) - T_{34,45}(\mathbf{0})]}{T_5(\mathbf{0})}$$

$$= \frac{x_{11}(G'_e G_f + G_f G_2 + G'_e G_2)}{G_5(G'_e G_f + G'_e G_2 + G_2 G_f)} = 42.88407 \qquad (10.151b)$$

$$F_{12} = \frac{x_{22}[T_{53,25}(\mathbf{0}) - T_{55,23}(\mathbf{0})]}{T_5(\mathbf{0})}$$

$$= \frac{x_{22}(0 - G_f G_5)}{G_5(G'_e G_f + G'_e G_2 + G_2 G_f)} = -2.06529 \qquad (10.151c)$$

$$F_{22} = 1 + \frac{x_{22}[T_{55,24}(\mathbf{0}) - T_{54,25}(\mathbf{0})]}{T_5(\mathbf{0})}$$

$$= 1 + \frac{x_{22}(0 - 0)}{G_5(G'_e G_f + G'_e G_2 + G_2 G_f)} = 1 \qquad (10.151d)$$

confirming (7.37), as previously computed in Sec. 7.2.1 of Chapter 7.

Alternatively, F_{11} can be computed by using the digraph of Fig. 10.17 and applying (10.147):

$$F_{11} = \frac{T_m(\mathbf{X}^{11})}{T_m(\mathbf{0})} = \frac{T_5(\mathbf{X}^{11})}{T_5(\mathbf{0})} = \frac{T_4(\mathbf{X}^{11})}{T_5(\mathbf{0})}$$

$$= \frac{\begin{array}{c} G'_e G_5 G_f + x_{11} G_5 G_f + G'_e G_5 G_2 \\ + x_{11} G_5 G_2 + G_5 G_f G_2 \end{array}}{G_5(G'_e G_f + G'_e G_2 + G_2 G_f)}$$

$$= 5.13058 \qquad (10.152)$$

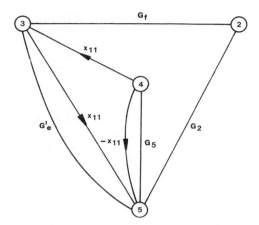

Figure 10.17 The digraph used to compute the numerator of the diagonal element F_{11} of the return difference matrix with respect to the controlling parameters of the network of Fig. 10.15.

Suppose that the two controlled current sources in Fig. 10.15 are characterized by the equation

$$\mathbf{I}_\alpha = \begin{bmatrix} I_a \\ I_b \end{bmatrix} = \begin{bmatrix} 0.0455 & 0.01 \\ 0.02 & 0.0455 \end{bmatrix} \begin{bmatrix} V_{13} \\ V_{45} \end{bmatrix} = \mathbf{X}\mathbf{V}_\beta \qquad (10.153)$$

To compute the return difference matrix $\mathbf{F}(\mathbf{X})$ using (10.144) and (10.145), we still can use the graph of Fig. 10.16 with the same labeling as indicated in (10.150). The results are given by

$$F_{11} = 1 + \frac{x_{11}[T_{33,45}(\mathbf{0}) - T_{35,43}(\mathbf{0})] + x_{21}[T_{53,25}(\mathbf{0}) - T_{55,23}(\mathbf{0})]}{T_5(\mathbf{0})} = 4.22276$$

$$(10.154a)$$

$$F_{21} = \frac{x_{11}[T_{35,44}(\mathbf{0}) - T_{34,45}(\mathbf{0})] + x_{21}[T_{55,24}(\mathbf{0}) - T_{54,25}(\mathbf{0})]}{T_5(\mathbf{0})} = 42.88407$$

$$(10.154b)$$

$$F_{12} = \frac{x_{12}[T_{33,45}(\mathbf{0}) - T_{35,43}(\mathbf{0})] + x_{22}[T_{53,25}(\mathbf{0}) - T_{55,23}(\mathbf{0})]}{T_5(\mathbf{0})} = -1.15747$$

$$(10.154c)$$

$$F_{22} = 1 + \frac{x_{12}[T_{35,44}(\mathbf{0}) - T_{34,45}(\mathbf{0})] + x_{22}[T_{55,24}(\mathbf{0}) - T_{54,25}(\mathbf{0})]}{T_5(\mathbf{0})} = 10.42507$$

$$(10.154d)$$

10.8.2 The Null Return Difference Matrix

The null return difference matrix is found to be very useful in measurement situations and in the computation of the sensitivity for the multiple-loop feedback amplifiers. In the following, we present topological formulas for the evaluation of its elements.

The null return difference matrix $\hat{\mathbf{F}}(\mathbf{X})$, as defined in (10.134), is the return difference matrix with respect to $\mathbf{X}$ when the input excitation I_s is adjusted so that the output voltage V_o of Fig. 10.12 (I_s is not explicitly shown) is identically zero. Thus, $\hat{\mathbf{A}}$ is the matrix relating the controlling voltage vector $\mathbf{V}_\beta$ to the controlled current vector $\mathbf{I}_\alpha$ when input excitation I_s is adjusted so that the output voltage V_o is identically zero. Under this situation, to compute the jth column elements of $\hat{\mathbf{A}}\mathbf{X}$, we replace the controlled current sources $I_{\alpha k}$ ($k = 1, 2, \ldots, q$) in Fig. 10.10 by the independent current sources of strengths x_{kj}, as depicted in Fig. 10.12, and then adjust the input excitation I_s so that the output voltage V_o is identically zero. The voltage $\tilde{V}_{\beta i}$ appearing at ith controlling branch is the (i, j)th element of $\hat{\mathbf{A}}\mathbf{X}$. Let the desired input excitation be designated by I_{sj} ($j = 1, 2, \ldots, p$). Applying the principle of superposition, the output voltage V_o is given by

$$V_o = I_s \frac{Y_{rp,sq}(\mathbf{0})}{Y_{uv}(\mathbf{0})} + \sum_{k=1}^{q} \frac{x_{kj} Y_{d_k p, c_k q}}{Y_{uv}(\mathbf{0})} \tag{10.155}$$

Setting $V_o = 0$ yields

$$I_s = I_{sj} = -\sum_{k=1}^{q} x_{kj} \frac{Y_{d_k p, c_k q}(\mathbf{0})}{Y_{rp,sq}(\mathbf{0})}, \quad j = 1, 2, \ldots, p \tag{10.156}$$

Appealing to (10.139), the desired input excitation can be computed by the formulas

$$I_{sj} = \sum_{k=1}^{q} x_{kj} \frac{T_{d_k q, c_k p}(\mathbf{0}) - T_{d_k p, c_k q}(\mathbf{0})}{T_{rp,sq}(\mathbf{0}) - T_{rq,sp}(\mathbf{0})} \tag{10.157}$$

The voltage appearing at the ith controlling branch is found to be

$$\tilde{V}_{\beta i} = I_{sj} \frac{Y_{ra_i, sb_i}(\mathbf{0})}{Y_{uv}(\mathbf{0})} + \sum_{k=1}^{q} \frac{x_{kj} Y_{d_k a_i, c_k b_i}}{Y_{uv}(\mathbf{0})} \tag{10.158}$$

Let

$$\hat{\mathbf{F}}(\mathbf{X}) = [\hat{F}_{ij}] \tag{10.159}$$

Then from (10.158) in conjunction with (10.138) and (10.139), we obtain

$$\hat{F}_{ij} = \frac{\sum_{k=0}^{q} x_{kj}[T_{d_k b_i, c_k a_i}(\mathbf{0}) - T_{d_k a_i, c_k b_i}(\mathbf{0})]}{T_m(\mathbf{0})}, \quad i \neq j$$

(10.160)

$$\hat{F}_{ii} = 1 + \frac{\sum_{k=0}^{q} x_{ki}[T_{d_k b_i, c_k a_i}(\mathbf{0}) - T_{d_k a_i, c_k b_i}(\mathbf{0})]}{T_m(\mathbf{0})}$$

(10.161)

where $x_{0j} = I_{sj}$, $c_0 = s$, and $d_0 = r$.

In the special situation where $\mathbf{X}$ is diagonal, (10.160) and (10.161) reduce to

$$\hat{F}_{ij} = \frac{\begin{aligned}&I_{sj}[T_{rb_i, sa_i}(\mathbf{0}) - T_{ra_i, sb_i}(\mathbf{0})]\\&+x_{jj}[T_{d_j b_i, c_j a_i}(\mathbf{0}) - T_{d_j a_i, c_j b_i}(\mathbf{0})]\end{aligned}}{T_m(\mathbf{0})}, \quad i \neq j$$

(10.162)

and from (10.156) and (10.158)

$$\begin{aligned}\hat{F}_{ii} &= 1 - \frac{I_{si} Y_{ra_i, sb_i}(\mathbf{0}) + x_{ii} Y_{d_i a_i, c_i b_i}(\mathbf{0})}{Y_{uv}(\mathbf{0})}\\&= 1 - \frac{x_{ii}[Y_{d_i a_i, c_i b_i}(\mathbf{0})Y_{rp, sq}(\mathbf{0}) - Y_{d_i p, c_i q}(\mathbf{0})Y_{ra_i, sb_i}(\mathbf{0})]}{Y_{rp, sq}(\mathbf{0})Y_{uv}(\mathbf{0})}\\&= 1 + \frac{x_{ii}\dot{Y}_{rp, sq}(\mathbf{0})}{Y_{rp, sq}(\mathbf{0})}\end{aligned}$$

(10.163)

where $\dot{Y}_{rp, sq}(\mathbf{0})$ denotes the partial derivative of $Y_{rp, sq}(\mathbf{X})$ with respect to the element x_{ii} evaluated at $\mathbf{X} = \mathbf{0}$, i.e.

$$\dot{Y}_{rp, sq}(\mathbf{0}) = \left.\frac{\partial Y_{rp, sq}(\mathbf{X})}{\partial x_{ii}}\right|_{\mathbf{X}=\mathbf{0}}$$

(10.164)

The third line of (10.163) follows directly from the identity given in (4.144):

$$\begin{aligned}Y_{uv}(\mathbf{X})\dot{Y}_{rp, sq}(\mathbf{X}) &= Y_{ra_i, sb_i}(\mathbf{X})Y_{d_i p, c_i q}(\mathbf{X})\\&\quad -Y_{d_i a_i, c_i b_i}(\mathbf{X})Y_{rp, sq}(\mathbf{X})\end{aligned}$$

(10.165)

Using $\mathbf{X}^{ii}$ defined (10.147), it is not difficult to show that $Y_{rp, sq}(\mathbf{X}^{ii})$ can be expanded as

$$Y_{rp, sq}(\mathbf{X}^{ii}) = Y_{rp, sq}(\mathbf{0}) + x_{ii}\dot{Y}_{rp, sq}(\mathbf{0})$$

(10.166)

Substituting (10.166) in (10.163) yields

$$\hat{F}_{ii} = \frac{Y_{rp,sq}(\mathbf{X}^{ii})}{Y_{rp,sq}(\mathbf{0})} = \frac{T_{rp,sq}(\mathbf{X}^{ii}) - T_{rq,sp}(\mathbf{X}^{ii})}{T_{rp,sq}(\mathbf{0}) - T_{rq,sp}(\mathbf{0})} \qquad (10.167)$$

We illustrate the above results by the following example.

Example 10.6 We use the same voltage-series feedback amplifier of Fig. 10.13, as considered in Example 10.5. Assume that $\mathbf{X}$ of (10.149) is the matrix of interest. Since $\mathbf{X}$ is diagonal and contains all the controlling parameters of the controlled sources, (10.162) applies and required digraph for (10.162) is simply the graph representing the equivalent network of Fig. 10.15 after the independent voltage source V_s and the controlled current sources I_a and I_b have been removed. The resulting graph is shown in Fig. 10.18 with

$$a_1 = 5, \quad b_1 = 3, \quad c_1 = 4, \quad d_1 = 3, \quad r = 1, \quad p = 2 \qquad (10.168a)$$
$$a_2 = 4, \quad b_2 = 5, \quad c_2 = 2, \quad d_2 = 5, \quad s = 5, \quad q = 5 \qquad (10.168b)$$

From (10.162) in conjunction with (10.157), we obtain

$$\hat{F}_{12} = \frac{I_{s2}[T_{13,51}(\mathbf{0}) - T_{11,53}(\mathbf{0})] + x_{22}[T_{53,21}(\mathbf{0}) - T_{51,23}(\mathbf{0})]}{T_5(\mathbf{0})}$$
$$= \frac{I_{s2}G_5(-G_eG_f - G_eG_2 - G_2G_f) - x_{22}G_5G_1G_f}{T_5(\mathbf{0})}$$
$$= -2402.264 \qquad (10.169)$$

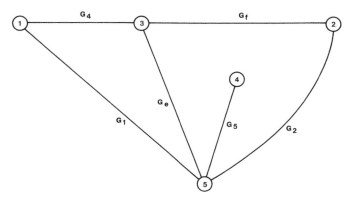

Figure 10.18 The graph used to compute the elements of the null return difference matrix with respect to the controlling parameters of the network of Figure 10.15.

where

$$I_{s2} = x_{22} \frac{T_{55,22}(\mathbf{0}) - T_{52,25}(\mathbf{0})}{T_{12,55}(\mathbf{0}) - T_{15,52}(\mathbf{0})}$$

$$= x_{22} \frac{G_5[(G_f + G_e)(G_1+G_4) + G_1 G_4]}{G_5 G_4 G_f}$$

$$= 2.25691 \tag{10.170}$$

$$T_5(\mathbf{0}) = G_5\{G_1[(G_4 + G_e)(G_2+G_f) + G_2 G_f]$$

$$+ G_4 G_e (G_2 + G_f) + G_2 G_4 G_f\}$$

$$= 4.28764 \cdot 10^{-12} \tag{10.171}$$

and

$$\hat{F}_{21} = \frac{I_{s1}[T_{15,54}(\mathbf{0}) - T_{14,55}(\mathbf{0})] + x_{11}[T_{35,44}(\mathbf{0}) - T_{34,45}(\mathbf{0})]}{T_5(\mathbf{0})}$$

$$= \frac{I_{s1} 0 + x_{11} T_5(\mathbf{0})/G_5}{T_5(\mathbf{0})} = \frac{x_{11}}{G_5} = 42.88407 \tag{10.172}$$

To compute $\hat{F}_{11}$, we use Fig. 10.18 for the denominator of (10.167) and Fig. 10.19 for its numerator. The digraph of Fig. 10.19 is obtained from that of Fig. 10.18 by

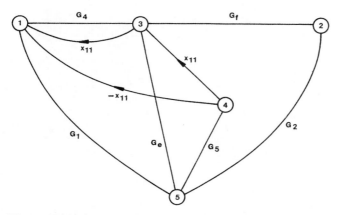

Figure 10.19 The digraph used to compute the numerator of diagonal element $\hat{F}_{11}$ of the null return difference matrix for the network of Fig. 10.15.

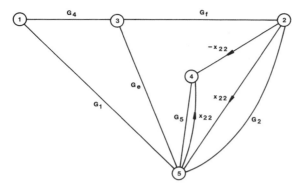

Figure 10.20 The digraph used to compute the numerator of the diagonal element $\hat{F}_{22}$ of the null return difference matrix for the network of Figure 10.15.

inserting the associated digraph of the controlled current source I_a. This gives

$$\hat{F}_{11} = \frac{T_{12,55}(\mathbf{X}^{11}) - T_{15,52}(\mathbf{X}^{11})}{T_{12,55}(\mathbf{0}) - T_{15,52}(\mathbf{0})} = \frac{G_5(G_4 + x_{11})G_f}{G_5 G_4 G_f}$$

$$= 1 + \frac{x_{11}}{G_4} = 51.05501 \tag{10.173}$$

In fact, the denominator of $\hat{F}_{11}$ can be obtained from its numerator simply by setting x_{11} to zero. Finally, for $\hat{F}_{22}$ we use the digraph of Fig. 10.20 for the numerator of (10.167). The digraph of Fig. 10.20 is obtained from that of Fig. 10.18 by inserting the associated digraph of the controlled current source I_b. This gives

$$\hat{F}_{22} = \frac{T_{12,55}(\mathbf{X}^{22}) - T_{15,52}(\mathbf{X}^{22})}{T_{12,55}(\mathbf{0}) - T_{15,52}(\mathbf{0})} = \frac{G_5 G_4 G_f}{G_5 G_4 G_f} = 1 \tag{10.174}$$

These results are consistent with those obtained earlier in (7.80) of Chapter 7.

As in (10.173), the denominator of $\hat{F}_{22}$ can be obtained from the numerator by setting x_{22} to zero.

Alternatively, $\hat{F}_{11}$ and $\hat{F}_{22}$ can be computed by using only the graph of Fig. 10.18 and formulas (10.157) and (10.161), as follows:

$$\hat{F}_{11} = 1 + \frac{I_{s1}[T_{13,51}(\mathbf{0}) - T_{11,53}(\mathbf{0})] + x_{11}[T_{33,41}(\mathbf{0}) - T_{31,43}(\mathbf{0})]}{T_5(\mathbf{0})}$$

$$= 1 + \frac{I_{s1}G_5(-G_e G_f - G_e G_2 - G_2 G_f) + x_{11}G_1 G_5(G_2 + G_f)}{T_5(\mathbf{0})}$$

$$= 1 + \frac{x_{11}}{G_4} = 51.05501 \tag{10.175}$$

where $T_5(\mathbf{0})$ is given by (10.171) and

$$I_{s1} = x_{11}\frac{T_{35,42}(\mathbf{0}) - T_{32,45}(\mathbf{0})}{T_{12,55}(\mathbf{0}) - T_{15,52}(\mathbf{0})} = -x_{11}\frac{G_5(G_1+G_4)G_f}{G_4 G_5 G_f}$$

$$= -x_{11}\left(1 + \frac{G_1}{G_4}\right) = -0.046902 \tag{10.176}$$

and

$$\hat{F}_{22} = 1 + \frac{I_{s2}[T_{15,54}(\mathbf{0}) - T_{14,55}(\mathbf{0})] + x_{22}[T_{55,24}(\mathbf{0}) - T_{54,25}(\mathbf{0})]}{T_5(\mathbf{0})}$$

$$= 1 + \frac{I_{s2}0 - x_{22}0}{T_5(\mathbf{0})} = 1 \tag{10.177}$$

This confirms (10.173) and (10.174). Observe that (10.176) and (10.177) are much more involved than those of (10.173) and (10.174). However, (10.176) and (10.177) require only the graph of Fig. 10.18 while (10.173) and (10.174) need the digraphs of Figs. 10.19 and 10.20.

10.9 SUMMARY

In the foregoing, we have shown that many of the invariant characters of network determinants can be derived from graph-theoretic considerations rather than from physical arguments. Thus, they are valid not only for the electrical networks, but also for other systems. Since the cofactors in general are not invariant with respect to the transformations of either circuits or cutsets, their values can only be stated in terms of the values of some precisely defined quantities called the generalized cofactors. It has been shown that although the generalized network determinants and their generalized cofactors are not invariant for a general system of circuits and cutsets, they are related by a real constant depending only upon the two choices of circuits or cutsets, and that for the most common and important types of representations such as the fundamental systems of circuits and cutsets, the circuits formed by the windows of a planar graph, and the set of incidence cutsets, they all are invariant. It has also been shown that the ratio of the determinant of a generalized network matrix and one of its generalized cofactors or the ratio of two of its generalized cofactors is invariant with respect to the general transformations of the reference frame. Thus, in particular, the network functions are invariant with respect to the general transformations of the reference frame since they can always be expressed as the ratios of the above types. In other words, we have shown that many of the invariant characters of a network are consequences of the network topology rather than the characteristics of the branches of the network.

We have indicated that, although the general return difference in general is not invariant under the general transformation of the reference frame, it is invariant for the most common and important types of feedback networks, in which the controlled sources are not cyclically coupled. We also derived conditions that are both necessary and sufficient for the invariance of the general return difference. These conditions are stated in terms of the nonsingular branch-immittance matrix. This can always be done if the network possesses a unique solution, since by applying Thévenin and Norton theorems and the Blakesley transformation if necessary, we can consider the equivalent modified networks N_c and N_m in which each controlled current source is connected in parallel with a finite nonzero admittance and each controlled voltage source is connected in series with a finite nonzero impedance.

We have also defined a related general return difference function called the general null return difference which was shown to be consistent with the physical interpretation of the ordinary null return difference used extensively in the literature. The significance is that, unlike the ordinary null return difference, which varies even with respect to the transformations from a system of basis of circuits or cutsets to another, the general null return difference defined here possesses all the properties and the general return difference. Thus, it is invariant under the general transformations of the reference frame for nearly all practical feedback networks. Furthermore, no specific details on the formulation of the network equations are required in the definition.

Since it is well known that the relative sensitivity function, in particular the sensitivity function, is equal to the difference of the reciprocals of the general return difference and the general null return difference, the relative sensitivity varies in a similar fashion as the general return difference under the general transformations of the reference frame. However, the driving-point impedance of a feedback network is, as expected, invariant under the general transformations of the reference frame, since it can be expressed in terms of the ratio of the general null return difference and the general return difference, which is known as Blackman's formula [see Blackman (1943)].

We have demonstrated that the term *network determinant* defined in terms of the determinant of the coefficient matrix of the primary systems of equations can also be used to define the general return difference, yielding the same properties as those based on the loop and cutset formulations, which are known as the secondary systems of equations.

We have indicated that the determinants of the generalized network matrices and their generalized cofactors, although they are in general not invariant with respect to the choice of the incidence functions of the graph, are invariant for general electrical networks.

Finally, we have presented topological formulas for the evaluation of the elements of the return difference matrix and the null return difference matrix of a multiple-loop feedback network in terms of the sums of the directed-tree and directed-two-tree admittance products. The elements of interest are either the ones that are inherently sensitive to variation or the ones whose effect on the over-all amplifier performance is of paramount importance to the designers. In most situations, we choose the controlling parameters of all the controlled sources as the elements of interest. In such a case, the associated digraph is simply the graph representing the feedback network after the independent source and the controlled sources have been removed. The elements of the return difference and null return difference matrices can be expressed in terms of the tree-admittance and two-tree-admittance products in this graph. The significance of this approach is that it not only provides a short-cut for the evaluation of these matrices, but also gives an insight into the behavior of the feedback system under consideration. Furthermore, if a single element or a group of elements of the return difference matrix or the null return difference matrix is required, it can be computed directly from the associated graph without the necessity of first computing the transfer-impedance matrices $\mathbf{A}$, $\mathbf{B}$, $\mathbf{C}$ and D of various orders. This is a real saving!

We remark that using the established identities, we have shown that the topological formulas of a network can easily be derived from one another. Thus, the results provide an insight into and a unification of many of the existing results in the literature.

BIBLIOGRAPHY

Aitken, A. C.: "Determinants and Matrices," 9th ed., New York: Interscience, 1962.

Blackman, R. B.: Effect of Feedback on Impedance, *Bell Syst. Tech. J.*, vol. 22, pp. 268–277, 1943.

Brown, D. P.: Topological Properties of Resistance Matrices, *SIAM J. Appl. Math.*, vol. 16, pp. 387–394, 1968.

Cederbaum, I.: Invariance and Mutual Relations of Electrical Network Determinants, *J. Math. Phys.*, vol. 35, pp. 236–244, 1956a.

Cederbaum, I.: On Network Determinants, *Proc. IRE*, vol. 44, pp. 258–259, 1956b.

Chen, W. K.: Graph-Theoretic Considerations on the Invariance and Mutual Relations of the Determinants of the Generalized Network Matrices and Their Generalized Cofactors, *Quart. J. Math. Oxford* (2), vol. 21, no. 84, pp. 459–479, 1970.

Chen, W. K.: Graph-Theoretic Considerations on the Invariance of Return Difference, *J. Franklin Inst.*, vol. 298, no. 2, pp. 81–100, 1974.

Chen, W. K.: Indefinite-Admittance Matrix Formulation of Feedback Amplifier Theory, *IEEE Trans. Circuits and Systems*, vol. CAS-23, no. 8, pp. 498–505, 1976a.

Chen, W. K.: "Applied Graph Theory: Graphs and Electrical Networks," 2nd rev. ed., New York: American Elsevier, and Amsterdam: North-Holland, 1976b.

Chen, W. K.: The Hybrid Matrix in Linear Multiple-Loop Feedback Networks, *IEEE Trans. Circuits and Systems*, vol. CAS-24, pp. 469–474, 1977.

Chen, W. K.: On Second-Order Cofactors and Null Return Difference in Feedback Amplifier Theory, *Int. J. Circuit Theory Applications*, vol. 6, pp. 305–312, 1978.

Chen, W. K.: Topological Evaluation of Feedback Matrices in Multiple-Loop Feedback Amplifiers, *J. Franklin Inst.*, vol. 308, no. 2, pp. 125–139, 1979.

Chen, W. K. and H. M. Elsherif: Determinant of the Null Return-Difference Matrix, *Electron. Lett.*, vol. 13, pp. 306–307, 1977.

Hakim, S. S.: Aspects of Return-Difference Evaluation in Transistor Feedback Amplifiers, *Proc. IEE (London)*, vol. 112, pp. 1700–1704, 1965.

Hoskins, R. F.: Definition of Loop Gain and Return Differences in Transistor Feedback Amplifiers, *Proc. IEE (London)*, vol. 112, pp. 1995–2000, 1965.

Jeans, J.: "The Mathematical Theory of Electricity and Magnetism," Cambridge, UK: Cambridge University Press, 1948.

Kuh, E. S. and R. A. Rohrer: "Theory of Linear Active Networks," San Francisco, Calif.: Holden Day, 1967.

Malik, N. R. and H. W. Hale: Relationships between Nonzero Determinants Formed from Vertex and Circuit Matrices, *IEEE Trans. Circuit Theory*, vol. CT-13, p. 196, 1966.

Nerode, A. and H. Shank: An Algebraic Proof of Kirchhoff's Network Theorem, *Amer. Math. Monthly*, vol. 68, pp. 244–247, 1961.

Okada, S.: On Node and Mesh Determinants, *Proc. IRE*, vol. 43, p. 1527, 1955.

Percival, W. S.: Improved Matrix and Determinant Methods for Solving Networks, *Proc. IEE (London)*, vol. 101, pp. 258–265, 1954.

Saltzer, C.: The Second Fundamental Theorem of Electrical Networks, *Quart. Appl. Math.*, vol. 11, pp. 119–123, 1953.

Sandberg, I. W.: On the Theory of Linear Multi-Loop Feedback Systems, *Bell Syst. Tech. J.*, vol. 42, pp. 355–382, 1963.

Seshu, S.: The Mesh Counterpart of Shekel's Theorem, *Proc. IRE*, vol. 43, p. 342, 1955.

Seshu, S. and M. B. Reed: "Linear Graphs and Electrical Networks," Reading, Mass.: Addison-Wesley, 1961.

Shekel, J.: Two Network Theorems Concerning Change of Voltage Reference Terminal, *Proc. IRE*, vol. 42, p. 1125, 1954.

Slepian, P.: "Mathematical Foundations of Network Analysis," p. 46 and pp. 177–186, New York: Springer-Verlag, 1968.

Tasny-Tschiassny, L.: The Return Difference Matrix in Linear Networks, *Proc. IEE (London)*, vol. 100, pt. IV, pp. 39–46, 1953.

Tsang, N. F.: On Electrical Network Determinants, *J. Math. Phys.*, vol. 33, pp. 185–193, 1954.

THE INDEFINITE-IMPEDANCE MATRIX FORMULATION OF FEEDBACK AMPLIFIER THEORY

The indefinite-admittance matrix was introduced in Chapter 2, and its applications to the feedback amplifier theory were taken up in Chapters 4 and 5. We indicated that Bode's concept of return difference plays an important role in the design of feedback amplifiers. Among the many important properties, the return difference is a generalization of the concept of the feedback factors of ideal feedback model, the sensitivity function of the amplifier is closely related to the return difference, and the return difference is basic to the study of the stability of the feedback system and to the determination of its transmission and driving-point properties. We showed that the return difference and the null return difference can elegantly and compactly expressed in terms of the first- and second-order cofactors of the elements of the indefinite-admittance matrix. However, the dual concept, the indefinite-impedance matrix is rarely considered. Perhaps this is due to the fact that measuring the branch voltage is easier than measuring the branch current. However, advances in integrated op-amp circuits have made it possible to measure the branch current on line without opening any branch.

The indefinite-impedance matrix was first introduced in Sec. 9.5.2 of Chapter 9 for the loop formulation of return difference. To simplify the formulation, we only considered connected planar networks, and obtained the dual results on the summations of the reciprocals of the return differences and the products of transimpedances and transfer admittances. In the present chapter, we shall study the subject further and extend the results to nonplanar networks.

11.1 THE INDEFINITE-IMPEDANCE MATRIX

We begin by considering a connected planar network N of e branches $b_x (x = 1, 2, \ldots, e)$ and of nullity m. Assume that each branch b_x has a nonzero impedance z_x. Since a one-port impedance or a transformer can be equivalently represented by current-controlled voltage sources, without loss of generality, we may assume that all branches of N are current-controlled voltage sources

$$V_x = z_x (I_{j_1} - I_{j_2}), \quad x = 1, 2, \ldots, e \tag{11.1}$$

as in (9.125), where I_u ($u = 1, 2, \ldots, m + 1$) are the mesh currents. As before, all the mesh currents are oriented clockwise except the outer mesh I_{m+1} which is counterclockwise. For $j = x$

$$V_x = z_x (I_{x_1} - I_{x_2}) \tag{11.2}$$

and z_x becomes the one-port impedance of branch b_x. The matrix $\mathbf{Z}$ relating the mesh-current vector $\mathbf{J}$ and the mesh voltage-source vector $\mathbf{E}$

$$\begin{bmatrix} z_{11} & z_{12} & \cdots & z_{1\tilde{m}} \\ z_{21} & z_{22} & \cdots & z_{2\tilde{m}} \\ \vdots & \vdots & \ddots & \vdots \\ z_{\tilde{m}1} & z_{\tilde{m}2} & \cdots & z_{\tilde{m}\tilde{m}} \end{bmatrix} \begin{bmatrix} j_1 \\ j_2 \\ \vdots \\ j_{\tilde{m}} \end{bmatrix} = \begin{bmatrix} E_1 \\ E_2 \\ \vdots \\ E_{\tilde{m}} \end{bmatrix} \tag{11.3a}$$

is called the *indefinite-impedance matrix*, which is square and of order $\tilde{m} = m + 1$, or more compactly

$$\mathbf{ZJ} = \mathbf{E} \tag{11.3b}$$

In general, the impedance z_x enters the indefinite-impedance matrix $\mathbf{Z}$ in a rectangular pattern, as in (9.95), with z_x appearing at the entries (x_1, j_1) and (x_2, j_2) and $-z_x$ at (x_1, j_2) and (x_2, j_1), where branch b_x is traversed by mesh currents I_{x_1} and I_{x_2} as shown in (9.128). These rules remain valid for coupled inductors if they are transformed into two controlled voltage sources.

As an example, consider the planar network of Fig. 11.1a, the mutual inductance of which can be equivalently represented by two current-controlled voltage sources as shown in Fig. 11.1b. By applying the above rules as depicted

Figure 11.1 (a) A planar network with mutual inductance M and an r-type current-controlled voltage source. (b) The equivalent representation of the mutual inductance M in (a) by two current-controlled voltage source.

in (9.128), the indefinite-impedance matrix of the network of Fig. 11.1a can be written down by inspection as

$$
\mathbf{Z} = \begin{bmatrix}
R_3 + L_1 s + \dfrac{1}{C_4 s} & M s - \dfrac{1}{C_4 s} & -R_3 - L_1 s - M s \\[2ex]
r + M s - \dfrac{1}{C_4 s} & R_5 + L_2 s + \dfrac{1}{C_4 s} & -R_5 - r - L_2 s - M s \\[2ex]
-R_3 - r - L_1 s - M s & -R_5 - L_2 s - M s & \begin{array}{c} r + R_3 + R_5 + L_1 s \\ + L_2 s + 2 M s \end{array}
\end{bmatrix}
$$

$$(11.4)$$

As the rules ensure $\mathbf{Z}$ to be an equicofactor matrix, this property can be used to reduce the labor in network computations. For example, in order to obtain a first-order cofactor of an element of $\mathbf{Z}$, we can choose the most convenient one to calculate.

11.1.1 Reduction of the Order of Z

In a given network N, if some of the meshes are not available for external excitation by independent sources, these meshes are said to be *inaccessible* and can be suppressed by setting the corresponding voltage-source subvector to zero. The resulting indefinite-impedance matrix can be partitioned from (11.3) as

$$
\mathbf{ZJ} = \begin{bmatrix} \mathbf{Z}_{11} & \mathbf{Z}_{12} \\ \mathbf{Z}_{21} & \mathbf{Z}_{22} \end{bmatrix} \begin{bmatrix} \mathbf{J}_1 \\ \mathbf{J}_2 \end{bmatrix} = \begin{bmatrix} \mathbf{E}_1 \\ \mathbf{0} \end{bmatrix}
$$

$$(11.5)$$

where $\mathbf{Z}_{22}$ and $\mathbf{J}_2$ correspond to the meshes being suppressed, and $\mathbf{Z}_{11}$ and $\mathbf{J}_1$ correspond to the rest. From the second equation of (11.5), we solve for $\mathbf{J}_2$ and

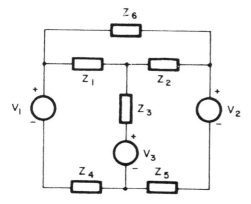

Figure 11.2 A planar network containing a voltage-source cutset.

substitute the result into the first to obtain

$$(\mathbf{Z}_{11} - \mathbf{Z}_{12}\mathbf{Z}_{22}^{-1}\mathbf{Z}_{21})\mathbf{J}_1 = \mathbf{E}_1 \tag{11.6}$$

The resulting reduced indefinite-impedance matrix becomes

$$\mathbf{Z}_{\mathrm{red}} = \mathbf{Z}_{11} - \mathbf{Z}_{12}\mathbf{Z}_{22}^{-1}\mathbf{Z}_{21} \tag{11.7}$$

As an example, consider the network of Fig. 11.2 containing three voltage sources forming a cutset. The indefinite impedance matrix $\mathbf{Z}$ is found to be

$$\mathbf{Z} = \begin{bmatrix} z_1 + z_2 + z_6 & -z_1 & -z_2 & -z_6 \\ -z_1 & z_1 + z_3 + z_4 & -z_3 & -z_4 \\ -z_2 & -z_3 & z_2 + z_3 + z_5 & -z_5 \\ -z_6 & -z_4 & -z_5 & z_4 + z_5 + z_6 \end{bmatrix} \tag{11.8}$$

Suppose that we wish to suppress the mesh corresponding to first row and first column of (11.8), and the resulting matrix is obtained from (11.7) as

$$\mathbf{Z}_{\mathrm{red}} = \begin{bmatrix} z_1 + z_3 + z_4 & -z_3 & -z_4 \\ -z_3 & z_2 + z_3 + z_5 & -z_5 \\ -z_4 & -z_5 & z_4 + z_5 + z_6 \end{bmatrix}$$

$$- \frac{1}{z_1 + z_2 + z_6} \begin{bmatrix} -z_1 \\ -z_2 \\ -z_3 \end{bmatrix} \begin{bmatrix} -z_1 & -z_2 & -z_6 \end{bmatrix}$$

$$
= \begin{bmatrix}
z_3 + z_4 + \dfrac{z_1 z_2 + z_1 z_6}{z_1 + z_2 + z_6} & -\left(z_3 + \dfrac{z_1 z_2}{z_1 + z_2 + z_6}\right) & -\left(z_4 + \dfrac{z_1 z_6}{z_1 + z_2 + z_6}\right) \\[2ex]
-\left(z_3 + \dfrac{z_1 z_2}{z_1 + z_2 + z_6}\right) & z_3 + z_5 + \dfrac{z_1 z_2 + z_2 z_6}{z_1 + z_2 + z_6} & -\left(z_5 + \dfrac{z_2 z_6}{z_1 + z_2 + z_6}\right) \\[2ex]
-\left(z_4 + \dfrac{z_1 z_6}{z_1 + z_2 + z_6}\right) & -\left(z_5 + \dfrac{z_2 z_6}{z_1 + z_2 + z_6}\right) & z_4 + z_5 + \dfrac{z_1 z_6 + z_2 z_6}{z_1 + z_2 + z_6}
\end{bmatrix}
$$

$$(11.9)$$

11.1.2 Calculations of Network Functions

Consider a linear time-invariant network consisting of *RLC* elements and *r*-type current-controlled voltage sources as shown in Fig. 11.3. The network N is excited by the voltage source V_{in} and terminated in the impedance Z_L. Assume that the meshes s and r contain the source V_{in} and the other two meshes p and q contain the load Z_L. Their directions are as shown in Fig. 11.3. The network together with its terminations is denoted by $\tilde{N}$.

Suppose that we wish to compute the transfer admittance $Y_H \equiv y_{rp,sq}$ of Fig. 11.3 defined as

$$
Y_H \equiv y_{rp,sq} = \frac{I_o}{V_{in}} \tag{11.10}
$$

For simplicity, let $J_q = 0$ and delete the sth mesh equation from the equations (11.3b) with r' and p' denoting the new row indices of the meshes r and p after the deleting of the sth mesh equation, respectively. This results in

$$
\mathbf{Z}_{sq}\mathbf{J} = \mathbf{E} \tag{11.11}
$$

where $\mathbf{Z}_{sq}$ is the definite impedance matrix relating vectors

$$
\mathbf{J} = \left[J_1, J_2, \ldots, J_{q-1}, J_{q+1}, \ldots, J_m\right]' \tag{11.12a}
$$

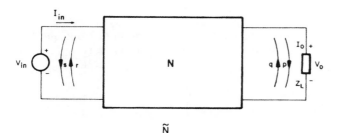

$$\tilde{N}$$

Figure 11.3 An *RLC-r* network N excited by the voltage sources V_{in} and terminated in the impedance Z_L with the complete network being denoted by $\tilde{N}$.

and

$$\mathbf{E} = \left[0, 0, \ldots, 0, E_{r'} = V_{in}, 0, \ldots, 0 \right]' \qquad (11.12b)$$

with the prime denoting the matrix transpose.

Solving (11.11) for J_p gives

$$J_p = I_o = \frac{Z_{rp,sq}}{Z_{sq}} V_{in} \qquad (11.13)$$

or, invoking the equicofactor property of $\mathbf{Z}$, we have

$$Y_H = y_{rp,sq} \equiv \frac{I_o}{V_{in}} = \frac{Z_{rp,sq}}{Z_{sq}} = \frac{Z_{rp,sq}}{Z_{nn}} \qquad (11.14)$$

where Z_{sq} and $Z_{rp,sq}$ are the first- and second-order cofactors of the elements of $\mathbf{Z}$, respectively. When $r = p$ and $s = q$, we obtain the driving-point admittance

$$Y_{in} \equiv \frac{I_{in}}{V_{in}} = \frac{Z_{rr,ss}}{Z_{nn}} \qquad (11.15)$$

From (11.14) and (11.15), the current gain $A_I \equiv h_{rp,sq}$ and voltage gain $A_V \equiv g_{rp,sq}$ are found to be

$$A_I \equiv h_{rp,sq} \equiv \frac{I_o}{I_{in}} = \frac{Z_{rp,sq}}{Z_{rr,ss}} \qquad (11.16)$$

$$A_V \equiv g_{rp,sq} \equiv \frac{V_o}{V_{in}} = \frac{Z_{rp,sq}}{Z_{nn}} Z_L \qquad (11.17)$$

11.1.3 Calculations of Return Difference and Null Return Difference

As in (9.132), the return difference $F_k(x)$ with respect to a network element x for a general reference value k can be defined as the ratio of two functional values assumed by the first-order cofactor $Z_{uv}(x)$ of the uth row and vth column element of the indefinite-impedance matrix explicitly written as $\mathbf{Z} = \mathbf{Z}(x)$ under the condition that the element x assumes its nominal value and the condition that the element x assumes the value k:

$$F_k(x) = \frac{Z_{uv}(x)}{Z_{uv}(k)} \qquad (11.18)$$

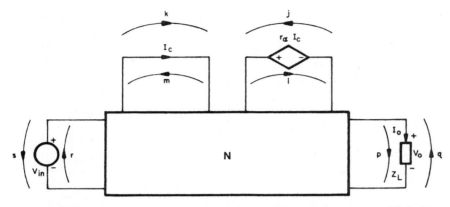

Figure 11.4 A linear time-invariant network N with its exciting source and a current-controlled voltage source shown explicitly.

As before, if $k = 0$,

$$F_0(x) \equiv F(x) = \frac{Z_{uv}(x)}{Z_{uv}(0)} \tag{11.19}$$

is the ordinary return difference.

Recall that the null return difference is regarded as the return difference evaluated under the condition that the input is adjusted to give zero output. Here, in order to exhibit the role of a current-controlled voltage source, this source is explicitly as shown in Fig. 11.4.

First, we consider the null return difference $\hat{F}(r_\alpha)$ with respect to the controlling parameter r_α of the controlled voltage source $V = r_\alpha I_c$ for the zero reference value, where I_c is the controlling current. Recall that $\hat{F}(r_\alpha)$ is defined as one minus the returned current I_c flowing through the controlling branch of the controlled source when the controlled source is replaced by an independent voltage source of r_α volts and when the input voltage source of N is adjusted so that its output current I_o is identically zero.

Applying superposition principle and using (11.14), we obtain

$$I_o = \frac{Z_{rp,sq}(0)}{Z_{nn}(0)} V_{in} + \frac{Z_{jp,iq}(0)}{Z_{nn}(0)} r_\alpha \tag{11.20}$$

and

$$I_c = \frac{Z_{rk,sm}(0)}{Z_{nn}(0)} V_{in} + \frac{Z_{jk,im}(0)}{Z_{nn}(0)} r_\alpha \tag{11.21}$$

Setting $I_o = 0$ in (11.20) yields

$$V_{in} = -\frac{Z_{jp,iq}(0)}{Z_{rp,sq}(0)} r_\alpha \tag{11.22}$$

Under this condition, (11.21) becomes

$$I_c = \frac{r_\alpha}{Z_{rp,sq}(0)} \cdot \frac{1}{Z_{nn}(0)} [Z_{rp,sq}(0) Z_{jk,im}(0) - Z_{rk,sm}(0) Z_{jp,iq}(0)] \tag{11.23}$$

Since $\mathbf{Z}$ is an equicofactor matrix, the identity (4.144) of Chapter 4 applies:

$$Z_{uv} \dot{Z}_{rp,sq} = Z_{ra,sb} Z_{dp,cq} - Z_{rp,sq} Z_{da,cb} \tag{11.24}$$

where $\dot{Z}_{rp,sq}$ denotes the partial derivative of the function $Z_{rp,sq}$ with respect to the controlling parameter $x = r_\alpha$. Since $\mathbf{Z}(0)$ remains to be an equicofactor matrix, the identity (11.24) remains valid when r_α is set to zero:

$$Z_{uv}(0) \dot{Z}_{rp,sq}(0) = Z_{ra,sb}(0) Z_{dp,cq}(0) - Z_{rp,sq}(0) Z_{da,cb}(0) \tag{11.25a}$$

or, equivalently, by relabeling some of the subscripts, (11.25a) can be rewritten as

$$-Z_{uv}(0) \dot{Z}_{rp,sq}(0) = Z_{rp,sq}(0) Z_{jk,im}(0) - Z_{rk,sm}(0) Z_{jp,iq}(0) \tag{11.25b}$$

Substituting this in (11.23) gives the returned current

$$I_c = -\frac{r_\alpha}{Z_{rp,sq}(0)} \cdot \frac{Z_{uv}(0) \dot{Z}_{rp,sq}(0)}{Z_{nn}(0)} = -r_\alpha \frac{\dot{Z}_{rp,sq}(0)}{Z_{rp,sq}(0)} \tag{11.26}$$

As before, the *null return ratio* $\hat{T}$ with respect to the controlling parameter $x = r_\alpha$ is defined as the negative of the returned current I_c or algebraically as

$$\hat{T} = r_\alpha \frac{\dot{Z}_{rp,sq}(0)}{Z_{rp,sq}(0)} \tag{11.27}$$

Then we have

$$1 + \hat{T} = 1 - I_c = 1 + r_\alpha \frac{\dot{Z}_{rp,sq}(0)}{Z_{rp,sq}(0)} = \frac{Z_{rp,sq}(0) + r_\alpha \dot{Z}_{rp,sq}(0)}{Z_{rp,sq}(0)}$$

$$= \frac{Z_{rp,sq}(r_\alpha)}{Z_{rp,sq}(0)} = \hat{F}(r_\alpha) \tag{11.28}$$

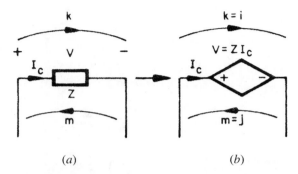

k

+   V   −

I_c

Z

m

k = i

V = Z I_c

I_c

+   −

m = j

(a)   (b)

Figure 11.5 (a) A one-port impedance z and (b) its equivalent representation as a current-controlled voltage source.

showing that the null return difference is equal to one minus the returned current I_c, or

$$\hat{F}(r_\alpha) = \frac{Z_{rp,sq}(r_\alpha)}{Z_{rp,sq}(0)} \tag{11.29}$$

If the element of interest is a one-port impedance z, it can be considered as a current-controlled voltage source $V = zI_c$, as shown in Fig. 11.5. Thus, the null return difference $\hat{F}(x)$ with respect to any element x for the zero reference value can again be written as

$$\hat{F}(x) = \frac{Z_{rp,sq}(x)}{Z_{rp,sq}(0)} \tag{11.30}$$

In the case of the *general null return difference*, we replace the current-controlled voltage source of Fig. 11.4 by an independent voltage source of $r'_\alpha = r_\alpha - k$ volts in series with a current-controlled voltage source $V = kI_c$. The resulting indefinite-impedance matrix is simply $\mathbf{Z}(k)$. As before, applying the principle of superposition, the output current I_o at the load of Fig. 11.4 can be expressed as

$$I_o = \frac{Z_{rp,sq}(k)}{Z_{nn}(k)} V_{in} + r'_\alpha \frac{Z_{jp,iq}(k)}{Z_{nn}(k)} \tag{11.31}$$

Setting $I_o = 0$ gives

$$V_{in} = -r'_\alpha \frac{Z_{jp,iq}(k)}{Z_{rp,sq}(k)} \tag{11.32}$$

which is the input excitation that will make the output of the amplifier identically zero. With this input voltage, the returned current I_c of Fig. 11.4 is given by

$$
\begin{aligned}
I_c &= \frac{Z_{rk,sm}(k)}{Z_{nn}(0)} V_{in} + r'_\alpha \frac{Z_{jk,im}(k)}{Z_{nn}(0)} \\
&= -r'_\alpha \frac{Z_{rk,sm}(k) Z_{jp,iq}(k)}{Z_{nn}(0) Z_{rp,sq}(k)} + r'_\alpha \frac{Z_{jk,im}(k)}{Z_{nn}(0)} \\
&= r'_\alpha \frac{Z_{jk,im}(k) Z_{rp,sq}(k) - Z_{rk,sm}(k) Z_{jp,iq}(k)}{Z_{nn}(0) Z_{rp,sq}(k)}
\end{aligned}
\tag{11.33}
$$

From (11.25b), we have

$$
-Z_{uv}(0)\dot{Z}_{rp,sq} = Z_{rp,sq}(k) Z_{jk,im}(k) - Z_{rk,sm}(k) Z_{jp,iq}(k)
\tag{11.34}
$$

We remark that, like $\dot{Z}_{rp,sq}$, the functions $Z_{rp,sq}$, $Z_{jk,im}$, $Z_{rk,sm}$ and $Z_{jp,iq}$ are independent of k, and thus the argument k need not be exhibited explicitly. It is expressed here merely to conform with (11.33). Substituting (11.34) in (11.33) yields

$$
I_c = -r'_\alpha \frac{Z_{uv}(0)\dot{Z}_{rp,sq}(k)}{Z_{nn}(0) Z_{rp,sq}(k)} = -r'_\alpha \frac{\dot{Z}_{rp,sq}(k)}{Z_{rp,sq}(k)}
\tag{11.35}
$$

Like (11.27), define the *general null return ratio* $\hat{T}_k$ with respect to r_α as being the negative of the returned current I_c or algebraically as

$$
\hat{T}_k = r'_\alpha \frac{\dot{Z}_{rp,sq}(k)}{Z_{rp,sq}(k)}
\tag{11.36}
$$

Then we have

$$
1 + \hat{T}_k = 1 - I_c = 1 + r'_\alpha \frac{\dot{Z}_{rp,sq}(k)}{Z_{rp,sq}(k)} = \frac{Z_{rp,sq}(k) + r'_\alpha \dot{Z}_{rp,sq}(k)}{Z_{rp,sq}(k)}
\tag{11.37}
$$

Since

$$
Z_{rp,sq}(r_\alpha) = Z_{rp,sq}(0) + r_\alpha \dot{Z}_{rp,sq}(k) = Z_{rp,sq}(k) + r'_\alpha \dot{Z}_{rp,sq}(k)
\tag{11.38}
$$

which when substituted in (11.37) gives

$$
1 + \hat{T}_k = \frac{Z_{rp,sq}(r_\alpha)}{Z_{rp,sq}(k)} = \hat{F}_k(r_\alpha)
\tag{11.39}
$$

Thus, the general null return difference can again be interpreted as one minus the returned current I_c in Fig. 11.4.

11.1.4 The Relations between Network Functions and Return Difference and Null Return Difference

In this section, we show how the four types of network functions discussed in Sec. 11.1.2 are related to the return difference $F_k(x)$ and null return difference $\hat{F}_k(x)$. For the transfer admittance written as $Y_T(x) = y_{rp,sq}(x)$, we obtain from (11.14)

$$\frac{Y_T(x)}{Y_T(k)} = \frac{Z_{rp,sq}(x)}{Z_{rp,sq}(k)} \cdot \frac{Z_{nn}(k)}{Z_{nn}(x)} = \frac{\hat{F}_k(x)}{F_k(x)} \tag{11.40}$$

For the driving-point admittance Y_{in}, we have from (11.15) or (11.40)

$$\frac{Y_{in}(x)}{Y_{in}(k)} = \frac{F_k(\text{input open-circuited})}{F_k(\text{input short-circuited})} \tag{11.41}$$

For voltage gain A_V, two cases are distinguished. In case 1, the element of interest x is not Z_L. Then we have from (11.17)

$$\frac{A_V(x)}{A_V(k)} = \frac{\hat{F}_k(x)}{F_k(x)} \tag{11.42}$$

In case 2, the element of interest is Z_L. Then from (11.17)

$$\frac{A_V(x)}{A_V(k)} = \frac{\hat{F}_k(Z_L)}{F_k(Z_L)} \cdot \frac{Z_L}{k} \tag{11.43}$$

For the current gain A_I, we obtain the relation from (11.16) as

$$\frac{A_I(x)}{A_I(k)} = \frac{F_k(\text{output open-circuited})}{F_k(\text{input open-circuited})} \tag{11.44}$$

These relations are useful in measurement situations. For example, if k is set to zero in (11.41), the resulting equation becomes the well-known Blackman's formula first given by Blackman (1943).

11.1.5 Relations between Sensitivities and Return Difference and Null Return Difference

The relative sensitivity of a network function H with respect to an element x with reference value k is defined as

$$S_{x'}^H = \frac{\partial \ln H}{\partial \ln x'} \tag{11.45}$$

where $x' = x - k$. When $H = Y_H$, the relative sensitivity function becomes

$$S_{x'}^{Y_H} = x' \left[\frac{\partial \ln \left(Z_{rp,sq}/Z_{nn} \right)}{\partial x'} \right] = \frac{1}{Z_{rp,sq}} \cdot x' \frac{\partial Z_{rp,sq}}{\partial x'} - \frac{1}{Z_{nn}} \cdot x' \frac{\partial Z_{nn}}{\partial x'}$$

(11.46)

Referring to (4.149*b*) of Chapter 4, we have

$$x \frac{\partial Z_{rp,sq}}{\partial x} = Z_{rp,sq}(x) - Z_{rp,sq}(0)$$

(11.47)

Substituting x' for x, (11.47) becomes

$$x' \frac{\partial Z_{rp,sq}}{\partial x'} = Z_{rp,sq}(x) - Z_{rp,sq}(k)$$

(11.48)

Likewise, we have

$$x' \frac{\partial Z_{nn}}{\partial x'} = Z_{nn}(x) - Z_{nn}(k)$$

(11.49)

Substituting (11.48) and (11.49) in (11.46) yields

$$S_{x'}^{Y_H} = \frac{Z_{rp,sq}(x) - Z_{rp,sq}(k)}{Z_{rp,sq}(x)} - \frac{Z_{nn}(x) - Z_{nn}(k)}{Z_{nn}(x)}$$

(11.50)

or

$$S_{x'}^{Y_H} = \frac{1}{F_k(x)} - \frac{1}{\hat{F}_k(x)}$$

(11.51)

We remark that if $H = Y_{in}$, we obtain the same expression as (11.50), where the output variable is the input current at the input port instead of the output current at the output port. This is equivalent to saying that

$$S_{x'}^{Y_{in}} = \frac{1}{F_k(\text{input short-circuited})} - \frac{1}{F_k(\text{input open-circuited})}$$

(11.52)

If $H = A_V$, two cases are considered. In case 1, the element x of interest is not Z_L. Then

$$S_{x'}^{A_V} = \frac{1}{F_k(x)} - \frac{1}{\hat{F}_k(x)}$$

(11.53)

In case 2, the element x of interest is Z_L. Then

$$S_{x'}^{A_V} = 1 - \frac{k}{Z_L} + \frac{1}{F_k(Z_L)} - \frac{1}{\hat{F}_k(Z_L)}$$

(11.54)

Finally, if $H = A_I$, the result is

$$S_{x'}^{A_I} = \frac{1}{F_k(\text{input open-circuited})} - \frac{1}{F_k(\text{output open-circuited})} \tag{11.55}$$

When the reference value k assumes the zero value, (11.51)–(11.55) give the relationship between the sensitivities S_x^H and $F_k(x)$ and $\hat{F}_k(x)$.

11.1.6 Illustrative Examples

Consider the network of Fig. 11.6, the mesh currents of which are as indicated. Using the rules as outlined earlier, the indefinite-impedance matrix can be written down by inspection as

$$\mathbf{Z} = \begin{bmatrix} z_1 + z_2 & -z_2 & -z_1 \\ -z_2 & z_2 + z_L & -z_L \\ -z_1 & -z_L & z_1 + z_L \end{bmatrix} \tag{11.56}$$

Using the formulas derived in the proceeding sections, the desired network function are found to be

$$Y_H = \frac{I_o}{V_{in}} = \frac{Z_{12,33}}{Z_{33}} = \frac{z_2}{z_1 z_2 + z_1 z_L + z_2 z_L} \tag{11.57a}$$

$$Y_{in} = \frac{I_i}{V_{in}} = \frac{Z_{11,33}}{Z_{33}} = \frac{z_2 + z_L}{z_1 z_2 + z_1 z_L + z_2 z_L} \tag{11.57b}$$

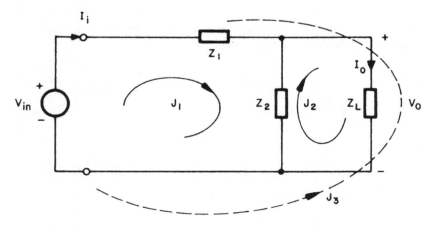

Figure 11.6 An network used for illustration.

Figure 11.7 Network used to calculate the return difference $F_k(Z_1)$.

$$A_V = \frac{V_o}{V_{in}} = Z_L \frac{Z_{12,33}}{Z_{33}} = \frac{z_2 z_L}{z_1 z_2 + z_1 z_L + z_2 z_L} \qquad (11.57c)$$

$$A_I = \frac{I_o}{I_i} = \frac{Z_{12,33}}{Z_{11,33}} = \frac{z_2}{z_2 + z_L} \qquad (11.57d)$$

To compute the general return difference, we use (11.18) and Fig. 11.7, and obtain

$$F_k(z_1) = \frac{Z_{uv}(z_1)}{Z_{uv}(k)} = \frac{Z_{33}(z_1)}{Z_{33}(k)} = (1 - I_c)]_{V_{in}=0}$$

$$= 1 + \frac{z_1'(z_2 + z_L)}{z_2 z_L + k(z_2 + z_L)} = \frac{z_2 z_L + z_1(z_2 + z_L)}{z_2 z_L + k(z_2 + z_L)} \qquad (11.58a)$$

where $z_1' = z_1 - k$. Likewise, from Fig. 11.8 with $z_2' = z_2 - k$ we have

$$F_k(z_2) = \frac{Z_{uv}(z_2)}{Z_{uv}(k)} = \frac{Z_{33}(z_2)}{Z_{33}(k)} = (1 - I_c)]_{V_{in}=0}$$

$$= 1 + \frac{z_2'}{k + \frac{z_1 z_L}{z_1 + z_L}} = \frac{z_1 z_L + z_2(z_1 + z_L)}{z_1 z_L + k(z_1 + z_L)} \qquad (11.58b)$$

Finally, by replacing z_2 and z_L by z_L and z_2, respectively, in (11.58b) we obtain

$$F_k(z_L) = \frac{Z_{33}(z_L)}{Z_{33}(k)} = \frac{z_1 z_2 + z_L(z_1 + z_2)}{z_1 z_2 + k(z_1 + z_2)} \qquad (11.59)$$

Figure 11.8 Network used to calculate the return difference $F_k(Z_2)$.

Figure 11.9 Network used to calculate the return difference $\hat{F}_k(z_1)$.

To compute the null return difference $\hat{F}_k(z_1)$, we refer to Fig. 11.9. If I_o is the output current, then

$$\hat{F}_k(z_1) = \frac{Z_{12,33}(z_1)}{Z_{12,33}(k)} = (1 - I_c)]_{I_o=0} = 1 \tag{11.60}$$

where V_{in} was adjusted so that $I_o = 0$. On the other hand, if I_i is the output variable, then

$$\hat{F}_k(z_1) = \frac{Z_{11,33}(z_1)}{Z_{11,33}(k)} = (1 - I_c)]_{I_i=0} = 1 \tag{11.61}$$

where V_{in} was adjusted so that $I_i = 0$. Likewise, to compute the null return difference $\hat{F}_k(z_2)$, we refer to Fig. 11.10. If I_o is the output current, then

$$\hat{F}_k(z_2) = \frac{Z_{12,33}(z_2)}{Z_{12,33}(k)} = (1 - I_c)_{I_o=0} = \frac{z_2}{k} \tag{11.62}$$

Figure 11.10 Network used to calculate the return difference $\hat{F}_k(z_2)$.

Figure 11.11 Network used to calculate the return difference $\hat{F}_k(z_L)$.

where

$$V_{in} = z_2' - \frac{z_2'(z_1 + k)}{k} \tag{11.63}$$

If I_i is the output variable, then

$$\hat{F}_k(z_2) = \frac{Z_{11,33}(z_2)}{Z_{11,33}(k)} = (1 - I_c)]_{I_i=0} = 1 + \frac{z_2'}{k + z_L} = \frac{z_2 + z_L}{k + z_L} \tag{11.64}$$

For $\hat{F}_k(z_L)$, we use Fig. 11.11 and obtain

$$\hat{F}_k(z_L) = \frac{Z_{12,33}(z_L)}{Z_{12,33}(k)} = (1 - I_c)]_{I_o=0} = 1 \tag{11.65}$$

$$\hat{F}_k(z_L) = \frac{Z_{11,33}(z_L)}{Z_{11,33}(k)} = (1 - I_c)]_{I_i=0}$$

$$= 1 + \frac{z_1'}{k + z_2} = \frac{z_L + z_2}{k + z_2} \tag{11.66}$$

We next verify the relations between network functions and the general return differences and null return differences, as follows:

$$
\begin{aligned}
\frac{Y_{in}(z_2)}{Y_{in}(k)} &= \frac{(z_2 + z_L)(z_1 k + k z_L + z_1 z_L)}{(k + z_L)(z_1 z_2 + z_2 z_L + z_1 z_L)} \\
&= \frac{F_k(\text{input open-circuited})}{F_k(\text{input short-circuited})}
\end{aligned}
\tag{11.67}
$$

$$
\frac{A_V(z_L)}{A_V(k)} = \frac{z_L}{k} \cdot \frac{z_1 z_2 + (z_1 + z_2)k}{z_1 z_2 + (z_1 + z_2)z_L} = \frac{z_L}{k} \cdot \frac{\hat{F}_k(z_L)}{F_k(z_L)}
\tag{11.68}
$$

$$
\frac{A_I(z_2)}{A_I(k)} = \frac{z_2(k + z_L)}{k(z_2 + z_L)} = \frac{F_k(\text{output open-circuited})}{F_k(\text{input open-circuited})}
\tag{11.69}
$$

Finally, we confirm the relationships between sensitivity functions and the general return differences and null return differences, as follows:

$$
\begin{aligned}
S_{z_L'}^{A_V} &= \frac{z_L'}{A_V} \cdot \frac{\partial A_V}{\partial z_L'} = 1 - \frac{k}{z_L} + \frac{1}{F_k(z_L)} - \frac{1}{\hat{F}_k(z_L)} \\
&= \frac{z_1 z_2 (z_L - k)}{z_L (z_1 z_2 + z_1 z_L + z_2 z_L)}
\end{aligned}
\tag{11.70}
$$

$$
\begin{aligned}
S_{z_1'}^{A_V} &= \frac{z_1'}{A_V} \cdot \frac{\partial A_V}{\partial z_1'} = \frac{1}{F_k(\text{input short-circuited})} \\
&\quad - \frac{1}{F_k(\text{output open-circuited})} = \frac{-z_1'(z_2 + z_L)}{z_1 z_2 + z_1 z_L + z_2 z_L}
\end{aligned}
\tag{11.71}
$$

$$
\begin{aligned}
S_{z_2'}^{A_I} &= \frac{z_2'}{A_I} \cdot \frac{\partial A_I}{\partial z_2'} = \frac{1}{F_k(\text{input open-circuited})} \\
&\quad - \frac{1}{F_k(\text{output open-circuited})} = \frac{z_L(z_2 - k)}{z_2(z_2 + z_L)}
\end{aligned}
\tag{11.72}
$$

11.2 EXTENSION TO NONPLANAR NETWORKS

In a planar network or graph, a mesh is a circuit formed by the boundary edges of a region. As indicated in the preceding section, we assume that all the interior meshes are oriented clockwise and the exterior mesh counterclockwise. Then each edge of the planar graph is traversed by exactly two meshes oriented in the opposite directions. This result is also found to be sufficient in characterizing a planar graph by MacLane (1937). We state MacLane's result as follows.

Theorem 11.1 A graph is planar if and only if every block of the graph with at least three nodes has a basis circuit set and one additional circuit such that every edge occurs in exactly two of these circuits.

A *loop* is an oriented circuit. A *loop set* is either a loop or an edge-disjoint union of loops. The question arises as to whether or not it is possible to assign a set of m loop sets to a nonplanar graph of nullity m and one additional loop set, so that the number of loop sets traversing in one direction through any edge equals the number of loop sets traversing in the opposite direction of the edge. Furthermore, any m of these $m + 1$ loop sets are linearly independent. The answer to this question is affirmative. In the present section, we give a constructive proof of this result. The proof itself serves as an algorithm for the enumeration of the desired loop sets. This result is based on the work of Chen and Lu (1991).

We first state the main result of this section as follows.

Theorem 11.2 It is possible to assign $m + 1$ loop sets to a nonplanar graph of nullity m, so that the number of loop sets traversing in one direction through any edge equals that traversing in the opposite direction of the edge. Furthermore, any m of these $m + 1$ loop sets are linearly independent.

PROOF. We prove this result by induction over the number of edges of a given nonseparable graph G of nullity m. The theorem is certainly true for a graph containing one or two edges because it is true for any planar graph. Assume that the assertion is true for any graph containing e edges $b_x (x = 1, 2, \ldots, e)$ or less and of nullity m. We show that it is also true for any graph G' containing $e + 1$ edges and of nullity m'.

To this end, let G be the graph of nullity m obtained from G' by the removal of an edge b_{p_0}. By induction hypothesis, $m + 1$ loop sets can be assigned to G so that the number of loop sets traversing in one direction through any edge equals that traversing in the opposite direction of the edge. Let

$$L_{p_1}, L_{p_2}, \ldots, L_{p_{m+1}} \tag{11.73}$$

be these $m + 1$ loop sets in G. Since G' is nonseparable, $m' = m + 1$. To complete the proof, we must construct $m + 2$ such loop set in G'.

Suppose that we insert the removed edge b_{p_0} back to G to yield the original graph G'. In G' let

$$L_{p_0} = b_{p_0} b_{p_1} \ldots b_{p_u} = (\alpha_0, \alpha_1)(\alpha_1, \alpha_2) \ldots (\alpha_u, \alpha_0) \tag{11.74}$$

be a newly chosen loop set containing the edge b_{p_0}, where

$$b_{p_v} = (\alpha_v, \alpha_{v+1}), \quad v = 0, 1, 2, \ldots, u \tag{11.75}$$

denotes the edge with endpoints α_v and α_{v+1}, and $\alpha_{u+1} = \alpha_0$. We now consider the loop L_{p0} to be the $(m+2)$th loop set in G'. For our purposes, we assume that this $(m+2)$th loop set travers the edge $b_{pV} = (\alpha_v, \alpha_{v+1})$ from α_v to α_{v+1}. Observe that the addition of this new loop set tilts the balance of the number of loop set at the edges b_{p_v} of L_{p0} in G' in that the number of loop sets traversing in one direction through an edge b_{p_v} is no longer equal to that traversing in the opposite direction through b_{p_v}. To rebalance them, we resort the following operations.

Let L_{p_v} be one of the loop sets traversing in the direction from α_v to α_{v+1} through the edge b_{p_v}, as shown in Fig. 11.12, where $L_{p_v} \neq L_{p0}$. To rebalance the loop sets at the edges $b_{p0}, b_{p1}, \ldots, b_{p_u}$, we replace the loop sets L_{p_v} by the loop set $L'_{p_v} = L_{p0} \oplus L_{p_v}$ or

$$L'_{p_v} = (\alpha_0, \alpha_u)(\alpha_u, \alpha_{u-1})\ldots(\alpha_{v+2}, \alpha_{v+1})(\alpha_{v+1}, \beta_1)$$
$$\ldots(\beta_t, \alpha_v)(\alpha_v, \alpha_{v-1})\ldots(\alpha_2, \alpha_1)(\alpha_1, \alpha_0) \qquad (11.76)$$

as depicted in Fig. 11.13, because the ring sum of two loop sets is a loop set, where

$$L_{p_v} = (\alpha_v, \alpha_{v+1})(\alpha_{v+1}, \beta_1)(\beta_1, \beta_2)\ldots(\beta_{t-1}, \beta_t)(\beta_t, \alpha_v) \qquad (11.77)$$

is shown in Fig. 11.12. Observe that each of the edges $b_{p0}, b_{p1}, \ldots, b_{p_u}$ of L_{p0} is now balanced in that the number of loop sets traversing in one direction through any edge equals that traversing in the opposite direction of the edge. Furthermore, with the exception of the edge b_{p_v}, the edges traversed by the loop set L_{p_v} are now traversed by the loop set L'_{p_v} in the same direction as that of L_{p_v}. The edge b_{p_v},

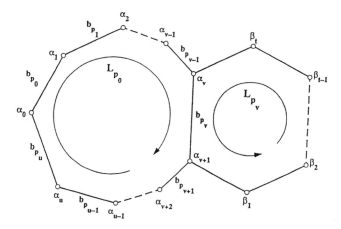

Figure 11.12 A subgraph of the graph G' showing the loop sets L_{p0} and L_{p_v}.

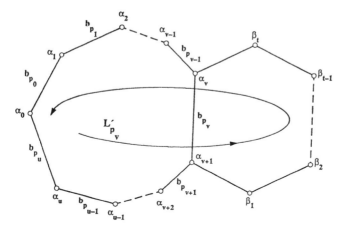

Figure 11.13 A subgraph of the graph G' showing the loop sets L'_{p_v}.

previously traversed by L_{p_v} from α_v to α_{v+1}, is now traversed by L_{p_0} from α_v to α_{v+1}. If there are other common edges between L_{p_0} and L_{p_v}, the same procedure applies.

Now consider the set of $(m+2)$ loop sets

$$L_{p_0}, L_{p_1}, L_{p_2}, \ldots, L_{p_{v-1}}, L'_{p_v}, L_{p_{v+1}}, \ldots, L_{p_m}, L_{p_{m+1}} \qquad (11.78)$$

of the graph G'. The number of these loop sets traversing in one direction through any edge of G' equals the number of these loop sets traversing in the opposite direction of the edge. It remains to be shown that any $m+1$ of these loop sets are linearly independent.

From induction hypothesis, any m of the $m+1$ loop sets $L_{p_1}, L_{p_2}, \ldots, L_{p_{m+1}}$ with L'_{p_v} replacing L_{p_v} are linearly independent. Since L_{p_0} is a loop set containing the edge b_{p_0} and since b_{p_0} is not contained in any of the loop sets $L_{p_1}, L_{p_2}, \ldots, L_{p_{m+1}}$ except in L'_{p_v}, it follows that any $m+1$ of the $m+2$ loop sets $L_{p_0}, L_{p_1}, L_{p_2}, \ldots, L_{p_{m+1}}$ of (11.78) are linearly independent, because there is at least one edge in L'_{p_v} that is not contained in L_{p_0}. Also from (11.74), (11.76) and (11.77), we see that the loop set L'_{p_v} is the ring sum of the loop sets L_{p_0} and L_{p_v}, or

$$L'_{p_v} = L_{p_0} \oplus L_{p_v} \qquad (11.79)$$

Thus, any $m+1$ of the $m+2$ loop sets of (11.78) are linearly independent. This completes the proof of the theorem.

11.3 EXTENSION OF A DUAL THEOREM ON THE SUMMATION OF RETURN DIFFERENCES

In Sec. 9.5 of Chapter 9, we presented theorems on return differences by showing that, based on the indefinite-admittance matrix approach, the sum of the reciprocals of the return differences with respect to all network elements equals the nullity of the network, and that, in terms of the indefinite-impedance matrix, the sum of the reciprocals of the return differences with respect to all network elements of a planar network equals the rank of the network. In this section, we extend these results on return differences for a planar network to any network, planar or nonplanar. Similar results on the summation of driving-point and transfer admittances are obtained for planar or nonplanar networks containing active and coupled elements. These results are based on the work of Wang and Chen (1991).

Let N be a network of e branches $b_x (x = 1, 2, \ldots, e)$, and of nullity m where each branch b_x has a nonzero impedance z_x. Since a one-port impedance or a transformer can be represented equivalently by current-controlled voltage sources, without loss of generality we may assume that all branches of N are current-controlled voltage sources

$$V_x = z_x \left(\sum_{k=1}^{w_j} i_{j_k} - \sum_{h=w_j+1}^{2w_j} i_{j_h} \right) \tag{11.80}$$

where $i_{j_k} (k = 1, 2, \ldots, w_j)$ are the loop currents traversing in one direction through b_j, $i_{j_h} (h = w_j + 1, w_j + 2, \ldots, 2w_j)$ are the loop currents traversing in the opposite direction through b_j, and w_j is a positive integer. Here, we assume that $m + 1$ loop currents are assigned in N, so that the number of loop currents traversing in one direction through a branch equals the number of loop currents traversing in the opposite direction through the branch, where any m of these loop currents are linearly independent. That this is always possible was shown in the preceding section. Figure 11.14 is an example of such an assignment. Another such example was given in Fig. 9.32 of Chapter 9.

The matrix $\mathbf{Z}$ relating the loop-current vector $\mathbf{J}$ and the loop voltage-source vector $\mathbf{E}$

$$\mathbf{ZJ} = \mathbf{E} \tag{11.81}$$

is the *indefinite-impedance matrix*, which is square and of order $m + 1$.

For $w_x = w_j = 1$, $V_x = z_x(i_{j_1} - i_{j_2})$ and the impedance z_x enters the indefinite-impedance matrix $\mathbf{Z}$ in a rectangular pattern with z_x appearing at the entries (x_1, j_1) and (x_2, j_2) and $-z_x$ at (x_1, j_2) and (x_2, j_1). For $w_x =$

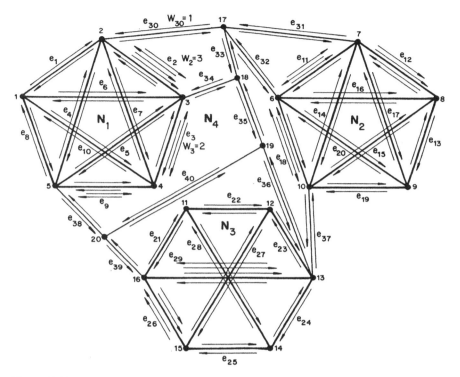

Figure 11.14 A nonplanar network showing the assignment of loop currents.

$w_j = 2$, $V_x = z_x(i_{j_1} + i_{j_2} - i_{j_3} - i_{j_4})$ and the impedance z_x enters the indefinite-impedance matrix **Z** in a rectangular pattern with z_x appearing at the entries $(x_1, j_1), (x_1, j_2), (x_2, j_1), (x_2, j_2), (x_3, j_3), (x_3, j_4), (x_4, j_3)$ and (x_4, j_4) and $-z_x$ at $(x_1, j_3), (x_1, j_4), (x_2, j_3), (x_2, j_4), (x_3, j_1), (x_3, j_2), (x_4, j_1)$ and (x_4, j_2). In general, when w_x and w_j are any positive integers with

$$V_x = z_x \left(\sum_{k=1}^{w_j} i_{j_k} - \sum_{h=w_j+1}^{2w_j} i_{j_h} \right) \qquad (11.82)$$

the impedance z_x enters the indefinite-impedance matrix **Z** in the form

$$z_x \begin{bmatrix} \mathbf{A}_{w_x w_j} & -\mathbf{A}_{w_x w_j} \\ -\mathbf{A}_{w_x w_j} & \mathbf{A}_{w_x w_j} \end{bmatrix} \qquad (11.83)$$

where $\mathbf{A}_{w_x w_j}$ is a square matrix of order $w_x \times w_j$ consisting only of 1's. Recall that branch b_x is traversed by loop currents $i_{x_k} (k = 1, 2, \ldots, w_x)$ in one direction and by loop currents $i_{x_h} (h = w_x + 1, w_x + 2, \ldots, 2w_x)$ in the opposite direction. Thus, for $j = x$ we obtain

$$V_x = z_x \left(\sum_{k=1}^{w_x} i_{x_k} - \sum_{h=w_x+1}^{2w_x} i_{x_h} \right) \tag{11.84}$$

Notice that z_x now becomes the one-port impedance of branch b_x.

Following (9.129) and (9.130), denote by Z_{ij} and $Z_{rp,sq}$ the first- and second-order cofactors of the elements of the indefinite-impedance matrix $\mathbf{Z}$, respectively. As before, it can be shown that $\mathbf{Z}$ is an equicofactor matrix:

$$Z_{ij} = Z_{uv} \tag{11.85}$$

for all i, j, u and v. Recall that the *return difference* $F(x)$ with respect to a network element x is defined as the ratio of the two functional values assumed by the loop determinant under the condition that the element x assumes its nominal value and the condition that the element x assumes the zero value:

$$F(x) = \frac{Z_{uv}(x)}{Z_{uv}(0)} \tag{11.86}$$

where again we write $\mathbf{Z} = \mathbf{Z}(x)$ to emphasize the importance of x, even though it is also a function of the complex-frequency variable s.

Theorem 11.3 Let N be a network of e branches and of rank r. Then

$$\sum_{x=1}^{e} \frac{1}{F(z_x)} = r \tag{11.87}$$

PROOF. Without loss of generality, assume that N is an n-node connected network of nullity m, the indefinite-impedance matrix of which is $\mathbf{Z}$. We expand the first-order cofactor $Z_{uv}(z_x)$ along each column (row) of $\mathbf{Z}_{uv}(z_x)$ and then add the resulting expansions. This gives

$$\sum_{x=1}^{e} z_x (Z_{x_1 j_1, uv} + Z_{x_2 j_2, uv} - Z_{x_1 j_2, uv} - Z_{x_2 j_1, uv}) = m Z_{uv} \tag{11.88}$$

for $V_x = z_x(i_{j_1} - i_{j_2})$, and

$$\sum_{x=1}^{e} z_x (Z_{x_1 j_1,uv} + Z_{x_1 j_2,uv} + Z_{x_2 j_1,uv} + Z_{x_2 j_2,uv} + Z_{x_3 j_3,uv}$$

$$+ Z_{x_3 j_4,uv} + Z_{x_4 j_3,uv} + Z_{x_4 j_4,uv} - Z_{x_1 j_3,uv} - Z_{x_1 j_4,uv} - Z_{x_2 j_3,uv}$$

$$- Z_{x_2 j_4,uv} - Z_{x_3 j_1,uv} - Z_{x_3 j_2,uv} - Z_{x_4 j_1,uv} - Z_{x_4 j_2,uv})$$

$$= \sum_{x=1}^{e} z_x [(Z_{x_1 j_1,uv} + Z_{x_3 j_3,uv} - Z_{x_1 j_3,uv} - Z_{x_3 j_1,uv})$$

$$+ (Z_{x_2 j_2,uv} + Z_{x_4 j_4,uv} - Z_{x_2 j_4,uv} - Z_{x_4 j_2,uv})$$

$$+ (Z_{x_1 j_2,uv} + Z_{x_3 j_4,uv} - Z_{x_1 j_4,uv} - Z_{x_3 j_2,uv})$$

$$+ (Z_{x_2 j_1,uv} + Z_{x_4 j_3,uv} - Z_{x_2 j_3,uv} - Z_{x_4 j_1,uv})]$$

$$= m Z_{uv} \tag{11.89}$$

for $V_x = z_x(i_{j_1} + i_{j_2} - i_{j_3} - i_{j_4})$. From (4.148) we have

$$Z_{rp,sq} = Z_{rp,uv} + Z_{sq,uv} - Z_{rq,uv} - Z_{sp,uv} \tag{11.90}$$

for all $r, p, s, q, u, v \leq m + 1$. In general, we can write for (11.80) the relation

$$\sum_{x=1}^{e} z_x \left(\sum_{p,q=1}^{w_x} Z_{x_p j_q, x_{(p+w_x)} j_{(q+w_x)}} \right) = m Z_{uv} \tag{11.91}$$

Also, Z_{uv} can be expanded as

$$Z_{uv} = Z_{uv}(z_x) = Z_{uv}(0) + z_x \sum_{p,q=1}^{w_x} Z_{x_p j_q, x_{(p+w_x)} j_{(q+w_x)}} \tag{11.92}$$

or

$$\sum_{x=1}^{e} Z_{uv}(z_x) - \sum_{x=1}^{e} Z_{uv}(0) = \sum_{x=1}^{e} z_x \sum_{p,q=1}^{w_x} Z_{x_p j_q, x_{(p+w_x)} j_{(q+w_x)}} \tag{11.93}$$

Combining this with (11.91), we obtain

$$e Z_{uv}(z_x) - \sum_{x=1}^{e} Z_{uv}(0) = m Z_{uv}(z_x) \tag{11.94}$$

Dividing both side by $Z_{uv}(z_x)$ gives

$$\sum_{x=1}^{e} \frac{Z_{uv}(0)}{Z_{uv}(z_x)} = e - m = r \tag{11.95}$$

Equation (11.87) follows. This completes the proof of the theorem.

Assume that N is excited by a voltage source V_{in}, the output current of which is I_o. If V_{in} is traversed by the loop currents i_{x_1} and i_{x_2} and by I_o by i_{j_1} and i_{j_2} as shown in Fig. 11.15, the transfer admittance becomes

$$y_{x_1 j_1, x_2 j_2} \equiv \frac{I_o}{V_{in}} = \frac{Z_{x_1 j_1, x_2 j_2}}{Z_{uv}} \tag{11.96}$$

Thus, if V_{in} is traversed by the loop currents $i_{x_1}, i_{x_2}, i_{x_3}$ and i_{x_4} and by I_o by $i_{j_1}, i_{j_2}, i_{j_3}$ and i_{j_4} as shown in Fig. 11.16, the transfer admittance becomes

$$y_{x_1 j_1, x_2 j_2, x_3 j_3, x_4 j_4} \equiv \frac{I_o}{V_{in}} = \frac{Z_{x_1 j_1, x_3 j_3} + Z_{x_2 j_2, x_4 j_4} + Z_{x_1 j_2, x_3 j_4} + Z_{x_2 j_1, x_4 j_3}}{Z_{uv}}$$

$$\tag{11.97}$$

Figure 11.15 The network used to define transfer admittance $y_{x_1 j_1, x_2 j_2}$.

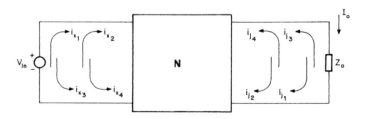

Figure 11.16 The network used to define transfer admittance $y_{x_1 j_1, x_2 j_2, x_3 j_3, x_4 j_4}$.

In general, for $w_x = w_j$ the transfer admittance is given by

$$y_{x_1 j_1, x_2 j_2, \ldots, x_{2w_x} j_{2w_x}} \equiv \frac{I_o}{V_{in}} = \frac{\sum_{p,q=1}^{w_x} Z_{x_p j_q, x_{(p+w_x)} j_{(q+w_x)}}}{Z_{uv}} \tag{11.98}$$

Substituting this in (11.91) yields the following theorem.

Theorem 11.4 If the nullity of an e-edge network is m, then

$$\sum_{x=1}^{e} z_x y_{x_1 j_1, x_2 j_2, \ldots, x_{2w_x} j_{2w_x}} = m \tag{11.99}$$

In particular, when $x = j$, $y_{x_1 j_1, x_2 j_2, \ldots, x_{2w_x} j_{2w_x}}$ denotes the driving-point admittance of N and Theorem 11.4 becomes the following corollary.

Corollary 11.1 For an e-branch RLC network of nullity m with no coupling among the elements,

$$\sum_{x=1}^{e} z_x y_{x_1 x_1, x_2 x_2, \ldots, x_{2w_x} x_{2w_x}} = m \tag{11.100}$$

An illustration of the above results was given in Chapter 9 as Example 9.17, where it was used to demonstrate without justification that Theorems 9.11 and 9.12 remain valid for nonplanar networks. The nonplanar graph associated with the nonplanar network of Fig. 9.32 is shown in Fig. 11.17.

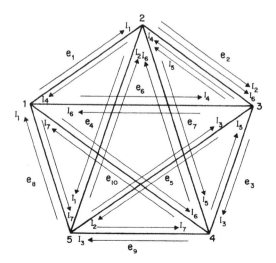

Figure 11.17 The nonplanar graph associated with the nonplanar network of Fig. 9.32.

11.4 DUAL TOPOLOGICAL THEOREMS OF LINEAR ACTIVE NETWORKS

In this section, we show that, in a linear active network with current-controlled voltage sources (CCVS), the sum of the products of the impedances of the tree branches and their associated transfer admittances plus the sum of the products of the controlling parameters of CCVS and their associated transfer admittances equals the sum of the products of the admittances of the cotree links and their associated transfer impedances. In a linear active network with voltage-controlled current sources (VCCS), the sum of the products of the admittances of cotree links and their associated transfer impedances plus the sum of the products of the controlling parameters of VCCS and their associated transfer impedances equals the sum of the products of the impedances of tree branches and their associated transfer admittances. These dual topological theorems and their corollaries bring to light important topological relations in linear active networks.

11.4.1 Preliminary Considerations and Definitions

Referring to Fig. 11.18, let N be a network of e branches $b_x (x = 1, 2, \ldots, e)$ and of nullity m, where each branch b_x has a nonzero impedance z_x, where $i_{j_k} (k = 1, 2, \ldots, w_j)$ are the loop currents traversing in one direction through b_j, and $i_{j_h} (h = w_j + 1, w_j + 2, \ldots, 2w_j)$ are the loop currents traversing in the opposite direction through b_j, where w_j is a positive integer. We assume that $m + 1$ loop currents are assigned in N, so that the number of loop currents traversing in one direction through a branch equals the number of loop currents traversing in the opposite direction through the branch, where any m of these loop currents are linearly independent. That this is always possible was shown in the Sec. 11.2. Likewise, let $i_{x_k} (k = 1, 2, \ldots, w_x)$ be the loop currents traversing in one direction through b_x, and $i_{x_h} (h = w_x + 1, w_x + 2, \ldots, 2w_x)$ be the loop currents traversing in the opposite direction through b_x, where w_x is a positive integer. As

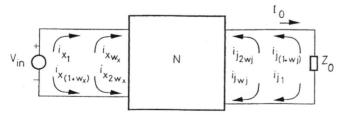

Figure 11.18 Network used to define transfer admittance T_{xj}^{y}.

shown in (11.98), the transfer admittance, now rewritten as T_{xj}^y, can be expressed compactly as

$$T_{xj}^y \equiv y_{x_1 j_1, x_2 j_2, \dots, x_{2w_x} j_{2w_x}} \equiv \frac{I_o}{V_{in}} = \frac{\sum_{p,q=1}^{w_x} Z_{x_p j_q, x_{(p+w_x)} j_{(q+w_x)}}}{Z_{uv}} \qquad (11.101)$$

From Theorem 11.4, we have

$$\sum_{x=1}^{e} z_x T_{xj}^y = m \qquad (11.102)$$

In particular, when $x = j$, T_{xx}^y denotes the driving-point admittance of N, as shown in Fig. 11.19. For an RLC network with no coupling among the elements, the above result becomes

$$\sum_{x=1}^{e} z_x T_{xx}^y = m \qquad (11.103)$$

Now, referring to Fig. 11.20, assume that a current source is connected between any two nodes x_1 and x_2, so that a current I_s is injected into the x_1 node and at the

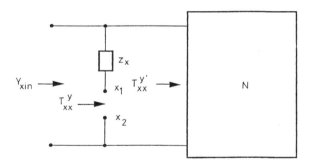

Figure 11.19 Network used to define transfer admittance T_{xx}^y.

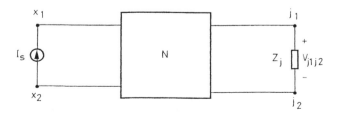

Figure 11.20 Network used to define transfer impedance T_{xj}^z.

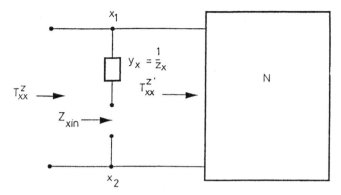

Figure 11.21 Network used to define driving-point impedance T_{xx}^z.

same time is extracted from the x_2 node. The transfer impedance denoted by T_{xj}^z between the node pairs x_1, x_2 and j_1, j_2 is defined as in (2.89) by the relation

$$-T_{xj}^z \equiv z_{x_1j_1,x_2j_2} \equiv \frac{V_{j_1j_2}}{I_s} \tag{11.104}$$

According to (2.94), (11.104) can be compactly expressed in terms of the first-and second-order cofactors of the elements of the indefinite-admittance matrix $\mathbf{Y}$ of N as

$$-T_{xj}^z \equiv z_{x_1j_1,x_2j_2} = \frac{Y_{x_1j_1,x_2j_2}}{Y_{uv}} \tag{11.105}$$

Appealing to Theorem 9.10 or (9.115) of Chapter 9, we have

$$\sum_{x=1}^{e} y_x T_{xj}^z = r \tag{11.106}$$

where $y_x = 1/z_x$. In particular, when $x_1 = j_1$ and $x_2 = j_2$, T_{xx}^z denotes the driving-point impedance of N across the nodes x_1 and x_2, as shown in Fig. 11.21. For an RLC network with no coupling among the elements, (11.106) becomes

$$\sum_{x=1}^{e} y_x T_{xx}^z = r \tag{11.107}$$

11.4.2 Topological Theorems

As shown in Fig. 11.19, Y_{xin} is the input admittance looking into the two input terminals. Likewise, Z_{xin} is the input impedance looking into the two input terminals as indicated in Fig. 11.21. From these two networks, we obtain

$$T_{xx}^z = \frac{1}{Y_{xin}} = \frac{1}{y_x + T_{xx}^{y'}} \tag{11.108}$$

$$T_{xx}^y = \frac{1}{Z_{xin}} = \frac{1}{z_x + T_{xx}^{z'}} \tag{11.109}$$

yielding

$$T_{xx}^y = y_x(1 - y_x T_{xx}^z) \tag{11.110}$$

$$T_{xx}^z = z_x(1 - z_x T_{xx}^y) \tag{11.111}$$

Theorem 11.5 Let N be a network of e branches and of nullity m, where branch e_x is a passive branch with nonzero impedance $z_x = 1/y_x\,(x = 1, 2, \ldots, b)$, and let e_α be the controlled branch of a CCVS, the controlling parameter of which is $\alpha_p\,(p = 1 + b, 2 + b, \ldots, e)$. Then for a tree t containing all the CCVS

$$\sum_{(t)} z_x T_{xx}^y + \sum_{(e_\alpha)} \alpha_p T_{xj}^y = \sum_{(\bar{t})} y_x T_{xx}^z \tag{11.112}$$

where the summations $\Sigma_{(t)}$, $\Sigma_{(\bar{t})}$ and $\Sigma_{(e_\alpha)}$ are taken over all the passive tree branches of a tree t, all the passive cotree links of the cotree $\bar{t}$ of the tree t, and all the controlled branches e_α, respectively.

Proof. According to (11.102), we have

$$\sum_{x=1}^{e} z_x T_{xj}^y = \sum_{(t)} z_x T_{xj}^y + \sum_{(\bar{t})} z_x T_{xj}^y + \sum_{(e_\alpha)} \alpha_p T_{xj}^y = m \tag{11.113}$$

Since the summation $\Sigma_{(t)}$ is taken over all the passive tree branches of a tree t and since there are no active cotree links of the cotree $\bar{t}$ of the tree t, (11.113) becomes

$$\sum_{(t)} z_x T_{xx}^y + \sum_{(\bar{t})} z_x T_{xx}^y + \sum_{(e_\alpha)} \alpha_p T_{xj}^y = m \tag{11.114}$$

By rearranging the terms of the last equality yields

$$\sum_{(t)} z_x T_{xx}^y + \sum_{(e_\alpha)} \alpha_p T_{xj}^y = \sum_{(\bar{t})} (1 - z_x T_{xx}^y) \tag{11.115}$$

Combining (11.111) and (11.115) completes the proof of the theorem.

It was shown by Chen and Chan (1974) that if a linear active network is uniquely solvable, all the voltage sources or dependent and independent voltage sources can be made tree branches of a tree t. Therefore in a network with CCVS the sum of the products of the passive impedances of tree branches and their driving-point admittance plus the sum of the products of the controlling parameters of the CCVS and their associated transfer admittance equals the sum of the products of the admittances of the cotree links and their driving-point impedances.

Theorem 11.6 Let N be a network of e branches and of rank r, where branch e_x is a passive branch with nonzero impedance $z_x = 1/y_x (x = 1, 2, \ldots, d)$, and let e_β be the controlled branch of a VCCS, the controlling parameter of which is $\beta_q (q = 1 + d, 2 + d, \ldots, e)$. Then for a cotree $\bar{t}$ containing all the VCCS

$$\sum_{(\bar{t})} y_x T_{xx}^z + \sum_{(e_\beta)} \beta_q T_{xj}^z = \sum_{(t)} z_x T_{xx}^y \tag{11.116}$$

where the summations $\Sigma_{(t)}$, $\Sigma_{(\bar{t})}$ and $\Sigma_{(e_\beta)}$ are taken over all the passive tree branches of a tree t, all the passive cotree links of the cotree $\bar{t}$ of the tree t, and all the controlled branches e_β, respectively.

PROOF. According to (11.106), we have

$$\sum_{x=1}^{e} y_x T_{xj}^z = \sum_{(t)} y_x T_{xj}^z + \sum_{(\bar{t})} y_x T_{xj}^z + \sum_{(e_\beta)} \beta_q T_{xj}^z = r \tag{11.117}$$

Since no active elements are included in $\Sigma_{(t)}$ and since the summation $\Sigma_{(\bar{t})}$ is take over all the passive cotree links of the cotree $\bar{t}$ of the tree t, (11.117) becomes

$$\sum_{(t)} y_x T_{xx}^z + \sum_{(\bar{t})} y_x T_{xx}^z + \sum_{(e_\beta)} \beta_q T_{xj}^z = r \tag{11.118}$$

By rearranging the terms of the last equality yields

$$\sum_{(\bar{t})} y_x T_{xx}^z + \sum_{(e_\beta)} \beta_q T_{xj}^z = \sum_{(t)} (1 - y_x T_{xx}^z) \tag{11.119}$$

Combining this with (11.110), (11.116) follows.

Likewise, if a linear active network is uniquely solvable, all the controlled current sources or dependent and independent current sources can be made cotree links of a cotree $\bar{t}$. Therefore in a network with VCCS the sum of the products of the admittances of cotree links and their driving-point impedances plus the sum of the products of the controlling parameters of the VCCS and their associated transfer impedances equals the sum of the products of the impedances of the tree branches and their driving-point admittances.

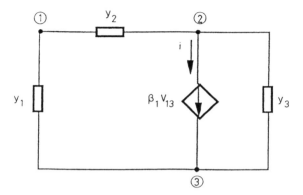

Figure 11.22 Network used to illustrate the theorems.

Corollary 11.2 For an e-branch RLC network of nullity m and of rank r with no coupling among the elements,

$$\sum_{(t)} z_x T_{xx}^y = \sum_{(\bar{t})} y_x T_{xx}^z \qquad (11.120)$$

Example 11.1 An active 3-node and 4-branch network is shown in Fig. 11.22. We use this network to verify (11.116) of Theorem 11.6. The indefinite-admittance matrix $\mathbf{Y}$ is found to be

$$\mathbf{Y} = \begin{bmatrix} y_1 + y_2 & -y_2 & -y_1 \\ -y_2 + \beta_1 & y_2 + y_3 & -y_3 - \beta_1 \\ -y_1 - \beta_1 & -y_3 & y_1 + y_3 + \beta_1 \end{bmatrix} \qquad (11.121)$$

from which

$$T_{xx}^{z_1} = \frac{Y_{11,33}}{Y_{33}} = \frac{y_2 + y_3}{y_1 y_2 + y_2 y_3 + y_1 y_3 + y_2 \beta_1} \qquad (11.122a)$$

$$T_{xx}^{z_2} = \frac{Y_{11,22}}{Y_{33}} = \frac{y_1 + y_3 + \beta_1}{y_1 y_2 + y_2 y_3 + y_1 y_3 + y_2 \beta_1} \qquad (11.122b)$$

$$T_{xx}^{z_3} = \frac{Y_{22,33}}{Y_{33}} = \frac{y_1 + y_2}{y_1 y_2 + y_2 y_3 + y_1 y_3 + y_2 \beta_1} \qquad (11.122c)$$

$$T_{xj}^{z\beta_1} = -\frac{Y_{23,31}}{Y_{33}} = \frac{y_2}{y_1 y_2 + y_2 y_3 + y_1 y_3 + y_2 \beta_1} \qquad (11.122d)$$

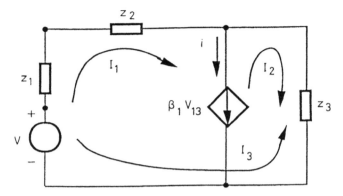

Figure 11.23 Network used to compute the driving-point admittance $T_{xx}^{y_1}$.

From (11.110) and (11.111) and Kirchhoff's current law, we can compute the driving-point admittances of the passive branches. For $T_{xx}^{y_1}$ we connect a voltage source V in series with the z_1 branch as shown in Fig. 11.23. The loop equations become

$$z_1 (I_1 - I_3) + z_2 (I_1 - I_3) + z_3 (I_2 - I_3) = V \qquad (11.123a)$$

$$i = \beta_1 V_{13} = (I_1 - I_2) = \beta_1 [V - z_1 (I_1 - I_3)] \qquad (11.123b)$$

$$(I_1 - I_3) = (I_1 - I_2) + (I_2 - I_3) \qquad (11.123c)$$

From (11.123), we obtain

$$z_1 (I_1 - I_3) + z_2 (I_1 - I_3) + z_3 [(I_1 - I_3) - (I_1 - I_2)] = V \qquad (11.124a)$$

$$(I_1 - I_3) (z_1 + z_2 + z_3) + z_3 [\beta_1 z_1 (I_1 - I_3) - \beta_1 V] = V \qquad (11.124b)$$

$$T_{xx}^{y_1} = \frac{I_1 - I_3}{V} = \frac{1 + \beta_1 z_3}{z_1 + z_2 + z_3 + z_1 z_3 \beta_1} \qquad (11.125a)$$

$$T_{xx}^{y_2} = \frac{1}{z_1 + z_2 + z_3 + z_1 z_3 \beta_1} \qquad (11.125b)$$

$$T_{xx}^{y_3} = \frac{1 + \beta_1 z_1}{z_1 + z_2 + z_3 + z_1 z_3 \beta_1} \qquad (11.125c)$$

yielding

$$y_1 T_{xx}^{z_1} + \beta_1 T_{xj}^{z\beta_1} = z_2 T_{xx}^{y_2} + z_3 T_{xx}^{y_3} \qquad (11.126a)$$

$$y_2 T_{xx}^{z_2} + \beta_1 T_{xj}^{z\beta_1} = z_1 T_{xx}^{y_1} + z_3 T_{xx}^{y_3} \qquad (11.126b)$$

$$y_3 T_{xx}^{z_3} + \beta_1 T_{xj}^{z\beta_1} = z_1 T_{xx}^{y_1} + z_2 T_{xx}^{y_2} \qquad (11.126c)$$

11.5 LOOP-IMPEDANCE MATRIX FORMULATION

In the foregoing, we used the indefinite-impedance matrix to compute the network functions, return difference and null return difference. In this section, we show that loop-impedance matrix can also be used for these purposes. The material is based on the work of Lan and Chen (1985).

11.5.1 Preliminary Considerations and Definition

Let G be a directed graph of n nodes, b branches and of nullity m and let its basis circuit matrix be $\mathbf{B} = [b_{ij}]$. For an arbitrary given matrix $\mathbf{W}$ of order b, the scalar quantity

$$M_{uv}(\mathbf{B}) = \sum_{i=1}^{n} \sum_{j=1}^{n} (-1)^{i+j} b_{iu} b_{jv} \det \mathbf{B}_{-i} \mathbf{W} \mathbf{B}'_{-j} \qquad (11.127)$$

as given in (10.61a) of Definition 10.5 of Chapter 10, is called the generalized cofactor of the elements of the matrix $\mathbf{BWB}'$. As mentioned before, the generalized cofactor, in general, is not invariant with respect to the choice of the circuit matrix $\mathbf{B}$, but it is invariant when $\mathbf{B}$ is the fundamental circuit matrix $\mathbf{B}_f$. If $\mathbf{W} = \mathbf{Z}_b = [z_{ij}]$ is the branch-impedance matrix of a network N, the matrix triple produce $\mathbf{BZ}_b\mathbf{B}' = \mathbf{Z}_m$ is called the *loop-impedance matrix* of N.

As in the case of the indefinite-admittance matrix or the indefinite-impedance matrix, we can also define the *second-order generalized cofactor* of the elements of the loop-impedance matrix $\mathbf{BZ}_b\mathbf{B}' = \mathbf{Z}_m$ similarly as

$$M_{rs,uv}(\mathbf{B}) = \sum_{i=1}^{m} \sum_{j=1}^{m} \sum_{k=1}^{m} \sum_{l=1}^{m} (-1)^{i+j+k+l} \mathrm{sgn}(i-k)\mathrm{sgn}(j-l)$$

$$\times b_{ir} b_{js} b_{ku} b_{lv} \det \mathbf{B}_{-ik} \mathbf{W} \mathbf{B}'_{-jl} \qquad (11.128)$$

for $r, s, u, v = 1, 2, \ldots, b$. If the edges e_r, e_s, e_u and e_v of G corresponding to the columns r, s, u and v in $\mathbf{B}$ are contained only in the circuits I_k, I_h, I_p and I_q

corresponding to the rows k, h, p and q in $\mathbf{B}$ and oriented in the same directions as the circuits, then

$$M_{rs,uv}(\mathbf{B}) = (-1)^{k+h+p+q}\operatorname{sgn}(k-p)\operatorname{sgn}(h-q)\det \mathbf{B}_{-kp}\mathbf{W}\mathbf{B}'_{-hq} \quad (11.129)$$

becomes the second-order cofactor of the (k, h)-element and the (p, q)-element of $\mathbf{Z}_m$.

We now proceed to establish the following useful identity:

$$M_{rs}(\mathbf{B})M_{uv}(\mathbf{B}) - M_{rv}(\mathbf{B})M_{us}(\mathbf{B}) = (\det \mathbf{B}\mathbf{Z}_b\mathbf{B}')M_{rs,uv}(\mathbf{B}) \quad (11.130)$$

PROOF. Let $\mathbf{B}$ be a fundamental circuit matrix. From (11.127), we obtain

$$M_{rs}(\mathbf{B}) = \sum_{i=1}^{m}\sum_{j=1}^{m}(-1)^{i+j} b_{ir}b_{js}\det \mathbf{B}_{-i}\mathbf{Z}_b\mathbf{B}'_{-j} \quad (11.131a)$$

$$M_{uv}(\mathbf{B}) = \sum_{i=1}^{m}\sum_{j=1}^{m}(-1)^{i+j} b_{iu}b_{jv}\det \mathbf{B}_{-i}\mathbf{Z}_b\mathbf{B}'_{-j} \quad (11.131b)$$

$$M_{rv}(\mathbf{B}) = \sum_{i=1}^{m}\sum_{j=1}^{m}(-1)^{i+j} b_{ir}b_{jv}\det \mathbf{B}_{-i}\mathbf{Z}_b\mathbf{B}'_{-j} \quad (11.131c)$$

$$M_{us}(\mathbf{B}) = \sum_{i=1}^{m}\sum_{j=1}^{m}(-1)^{i+j} b_{iu}b_{js}\det \mathbf{B}_{-i}\mathbf{Z}_b\mathbf{B}'_{-j} \quad (11.131d)$$

Next, compute the products

$$M_{rs}(\mathbf{B})M_{uv}(\mathbf{B}) = \sum_{i=1}^{m}\sum_{j=1}^{m}\sum_{k=1}^{m}\sum_{l=1}^{m} b_{ir}b_{js}b_{ku}b_{lv}\Delta_{ij}\Delta_{kl} \quad (11.132a)$$

$$M_{rv}(\mathbf{B})M_{us}(\mathbf{B}) = \sum_{i=1}^{m}\sum_{j=1}^{m}\sum_{k=1}^{m}\sum_{l=1}^{m} b_{ir}b_{js}b_{ku}b_{lv}\Delta_{il}\Delta_{kj} \quad (11.132b)$$

where

$$\Delta_{ij} (-1)^{i+j}\det \mathbf{B}_{-i}\mathbf{Z}_b\mathbf{B}'_{-j} \quad (11.133)$$

and their difference

$$M_{rs}(\mathbf{B})M_{uv}(\mathbf{B}) - M_{rv}(\mathbf{B})M_{us}(\mathbf{B})$$

$$= \sum_{i=1}^{m}\sum_{j=1}^{m}\sum_{k=1}^{m}\sum_{l=1}^{m} b_{ir}b_{js}b_{ku}b_{lv}(\Delta_{ij}\Delta_{kl} - \Delta_{il}\Delta_{kj}) \quad (11.134)$$

Refer to the well-known identity as given in (4.145) of Chapter 4

$$\Delta \Delta_{ab,cd} = \Delta_{ab}\Delta_{cd} - \Delta_{ad}\Delta_{cb} \tag{11.135}$$

where $\Delta_{ab,cd}$ is the abth cofactor of Δ_{cd}, which in turn is the cdth cofactor of the determinant Δ. In our case, $\Delta = \det \mathbf{Z}_m$, Δ_{ij} is the cofactor of the (i, j)-element of $\mathbf{Z}_m$, and $\Delta_{ab,cd}$ denotes the determinant of the submatrix obtained from $\mathbf{Z}_m$ by deleting rows a and c and columns b and d and prefixing with the sign.

$$(-1)^{a+b+c+d}\mathrm{sgn}(a - c)\mathrm{sgn}(b - d) \tag{11.136}$$

Appealing to this identity (11.135), (11.134) becomes

$$M_{rs}(\mathbf{B})M_{uv}(\mathbf{B}) - M_{rv}(\mathbf{B})M_{us}(\mathbf{B})$$

$$= \Delta \sum_{i=1}^{m}\sum_{j=1}^{m}\sum_{k=1}^{m}\sum_{l=1}^{m} b_{ir}b_{js}b_{ku}b_{lv}\,(-1)^{i+j+k+l}$$

$$\times \mathrm{sgn}(i - k)\mathrm{sgn}(j - l)\det \mathbf{B}_{-ik}\mathbf{W}\mathbf{B}'_{-jl}$$

$$= \Delta M_{rs,uv}(\mathbf{B}) = \left(\det \mathbf{B}\mathbf{Z}_b\mathbf{B}'\right)M_{rs,uv}(\mathbf{B})$$

$$= (\det \mathbf{Z}_m)\,M_{rs,uv}(\mathbf{B}) \tag{11.137}$$

This completes the proof of (11.130).

From (11.130), we have

$$M_{rs,uv}(\mathbf{B}) = \frac{M_{rs}(\mathbf{B})M_{uv}(\mathbf{B}) - M_{rv}(\mathbf{B})M_{us}(\mathbf{B})}{\det \mathbf{Z}_m} \tag{11.138}$$

Since $\mathbf{B}$ is a fundamental circuit matrix, the determinant of $\mathbf{Z}_m$ and the generalized cofactors of the elements of $\mathbf{Z}_m$ are invariant with respect to the transformation from one such system to another. Thus, from (11.138) we know that the second-order generalized cofactor of the elements of the matrix $\mathbf{B}\mathbf{Z}_b\mathbf{B}'$ is invariant with respect to the same transformation.

11.5.2   The Calculation of Network Functions

Consider a linear time-invariant RLCM network N containing the current controlled voltage sources as shown in Fig. 11.24. The network N whose network graph is G is excited by the voltage source V_u corresponding to the edge e_u and terminated at the output by the impedance z_v corresponding to the edge e_v.

respectively. Write $\mathbf{Z}_m = \mathbf{Z}_m(r_m)$ and $M_{uv}(\mathbf{B}) = M_{uv}(\mathbf{B}, r_m)$ to exhibit explicitly their dependence on r_m. Then from (11.144), we get

$$I = -r_m \frac{M_{uv}(\mathbf{B},0)}{\det \mathbf{Z}_m(0)} \qquad (11.149)$$

giving

$$1 - I = \frac{\det \mathbf{Z}_m(0) + r_m M_{uv}(\mathbf{B},0)}{\det \mathbf{Z}_m(0)} = \frac{\det \mathbf{Z}_m(r_m)}{\det \mathbf{Z}_m(0)} \qquad (11.150)$$

after appealing to the following relation

$$\det \mathbf{Z}_m(r_m) = \det \mathbf{Z}_m(0) + r_m M_{uv}(\mathbf{B}, 0) \qquad (11.151)$$

a proof of which is given in the Appendix at the end of Sec. 11.5. Thus, we have

$$F(r_m) = \frac{\det \mathbf{Z}_m(r_m)}{\det \mathbf{Z}_m(0)} = 1 - I \qquad (11.152)$$

showing that the return difference with respect to the controlling parameter r_m is one minus the returned current through the controlling branch of the controlled source when the controlled voltage source is replaced by the independent voltage source of r_m volts and the input voltage is set to zero. When x represents a one-port impedance of branch p, the return difference $F(x)$ with respect to x becomes

$$F(x) = \frac{\det \mathbf{Z}_m(x)}{\det \mathbf{Z}_m(0)} = \frac{\det \mathbf{Z}_m(0) + x M_{pp}(\mathbf{B},0)}{\det \mathbf{Z}_m(0)} = 1 + \frac{x M_{pp}(\mathbf{B},0)}{\det \mathbf{Z}_m(0)}$$

$$= 1 + x T_{pp}^y(0) = 1 + \frac{x}{T_{pp}^z(0)} = \frac{x + T_{pp}^z(0)}{T_{pp}^z(0)} = \frac{T_{pp}^z(x)}{T_{pp}^z(0)} \qquad (11.153)$$

where $M_{pp}(\mathbf{B}) \equiv M_{pp}(\mathbf{B},x)$. In other words, the return difference $F(x)$ with respect to the one-port impedance x is equal to the ratio of total impedance between the two nodes obtained by opening the one end of the branch p corresponding to impedance x when x assumes its normal value to the impedance when x assumes the value zero.

As in (11.18), the loop-based null return difference $\hat{F}_k^m(x)$ of a feedback amplifier with respect to an element x for the reference value k can be defined as

$$\hat{F}_k^m(x) = \frac{M_{rs}(\mathbf{B},x)}{M_{rs}(\mathbf{B},k)} \qquad (11.154)$$

where e_r and e_s are the input and output edges of the feedback amplifier, respectively. When $k = 0$, $\hat{F}_0^m(x) \equiv \hat{F}^m(x)$ becomes the usual null return difference.

Refer to Fig. 11.25. When $x = r_m$, we show that, like the return difference, $\hat{F}^m(r_m)$ can be interpreted as one minus the return current I through the controlling branch when the controlled voltage source is replaced by an independent voltage source of r_m volts and when the input voltage V_r is adjusted so that its output current I_s is identically zero.

Applying the superposition principle and (11.144), we obtain

$$I_s = \frac{-M_{us}(\mathbf{B}, 0)}{\det \mathbf{Z}_m(0)} r_m + \frac{-M_{rs}(\mathbf{B}, 0)}{\det \mathbf{Z}_m(0)} V_r = 0 \tag{11.155}$$

giving

$$V_r = -\frac{M_{us}(\mathbf{B}, 0)}{M_{rs}(\mathbf{B}, 0)} r_m \tag{11.156}$$

and the returned current

$$\begin{aligned} I &= \frac{-M_{uv}(\mathbf{B}, 0)}{\det \mathbf{Z}_m(0)} r_m + \frac{-M_{rv}(\mathbf{B}, 0)}{\det \mathbf{Z}_m(0)} V_r \\ &= \frac{-r_m [M_{rs}(\mathbf{B}, 0) M_{uv}(\mathbf{B}, 0) - M_{rv}(\mathbf{B}, 0) M_{us}(\mathbf{B}, 0)]}{[\det \mathbf{Z}_m(0)] M_{rs}(\mathbf{B}, 0)} \end{aligned} \tag{11.157}$$

Appealing to (11.138) yields

$$1 - I = \frac{M_{rs}(\mathbf{B}, 0) + r_m M_{rs,uv}(\mathbf{B}, 0)}{M_{rs}(\mathbf{B}, 0)} = \frac{M_{rs}(\mathbf{B}, r_m)}{M_{rs}(\mathbf{B}, 0)} = \hat{F}^m(r_m) \tag{11.158}$$

confirming our earlier assertion, or

$$\hat{F}^m(x) = \frac{M_{rs}(\mathbf{B}, x)}{M_{rs}(\mathbf{B}, 0)} \tag{11.159}$$

From (11.148) and (11.154), we again establish a relation between $F_k(x)$ and $\hat{F}_k^m(x)$ as follows:

$$\frac{\hat{F}_k^m(x)}{F_k(x)} = \frac{M_{rs}(\mathbf{B}, x)}{M_{rs}(\mathbf{B}, k)} \cdot \frac{\det \mathbf{Z}_m(k)}{\det \mathbf{Z}_m(x)} = \frac{T_{rs}^y(x)}{T_{rs}^y(k)} \tag{11.160}$$

where $T_{rs}^y(x)$ is the transfer admittance between input voltage V_r and output current I_s and $T_{rs}^y(k)$ is the same transfer admittance under the condition that the element x of N assumes the reference value k.

11.5.4 Relations between Sensitivities and Return Difference and Null Return Difference

As indicated in Sec. 5.1 of Chapter 5, the sensitivity function of a network function $w(x)$ with respect to an element x is defined as

$$S_x^w = \frac{\partial \ln w}{\partial \ln x} = \frac{x}{w} \cdot \frac{\partial w}{\partial x} \tag{11.161}$$

Refer to Fig. 11.25. If $w(x)$ represents either the voltage gain V_s/V_r or the transfer admittance function i_s/V_r, then from (11.146) and (11.144)

$$w(x) = G_{rs}(x) \equiv \frac{V_s}{V_r} = -\frac{M_{rs}(\mathbf{B},x)}{\det \mathbf{Z}_m(x)} z_s \tag{11.162}$$

$$w(x) = T_{rs}^y(x) \equiv \frac{i_s}{V_r} = -\frac{M_{rs}(\mathbf{B},x)}{\det \mathbf{Z}_m(x)} \tag{11.163}$$

From (11.151) and (11.158), we know that

$$\det \mathbf{Z}_m(x) = \det \mathbf{Z}_m(0) + x M_{pq}(\mathbf{B},x) \tag{11.164}$$

$$M_{rs}(\mathbf{B},x) = M_{rs}(\mathbf{B},0) + x M_{rs,pq}(\mathbf{B},0) \tag{11.165}$$

where the choice of edges e_p and e_q depends on the element x. When x represent a one-port element corresponding to the edge e_p, then $p = q$. When x represents the controlling parameter r_m of a controlled voltage source, e_p and e_q are the edges corresponding to the controlled branch and the controlling branch, respectively. Now we compute the sensitivity function S_x^w. From (11.164) and (11.165), we have

$$\frac{\partial w}{\partial x} = -\frac{M_{rs,pq}(\mathbf{B},0)[\det \mathbf{Z}_m(x)] - M_{pq}(\mathbf{B},0)M_{rs}(\mathbf{B},x)}{[\det \mathbf{Z}_m(x)]^2} \tag{11.166}$$

$$\begin{aligned}
S_x^w &= \frac{x}{w} \cdot \frac{\partial w}{\partial x} \\
&= \frac{x M_{rs,pq}(\mathbf{B},0)\det \mathbf{Z}_m(x) - x M_{pq}(\mathbf{B},0)M_{rs}(\mathbf{B},x)}{[\det \mathbf{Z}_m(x)] M_{rs}(\mathbf{B},x)} \\
&= \frac{[M_{rs}(\mathbf{B},x) - M_{rs}(\mathbf{B},0)]\det \mathbf{Z}_m(x) - M_{rs}(\mathbf{B},x)}{[\det \mathbf{Z}_m(x)] M_{rs}(\mathbf{B},x)}
\end{aligned}$$

$$= \frac{M_{rs}(\mathbf{B},x)\det \mathbf{Z}_m(0) - M_{rs}(\mathbf{B},0)\det \mathbf{Z}_m(x)}{[\det \mathbf{Z}_m(x)]\,M_{rs}(\mathbf{B},x)}$$

$$= \frac{\det \mathbf{Z}_m(0)}{\det \mathbf{Z}_m(x)} - \frac{M_{rs}(\mathbf{B},0)}{M_{rs}(\mathbf{B},x)} \tag{11.167}$$

yielding from (11.148) and (11.154)

$$S_x^w = \frac{1}{F(x)} - \frac{1}{\hat{F}^m(x)} \tag{11.168}$$

Since

$$\frac{\hat{F}^m(x)}{F(x)} = \frac{w(x)}{w(0)} \tag{11.169}$$

we have

$$S_x^w = \frac{1}{F(x)}\left[1 - \frac{w(0)}{w(x)}\right] \tag{11.170}$$

For $w(0) = 0$,

$$S_x^w = \frac{1}{F(x)} \tag{11.171}$$

In this case or when $|w(x)| \gg |w(0)|$, the sensitivity is equal to or approximately equal to the reciprocal of the return difference.

11.5.5 Illustrative Example

Consider a linear time-invariant RLC-r_m type network N and its associated directed graph G as shown in Figs. 11.26 and 11.27, respectively. The network is excited by the voltage source V_{in} corresponding to branch 1 and at a particular frequency. The values of the components are given by

$$z_2 = 1\,\Omega, \quad z_3 = 2\,\Omega, \quad z_4 = j2\,\Omega, \quad z_5 = 4\,\Omega, \quad z_7 = j2\,\Omega \tag{11.172a}$$

$$z_8 = -j2\,\Omega, \quad z_9 = 1\,\Omega, \quad z_{10} = 2\,\Omega, \quad r_m = 2\,\Omega \tag{11.172b}$$

Now, we compute the network functions and return difference and other related quantities. First, choose a tree composed of the branches e_4, e_5, e_7 and e_9 of G and

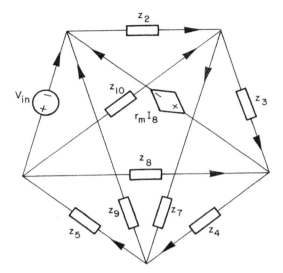

Figure 11.26 A linear time-Invariant RLC-r_m type network N.

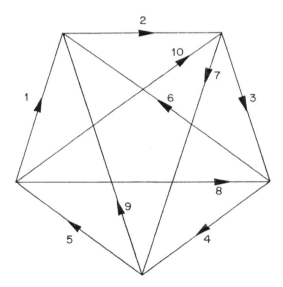

Figure 11.27 The associated directed graph G of the network N of Fig. 11.26.

write the fundamental circuit matrix **B** of G as

$$
\mathbf{B} =
\begin{array}{c}
\begin{array}{cccccccccc} 1 & 2 & 3 & \; & 4 & 5 & 6 & \; & 7 & 8 & \; & 9 & 10 \end{array} \\
\begin{bmatrix}
1 & 0 & 0 & & 0 & 1 & 0 & & 0 & 0 & & -1 & 0 \\
0 & 1 & 0 & & 0 & 0 & 0 & & 1 & 0 & & 1 & 0 \\
0 & 0 & 1 & & 1 & 0 & 0 & & -1 & 0 & & 0 & 0 \\
0 & 0 & 0 & & -1 & 0 & 1 & & 0 & 0 & & -1 & 0 \\
0 & 0 & 0 & & 1 & 1 & 0 & & 0 & 1 & & 0 & 0 \\
0 & 0 & 0 & & 0 & 1 & 0 & & 1 & 0 & & 0 & 1
\end{bmatrix}
\end{array}
\tag{11.173}
$$

The branch-impedance matrix $\mathbf{Z}_b$ and the loop-impedance matrix $\mathbf{Z}_m$ are found to be

$$
\mathbf{Z}_b =
\begin{array}{c}
\begin{array}{cccccccccc} 1 & 2 & 3 & 4 & 5 & 6 & 7 & 8 & 9 & 10 \end{array} \\
\begin{bmatrix}
0 & 0 & 0 & 0 & 0 & 0 & 0 & 0 & 0 & 0 \\
0 & z_2 & 0 & 0 & 0 & 0 & 0 & 0 & 0 & 0 \\
0 & 0 & z_3 & 0 & 0 & 0 & 0 & 0 & 0 & 0 \\
0 & 0 & 0 & z_4 & 0 & 0 & 0 & 0 & 0 & 0 \\
0 & 0 & 0 & 0 & z_5 & 0 & 0 & 0 & 0 & 0 \\
0 & 0 & 0 & 0 & 0 & 0 & 0 & r_m & 0 & 0 \\
0 & 0 & 0 & 0 & 0 & 0 & z_7 & 0 & 0 & 0 \\
0 & 0 & 0 & 0 & 0 & 0 & 0 & z_8 & 0 & 0 \\
0 & 0 & 0 & 0 & 0 & 0 & 0 & 0 & z_9 & 0 \\
0 & 0 & 0 & 0 & 0 & 0 & 0 & 0 & 0 & z_{10}
\end{bmatrix}
\end{array}
\tag{11.174}
$$

$$
\begin{aligned}
\mathbf{B}\mathbf{Z}_b\mathbf{B}' &=
\begin{bmatrix}
z_5 + z_9 & -z_9 & 0 & z_9 & z_5 & z_5 \\
-z_9 & z_2 + z_7 + z_9 & -z_7 & -z_9 & 0 & z_7 \\
0 & -z_7 & z_3 + z_4 + z_7 & -z_4 & z_4 & -z_7 \\
z_9 & -z_9 & -z_4 & z_4 + z_9 & r_m + z_4 & 0 \\
z_5 & 0 & z_4 & -z_4 & z_4 + z_5 + z_8 & z_5 \\
z_5 & z_7 & -z_7 & 0 & z_5 & z_5 + z_7 + z_{10}
\end{bmatrix} \\[2mm]
&=
\begin{bmatrix}
5 & -1 & 0 & 1 & 4 & 4 \\
-1 & 2 + j2 & -j2 & -1 & 0 & j2 \\
0 & -j2 & 2 & j2 & -j2 & -j2 \\
1 & -1 & j2 & 1 - j2 & 2 + j2 & 0 \\
4 & 0 & -j2 & j2 & 4 - j4 & 4 \\
4 & j2 & -j2 & 0 & 4 & 6 + j2
\end{bmatrix}
\end{aligned}
\tag{11.175}
$$

Suppose that z_7 is the output branch. We compute the transfer admittance T_{17}^y. From (11.144), we obtain

$$T_{uv}^y = T_{17}^y = -\frac{M_{17}(\mathbf{B})}{\det \mathbf{Z}_m} \tag{11.176}$$

where

$$\det \mathbf{Z}_m = 272 - j432 \tag{11.177a}$$

$$M_{17}(\mathbf{B}) = -\det \mathbf{B}_{-1}\mathbf{Z}_b(0)\mathbf{B}_{-2}' - \det \mathbf{B}_{-1}\mathbf{Z}_b(0)\mathbf{B}_{-3}'$$
$$- \det \mathbf{B}_{-1}\mathbf{Z}_b(0)\mathbf{B}_{-6}' = 40 + j64 \tag{11.177b}$$

yielding the transfer admittance

$$T_{17}^y = -\frac{40 + j64}{272 - j432} = -0.148\angle 115.6° \tag{11.178}$$

From (11.145), the driving-point admittance is computed by

$$T_{rr}^y = T_{11}^y = \frac{M_{11}(\mathbf{B})}{\det \mathbf{Z}_m} = \frac{280 - j312}{272 - j432} = 0.8212\angle 9.71° \tag{11.179}$$

where

$$M_{rr}(\mathbf{B}) = M_{11}(\mathbf{B}) = \det \mathbf{B}_{-1}\mathbf{Z}_b(0)\mathbf{B}_{-1}' = 280 - j312 \tag{11.180}$$

When r_m is zero, with $M_{rs}(\mathbf{B}) = M_{rs}(\mathbf{B},r_m)$ we have

$$\det \mathbf{Z}_m(0) = -80 - j352 \tag{11.181a}$$

$$M_{rs}(\mathbf{B},0) = M_{17}(\mathbf{B},0) = 8 + j16 \tag{11.181b}$$

$$M_{uv}(\mathbf{B},0) = M_{68}(\mathbf{B},0) = -\det \mathbf{B}_{-4}\mathbf{Z}_b(0)\mathbf{B}_{-5}' = 176 - j40 \tag{11.181c}$$

$$M_{rv}(\mathbf{B},0) = M_{18}(\mathbf{B},0) = \det \mathbf{B}_{-1}\mathbf{Z}_b(0)\mathbf{B}_{-5}' = -176 + j40 \tag{11.181d}$$

$$M_{us}(\mathbf{B},0) = M_{67}(\mathbf{B},0)$$
$$= -\det \mathbf{B}_{-4}\mathbf{Z}_b(0)\mathbf{B}_{-2}' + \det \mathbf{B}_{-4}\mathbf{Z}_b(0)\mathbf{B}_{-3}' + \det \mathbf{B}_{-4}\mathbf{Z}_b(0)\mathbf{B}_{-6}'$$
$$= 40 - j48 \tag{11.181e}$$

$$M_{rs,uv}(\mathbf{B},0) = M_{17,68}(\mathbf{B},0)$$
$$= \det \mathbf{B}_{-14}\mathbf{Z}_b(0)\mathbf{B}_{-25}' + \det \mathbf{B}_{-14}\mathbf{Z}_b(0)\mathbf{B}_{-35}' - \det \mathbf{B}_{-14}\mathbf{Z}_b(0)\mathbf{B}_{-65}'$$
$$= 96 - j8 - 40 + j16 - 40 + j16 = 16 + j24 \tag{11.181f}$$

giving

$$M_{rs}(\mathbf{B},0) + r_m M_{rs,uv}(\mathbf{B},0) = 8 + j16 + 2(16 + j24)$$

$$= 40 + j64 = M_{rs}(\mathbf{B}) \qquad (11.182)$$

confirming (11.177b) and from (11.181a) and (11.181c)

$$\det \mathbf{Z}_m(0) + r_m M_{uv}(\mathbf{B},0) = -80 - j352 + 2(176 - j40)$$

$$= 272 - j432 = \det \mathbf{Z}_m(r_m) \qquad (11.183)$$

confirming (11.151). From (11.181) and (11.183), we have

$$M_{rs}(\mathbf{B},0)M_{uv}(\mathbf{B},0) - M_{rv}(\mathbf{B},0)M_{us}(\mathbf{B},0) = 7168 - j7552 \qquad (11.184a)$$

$$M_{rs,uv}(\mathbf{B},0)\det \mathbf{Z}_m(0) = 7168 - j7552 \qquad (11.184b)$$

confirming (11.130).

The return difference $F(x)$ and the null return difference $\hat{F}^m(r_m)$ with respect to the element r_m are found to be

$$F(r_m) = \frac{\det \mathbf{Z}_m(r_m)}{\det \mathbf{Z}_m(0)} = \frac{272 - j432}{-80 - j352} = 1.414\angle + 45° \qquad (11.185a)$$

$$\hat{F}^m(r_m) = \frac{M_{rs}(\mathbf{B},r_m)}{M_{rs}(\mathbf{B},0)} = \frac{M_{17}(\mathbf{B},r_m)}{M_{17}(\mathbf{B},0)} = \frac{40 + j64}{8 + j16} = 4.22\angle - 5.43° \qquad (11.185b)$$

Finally, we compute the sensitivity function S_x^w. If S_x^w represents the sensitivity function of the transfer admittance T_{17}^y with respect to the controlling parameter r_m, then from (11.168) we obtain

$$S_{r_m}^{T_{17}^y} = \frac{1}{F(r_m)} - \frac{1}{\hat{F}^m(r_m)} = \frac{1}{1.414\angle + 45°} - \frac{1}{4.22\angle - 5.63°} = 0.545\angle 61° \qquad (11.186)$$

Appendix: Outline of a Proof of (11.151)

PROOF.

$$\det \mathbf{Z}_m(r_m) = \det \mathbf{Z}_m(0) + r_m M_{uv}(\mathbf{B},0) \qquad (11.187)$$

Observe that in the matrix triple product $\mathbf{B}\mathbf{Z}_b\mathbf{B}'$, the controlling parameter r_m is in the ith row and jth column position, where b_{iu} and b_{jv} are both nonzero for

$i, j = 1, 2, \ldots, m$, and it appears in the uth row and vth column position of the branch-impedance matrix $\mathbf{Z}_b$. Let

$$r_{ij} = b_{iu} b_{jv} r_m \quad (i, j = 1, 2, \ldots, m) \tag{11.188}$$

$$\mathbf{R} = [r_{ij}] = [\mathbf{r}_1, \mathbf{r}_2, \ldots, \mathbf{r}_m] \tag{11.189}$$

$$\mathbf{Z}_m = [\mathbf{z}_{m1}, \mathbf{z}_{m2}, \ldots, \mathbf{z}_{mm}] \tag{11.190}$$

where $\mathbf{r}_i$ and $\mathbf{z}_{mi}$ are the ith column vectors of the matrices $\mathbf{R}$ and $\mathbf{Z}_m$, respectively. Denote by $\mathbf{z}_{mi}^0$ the column vector $\mathbf{z}_{mi}$ when $r_m = 0$. Then we have

$$\mathbf{Z}_m = [\mathbf{z}_{m1}^0 + \mathbf{r}_1, \mathbf{z}_{m2}^0 + \mathbf{r}_2, \ldots, \mathbf{z}_{mm}^0 + \mathbf{r}_m] \tag{11.191}$$

Assume that $\mathbf{r}_1 \neq \mathbf{0}$. We get

$$\det \mathbf{Z}_m = \det[\mathbf{z}_{m1}^0, \mathbf{z}_{m2}, \mathbf{z}_{m3}, \ldots, \mathbf{z}_{mm}] + \det[\mathbf{r}_1, \mathbf{z}_{m2}, \mathbf{z}_{m3}, \ldots, \mathbf{z}_{mm}] \tag{11.192}$$

Since $\mathbf{r}_1$ is either equal to $\mathbf{r}_j$ or $-\mathbf{r}_j$ for any $\mathbf{r}_j \neq \mathbf{0}$, by column operation we have

$$\det[\mathbf{r}_1, \mathbf{z}_{m2}, \mathbf{z}_{m3}, \ldots, \mathbf{z}_{mm}] = \det[\mathbf{r}_1, \mathbf{z}_{m2}^0, \mathbf{z}_{m3}^0, \ldots, \mathbf{z}_{mm}^0] \tag{11.193}$$

In a similar way, (11.192) can be expanded as

$$
\begin{aligned}
\det \mathbf{Z}_m &= \det[\mathbf{z}_{m1}^0, \mathbf{z}_{m2}, \mathbf{z}_{m3}, \ldots, \mathbf{z}_{mm}] + \det[\mathbf{r}_1, \mathbf{z}_{m2}^0, \mathbf{z}_{m3}^0, \ldots, \mathbf{z}_{mm}^0] \\
&= \det[\mathbf{z}_{m1}^0, \mathbf{z}_{m2}^0, \mathbf{z}_{m3}, \ldots, \mathbf{z}_{mm}] + \det[\mathbf{r}_1, \mathbf{z}_{m2}^0, \mathbf{z}_{m3}^0, \ldots, \mathbf{z}_{mm}^0] \\
&\quad + \det[\mathbf{z}_{m1}^0, \mathbf{r}_2, \mathbf{z}_{m3}, \mathbf{z}_{m4}, \ldots, \mathbf{z}_{mm}] \\
&= \det[\mathbf{z}_{m1}^0, \mathbf{z}_{m2}^0, \mathbf{z}_{m3}, \ldots, \mathbf{z}_{mm}] + \det[\mathbf{r}_1, \mathbf{z}_{m2}^0, \mathbf{z}_{m3}^0, \ldots, \mathbf{z}_{mm}^0] \\
&\quad + \det[\mathbf{z}_{m1}^0, \mathbf{r}_2, \mathbf{z}_{m3}^0, \mathbf{z}_{m4}^0, \ldots, \mathbf{z}_{mm}^0] \\
&= \det[\mathbf{z}_{m1}^0, \mathbf{z}_{m2}^0, \ldots, \mathbf{z}_{mm}^0] + \det[\mathbf{r}_1, \mathbf{z}_{m2}^0, \ldots, \mathbf{z}_{mm}^0] \\
&\quad + \det[\mathbf{z}_{m1}^0, \mathbf{r}_2, \mathbf{z}_{m3}^0, \ldots, \mathbf{z}_{mm}^0 + \cdots + \det[\mathbf{z}_{m1}^0, \ldots, \mathbf{z}_{m(m-1)}^0, \mathbf{r}_m] \\
&= \det \mathbf{Z}_m(0) + \sum_{i=1}^{m} \sum_{j=1}^{m} (-1)^{i+j} b_{iu} b_{jv} (\det \mathbf{B}_{-i} \mathbf{Z}_b(0) \mathbf{B}'_{-j}) r_m \\
&= \det \mathbf{Z}_m(0) + r_m M_{uv}(\mathbf{B}, 0) \tag{11.194}
\end{aligned}
$$

This completes the proof of the identity.

11.6 SUMMARY

We began this chapter by defining the indefinite-impedance matrix for a planar network and then by showing its applications to network analysis. These results were extended to nonplanar networks by demonstrating that it is always possible to assign $m + 1$ loop sets in a graph, planar or nonplanar, of nullity m, so that the number of loop sets traversing in one direction through any edge equals the number of loop sets traversing in the opposite direction of the edge. Furthermore, any m of these $m + 1$ loop sets are linearly independent. A procedure was given for the enumeration of these loop sets. We established the duality between the indefinite-impedance matrix and the indefinite-admittance matrix. In practical applications, they complement each other, each having its own advantages and disadvantages. Because the indefinite-impedance matrix is also an equicofactor matrix, the formulation is applicable to the generation of symbolic network functions as discussed in Chen (1965) and Alderson and Lin (1973).

As a result, we extended and showed that the sum of the reciprocals of the return differences with respect to all network elements of any network, planar or nonplanar, equals the rank of the network. Similar results on the summation of driving-point and transfer admittances are also extended to any network, planar or nonplanar, networks. We showed that in a uniquely solvable linear network, the sum of the products of the impedances of the tree branches including the controlling parameters of voltage sources, and the their associated transfer admittances equals the sum of the products of the admittances of the cotree links including the controlling parameters of current sources, and their associated transfer impedances.

Finally, we derived a useful identity about the loop-impedance matrix and its generalized cofactors, and established the loop-impedance matrix formulas for the feedback amplifier theory. They are very similar to those derived by means of the indefinite-admittance matrix. It is easy to see that similar formulas for the calculation of the network functions, return difference and sensitivities by means of the cutset-admittance matrix can be derived. In most cases, these matrices can be written down directly from the network by inspection. For networks whose rank is greater than its nullity, the formulas given here are more convenient. As a rule, we should choose the input and output edges and as many edges of interest as possible as the links of a tree in order to simplify the calculations. For a planar network, if we select all interior meshes as the basic circuits and add a new row corresponding to the outer mesh to the circuit matrix, the indefinite-impedance matrix formed by these meshes can be defined, resulting in formulas dual to those derived by the indefinite-admittance matrix.

BIBLIOGRAPHY

Alderson, G. E. and P. M. Lin: Computer Generation of Symbolic Network Functions — A New Theory and Implementation, *IEEE Trans. Circuit Theory*, vol. CT-20, pp. 48–56, 1973.

Blackman, R. B.: Effect of Feedback on Impedance, *Bell Syst. Tech. J.*, vol. 22, pp. 268–277, 1943.

Bode, H. W.: "Network Analysis and Feedback Amplifier Design," Princeton, NJ: Van Nostrand, 1959.

Cederbaum, I.: Invariance and Mutual Relations of Electrical Network Determinants, *J. Math. Phys.*, vol. 35, pp. 236–244, 1956a.

Cederbaum, I.: On Network Determinants, *Proc. IRE*, vol. 44, pp. 258–259, 1956b.

Chen, W. K.: Topological Analysis for Active Networks, *IEEE Trans. Circuit Theory*, vol. CT-12, pp. 85–91, 1965.

Chen, W. K.: A Generalization of the Equicofactor Matrix, *IEEE Trans. Circuit Theory*, vol. CT-13, pp. 440–442, 1966.

Chen, W. K.: Graph-Theoretic Considerations on the Invariance and Mutual Relations of the Determinants of the Generalized Network Matrices and Their Generalized Cofactors, *Quart. J. Math. Oxford* (2), vol. 21, no. 84, pp. 459–479, 1970.

Chen, W. K.: On Equicofactor and Indefinite-Admittance Matrices, *Matrix Tensor Quart.*, vol. 23, pp. 26–28, 1972.

Chen, W. K.: Graph-Theoretic Considerations on the Invariance of Return Difference, *J. Franklin Inst.*, vol. 298, no. 2, pp. 81–100, 1974.

Chen, W. K.: Indefinite-Admittance Matrix Formulation of Feedback Amplifier Theory, *IEEE Trans. Circuit and Systems*, vol. CAS-23, no. 8, pp. 498–505, 1976a.

Chen, W. K.: "Applied Graph Theory: Graphs and Electrical Networks," 2nd rev. ed., New York: American Elsevier, and Amsterdam: North-Holland, 1976b.

Chen, W. K.: On Second-Order Cofactors and Null Return Difference in Feedback Amplifier Theory, *Int. J. Circuit Theory Applications*, vol. 6, pp. 305–312, 1978.

Chen, W. K.: Topological Evaluation of Feedback Matrices in Multiloop Feedback Amplifiers, *J. Franklin Inst.*, vol. 308, pp. 125–139, 1979.

Chen, W. K.: A Theorem on the Summation of Return Differences and Some Consequences, *Proc. IEEE*, vol. 72, pp. 396–397, 1984.

Chen, W. K.: A Dual Theorem on the Summation of Return Differences, *Proc. IEEE*, vol. 73, pp. 157–159, 1985.

Chen, W. K. and F. N. T. Chan: On the Unique Solvability of Linear Active Networks, *IEEE Trans. Circuits and Systems*, vol. CAS-21, pp. 26–35, 1974.

Chen, W. K. and H. M. Elsherif: Determinant of the Null Return-Difference Matrix, *Electron. Lett.*, vol. 13, pp. 306–307, 1977.

Chen, W. K. and J. Lu: On the Assignment of Loop Sets to Graphs, *IEEE Trans. Circuits and Systems*, vol. CAS-38, pp. 1090–1092, 1991.

Chen, W. K. and M. D. Tong: On Indefinite Impedance Matrix, *IEEE Trans. Circuits and Systems*, vol. CAS-32, pp. 840–847, 1985.

Hajj, I. N.: A Note on a Theorem on the Summation of Driving-Point and Transfer Network Impedances, *Proc. IEEE*, vol. 72, pp. 395–396, 1984.

Hakim, S. S.: Multiple-Loop Feedback Circuits, *Proc. IEE (London)*, vol. 110, pp. 1955–1959, 1963.

Hakim, S. S.: Aspects of Return-Difference Evaluation in Transistor Feedback Amplifiers, *Proc. IEE (London)*, vol. 112, pp. 1700–1704, 1965.

Hoskins, R. F.: Definition of Loop Gain and Return Differences in Transistor Feedback Amplifiers, *Proc. IEE (London)*, vol. 112, pp. 1995–2000, 1965.

Kuh, E. S.: Some Results in Linear Multiple Loop Feedback Systems, *Proc. Allerton. Conf. on Circuit and System Theory*, vol. 1, pp. 471–487, 1963.

Lan, J. L. and W. K. Chen: On Loop-Impedance Matrix Formulation of Feedback Amplifier Theory, *J. Franklin Inst.*, vol. 320, no. 1, pp. 1–14, 1985.

MacLane, S.: A Structural Characterization of Planar Combinatorial Graphs, *Duke Math. J.*, vol. 3, pp. 460–472, 1937.

Maclean, D. J. H.: Improved Methods of Assessing Feedback in Wideband Multiloop Amplifiers, *IEEE Trans. Circuits and Systems*, vol. CAS-27, pp. 779–792, 1980.

MacFarlance, A. G. J.: Return-Difference and Return-Ratio Matrices and Their Use in Analysis and Design of Multivariable Feedback Control Systems, *Proc. IEE (London)*, vol. 117, pp. 2037–2049, 1970.

MacFarlance, A. G. J.: Return-Difference and Return-Ratio Matrices and Their Use in Analysis and Design of Multivariable Feedback Control Systems, *Proc. IEE (London)*, vol. 118, pp. 946–947, 1971.

Malik, N. R. and H. W. Hale: Relationships between Nonzero Determinants Formed from Vertex and Circuit Matrices, *IEEE Trans. Circuit Theory*, vol. CT-13, p. 196, 1966.

Nerode, A. and H. Shank: An Algebraic Proof of Kirchhoff's Network Theorems, *Amer. Math. Monthly*, vol. 68, pp. 244–247, 1961.

Okada, S.: On Node and Mesh Determinants, *Proc. IRE*, vol. 43, p. 1527, 1955.

Percival, W. S.: Improved Matrix and Determinant Methods for Solving Networks, *Proc. IEE (London)*, vol. 101, pp. 258–265, 1954.

Saltzer, C.: The Second Fundamental Theorem of Electrical Networks, *Quart. Appl. Math.*, vol. 11, pp. 119–123, 1953.

Sandberg, I. W.: On the Theory of Linear Multi-Loop Feedback Systems, *Bell Syst. Tech. J.*, vol. 42, pp. 355–382, 1963.

Seshu, S.: The Mesh Counterpart of Shekel's Theorem, *Proc. IRE*, vol. 43, p. 342, 1955.

Sharpe, G. E. and B. Spain: On the Solution of Networks by Means of the Equicofactor Matrix, *IEEE Trans. Circuit Theory*, vol. CT-7, pp. 230–239, 1960.

Shekel, J.: Two Network Theorems Concerning Change of Voltage Reference Terminal, *Proc. IRE*, vol. 42, p. 1125, 1954.

Slepian, P.: "Mathematical Foundations of Network Analysis," p. 46 and pp. 177–186, New York: Springer-Verlag, 1968.

Tasny-Tschiassny, L.: The Return Difference Matrix in Linear Networks, *Proc. IEE (London)*, vol. 100, pt. IV, pp. 39–46, 1953.

Thulasiraman, K., R. Jayakumar and M. N. S. Swamy: Graph-Theoretic Proof of a Network Theorem and Some Consequences, *Proc. IEEE*, vol. 71, pp. 771–772, 1983.

Tong, M. D. and Y. H. Zhu: "Symbolic Network Functions and Indefinite Admittance Matrix," Beijing, China: Higher Education Publishing House of China, 1983.

Tsang, N. F.: On Electrical Network Determinants, *J. Math. Phys.*, vol. 33, pp. 185–193, 1954.

Wang, H. Y. and W. K. Chen: A General Theorem on the Summation of Return Differences, *Proc. IEEE Int. Symp. Circuits and Systems*, pp. 160–164, 1988.

Wang, H. Y. and W. K. Chen: An Extension of a Dual Theorem on the Summation of Return Differences, *IEEE Trans. Circuits and Systems*, vol. CAS-38, pp. 1110–1113, 1991.

Wang, H. Y. and W. K. Chen: Dual Topological Theorems of Linear Active Networks, *IEE Proc.- Circuits Devices Syst.*, vol. 143, pp. 225–228, 1996.

HERMITIAN FORMS

Hermitian forms and their properties are reviewed briefly in this appendix.

Let

$$\mathbf{A} = [a_{ij}] \tag{I.1}$$

be a hermitian matrix of order n and let $\mathbf{X} = [X_j]$ be a complex n-vector. Then the scalar expression

$$\mathbf{X}^*\mathbf{AX} = \sum_{i=1}^{n}\sum_{j=1}^{n} a_{ij}\overline{x}_i x_j \tag{I.2}$$

where

$$\mathbf{X}^* = \overline{X}' \tag{I.3}$$

$$\overline{X} = [\overline{x}_j] \tag{I.4}$$

and $\overline{x}_j$ denotes the complex conjugate of x_j, is called a *hermitian form*. The matrix $\mathbf{A}$ is referred to as the *matrix of the hermitian form* (I.2). Even though $\mathbf{A}$ and $\mathbf{X}$ are complex, the hermitian form $\mathbf{X}^*\mathbf{AX}$ is real. To see this, we take the conjugate of $\mathbf{X}^*\mathbf{AX}$. Since $\mathbf{X}^*\mathbf{AX}$ is a scalar, $(\mathbf{X}^*\mathbf{AX})' = \mathbf{X}^*\mathbf{AX}$. Hence, since $\mathbf{A}$ is hermitian, we have

$$\overline{\mathbf{X}^*\mathbf{AX}} = (\mathbf{X}^*\mathbf{AX})^* = \mathbf{X}^*\mathbf{A}^*\mathbf{X} = \mathbf{X}^*\mathbf{AX} \tag{I.5}$$

showing that $\mathbf{X}^*\mathbf{AX}$, being equal to its own conjugate, is real for every choice of $\mathbf{X}$. However, for a given $\mathbf{A}$, the sign associated with such a form normally depends on the values of $\mathbf{X}$. It may happen that, for some $\mathbf{A}$, its hermitian form remains of one sign, independent of the values of $\mathbf{X}$. Such forms are called *definite*. Since definiteness of a hermitian form must be an inherent property of its matrix $\mathbf{A}$, it is natural then to refer to the matrix $\mathbf{A}$ as *definite*. We now consider two subclasses of the class of definite hermitian matrices.

Definition I.1: Positive definite matrix An $n \times n$ hermitian matrix $\mathbf{A}$ is called a *positive definite matrix* if

$$\mathbf{X}^*\mathbf{A}\mathbf{X} > 0 \tag{I.6}$$

for all complex n-vectors $\mathbf{X} \neq 0$.

Definition I.2: Nonnegative definite matrix An $n \times n$ hermitian matrix $\mathbf{A}$ is called a *nonnegative definite matrix* if

$$\mathbf{X}^*\mathbf{A}\mathbf{X} \geqslant 0 \tag{I.7}$$

for all complex n-vectors $\mathbf{X}$.

Evidently, a positive definite matrix is also nonnegative definite. Very often, a nonnegative definite matrix is also called a *positive semidefinite matrix*, but we must bear in mind that some people define a positive semidefinite matrix $\mathbf{A}$ as one that satisfies (I.7) for all $\mathbf{X}$, provided that there is at least one $\mathbf{X} \neq \mathbf{0}$ for which the equality holds. In the latter case, it is evident that positive definiteness and positive semi-definiteness are mutually exclusive. Together, they form the class of nonnegative definite matrices. Thus, care must be taken to ensure the proper interpretation of the term "positive semidefiniteness."

In a matrix $\mathbf{A}$ of order n, define a *principal minor* of order r to be the determinant of a submatrix consisting of the rows $i_1, i_2, \ldots, i_r$, and the columns $i_1, i_2, \ldots, i_r$. The *leading principal minor* of order r is the determinant of the submatrix consisting of the first r rows and the first r columns. We now present Sylvester's criteria for positive definiteness and nonnegative definiteness of a hermitian matrix.

Theorem I.1 A hermitian matrix is positive definite if and only if all of its leading principal minors are positive. A hermitian matrix is nonnegative definite if and only if all of its principal minors are nonnegative.

We remark that in testing for positive definiteness, the positiveness of all of its leading principal minors also implies the positiveness of all of its principal minors. However, in testing for nonnegative definiteness, the nonnegativeness of all of its leading principal minors does not necessarily imply the nonnegativeness of all of its principal minors.

An $n \times n$ matrix $\mathbf{U}$ is said to be *unitary* if and only if it has the property that $\mathbf{U}^*\mathbf{U} = \mathbf{1}_n$. We write

$$\mathbf{D}(\lambda_1, \lambda_2, \ldots, \lambda_n) = \begin{bmatrix} \lambda_1 & 0 & \cdots & 0 \\ 0 & \lambda_2 & \cdots & 0 \\ & & \cdots & \\ 0 & 0 & \cdots & \lambda_n \end{bmatrix} \tag{I.8}$$

We can prove the following theorem by induction over the order of $\mathbf{A}$.

Theorem I.2 Let $\mathbf{A}$ be a hermitian matrix of order n. Then there exists a unitary matrix $\mathbf{U}$ such that $\mathbf{U}^*\mathbf{AU}$ is a diagonal matrix whose diagonal elements are the eigenvalues of $\mathbf{A}$; that is,

$$\mathbf{U}^*\mathbf{AU} = \mathbf{D}(\lambda_1, \lambda_2, \ldots, \lambda_n) \tag{I.9}$$

For a hermitian form (I.2), Theorem I.2 implies

Theorem I.3 By a suitable unitary transformation $\mathbf{X} = \mathbf{UY}$, the hermitian form $\mathbf{X}^*\mathbf{AX}$ can be reduced to the form

$$\mathbf{X}^*\mathbf{AX} = \mathbf{Y}^*\mathbf{D}(\lambda_1, \lambda_2, \ldots, \lambda_n)\mathbf{Y} = \sum_{j=1}^{n} \lambda_j \overline{y}_j y_j \tag{I.10}$$

where the λ's are the eigenvalues of the hermitian matrix $\mathbf{A}$ and y_j is the jth row element of the n-vector $\mathbf{Y}$.

It is possible to use other nonsingular transformations to reduce a given hermitian form to the diagonal form. No matter by what nonsingular transformation, a hermitian form can be reduced to a standard form given in the following theorem.

Theorem I.4 Irrespective of the nonsingular transformation used, a given hermitian form can be reduced to a form

$$d_1\overline{y}_1 y_1 + d_2\overline{y}_2 y_2 + \cdots + d_p\overline{y}_p y_p - d_{p+1}\overline{y}_{p+1} y_{p+1} - \cdots - d_r\overline{y}_r y_r \tag{I.11}$$

where the d's are all positive and the integers p and r remain invariant.

The number r of terms in (I.11) is called the *rank* of the form, and the number p of positive terms is referred to as the *index* of the form. The *signature* of the form is then defined to be $2p - r$. Evidently, the rank of the matrix of the hermitian form is equal to the rank of the form. Thus, we can state that a hermitian matrix is positive definite if and only if its order, rank, and the index of its hermitian form are equal, and that it is nonnegative definite if and only if the rank and index of its hermitian form are equal.

Theorem I.5 A hermitian matrix $\mathbf{A}$ is positive definite if and only if any one of the following conditions is satisfied:

1. $\mathbf{B}^*\mathbf{A}\mathbf{B}$ is positive definite for arbitrary and nonsingular $\mathbf{B}$.
2. $\mathbf{A}^q$ is positive definite for every integer q.
3. There exists a nonsingular matrix $\mathbf{B}$ such that $\mathbf{A} = \mathbf{B}^*\mathbf{B}$.

CONVERSION CHART FOR TWO-PORT PARAMETERS

$$\Delta_x = x_{11}x_{22} - x_{12}x_{21}$$

To \ From	**Z**		**Y**		**T**		**T**$^{-1}$		**H**		**H**$^{-1}$	
Z	z_{11}	z_{12}	$\frac{y_{22}}{\Delta_y}$	$-\frac{y_{12}}{\Delta_y}$	$\frac{A}{C}$	$\frac{\Delta_T}{C}$	$\frac{D'}{C'}$	$\frac{1}{C'}$	$\frac{\Delta_h}{h_{22}}$	$\frac{h_{12}}{h_{22}}$	$\frac{1}{g_{11}}$	$-\frac{g_{12}}{g_{11}}$
	z_{21}	z_{22}	$-\frac{y_{21}}{\Delta_y}$	$\frac{y_{11}}{\Delta_y}$	$\frac{1}{C}$	$\frac{D}{C}$	$\frac{\Delta_{T'}}{C'}$	$\frac{A'}{C'}$	$-\frac{h_{21}}{h_{22}}$	$\frac{1}{h_{22}}$	$\frac{g_{21}}{g_{11}}$	$\frac{\Delta_g}{g_{11}}$
Y	$\frac{z_{22}}{\Delta_z}$	$-\frac{z_{12}}{\Delta_z}$	y_{11}	y_{12}	$\frac{D}{B}$	$-\frac{\Delta_T}{B}$	$\frac{A'}{B'}$	$-\frac{1}{B'}$	$\frac{1}{h_{11}}$	$-\frac{h_{12}}{h_{11}}$	$\frac{\Delta_g}{g_{22}}$	$\frac{g_{12}}{g_{22}}$
	$-\frac{z_{21}}{\Delta_z}$	$\frac{z_{11}}{\Delta_z}$	y_{21}	y_{22}	$-\frac{1}{B}$	$\frac{A}{B}$	$-\frac{\Delta_{T'}}{B'}$	$\frac{D'}{B'}$	$\frac{h_{21}}{h_{11}}$	$\frac{\Delta_h}{h_{11}}$	$-\frac{g_{21}}{g_{22}}$	$\frac{1}{g_{22}}$
T	$\frac{z_{11}}{z_{21}}$	$\frac{\Delta_z}{z_{21}}$	$-\frac{y_{22}}{y_{21}}$	$-\frac{1}{y_{21}}$	A	B	$\frac{D'}{\Delta_{T'}}$	$\frac{B'}{\Delta_{T'}}$	$-\frac{\Delta_h}{h_{21}}$	$-\frac{h_{11}}{h_{21}}$	$\frac{1}{g_{21}}$	$\frac{g_{22}}{g_{21}}$
	$\frac{1}{z_{21}}$	$\frac{z_{22}}{z_{21}}$	$-\frac{\Delta_y}{y_{21}}$	$-\frac{y_{11}}{y_{21}}$	C	D	$\frac{C'}{\Delta_{T'}}$	$\frac{A'}{\Delta_{T'}}$	$-\frac{h_{22}}{h_{21}}$	$-\frac{1}{h_{21}}$	$\frac{g_{11}}{g_{21}}$	$\frac{\Delta_g}{g_{21}}$
T^{-1}	$\frac{z_{22}}{z_{12}}$	$\frac{\Delta_z}{z_{12}}$	$-\frac{y_{11}}{y_{12}}$	$-\frac{1}{y_{12}}$	$\frac{D}{\Delta_T}$	$\frac{B}{\Delta_T}$	A'	B'	$\frac{1}{h_{12}}$	$\frac{h_{11}}{h_{12}}$	$-\frac{\Delta_g}{g_{12}}$	$-\frac{g_{22}}{g_{12}}$
	$\frac{1}{z_{12}}$	$\frac{z_{11}}{z_{12}}$	$-\frac{\Delta_y}{y_{12}}$	$-\frac{y_{22}}{y_{12}}$	$\frac{C}{\Delta_T}$	$\frac{A}{\Delta_T}$	C'	D'	$\frac{h_{22}}{h_{12}}$	$\frac{\Delta_h}{h_{12}}$	$-\frac{g_{11}}{g_{12}}$	$-\frac{1}{g_{12}}$
H	$\frac{\Delta_z}{z_{22}}$	$\frac{z_{12}}{z_{22}}$	$\frac{1}{y_{11}}$	$-\frac{y_{12}}{y_{11}}$	$\frac{B}{D}$	$\frac{\Delta_T}{D}$	$\frac{B'}{A'}$	$\frac{1}{A'}$	h_{11}	h_{12}	$\frac{g_{22}}{\Delta_g}$	$-\frac{g_{12}}{\Delta_g}$
	$-\frac{z_{21}}{z_{22}}$	$\frac{1}{z_{22}}$	$\frac{y_{21}}{y_{11}}$	$\frac{\Delta_y}{y_{11}}$	$-\frac{1}{D}$	$\frac{C}{D}$	$-\frac{\Delta_{T'}}{A'}$	$\frac{C'}{A'}$	h_{21}	h_{22}	$-\frac{g_{21}}{\Delta_g}$	$\frac{g_{11}}{\Delta_g}$
H^{-1}	$\frac{1}{z_{11}}$	$-\frac{z_{12}}{z_{11}}$	$\frac{\Delta_y}{y_{22}}$	$\frac{y_{12}}{y_{22}}$	$\frac{C}{A}$	$-\frac{\Delta_T}{A}$	$\frac{C'}{D'}$	$-\frac{1}{D'}$	$\frac{h_{22}}{\Delta_h}$	$-\frac{h_{12}}{\Delta_h}$	g_{11}	g_{12}
	$\frac{z_{21}}{z_{11}}$	$\frac{\Delta_z}{z_{11}}$	$-\frac{y_{21}}{y_{22}}$	$\frac{1}{y_{22}}$	$\frac{1}{A}$	$\frac{B}{A}$	$\frac{\Delta_{T'}}{D'}$	$\frac{B'}{D'}$	$-\frac{h_{21}}{\Delta_h}$	$\frac{h_{11}}{\Delta_h}$	g_{21}	g_{22}

OUTLINE OF A DERIVATION OF EQ. (7.224)

Let $\mathbf{M}$ denote the coefficient matrix of (7.221). We first rewrite the matrix $\mathbf{M}$ as the product of two matrices and then evaluate its determinant:

$$
\det \mathbf{M} = \det \begin{bmatrix} \mathbf{1}_m & \mathbf{0} & \mathbf{0} & \mathbf{0} \\ \mathbf{0} & \mathbf{1}_r & \mathbf{0} & \mathbf{0} \\ \mathbf{0} & \mathbf{0} & \mathbf{1}_r & \mathbf{0} \\ \mathbf{0} & \mathbf{0} & \mathbf{0} & -\mathbf{H}_{\bar{t}\bar{t}} \end{bmatrix} \begin{bmatrix} -\mathbf{Q}'_{\bar{t}} & \mathbf{1}_m & \mathbf{0} & \mathbf{0} \\ \mathbf{0} & \mathbf{0} & \mathbf{1}_r & \mathbf{Q}_{\bar{t}} \\ \mathbf{H}_{tt} & \mathbf{0} & -\mathbf{1}_r & \mathbf{H}_{t\bar{t}} \\ -\mathbf{H}_{\bar{t}\bar{t}}^{-1}\mathbf{H}_{\bar{t}t} & \mathbf{H}_{\bar{t}\bar{t}}^{-1} & \mathbf{0} & -\mathbf{1}_m \end{bmatrix}
$$

$$
= (-1)^m \det \mathbf{H}_{\bar{t}\bar{t}} \det \begin{bmatrix} -\mathbf{Q}'_{\bar{t}} & \mathbf{1}_m & \mathbf{0} & \mathbf{0} \\ \mathbf{0} & \mathbf{0} & \mathbf{1}_r & \mathbf{Q}_{\bar{t}} \\ \mathbf{H}_{tt} - \mathbf{H}_{t\bar{t}}\mathbf{H}_{\bar{t}\bar{t}}^{-1}\mathbf{H}_{\bar{t}t} & \mathbf{H}_{t\bar{t}}\mathbf{H}_{\bar{t}\bar{t}}^{-1} & -\mathbf{1}_r & \mathbf{0} \\ -\mathbf{H}_{\bar{t}\bar{t}}^{-1}\mathbf{H}_{\bar{t}t} & \mathbf{H}_{\bar{t}\bar{t}}^{-1} & \mathbf{0} & -\mathbf{1}_m \end{bmatrix}
$$

$$
= (-1)^m \det \mathbf{H}_{\bar{t}\bar{t}} \det \begin{bmatrix} \mathbf{B}_f & \mathbf{0} \\ \mathbf{0} & \mathbf{Q}_f \\ \mathbf{Y}_b & -\mathbf{1}_b \end{bmatrix} \tag{III.1}
$$

where $b = r + m$ and $\mathbf{Y}_b$ is defined by (7.226). It should be noted that in deriving (III.1) we have implicitly assumed that the network possesses the admittance representation, which would require that $\mathbf{H}_{\bar{t}\bar{t}}$ be nonsingular. To continue our derivation, we apply the elementary row operations to the matrices in (III.1),

yielding

$$
\det \begin{bmatrix} \mathbf{B}_f & \mathbf{0} \\ \mathbf{0} & \mathbf{Q}_f \\ \mathbf{Y}_b & -\mathbf{1}_b \end{bmatrix}
$$

$$
= \det \begin{bmatrix} -\mathbf{Q}_{\bar{t}}' & \mathbf{1}_m & \mathbf{0} & \mathbf{0} \\ \mathbf{H}_{tt} - \mathbf{H}_{t\bar{t}}\mathbf{H}_{\bar{t}\bar{t}}^{-1}\mathbf{H}_{\bar{t}t} - \mathbf{Q}_{\bar{t}}\mathbf{H}_{\bar{t}\bar{t}}^{-1}\mathbf{H}_{\bar{t}t} & \mathbf{H}_{t\bar{t}}\mathbf{H}_{\bar{t}\bar{t}}^{-1} + \mathbf{Q}_{\bar{t}}\mathbf{H}_{\bar{t}\bar{t}}^{-1} & \mathbf{0} & \mathbf{0} \\ \mathbf{H}_{tt} - \mathbf{H}_{t\bar{t}}\mathbf{H}_{\bar{t}\bar{t}}^{-1}\mathbf{H}_{\bar{t}t} & \mathbf{H}_{t\bar{t}}\mathbf{H}_{\bar{t}\bar{t}}^{-1} & -\mathbf{1}_r & \mathbf{0} \\ -\mathbf{H}_{\bar{t}\bar{t}}^{-1}\mathbf{H}_{\bar{t}t} & \mathbf{H}_{\bar{t}\bar{t}}^{-1} & \mathbf{0} & -\mathbf{1}_m \end{bmatrix}
$$

$$
= (-1)^b \det \begin{bmatrix} \mathbf{0} & \mathbf{1}_m \\ \mathbf{H}_{tt} - \mathbf{H}_{t\bar{t}}\mathbf{H}_{\bar{t}\bar{t}}^{-1}\mathbf{H}_{\bar{t}t} - \mathbf{Q}_{\bar{t}}\mathbf{H}_{\bar{t}\bar{t}}^{-1}\mathbf{H}_{\bar{t}t} \\ +\mathbf{H}_{t\bar{t}}\mathbf{H}_{\bar{t}\bar{t}}^{-1}\mathbf{Q}_{\bar{t}}' + \mathbf{Q}_{\bar{t}}\mathbf{H}_{\bar{t}\bar{t}}^{-1}\mathbf{Q}_{\bar{t}}' & \mathbf{H}_{t\bar{t}}\mathbf{H}_{\bar{t}\bar{t}}^{-1} + \mathbf{Q}_{\bar{t}}\mathbf{H}_{\bar{t}\bar{t}}^{-1} \end{bmatrix}
$$

$$
= (-1)^{mr+b} \det (\mathbf{H}_{tt} - \mathbf{H}_{t\bar{t}}\mathbf{H}_{\bar{t}\bar{t}}^{-1}\mathbf{H}_{\bar{t}t} - \mathbf{Q}_{\bar{t}}\mathbf{H}_{\bar{t}\bar{t}}^{-1}\mathbf{H}_{\bar{t}t} + \mathbf{Q}_{\bar{t}}\mathbf{H}_{\bar{t}\bar{t}}^{-1}\mathbf{Q}_{\bar{t}}'
$$

$$
+ \mathbf{H}_{t\bar{t}}\mathbf{H}_{\bar{t}\bar{t}}^{-1}\mathbf{Q}_{\bar{t}}')
$$

$$
= (-1)^{mr+b} \det \mathbf{Q}_f \mathbf{Y}_b \mathbf{Q}_f' \tag{III.2}
$$

where $\mathbf{Q}_f$ is defined in (7.219a). Combining (III.1) and (III.2) gives

$$
\det \mathbf{M} = (-1)^{m+mr+b} \det \mathbf{H}_{\bar{t}\bar{t}} \det \mathbf{Q}_f \mathbf{Y}_b \mathbf{Q}_f'
$$

$$
= (-1)^{mr+r} \det \mathbf{H}_{\bar{t}\bar{t}} \det \mathbf{Q}_f \mathbf{Y}_b \mathbf{Q}_f' \tag{III.3}
$$

Since (7.222) is obtained from (7.221) by using only the elementary row operations to eliminate the variables $\mathbf{V}_{\bar{t}}$ and $\mathbf{I}_t$, it is not difficult to show that the determinants of the coefficient matrices of (7.221) and (7.222) are related by the equation

$$
\det \mathbf{M} = (-1)^{rb} \det \mathbf{H} \tag{III.4}
$$

Substituting (III.4) in (III.3), we obtain

$$
\det \mathbf{H} = \det \mathbf{H}_{\bar{t}\bar{t}} \det \mathbf{Q}_f \mathbf{Y}_b \mathbf{Q}_f' \tag{III.5}
$$

In a similar manner, if the feedback network possesses the impedance representation, $\mathbf{H}_{tt}$ is nonsingular and the determinant of $\mathbf{H}$ is related to

$\mathbf{B}_f \mathbf{Z}_b \mathbf{B}'_f$ by

$$\det \mathbf{H} = \det \mathbf{H}_{tt} \, \det \mathbf{B}_f \mathbf{Z}_b \mathbf{B}'_f \qquad (\text{III}.6)$$

where the fundamental circuit matrix $\mathbf{B}_f$ is defined in (7.219b), and the branch-impedance matrix $\mathbf{Z}_b$ is given by (7.225).

By appealing to (5.201) and (5.202), (III.5) and (III.6) can be expressed as

$$\det \mathbf{H} = \lambda_{hc}(\det \mathbf{H}_{\bar{t}\bar{t}}) \Delta_c \qquad (\text{III}.7a)$$

$$\det \mathbf{H} = \lambda_{hm}(\det \mathbf{H}_{tt}) \Delta_m \qquad (\text{III}.7b)$$

where λ_{hc} and λ_{hm} are real constants, depending only on the choices of the basis cutsets and loops.

SYMBOL INDEX

The symbols that occur most often are listed here, separated into four categories: Roman letters, Greek letters, matrices and vector, and others.

Roman letters

$\bar{a}_{ij}$, complex conjugate of a_{ij}, 29, 429

am, ammeter branch, 721

c, number of components, 704

cs, current source branch, 721

$\varepsilon(t)$, energy stored at time t, 9

$f(g)$, product of the weights associated with the edges of the subgraph g, 667

$F(x)$, $F_0(x)$, return difference, 265, 272, 479, 494

$\hat{F}(x)$, $\hat{F}_0(x)$, null return difference, 279, 490, 494

$F_k(x)$, general return difference, 321

$\hat{F}_k(x)$, general null return difference, 326

$\mathcal{G}$, transducer power gain, 151

$\mathcal{G}_a$, available power gain, 151

g_{ij}, g-parameters, 148

$G(N)$, associated digraph of an active network N, 667, 671

$\mathcal{G}_0$, maximum stable power gain, 203

$\mathcal{G}_p$, power gain, 151

$g_{rp,sq}$, voltage gain, 106

h_{ij}, general hybrid parameters or h-parameters, 39, 148

$H_{rp,sq}$, transmission, 720

$i_k(t)$, kth port or terminal current, 10, 73

$I_k(s)$, Laplace transform of $i_k(t)$, 23, 72

Im, imaginary part of, 38

k, general reference value, 300, 321

k_{ij}, general hybrid parameters of a two-port network, 148

KCL, Kirchoff's current law, 561

KVL, Kirchoff's voltage law, 561

LHS, left half of the complex frequency s-plane, 29, 179, 382

$m(g)$, nullity of the graph g, 672

M_i, terminating immittance at port i, 211

M_{ii}, driving-point immittance at port i, 187

N, network or n-port network, 2

op-amp, operational amplifier, 115

p, linear differential operator, 603

p_{rs}^p, path between r and s, 720

Re, real part of, 29

$r(g)$, rank of the graph g, 568

RHS, right half of the complex frequency s-plane, 29, 42

$s = \sigma + j\omega$, complex frequency, 39, 41

$S(x)$, sensitivity function, 153, 264

$\mathcal{S}(s_k; x)$, root sensitivity, 383

$S'(x')$, relative sensitivity function, 336

$\bar{T}$, cotree with respect to tree T, 580

T, T_0, return ratio, 267, 272

$\hat{T}$, $\hat{T}_0$, null return ratio, 279–280

T_k, general return ratio or directed tree with reference node k, 666–667

$\hat{T}_k$, general null return ratio, 326

SUBJECT INDEX